THE ONE-DIMENSIONAL HUBBARD MODEL

The description of a solid at a microscopic level is complex, involving the interaction of a huge number of its constituents, such as ions or electrons. It is impossible to solve the corresponding many-body problems analytically or numerically, although much insight can be gained from the analysis of simplified models. An important example is the Hubbard model, which describes interacting electrons in narrow energy bands, and which has been applied to problems as diverse as high-T_c superconductivity, band magnetism and the metal-insulator transition.

Remarkably, the one-dimensional Hubbard model can be solved exactly using the Bethe ansatz method. The resulting solution has become a laboratory for theoretical studies of non-perturbative effects in strongly correlated electron systems. Many methods devised to analyse such effects have been applied to this model, both to provide complementary insight into what is known from the exact solution and as an ultimate test of their quality.

This book presents a coherent, self-contained account of the exact solution of the Hubbard model in one dimension. The early chapters develop a self-contained introduction to Bethe's ansatz and its application to the one-dimensional Hubbard model, and will be accessible to beginning graduate students with a basic knowledge of quantum mechanics and statistical mechanics. The later chapters address more advanced topics, and are intended as a guide for researchers to some of the more recent scientific results in the field of integrable models.

The authors are distinguished researchers in the field of condensed matter physics and integrable systems, and have contributed significantly to the present understanding of the one-dimensional Hubbard model. FABIAN ESSLER is a University Lecturer in Condensed Matter Theory at Oxford University. HOLGER FRAHM is Professor of Theoretical Physics at the University of Hannover. FRANK GÖHMANN is a Lecturer at Wuppertal University, Germany. ANDREAS KLÜMPER is Professor of Theoretical Physics at Wuppertal University. VLADIMIR KOREPIN is Professor at the Yang Institute for Theoretical Physics, State University of New York at Stony Brook, and author of *Quantum Inverse Scattering Method and Correlation Functions* (Cambridge, 1993).

THE ONE-DIMENSIONAL HUBBARD MODEL

FABIAN H. L. ESSLER
Oxford University

HOLGER FRAHM
University of Hannover

FRANK GÖHMANN
Wuppertal University

ANDREAS KLÜMPER
Wuppertal University

VLADIMIR E. KOREPIN
State University of New York at Stony Brook

CAMBRIDGE
UNIVERSITY PRESS

CAMBRIDGE UNIVERSITY PRESS
Cambridge, New York, Melbourne, Madrid, Cape Town, Singapore,
São Paulo, Delhi, Dubai, Tokyo, Mexico City

Cambridge University Press
The Edinburgh Building, Cambridge CB2 8RU, UK

Published in the United States of America by Cambridge University Press, New York

www.cambridge.org
Information on this title: www.cambridge.org/9780521143943

First published 2005
First paperback printing 2010

A catalogue record for this publication is available from the British Library

Library of Congress Cataloguing in Publication data

The one-dimensional Hubbard model / F. H. L. Essler . . . [*et al.*].
p.   cm.
Includes bibliographical references and index.
ISBN 0 521 80262 8
1. Hubbard model.   2. Quantum theory.   3. Statistical mechanics.   I. Essler, F. H. L., 1965–
QC176.8.E4E89 2005
530.4′1–dc22   2004051917

ISBN 978-0-521-80262-8 Hardback
ISBN 978-0-521-14394-3 Paperback

Contents

Preface

On account of Lieb and Wu's 1968 Bethe ansatz solution, the one-dimensional Hubbard model has become a laboratory for theoretical studies of non-perturbative effects in strongly correlated electron systems. Many of the tools available for the analysis of such systems have been applied to this model, both to provide complementary insights to what is known from the exact solution or as an ultimate test of their quality. In parallel, due to the synthesis of new quasi one-dimensional materials and the refinement of experimental techniques, the one-dimensional Hubbard model has evolved from a toy model to a paradigm of experimental relevance for strongly correlated electron systems.

Due to the ongoing efforts to improve our understanding of one-dimensional correlated electron systems, there exists a large number of review articles and books covering various aspects of the general theory, as well as the Bethe ansatz and field theoretical methods. A collection of these works is listed in the General Bibliography below.

Still we felt – and many of our colleagues shared this view – that there would be a need for a coherent account of all of these aspects in a unified framework and from the perspective of the one-dimensional Hubbard model, which, moreover, would be accessible to beginners in the field. This motivated us to write this volume. It is intended to serve both as a textbook and as a monograph. The first chapters are supposed to provide a self-contained introduction to Bethe's ansatz and its application to the one-dimensional Hubbard model, accessible to beginning graduate students with only a basic knowledge of Quantum Mechanics and Statistical Mechanics. The later chapters address more advanced issues and are intended to guide the interested researcher to some of the more recent scientific developments in the field of integrable models.

Although this book concentrates on the one-dimensional Hubbard model, we would like to stress that the methods used in its solution are general in the sense that they apply equally well to other integrable models, some of which we actually deal with in passing. In fact, the application of Bethe's ansatz to the Hubbard model is more involved than in other cases. We expect the reader who has mastered the solution of the Hubbard model to be able to apply his/her knowledge readily to other integrable theories.

This volume does not pretend to cover its subject completely. Rather, we attempted to find a balance between being didactic and being comprehensive. Our selection of material was necessarily governed by our predispositions. We apologize if we have failed to cover important issues adequately.

Ultimately this book originates in the many collaborations between the authors over the last ten years, which are documented in the reference section at the end of the book. Although the material presented has matured in the discussions between us, it is not difficult to infer from our different styles which author bears primary responsibility for which chapter, namely FG for chapters 2, 3, 11, 12, 14, 15, FHLE for chapters 4–7, 10 and 17, HF for chapters 8 and 9, AK for chapter 13, VEK for chapter 16, and FG and FHLE jointly for chapter 1.

Throughout this project and in many fruitful collaborations before we have benefitted immeasurably from numerous discussions with our colleagues and friends A. M. Tsvelik, N. d'Ambrumenil, T. Deguchi, H. Fehske, F. Gebhard, F. D. M. Haldane, V. I. Inozemtsev, A. R. Its, E. Jeckelmann, G. Jüttner, N. Kawakami, R. M. Konik, E. H. Lieb, S. Lukyanov, M. J. Martins, S. Murakami, A. A. Nersesyan, K. Schoutens, H. Schulz, M. Shiroishi, F. Smirnov, J. Suzuki, M. Takahashi, M. Wadati, A. Weisse and J. Zittartz. Special thanks are due to Andreas Schadschneider for discussions and his constructive criticism after reading the entire manuscript. We are grateful to M. Bortz, A. Fledderjohann, M. Karbach, P. Boykens, A. Grage, M. Hartung, R. M. Konik and A. Seel for proofreading parts of the manuscript and helpful comments.

Despite the joint efforts of many dear friends we do not expect the first edition of such a thick volume to be free of misprints. We plan to keep a record of all misprints brought to our knowledge on our personal websites.

We thank the Physics Departments at Brookhaven National Laboratory and the Universities of Bayreuth, Dortmund, Hannover, Stony Brook, Warwick and Wuppertal for providing stimulating environments during the course of writing this book.

FHLE acknowledges support by the Department of Energy under contract DE-AC02-98 CH10886.

General bibliography

Books

R. J. Baxter, *Exactly Solved Models in Statistical Mechanics* (London: Academic Press, 1982).

J. L. Cardy, *Scaling and Renormalization in Statistical Physics* (Cambridge: Cambridge University Press, 1996).

P. Di Francesco, P. Mathieu and D. Sénéchal, *Conformal Field Theory* (New York: Springer Verlag, 1997).

E. Fradkin, *Field Theories of Condensed Matter Systems* (Reading, Mass.: Addison Wesley, 1991).

P. Fulde, *Electron Correlations in Molecules and Solids* (Berlin: Springer Verlag, 1991).

M. Gaudin, *La fonction de l'onde de Bethe pour les modèles exacts de la méchanique statistique* (Paris: Masson, 1983).

F. Gebhard, *The Mott Metal-Insulator Transition* (Berlin: Springer Verlag, 1997).

A. O. Gogolin, A. A. Nersesyan and A. M. Tsvelik, *Bosonization and Strongly Correlated Systems* (Cambridge: Cambridge University Press, 1998).

I. A. Iziumov and I. N. Skriabin, *Statistical Mechanics of Magnetically Ordered Systems* (New York: Consultants Bureau, 1988).

V. E. Korepin, N. M. Bogoliubov and A. G. Izergin, *Quantum Inverse Scattering Method and Correlation Functions* (Cambridge: Cambridge University Press, 1993).

S. Sachdev, *Quantum Phase Transitions* (Cambridge: Cambridge University Press, 2000).

F. Smirnov, *Form Factors in Completely Integrable Models of Quantum Field Theory* (Singapore: World Scientific, 1992).

B. Sutherland, *Beautiful Models: 70 Years of Exactly Solvable Quantum Many–Body Problems* (Singapore: Worlds Scientific, 2004).

M. Takahashi, *Thermodynamics of One-Dimensional Solvable Models* (Cambridge: Cambridge University Press, 1999).

A. M. Tsvelik, *Quantum Field Theory in Condensed Matter Physics* (Cambridge: Cambridge University Press, 1995; 2nd edition 2003).

K. Yosida, *Theory of Magnetism* (Berlin: Springer-Verlag, 1996).

Review articles

I. Affleck, Field theory methods and quantum critical phenomena, in E. Brézin and J. Zinn-Justin, eds., *champs, cordes et phénomène critique* (Amsterdam: North-Holland, 1990), 563–640. Les Houches, Session XLIX, 28 June–5 August 1988.

N. Andrei, Integrable models in condensed matter physics, preprint, cond-mat/9408101.

N. Andrei, K. Furuya and J. H. Lowenstein, Solution of the Kondo problem, *Rev. Mod. Phys.* **55** (1983) 331.

J. L. Cardy, Conformal invariance and statistical mechanics, in E. Brézin and J. Zinn-Justin, eds., *champs, cordes et phénomène critique* (Amsterdam: North-Holland, 1990), 169–245. Les Houches, Session XLIX, 28 June–5 August 1988.

T. Deguchi, F. H. L. Essler, F. Göhmann, A. Klümper, V. E. Korepin and K. Kusakabe, Thermodynamics and excitations of the one-dimensional Hubbard model, *Phys. Rep.* **331** (2000) 197.

P. Ginsparg, Applied conformal field theory, in E. Brézin and J. Zinn-Justin, eds., *Champs, cordes et phénomène critique* (Amsterdam: North-Holland, 1990), 1–168. Les Houches, Session XLIX, 28 June–5 August 1988.

P. P. Kulish and E. K. Sklyanin, Quantum spectral transform method – recent developments, in *Lecture Notes in Physics 151* (Berlin: Springer Verlag, 1982), 61–119.

P. P. Kulish and E. K. Sklyanin, Solutions of the Yang-Baxter equation, *J. Soviet Math.* **19** (1982) 1596.

E. H. Lieb, The Hubbard model – some rigorous results and open problems, in D. Iagolnitzer, ed., *Proceedings of the XIth International Congress of Mathematical Physics, Paris 1994* (Cambridge, Mass.: International Press, 1995), 392–412.

H. Saleur, Lectures on non-perturbative quantum field theories and quantum impurity problems, in A. Comtet, T. Jolicoeur, S. Ouvry and F. David, eds., *Aspects topologiques de la physique en basse dimension* (Berlin: Springer Verlag, 1999). Les Houches, Session LXIX, 7–31 July 1998, 473–550.

H. Saleur, Lectures on non-perturbative quantum field theories and quantum impurity problems: part II, preprint, cond-mat/0007309 (2000).

P. Schlottmann, Exact results for highly correlated electron systems in one dimension, *Int. J. Mod. Phys. B* **11** (1997) 355.

H. J. Schulz, Interacting fermions in one dimension: from weak to strong correlation, in V. J. Emery, ed., *Correlated Electron Systems, Vol. IX* (Singapore: World Scientific, 1993), 199–241. Jerusalem Winter School for Theoretical Physics, 30 Dec. 1991–8 Jan. 1992.

H. J. Schulz, G. Cuniberti and P. Pieri, Fermi liquids and Luttinger liquids, in G. Morandi *et al.* eds, *Field Theories for Low-Dimensional Condensed Matter Systems* (Berlin: Springer Verlag, 2000). Chia Laguna Summer School, September 1997, preprint cond-mat/9807366.

B. Sutherland, An introduction to the Bethe ansatz, in *Lecture Notes in Physics 242* (Berlin: Springer Verlag, 1985), 1–95.

H. Tasaki, The Hubbard model – an introduction and selected rigorous results, *J. Phys.: Condens. Matter* **10** (1998) 4353.

A. M. Tsvelick and P. B. Wiegmann, Exact results in the theory of magnetic alloys, *Adv. Phys.* **32** (1983) 453.

Reprint volumes

M. Jimbo, *Yang-Baxter Equation in Integrable Systems* (Singapore: World Scientific, 1989).

V. E. Korepin and F. H. L. Essler, *Exactly Solvable Models of Strongly Correlated Electrons*, vol. XVIII of Advanced Series in Mathematical Physics (Singapore: World Scientific, 1994).

A. Montorsi, *The Hubbard Model* (Singapore: World Scientific, 1992).

M. Rasetti, *The Hubbard Model, Recent Results* (Singapore: World Scientific, 1991).

Instead of a reading guide

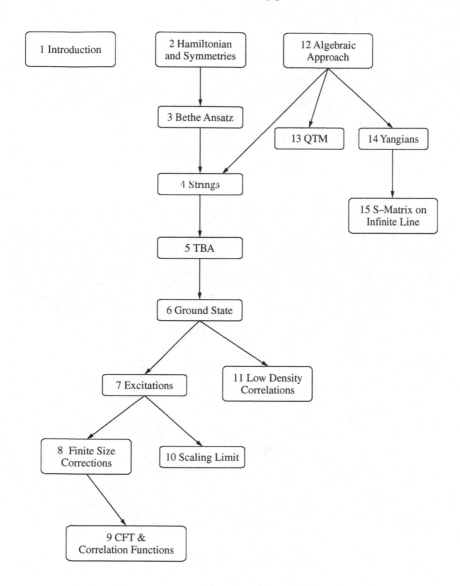

The figure shows the logical interdependence of the chapters and may serve the reader to find individual paths through this book. Chapters 16 and 17 have the character of appendices and are logically independent from the remaining part of the book.

1

Introduction

The purpose of this opening chapter is threefold: to introduce the Hubbard model, to discuss its origin and significance and to give a brief summary of its history. Rather than beginning with more general and historical considerations we will start with a concrete albeit somewhat technical discussion of how the Hubbard model arises as an effective description of electronic degrees of freedom in solids.

1.1 On the origin of the Hubbard model

The Hubbard model is named after John Hubbard, who in a series of influential articles [201–206] introduced[1] the Hamiltonian in order to model electronic correlations in narrow energy bands and proposed a number of approximate treatments of the associated many-body problem. Our following discussion of how the Hubbard Hamiltonian arises in an approximate description of interacting electrons in a solid loosely parallels Hubbard's original work. We will assume that the reader is familiar with the basic concepts of solid state theory (see e.g. [25, 509]) and with the formalism of second quantization (e.g. [283]). For further reading we refer to the original literature [188, 201, 233] and to the monographs [27, 158, 498].

A solid consists of ions and electrons condensed in a three-dimensional crystalline structure. Since the ions are much heavier than the electrons, it is often a good phenomenological starting point for the exploration of the electronic properties of solids to think of the ions as forming a *static* lattice.[2] In this approximation the dynamics of the electrons is governed by the Hamiltonian

$$H = \sum_{i=1}^{N} \left(\frac{\mathbf{p}_i^2}{2m} + V_I(\mathbf{x}_i) \right) + \sum_{1 \leq i < j \leq N} V_C(\mathbf{x}_i - \mathbf{x}_j), \qquad (1.1)$$

where N is the number of electrons, $V_I(\mathbf{x})$ is the periodic potential of the ions and

$$V_C(\mathbf{x}) = \frac{e^2}{\| \mathbf{x} \|} \qquad (1.2)$$

is the Coulomb repulsion among the electrons.

---

[1] The Hubbard model was independently introduced by Gutzwiller [188] and Kanamori [233] around the same time.
[2] This can be further justified within the Born-Oppenheimer approximation [68].

In spite of the drastic approximation we made by assuming a static lattice the Hamiltonian (1.1) is far too complicated to be solved exactly. It still bears all the difficulties of a generic many-body system. Much of the success of solid state theory derives from efficient 'mean-field' one-particle approximations to (1.1). On a technical level these approximations are based on adding an auxiliary potential $V_A(\mathbf{x})$ to the one-particle piece of the Hamiltonian (1.1) and then subtracting it again in the two-body part, i.e. we may write

$$H = \sum_{i=1}^{N} \left(\frac{\mathbf{p}_i^2}{2m} + V(\mathbf{x}_i) \right) + \sum_{1 \le i < j \le N} U(\mathbf{x}_i, \mathbf{x}_j), \tag{1.3}$$

where we introduced effective one- and two-body potentials $V(\mathbf{x})$ and $U(\mathbf{x}, \mathbf{y})$ as

$$V(\mathbf{x}) = V_I(\mathbf{x}) + V_A(\mathbf{x}), \tag{1.4a}$$

$$U(\mathbf{x}, \mathbf{y}) = V_C(\mathbf{x} - \mathbf{y}) - \tfrac{1}{N-1} \big(V_A(\mathbf{x}) + V_A(\mathbf{y}) \big). \tag{1.4b}$$

Mean-field approximations to H amount to simply setting $U(\mathbf{x}, \mathbf{y})$ equal to zero. In order for this to be sensible, the auxiliary potential needs to be chosen in such a way, that the matrix elements of the effective two-body potential $U(\mathbf{x}, \mathbf{y})$ between the eigenstates of the one-particle Hamiltonian

$$h_1(\mathbf{x}, \mathbf{p}) = \frac{\mathbf{p}^2}{2m} + V(\mathbf{x}) \tag{1.5}$$

become small. Even in circumstances when this cannot be achieved, the two-body interaction $U(\mathbf{x}, \mathbf{y})$ may still be considerably reduced in range and magnitude compared to the full Coulomb interaction $V_C(\mathbf{x} - \mathbf{y})$.

The physical idea behind the introduction of the auxiliary potential $V_A(\mathbf{x})$ may be formulated as follows. Let us assume we have a large number N of electrons in the ground state Ψ_0 of the Hamiltonian H. If we insert an additional electron locally into the system, what potential does it feel? Superimposed on the periodic potential of the ions it feels the electro-static potential which stems from the ground state density $|\Psi_0(\mathbf{x}_1, \ldots, \mathbf{x}_N)|^2$ of the other electrons. This potential is periodic with the same periods as the ionic potential. It is, however, of the opposite sign and therefore screens the attractive interaction of the ions. Of course, this picture is only approximately correct because the additional electron itself causes a change in the ground state density. Still, we may imagine that (again because of the screening) the effect of the additional electron is only local and therefore small.

We now wish to 'second-quantize' the Hamiltonian (1.3) in a suitable basis of states. In order to construct this basis we consider eigenstates of the one-particle Hamiltonian h_1. Since the one-body potential $V(\mathbf{x})$ in (1.4a) is periodic, the eigenfunctions of h_1 are Bloch functions (see e.g. [25, 509]), i.e., they are of the form

$$\varphi_{\alpha \mathbf{k}}(\mathbf{x}) = e^{i\mathbf{k} \cdot \mathbf{x}} u_{\alpha \mathbf{k}}(\mathbf{x}). \tag{1.6}$$

Here $u_{\alpha \mathbf{k}}(\mathbf{x})$ has the periodicity of the lattice, $\mathbf{k}$ is the quasi momentum and α the band index. The quasi momentum vector $\mathbf{k}$ runs over the first Brillouin zone. Being eigenfunctions of the one-particle Hamiltonian h_1,

$$h_1 \varphi_{\alpha\mathbf{k}}(\mathbf{x}) = \varepsilon_{\alpha\mathbf{k}} \varphi_{\alpha\mathbf{k}}(\mathbf{x}), \tag{1.7}$$

the functions $\varphi_{\alpha\mathbf{k}}(\mathbf{x})$ constitute a basis of one-particle states.

A complementary one-particle basis is provided by the Wannier functions [25, 509] $\phi_\alpha(\mathbf{x} - \mathbf{R}_i)$, where $\mathbf{R}_i$ is a lattice vector and $\phi_\alpha(\mathbf{x})$ is defined as

$$\phi_\alpha(\mathbf{x}) = \frac{1}{\sqrt{L}} \sum_\mathbf{k} \varphi_{\alpha\mathbf{k}}(\mathbf{x}). \tag{1.8}$$

Here L denotes the number of ions. The Wannier functions $\phi_\alpha(\mathbf{x} - \mathbf{R}_i)$ are centred around $\mathbf{R}_i$. They are lattice analogues of atomic wave functions and have the advantage of being mutually orthogonal for different band and site indices α and i. The Bloch functions are expressed in terms of the Wannier functions by means of Fourier inversion,

$$\varphi_{\alpha\mathbf{k}}(\mathbf{x}) = \frac{1}{\sqrt{L}} \sum_i e^{i\mathbf{k}\cdot\mathbf{R}_i} \phi_\alpha(\mathbf{x} - \mathbf{R}_i). \tag{1.9}$$

Let us introduce creation operators $c_{\alpha\mathbf{k},a}^\dagger$ of electrons of spin a in Bloch states $\varphi_{\alpha\mathbf{k}}(\mathbf{x})$. We further introduce their Fourier transforms

$$c_{\alpha i,a}^\dagger = \frac{1}{\sqrt{L}} \sum_\mathbf{k} e^{-i\mathbf{k}\cdot\mathbf{R}_i} c_{\alpha\mathbf{k},a}^\dagger. \tag{1.10}$$

Then, using (1.9), we may express the field operator, which creates an electron of spin a at position $\mathbf{x}$, in two different ways,

$$\Psi_a^\dagger(\mathbf{x}) = \sum_{\alpha\mathbf{k}} \varphi_{\alpha\mathbf{k}}^*(\mathbf{x}) c_{\alpha\mathbf{k},a}^\dagger = \sum_{\alpha i} \phi_\alpha^*(\mathbf{x} - \mathbf{R}_i) c_{\alpha i,a}^\dagger. \tag{1.11}$$

Here the asterisk denotes complex conjugation.

Finally, the general formula (see [283]) relating first and second quantized formalisms

$$H = \sum_{a=\uparrow,\downarrow} \int dx^3 \, \Psi_a^\dagger(\mathbf{x}) h_1 \Psi_a(\mathbf{x})$$
$$+ \frac{1}{2} \sum_{a,b=\uparrow,\downarrow} \int dx^3 dy^3 \, \Psi_a^\dagger(\mathbf{x}) \Psi_b^\dagger(\mathbf{y}) U(\mathbf{x},\mathbf{y}) \Psi_b(\mathbf{y}) \Psi_a(\mathbf{x}), \tag{1.12}$$

enables us to express the Hamiltonian (1.1) in second quantized form in the basis of Wannier states,

$$H = \sum_{\alpha,i,j,a} t_{ij}^\alpha c_{\alpha i,a}^\dagger c_{\alpha j,a} + \frac{1}{2} \sum_{\substack{\alpha,\beta,\gamma,\delta \\ i,j,k,l}} \sum_{a,b} U_{ijkl}^{\alpha\beta\gamma\delta} c_{\alpha i,a}^\dagger c_{\beta j,b}^\dagger c_{\gamma k,b} c_{\delta l,a}. \tag{1.13}$$

Here the hopping matrix elements t_{ij}^α are given by

$$t_{ij}^\alpha = \int dx^3 \, \phi_\alpha^*(\mathbf{x} - \mathbf{R}_i) h_1 \phi_\alpha(\mathbf{x} - \mathbf{R}_j) = \frac{1}{L} \sum_\mathbf{k} e^{i\mathbf{k}\cdot(\mathbf{R}_i - \mathbf{R}_j)} \varepsilon_{\alpha\mathbf{k}}. \tag{1.14}$$

Similarly, the interaction parameters $U_{ijkl}^{\alpha\beta\gamma\delta}$ are expressed as 'overlap integrals'

$$U_{ijkl}^{\alpha\beta\gamma\delta} = \int dx^3 dy^3 \; \phi_\alpha^*(\mathbf{x} - \mathbf{R}_i)\phi_\beta^*(\mathbf{y} - \mathbf{R}_j)U(\mathbf{x}, \mathbf{y})\phi_\gamma(\mathbf{y} - \mathbf{R}_k)\phi_\delta(\mathbf{x} - \mathbf{R}_l). \quad (1.15)$$

We note that H in equation (1.13) is still completely equivalent to the first quantized Hamiltonian (1.1). An optimal choice of the Wannier functions $\phi_\alpha(\mathbf{x})$ through an optimal choice of the auxiliary potential $V_A(\mathbf{x})$ minimizes the influence of the mutual Coulomb interaction. When the interaction parameters are small compared to the hopping matrix elements, they can be set equal to zero in a first approximation, and can later be taken into account by perturbation theory. This is the realm of band theory.

The Hubbard model is obtained from (1.13) when the interaction parameters are no longer negligible, but their range is still very small, i.e., when the intra-atomic Coulomb interaction $U_{iiii}^{\alpha\beta\gamma\delta}$ is large compared to the inter atomic interaction parameters and, at the same time, cannot be neglected compared to the hopping matrix elements. This situation is believed to be characteristic for transition and rare earth metals.

When the Fermi surface lies inside a single conduction band, say $\alpha = 1$, it is sometimes justified to 'project' the multi-band Hamiltonian onto an effective one-band model. Let us imagine a situation where the interband interactions are weak and at the same time all bands except the $\alpha = 1$ conduction band are far away from the Fermi level. As long as we are interested only in energies in the vicinity of the Fermi level, the main effect of the high energy bands is to change the hopping and interaction parameters of the electrons in the conduction band. Then we may replace the multi-band Hamiltonian (1.13) by a one-band model with effective parameters t_{ij} and U

$$H = \sum_{ij} t_{ij} c_{i,a}^\dagger c_{j,a} + \frac{U}{2} \sum_i c_{i,a}^\dagger c_{i,b}^\dagger c_{i,b} c_{i,a} \,. \qquad (1.16)$$

Whereas the hopping matrix elements can usually be determined accurately in the framework of density-functional theory (see e.g. [106]), the effective interaction parameter U is much more difficult to estimate and is perhaps best fixed by comparing theoretical predictions to experimental results.

It has to be said that the Hamiltonian (1.16) is not expected to describe the transition or rare earth metals quantitatively, since the interaction between overlapping bands is important in both cases. The Hamiltonian (1.16) is most appropriately regarded as an effective Hamiltonian that is believed to capture, at least qualitatively, some of the electronic features of the transition metals.

A further simplification of the Hamiltonian, which is compatible with the assumption that the Wannier functions $\phi_\alpha(\mathbf{x} - \mathbf{R}_i)$ are strongly localized around $\mathbf{R}_i$ is the tight-binding approximation, where one retains only hopping matrix elements between nearest neighbours. Then, upon introducing the particle number operators $n_{i\uparrow} = c_{i\uparrow}^\dagger c_{i\uparrow}$ and $n_{i\downarrow} = c_{i\downarrow}^\dagger c_{i\downarrow}$, the Hamiltonian (1.16) reduces to

$$H = -t \sum_{\langle i,j \rangle} c_{i,a}^{\dagger} c_{j,a} + U \sum_{i} n_{i\uparrow} n_{i\downarrow} . \tag{1.17}$$

Here the symbol $\langle i, j \rangle$ denotes summation over ordered pairs of nearest neighbours. We have assumed isotropic hopping of strength $-t$ between nearest neighbours and have suppressed the on-site terms t_{ii}, since they may be absorbed into the chemical potential in a grand canonical description of the model.

The one-dimensional version of the Hamiltonian (1.17) is easily identified with the Hubbard Hamiltonian (2.1) of Chapter 2 which is the actual starting point of this book. Its peculiar charm is certainly due to the fact that, in spite of its simplicity, it cannot be reduced to a one-particle theory and for this reason shows a rich spectrum of physical phenomena.

1.2 The Hubbard model – a paradigm in condensed matter physics

One of the most successful descriptions of electrons in solids is band theory. It is based on reducing many-body interactions to an effective one-body description, i.e., on neglecting the two-body potential $U(\mathbf{x}, \mathbf{y})$ in equation (1.4b) or equivalently the interaction parameters $U_{ijkl}^{\alpha\beta\gamma\delta}$ in equation (1.13). However, there are various situations of physical interest where band theory fails by construction. Arguably the most prominent example are Mott insulators: these have an odd number of valence electrons per elementary cell and yet are insulating, in contradiction with predictions of band theory.

One of the main motivations for studying the Hubbard model is that it is the simplest generalization beyond the band theory description of solids, yet still appears to capture the gross physical features of many systems characterized by more general interaction parameters in (1.13). The Hubbard model has been used in attempts to describe

 (i) the electronic properties of solids with narrow bands,
 (ii) band magnetism in iron, cobalt, nickel,
 (iii) the Mott metal-insulator transition,
 (iv) electronic properties of high-T_c cuprates in the normal state.

Despite its apparent simplicity, no fully consistent treatment of the Hubbard model is available in general. However, there are two cases in which one is more fortunate and many properties are calculable, namely the extremes of lattice coordination numbers two and infinity.[3] One might naively expect that the latter case can be easily understood by means of a mean-field approximation. Surprisingly, there is a particular way [325] of performing the limit of infinite lattice 'dimension' $D \to \infty$, in which the behaviour of the Hubbard model does not become mean-field like, but the model remains tractable. A striking result obtained in this approach is an understanding of the Mott transition between a paramagnetic metal and a correlated insulator. For details we refer the interested reader to the review article [163] and to the monograph [158].

---

[3] Notice that a few rigorous results [294, 452] also hold in the general case of arbitrary lattice dimension.

Here we are concerned with the first case, which corresponds to a one-dimensional lattice. The one-dimensional Hubbard model has a distinctive feature: it is 'integrable'. This essentially means that many physical properties can be determined exactly.[4] Integrable models are rather special and occur mostly in $D = 1$. Integrability is a fragile property: adding extra terms to an integrable Hamiltonian will in general break it. This fact is a frequent source of criticism. However, often the physics of a given problem is more robust than the mathematics: adding small perturbations to a given Hamiltonian does not necessarily lead to any dramatic changes in the physical properties. This point can be made more precise in the context of universality classes of critical behaviour (see e.g. [76]). By the same token the 1D Hubbard model may yet prove to be directly relevant for the description of experiments.

In our view, the central importance of integrable models rests with the fact that they constitute paradigms of diverse physical phenomena, which can be understood and characterized in their entirety. They allow us to study many-body physics beyond the restrictions of perturbation theory or intuitive non-systematic approximations. Their analysis permits us to develop an intuitive understanding of 'non-perturbative' effects. Last but not least, integrable models provide benchmarks for the development of approximate and numerical methods.

1.2.1 Integrable models

The history of exactly solvable many-body quantum systems traces back to H. Bethe's 1931 article [60] on the spin-$\frac{1}{2}$ Heisenberg chain in the early days of Quantum Theory. Bethe constructed the many-body wave functions and reduced the problem of calculating the spectrum of the Hamiltonian to solving a set of N coupled algebraic equations ('Bethe ansatz equations'), where N is the number of overturned spins. In this way a problem of exponential complexity is reduced to one of polynomial complexity. Bethe's work provides an explicit answer for the ground state properties and excitations of the ferromagnetic Heisenberg model. The energy per lattice site in the antiferromagnetic ground state was calculated by L. Hulthén in [207]. Hulthén recognized that in the thermodynamic limit the ground state can be characterized by the solution of a linear integral equation.

At the time Bethe's work was considered to be a fascinating but mostly academic exercise and it was hoped that it might serve as a stepping stone on the path to a solution of Heisenberg models on two- and three-dimensional lattices. Since then, Bethe's work has remained a constant source of inspiration for generations of researchers in theoretical and mathematical physics. It marked the beginning of a new branch of mathematical physics, the theory of exactly solvable quantum systems.

The next milestone was reached in 1944 [350] with L. Onsager's solution of the two-dimensional Ising model, which is based on an infinite-dimensional symmetry algebra ('Onsager Algebra') and a transfer matrix approach. Furthermore, the 'star-triangle relation', which played a crucial role in many subsequent developments, was mentioned for the first

---

[4] A precise definition of integrability is given in Chapter 12.

time. Onsager's work had a lasting impact on the microscopic foundations of the general theory of phase transitions. It established that singularities in the free energy can be obtained by a direct calculation of the partition function of a microscopic model. Furthermore, the solution demonstrated that critical exponents need not be mean-field like.

New applications of Bethe's ansatz were discovered during the 1960s, starting with the work [296] of E. H. Lieb and W. Liniger on the Bose gas with delta-function interactions. The extension of Bethe's ansatz to problems in statistical mechanics was achieved by E. H. Lieb in his solution of three archetypical cases of the six-vertex model (ice, KDP, F) [290–292]. The general case was solved shortly after by B. Sutherland [422] (for a review see [299]).

The generalization of Bethe's ansatz to models with internal degrees of freedom like spin proved to be very hard, because scattering involves changes of the internal states of the scatterers. This problem was eventually solved by C. N. Yang [493] and M. Gaudin [154] by means of what is nowadays called 'nested Bethe ansatz'. The condition for the applicability of the nested Bethe ansatz is the consistent factorization of multi-particle scattering processes into two-particle ones. Consistency requires the two-particle scattering matrices to fulfill certain algebraic equations, the 'Yang-Baxter Equations'.

In 1969 C. N. Yang and C. P. Yang [496] showed that Bethe's ansatz allows for the calculation of finite temperature properties of the delta-function Bose gas. This astonishing result is the first exact treatment of the thermodynamics of an interacting many-body quantum system.

Starting from the observation that the eigenstates of the transfer matrix of the six-vertex model are independent of one of the parameters, R. J. Baxter realized that there must be an entire family of commuting transfer matrices. He discovered a simple explanation of this remarkable fact by showing that it follows from ternary relations for the local Boltzmann weights [45]. These relations are a sufficient condition for the solvability of the model and are identical to the Yang-Baxter Equations obtained previously by Yang in his construction of the nested Bethe ansatz. Using his insights, Baxter realized the equivalence of the Yang-Baxter Equations to Onsager's star-triangle relations. He also discovered that the Boltzmann weights of an eight-vertex model satisfy the Yang-Baxter Equations, establishing solvability in the sense of the existence of a family of commuting transfer matrices [43,44]. Interestingly the model cannot be solved by Bethe's ansatz. However, Baxter managed to develop novel methods for the calculation of partition functions as well as one-point functions [45].

The role of the Yang-Baxter Equations as the defining structure of integrable models was emphasized by L. D. Faddeev, E. K. Sklyanin and L. Takhtajan and other members of the St Petersburg branch of the Stekhlov Mathematical Institute. They established a relation between quantum many-body models solved by Bethe's ansatz and classical integrable evolution equations [404,410,411]. Building on this connection, they initiated a systematic search for solutions of the Yang-Baxter Equations [274,276] and developed a programme for the solution of integrable models they called the 'Quantum Inverse Scattering Method'. An important element of this method is the algebraization of the construction of eigenstates of the transfer matrix [132,445]. The developments initiated by the Stekhlov group culminated in the advent of 'Quantum Groups' [107,225].

1.2.2 Bethe ansatz solution of the Hubbard model

The history of the one-dimensional Hubbard model as an exactly solvable model began in 1968 with E. H. Lieb and F. Y. Wu's article [298]. Lieb and Wu discovered that Bethe's ansatz can be applied to the Hubbard model and reduced the spectral problem of the Hamiltonian to solving a set of algebraic equations, nowadays known as the Lieb-Wu equations (see Chapter 3). They succeeded in calculating the ground state energy and demonstrated that the Hubbard model undergoes a Mott metal-insulator transition at half filling (one electron per site) with critical interaction strength $U = 0$ (Chapter 6).[5]

In the 35 years since Lieb and Wu's fundamental work appeared in print there have been hundreds of publications on the subject. It is clearly an impossible feat to do all of them justice within the confines of this short introduction. Hence we will constrain the following discussion to a small selection of works, which in our very personal and subjective view are of particular importance.

In 1972 M. Takahashi [435] proposed a classification of the solutions of the Lieb-Wu equations in terms of a 'string hypothesis' (see Chapter 4). He employed this hypothesis to replace the Lieb-Wu equations by simpler ones and then proceeded to derive a set of non-linear integral equations, which determine the Gibbs free energy of the Hubbard model (Chapter 5). These integral equations are known as thermodynamic Bethe ansatz (TBA) equations. Solving them in the limit of small temperatures Takahashi calculated the specific heat [436]. Later on a more complete picture of the thermodynamics of the Hubbard model was obtained from numerical solutions of the TBA equations [240, 469].

In fact, Takahashi's equations, in conjunction with the TBA equations, can be used to calculate any physical quantity that pertains to the energy spectrum of the Hubbard model. In particular, the dispersion curves of all elementary excitations can be obtained from the TBA equations in the limit $T \to 0$ [95]. Constraints on the quantum numbers in Takahashi's equations imply certain selection rules that determine the allowed combinations of elementary excitations and therefore the physical excitation spectrum [95]. Historically, the pioneering works in which ground state properties [298, 396, 429, 492] and the excitation spectrum [88, 258, 352, 481, 482, 485, 486]) were determined followed a different approach. In our view the thermodynamic Bethe ansatz is perhaps the most systematic approach for studying the ground state and the physical excitation spectrum of the one-dimensional Hubbard model and will serve as the basis of the corresponding Chapters 6 and 7 of this book.

Takahashi's equations may also serve as starting point for the calculation of the scattering matrix of the elementary excitations. For the half-filled Hubbard model in vanishing magnetic field the S-matrix was calculated in [120, 121]. It was shown that the excitation spectrum at half filling is given by scattering states of four elementary excitations: holon and antiholon with spin 0 and charge $\pm e$ and charge neutral spinons with spin up or down respectively. This is remarkable, since away from half filling, or at finite magnetic field, the number of elementary excitations is infinite [95]. It was further shown in [120, 121] that

---

[5] The ground state of the half-filled Hubbard model is metallic for $U = 0$ but Mott-insulating for all $U > 0$.

the four particles can only be excited in SO(4) multiplets (for the SO(4) symmetry of the model see Chapters 2 and 3).

A new chapter in the analysis of integrable models was opened with the advent of Conformal Field Theory [51]. In conformally invariant one-dimensional quantum systems the critical exponents governing the power-law decay of correlation functions are directly related to the energy levels in a large finite volume [6, 62, 74, 75]. Fortuitously in integrable theories the finite-size corrections to the energies of the ground state and low lying excited states can be determined from the Bethe ansatz [470]. The relations derived in [6, 62, 75] have been applied to numerous integrable models, including the attractive Hubbard model [65], which are conformally invariant in the low energy limit. Systematic studies of the finite size corrections in the spectrum of the Hubbard model were performed in the late eighties starting with work on the half-filled case [489]. The generic finite size spectrum away from half filling in finite magnetic fields was eventually obtained by F. Woynarovich [487]. In [140, 141] the aforementioned relation between the finite-size spectrum and the asymptotic behaviour of correlation functions was extended to models with several critical degrees of freedom and then utilized to calculate the critical exponents of general two-point correlation functions for the repulsive Hubbard model using Woynarovich's results (see Chapter 9). This was a breakthrough in the understanding of correlations in interacting one-dimensional quantum systems. The method has subsequently been applied to many other integrable models.

When considering a finite volume, an issue arises which we have hitherto ignored, namely the boundary conditions imposed on the system. In Lieb and Wu's work periodic boundary conditions were chosen. In 1985, the Hubbard model with reflecting ends was solved by H. Schulz [380] by means of a method introduced by M. Gaudin [155] for the δ-function Bose gas enclosed in a box and the open spin-$\frac{1}{2}$ Heisenberg chain.

The 'conformal approach' provides information on the large-distance/low-energy behaviour of correlation functions in the Hubbard model for band fillings strictly larger than zero and strictly less than one, for all positive values of U and low temperatures. In the Mott insulating phase at half filling the conformal approach is not applicable. However, in the small-U and scaling limits [324, 490, 491] methods of integrable quantum field theory can be employed to determine dynamical correlation functions at low energies [89, 105, 128, 129, 224] (Chapter 10). Another phase where the conformal approach is not applicable is the so-called gas phase characterized by sufficiently negative chemical potential and hence describing a correlated band insulator. The large-distance asymptotics of finite temperature correlation functions in the gas phase can be obtained exactly by a Bethe ansatz based approach [173, 176] (see Chapter 11). In the strong coupling limit $U \to \infty$ at zero temperature it is possible to obtain information about correlation functions at all energy scales by a combination of analytical and numerical techniques [153, 160, 344, 345, 353–358] for all band fillings.

The Bethe ansatz solution also supplies the coordinate wave functions of eigenstates of the Hamiltonian. An explicit representation for the wave functions was given by F. Woynarovich [481]. Only much later [122, 125] it was proven that the corresponding eigenstates are highest weight states with respect to the SO(4) symmetry [197, 497] of the Hamiltonian (Chapter 3). Combining this result with the implications of the string hypothesis it became possible to give a completeness argument in [123] (Chapter 4). Unfortunately the Bethe ansatz wave functions appear to be too complicated to allow for a direct calculation of correlation functions. Even an expression for the norms (Chapter 3) of the eigenstates has only been conjectured [175] and is still awaiting a proof.

In 1986 B. S. Shastry opened up a new way for studying the Hubbard model by placing it into the framework of the quantum inverse scattering method. Using a Jordan-Wigner transformation he mapped the Hubbard model to a spin model and then demonstrated that the resulting spin Hamiltonian commutes with the transfer matrix of a related covering vertex model [392]. In [391] Shastry first obtained the R-matrix of the spin model, thus embedding it into the general classification of 'integrable models' (see Chapter 12). Alternative derivations were obtained in [393] and in [348, 349, 475]. The latter references also include a formulation in terms of fermions that applies more directly to the Hubbard model.

It then took about ten years before Shastry's construction was really utilized. In [401] it was shown that Shastry's R-matrix satisfies the Yang-Baxter equation. An algebraic Bethe ansatz for the Hubbard model was constructed in [320, 371] and expressions for the eigenvalues of the transfer matrix of the two-dimensional statistical covering model were obtained (see also [499]). This result was of crucial importance for the quantum transfer matrix approach to the thermodynamics [232] of the Hubbard model (Chapter 13). This approach allows for a drastically simplified description of the thermodynamics in terms of the solution of a finite set of nonlinear integral equations, rather than the infinite set originally obtained by Takahashi in 1972 [435]. Within the quantum transfer matrix approach thermodynamic quantities can be calculated numerically with a very high precision. The approach can be extended to the calculation of correlation lengths at finite temperature [459, 465].

Another important algebraic result, which was unrelated to Shastry's work at first sight, was the discovery of a quantum group symmetry of the Hubbard model on the infinite line (see Chapter 14): the Hamiltonian is invariant under the direct sum of two Y(sl(2)) Yangians [462] (see also [172]). The relation of these Yangians to Shastry's R-matrix and the implications of one of these Yangians for the structure of the bare excitations (Chapter 15) were clarified in [335, 336].

1.2.3 The one-dimensional Hubbard model and experiments

The one-dimensional Hubbard model has been of immense conceptual value in facilitating the interpretation of experiments on quasi one-dimensional materials. Although it is not strictly a perfect model for any existing material, many of its qualitative features seem to be realized in nature. At present there is a sizeable list of materials, for which the electronic degrees of freedom are believed to be described by 'Hubbard-like' Hamiltonians. Examples are the chain cuprates Sr_2CuO_3 [150, 340] and $SrCuO_2$ [245, 246], organic conductors such as the Bechgaard salts [69, 386] or TTF-TCNQ [85] and π-conjugated polymers [35, 135] like polydiacetylene. However, in all these cases the appropriate electronic Hamiltonians differ significantly from a simple one-band Hubbard model. In the chain cuprates several bands of electrons need to be taken into account whereas organic conductors and polymers often have a tendency towards dimerization. Furthermore, it is usually not a good approximation to replace the Coulomb repulsion by a simple on-site Hubbard interaction. Nevertheless it is perhaps only a question of time until a material is discovered, which at least in some regime of temperatures and energies can be described in terms of a one-band Hubbard model.

1.3 External fields

In the presence of an external electro-magnetic field the Hubbard Hamiltonian has to be modified. Below we derive the modifications required in case of the one-band Hubbard model in the tight-binding approximation. By considering the reaction to a small external field we can also obtain a formula for the zero-temperature conductivity of the periodic Hubbard chain. This is presented in appendix 1.A.1.

1.3.1 External fields in three dimensions

Our starting point is electrons on a three-dimensional lattice. An external, time-dependent electro-magnetic field modifies the individual particle momenta and therefore affects only the one-particle part h_1 (1.5) of the Hamiltonian,

$$h_1(t) \longrightarrow \frac{1}{2m} \left(p^\alpha + \frac{e A^\alpha(\mathbf{x}, t)}{c} \right)^2 - e\, \Phi(\mathbf{x}, t) + V(\mathbf{x}) - 2\mu_B \mathbf{B} \cdot \mathbf{S}. \qquad (1.18)$$

Here A^α, $\alpha = x, y, z$ denote the components of the vector potential of the external field, Φ is its scalar potential, $\mathbf{B} = \text{rot}\, \mathbf{A}$ is the magnetic field, μ_B the Bohr magneton and $\mathbf{S}$ the spin operator. The magnetic field term in (1.18) acts only on the spin part of the wave function and therefore can be treated separately. In order to keep the following discussion simple we disregard it for the time being and restore it in the end. We are free to choose the gauge

$$\Phi(\mathbf{x}, t) = 0. \qquad (1.19)$$

Let $c_{j,a}^\dagger, c_{j,a}$ be creation and annihilation operators of electrons in Wannier states $\phi(\mathbf{x} - \mathbf{R}_j)$. We suppress the band index from the very beginning, since we are interested only in the one-band model. The modified one-particle Hamiltonian (1.18) leads to modified hopping matrix elements

$$
\begin{aligned}
t_{ij}(t) &= \int dx^3 \, \phi^*(\mathbf{x} - \mathbf{R}_i) \, h_1(t) \phi(\mathbf{x} - \mathbf{R}_j) \\
&= \int dx^3 \, \phi^*(\mathbf{x} - \mathbf{R}_i) \left[\frac{1}{2m} \left(p^\alpha + \frac{eA^\alpha}{c} \right)^2 + V(\mathbf{x}) \right] \phi(\mathbf{x} - \mathbf{R}_j) \\
&= \int dx^3 \, \phi^*(\mathbf{x} - \mathbf{R}_i) e^{-ie\lambda/c} \cdot \left[\frac{1}{2m} \left(p^\alpha + \frac{e(A^\alpha - \partial_\alpha \lambda)}{c} \right)^2 + V(\mathbf{x}) \right] e^{ie\lambda/c} \phi(\mathbf{x} - \mathbf{R}_j).
\end{aligned}
\tag{1.20}
$$

Here we used (1.14) and (1.19) in the first and second equality. The third equality holds trivially for any differentiable complex valued function $\lambda(\mathbf{x}, t)$. The particular choice

$$
\lambda(\mathbf{x}, t) = \int_{\mathbf{x}_0}^{\mathbf{x}} dy_\alpha \, A^\alpha(\mathbf{y}, t),
\tag{1.21}
$$

where $\mathbf{x}_0$ is an arbitrary fixed point, along with the definition

$$
\tilde{\phi}(\mathbf{x} - \mathbf{R}_j) = e^{ie\lambda(\mathbf{x}, t)/c} \phi(\mathbf{x} - \mathbf{R}_j)
\tag{1.22}
$$

transforms the hopping matrix elements into

$$
t_{ij}(t) = \int dx^3 \, \tilde{\phi}^*(\mathbf{x} - \mathbf{R}_i) \left[\frac{\mathbf{p}^2}{2m} + V(\mathbf{x}) \right] \tilde{\phi}(\mathbf{x} - \mathbf{R}_j).
\tag{1.23}
$$

Thus, the modified hopping matrix elements $t_{ij}(t)$ are of the same form as the hopping matrix elements t_{ij} in the absence of an external field. The field dependence has been absorbed into the redefinition (1.22) of the Wannier functions.

Let us now assume that the Wannier functions $\phi(\mathbf{x} - \mathbf{R}_j)$ are strongly localized around $\mathbf{R}_j$ and that the vector potential A^α is slowly varying on an atomic scale. Then the approximation

$$
\tilde{\phi}(\mathbf{x} - \mathbf{R}_j) = e^{ie\lambda(\mathbf{R}_j, t)/c} \phi(\mathbf{x} - \mathbf{R}_j)
\tag{1.24}
$$

is justified. Furthermore, the tight-binding approximation will be valid, and we may retain only hopping matrix elements between neighbouring lattice sites. As a result we arrive at the following modification of the Hamiltonian (1.17),

$$
H(t) = -t_0 \sum_{\langle i,j \rangle} e^{-ie(\lambda(\mathbf{R}_i, t) - \lambda(\mathbf{R}_j, t))/c} c_{i,a}^\dagger c_{j,a} + U \sum_i n_{i\uparrow} n_{i\downarrow}.
\tag{1.25}
$$

The phases in the first term on the right are called Peierls phases. Here we wrote t_0 instead of t for the hopping matrix element between neighbouring lattice sites in order to distinguish it from the time variable. We note that in equation (1.25) we have discarded the coupling of the electron spins to the magnetic field $\mathbf{B} = \mathrm{rot}\,\mathbf{A}$. It may be added at any later stage.

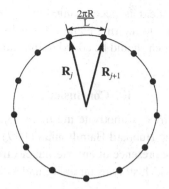

Fig. 1.1. The one-dimensional Hubbard model on a ring.

1.3.2 External fields in one dimension

Let us now specialize to the case where the system is a ring of radius R with L equidistantly spaced lattice sites (see figure 1.1). We may choose the centre of the ring as the origin of our coordinate system. In this one-dimensional geometry the Hamiltonian (1.25) turns into

$$H(t) = -t_0 \sum_{j=1}^{L} \left(e^{i\lambda_{j,j+1}} c_{j,a}^{\dagger} c_{j+1,a} + e^{-i\lambda_{j,j+1}} c_{j+1,a}^{\dagger} c_{j,a} \right) + U \sum_i n_{i\uparrow} n_{i\downarrow} , \qquad (1.26)$$

where $\lambda_{j,j+1}$ is the integral

$$\lambda_{j,j+1} = \frac{e}{c} \int_{\mathbf{R}_j}^{\mathbf{R}_{j+1}} dx_{\alpha} \, A^{\alpha}(\mathbf{x}, t) \qquad (1.27)$$

and periodic boundary conditions on the $\lambda_{j,j+1}$ are assumed.

In the appendix we shall consider the response to a spatially homogeneous electric field. For our one-dimensional periodic system this means that the electric field is of the form $\mathbf{E}(\mathbf{x}, t) = -E(\rho, t)\mathbf{e}_{\varphi}$, where $\mathbf{e}_{\varphi}$ is the unit vector along the ring and ρ is the distance of the point $\mathbf{x}$ from the axis perpendicular to the ring plane through the centre of the ring, say, the z-axis. Since in the gauge (1.19)

$$\mathbf{E}(\mathbf{x}, t) = -\frac{1}{c} \partial_t \mathbf{A}(\mathbf{x}, t) , \qquad (1.28)$$

we may take the vector potential to be of the form

$$\mathbf{A}(\mathbf{x}, t) = -A(\rho, t)\mathbf{e}_{\varphi} . \qquad (1.29)$$

It follows that all phases $\lambda_{j,j+1}$ in (1.27) are equal to

$$\lambda(t) = \frac{ea_0}{c} A(t) , \qquad (1.30)$$

where we introduced the shorthand notation $A(t) = A(R, t)$ and the lattice spacing $a_0 = \frac{2\pi R}{L}$.

We note that the electric field is accompanied by a homogeneous magnetic field $\mathbf{B} = \text{rot}\,\mathbf{A} = B\mathbf{e}_z$ perpendicular to the ring plane. This field couples to the electron spins through a term $-2\mu_B B S^z$, which should be added to the Hamiltonian (1.26).

1.4 Conclusions

In this introduction we have tried to motivate the main object of this book which is the one-dimensional version of the Hubbard Hamiltonian (1.17). We further showed how the Hamiltonian is modified in the presence of an external electro-magnetic field (see (1.26)), and we gave a short review of its history as far as we understand it.

Appendices to Chapter 1

1.A Response to external fields

As an important example of the response of the Hubbard model to an external perturbation we show in this appendix how an electro-magnetic field induces a current into the one-dimensional Hubbard model in the ring geometry of section (1.3). By consistently linearizing in the amplitude of the applied field and using the Born approximation for the time evolution operator in the interaction picture we obtain a linear relation between applied field and induced current. The Fourier transform of the kernel of the linear operator connecting field and current is the AC conductivity $\sigma(\omega)$.

In Chapter 10 we shall also consider the linear response to other types of perturbations. The general scheme of expressing measurable quantities like transport coefficients through correlation functions will always be similar to the example below. For further reading we recommend standard text books on many-body solid-state theory like e.g. [315].

1.A.1 The current operator

The operator

$$j_{m,\uparrow} = -it_0\left(e^{i\lambda(t)}c_{m,\uparrow}^\dagger c_{m+1,\uparrow} - e^{-i\lambda(t)}c_{m+1,\uparrow}^\dagger c_{m,\uparrow}\right) \tag{1.A.1}$$

can be interpreted as the operator of the mass current of up-spin electrons per lattice site. It satisfies the continuity equation

$$\dot{n}_{m,\uparrow} + j_{m,\uparrow} - j_{m-1,\uparrow} = 0, \tag{1.A.2}$$

where $\dot{n}_{m,\uparrow}$ is the operator for the change of particle density,

$$\dot{n}_{m,\uparrow} = i[H, n_{m,\uparrow}]. \tag{1.A.3}$$

Similar equations hold for down-spin electrons.

15

The mass current per lattice site is $j_m = j_{m,\uparrow} + j_{m,\downarrow}$. Summation over all site indices yields the current operator

$$J(t) = -it_0 \sum_{m=1}^{L} \left(e^{i\lambda(t)} c_{m,a}^{\dagger} c_{m+1,a} - e^{-i\lambda(t)} c_{m+1,a}^{\dagger} c_{m,a} \right). \tag{1.A.4}$$

The corresponding electric current is

$$J_{el}(t) = -ea_0 J(t). \tag{1.A.5}$$

Hence the electric current per unit volume[1] is given by

$$j_{el}(t) = -\frac{e}{a_0^2} J(t). \tag{1.A.6}$$

We are interested in the response of the Hubbard model to a small electric field. Thus, we shall assume that λ is a small quantity, and the effect of a small electric field is properly taken into account by retaining only the linear terms in the expansion of the Hamiltonian (1.26) and the current operator (1.A.4) in λ. We obtain

$$J(t) = J - \lambda(t)H_0, \tag{1.A.7}$$

$$H(t) = H + \lambda(t)J, \tag{1.A.8}$$

where J and H are the current operator (1.A.4) and the Hamiltonian (1.26) in zero external field $\lambda(t) = 0$, respectively, and H_0 is the Hubbard Hamiltonian for vanishing interaction $U = 0$ which later will be called the 'tight-binding Hamiltonian' (see equation (2.12)).

1.A.2 Linear response

In order to obtain an expression for the conductivity, we shall calculate the expectation value of the current $J(t)$ in a state, which develops from the ground state $|\psi_0\rangle$ of the unperturbed Hamiltonian H under the influence of the small perturbation $\lambda(t)$. We calculate this expectation value in first order time dependent perturbation theory and to linear order in $\lambda(t)$. To make sure that the system evolves from the ground state of the unperturbed Hamiltonian we have to require the perturbation to be switched off for $t \to -\infty$,

$$\lim_{t \to -\infty} \lambda(t) = 0. \tag{1.A.9}$$

The time evolution operator U of the unperturbed Hamiltonian H is the solution of the initial value problem

$$i\partial_t U = HU, \quad \lim_{t \to -\infty} U(t) = \mathrm{id}, \tag{1.A.10}$$

---

[1] The electric current is conventionally normalized to a 3d unit volume, even in 1d systems.

where id is the identity operator. For arbitrary states $|\psi\rangle$ and operators X let

$$|\psi_t\rangle = U^{-1}(t)|\psi\rangle\,, \tag{1.A.11}$$

$$X_t = U^{-1}(t)XU(t)\,. \tag{1.A.12}$$

Then for $|\psi\rangle$ a solution of the Schrödinger equation $i\partial_t|\psi\rangle = H(t)|\psi\rangle$ we find

$$i\partial_t|\psi_t\rangle = \lambda(t)J_t|\psi_t\rangle\,. \tag{1.A.13}$$

Here we have used (1.A.8) and (1.A.10). The time evolution operator $\hat{U}$ corresponding to the reduced Schrödinger equation (1.A.13) is determined by the initial value problem

$$i\partial_t\hat{U} = \lambda(t)J_t\hat{U}\,, \qquad \lim_{t\to-\infty}\hat{U}(t) = \mathrm{id}\,, \tag{1.A.14}$$

which may equivalently be written as an integral equation,

$$\hat{U}(t) = \mathrm{id} - i\int_{-\infty}^{t} dt'\,\lambda(t')J_{t'}\hat{U}(t')\,. \tag{1.A.15}$$

This equation can be solved by iteration. The solution to linear order, which is sometimes called the 'Born approximation', is

$$\hat{U}(t) = \mathrm{id} - i\int_{-\infty}^{t} dt'\,\lambda(t')J_{t'}\,. \tag{1.A.16}$$

The latter result will be sufficient for our purpose of calculating the linear response of the Hubbard model to a small perturbation.

First note that by (1.A.10), (1.A.12) and (1.A.14) the product $U\hat{U}$ is the time evolution operator of the time dependent Hamiltonian (1.A.8),

$$i\partial_t U\hat{U} = (H + \lambda(t)J)U\hat{U}\,, \qquad \lim_{t\to-\infty}U(t)\hat{U}(t) = \mathrm{id}\,. \tag{1.A.17}$$

It follows for a state $|\psi(t)\rangle$, evolving from the ground state $|\psi_0\rangle$ of the unperturbed Hamiltonian, that

$$\begin{aligned}
\langle\psi(t)|J(t)|\psi(t)\rangle &= \langle\psi_0|\hat{U}^{-1}J_t(t)\hat{U}|\psi_0\rangle \\
&= \langle\psi_0|J|\psi_0\rangle - \langle\psi_0|H_0|\psi_0\rangle\lambda(t) \\
&\quad + i\int_{-\infty}^{t} dt'\,\langle\psi_0|[J_{t'}, J_t]|\psi_0\rangle\lambda(t') + \mathcal{O}(\lambda^2)\,.
\end{aligned} \tag{1.A.18}$$

Here we have used (1.A.7) and (1.A.16). Next, it is necessary to expand the ground state mean value $\langle\psi_0|[J_{t'}, J_t]|\psi_0\rangle$ in terms of form factors by inserting a complete set of eigenstates $|\psi_n\rangle$ of the unperturbed Hamiltonian H. Let us denote its eigenvalues by E_n with $E_{n+1} \geq E_n$ and E_0 being the ground state energy. We further introduce the abbreviation $\omega_n = E_n - E_0$. Then

$$\langle\psi_0|[J_{t'}, J_t]|\psi_0\rangle = 2i\sum_{n>0}\sin(\omega_n(t - t'))|\langle\psi_n|J|\psi_0\rangle|^2\,. \tag{1.A.19}$$

Inserting (1.A.19) into (1.A.18) we obtain

$$\langle \psi(t)|J(t)|\psi(t)\rangle - \langle \psi_0|J|\psi_0\rangle$$

$$= -\langle \psi_0|H_0|\psi_0\rangle \lambda(t) - 2 \int_{-\infty}^{t} dt' \sum_{n>0} |\langle \psi_n|J|\psi_0\rangle|^2 \sin(\omega_n(t-t'))\lambda(t')$$

$$= -\langle \psi_0|H_0|\psi_0\rangle \lambda(t) - 2 \sum_{n>0} \frac{|\langle \psi_n|J|\psi_0\rangle|^2}{\omega_n} \lambda(t)$$

$$+ 2 \int_{-\infty}^{t} dt' \sum_{n>0} \frac{|\langle \psi_n|J|\psi_0\rangle|^2}{\omega_n} \cos(\omega_n(t-t')) \partial_{t'}\lambda(t')$$

$$= 2 \int_{-\infty}^{t} dt' \left[-\frac{1}{2}\langle \psi_0|H_0|\psi_0\rangle - \sum_{n>0} \frac{|\langle \psi_n|J|\psi_0\rangle|^2}{\omega_n} \right.$$

$$\left. + \sum_{n>0} \frac{|\langle \psi_n|J|\psi_0\rangle|^2}{\omega_n} \cos(\omega_n(t-t')) \right] \partial_{t'}\lambda(t') . \tag{1.A.20}$$

We finally observe that

$$\partial_t \lambda(t) = -ea_0 E(R,t), \tag{1.A.21}$$

where $E(R,t)$ is the electric field (see equation (1.28) and above). Thus, to linear order, the electric field and the electric current per unit volume (1.A.6) are related by the equation

$$\langle \psi(t)|j_{el}(t)|\psi(t)\rangle - \langle \psi_0|j_{el}|\psi_0\rangle = \int_{-\infty}^{\infty} dt' \, \sigma(t-t')E(R,t'), \tag{1.A.22}$$

where

$$\sigma(t) = \frac{2e^2}{a_0} \theta_H(t) \left[-\frac{1}{2}\langle \psi_0|H_0|\psi_0\rangle - \sum_{n>0} \frac{|\langle \psi_n|J|\psi_0\rangle|^2}{\omega_n} \right.$$

$$\left. + \sum_{n>0} \frac{|\langle \psi_n|J|\psi_0\rangle|^2}{\omega_n} \cos(\omega_n t) \right]. \tag{1.A.23}$$

The function $\theta_H(t)$ is the step function, being equal to one for $t > 0$ and equal to zero for $t < 0$. It is clear from the form of equation (1.A.22) that $\sigma(t)$ is the conductivity. Note that the causality postulate $\sigma(t) = \theta_H(t)\sigma(t)$ of classical electrodynamics comes out as a result of our quantum calculation (and the particular way of switching on the perturbation).

1.A.3 Optical conductivity, Drude weight and f-sum rule

Transport experiments measure the Fourier transform

$$\sigma(\omega) = \int_{-\infty}^{\infty} dt \, \sigma(t)e^{i\omega t} = \int_{0}^{\infty} dt \, \sigma(t)e^{i\omega t} \tag{1.A.24}$$

of the conductivity $\sigma(t)$. The function $\sigma(\omega)$ is called the optical conductivity. From the second equation (1.A.24) it is clear that $\sigma(\omega)$ exists in the upper complex half plane as an analytic function of ω. This fact is crucial for proving the Kramers-Kronig relation (see e.g. [220]). The optical conductivity can be continued to the real axis as a generalized function,

$$\sigma(\omega) = \frac{2e^2}{a_0} \Bigg[\left(-\frac{1}{2}\langle\psi_0|H_0|\psi_0\rangle - \sum_{n>0} \frac{|\langle\psi_n|J|\psi_0\rangle|^2}{\omega_n} \right) \frac{i}{\omega + i0}$$
$$+ \sum_{n>0} \frac{|\langle\psi_n|J|\psi_0\rangle|^2}{2\omega_n} \left(\frac{i}{\omega + \omega_n + i0} + \frac{i}{\omega - \omega_n + i0} \right) \Bigg] \quad (1.A.25)$$

for real ω.

By application of the Plemelj formula,

$$\frac{1}{\omega + i0} = -\pi i \delta(\omega) + \text{p.v.} \frac{1}{\omega}, \quad (1.A.26)$$

we may separate the optical conductivity into its real and imaginary parts,

$$\text{Re}(\sigma(\omega)) = 2\pi \frac{e^2}{a_0} \Bigg[\left(-\frac{1}{2}\langle\psi_0|H_0|\psi_0\rangle - \sum_{n>0} \frac{|\langle\psi_n|J|\psi_0\rangle|^2}{\omega_n} \right) \delta(\omega)$$
$$+ \sum_{n>0} \frac{|\langle\psi_n|J|\psi_0\rangle|^2}{2\omega_n} \left(\delta(\omega + \omega_n) + \delta(\omega - \omega_n) \right) \Bigg], \quad (1.A.27a)$$
$$\text{Im}(\sigma(\omega)) = \frac{2e^2}{a_0} \Bigg[\left(-\frac{1}{2}\langle\psi_0|H_0|\psi_0\rangle - \sum_{n>0} \frac{|\langle\psi_n|J|\psi_0\rangle|^2}{\omega_n} \right) \text{p.v.} \frac{1}{\omega}$$
$$+ \sum_{n>0} \frac{|\langle\psi_n|J|\psi_0\rangle|^2}{\omega_n} \text{p.v.} \frac{\omega}{\omega^2 - \omega_n^2} \Bigg]. \quad (1.A.27b)$$

The prefactor of $\delta(\omega)$ on the right hand side of (1.A.27a) is the so-called Drude weight,

$$D = -\frac{1}{2}\langle\psi_0|H_0|\psi_0\rangle - \sum_{n>0} \frac{|\langle\psi_n|J|\psi_0\rangle|^2}{\omega_n}. \quad (1.A.28)$$

Integrating (1.A.27a) over ω we obtain the well-known [36] f-sum rule

$$\int_{-\infty}^{\infty} d\omega \, \text{Re}(\sigma(\omega)) = -\pi \frac{e^2}{a_0} \langle\psi_0|H_0|\psi_0\rangle. \quad (1.A.29)$$

In the context of the one-dimensional Hubbard model (1.A.27a) and (1.A.28) were considered by Shastry and Sutherland and by Zvyagin [395, 510].

2

The Hubbard Hamiltonian and its symmetries

The Hubbard model is a model of itinerant, interacting electrons on a lattice. The structure and the dimension of the lattice influence its features. This book deals with the special case when the lattice is one-dimensional, since only then an exact solution is known. Only in one dimension we have the rare opportunity to get a deep and rigorous insight into the rich structure of an interacting many-body system.

In this chapter we work out the basic properties of the Hubbard Hamiltonian. We try to do this rather explicitly and in an elementary way that requires no preknowledge of solid state physics. Although we restrict ourselves from the very beginning to the one-dimensional model, the results presented here readily generalize to (bipartite) lattices of arbitrary dimension. The peculiarities of the one-dimensional model will be explored in the remaining part of the book.

2.1 The Hamiltonian

As with many other models in physics the term 'Hubbard model' is used somewhat freely in the literature. Many variants and generalizations of Hubbard's original Hamiltonian (1.17) have been considered over the years. In what follows, however, we shall be rather specific: by 'Hubbard model' we shall mean the one-dimensional one-band electronic model with nearest-neighbour hopping defined by the Hamiltonian [298]

$$H = -t \sum_{j=1}^{L} \sum_{a=\uparrow,\downarrow} (c_{j,a}^{\dagger} c_{j+1,a} + c_{j+1,a}^{\dagger} c_{j,a}) + U \sum_{j=1}^{L} n_{j\uparrow} n_{j\downarrow}. \tag{2.1}$$

Here $c_{j,a}^{\dagger}$ and $c_{j,a}$ are creation and annihilation operators of electrons of spin a ($a = \uparrow$ or $a = \downarrow$) localized in an orbital at site j of a one-dimensional lattice, and $n_{j,a} = c_{j,a}^{\dagger} c_{j,a}$. U and t are real numbers, which set the energy scale and fix the relative strength of the two sums that contribute to the Hamiltonian. We impose periodic boundary conditions on the operators, $c_{L+1,a} = c_{1,a}$. Due to this definition the Hamiltonian is invariant under cyclic permutations of the lattice sites, or, equivalently, under lattice translations on a ring of L sites. Different kinds of boundary conditions are discussed below in Chapter 8.3.

The operators $c_{j,a}^{\dagger}$ and $c_{j,a}$ are canonical Fermi operators. They satisfy the anticommutation relations

$$\{c_{j,a}, c_{k,b}\} = \{c_{j,a}^{\dagger}, c_{k,b}^{\dagger}\} = 0, \tag{2.2a}$$

$$\{c_{j,a}, c_{k,b}^{\dagger}\} = \delta_{jk}\delta_{ab} \tag{2.2b}$$

for $j, k = 1, \ldots, L$ and $a, b = \uparrow, \downarrow$. The creation operators $c_{j,a}^{\dagger}$ generate the space of states $\mathcal{H}^{(L)}$ of the Hubbard model by acting on the empty lattice (or 'vacuum state') $|0\rangle$ defined by the condition

$$c_{j,a}|0\rangle = 0, \quad j = 1, \ldots, L, \quad a = \uparrow, \downarrow. \tag{2.3}$$

Let us introduce row vectors of electron and spin coordinates, $\mathbf{x} = (x_1, \ldots, x_N)$ and $\mathbf{a} = (a_1, \ldots, a_N)$ with $x_j \in \{1, \ldots, L\}$ and $a_j = \uparrow, \downarrow$. The space of states of the Hubbard model is spanned by all linear combinations of the so-called Wannier states

$$|\mathbf{x}, \mathbf{a}\rangle = c_{x_N,a_N}^{\dagger} \ldots c_{x_1,a_1}^{\dagger} |0\rangle. \tag{2.4}$$

We fancy these states as states where electrons of spin a_j are located in atomic orbitals at lattice sites x_j, $j = 1, \ldots, N$. We say the sites x_j are occupied by electrons.

The number of linearly independent Wannier states is necessarily finite, since according to (2.2a), creation operators at different sites or with different spin indices anticommute, and $(c_{j,a}^{\dagger})^2 = 0$. A basis $\mathcal{B}$ of the space of states is obtained by ordering the Fermi operators in (2.4). We may choose for instance

$$\mathcal{B} = \left\{ |\mathbf{x}, \mathbf{a}\rangle \in \mathcal{H}^{(L)} \,\middle|\, \begin{array}{l} N = 0, \ldots, 2L \\ x_{j+1} \geq x_j, \; a_{j+1} > a_j \text{ if } x_{j+1} = x_j \end{array} \right\}, \tag{2.5}$$

where by convention $\uparrow < \downarrow$, and $N = 0$ corresponds to the vacuum state $|0\rangle$. The basis $\mathcal{B}$ is called the Wannier basis.

The number of all linearly independent vectors of the form (2.4) for a fixed number of particles N is equal to $\binom{2L}{N}$. Thus, the dimension of the space of states $\mathcal{H}^{(L)}$ is

$$\dim \mathcal{H}^{(L)} = \sum_{N=0}^{2L} \binom{2L}{N} = 4^L. \tag{2.6}$$

The same number follows more easily from the fact that the four states

$$|0\rangle, \quad c_{j,\uparrow}^{\dagger}|0\rangle, \quad c_{j,\downarrow}^{\dagger}|0\rangle, \quad c_{j,\uparrow}^{\dagger}c_{j,\downarrow}^{\dagger}|0\rangle \tag{2.7}$$

are associated with every lattice site. These states correspond to an empty site, a site occupied by one electron with spin up or down, or a doubly occupied site, respectively. Since $(c_{j,a}^{\dagger})^2 = 0$, electrons of the same spin cannot occupy the same lattice site. This is the Pauli principle which is built-in into the definition of the Fermi operators.

The operator $n_{j,a} = c_{j,a}^{\dagger}c_{j,a}$ is the local particle number operator for electrons of spin a at site j. Let us recall why this name is justified. Due to (2.2) and (2.3) it follows that

$$[n_{j,a}, c^\dagger_{k,b}] = \delta_{jk}\delta_{ab}c^\dagger_{k,b}, \quad n_{j,a}|0\rangle = 0, \tag{2.8}$$

and therefore

$$n_{j,a}|\mathbf{x}, \mathbf{a}\rangle = \sum_{k=1}^{N} \delta_{j,x_k}\delta_{a,a_k}|\mathbf{x}, \mathbf{a}\rangle. \tag{2.9}$$

Thus, $n_{j,a}|\mathbf{x}, \mathbf{a}\rangle = |\mathbf{x}, \mathbf{a}\rangle$, if site j is occupied by an electron of spin a, and zero otherwise.

A first interpretation of the Hubbard model can be obtained by considering separately the two contributions that make up the Hamiltonian (2.1). For $t = 0$ or $U = 0$ it can be diagonalized and understood by elementary means. For $t = 0$ the Hamiltonian reduces to $H = UD$, where

$$D = \sum_{j=1}^{L} n_{j\uparrow}n_{j\downarrow}. \tag{2.10}$$

Using (2.9) we can calculate the action of D on a state $|\mathbf{x}, \mathbf{a}\rangle$,

$$
\begin{aligned}
D|\mathbf{x}, \mathbf{a}\rangle &= \sum_{k,l=1}^{N} \delta_{x_k,x_l}\delta_{\uparrow,a_k}\delta_{\downarrow,a_l}|\mathbf{x}, \mathbf{a}\rangle \\
&= \sum_{1\le k<l\le N} \delta_{x_k,x_l}(\delta_{\uparrow,a_k}\delta_{\downarrow,a_l} + \delta_{\downarrow,a_k}\delta_{\uparrow,a_l})|\mathbf{x}, \mathbf{a}\rangle \\
&= \sum_{1\le k<l\le N} \delta_{x_k,x_l}(\delta_{\uparrow,a_k} + \delta_{\downarrow,a_k})(\delta_{\uparrow,a_l} + \delta_{\downarrow,a_l})|\mathbf{x}, \mathbf{a}\rangle \\
&= \sum_{1\le k<l\le N} \delta_{x_k,x_l}|\mathbf{x}, \mathbf{a}\rangle.
\end{aligned}
\tag{2.11}
$$

Here we used $\delta_{\uparrow,a_k}\delta_{\downarrow,a_k} = 0$ in the second equation and the Pauli principle in the third equation. As we learn from (2.11) every state $|\mathbf{x}, \mathbf{a}\rangle$ is an eigenstate of the operator D. Thus, D is diagonal in the Wannier basis. The limit $t \to 0$ of the Hubbard Hamiltonian (2.1) is called the atomic limit, because the eigenstate $|\mathbf{x}, \mathbf{a}\rangle$ describes electrons localized at the sites $x_1, \ldots, x_N$, which are identified with the loci of the atomic orbitals the electrons may occupy.

The meaning of the operator D is evident from equation (2.11). D counts the number of doubly occupied sites in the state $|\mathbf{x}, \mathbf{a}\rangle$. The contribution of the term UD to the energy is non-negative for positive U and increases with the number of doubly occupied sites. This can be viewed as on-site repulsion among the electrons. Negative U on the other hand, means on-site attraction. Hence, it is natural to refer to D as to the operator of the on-site interaction.

In the other extreme, when $U = 0$, the Hamiltonian (2.1) turns into

$$H_0 = -t \sum_{j=1}^{L} \sum_{a=\uparrow,\downarrow} (c^\dagger_{j,a}c_{j+1,a} + c^\dagger_{j+1,a}c_{j,a}). \tag{2.12}$$

This is called the tight-binding Hamiltonian. Like every translationally invariant one-body Hamiltonian it can be diagonalized by discrete Fourier transformation. Let us define

$$\tilde{c}^\dagger_{k,a} = \frac{1}{\sqrt{L}} \sum_{j=1}^{L} e^{i\phi kj}\, c^\dagger_{j,a}\,, \quad k = 0, \ldots, L-1, \tag{2.13}$$

where $\phi = 2\pi/L$. Then, by Fourier inversion

$$c^\dagger_{j,a} = \frac{1}{\sqrt{L}} \sum_{k=0}^{L-1} e^{-i\phi jk}\, \tilde{c}^\dagger_{k,a}\,, \quad j = 1, \ldots, L. \tag{2.14}$$

Equation (2.14) is readily verified by inserting (2.13) into the right hand side and using the geometric sum formula. Clearly, $\tilde{c}^\dagger_{k+L,a} = \tilde{c}^\dagger_{k,a}$. Insertion of (2.14) into (2.12) leads to

$$H_0 = -2t \sum_{k=0}^{L-1} \sum_{a=\uparrow,\downarrow} \cos(\phi k)\tilde{n}_{k,a}\,, \tag{2.15}$$

where $\tilde{n}_{k,a} = \tilde{c}^\dagger_{k,a}\tilde{c}_{k,a}$.

The Fourier transformation leaves the canonical anticommutation relations (2.2) invariant,

$$\{\tilde{c}_{j,a}, \tilde{c}_{k,b}\} = \{\tilde{c}^\dagger_{j,a}, \tilde{c}^\dagger_{k,b}\} = 0\,, \tag{2.16a}$$

$$\{\tilde{c}_{j,a}, \tilde{c}^\dagger_{k,b}\} = \delta_{jk}\delta_{ab}\,. \tag{2.16b}$$

A transformation with this property is called canonical. Applying (2.13) to the empty lattice state $|0\rangle$, we obtain the analogue of (2.3),

$$\tilde{c}_{k,a}|0\rangle = 0\,, \quad k = 0, \ldots, L-1, \quad a = \uparrow, \downarrow\,. \tag{2.17}$$

Thus, acting with the creation operators $\tilde{c}^\dagger_k$ on the empty lattice $|0\rangle$ we obtain an alternative basis $\tilde{\mathcal{B}}$. Let us introduce the row vectors $\mathbf{q} = (q_1, \ldots, q_N) = \phi(k_1, \ldots, k_N)$ and the states

$$|\mathbf{q}, \mathbf{a}\rangle = \tilde{c}^\dagger_{k_N,a_N} \ldots \tilde{c}^\dagger_{k_1,a_1}|0\rangle\,. \tag{2.18}$$

These states are eigenstates of the lattice momentum operator with eigenvalue $\left(\sum_{j=1}^{N} q_j\right) \bmod 2\pi$, as we shall see in the following section on symmetries. The set

$$\tilde{\mathcal{B}} = \left\{ |\mathbf{q}, \mathbf{a}\rangle \in \mathcal{H}^{(L)} \,\middle|\, \begin{array}{l} N = 0, \ldots, 2L \\ q_{j+1} \geq q_j,\ a_{j+1} > a_j\ \text{if}\ q_{j+1} = q_j \end{array} \right\} \tag{2.19}$$

is a basis of $\mathcal{H}^{(L)}$. This basis is sometimes called the Bloch basis. Electrons in Bloch states $|\mathbf{q}, \mathbf{a}\rangle$ are delocalized, but have definite momenta $q_1, \ldots, q_N$.

By virtue of (2.16), the analogues of (2.8) and (2.9) are satisfied by $\tilde{n}_{j,a}$ and $\tilde{c}^\dagger_{k,b}$. It follows that

$$H_0|\mathbf{q}, \mathbf{a}\rangle = -2t \sum_{j=1}^{N} \cos(q_j)|\mathbf{q}, \mathbf{a}\rangle\,. \tag{2.20}$$

Thus, the tight-binding Hamiltonian H_0 is diagonal in the Bloch basis. It describes non-interacting band electrons in a cosine-shaped band of width $4t$.

The tight-binding Hamiltonian H_0 and the operator D which counts the number of doubly occupied sites do not commute. Therefore the Hubbard Hamiltonian can neither be diagonal in the Bloch basis nor in the Wannier basis. The construction of its eigenstates for general t and U will be the subject of the next chapter. The physics of the Hubbard model may be understood as arising from the competition between the two contributions, H_0 and D, to the Hamiltonian (2.1). The tight-binding contribution H_0 prefers to delocalize the electrons, while the on-site interaction D favours localization. The ratio

$$u = \frac{U}{4t} \tag{2.21}$$

is a measure for the relative contribution of both terms and is the intrinsic, dimensionless coupling constant of the Hubbard model.

For the purpose of this book it is natural to measure energies in units of t. This is equivalent to setting $t = 1$. Then the Hamiltonian (2.1) turns into

$$H = -\sum_{j=1}^{L}\sum_{a=\uparrow,\downarrow}(c_{j,a}^{\dagger}c_{j+1,a} + c_{j+1,a}^{\dagger}c_{j,a}) + 4u\sum_{j=1}^{L}n_{j\uparrow}n_{j\downarrow}. \tag{2.22}$$

Later on we shall also discuss the influence of an external magnetic field B coupled to the spins of the electrons and of a chemical potential μ. Then the Hamiltonian has to be modified as

$$H_{\mu,B} = H - \mu\hat{N} - 2BS^z, \tag{2.23}$$

where we have introduced the particle number operator

$$\hat{N} = \sum_{j=1}^{L}(n_{j,\uparrow} + n_{j,\downarrow}) \tag{2.24}$$

and the operator

$$S^z = \tfrac{1}{2}\sum_{j=1}^{L}(n_{j,\uparrow} - n_{j,\downarrow}) \tag{2.25}$$

of the z-component of the total spin. H and $H_{\mu,B}$ have the same set of eigenstates since the particle number and the z-component of the total spin are conserved,

$$[H, \hat{N}] = [H, S^z] = 0 \tag{2.26}$$

and since $[\hat{N}, S^z] = 0$.

This can be seen as follows: let us introduce the particle number operators for up- and down-spin electrons, respectively,

$$\hat{N}_a = \sum_{j=1}^{L}n_{j,a}, \quad a = \uparrow,\downarrow. \tag{2.27}$$

Summation of equation (2.8) over j yields

$$[\hat{N}_a, c_{k,b}^\dagger] = \delta_{ab} c_{k,b}^\dagger, \qquad [\hat{N}_a, c_{k,b}] = -\delta_{ab} c_{k,b}. \qquad (2.28)$$

Here the second equation is the Hermitian conjugate of the first one. It follows that

$$[\hat{N}_a, c_{j,b}^\dagger c_{k,b}] = [\hat{N}_a, c_{j,b}^\dagger] c_{k,b} + c_{j,b}^\dagger [\hat{N}_a, c_{k,b}] = 0. \qquad (2.29)$$

From the latter equation we conclude that

$$[H, \hat{N}_a] = 0, \qquad a = \uparrow, \downarrow. \qquad (2.30)$$

The numbers of up- and down-spin electrons are separately conserved. Now $\hat{N} = \hat{N}_\uparrow + \hat{N}_\downarrow$, $S^z = \frac{1}{2}(\hat{N}_\uparrow - \hat{N}_\downarrow)$, and we obtain (2.26).

We shall denote the conserved number of down spin electrons by M and the conserved total number of electrons by N. The value of the z-component of the total spin for a state with M down-spin electrons is $N/2 - M$. In Chapter 3 we shall diagonalize H for fixed values of M and N.

Because of the particle number conservation, we may add a term $-2u\hat{N} + uL$ to the Hamiltonian (2.22) without affecting its eigenfunctions. The resulting expression

$$H = -\sum_{j=1}^{L} \sum_{a=\uparrow,\downarrow} (c_{j,a}^\dagger c_{j+1,a} + c_{j+1,a}^\dagger c_{j,a}) + u \sum_{j=1}^{L} (1 - 2n_{j\uparrow})(1 - 2n_{j\downarrow}) \qquad (2.31)$$

will turn out to be particularly convenient for our further discussion. As we shall see below (2.31) is of higher symmetry than (2.22), if L is even.

2.2 Symmetries

The one-dimensional Hubbard model has many symmetries. Some of them, like the translational symmetry or the symmetry under spin flips, are obvious and common. Apart from the obvious symmetries, however, there are others which are rather unusual.

We distinguish symmetries which are independent of the coupling constant u, from symmetries which, on the contrary, depend on u. The existence of the latter type of symmetries relates to the fact that the one-dimensional Hubbard model is exactly solvable. This symmetry type includes an Abelian symmetry generated by a series of mutually commuting higher conserved operators and the non-Abelian so-called Yangian symmetry (see Chapters 12 and 14).

Here we concentrate on the u-independent symmetries. In the context of the one-dimensional Hubbard model they were systematically studied by Heilmann and Lieb [197] as early as 1971. Later, in a period of renewed interest in the Hubbard model it was observed [361, 494, 497] that much of the analysis of Heilmann and Lieb carries over to an arbitrary dimension of the lattice. The symmetries considered by Heilmann and Lieb are of three kinds: spatial symmetries related to the lattice, symmetries connected to the spin and symmetries associated with special features of the Hubbard model.

2.2.1 Permutations

Since the Hubbard model is defined on a lattice, the set of all possible spatial transformations is equal to the set of all permutations of site indices. These permutations form the symmetric group $\mathfrak{S}^L$. Our first goal is therefore to construct a faithful representation of the symmetric group in terms of Fermi operators. For this purpose it is sufficient to construct representations of the elementary transpositions, which generate the symmetric group.

Let us start with spinless fermions, $\{c_j, c_k\} = \{c_j^\dagger, c_k^\dagger\} = 0$, $\{c_j, c_k^\dagger\} = \delta_{jk}$, on a one-dimensional lattice of L sites. Let

$$P_{ij} = 1 - (c_i^\dagger - c_j^\dagger)(c_i - c_j). \tag{2.32}$$

It is not difficult to see that P_{ij} permutes fermions. We have the obvious identities

$$P_{ij} = P_{ij}^\dagger, \quad P_{ij} = P_{ji}. \tag{2.33}$$

Use of the fundamental anticommutators for the fermions leads to

$$P_{ij}c_i = c_i + (c_i^\dagger - c_j^\dagger)c_j(c_i - c_j) = c_i + (-1 + c_j c_j^\dagger - c_j c_i^\dagger)(c_i - c_j) = c_j P_{ij}. \tag{2.34}$$

It then follows from (2.33) that

$$P_{ij}c_j = c_i P_{ij}, \quad P_{ij}c_i^\dagger = c_j^\dagger P_{ij}, \quad P_{ij}c_j^\dagger = c_i^\dagger P_{ij}. \tag{2.35}$$

Thus, the operators P_{ij} induce the action of transpositions on the site indices of the Fermi operators.

Let us show that the P_{ij} generate a representation of the symmetric group. First of all, we obtain from (2.34) and (2.35) that

$$P_{ij}P_{jk} = P_{ik}P_{ij} = P_{jk}P_{ik}, \quad i \neq j \neq k \neq i. \tag{2.36}$$

A short calculation similar to the one in equation (2.34) shows that

$$P_{ij}P_{ij} = 1. \tag{2.37}$$

Finally, we have the obvious identity

$$[P_{ij}, P_{kl}] = 0 \tag{2.38}$$

for $i, j \neq k, l$. The relations (2.36)–(2.38) are a possible choice of defining relations for the symmetric group.

Formally, the site indices i, j, k in the preceding calculations are just labels. We may replace them by more complicated labels without spoiling the validity of our results. The replacement $j \to ja$, where j is a site index and $a = \uparrow, \downarrow$ is a spin index leads to

$$P_{ia,jb} = 1 - (c_{i,a}^\dagger - c_{j,b}^\dagger)(c_{i,a} - c_{j,b}). \tag{2.39}$$

These transposition operators were introduced by Heilmann and Lieb [197] in their analysis of the symmetries of the Hubbard model.[1] They describe simultaneous transpositions of site and spin indices, or, in more physical terms, they interchange electrons in Wannier orbitals.

---

[1] Unfortunately, there is a typo in their definition, equation (8) of their paper.

2.2.2 Spatial symmetries

We may imagine the L Wannier orbitals of the one-dimensional Hubbard model as forming a regular polygon with L edges and corners. The spatial symmetries of the Hubbard model are then the symmetries of this polygon. They are generated by a rotation through $2\pi/L$ and by an arbitrary reflection which maps the polygon onto itself. The corresponding symmetry operators are the shift operator and the parity operator.

The shift operator is a representation of the generator of the cyclic subgroup of order L of the symmetric group [197]. For spinless fermions we define

$$\hat{U}_n = P_{n-1n} \dots P_{23} P_{12}, \quad n = 2, \dots, L.$$ (2.40)

Using equation (2.34) we can readily verify that

$$\hat{U}_L c_j = \begin{cases} c_{j-1} \hat{U}_L & j - 2, \dots, L \\ c_L \hat{U}_L & j = 1. \end{cases}, \quad \text{if}$$ (2.41)

This means that $\hat{U}_L$ is acting as a left shift operator on the elementary Fermi operators. Now, (2.40) implies that

$$\hat{U}_L^\dagger = P_{12} \dots P_{L-1L}.$$ (2.42)

It follows from equation (2.37) that $\hat{U}_L$ is unitary, $\hat{U}_L \hat{U}_L^\dagger = \hat{U}_L^\dagger \hat{U}_L = 1$. The operator $\hat{U}_L^\dagger$ generates a shift to the right by one lattice site.

To realize the shift operator for electrons, we have to attach a spin label to the above operators. We define shift operators $U_{L\uparrow}$ and $U_{L\downarrow}$ for up- and down-spin electrons by replacing $P_{j,j+1}$ in equation (2.40) with $P_{j\uparrow,j+1\uparrow}$ and $P_{j\downarrow,j+1\downarrow}$, respectively. We observe that $[\hat{U}_{L\uparrow}, c_{j\downarrow}] = [\hat{U}_{L\downarrow}, c_{j\uparrow}] = 0$. Thus,

$$\hat{U} = \hat{U}_{L\uparrow} \hat{U}_{L\downarrow}$$ (2.43)

is the left shift operator for electrons.

Using the definition (2.13) of $\tilde{c}_{k,a}$ we obtain

$$\hat{U} \tilde{c}_{k,a}^\dagger = e^{i\phi k} \tilde{c}_{k,a}^\dagger \hat{U}.$$ (2.44)

It follows that $\hat{U}$ acts diagonally on the basis $\tilde{B}$ of Bloch states, equation (2.19),

$$\hat{U} |\mathbf{q}, \mathbf{a}\rangle = e^{i\phi(k_1 + \dots + k_N)} |\mathbf{q}, \mathbf{a}\rangle.$$ (2.45)

Obviously, the Hubbard Hamiltonian (2.31) is invariant under a change of site indices of the Fermi operators from j to $L - j + 1$. The corresponding parity operator R_L can be conveniently expressed in terms of the operators $\hat{U}_n$, equation (2.40), as the ordered product

$$R_L = \hat{U}_2 \dots \hat{U}_L.$$ (2.46)

Equivalently, R_L may be written as

$$R_L = \prod_{j=1}^{[L/2]} P_{j,L-j+1},$$ (2.47)

where $[L/2]$ denotes the integer part of $L/2$, which is $L/2$ for even L and $(L-1)/2$ for odd L. From the form (2.47) of the parity operator and from (2.37) it is clear that $R_L^2 = \mathrm{id}$ and $R_L = R_L^\dagger$. Thus, R_L is unitary and Hermitian. We may again define a parity operator for electrons $R = R_{L\uparrow} R_{L\downarrow}$ after attaching spin labels $\uparrow$ and $\downarrow$ to the operators in (2.47).

2.2.3 The momentum operator

Observables in quantum mechanics are described by Hermitian operators. For this reason we would like to define a Hermitian momentum operator that generates the shifts on the lattice. Generically, the momentum operator is defined as the generator of infinitesimal spatial shifts. This definition, however, does not work on a lattice. Alternatively, following [178], we may try to define a lattice momentum operator $\hat{\Pi}$ by the following three requirements.

$$e^{i\hat{\Pi}} = \hat{U}, \qquad (2.48a)$$

$$\hat{\Pi} = \hat{\Pi}^\dagger, \qquad (2.48b)$$

$$[H, \hat{\Pi}] = 0. \qquad (2.48c)$$

Note that the choice

$$\Pi = \phi \sum_{k=1}^{L-1} k\, \tilde{c}_k^\dagger \tilde{c}_k, \qquad (2.49)$$

which is often found in the literature and acts diagonally on the basis $\widetilde{B}$ of Bloch states, does not satisfy (2.48c). It is *not* a conserved quantity for the Hubbard model. A way out of this dilemma comes from the fact that the condition (2.48a) fixes the momentum only modulo 2π.

For $\alpha \in \mathbb{C}$ let us define

$$g(\alpha) := \sum_{k=0}^{L-1} \mathrm{i}e^{-\mathrm{i}\phi k\alpha} = \mathrm{i}\,\frac{1 - e^{-\mathrm{i}\phi L\alpha}}{1 - e^{-\mathrm{i}\phi\alpha}}. \qquad (2.50)$$

Here we applied the geometric sum formula to get the second equation. It follows from (2.50) that

$$g'(m) = \sum_{k=1}^{L-1} \phi k e^{-\mathrm{i}\phi km}, \qquad m = 1, \ldots, L. \qquad (2.51)$$

Thus, by Fourier inversion,

$$\phi k = \frac{1}{L} \sum_{m=1}^{L} g'(m) e^{\mathrm{i}\phi km}, \qquad k = 0, \ldots, L-1. \qquad (2.52)$$

For $m = 1, \ldots, L-1$ the coefficients $g'(m)$ are obtained by differentiating the right hand side of (2.50). We find $g'(m) = \phi L/(e^{-\mathrm{i}\phi m} - 1)$. We further read off from (2.51) that

$g'(L) = \phi L(L-1)/2$. Thus,

$$\phi k = \phi \sum_{m=1}^{L-1} \left(\frac{1}{2} + \frac{e^{i\phi km}}{e^{-i\phi m} - 1} \right), \quad k = 0, \dots, L-1. \tag{2.53}$$

The right hand side of this equation is a Fourier sum which periodically extends to all integers $k \in \mathbb{Z}$. Setting $x = \phi k$ we see that it defines the 'saw tooth function' $f(x) = x \bmod 2\pi$ on the set $\phi \mathbb{Z}$.

By hypothesis the spectrum of the momentum operator $\hat{\Pi}$ we are looking for is contained in $\phi \mathbb{Z}$ (see condition (2.48a)). Let us assume we are given a momentum operator which satisfies (2.48). Then, substituting $\hat{\Pi}$ for ϕk in (2.53) leads to a restriction of $\hat{\Pi}$ modulo 2π. Because of our condition (2.48a), we obtain

$$\hat{\Pi} = \phi \sum_{m=1}^{L-1} \left(\frac{1}{2} + \frac{\hat{U}^m}{e^{-i\phi m} - 1} \right). \tag{2.54}$$

Thus, no matter what the actual form of the operator $\hat{\Pi}$ is, its restriction modulo 2π leads to something known: the right hand side of (2.54) is a polynomial in the shift operator $\hat{U}$.

We may therefore take (2.54) as the definition of the momentum operator. Then we must verify that $\hat{\Pi}$ defined in this way indeed satisfies (2.48). First of all, using (2.45) and (2.53), we obtain

$$\hat{\Pi} |\mathbf{q}, \mathbf{a}\rangle = \phi \sum_{m=1}^{L-1} \left(\frac{1}{2} + \frac{e^{i\phi(k_1 + \dots + k_N)m}}{e^{-i\phi m} - 1} \right) |\mathbf{q}, \mathbf{a}\rangle = (\phi(k_1 + \dots + k_N) \bmod 2\pi) |\mathbf{q}, \mathbf{a}\rangle. \tag{2.55}$$

Using (2.45) and (2.55) we conclude that

$$e^{i\hat{\Pi}} |\mathbf{q}, \mathbf{a}\rangle = e^{i\phi(k_1 + \dots + k_N)} |\mathbf{q}, \mathbf{a}\rangle = \hat{U} |\mathbf{q}, \mathbf{a}\rangle \tag{2.56}$$

for all $|\mathbf{q}, \mathbf{a}\rangle \in \tilde{\mathcal{B}}$. Thus, (2.48a) is satisfied. To verify condition (2.48b) we use the unitarity of $\hat{U}$ and the fact that $\hat{U}^L = 1$. (2.48c) is satisfied, because the Hubbard Hamiltonian commutes with $\hat{U}$.

2.2.4 More discrete symmetries

We would like to consider two more discrete transformations: the spin flip and the so-called Shiba transformation. Both are useful to restrict the ranges of N and M, the numbers of electrons and down spins, and will be utilized in the next chapter, where we diagonalize the one-dimensional Hubbard Hamiltonian. Moreover, the invariance of the Hubbard Hamiltonian, modulo sign of the coupling, under the Shiba transformations is the reason for the appearance of a second su(2) symmetry besides the more or less obvious rotational symmetry.

The Hubbard Hamiltonian is invariant under the reversal of all spins, caused by a similarity transformation with the operator

$$J^{(s)} = \prod_{j=1}^{L} P_{j\uparrow,j\downarrow}.$$ (2.57)

This transformation maps the eigenstates with M down-spin electrons and $N - M$ up-spin electrons one-to-one onto the eigenstates with M up-spin electrons and $N - M$ down-spin electrons. Thus, the z-component of the total spin changes its sign. As a consequence we may restrict ourselves to non-negative values $N/2 - M$ of S^z, when we diagonalize the Hubbard Hamiltonian in Chapter 3. It is clear from the definition of the transposition operators (2.39) that the transformation (2.57) leaves the empty lattice state $|0\rangle$ invariant.

Let us now consider a lattice with an even number of sites. We define the operators

$$J_a^{(sh)} = (c_{L,a}^\dagger - c_{L,a})(c_{L-1,a}^\dagger + c_{L-1,a}) \ldots (c_{2,a}^\dagger - c_{2,a})(c_{1,a}^\dagger + c_{1,a}),$$ (2.58)

$a = \uparrow, \downarrow$. Note that the signs in the bracket on the right hand side alternate and are plus for odd lattice sites and minus for even lattice sites. The operators (2.58) generate a particle-hole transformation on spin species a, accompanied by a change of sign on every second lattice site. We obtain, for instance, $[J_\downarrow^{(sh)}, c_{j,\uparrow}] = 0$ and

$$J_\downarrow^{(sh)} c_{j,\downarrow} \left(J_\downarrow^{(sh)}\right)^\dagger = (-1)^j c_{j,\downarrow}^\dagger.$$ (2.59)

Clearly, for an even number of lattice sites, the tight-binding part of the Hubbard Hamiltonian (2.31) is invariant under the transformations generated by $J_a^{(sh)}$, $a = \uparrow, \downarrow$, while the interaction part changes its sign. Thus, $H(u)$ is mapped to $H(-u)$. The empty lattice is mapped to

$$J_a^{(sh)}|0\rangle = c_{L,a}^\dagger \ldots c_{1,a}^\dagger |0\rangle,$$ (2.60)

which is the fully spin polarized half-filled band state.

The transformation (2.59) is often called the Shiba transformation but was probably first obtained in [197, 298]. Of course, it is possible to define the Shiba transformation for an odd number of lattice sites. In that case, however, the tight-binding part of the Hubbard Hamiltonian (2.31) is not invariant under the Shiba transformation anymore. Due to the periodic boundary conditions the two odd lattice sites 1 and L are nearest neighbours, and the terms $c_1^\dagger c_L$ and $c_L^\dagger c_1$ pick up a minus sign. This may be avoided by switching to open boundaries. A Hubbard Hamiltonian invariant under a Shiba transformation can always be defined on a bipartite lattice with appropriate boundary conditions. A bipartite lattice is a union of two complementary sublattices Γ_1 and Γ_2, where each lattice site in Γ_1 has nearest neighbours only in Γ_2 and vice versa.

For a lattice with an even number of sites let us perform Shiba transformations for the up- and for the down spins. Then the Hamiltonian is not altered, since the sign of the coupling is switched twice, but the empty lattice state is mapped onto a state with all sites doubly occupied. Thus, all eigenstates of the Hubbard Hamiltonian (2.31) with N electrons are mapped onto eigenstates with $2L - N$ electrons, and we may restrict ourselves to $N \leq L$, when we diagonalize the Hubbard Hamiltonian in Chapter 3.

The spin flip and the Shiba transformations affect the particle number operator and the operator of the z-component of the total spin in a non-trivial way,

$$J^{(s)} \hat{N} J^{(s)} = \hat{N}, \qquad\qquad J^{(s)} S^z J^{(s)} = -S^z, \qquad\qquad (2.61a)$$

$$J_\downarrow^{(sh)} \hat{N} (J_\downarrow^{(sh)})^\dagger = L + 2S^z, \qquad J_\downarrow^{(sh)} 2S^z (J_\downarrow^{(sh)})^\dagger = \hat{N} - L, \qquad (2.61b)$$

$$J_\uparrow^{(sh)} \hat{N} (J_\uparrow^{(sh)})^\dagger = L - 2S^z, \qquad J_\uparrow^{(sh)} 2S^z (J_\uparrow^{(sh)})^\dagger = L - \hat{N}. \qquad (2.61c)$$

This has immediate implications for the Gibbs free energy per lattice site

$$f(\mu, B, T, u) = -\frac{T}{L} \ln\left(\text{tr}\left\{ \exp\left(-\frac{H(u) - \mu\hat{N} - 2BS^z}{T} \right) \right\} \right) \qquad (2.62)$$

which determines the equilibrium thermodynamic properties of the Hubbard model as a function of the chemical potential μ, the magnetic field B and the temperature T and parametrically depends on the coupling u. Using the mutual commutativity of the operators H, $\hat{N}$ and S^z and the invariance of the trace of a product of matrices under cyclic permutations of the matrices we conclude with (2.61) that

$$f(\mu, B, T, u) = f(\mu, -B, T, u) \qquad (2.63a)$$

$$= f(B, \mu, T, -u) - \mu + B \qquad (2.63b)$$

$$= f(-B, -\mu, T, -u) - \mu - B. \qquad (2.63c)$$

Combining the latter equations we further obtain

$$f(\mu, B, T, u) + \mu = f(-\mu, B, T, u) - \mu. \qquad (2.64)$$

Finally we note for later reference the transformation formulae

$$J_\uparrow^{(sh)} J_\downarrow^{(sh)} \hat{N} (J_\downarrow^{(sh)})^\dagger (J_\uparrow^{(sh)})^\dagger = 2L - \hat{N}, \qquad (2.65a)$$

$$J_\uparrow^{(sh)} J_\downarrow^{(sh)} S^z (J_\downarrow^{(sh)})^\dagger (J_\uparrow^{(sh)})^\dagger = -S^z \qquad (2.65b)$$

for a simultaneous application of the two Shiba transformations to the particle number operator and to the operator of the z-component of the total spin which follow from (2.61b) and (2.61c).

2.2.5 SO(4) symmetry

We saw in Section 2.1 that the Hubbard Hamiltonian (2.31) conserves the z-component S^z of the total spin and the particle number $\hat{N}$. Both operators generate U(1) transformations. The operator S^z is the generator of rotations about the z-axis, the particle number operator $\hat{N}$ generates the global gauge transformations. We shall now define the operators S^x and S^y of the x- and y-components of the total spin. S^x, S^y and S^z combine into a representation of the Lie algebra su(2) which generates the group SU(2) of rotations in spin space. We shall show that the Hubbard Hamiltonian commutes with S^x, S^y and S^z and thus is fully rotationally invariant. For an even number of lattice sites, we shall see that the particle number is part of another, hidden su(2) symmetry which has its origin in the invariance modulo sign of the coupling of the Hubbard Hamiltonian under the Shiba transformation (2.59). This non-Abelian extension of the gauge symmetry is called the η-pairing symmetry.

We define the operators of the components of the total spin as

$$S^\alpha = \tfrac{1}{2} \sum_{j=1}^{L} \sum_{a,b=1}^{2} c_{j,a}^\dagger (\sigma^\alpha)_b^a c_{j,b}, \quad \alpha = x, y, z. \tag{2.66}$$

In the second summation we identified 1 with ↑ and 2 with ↓. The matrices σ^α are the Pauli matrices

$$\sigma^x = \begin{pmatrix} 0 & 1 \\ 1 & 0 \end{pmatrix}, \quad \sigma^y = \begin{pmatrix} 0 & -i \\ i & 0 \end{pmatrix}, \quad \sigma^z = \begin{pmatrix} 1 & 0 \\ 0 & -1 \end{pmatrix}. \tag{2.67}$$

They form a basis of the fundamental representation of the Lie algebra su(2) and satisfy the commutation relations

$$[\sigma^\alpha, \sigma^\beta] = 2i\varepsilon^{\alpha\beta\gamma}\sigma^\gamma, \quad \alpha, \beta = x, y, z, \tag{2.68}$$

where $\varepsilon^{\alpha\beta\gamma}$ is the totally antisymmetric tensor. As was claimed above the spin operators generate a representation of su(2),

$$[S^\alpha, S^\beta] = i\varepsilon^{\alpha\beta\gamma} S^\gamma, \quad \alpha, \beta = x, y, z, \tag{2.69}$$

and commute with the Hubbard Hamiltonian,

$$[H, S^\alpha] = 0, \quad \alpha = x, y, z. \tag{2.70}$$

It is not difficult to verify (2.69) and (2.70). One has to use the fundamental anticommutation relations (2.2) in the calculation of various commutators. As a convenient means to deal with all of these commutators simultaneously we introduce so-called current operators [172]. We first of all define a 2×2 operator valued matrix S_{jk} with matrix elements

$$S_{jk}{}_a{}^b = c_{j,a}^\dagger c_{k,b} \tag{2.71}$$

for $a, b = 1, 2$. Using the fundamental anticommutation relations (2.2) we obtain the commutators

$$[S_{jk}{}_a{}^b, S_{lm}{}_c{}^d] = \delta_{kl}\delta_c^b S_{jm}{}_a{}^d - \delta_{mj}\delta_a^d S_{lk}{}_c{}^b. \tag{2.72}$$

We further introduce the projections of S_{jk} onto a gl(2) basis consisting of the Pauli matrices and the 2×2 unit matrix I_2,

$$S_{jk}^\alpha = \mathrm{tr}(\sigma^\alpha S_{jk}), \quad S_{jk}^0 = \mathrm{tr}(S_{jk}). \tag{2.73}$$

These are the above mentioned current operators. As a consequence of (2.72), we obtain the commutators

$$[S^0_{jk}, S^0_{lm}] = \delta_{kl} S^0_{jm} - \delta_{mj} S^0_{lk} , \tag{2.74a}$$

$$[S^0_{jk}, S^\alpha_{lm}] = \delta_{kl} S^\alpha_{jm} - \delta_{mj} S^\alpha_{lk} , \tag{2.74b}$$

$$[S^\alpha_{jk}, S^\beta_{lm}] = \delta^{\alpha\beta} \left(\delta_{kl} S^0_{jm} - \delta_{mj} S^0_{lk} \right) + i\varepsilon^{\alpha\beta\gamma} \left(\delta_{kl} S^\gamma_{jm} + \delta_{mj} S^\gamma_{lk} \right) . \tag{2.74c}$$

Equation (2.74a) is obtained by setting $a = b$ and $c = d$ in (2.72) and summing over a and c. For (2.74b) we first multiply (2.72) by a Pauli matrix and then take the traces. Similarly, to obtain (2.74c) we have to multiply by two Pauli matrices and have to use the identity $\sigma^\alpha \sigma^\beta = \delta^{\alpha\beta} + i\varepsilon^{\alpha\beta\gamma}\sigma^\gamma$.

We note that $S^\alpha_j = \frac{1}{2} S^\alpha_{jj}$ is a local spin operator ('spin density operator') and that $S^0_j = S^0_{jj} = n_{j,\uparrow} + n_{j,\downarrow}$ is the local particle number operator. Putting $j = k$ and $l = m$ in (2.74c) and multiplying by $\frac{1}{4}$ leads to the commutators

$$[S^\alpha_j, S^\beta_l] = \delta_{jl} i\varepsilon^{\alpha\beta\gamma} S^\gamma_j \tag{2.75}$$

for local spin operators. From this equation we obtain (2.69) by summation over j and l.

The Hamiltonian (2.31) expressed in terms of the operators S^0_{jk} reads

$$H = -\sum_{j=1}^{L} \left(S^0_{jj+1} + S^0_{j+1j} - 2u(S^0_{jj} - 1)^2 + u \right) . \tag{2.76}$$

To obtain the squared bracket one has to use $n^2_{j,\uparrow} = c^\dagger_{j,\uparrow} c_{j,\uparrow} c^\dagger_{j,\uparrow} c_{j,\uparrow} = c^\dagger_{j,\uparrow} c_{j,\uparrow} (1 - c_{j,\uparrow} c^\dagger_{j,\uparrow}) = n_{j,\uparrow}$ and $n^2_{j,\downarrow} = n_{j,\downarrow}$. Using (2.74b) we obtain

$$[S^0_{jk}, S^\alpha_l] = \frac{1}{2}(\delta_{kl} S^\alpha_{jl} - \delta_{lj} S^\alpha_{lk}) . \tag{2.77}$$

Summation over l yields $[S^0_{jk}, S^\alpha] = 0$ which, together with (2.76), implies the conservation of the total spin, equation (2.70).

We shall usually use the ladder operators $S^\pm = S^x \pm iS^y$ instead of S^x and S^y. They have the explicit form

$$S^+ = \sum_{j=1}^{L} c^\dagger_{j,\uparrow} c_{j,\downarrow} , \quad S^- = \sum_{j=1}^{L} c^\dagger_{j,\downarrow} c_{j,\uparrow} \tag{2.78}$$

and obey the commutation relations

$$[S^z, S^\pm] = \pm S^\pm , \quad [S^+, S^-] = 2S^z , \tag{2.79}$$

which follow from (2.69).

Let us now turn to the η-pairing symmetry. It has its origin in the invariance modulo the sign of the coupling of the Hubbard Hamiltonian (2.31) under the Shiba transformation

(2.59). Let us apply the Shiba transformation (2.59) to the spin operators $S^\pm$, S^z. Then

$$J_\downarrow^{(sh)} S^+ \left(J_\downarrow^{(sh)} \right)^\dagger = \sum_{j=1}^{L} (-1)^j c_{j,\uparrow}^\dagger c_{j,\downarrow}^\dagger = -\eta^+ , \tag{2.80a}$$

$$J_\downarrow^{(sh)} S^- \left(J_\downarrow^{(sh)} \right)^\dagger = \sum_{j=1}^{L} (-1)^j c_{j,\downarrow} c_{j,\uparrow} = -\eta^- , \tag{2.80b}$$

$$J_\downarrow^{(sh)} S^z \left(J_\downarrow^{(sh)} \right)^\dagger = \tfrac{1}{2} \sum_{j=1}^{L} (n_{j,\uparrow} + n_{j,\downarrow} - 1) = \tfrac{1}{2}(\hat{N} - L) = \eta^z . \tag{2.80c}$$

We introduced the minus signs in (2.80a) and (2.80b) in order to match the conventions in the literature. Application of the Shiba transformation to (2.79) yields the su(2) commutation relations

$$[\eta^z, \eta^\pm] = \pm \eta^\pm , \quad [\eta^+, \eta^-] = 2\eta^z \tag{2.81}$$

for the η-pairing operators. Let us define the analogues of S^x and S^y,

$$\eta^x = \tfrac{1}{2}(\eta^+ + \eta^-) , \quad \eta^y = -\tfrac{i}{2}(\eta^+ - \eta^-) , \tag{2.82}$$

which due to (2.81) satisfy

$$[\eta^\alpha, \eta^\beta] = i\varepsilon^{\alpha\beta\gamma} \eta^\gamma , \quad \alpha, \beta = x, y, z . \tag{2.83}$$

The invariance of the Hubbard Hamiltonian (2.31) under the η-pairing symmetry follows from $[H(-u), S^\alpha] = 0$ by application of the Shiba transformation,

$$[H, \eta^\alpha] = 0 , \quad \alpha = x, y, z . \tag{2.84}$$

A trivial yet important implication of (2.70) and (2.84) that will be useful later on is that the Hubbard Hamiltonian (2.31) also commutes with the Casimir operators $(S^\alpha)^2$ and $(\eta^\alpha)^2$.

To get a full understanding of the symmetries of the Hubbard Hamiltonian we still have to consider the mutual commutators of the two sets of generators S^α and η^β. We claim that

$$[S^\alpha, \eta^\beta] = 0 , \quad \alpha, \beta = x, y, z . \tag{2.85}$$

Because of the antisymmetry, this is a set of six independent equations to be verified. We may, for instance, start with $c_{j,\uparrow}^\dagger c_{j,\downarrow} c_{k,\downarrow}^\dagger c_{k,\uparrow}^\dagger = c_{k,\downarrow}^\dagger c_{k,\uparrow}^\dagger c_{j,\uparrow}^\dagger c_{j,\downarrow}$, which implies that $[S^+, \eta^+] = 0$. Reversing all spins we obtain $[S^-, \eta^+] = 0$. Then, because of (2.79), also $[S^z, \eta^+] = 0$. Hence, η^+ is invariant under rotations. But η^- is the Hermitian conjugate of η^+, so it is invariant under rotations as well. Finally, because of (2.81), the same is true for η^z, and we have proved (2.85).

So far we have shown that the Hubbard Hamiltonian (2.31) commutes with the direct sum of two representations of su(2). We have to keep in mind that the invariance under the η-pairing symmetry, equation (2.84), holds only for an even number of lattice sites L. This fact imposes restrictions on joint irreducible representations of spin and η-spin realized on eigenstates of the Hubbard Hamiltonian. From the definitions (2.25), (2.27) and (2.80c) we obtain

$$S^z + \eta^z = \hat{N}_\uparrow - L/2 \,. \tag{2.86}$$

Thus, $S^z + \eta^z$ when acting on a joint highest weight state takes only integer eigenvalues (see Section 3.4). Spin and η-spin are either both integer or both half odd integer.

It follows from (2.80c) and (2.81) that η^+ and η^- do not preserve the number of particles,

$$[\hat{N}, \eta^\pm] = \pm 2\eta^\pm \,. \tag{2.87}$$

The operator η^+ creates a so-called η-pair in an eigenstate of the Hamiltonian. η^- is the corresponding annihilation operator. As an immediate consequence of the definitions (2.80a), (2.80b) we obtain

$$\{\hat{U}, \eta^\pm\} = 0 \,. \tag{2.88}$$

Thus, η-pairs have lattice momentum π.

The full SO(4) $\cong$ SU(2) $\times$ SU(2)/$\mathbb{Z}_2$ symmetry is only realized for the Hubbard Hamiltonian of the form (2.31). Adding a magnetic field term $-2BS^z$ breaks the rotational invariance, while the η-pairing invariance is preserved. Adding a chemical potential term $-\mu\hat{N}$, on the other hand, breaks the η-pairing symmetry, but preserves the invariance under rotations. The chemical potential μ plays the same role of a symmetry breaking field for the η-spin as the magnetic field B for the spin.

2.3 Conclusions

In this introductory chapter we gave a thorough description of the one-dimensional Hubbard model and considered its basic properties and symmetries. The features of the Hubbard model as far as considered here are not peculiar to the one-dimensional model. They readily generalize to higher dimensional bipartite lattices.

Appendix 2.A is devoted to the strong coupling limit of the Hubbard model, which gives rise to a number of other prominent models in solid state theory, like the isotropic Heisenberg model or the t-J model. In appendix 2.B we consider several possible continuum limits of the Hubbard Hamiltonian. The reader who is less interested in or already familiar with the limiting cases of the Hubbard model may skip the appendices and proceed directly to Chapter 3.

Appendices to Chapter 2

2.A The strong coupling limit

In this section we study the large U perturbation theory of the Hamiltonian

$$H = T + UD, \qquad (2.A.1)$$

where

$$T = \sum_{j,k=1}^{L} t_{jk} c_{j,a}^{\dagger} c_{k,a}, \qquad D = \sum_{j=1}^{L} n_{j,\uparrow} n_{j,\downarrow}. \qquad (2.A.2)$$

On the right hand side of the first equation (2.A.2) implicit summation over $a = \uparrow, \downarrow$ is understood. We shall assume that $t_{jj} = 0$, $t_{kj} = t_{jk}^{*}$ and $U > 0$. The discussion of the attractive case, $U < 0$, is very similar and is left as an exercise to the reader. The Hamiltonian (2.A.1) is a slight generalization of the Hubbard Hamiltonian (2.1). It turns into (2.1) for an appropriate choice of the hopping matrix elements t_{jk}.

Let us assume that $|t_{jk}| \ll U$ for $j, k = 1, \ldots, L$. Then T can be considered as a small perturbation to UD. As we have seen in Section 2.1 the operator D acts diagonally on the basis $\mathcal{B}$ of Wannier states. It counts the number of doubly occupied states. Its eigenvalues are thus $n = 0, 1, \ldots, L$. Let us denote the projectors onto the corresponding eigenspaces $\mathcal{H}_n$ by P_n. The Hilbert space $\mathcal{H}^{(L)}$ of our model (2.A.1) decomposes into $\mathcal{H}^{(L)} = \mathcal{H}_0 \oplus \cdots \oplus \mathcal{H}_L$, and D has the spectral decomposition

$$D = \sum_{n=1}^{L} n P_n. \qquad (2.A.3)$$

The Hamiltonian (2.A.1) conserves the number of electrons N. In the following we will be interested in the case when $N \leq L$. In this case the space of N-particle eigenstates has non-zero overlap with the eigenspace $\mathcal{H}_0$ of P_0, which is spanned by the degenerate ground states of D. The number of these ground states for a fixed number of electrons N is $2^N \binom{L}{N}$. Hence, the total number of Wannier states with no site doubly occupied is

$$\dim \mathcal{H}_0 = 3^L. \qquad (2.A.4)$$

The perturbation T will partially lift this large degeneracy. The lowest energy level of UD

will split into many levels, which are expected, however, to be well separated from the first excited level of UD as long as the condition $|t_{jk}| \ll U$ holds. The splitting of the ground state of UD (in second order perturbation theory) will be described by an effective Hamiltonian, the so-called t-J-Hamiltonian, whose derivation is the subject of the following subsections.

The t-J-Hamiltonian is the most general effective strong coupling approximation to the Hamiltonian (2.A.1). Upon further restriction of the parameters it generates a number of prominent models in solid state theory. At half-filling ($N = L$) it turns into the Hamiltonian of the isotropic spin-$\frac{1}{2}$ Heisenberg chain. The restriction to first order perturbation theory leads to the so-called t-0 model. Keeping only the two-site terms another important variant of the t-J model is obtained.

Our approach here is quite standard and has earlier been described in other textbooks (see e.g. [27, 152]). For further reading we recommend Takahashi's article [437], where the strong coupling perturbation theory for the half-filled Hubbard model has been carried out to higher orders and where references to the older literature can be found.

2.A.1 Projectors

As an input for the perturbative treatment of H we will need expressions for the projectors P_n in terms of fermions. A fermionic representation of the projector P_n can be obtained from the generating function

$$G(\alpha) = \prod_{j=1}^{L}(1 - \alpha \, n_{j,\uparrow}n_{j,\downarrow}). \tag{2.A.5}$$

Let us consider the action of $G(\alpha)$ on Wannier states $|\mathbf{x}, \mathbf{a}\rangle$. A calculation similar to the one presented in equation (2.11) of Chapter 2.1 gives

$$n_{j,\uparrow}n_{j,\downarrow}|\mathbf{x}, \mathbf{a}\rangle = \sum_{1 \leq k < l \leq N} \delta_{j,x_k}\delta_{j,x_l}|\mathbf{x}, \mathbf{a}\rangle = \begin{cases} |\mathbf{x}, \mathbf{a}\rangle & \text{if site } j \text{ is doubly occupied,} \\ 0 & \text{else.} \end{cases} \tag{2.A.6}$$

We conclude that

$$G(\alpha)|\mathbf{x}, \mathbf{a}\rangle = (1 - \alpha)^n |\mathbf{x}, \mathbf{a}\rangle , \tag{2.A.7}$$

where n is the number of doubly occupied sites in the state $|\mathbf{x}, \mathbf{a}\rangle$, i.e., $D|\mathbf{x}, \mathbf{a}\rangle = n|\mathbf{x}, \mathbf{a}\rangle$. It follows from (2.A.7) that

$$\frac{(-1)^k}{k!} \partial_\alpha^k G(\alpha)\Big|_{\alpha=1} |\mathbf{x}, \mathbf{a}\rangle = \delta_{kn}|\mathbf{x}, \mathbf{a}\rangle \tag{2.A.8}$$

and thus,

$$P_n = \frac{(-1)^n}{n!} \partial_\alpha^n G(\alpha)\Big|_{\alpha=1}. \tag{2.A.9}$$

Let us note that the latter equation is naturally extended to $n = 0$, since (2.A.7) implies $P_0 = G(1)$.

2.A.2 Second order perturbation theory around an energy level

Consider any Hamiltonian H acting on a Hilbert space $\mathcal{H}$. Let P be a projection on a subspace $P\mathcal{H}$ of $\mathcal{H}$ and $Q = 1 - P$. $|\psi\rangle$ is a solution of the Schrödinger equation

$$H|\psi\rangle = E|\psi\rangle \tag{2.A.10}$$

with eigenvalue E, if and only if

$$PHP|\psi\rangle + PHQ|\psi\rangle = EP|\psi\rangle, \tag{2.A.11a}$$

$$QHP|\psi\rangle + QHQ|\psi\rangle = EQ|\psi\rangle. \tag{2.A.11b}$$

Let us solve equation (2.A.11b) for $Q|\psi\rangle$ and insert the result into (2.A.11a). We obtain

$$\hat{H}(E)|\varphi\rangle = E|\varphi\rangle, \tag{2.A.12}$$

where

$$\hat{H}(E) = PH(1 + (E - QH)^{-1}QH)P \tag{2.A.13}$$

and $|\varphi\rangle = P|\psi\rangle$. Thus, $|\varphi\rangle \in P\mathcal{H}$ is a solution of the spectral problem (2.A.12) with eigenvalue E. Conversely, if $|\varphi\rangle \in P\mathcal{H}$ solves (2.A.12), the vector

$$|\psi\rangle = (1 + (E - QH)^{-1}QH)|\varphi\rangle \tag{2.A.14}$$

is a solution of the full stationary Schrödinger equation (2.A.10) with eigenvalue E.

Remark. We have assumed the spectra of H and QH be disjoint. This should be true for generic choice of P. Then, by construction, equation (2.A.12) contains the *complete* information about the spectrum of H and all eigenstates of H follow from (2.A.14). Since $\hat{H}(E)$ is nonlinear in E, there may be several eigenvalues belonging to the same eigenvector $|\varphi\rangle$. In the extreme case a single eigenvector of $\hat{H}(E)$ determines the complete spectrum of H. The reader may verify this statement with the toy example

$$H = \begin{pmatrix} 1 & 0 & 0 \\ 0 & 2 & 0 \\ 0 & 0 & 3 \end{pmatrix}, \quad P = \frac{1}{3}\begin{pmatrix} 1 & 1 & 1 \\ 1 & 1 & 1 \\ 1 & 1 & 1 \end{pmatrix}. \tag{2.A.15}$$

Equations (2.A.12) and (2.A.13) are a convenient starting point for the perturbation theory around a given degenerate energy level. Let us consider a Hamiltonian

$$H = H_0 + \lambda H_1 \tag{2.A.16}$$

composed of a contribution H_0 with known spectral decomposition

$$H_0 = \sum_n E_n P_n \tag{2.A.17}$$

and a perturbation H_1 coupled to H_0 by a coupling constant λ.

Replacing P by P_n in (2.A.13) we obtain

$$\hat{H}_n(E) = P_n H (1 + (E - Q_n H)^{-1} Q_n H) P_n . \qquad (2.A.18)$$

This operator acts non-trivially only on the degeneracy subspace corresponding to the nth energy level E_n of H_0. Nevertheless, it determines the complete spectrum and all eigenfunctions of H through (2.A.12) and (2.A.14). We may use the explicit form (2.A.17) of the unperturbed Hamiltonian H_0 to express $\hat{H}_n(E)$ entirely in terms of the projectors P_n, the unperturbed energy levels E_n and the perturbation H_1. A short calculation based on the elementary relations

$$P_m P_n = \delta_{mn} P_n , \qquad (2.A.19a)$$

$$P_n H_0 = H_0 P_n = E_n P_n , \qquad Q_n H_0 = H_0 Q_n = \sum_{m \ (m \neq n)} E_m P_m , \qquad (2.A.19b)$$

$$P_n H Q_n = \lambda P_n H_1 Q_n , \qquad Q_n H P_n = \lambda Q_n H_1 P_n \qquad (2.A.19c)$$

yields

$$\hat{H}_n(E) = \left[E_n + P_n H_1 \sum_{k=0}^{\infty} \lambda^{k+1} \left(\sum_{m \ (m \neq n)} \frac{P_m H_1}{E - E_m} \right)^k \right] P_n . \qquad (2.A.20)$$

Thus, the spectral problem (2.A.12) turns into

$$P_n H_1 \sum_{k=0}^{\infty} \lambda^{k+1} \left(\sum_{m \ (m \neq n)} \frac{P_m H_1}{E - E_m} \right)^k |\varphi\rangle = (E - E_n) |\varphi\rangle \qquad (2.A.21)$$

for $|\varphi\rangle \in P_n \mathcal{H}$.

So far we have achieved nothing but a reformulation of our original spectral problem (2.A.10). Now perturbation theory around the energy level E_n is the iterative solution of equation (2.A.21) in ascending orders of λ. For $\lambda = 0$ the left hand side of (2.A.21) vanishes and we have $E = E_n$. For small λ we expect corrections to E_n of the form

$$E = E_n + \lambda E_n^{(1)} + \lambda^2 E_n^{(2)} + \mathcal{O}(\lambda^3) . \qquad (2.A.22)$$

On the left hand side of equation (2.A.21) these corrections do not contribute to the first and second order terms in an expansion in λ. Thus, for the part of the spectrum of H, which evolves from the energy level E_n as the coupling λ is turned on, we obtain to quadratic order in λ the linear spectral problem

$$\left[P_n H_1 P_n + \lambda \sum_{m \ (m \neq n)} \frac{P_n H_1 P_m H_1 P_n}{E_n - E_m} \right] |\varphi\rangle = \frac{E - E_n}{\lambda} |\varphi\rangle . \qquad (2.A.23)$$

The term in square brackets is an effective Hamiltonian on the restricted Hilbert space $P_n \mathcal{H}$. It describes the splitting of the energy level E_n of the Hamiltonian H_0 under the influence of the small perturbation λH_1. The corresponding eigenstates of the full Hamiltonian H are

obtainable by application of the operator $1 + (E - Q_n H)^{-1} Q_n H P_n$ (see (2.A.14)) to the
eigenstates of (2.A.23). Another short calculation gives

$$(E - Q_n H)^{-1} Q_n H P_n = \lambda \cdot \sum_{\substack{m \ (m \neq n)}} \frac{P_m H_1 P_n}{E - E_m} + \lambda^2 \sum_{\substack{l,m \\ (l,m \neq n)}} \frac{P_l H_1 P_m H_1 P_n}{(E - E_l)(E - E_m)} + \mathcal{O}(\lambda^3).$$

(2.A.24)

2.A.3 The Hubbard model in the strong coupling limit

Let us now apply the formalism of the previous section to the Hamiltonian (2.A.1). We
rescale the Hamiltonian as

$$H/U = D + T/U.$$ (2.A.25)

Then, the energy has to be rescaled as well. H/U is of the form (2.A.16) with $H_0 = D$,
$H_1 = T$ and $\lambda = 1/U$. Hence, (2.A.23) applies for large U.

The operator D has eigenvalues $E_n = n$, for $n = 0, 1, \ldots, L$. Setting $n = 0$ in (2.A.23)
and inserting the above data we obtain

$$\left[P_0 T P_0 - \frac{1}{U} \sum_{m=1}^{L} \frac{P_0 T P_m T P_0}{m} \right] |\varphi\rangle = E |\varphi\rangle.$$ (2.A.26)

The operator on the left hand side is often called the t-J-Hamiltonian. We will denote
it by H_{t-J}. It describes the splitting of the lowest energy level $E_0 = 0$ of the Hubbard
Hamiltonian in the atomic limit, in the situation, when the hopping amplitudes are small
compared to the Hubbard interaction U. The projection operators P_m, $m = 0, \ldots, L$, are
given by equation (2.A.9). In particular,

$$P_0 = \prod_{j=1}^{L} (1 - n_{j,\uparrow} n_{j,\downarrow})$$ (2.A.27)

is the projection operator onto the space with no doubly occupied site.

The work that remains to be done is to express the t-J-Hamiltonian in a more convenient
and intuitively understandable form. First note the relation

$$n_{j,\uparrow} n_{j,\downarrow} P_0 = P_0 n_{j,\uparrow} n_{j,\downarrow} = 0$$ (2.A.28)

for $j = 1, \ldots, L$, which implies that

$$G(\alpha) T P_0 = \sum_{j,k=1}^{L} t_{jk} (1 - \alpha\, n_{j,\uparrow} n_{j,\downarrow}) c_{j,a}^{\dagger} c_{k,a} P_0.$$ (2.A.29)

Here the right hand side is linear in α. Thus, by application of (2.A.9), it follows that

$$\sum_{m=1}^{L} \frac{P_m T P_0}{m} = \sum_{j,k=1}^{L} t_{jk} n_{j,\uparrow} n_{j,\downarrow} c_{j,a}^{\dagger} c_{k,a} P_0,$$ (2.A.30)

and thus

$$H_{t-J} = P_0\left[\sum_{j,k=1}^{L} t_{jk}c_{j,a}^\dagger c_{k,a} - \frac{1}{U}\sum_{j,k,k',l=1}^{L} t_{jk}t_{k'l}c_{j,a}^\dagger c_{k,a}n_{k',\uparrow}n_{k',\downarrow}c_{k',b}^\dagger c_{l,b}\right]P_0 . \quad (2.A.31)$$

The second sum in this expression can be further simplified. Let us concentrate for a moment on the sum over k'. For $k' \neq j, k$ the term $n_{k',\uparrow}n_{k',\downarrow}$ can be commuted through $c_{j,a}^\dagger c_{k,a}$, and, by (2.A.28), the corresponding terms in the second sum in (2.A.31) vanish. The term with $k' = j$ also vanishes. For $j \neq k$ this follows from $c_{j,a}^\dagger n_{j,\uparrow}n_{j,\downarrow} = 0$ and for $j = k$ from $c_{j,a}^\dagger c_{j,a}n_{j,\uparrow}n_{j,\downarrow} = 2n_{j,\uparrow}n_{j,\downarrow}$ and again (2.A.28). Thus, the only non-vanishing contribution in the sum over k' comes from the term $k' = k$, and the second sum in (2.A.31) reduces to a triple sum over j, k, and l. Finally, in the remaining triple sum, we may commute the term $n_{k,\uparrow}n_{k,\downarrow}$ to the left and use once more equation (2.A.28). The resulting expression for the t-J-Hamiltonian is most conveniently expressed in terms of the operators S_{jk}^α and S_{jk}^0 introduced in equation (2.73),

$$H_{t-J} = P_0\left[\sum_{j,k=1}^{L} t_{jk}S_{jk}^0 + \frac{1}{2U}\sum_{j,k,l=1}^{L} t_{jk}t_{kl}(S_{jl}^\alpha S_{kk}^\alpha - S_{jl}^0 S_{kk}^0)\right]P_0 \quad (2.A.32a)$$

$$= P_0\left[\sum_{\substack{j,k=1\\j\neq k}}^{L} t_{jk}S_{jk}^0 + \frac{1}{2U}\sum_{\substack{j,k=1\\j\neq k}}^{L} |t_{jk}|^2(S_{jj}^\alpha S_{kk}^\alpha - S_{jj}^0 S_{kk}^0)\right.$$

$$\left.+ \frac{1}{2U}\sum_{\substack{j,k,l=1\\j\neq k\neq l\neq j}}^{L} t_{jk}t_{kl}(S_{jl}^\alpha S_{kk}^\alpha - S_{jl}^0 S_{kk}^0)\right]P_0 . \quad (2.A.32b)$$

More conventionally the t-J-Hamiltonian is written as

$$H_{t-J} = P_0\left[\sum_{\substack{j,k=1\\j\neq k}}^{L} t_{jk}c_{j,a}^\dagger c_{k,a} + \sum_{\substack{j,k=1\\j\neq k}}^{L} \frac{2|t_{jk}|^2}{U}\left(S_j^\alpha S_k^\alpha - \frac{n_j n_k}{4}\right)\right.$$

$$\left.+ \frac{1}{U}\sum_{\substack{j,k,l=1\\j\neq k\neq l\neq j}}^{L} t_{jk}t_{kl}\left(c_{j,a}^\dagger\sigma_{ab}^\alpha c_{l,b}S_k^\alpha - \frac{1}{2}c_{j,a}^\dagger c_{l,a}n_k\right)\right]P_0 , \quad (2.A.33)$$

where the $S_j^\alpha = \frac{1}{2}S_{jj}^\alpha$, $\alpha = x, y, z$, are local spin operators and $n_j = n_{j,\uparrow} + n_{j,\downarrow}$ is the local particle number operator. Yet another useful form of the t-J-Hamiltonian is obtained, when we move in (2.A.33) the operator P_0 on the left through the sums,

$$H_{t-J} = \sum_{\substack{j,k=1\\j\neq k}}^{L} t_{jk}c_{j,a}^\dagger c_{k,a}(1-n_j) + \sum_{\substack{j,k=1\\j\neq k}}^{L} \frac{2|t_{jk}|^2}{U}\left(S_j^\alpha S_k^\alpha - \frac{n_j n_k}{4}\right)$$

$$+ \frac{1}{U}\sum_{\substack{j,k,l=1\\j\neq k\neq l\neq j}}^{L} t_{jk}t_{kl}\left(c_{j,a}^\dagger\sigma_{ab}^\alpha c_{l,b}S_k^\alpha - \frac{1}{2}c_{j,a}^\dagger c_{l,a}n_k\right)(1-n_j) . \quad (2.A.34)$$

Here we left out the operator P_0 on the very right, which is justified, since H_{t-J} acts on the projected space $P_0 \mathcal{H}^{(L)}$.

It is easy to verify that the t-J-Hamiltonian conserves the number of particles and the total spin,

$$[H_{t-J}, \hat{N}] = [H_{t-J}, S^\alpha] = 0 \tag{2.A.35}$$

for $\alpha = x, y, z$.

Let us summarize our results. Using a kind of second order perturbation theory, which includes projection onto the restricted Hilbert space $\mathcal{H}_0 = P_0 \mathcal{H}^{(L)}$, we obtained an effective strong coupling approximation to the Hamiltonian (2.A.1). We presented this so-called t-J-Hamiltonian in three different forms (2.A.32)-(2.A.34). Note that our derivation did not rely on the one-dimensionality of the Hamiltonian (2.A.1). Results similar to (2.A.32)–(2.A.34) hold in arbitrary dimensions. In order to obtain the low lying eigenstates of the full Hamiltonian in second order perturbation theory one has to apply the operator (2.A.24) to the eigenstates of the t-J-Hamiltonian. In contrast to the eigenstates of the t-J-Hamiltonian the eigenstates of the full Hamiltonian in second order perturbation theory have a small probability of having doubly occupied sites.

Usually the t-J-Hamiltonian as presented in, say, (2.A.34) is not directly studied in the literature. It rather is considered in certain more specialized situations that will be discussed below.

2.A.4 Heisenberg spin chain and Mott transition

H_{t-J} conserves the number of electrons (see (2.A.35)). At half-filling, when the number of electrons N equals the number of lattice sites L, all eigenstates of H_{t-J} must be 'pure spin states' of the form $|a_1, \ldots, a_L\rangle = c^\dagger_{L,a_L} \cdots c^\dagger_{1,a_1} |0\rangle$. In these states every lattice site is occupied precisely by one electron. Then $(1 - n_j)|a_1, \ldots, a_L\rangle = 0$, and the Hamiltonian (2.A.34) reduces to

$$H_{Spin} = \sum_{\substack{j,k=1 \\ j \neq k}}^{L} \frac{2|t_{jk}|^2}{U} \left(S_j^\alpha S_k^\alpha - \tfrac{1}{4} \right) . \tag{2.A.36}$$

This is the isotropic spin-$\tfrac{1}{2}$ Heisenberg chain with exchange couplings $J_{jk} = 2|t_{jk}|^2/U$ between spins at sites j and k. For the special case of nearest-neighbour hopping $t_{jk} = -t(\delta_{j,k-1} + \delta_{j,k+1})$ the Hamiltonian (2.A.1) turns into the Hubbard Hamiltonian (2.1). The corresponding spin chain Hamiltonian (2.A.36) becomes

$$H_{Spin} = \frac{4t^2}{U} \sum_{j=1}^{L} \left(S_j^\alpha S_{j+1}^\alpha - \tfrac{1}{4} \right), \tag{2.A.37}$$

where periodic boundary conditions $S_{L+1}^\alpha = S_1^\alpha$ are implied. The sign of the exchange coupling is the same as the sign of the Hubbard interaction U. Thus, we have shown the following remarkable fact: *At half-filling and for strong repulsion $U \gg |t|$ the low-lying excitations of the Hubbard model are effectively described by the excitations of an antiferromagnetic Heisenberg chain.*

The spin Hamiltonian (2.A.37) is exactly solvable by Bethe ansatz and is in fact the model the Bethe ansatz was first applied to in Bethe's original article [60].

If we apply an external electro-magnetic field to the Hubbard model it appears as a complex phase in the hopping matrix elements, $t_{jk} \rightarrow t_{jk}e^{i\lambda_{jk}}$ (see Chapter 1.3), and the first and the third sum in (2.A.34) have to be modified. The second sum, however, containing merely the modulus $|t_{jk}|$, remains unchanged. At half-filling the effective Hamiltonian is again (2.A.36). This means that to second order strong coupling perturbation theory the Hubbard Hamiltonian does not couple to an external electro-magnetic field. Hence, its conductivity is zero. On the other hand, the tight-binding model, which is the weak coupling limit $U \rightarrow 0$ of the Hubbard model has a non-vanishing conductivity [159]. This hints that at a critical interaction U_c there is an interaction induced phase transition from a conducting phase to an insulating phase. Such kind of phase transition is called a Mott metal-insulator transition. Later we shall see that the critical coupling strength in case of the one-dimensional Hubbard model is $U_c = 0$.

2.A.5 Neglecting the three-site terms

Close to half-filling the mean values of the first and the third sum on the right hand side of (2.A.34) are still expected to be small. In particular, the third sum may be considered to be unimportant compared to the second sum. On the other hand, the third sum in (2.A.34) is of higher order in $1/U$ than the first sum. Thus, close to half-filling the main features of the Hamiltonian (2.A.34) are expected to be properly described by its simplified ver ion

$$H_{t-J}^{(2)} = \sum_{\substack{j,k=1 \\ j \neq k}}^{L} t_{jk} c_{j,a}^{\dagger} c_{k,a}(1 - n_j) + \sum_{\substack{j,k=1 \\ j \neq k}}^{L} \frac{2|t_{jk}|^2}{U} \left(S_j^\alpha S_k^\alpha - \frac{n_j n_k}{4} \right) . \qquad (2.A.38)$$

We denoted this Hamiltonian by $H_{t-J}^{(2)}$, since the three-site terms are neglected here. Specializing again to nearest-neighbour hopping and going back to the notation of equation (2.A.33) we obtain.

$$H_{t-J}^{(2)} = P_0 \left[-t \sum_{j=1}^{L} (c_{j,a}^{\dagger} c_{j+1,a} + c_{j+1,a}^{\dagger} c_{j,a}) + \frac{4t^2}{U} \sum_{j=1}^{L} \left(S_j^\alpha S_{j+1}^\alpha - \frac{n_j n_{j+1}}{4} \right) \right] P_0 .$$

$$(2.A.39)$$

In this form $H_{t-J}^{(2)}$ is most frequently encountered in the literature. Unfortunately, for the lack of better names, the Hamiltonian (2.A.39) is also called t-J-Hamiltonian.[1]

In general, the t-J-Hamiltonian (2.A.39) is not solvable by Bethe ansatz [377]. For the special values $4t/U = 2$ and $4t/U = 0$ of the coupling constant, however, Bethe ansatz

---

[1] A common short-hand notation in the literature is $J = 4t^2/U$, whence the name.

solutions exist [377]. At $4t/U = 2$ the model has higher global symmetry. The model becomes invariant under the action of the Lie super algebra gl(1|2). Therefore the t-J-Hamiltonian $H^{(2)}_{t-J}$ with $4t/U = 2$ is called the supersymmetric t-J-Hamiltonian.

The history of the supersymmetric t-J model as a solvable model is rather involved. Many authors contributed to its solution. For the reader who is interested in more details we recommend the article [119], where the algebraic Bethe ansatz solution was obtained and the role of the gl(1|2) symmetry has been stressed. The article [119] also contains a brief account of the history of the supersymmetric t-J model with many references to further original articles.

2.A.6 The t-0 model

For $4t/U = 0$ the Hamiltonian (2.A.39) turns into

$$H_{t-0} = -tP_0\left[\sum_{j=1}^{L}(c^\dagger_{j,a}c_{j+1,a} + c^\dagger_{j+1,a}c_{j,a})\right]P_0. \qquad (2.A.40)$$

This model is called the t-0 model or the restricted hopping model. It is also obtained from (2.A.33) in the case of nearest-neighbour hopping and for $U \to \infty$, i.e., when the strong coupling perturbation theory, which led to the t-J-Hamiltonian (2.A.33), is restricted to first order.

The t-0 model is interesting because of its extreme simplicity. From the point of view of its algebraic structure it is closely related to the XX spin chain [179,313]. It has been solved by the algebraic Bethe ansatz [179] and by coordinate Bethe ansatz [219]. The t-0 model is an interesting toy model for the study of quantum correlation functions. A determinant representation (see [270]) for the two-particle Green's functions was obtained in [219].

2.A.7 An overview over the strong coupling effective models related to the Hubbard model

We have become acquainted with the various descendants of the Hubbard model in the strong coupling limit $|t| \ll U$. A schematic picture of their interdependencies is given in figure 2.A.1.

First of all, the t-J-Hamiltonian H_{t-J}, which includes three-site terms, is obtained from the Hubbard Hamiltonian in second order perturbation theory in $|t|/U$. The Heisenberg spin chain is the effective Hamiltonian at half-filling, $N/L = 1$. Ignoring the three-site terms in H_{t-J}, we obtained the t-J-Hamiltonian $H^{(2)}_{t-J}$. We argued that the three-site terms might be non-essential at strong coupling and close to half filling. In figure 2.A.1 we tried to indicate the somewhat heuristic nature of our arguments by drawing a dashed arrow between H_{t-J} and $H^{(2)}_{t-J}$. The t-0-Hamiltonian follows from H_{t-J} (or $H^{(2)}_{t-J}$) in the limit $U \to \infty$ and corresponds to first order perturbation theory. Finally, we included the supersymmetric t-J-Hamiltonian $H^{(s)}_{t-J}$ into our scheme. It is equal to the t-J-Hamiltonian $H^{(2)}_{t-J}$ at a certain

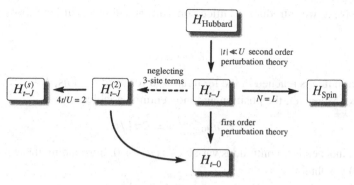

Fig. 2.A.1. The various models related to the strong coupling limit of the Hubbard model.

value, $4t/U = 2$, of the coupling and, like the Hubbard or Heisenberg chain, is solvable by Bethe ansatz.

Let us note that second order perturbation theory is not the only possible starting point for the derivation of the effective strong coupling Hamiltonian H_{t-J}. We choose this starting point here, since it is close to common textbook knowledge and rather universally applicable. For an alternative derivation of the strong coupling descendants of the Hubbard model, which uses more of the specific features of the Hubbard model, the reader is referred to [314].

2.B Continuum limits

The Hubbard model is a lattice model. Some of its peculiar features, like the existence of *two* su(2) symmetries, or the existence of certain kinds of bound states, which will be discussed in later chapters, can be directly attributed to the discreteness of the lattice. Quite generally, lattice models have a richer phenomenology than the continuum models they contain as limiting cases. Still, these continuum models describe certain aspects of the physics of the lattice model and may also be interesting on their own right.

How can we perform a continuum limit on Fermi operators? We shall start with certain formal manipulations and, in the sequel, try to give them a more rigorous meaning. Let us start with equations (2.13) and (2.14), which are the Fourier transformation formulae for switching from creation operators of electrons in Wannier states to creation operators of electrons in Bloch states. Shifting the limits of summation in (2.14) we obtain

$$\tilde{c}_{k,a}^\dagger = \frac{1}{\sqrt{L}} \sum_{n=1}^{L} e^{i\phi kn}\, c_{n,a}^\dagger \,, \quad k = k_-, \ldots, k_+, \tag{2.B.1a}$$

$$c_{n,a}^\dagger = \frac{1}{\sqrt{L}} \sum_{k=k_-}^{k_+} e^{-i\phi kn}\, \tilde{c}_{k,a}^\dagger \,, \quad n = 1, \ldots, L \,, \tag{2.B.1b}$$

where $k_+ = -k_- = (L-1)/2$ for L odd and $k_+ = -k_- + 1 = L/2$ for L even (recall that

$\phi = 2\pi/L$). Next, we introduce a lattice constant a_0 and the total length of the lattice ℓ, such that

$$\ell = L a_0 . \tag{2.B.2}$$

Then the continuum is reached in the limit $a_0 \to 0$ for fixed ℓ. Thus, $L \to \infty$, $k_- \to -\infty$ and $k_+ \to \infty$. Moreover, the position and momentum variables

$$x = n a_0 , \quad q_k = 2\pi k/\ell \tag{2.B.3}$$

become new, independent continuous variables as $a_0 \to 0$. Introducing these new variables into (2.B.1b) we obtain

$$\frac{c_{n,a}^\dagger}{\sqrt{a_0}} = \frac{1}{\sqrt{\ell}} \sum_{k=k_-}^{k_+} e^{-iq_k x} \, \tilde{c}_{k,a}^\dagger . \tag{2.B.4}$$

Let us fix x. Then the right hand side of (2.B.4) depends on a_0 only through the limits k_+ and k_- of summation, and we may formally define

$$\Psi_a^\dagger(x) = \lim_{a_0 \to 0} \left. \frac{c_{n,a}^\dagger}{\sqrt{a_0}} \right|_x = \frac{1}{\sqrt{\ell}} \sum_{k=-\infty}^{\infty} e^{-iq_k x} \, \tilde{c}_{k,a}^\dagger . \tag{2.B.5}$$

At first glance the sum on the right hand side of (2.B.5) is only a formal expression. If we restrict ourselves, however, to the space generated by the action of finitely many creation operators $\tilde{c}_{k,a}^\dagger$ on a Fock vacuum $|0\rangle$, this sum becomes perfectly sensible and gives meaning to the rather terrific expression $\lim_{a_0 \to 0}(c_{n,a}^\dagger/\sqrt{a_0})|_x$. Creation of finitely many particles on infinitely many sites means to consider a situation, where the density of particles N/L is zero. This is in agreement with our intuitive understanding. The continuum limit should work for low densities when all particles can be in states with small momenta. Small momenta correspond to wave lengths large compared to the lattice constant.

Fourier inversion of equation (2.B.5) gives

$$\tilde{c}_{k,a}^\dagger = \frac{1}{\sqrt{\ell}} \int_0^\ell dx \, e^{iq_k x} \, \Psi_a^\dagger(x) . \tag{2.B.6}$$

This equation can also be obtained from (2.B.1a) observing that

$$\sum_{n=1}^{L} a_0 \, f(n a_0) \xrightarrow{a_0 \to 0} \int_0^\ell dx \, f(x) . \tag{2.B.7}$$

For the fundamental anticommutators (2.2) we find

$$\left\{ c_{m,a}/\sqrt{a_0} , c_{n,b}^\dagger/\sqrt{a_0} \right\} = \delta_{a,b}\delta_{m,n}/a_0 . \tag{2.B.8}$$

Now,

$$\sum_{m=1}^{L} a_0 \, \frac{\delta_{m,n}}{a_0} \, f(m a_0) = f(n a_0) , \tag{2.B.9}$$

and thus, by comparison with (2.B.7),

$$\{\Psi_a(x), \Psi_b^\dagger(y)\} = \delta_{a,b}\delta(x-y). \tag{2.B.10}$$

Note that through (2.B.5) we can also define derivatives of the field operators $\Psi_a(x)$.

Let us now apply the above ideas to the Hubbard model. We start with the tight-binding Hamiltonian (2.12). Since

$$-\frac{1}{a_0}\left(c_{n,a}^\dagger c_{n+1,a} + c_{n+1,a}^\dagger c_{n,a}\right) \to -\left(\Psi_a^\dagger(x)\Psi_a(x+a_0) + \Psi_a^\dagger(x+a_0)\Psi_a(x)\right)$$

$$= -2\Psi_a^\dagger(x)\Psi_a(x) - a_0\partial_x\left(\Psi_a^\dagger(x)\Psi_a(x)\right) - \frac{a_0^2}{2}\partial_x^2\left(\Psi_a^\dagger(x)\Psi_a(x)\right)$$

$$+ a_0^2\left(\partial_x \Psi_a^\dagger(x)\right)\partial_x \Psi_a(x) + \mathcal{O}(a_0^3), \tag{2.B.11}$$

we find

$$H_0 \to -2\int_0^\ell dx\; \Psi_a^\dagger(x)\Psi_a(x) + a_0^2\int_0^\ell dx\; \left(\partial_x \Psi_a^\dagger(x)\right)\partial_x \Psi_a(x) + \mathcal{O}(a_0^3). \tag{2.B.12}$$

The operator appearing in the second order term in a_0 is the free Hamiltonian of the continuum model on a ring of length ℓ.

Because of the invariance of the tight-binding Hamiltonian under the Shiba transformation (2.59) for even L, four different continuum limits are possible,

$$\Psi_\uparrow^\dagger(x) = \lim_{a_0\to 0} \frac{c_{n,\uparrow}^\dagger}{\sqrt{a_0}}\bigg|_x, \qquad \Psi_\downarrow^\dagger(x) = \lim_{a_0\to 0} \frac{c_{n,\downarrow}^\dagger}{\sqrt{a_0}}\bigg|_x, \tag{2.B.13a}$$

$$\Psi_\uparrow^\dagger(x) = \lim_{a_0\to 0} \frac{c_{n,\uparrow}^\dagger}{\sqrt{a_0}}\bigg|_x, \qquad \Psi_\downarrow^\dagger(x) = \lim_{a_0\to 0} \frac{(-1)^n c_{n,\downarrow}}{\sqrt{a_0}}\bigg|_x, \tag{2.B.13b}$$

$$\Psi_\uparrow^\dagger(x) = \lim_{a_0\to 0} \frac{(-1)^n c_{n,\uparrow}}{\sqrt{a_0}}\bigg|_x, \qquad \Psi_\downarrow^\dagger(x) = \lim_{a_0\to 0} \frac{c_{n,\downarrow}^\dagger}{\sqrt{a_0}}\bigg|_x, \tag{2.B.13c}$$

$$\Psi_\uparrow^\dagger(x) = \lim_{a_0\to 0} \frac{(-1)^n c_{n,\uparrow}}{\sqrt{a_0}}\bigg|_x, \qquad \Psi_\downarrow^\dagger(x) = \lim_{a_0\to 0} \frac{(-1)^n c_{n,\downarrow}}{\sqrt{a_0}}\bigg|_x, \tag{2.B.13d}$$

which all lead to the same continuum Hamiltonian (2.B.12).

Let us consider the second equation (2.B.13b). For even L we have

$$\frac{(-1)^n c_{n,\downarrow}}{\sqrt{a_0}} = \frac{1}{\sqrt{\ell}}\sum_{k=k_-}^{k_+} e^{-iq_k x}\, \tilde{c}_{k-k_+,\downarrow}. \tag{2.B.14}$$

Thus, defining $\tilde{\tilde{c}}_{k,\downarrow} = \tilde{c}_{k-k_+,\downarrow}$, we formally obtain

$$\Psi_\downarrow^\dagger(x) = \frac{1}{\sqrt{\ell}}\sum_{k=-\infty}^{\infty} e^{-iq_k x}\, \tilde{\tilde{c}}_{k,\downarrow}. \tag{2.B.15}$$

In this case the long wave-length excitations of the continuum model correspond to short wave lengths in the lattice model. Recall that the Shiba transformation affects the vacuum (2.60). The (lattice) Fock vacuum associated with the limit (2.B.13b) is the fully spin-polarized half-filled band, $| \downarrow \rangle = c_{L,\downarrow}^\dagger \cdots c_{1,\downarrow}^\dagger |0\rangle$. Similar considerations are valid for the remaining cases (2.B.13c), (2.B.13d) with Fock vacua $| \uparrow \rangle = c_{L,\uparrow}^\dagger \cdots c_{1,\uparrow}^\dagger |0\rangle$, $| \uparrow\downarrow \rangle = c_{L,\uparrow}^\dagger c_{L,\downarrow}^\dagger \cdots c_{1,\uparrow}^\dagger c_{1,\downarrow}^\dagger |0\rangle$.

For the interaction part of the Hubbard Hamiltonian (2.31) we obtain

$$u(1 - 2n_{j,\uparrow})(1 - 2n_{j,\downarrow}) \rightarrow \pm u\left(1 - 2a_0 \Psi_a^\dagger(x)\Psi_a(x) + 2a_0^2 \Psi_a^\dagger(x)\Psi_b^\dagger(x)\Psi_b(x)\Psi_a(x)\right),$$
$$(2.B.16)$$

where the plus sign has to be taken for (2.B.13a) and (2.B.13d), and the minus sign for (2.B.13b) and (2.B.13c), respectively. The resulting continuum limits of the Hubbard Hamiltonian are

$$H \rightarrow \pm\frac{u\ell}{a_0} - 2(1 \pm u) \int_0^\ell dx\ \Psi_a^\dagger(x)\Psi_a(x)$$
$$+ a_0^2 \int_0^\ell dx\ \left[(\partial_x \Psi_a^\dagger(x))\partial_x \Psi_a(x) \pm \frac{2u}{a_0} \Psi_a^\dagger(x)\Psi_b^\dagger(x)\Psi_b(x)\Psi_a(x)\right] + \mathcal{O}(a_0^3).$$
$$(2.B.17)$$

Here the operator that appears in second order in a_0 is the well known Hamiltonian of the (electronic) non-linear Schrödinger equation. Setting $c = u/a_0$ we may define

$$H_{NLS} = \int_0^\ell dx\ \left[(\partial_x \Psi_a^\dagger(x))\partial_x \Psi_a(x) \pm 2c\ \Psi_a^\dagger(x)\Psi_b^\dagger(x)\Psi_b(x)\Psi_a(x)\right]. \qquad (2.B.18)$$

Our result has two different interpretations. First, the Hubbard Hamiltonian is a lattice regularization of the Hamiltonian (2.B.18) of the non-linear Schrödinger equation (NLS Hamiltonian). If we rescale the coupling of the Hubbard model as $u \rightarrow a_0 c$, we obtain

$$H_{NLS} = \lim_{a_0 \rightarrow 0} \frac{1}{a_0^2}\left\{H - \frac{u\ell}{a_0} + 2(1 + u)\hat{N}\right\}, \qquad (2.B.19)$$

where we choose the plus sign in (2.B.18) for definiteness. From equation (2.B.19) we can obtain the spectrum of the NLS Hamiltonian, once the spectrum of the Hubbard Hamiltonian is known. The NLS Hamiltonian for electrons is exactly solvable and historically was the first Hamiltonian diagonalized by the *nested* Bethe ansatz [493].

A second interpretation of (2.B.17) and (2.B.18) is the following. There are four different 'low density cases', (2.B.13), where the NLS Hamiltonian is a good approximation to the Hubbard Hamiltonian. The case (2.B.13a) will be discussed in Chapter 11, where we consider the so-called gas phase. Note that the Hubbard Hamiltonian at *finite* coupling u corresponds to the NLS Hamiltonian at *infinite* coupling $2u/a_0$.

Finally, let us investigate what happens to the two su(2) symmetries of the Hubbard Hamiltonian in the continuum limit. Using the limit (2.B.13a) the spin operators S^α (see

(2.66)) turn into

$$S^\alpha = \frac{1}{2} \int_0^\ell dx \ \Psi_a^\dagger(x) \sigma_{ab}^\alpha \Psi_b(x).$$

(2.B.20)

Similarly the operator $2\eta^z + L$ turns into

$$\hat{N} = \int_0^\ell dx \ \Psi_a^\dagger(x) \Psi_a(x).$$

(2.B.21)

For $\eta^+ = \sum_{n=1}^L (-1)^n c_{n,\downarrow}^\dagger c_{n,\uparrow}^\dagger$ and $\eta^- = \sum_{n=1}^L (-1)^n c_{n,\uparrow} c_{n,\downarrow}$, however, the limit (2.B.13a) does not make sense, since the factor of $(-1)^n$ appearing under the sums does not turn into a smooth function of x. In other words, the continuum limit (2.B.13a) destroys the η-pairing symmetry of the Hubbard Hamiltonian. Similarly, the limit (2.B.13b) exists for η^u, $\alpha = x, y, z$ and for S^z, but *not* for S^x and S^y.

3

The Bethe ansatz solution

The bare existence of this book is due to the amazing fact that the solution of the stationary Schrödinger equation for the one-dimensional Hubbard model can be reduced to a set of *algebraic* equations, which is tractable in the thermodynamic limit. These equations will be derived in this chapter. We will call them the Lieb-Wu equations to honour E. H. Lieb and F. Y. Wu, who first obtained them [298]. The derivation is based on a method, called the nested (coordinate) Bethe ansatz, which goes back to the seminal articles of C. N. Yang [493] and M. Gaudin [154], who generalized earlier work [60, 296] on exactly solvable models to models with internal degrees of freedom.

The roots of the Lieb-Wu equations parameterize the eigenvalues and the eigenstates of the Hamiltonian of the one-dimensional Hubbard model. They encode the complete information about the model. These roots are not explicitly known in the general N-particle case. In the thermodynamic limit ($N \to \infty$), however, only the distributions of the roots in the complex plane matter, and many physical quantities can be calculated as solutions of linear or non-linear Fredholm type integral equations. Moreover, it is sometimes possible to use the Lieb-Wu equations in an implicit way even for finite N, e.g., in the proof of the SO(4) highest weight properties of the eigenstates in Sections 3.D and 3.F of the appendix or in the calculation of their norm in Section 3.5.

The thermodynamic limit and the derivation of suitable integral equations will be the subject of the following chapters. Here we shall concentrate on the basic ideas of the nested Bethe ansatz and shall derive the Bethe ansatz wave functions and the Lieb-Wu equations. In Section 3.1 we take advantage of the particle number conservation. We define N-particle wave functions and the N-particle Schrödinger equation they ought to satisfy. We explain how periodic boundary conditions are dealt with within the Bethe ansatz, which is a rather subtle point of the method. Section 3.2 is devoted to a careful study of the two-particle problem. The presentation is aimed at the beginner. We show all details of the calculations. An essential part of what is necessary to understand the N-particle problem can be learned here. The results for the N-particle case are presented without derivation in Sections 3.3 and 3.4. The derivations have been placed in a number of appendices, where they are shown in great detail. In Section 3.5 we describe our conjecture about the norm of the N-particle eigenstates.

The Lieb-Wu equations (3.95), (3.96) and the expressions (3.97) for the energy and momentum eigenvalues are essential for the following chapters. They will be our starting point for the exploration of the physical properties of the Hubbard model.

3.1 The Hamiltonian in first quantization

Our goal in this chapter is to construct translationally invariant eigenstates of the Hubbard Hamiltonian, i.e., we are looking for solutions of the stationary Schrödinger equation,

$$H|\psi\rangle = E|\psi\rangle,\tag{3.1}$$

which at the same time solve the eigenvalue problem

$$\hat{U}|\psi\rangle = \omega|\psi\rangle\tag{3.2}$$

for the shift operator. We may work with one of the three forms (2.22), (2.23) or (2.31) of the Hubbard Hamiltonian. They mutually commute and thus have the same eigenfunctions. For the time being let us choose (2.22).

In this section we take advantage of the fact that H and $\hat{U}$ preserve the number of particles, $[H, \hat{N}] = [\hat{U}, \hat{N}] = 0$. Hence, we may consider the eigenvalue problems (3.1), (3.2) in the sectors of fixed numbers of particles $N = 0, 1, \ldots, 2L$. This corresponds to switching from second to first quantization, or from lattice quantum field theory to quantum mechanics on the lattice.

Let us consider an arbitrary N-particle state $|\psi\rangle \in \mathcal{H}^{(L)}$ and an N-particle Wannier state $|\mathbf{x}, \mathbf{a}\rangle \in \mathcal{H}^{(L)}$. By analogy with continuum models the amplitude

$$\psi(\mathbf{x}; \mathbf{a}) = \langle \mathbf{x}, \mathbf{a}|\psi\rangle\tag{3.3}$$

is called the (N-particle) wave function. $\psi(\mathbf{x}; \mathbf{a})$ is a complex function of the positions $\mathbf{x} = (x_1, \ldots, x_N)$ and spins $\mathbf{a} = (a_1, \ldots, a_N)$ of the electrons, and, by construction, is totally antisymmetric under exchange of electrons.

The action of the Hamiltonian on Wannier states $|\mathbf{x}, \mathbf{a}\rangle$ induces an action on the N-particle wave function $\psi(\mathbf{x}; \mathbf{a})$. To conveniently express this action we introduce row vectors $\mathbf{e}^\alpha$, $\alpha = 1, \ldots, N$, that have a one in column α and zeros elsewhere.

Let us first calculate the action of the tight-binding part (2.12) of the Hubbard Hamiltonian on a state $|\mathbf{x}, \mathbf{a}\rangle$. The elementary anticommutation relations (2.2) imply that

$$[c_{j,a}^\dagger c_{j\pm1,a}, c_{k,b}^\dagger] = c_{j,a}^\dagger \{c_{j\pm1,a}, c_{k,b}^\dagger\} = \delta_{j\pm1,k}\delta_{ab}c_{j,a}^\dagger,\tag{3.4}$$

for $j = 1, \ldots, L$. Here periodic boundary conditions on the Fermi operators are understood,

$$c_{0,b}^\dagger = c_{L,b}^\dagger, \quad c_{L+1,b}^\dagger = c_{1,b}^\dagger.\tag{3.5}$$

Using (3.4) we obtain

$$[H_0, c_{x_j,b}^\dagger] = -c_{x_j-1,b}^\dagger - c_{x_j+1,b}^\dagger\tag{3.6}$$

and thus,

$$H_0|\mathbf{x}, \mathbf{a}\rangle = -\sum_{j=1}^{N} \left(|(\mathbf{x} - \mathbf{e}^j) \bmod L, \mathbf{a}\rangle + |(\mathbf{x} + \mathbf{e}^j) \bmod L, \mathbf{a}\rangle \right), \qquad (3.7)$$

where by definition $\mathbf{x} \bmod L = (x_\alpha \bmod L)\mathbf{e}^\alpha$.

The action of the interaction part D of the Hubbard Hamiltonian on Wannier states was obtained in (2.11). Combining (3.7) and (2.11) and using the hermiticity of the Hamiltonian we find

$$\langle \mathbf{x}, \mathbf{a}|H|\psi\rangle = -\sum_{j=1}^{N} \left(\Delta_{j,L}^{-}\psi(\mathbf{x}; \mathbf{a}) + \Delta_{j,L}^{+}\psi(\mathbf{x}; \mathbf{a}) \right) + 4u \sum_{1 \le j < k \le N} \delta_{x_j, x_k} \psi(\mathbf{x}; \mathbf{a}), \qquad (3.8)$$

with the operators $\Delta_{j,L}^{\pm}$ being cyclic one-particle shift operators,

$$\Delta_{j,L}^{\pm}\psi(\mathbf{x}; \mathbf{a}) = \psi\big((\mathbf{x} \pm \mathbf{e}^j) \bmod L; \mathbf{a}\big), \qquad (3.9)$$

$j = 1, \ldots, N$. The periodic boundary conditions on the Fermi operators (3.5) translate into the property $\left(\Delta_{j,L}^{\pm}\right)^L = \mathrm{id}$, for $j = 1, \ldots, N$, of the one-particle shift operators.

Equation (3.8) suggests to introduce the cyclic N-particle Hubbard Hamiltonian

$$H_N^{(L)} = -\sum_{j=1}^{N} (\Delta_{j,L}^{+} + \Delta_{j,L}^{-}) + 4u \sum_{1 \le j < k \le N} \delta_{x_j, x_k} \qquad (3.10)$$

in 'coordinate representation'. It is a hermitian operator on the Hilbert space of complex functions on $\{1, \ldots, L\}^N$ endowed with the scalar product

$$\langle \psi, \varphi \rangle = \frac{1}{N!} \sum_{x_1, \ldots, x_N = 1}^{L} \sum_{a_1, \ldots, a_N = \uparrow, \downarrow} \overline{\psi(\mathbf{x}; \mathbf{a})} \varphi(\mathbf{x}; \mathbf{a}). \qquad (3.11)$$

Given an N-particle solution $|\psi\rangle$ of the eigenvalue problem (3.1) we have $\langle \mathbf{x}, \mathbf{a}|H|\psi\rangle = E\psi(\mathbf{x}; \mathbf{a})$, and it follows from (3.8) that $\psi(\mathbf{x}; \mathbf{a})$ solves the eigenvalue problem

$$(H_N^{(L)} - E)\psi(\mathbf{x}; \mathbf{a}) = 0. \qquad (3.12)$$

Thus, every N-particle solution of the eigenvalue problem (3.1) provides a solution of the eigenvalue problem (3.12) for the cyclic N-particle Hamiltonian (3.10).

The converse is also true: be $\psi(\mathbf{x}; \mathbf{a})$ a totally antisymmetric solution of the Schrödinger equation (3.12). Define the state

$$|\psi\rangle = \frac{1}{N!} \sum_{x_1, \ldots, x_N = 1}^{L} \sum_{a_1, \ldots, a_N = \uparrow, \downarrow} \psi(\mathbf{x}; \mathbf{a})|\mathbf{x}, \mathbf{a}\rangle. \qquad (3.13)$$

Then $\langle \mathbf{x}, \mathbf{a}|\psi\rangle = \psi(\mathbf{x}; \mathbf{a})$, and $|\psi\rangle$ satisfies (3.1). Equations (3.3) and (3.13) are mutually inverse. These formulae describe how to switch from second to first quantization and vice versa. A derivation of (3.13) is presented in appendix 3.A.

We have just seen that, instead of dealing with (3.1), we may equivalently solve the Schrödinger equation (3.12) for $N = 0, 1, \ldots, 2L$. Due to the cyclic nature of the one-particle shift operators $\Delta_{j,L}^{\pm}$, however, equation (3.12) is still hard to deal with. Instead of working directly with equation (3.12) the Bethe ansatz method proceeds in two separate steps: let us introduce (non-cyclic) shift operators $\Delta_j^{\pm}$,

$$\Delta_j^{\pm} \psi(\mathbf{x}; \mathbf{a}) = \psi(\mathbf{x} \pm \mathbf{e}^j; \mathbf{a}), \tag{3.14}$$

$j = 1, \ldots, N$, which necessarily act on functions $\psi(\mathbf{x}, \mathbf{a})$ defined on the infinite lattice. In terms of these operators we define the Hamiltonian

$$H_N = -\sum_{j=1}^N (\Delta_j^+ + \Delta_j^-) + 4u \sum_{1 \le j < k \le N} \delta_{x_j, x_k}. \tag{3.15}$$

Then the first step in the Bethe ansatz calculation is to solve the difference equation

$$(H_N - E)\psi(\mathbf{x}; \mathbf{a}) = 0 \tag{3.16}$$

on the infinite lattice, $\mathbf{x} \in \mathbb{Z}^N$. In a second step we require the solutions of (3.16) to satisfy the equations

$$(\Delta_j^{\pm} - \Delta_{j,L}^{\pm})\psi(\mathbf{x}; \mathbf{a}) = 0, \tag{3.17}$$

for $\mathbf{x} \in \{1, \ldots, L\}^N$ and $j = 1, \ldots, N$. Obviously, every solution of (3.16) and (3.17) provides a solution of (3.12). However, as we shall see below, the converse is not true. We shall call equation (3.16) 'the N-particle Schrödinger equation' and equation (3.17) 'the periodic boundary conditions'.

Let us also translate the equation (3.2) for the shift operator into its first quantized form. From equation (2.43) we deduce

$$\langle \mathbf{x}, \mathbf{a} | \hat{U} | \psi \rangle = \prod_{j=1}^N \Delta_{j,L}^+ \psi(\mathbf{x}; \mathbf{a}). \tag{3.18}$$

Note that there is no ordering required in the product on the right-hand side, since $[\Delta_{j,L}^+, \Delta_{k,L}^+] = 0$. We now proceed similarly as in case of the Schrödinger equation. Every eigenstate of $\hat{U}$ with eigenvalue ω leads to

$$(\hat{U}_N^{(L)} - \omega)\psi(\mathbf{x}; \mathbf{a}) = 0, \tag{3.19}$$

where $\hat{U}_N^{(L)}$ is the cyclic shift operator in 'coordinate representation',

$$\hat{U}_N^{(L)} = \prod_{j=1}^N \Delta_{j,L}^+. \tag{3.20}$$

Instead of looking for solutions of (3.19) we introduce the infinite interval counterpart

$$\hat{U}_N = \prod_{j=1}^N \Delta_j^+ \tag{3.21}$$

of the cyclic shift operator $\hat{U}_N^{(L)}$ and solve the difference equation

$$(\hat{U}_N - \omega)\psi(\mathbf{x}; \mathbf{a}) = 0 \tag{3.22}$$

subject to the periodic boundary conditions (3.17). Every solution of (3.22) that satisfies (3.17) is a solution of (3.19).

In the following section we discuss (3.16), (3.22) and (3.17) in the simplest non-trivial case of two particles. This will prepare us for the solution of the general N-particle case, which is presented in Section 3.3 and derived in detail in appendix 3.B. The two-particle case will help us to develop an intuitive understanding of the lattice model. We will learn, for instance, that on the lattice there may be bound states of electrons, even though the interaction is repulsive. Yang's Y-operators, which are an essential technical tool in the derivation of the solution of the N-particle problem, are easily introduced and understood within the context of the two-particle problem. Finally, we shall see in the next section that equations (3.16) and (3.17) are *not* equivalent to the N-particle eigenvalue problem (3.12). They are sufficient but not necessary. A solution of (3.12) exists, the so-called η-pair, which does not follow directly from (3.16), (3.17) but indirectly by acting with the operator η^+ on the empty lattice $|0\rangle$, which may be considered as a solution of (3.16), (3.17) in the sector $N = 0$. This situation, exemplified with the two-particle case, will turn out to be typical. The empty lattice is a so-called Bethe ansatz state, i.e., by definition a solution of (3.16), (3.17) of a certain form (see below). We shall see in Section 3.4 that all Bethe ansatz states are annihilated by η^- and that more eigenstates can be obtained by acting on Bethe ansatz states with η^+.

3.2 Solution of the two-particle problem

We start our discussion of the Schrödinger equation (3.16) with a thorough discussion of the two-particle case. In this case the Schrödinger equation reads

$$H_2\psi(x_1, x_2; a_1, a_2) = E\psi(x_1, x_2; a_1, a_2), \tag{3.23}$$

$$H_2 = -(\Delta_1^+ + \Delta_1^- + \Delta_2^+ + \Delta_2^-) + 4u\delta_{x_1,x_2}. \tag{3.24}$$

The equation (3.22) for the two-particle shift operator $\hat{U}_2 = \Delta_1^+ \Delta_2^+$ is

$$\hat{U}_2\psi(x_1, x_2; a_1, a_2) = \omega\psi(x_1, x_2; a_1, a_2). \tag{3.25}$$

In a first step we shall construct common solutions of (3.23) and (3.25), which are totally antisymmetric,

$$\psi(x_1, x_2; a_1, a_2) = -\psi(x_2, x_1; a_2, a_1). \tag{3.26}$$

After imposing the periodic boundary conditions (3.17) in a second step these solutions become two-particle, translationally invariant electronic wave functions corresponding to the Hamiltonian (2.22).

3.2.1 Separation of variables

Since the 'two-body potential' $4u\delta_{x_1,x_2}$ depends only on the difference $x_1 - x_2$, we expect the Schrödinger equation (3.23) to separate after introducing centre of mass coordinates $m = x_1 + x_2$ and relative coordinates $n = x_1 - x_2$. Let

$$\psi(x_1, x_2; a_1, a_2) = f(m)g(n), \qquad (3.27)$$

where we suppressed the spin variables on the right-hand side for brevity. Insertion of (3.27) into (3.23) leads to

$$\frac{f(m + 1) + f(m - 1)}{f(m)} = \frac{(4u\delta_{n,0} - E)g(n)}{g(n + 1) + g(n - 1)}. \qquad (3.28)$$

The left hand side of this equation depends only on m, the right hand side depends only on n. Hence it may be separated into

$$f(m + 1) + f(m - 1) = Cf(m), \qquad (3.29)$$
$$C(g(n + 1) + g(n - 1)) = (4u\delta_{n,0} - E)g(n), \qquad (3.30)$$

where C is independent of m and n. Equations (3.29) and (3.30) are the discrete analogues of the Schrödinger equation of a free particle and of a particle scattered by a delta-function potential, respectively.

3.2.2 The centre of mass motion

Equation (3.29) is a second order linear difference equation with constant coefficients. It has two fundamental solutions, $w^{\pm m}$, where $w + 1/w = C$. The general solution of equation (3.29) is a linear combination

$$f(m) = A^+ w^m + A^- w^{-m} \qquad (3.31)$$

of its fundamental solutions with two complex amplitudes A^+, A^-. Assuming $g(n)$ not to be identically vanishing, the equation (3.25) for the shift operator is equivalent to

$$f(m + 2) = \omega f(m). \qquad (3.32)$$

So (for $w \neq 1$), either A^+ or A^- in (3.31) must be zero, and we obtain, say, $f(m) = A^+ w^m$. It follows that

$$\omega = w^2. \qquad (3.33)$$

The amplitude A^+ may, in principle, be spin dependent. Since it merely appears as an overall factor in equation (3.27), however, we may assume without any loss of generality that only $g(n)$ depends on the spin variables and that

$$f(m) = w^m. \qquad (3.34)$$

3.2.3 The relative motion

Let us assume for a while that $C \neq 0$. The degenerate case $C = 0$ will be discussed later. Then, outside the origin, $n = 0$, equation (3.30) is a second order linear difference equation with constant coefficients. It has the general solution

$$g(n) = \begin{cases} A^{-+}z_-^n - A^{--}z_-^{-n} & n < 0 \\ A^{++}z_+^n - A^{+-}z_+^{-n} & n > 0, \end{cases} \tag{3.35}$$

where the minus signs have been introduced for later convenience. Let us insert (3.35) into (3.30). Assuming that $g(n)$ does not vanish identically we obtain

$$E = -\left(w + \frac{1}{w}\right)\left(z_+ + \frac{1}{z_+}\right) = -\left(w + \frac{1}{w}\right)\left(z_- + \frac{1}{z_-}\right). \tag{3.36}$$

Since $w^2 \neq -1$ by hypothesis, it follows that

$$z_+ + \frac{1}{z_+} = z_- + \frac{1}{z_-} \Leftrightarrow (z_+ - z_-)\left(1 - \frac{1}{z_+ z_-}\right) = 0$$

$$\Leftrightarrow z_+ = z_- \quad \text{or} \quad z_+ = \frac{1}{z_-}. \tag{3.37}$$

Thus, the general form of $g(n)$ must be

$$g(n) = \begin{cases} A^{-+}z^n - A^{--}z^{-n} & n < 0 \\ A^{++}z^n - A^{+-}z^{-n} & n > 0. \end{cases} \tag{3.38}$$

For $g(n)$ to be uniquely defined at $n = 0$ it is necessary and sufficient that

$$A^{+-} + A^{-+} = A^{++} + A^{--}. \tag{3.39}$$

Furthermore, $g(n)$ has to satisfy (3.30) at $n = 0$. This requirement leads to

$$\left(w + \frac{1}{w}\right)\left(g(1) + g(-1)\right) = (4u - E)g(0). \tag{3.40}$$

Here we insert (3.38) and then use (3.39). After a short calculation we obtain

$$\frac{1}{2}\left(wz - \frac{1}{wz} - \frac{w}{z} + \frac{z}{w}\right)(A^{++} + A^{+-} - A^{-+} - A^{--}) = 4u(A^{-+} - A^{--}). \tag{3.41}$$

We now introduce $k_1, k_2 \in \mathbb{C}$ such that

$$wz = e^{ik_1}, \quad \frac{w}{z} = e^{ik_2}, \tag{3.42}$$

and define

$$s_j = \sin k_j, \quad j = 1, 2. \tag{3.43}$$

Then, (3.41) turns into

$$(s_1 - s_2)(A^{++} + A^{+-} - A^{-+} - A^{--}) = 4iu(A^{--} - A^{-+}). \tag{3.44}$$

Equations (3.39) and (3.44) are two equations that restrict the choice of the amplitudes $A^{\alpha\beta}$. Further restrictions arise from the antisymmetry of the wavefunction (3.26), i.e., from the fact that we are dealing with fermions. In order to implement these restrictions we have to remember that the amplitudes $A^{\alpha\beta}$, $\alpha, \beta = \pm$, depend on the spin variables $a_1, a_2 = \uparrow, \downarrow$. The amplitudes $A^{\alpha\beta}$ are spinors with components $A^{\alpha\beta}_{a_1 a_2}$.

Let us introduce the permutation matrix Π that interchanges the spin variables,

$$\Pi^{b_1 b_2}_{a_1 a_2} = \delta^{b_1}_{a_2}\delta^{b_2}_{a_1}. \tag{3.45}$$

Then

$$(\Pi A^{\alpha\beta})_{a_1 a_2} = \Pi^{b_1 b_2}_{a_1 a_2} A^{\alpha\beta}_{b_1 b_2} = A^{\alpha\beta}_{a_2 a_1}. \tag{3.46}$$

Here we used implicit summation over doubly occurring indices. The wave function $\psi(x_1, x_2; a_1, a_2)$ must be totally antisymmetric. Since $f(m) = f(x_1 + x_2)$ is symmetric in the electron coordinates and does not depend on the spin variables, we must have $g(-n) = -\Pi g(n)$. This leads to

$$A^{++}z^{-n} - A^{+-}z^n = -\Pi A^{-+}z^n + \Pi A^{--}z^{-n} \tag{3.47}$$

for all $n < 0$. It follows that

$$A^{+-} = \Pi A^{-+}, \quad A^{++} = \Pi A^{--}. \tag{3.48}$$

Thus, we have obtained four conditions (equations (3.39), (3.44) and two equations (3.48)) on four amplitudes $A^{\alpha\beta}$. As we shall see below only three of these four conditions are independent, and we will remain with a single free amplitude.

From this point there are two ways to proceed in our calculation. The more elementary way is by assuming that the amplitudes $A^{\alpha\beta}$, $\alpha, \beta = \pm$, are either all symmetric or all antisymmetric in the spin variables. This means to assume that the two electrons are in a spin triplet or in a spin singlet state, which would be justified because the Hamiltonian H_2, equation (3.24), does not depend on the spin variables. This way of proceeding is, however, not much in the spirit of the Bethe ansatz calculation for the N-electron system, and is therefore not presented here, but left as an exercise to the reader. Instead, we will proceed in a different way, that will naturally lead us to the introduction of the so-called Y-operators of C. N. Yang. These operators play a crucial role in the generalization of the two-particle results to an arbitrary number of particles N.

Let us insert (3.48) into (3.39) and (3.44). We obtain

$$(1 + \Pi)A^{--} = (1 + \Pi)A^{-+} \tag{3.49}$$

and

$$\left(4iu + (s_1 - s_2)(1 - \Pi)\right)A^{--} = \left(4iu - (s_1 - s_2)(1 - \Pi)\right)A^{-+}. \tag{3.50}$$

The latter equation is easily solved for A^{--}. Using $\Pi^2 = 1$, we obtain

$$\left(4iu + (s_1 - s_2)(1 + \Pi)\right)\left(4iu + (s_1 - s_2)(1 - \Pi)\right) = 4iu\left(4iu + 2(s_1 - s_2)\right), \tag{3.51}$$

$$\left(4iu + (s_1 - s_2)(1 + \Pi)\right)\left(4iu - (s_1 - s_2)(1 - \Pi)\right) = 4iu\left(4iu + 2(s_1 - s_2)\Pi\right), \tag{3.52}$$

and thus,

$$A^{--} = \frac{2iu + (s_1 - s_2)\Pi}{2iu + (s_1 - s_2)} A^{-+}. \tag{3.53}$$

Equations (3.49) and (3.53) are *two* equations that connect A^{--} and A^{-+}. Are they compatible? The answer is yes. Since $\Pi^2 = 1$, we have $(1 + \Pi)\Pi = (1 + \Pi)$, and (3.53) implies (3.49).

Now we define the Y-operator

$$Y(\lambda) = \frac{2iu + \lambda\Pi}{2iu + \lambda}. \tag{3.54}$$

With the aid of this operator (3.53) can be written as

$$A^{--} = Y(s_1 - s_2)A^{-+}. \tag{3.55}$$

Using (3.48) and (3.55) we can express all amplitudes $A^{\alpha\beta}$ in (3.38) in terms of A^{-+}, and $g(n)$ takes the following form,

$$g(n) = \begin{cases} A^{-+}z^n - Y(s_1 - s_2)A^{-+}z^{-n} & n \leq 0 \\ Y(s_1 - s_2)\Pi A^{-+}z^n - \Pi A^{-+}z^{-n} & n \geq 0. \end{cases} \tag{3.56}$$

Finally, we may insert (3.34) and (3.56) into (3.27). Because of (3.42) we have

$$w^m z^n = e^{i(k_1 x_1 + k_2 x_2)}, \qquad w^m z^{-n} = e^{i(k_1 x_2 + k_2 x_1)}, \tag{3.57}$$

and thus,

$$\psi(x_1, x_2) = \begin{cases} A^{-+}e^{i(k_1 x_1 + k_2 x_2)} - Y(s_1 - s_2)A^{-+}e^{i(k_1 x_2 + k_2 x_1)} & x_1 \leq x_2 \\ Y(s_1 - s_2)\Pi A^{-+}e^{i(k_1 x_1 + k_2 x_2)} - \Pi A^{-+}e^{i(k_1 x_2 + k_2 x_1)} & x_2 \leq x_1. \end{cases} \tag{3.58}$$

By construction, $\psi(x_1, x_2)$ is a totally antisymmetric solution of the Schrödinger equation (3.23) for arbitrary $k_1, k_2 \in \mathbb{C}$, $k_1 + k_2 \neq \pi \bmod 2\pi$. The corresponding values of E and ω in (3.23) and (3.25) are (see (3.36), (3.33))

$$E = -2\cos(k_1) - 2\cos(k_2), \tag{3.59}$$

$$\omega = e^{i(k_1 + k_2)}. \tag{3.60}$$

The wavefunction $\psi(x_1, x_2)$ is determined by equation (3.58) up to an arbitrary function A^{-+} of the spin variables a_1, a_2, which accounts for the fact that the Hamiltonian (3.24) is spin independent. In general, A^{-+} is a linear combination of the spin singlet state

$$\varphi_a(a_1, a_2) = \delta_{a_1,\uparrow}\delta_{a_2,\downarrow} - \delta_{a_1,\downarrow}\delta_{a_2,\uparrow} \tag{3.61}$$

and the spin triplet states

$$\varphi_s(a_1, a_2) = \begin{cases} \delta_{a_1,\uparrow}\delta_{a_2,\uparrow} \\ \delta_{a_1,\uparrow}\delta_{a_2,\downarrow} + \delta_{a_1,\downarrow}\delta_{a_2,\uparrow} \\ \delta_{a_1,\downarrow}\delta_{a_2,\downarrow} \end{cases}. \tag{3.62}$$

These states span the space of states of two spins $\frac{1}{2}$ and correspond to the irreducible subspaces with total spin $S = 0$ and $S = 1$, respectively. Substituting φ_a or φ_s for A^{-+} in (3.58) and using that

$$\Pi\varphi_a = -\varphi_a, \quad \Pi\varphi_s = \varphi_s, \tag{3.63}$$

we obtain the joint spin singlet solution,

$$\psi(x_1, x_2) = \varphi_a \cdot \begin{cases} e^{i(k_1x_1+k_2x_2)} + \frac{s_1-s_2-2iu}{s_1-s_2+2iu} e^{i(k_1x_2+k_2x_1)} & x_1 \leq x_2 \\ \frac{s_1-s_2-2iu}{s_1-s_2+2iu} e^{i(k_1x_1+k_2x_2)} + e^{i(k_1x_2+k_2x_1)} & x_2 \leq x_1, \end{cases} \tag{3.64}$$

and spin triplet solutions,

$$\psi(x_1, x_2) = \varphi_s \cdot (e^{i(k_1x_1+k_2x_2)} - e^{i(k_1x_2+k_2x_1)}), \tag{3.65}$$

of the two-particle Schrödinger equation (3.23) and of the eigenvalue equation (3.25) for the shift operator. The spin-singlet state (3.64) is antisymmetric in the spin variables and symmetric in the electron coordinates while the situation is reversed for the spin-triplet states (3.65). They are symmetric in the spin variables and antisymmetric in the electron coordinates. For the latter reason two electrons never sit at the same site and therefore never feel the local interaction. This makes the triplet wave functions look like the wave functions of free electrons.

So far we have restricted our attention to the case $C \neq 0$, when equation (3.30) is non-degenerate. For $C = 0$ equation (3.30) turns into

$$(4u\delta_{n,0} - E)g(n) = 0. \tag{3.66}$$

This equation has non-trivial solutions in two cases, (a) for $E = 0$ and (b) for $E = 4u$. In case (a) we necessarily have $g(0) = 0$. In case (b), on the contrary, we must have $g(n) = 0$ for $n \neq 0$. In both cases $f(m)$ is given by (3.34) with $w^2 = -1$.

Let us consider case (a). In this case (3.38) is still a non-trivial solution of (3.66) for $n \neq 0$, albeit not the most general one. Furthermore, if (3.39) and (3.40) are satisfied for $E = 0$, then $g(n)$ according to (3.38) is a solution of (3.66), and all the calculations following (3.38) go through. We conclude that we may relax our restriction $k_1 + k_2 \neq \pi \mod 2\pi$. The function $\psi(x_1, x_2)$, equation (3.58), is a totally antisymmetric, translationally invariant solution of the two-particle Schrödinger equation for *arbitrary* $k_1, k_2 \in \mathbb{C}$.

In case (b) the function $g(n)$ is of the form $g(n) = A\delta_{n,0}$. Antisymmetry requires $g(-n) = -\Pi g(n)$. It follows that $\Pi A = -A$, and thus, without any loss of generality, $g(n) = \varphi_a\delta_{n,0} = \varphi_a\delta_{x_1,x_2}$. Inserting this result and $f(m) = (\pm i)^m$ into (3.27) we end up with

$$\psi(x_1, x_2) = \varphi_a \, (-1)^{x_1} \delta_{x_1, x_2} \, . \tag{3.67}$$

As we shall see in the next subsection, this solution is also contained as a limit in our general solution (3.58). It describes a bound state that is localized in the centre of mass frame.

Remark. It is clear from equation (3.56) that

$$X(s_1 - s_2) = Y(s_1 - s_2)\Pi \tag{3.68}$$

is the two-particle S-matrix, since (3.56) for $k_1, k_2 \in \mathbb{R}$ describes a scattering solution of the Schrödinger equation (3.23). The operators $X(\lambda)$ and $Y(\lambda)$ play a crucial role in the construction of the N-electron Bethe ansatz wavefunction (see appendix 3.B).

3.2.4 Eigenstates on the infinite interval

So far we have constructed the joint solutions (3.64), (3.65) of the two-particle Schrödinger equation (3.23) and the equation (3.25) for the shift operator. We have not yet specified the boundary conditions. By imposing periodic boundary conditions the Schrödinger equation (3.23) turns into an eigenvalue equation for the Hamiltonian H_2. The derivation of the spectrum and the eigenfunctions of H_2 under periodic boundary conditions is the main purpose of this section and will be completed in the next subsection. Here we deviate from our main line of reasoning and consider the Hamiltonian on the infinite interval.

Any solution of the two-particle Schrödinger equation (3.23) which is bounded for $x_1, x_2 \to \pm\infty$ is an eigenfunction of the Hamiltonian on the infinite interval. Hence, in order to find all eigenfunctions, we have to find all values of $k_1, k_2 \in \mathbb{C}$ for which our general solutions (3.64) and (3.65) are bounded. Clearly, this is the case for all real k_1, k_2. When k_1 and k_2 are real, the spin singlet and spin triplet states (3.64) and (3.65) describe the scattering of two electrons. The energy of the scattering states is given by (3.59). Note that for fixed k_1, k_2 the spin singlet state and the spin triplet states belong to degenerate energy eigenvalues. This degeneracy is peculiar of the system on the infinite interval and is lifted by periodic boundary conditions (see next subsection). Later, in Chapter 15, we shall see that the degeneracy is due to an additional so-called Yangian symmetry (see Chapter 14), which occurs in the thermodynamic limit.

Is it possible to have eigenstates with non-real k_1 and k_2? The answer is different for the spin-singlet and spin-triplet solutions, respectively. If k_1 or k_2 in (3.65) has non-zero imaginary part, the wave function is always unbounded. Thus, k_1 and k_2 must be real in the spin triplet state. Similarly, the spin singlet solution (3.64) is easily seen to be unbounded, if $s_1 - s_2 \neq 2iu$. For $s_1 - s_2 = 2iu$, however, the two scattering amplitudes in (3.64) vanish, and the spin singlet solution simplifies to

$$\psi(x_1, x_2) = \varphi_a \mathrm{e}^{i(k_1+k_2)(x_1+x_2)/2} \mathrm{e}^{-i(k_1-k_2)|x_1-x_2|/2} \, . \tag{3.69}$$

This solution is bounded, if and only if $\text{Im}(k_1 + k_2) = 0$ and $\text{Im}(k_1 - k_2) \leq 0$, and there is indeed a possibility to have non-real k_1 and k_2 of the form $k_1 = q_1 - i\kappa$, $k_2 = q_2 + i\kappa$ with $q_1, q_2 \in [0, 2\pi]$, $\kappa > 0$. Assuming $u \neq 0$, the condition $s_1 - s_2 = 2iu$ implies $q_1 = q_2$ and, using the shorthand notation $q = q_1 = q_2$,

$$\kappa(q) = -\text{arsinh}\left(\frac{u}{\cos q}\right). \tag{3.70}$$

The latter equation is compatible with $\kappa > 0$ for $\frac{\pi}{2} < q < \frac{3\pi}{2}$ and $u > 0$, or for $\frac{\pi}{2} < |q - \pi| < \pi$ and $u < 0$.

We thus have obtained the one-parameter family of eigenstates

$$\psi(x_1, x_2) = \varphi_a e^{iq(x_1+x_2)} e^{-\kappa(q)|x_1-x_2|} \tag{3.71}$$

of the Hubbard Hamiltonian on the infinite interval. $\psi(x_1, x_2)$ describes a bound state, which is moving with centre of mass momentum q. The interesting point about this state is that it exists even if the interaction between the electrons is repulsive ($u > 0$). The existence of a bound state for repulsive coupling is a lattice effect. In fact, in the continuum limit (see appendix 2.B), when we replace q by $a_0 q$, κ by $a_0 \kappa$, u by $a_0 c$ and take the limit $a_0 \to 0$ for fixed κ, q and c, equation (3.70) turns into $\kappa = -c$, and κ can only be positive for negative c. Hence, there are no bound states in the continuum model if the interaction is repulsive.

Let us note that the bound state solution (3.71) turns into (3.67) in the limit $q \to \frac{\pi}{2}$.

3.2.5 Periodic boundary conditions

Next, we want to single out those of the wave functions $\psi(x_1, x_2; a_1, a_2)$ that satisfy the periodic boundary conditions (3.17). In the two-particle case equations (3.17) reduce to four non-trivial conditions,

$$\psi(0, x_2; \mathbf{a}) = \psi(L, x_2; \mathbf{a}), \quad \psi(L + 1, x_2; \mathbf{a}) = \psi(1, x_2; \mathbf{a}), \tag{3.72a}$$

$$\psi(x_1, 0; \mathbf{a}) = \psi(x_1, L; \mathbf{a}), \quad \psi(x_1, L + 1; \mathbf{a}) = \psi(x_1, 1; \mathbf{a}). \tag{3.72b}$$

Because of the antisymmetry it is sufficient to consider one pair of equations. They are equivalent to one another. Suppose, for instance, that (3.72a) is satisfied; then

$$\psi(x_1, 0; a_1, a_2) = -\psi(0, x_1; a_2, a_1) = -\psi(L, x_1; a_2, a_1) = \psi(x_1, L; a_1, a_2), \tag{3.73}$$

which is the first equation (3.72b).

Let us insert the wave function (3.58) into the first equation (3.72a), and let $x_2 \in \{1, \ldots, L\}$. Using $Y(\lambda)Y(-\lambda) = 1$ we obtain

$$e^{ik_1 L}\left(Y(s_1 - s_2)\Pi - e^{-ik_1 L}\right) A^{-+} e^{ik_2 x_2}$$
$$-e^{ik_2 L} Y(s_1 - s_2)\left(Y(s_2 - s_1)\Pi - e^{-ik_2 L}\right) A^{-+} e^{ik_1 x_2} = 0. \tag{3.74}$$

A sufficient condition for (3.74) to be satisfied is

$$X(s_1 - s_2)A^{-+} = e^{-ik_1 L}A^{-+}, \tag{3.75a}$$

$$X(s_2 - s_1)A^{-+} = e^{-ik_2 L}A^{-+}. \tag{3.75b}$$

Equations (3.75) are also necessary, if $k_1 \neq k_2$, since, in this case, the functions $e^{ik_1 x_2}$ and $e^{ik_2 x_2}$ are linearly independent. Inserting the wave function (3.58) into the second equation (3.72a) we obtain an equation which is similar to (3.74) and leads again to (3.75). Thus, the equations (3.75) are sufficient for $\psi(x_1, x_2; a_1, a_2)$ to satisfy the periodic boundary conditions (3.72).

From the explicit form

$$X(s_1 - s_2) = \frac{s_1 - s_2 + 2iu\,\Pi}{s_1 - s_2 + 2iu} \tag{3.76}$$

of the two-particle S-matrix $X(s_1 - s_2)$ we see that every eigenvector of the permutation matrix Π is an eigenvector of $X(s_1 - s_2)$. The eigenvectors of Π are the spin singlet and spin triplet states (see (3.63)). Applying $X(s_1 - s_2)$ to (3.61) and (3.62) we obtain

$$X(s_1 - s_2)\varphi_a = \frac{s_1 - s_2 - 2iu}{s_1 - s_2 + 2iu}\,\varphi_a, \tag{3.77a}$$

$$X(s_1 - s_2)\varphi_s = \varphi_s. \tag{3.77b}$$

Hence, equations (3.75) are satisfied, if $A^{-+} = \varphi_a$ and if the 'charge momenta' k_1 and k_2 satisfy the quantization conditions

$$e^{ik_1 L} = \frac{s_1 - s_2 + 2iu}{s_1 - s_2 - 2iu}, \qquad e^{ik_2 L} = \frac{s_2 - s_1 + 2iu}{s_2 - s_1 - 2iu}. \tag{3.78}$$

Similarly, in the spin-triplet case $A^{-+} = \varphi_s$, we obtain the quantization conditions

$$e^{ik_1 L} = e^{ik_2 L} = 1. \tag{3.79}$$

The corresponding wave functions are given by (3.64) and (3.65), the energies and eigenvalues of the shift operator by (3.59) and (3.60).

The wave functions (3.64) and the wave functions (3.65) with $\varphi_s(a_1, a_2) = \delta_{a_1, \uparrow}\delta_{a_2, \uparrow}$ are the so-called Bethe-ansatz eigenfunctions of the two-particle Hubbard Hamiltonian in 'coordinate representation' (3.24). In the next section we shall generalize these wave functions to the many particle case. The solutions of equations (3.78) and (3.79) and the issue of completeness will be discussed in Chapter 4.

Remark. The alert reader may wonder how our general result in the next sections compares for $N = 2, M = 1$ with (3.78). In general, except for the charge momenta, we have a second set of quantum numbers that characterize the spin degrees of freedom. For $N = 2, M = 1$

there is one such quantum number λ, and equations (3.95) and (3.96) turn into

$$e^{ik_1 L} = \frac{\lambda - s_1 - iu}{\lambda - s_1 + iu}, \qquad e^{ik_2 L} = \frac{\lambda - s_2 - iu}{\lambda - s_2 + iu}, \qquad (3.80)$$

$$\frac{\lambda - s_1 - iu}{\lambda - s_1 + iu} \cdot \frac{\lambda - s_2 - iu}{\lambda - s_2 + iu} = 1 . \qquad (3.81)$$

To reproduce (3.78) from these equations one has to solve (3.81) for λ and has to insert the result, $\lambda = (s_1 + s_2)/2$, into (3.80).

3.2.6 The η-pair

An important issue in the context of the Bethe ansatz method is the question of completeness. We shall elaborate on this question in Chapter 4, where we are counting eigenstates. Here we want to point out a subtlety which is closely related to the completeness problem.

Consider the solution $\psi(x_1, x_2) = \varphi_a (-1)^{x_1} \delta_{x_1, x_2}$ (see equation (3.67)) of the two-particle Schrödinger equation. Clearly this solution is incompatible with the periodic boundary conditions (3.72). For we have, for instance, $\psi(0, L) = 0 \neq \psi(L, L) = \varphi_a(-1)^L$. Hence, this solution is outside the Bethe ansatz. On the other hand, the reader may easily verify that this solution is an eigenfunction of the cyclic N-particle Hubbard Hamiltonian $H_2^{(L)}$, equation (3.10):

$$(H_2^{(L)} \psi)(x_1, x_2) = 4u \, \psi(x_1, x_2) , \qquad (3.82)$$

if L is even. We learn from this example that the N-particle Schrödinger equation (3.16) and the periodic boundary conditions (3.17) do *not* determine all eigenfunctions of the cyclic N-particle Hubbard Hamiltonian $H_N^{(L)}$. We claim, however, that the states missed by the Bethe ansatz can be obtained by acting with symmetry operators on Bethe ansatz states (see Section 3.4). In the present case this statement means the following. Let us use equation (3.13) to switch back to the language of second quantization. We obtain

$$|\psi\rangle = \frac{1}{2} \sum_{x_1, x_2 = 1}^{L} \sum_{a_1, a_2 = \uparrow, \downarrow} (-1)^{x_1} \delta_{x_1, x_2} (\delta_{a_1, \uparrow} \delta_{a_2, \downarrow} - \delta_{a_1, \downarrow} \delta_{a_2, \uparrow}) c_{x_2, a_2}^{\dagger} c_{x_1, a_1}^{\dagger} |0\rangle$$

$$= \sum_{x_1 = 1}^{L} (-1)^{x_1} c_{x_1, \downarrow}^{\dagger} c_{x_1, \uparrow}^{\dagger} |0\rangle . \qquad (3.83)$$

Comparing the latter result with our definition (2.80a) of the operator η^+, we see that

$$|\psi\rangle = \eta^+ |0\rangle . \qquad (3.84)$$

The state $|\psi\rangle$ is obtained by acting with the operator η^+ on the Fock vacuum. $|\psi\rangle$ is not a Bethe ansatz state but $|0\rangle$ is. We shall see below that our example is typical. For even L all N-particle Bethe ansatz states (see Section 3.3) are lowest weight states with respect to the η-spin representation of su(2). Multiplets of eigenstates of the Hubbard Hamiltonian are thus generated by acting with η^+ on the Bethe ansatz states.

We should mention here that a similar statement holds for the spin su(2) symmetry. Our method of solving the N-particle generalization of the eigenvalue problem (3.75) will give us at first instance only those eigenstates which are highest weight with respect to spin. The missing states then can be obtained by acting with the ladder operator S^- on the highest weight states. Applying our general result to the two-particle problem, we would obtain only one of the spin triplet states, namely $\varphi_s(a_1, a_2) = \delta_{a_1,\uparrow}\delta_{a_2,\uparrow}$.

3.3 Many-particle wave functions and Lieb-Wu equations

Now we shall show how the results of the previous section generalize to an arbitrary number of electrons. We shall see that the general N-particle Bethe ansatz eigenstates of the Hubbard Hamiltonian (2.31) parametrically depend on two sets of quantum numbers: the charge momenta k_j, $j = 1, \ldots, N$, which also determine the total momentum and the energy of the Bethe ansatz states, and the spin rapidities λ_ℓ, $\ell = 1, \ldots, M$. The charge momenta and spin rapidities obey a set of coupled algebraic equations that basically arise as a consequence of the periodic boundary conditions. These algebraic equations are the Lieb-Wu equations mentioned in the introduction to this chapter.

The derivation of the Bethe ansatz eigenstates and the Lieb-Wu equations is rather lengthy and technical and is therefore presented separately in appendix 3.B. Here we concentrate on the mere description of the result which we recommend the readers to familiarize themselves with before studying its derivation.

3.3.1 The symmetric group

In this section and also in appendix 3.B where the results of this section are derived, we shall resort to some basic knowledge of the symmetric group. We therefore begin by recalling a few facts and introduce notations appropriate for our purposes. The symmetric group of order N, $\mathfrak{S}^N$, is the group of permutations of N distinct objects. In more mathematical terms we may define $\mathfrak{S}^N$ as the set of all one-to-one mappings of the set of numbers $\mathbb{Z}_N = \{1, \ldots, N\}$ onto itself.

A transposition $\Pi_{j,k}$ is a permutation that interchanges j and k and leaves all other elements of $\mathbb{Z}_N$ fixed. The symmetric group $\mathfrak{S}^N$ is generated by the transpositions $\Pi_{n,n+1}$, $n = 1, \ldots, N - 1$, of nearest neighbours. A realization of a permutation as a product of transpositions of nearest neighbours is, of course, not unique. But for a given permutation the number of factors is always either odd or even. Accordingly, a permutation is called odd or even. This defines the sign function or parity on $\mathfrak{S}^N$: $\text{sign}(Q) = 1$, if Q is even, and $\text{sign}(Q) = -1$, if Q is odd. Clearly $\text{sign}(PQ) = \text{sign}(P)\text{sign}(Q)$, $\text{sign}(Q^{-1}) = \text{sign}(Q)$, and $\text{sign}(\Pi_{j,k}) = -1$. It follows that the even permutations form a subgroup $\mathfrak{A}^N$ of $\mathfrak{S}^N$.

This subgroup is called the alternating group. Any transposition $\Pi_{j,k}$ generates a coset decomposition, $\mathfrak{S}^N = \mathfrak{A}^N \cup \Pi_{j,k}\mathfrak{A}^N = \mathfrak{A}^N \cup \mathfrak{A}^N \Pi_{j,k}$.

The symmetric group $\mathfrak{S}^N$ has a natural action on N-component row vectors,

$$\mathbf{x}Q = (x_{Q(1)}, \dots, x_{Q(N)}) = x_{Q(\alpha)}\mathbf{e}^\alpha = x_\alpha \mathbf{e}^{Q^{-1}(\alpha)}, \tag{3.85}$$

which defines a representation of $\mathfrak{S}^N$:

$$(\mathbf{e}^\alpha Q)P = \mathbf{e}^{Q^{-1}(\alpha)}P = \mathbf{e}^{P^{-1}Q^{-1}(\alpha)} = \mathbf{e}^{(QP)^{-1}(\alpha)} = \mathbf{e}^\alpha(QP). \tag{3.86}$$

For the Euclidean scalar product

$$\langle \mathbf{k}, \mathbf{x} \rangle = k_\alpha x_\alpha \tag{3.87}$$

we find that

$$\langle \mathbf{k}, \mathbf{x}Q \rangle = k_\alpha x_{Q(\alpha)} = k_{Q^{-1}(\alpha)} x_\alpha = \langle \mathbf{k}Q^{-1}, \mathbf{x} \rangle. \tag{3.88}$$

The matrix Q^{-1} is therefore equal to the transposed Q^t of Q, and the representation (3.85) is orthogonal with respect to the scalar product (3.87).

3.3.2 Many-particle wave functions and Lieb-Wu equations

The Bethe ansatz eigenstates of the Hubbard Hamiltonian (2.31) for N electrons and M down spins are characterized by two sets of quantum numbers $\{k_j\}_{j=1}^N$ and $\{\lambda_\ell\}_{\ell=1}^M$, $2M \leq N \leq L$. It is convenient to group these quantum numbers into row vectors $\mathbf{k} = (k_1, \dots, k_N)$ and $\boldsymbol{\lambda} = (\lambda_1, \dots, \lambda_M)$. The Bethe ansatz eigenstates may then be represented as (see (3.13))

$$|\psi_{\mathbf{k},\boldsymbol{\lambda}}\rangle = \frac{1}{N!} \sum_{x_1,\dots,x_N=1}^L \sum_{a_1,\dots,a_N=\uparrow,\downarrow} \psi(\mathbf{x};\mathbf{a}|\mathbf{k};\boldsymbol{\lambda})|\mathbf{x},\mathbf{a}\rangle, \tag{3.89}$$

where $\psi(\mathbf{x};\mathbf{a}|\mathbf{k};\boldsymbol{\lambda})$ is the N-particle Bethe ansatz wave function derived in appendix 3.B and described below.

The Bethe ansatz wave function depends on the relative ordering of the coordinates x_j. Any ordering is assigned to a permutation $Q \in \mathfrak{S}^N$ through the inequality

$$1 \leq x_{Q(1)} \leq x_{Q(2)} \leq \cdots \leq x_{Q(N)} \leq L. \tag{3.90}$$

The inequality (3.90) divides the configuration space of N electrons into $N!$ sectors, which can be labeled by the permutations Q. In sector Q the Bethe ansatz wave functions take the form

$$\psi(\mathbf{x};\mathbf{a}|\mathbf{k};\boldsymbol{\lambda}) = \sum_{P\in\mathfrak{S}^N} \text{sign}(PQ)\langle \mathbf{a}Q|\mathbf{k}P,\boldsymbol{\lambda}\rangle \, e^{i\langle \mathbf{k}P,\mathbf{x}Q\rangle} \tag{3.91}$$

with spin dependent amplitudes $\langle \mathbf{a}Q|\mathbf{k}P,\boldsymbol{\lambda}\rangle$ derived in appendices 3.B and 3.E. The amplitudes are again of 'Bethe ansatz form'. They are the Bethe ansatz wave functions of an inhomogeneous XXX spin chain, i.e.,

$$\langle \mathbf{a}Q | \mathbf{k}P, \boldsymbol{\lambda} \rangle = \sum_{R \in \mathfrak{S}^M} A(\lambda R) \prod_{\ell=1}^{M} F_{\mathbf{k}P}(\lambda_{R(\ell)}; y_\ell), \tag{3.92}$$

where

$$F_{\mathbf{k}}(\lambda; y) = \frac{2iu}{\lambda - \sin k_y + iu} \prod_{j=1}^{y-1} \frac{\lambda - \sin k_j - iu}{\lambda - \sin k_j + iu}, \tag{3.93}$$

and

$$A(\lambda) = \prod_{1 \le m < n \le M} \frac{\lambda_m - \lambda_n - 2iu}{\lambda_m - \lambda_n}. \tag{3.94}$$

In the above equations y_j denotes the position of the jth down spin in the sequence $a_{Q(1)}, \ldots, a_{Q(N)}$. The y's are thus 'coordinates of down spins on electrons'. If the number of down spins in the sequence $a_{Q(1)}, \ldots, a_{Q(N)}$ is different from M, the amplitude $\langle \mathbf{a}Q | \mathbf{k}P, \boldsymbol{\lambda} \rangle$ vanishes. An alternative expression for $\langle \mathbf{a}Q | \mathbf{k}P, \boldsymbol{\lambda} \rangle$ based on the terminology of the algebraic Bethe ansatz can be found in appendix 3.B.

The quantum numbers k_j, $j = 1, \ldots, N$, and λ_ℓ, $\ell = 1, \ldots, M$, may in general be complex. We call them charge momenta and spin rapidities, respectively. They have to be calculated from the Lieb-Wu equations (or 'periodic boundary conditions')

$$e^{ik_j L} = \prod_{\ell=1}^{M} \frac{\lambda_\ell - \sin k_j - iu}{\lambda_\ell - \sin k_j + iu}, \quad j = 1, \ldots, N, \tag{3.95}$$

$$\prod_{j=1}^{N} \frac{\lambda_\ell - \sin k_j - iu}{\lambda_\ell - \sin k_j + iu} = \prod_{\substack{m=1 \\ m \ne \ell}}^{M} \frac{\lambda_\ell - \lambda_m - 2iu}{\lambda_\ell - \lambda_m + 2iu}, \quad \ell = 1, \ldots, M. \tag{3.96}$$

Throughout this volume we will restrict our discussion to solutions of the Lieb-Wu equations which are finite and have a maximum number of charge momenta and spin rapidities fixed by the condition $2M \le N \le L$. These solutions will sometimes be called *regular* [125]. The restriction to regular solutions will become particularly important in Chapter 4 where the completeness problem is discussed.

The states (3.89) are joint eigenstates of the Hubbard Hamiltonian (2.31) and the momentum operator (2.54) with eigenvalues

$$E = -2 \sum_{j=1}^{N} \cos k_j + u(L - 2N), \quad P = \left[\sum_{j=1}^{N} k_j \right] \mod 2\pi. \tag{3.97}$$

Equations (3.95)–(3.97) are the most important result of this chapter. They determine the spectrum of the Hubbard Hamiltonian (2.31). They are the starting point of our exploration of the physical properties of the Hubbard model. In the following chapters these equations will be used to study the ground state and the elementary excitations of the Hubbard model in the thermodynamic limit. They will be used to study the thermodynamics of the Hubbard model. Moreover, finite size corrections to the thermodynamic limit will allow us to calculate the asymptotics of correlation functions within the so-called conformal approach.

3.4 Symmetry properties of wave functions and states

3.4.1 Symmetries under permutations

The eigenfunctions (3.91) have the following symmetry properties under permutations of electrons and quantum numbers:

$$\psi(\mathbf{x}R; \mathbf{a}R|\mathbf{k}; \boldsymbol{\lambda}) = \text{sign}(R)\psi(\mathbf{x}; \mathbf{a}|\mathbf{k}; \boldsymbol{\lambda}), \quad R \in \mathfrak{S}^N, \tag{3.98a}$$

$$\psi(\mathbf{x}; \mathbf{a}|\mathbf{k}R; \boldsymbol{\lambda}) = \text{sign}(R)\psi(\mathbf{x}; \mathbf{a}|\mathbf{k}; \boldsymbol{\lambda}), \quad R \in \mathfrak{S}^N, \tag{3.98b}$$

$$\psi(\mathbf{x}; \mathbf{a}|\mathbf{k}; \boldsymbol{\lambda}R) = \psi(\mathbf{x}; \mathbf{a}|\mathbf{k}; \boldsymbol{\lambda}), \quad R \in \mathfrak{S}^M, \tag{3.98c}$$

i.e., they are antisymmetric with respect to interchange of any two charge momenta k_j and symmetric with respect to interchange of any two spin rapidities λ_ℓ. They are antisymmetric with respect to simultaneous exchange of spin and space coordinates of two electrons, and hence respect the Pauli principle. Equation (3.98) are derived in appendix 3.B.

3.4.2 SO(4) multiplets

In appendix 3.D and appendix 3.F we work out the action of the spin operators and the η-spin operators on the Bethe ansatz states. It turns out that

$$S^+|\psi_{\mathbf{k},\boldsymbol{\lambda}}\rangle = 0, \tag{3.99}$$

$$S^z|\psi_{\mathbf{k},\boldsymbol{\lambda}}\rangle = \tfrac{1}{2}(N - 2M)|\psi_{\mathbf{k},\boldsymbol{\lambda}}\rangle. \tag{3.100}$$

The Bethe ansatz states are highest weight states of the total spin with highest weight $\frac{1}{2}(N - 2M)$. Similarly, for even length L of the lattice, they are lowest weight states of the η-spin,

$$\eta^-|\psi_{\mathbf{k},\boldsymbol{\lambda}}\rangle = 0, \tag{3.101}$$

$$\eta^z|\psi_{\mathbf{k},\boldsymbol{\lambda}}\rangle = \tfrac{1}{2}(N - L)|\psi_{\mathbf{k},\boldsymbol{\lambda}}\rangle. \tag{3.102}$$

We have seen in Section 2.2.5 that for even L both, spin and η-spin, are conserved and that spin and η-spin operators mutually commute. We conclude that the action of the ladder operators S^- and η^+ on a Bethe ansatz state generates a degenerate multiplet of eigenstates of dimension $(N - 2M + 1)(L - N + 1)$.

Instead of working with the spin operators $S^\pm$, S^z, let us introduce an alternative set of generators

$$\zeta = S^+, \quad \zeta^\dagger = S^-, \quad \zeta^z = -S^z. \tag{3.103}$$

These operators were used in [125], where equations (3.99)-(3.102) were first derived. They satisfy the commutation relations

$$[\zeta, \zeta^\dagger] = -2\zeta^z, \quad [\zeta^z, \zeta] = -\zeta, \quad [\zeta^z, \zeta^\dagger] = \zeta^\dagger. \tag{3.104}$$

For these alternative spin operators the spin multiplets are turned upside down, and the Bethe ansatz states are lowest weight states,

$$\zeta|\psi_{\mathbf{k},\lambda}\rangle = 0, \tag{3.105}$$

$$\zeta^z|\psi_{\mathbf{k},\lambda}\rangle = \tfrac{1}{2}(2M - N)|\psi_{\mathbf{k},\lambda}\rangle. \tag{3.106}$$

Every Bethe ansatz state $|\psi_{\mathbf{k},\lambda}\rangle$ generates the su(2)⊕su(2) multiplet of degenerate eigenstates

$$|\psi_{\mathbf{k},\lambda,\alpha,\beta}\rangle = (\zeta^\dagger)^\alpha (\eta^+)^\beta |\psi_{\mathbf{k},\lambda}\rangle, \tag{3.107}$$

$\alpha = 0, \ldots, N - 2M; \beta = 0, \ldots, L - N$. Recall that for the Bethe ansatz states $N \geq 2M$ and $L \geq N$. Thus, α and β are non-negative. Furthermore, we assume L to be even. This is necessary for the η-spin to be conserved (see section 2.2.5).

It was first noted in [497] that spin and η-spin in the multiplets (3.107) are not arbitrary. Since

$$(\zeta^z + \eta^z)|\psi_{\mathbf{k},\lambda}\rangle = (M - L/2)|\psi_{\mathbf{k},\lambda}\rangle \tag{3.108}$$

and L is even, the highest values of ζ^z and η^z must be either both odd or both even. Hence, not all possible su(2)⊕su(2) multiplets can occur for fixed even L. The multiplets $(0, 0)$, $(\frac{1}{2}, \frac{1}{2})$, $(0, 1)$, $(1, 0)$, ... are allowed, while, for instance, the multiplet $(0, \frac{1}{2})$ can not occur. This fact is called $\mathbb{Z}_2$-factorization. As a consequence the symmetry, when lifted to the group level, is $(SU(2) \times SU(2))/\mathbb{Z}_2 \cong SO(4)$ rather than $SU(2) \times SU(2)$.

We shall argue in Chapter 4 that the set (3.107) of all states $|\psi_{\mathbf{k},\lambda,\alpha,\beta}\rangle$ is complete. In other words, we argue that the Bethe ansatz together with the SO(4) symmetry provides all eigenstates of the Hubbard Hamiltonian.

3.5 The norm of the eigenfunctions

In this section we present a conjecture for the norm of the eigenstates (3.89) which was formulated in [175]. The proof of this conjecture is still an open problem. Judging from our experience with other Bethe ansatz solvable models, it seems likely that a proof will not be achieved within the coordinate Bethe ansatz method, and will rather rely on a better understanding of the algebraic Bethe ansatz for the Hubbard model (see Chapter 12).

To prepare for our formulation of the norm conjecture, we must first rewrite the Lieb-Wu equations in several different ways. In particular, we introduce a generating function for the logarithmic form of the Lieb-Wu equations, which plays a similar role as the Yang-Yang action [496] does in case of the Bose gas with delta-function interaction.

3.5.1 An action for the Lieb-Wu equations

Let us take the logarithm of the Lieb-Wu equations (3.95), (3.96),

$$k_j L - i \sum_{\ell=1}^{M} \ln \left(\frac{iu + \lambda_\ell - \sin k_j}{iu - \lambda_\ell + \sin k_j} \right) = 2\pi n_j^c \,, \tag{3.109}$$

$$i \sum_{j=1}^{N} \ln \left(\frac{iu + \lambda_\ell - \sin k_j}{iu - \lambda_\ell + \sin k_j} \right) - i \sum_{m=1}^{M} \ln \left(\frac{2iu + \lambda_\ell - \lambda_m}{2iu - \lambda_\ell + \lambda_m} \right) = 2\pi n_\ell^s \,. \tag{3.110}$$

Here the logarithm is defined in the cut complex plane, where the cut is along the real axis from $-\infty$ to zero. n_j^c in equation (3.109) is integer, if M is even and half odd integer, if M is odd. Similarly, n_ℓ^s in (3.110) is integer, if $N - M$ is odd, half odd integer, if $N - M$ is even.

In order to formulate the norm conjecture we shall introduce certain functions connected to the logarithmic form (3.109), (3.110) of the Lieb-Wu equations. We shall start with the definition

$$\Theta_u(x) = i \int_0^x dy \, \ln \left(\frac{iu + y}{iu - y} \right) \,. \tag{3.111}$$

In terms of this function the Lieb-Wu equations (3.109), (3.110) read

$$k_j L - \sum_{\ell=1}^{M} \Theta_u'(\lambda_\ell - \sin k_j) - 2\pi n_j^c = 0 \,, \tag{3.112}$$

$$\sum_{j=1}^{N} \Theta_u'(\lambda_\ell - \sin k_j) - \sum_{m=1}^{M} \Theta_{2u}'(\lambda_\ell - \lambda_m) - 2\pi n_\ell^s = 0 \,. \tag{3.113}$$

The primes denote derivatives with respect to the argument. The left hand side of these equations can be easily integrated with respect to the variables $\sin k_j$ and λ_ℓ, yielding the 'action'

$$\begin{aligned} S(\mathbf{k}; \boldsymbol{\lambda}) = & \sum_{j=1}^{N} (k_j \sin k_j + \cos k_j) L \\ & + \sum_{j=1}^{N} \sum_{\ell=1}^{M} \Theta_u(\lambda_\ell - \sin k_j) - \frac{1}{2} \sum_{\ell,m=1}^{M} \Theta_{2u}(\lambda_\ell - \lambda_m) \\ & - 2\pi \sum_{j=1}^{N} n_j^c \sin k_j - 2\pi \sum_{\ell=1}^{M} n_\ell^s \lambda_\ell \,. \end{aligned} \tag{3.114}$$

Thus, introducing the abbreviation $s_j = \sin k_j$, we can write the Lieb-Wu equations as extremum condition for S,

$$\frac{\partial S}{\partial s_j} = 0 \,, \quad j = 1, \ldots, N \,, \quad \frac{\partial S}{\partial \lambda_\ell} = 0 \,, \quad \ell = 1, \ldots, M \,. \tag{3.115}$$

We shall use the action S below in order to formulate the norm conjecture. A similar action was first introduced by Yang and Yang in the context of the Bose gas with delta-function interaction [496].

There is an interesting alternative way to write the Lieb-Wu equations. Let us define

$$\chi_j = k_j - \frac{1}{L} \sum_{\ell=1}^{M} \Theta'_u(\lambda_\ell - \sin k_j) - \frac{M\pi}{L}, \qquad (3.116)$$

$$\varphi_l = \frac{1}{N} \sum_{j=1}^{N} \Theta'_u(\lambda_\ell - \sin k_j) - \frac{1}{N} \sum_{m=1}^{M} \Theta'_{2u}(\lambda_\ell - \lambda_m) - \frac{(N-M+1)\pi}{N}, \qquad (3.117)$$

where $j = 1, \ldots, N$ and $\ell = 1, \ldots, M$. In terms of these new variables the Lieb-Wu equations (3.95), (3.96) become

$$e^{i\chi_j L} = 1, \quad j = 1, \ldots, N, \quad e^{i\varphi_\ell N} = 1, \quad \ell = 1, \ldots, M. \qquad (3.118)$$

This suggests to interpret the χ_j and φ_ℓ as the momenta of charge and spin degrees of freedom.

3.5.2 The norm formula

The square of the norm of the wave function (3.91) is by definition (see appendix 3.A)

$$\|\psi\|^2 = \langle \psi_{\mathbf{k},\lambda} | \psi_{\mathbf{k},\lambda} \rangle = \frac{1}{N!} \sum_{x_1,\ldots,x_N=1}^{L} \sum_{a_1,\ldots,a_N=\uparrow,\downarrow} |\psi(\mathbf{x}; \mathbf{a}|\mathbf{k}; \lambda)|^2. \qquad (3.119)$$

Here we have to insert the explicit expressions (3.91)-(3.94) on the right-hand side. After carrying out the trivial summation over the spin orientations we are left with a sum over all electron coordinates and a double sum over the symmetric group. These sums are hard to evaluate, in particular, because the charge momenta and spin rapidities are not arbitrary, but must satisfy the Lieb-Wu equations. This is the reason why at the present time we can only offer a conjecture for the square of the norm of the Hubbard wave function (3.91). The following formula was proposed in [175],

$$\|\psi\|^2 = (-1)^{M'}(2u)^M \prod_{j=1}^{N} \cos k_j \cdot \prod_{1 \leq j < k \leq M} \left(1 + \frac{4u^2}{(\lambda_j - \lambda_k)^2}\right) \cdot \det \begin{pmatrix} \dfrac{\partial^2 S}{\partial s^2} & \dfrac{\partial^2 S}{\partial s \partial \lambda} \\[2mm] \dfrac{\partial^2 S}{\partial \lambda \partial s} & \dfrac{\partial^2 S}{\partial \lambda^2} \end{pmatrix}. \qquad (3.120)$$

Here M' is the number of pairs of complex conjugated k_j's in a given solution of the Lieb-Wu equations. The determinant on the right-hand side of (3.120) is the determinant of an

$(N + M) \times (N + M)$-matrix. This matrix consists of four blocks with matrix elements

$$\left(\frac{\partial^2 S}{\partial s^2}\right)_{mn} = \frac{\partial^2 S}{\partial s_m \partial s_n} = \delta_{m,n}\left\{\frac{L}{\cos k_n} + \sum_{l=1}^{M}\frac{2u}{u^2 + (\lambda_l - s_n)^2}\right\},$$

$$m, n = 1, \ldots, N,\tag{3.121}$$

$$\left(\frac{\partial^2 S}{\partial \lambda \partial s}\right)_{mn} = \left(\frac{\partial^2 S}{\partial s \partial \lambda}\right)_{nm} = \frac{\partial^2 S}{\partial \lambda_m \partial s_n} = -\frac{2u}{u^2 + (\lambda_m - s_n)^2},$$

$$m = 1, \ldots, M, \quad n = 1, \ldots, N,\tag{3.122}$$

$$\left(\frac{\partial^2 S}{\partial \lambda^2}\right)_{mn} = \frac{\partial^2 S}{\partial \lambda_m \partial \lambda_n} = \delta_{m,n}\left\{\sum_{j=1}^{N}\frac{2u}{u^2 + (\lambda_n - s_j)^2} - \sum_{l=1}^{M}\frac{4u}{(2u)^2 + (\lambda_n - \lambda_l)^2}\right\}$$

$$+ \frac{4u}{(2u)^2 + (\lambda_m - \lambda_n)^2}, \quad m, n = 1, \ldots, M.\tag{3.123}$$

The norm is thus proportional to the Hessian determinant of the action S regarded as a function of the charge momenta k_j and spin rapidities λ_ℓ. Recalling the formulation (3.115) of the Lieb-Wu equations we see that a solution is non-degenerate (locally unique for fixed values of n_j^c and n_ℓ^s), if the norm of the corresponding wave function does not vanish. In other words, all eigenstates of the Hubbard Hamiltonian (2.31) correspond to non-degenerate solutions of the Lieb-Wu equations.

Another interesting form of the norm formula is obtained by expressing the Hessian determinant in equation (3.120) in terms of the momenta of elementary charge and spin excitations χ_j and φ_l,

$$\|\psi\|^2 = L^N N^M (-1)^{M'} (2u)^M \cdot \prod_{1 \le j < k \le M}\left(1 + \frac{4u^2}{(\lambda_j - \lambda_k)^2}\right) \cdot \frac{\partial(\chi_1, \ldots, \chi_N; \varphi_1, \ldots, \varphi_M)}{\partial(k_1, \ldots, k_N; \lambda_1, \ldots, \lambda_M)}.$$

$$\tag{3.124}$$

The norm is proportional to the Jacobian of the transformation from the set of charge momenta and spin rapidities k_j, λ_ℓ to the set of momenta of charge and spin degrees of freedom χ_j, φ_ℓ.

Let us list the arguments in support of (3.120):

(i) The experience with other Bethe ansatz solvable models [268, 373, 409, 448] shows that norm formulae for Bethe ansatz wave functions are generically of the form (3.120).

(ii) It is easy to see that (3.120) is valid for $M = 0$ and arbitrary N.

(iii) (3.120) was verified for $M = 1$ and $N = 2, 3$. The calculation is lengthy and involves highly non-trivial cancellations based on the Lieb-Wu equations.

(iv) (3.120) was verified for arbitrary N and M in the limit $u \to \infty$. This limit requires rescaling of the spin rapidities, $\lambda_j = 2u\tilde{\lambda}_j$. The remaining non-trivial factor in the expression for the norm reduces to the case of the XXX spin-$\frac{1}{2}$ chain, and the result of [268] can be applied.

(v) For arbitrary N and M the leading order term of the norm in the large L limit can be calculated. This term is proportional to L^N and fixes the prefactor in (3.120).

3.6 Conclusions

We have mapped the problem of solving the Schrödinger equation for the Hubbard model to the problem of solving the Lieb-Wu equations (3.95), (3.96). The Lieb-Wu equations are of primary importance for our further investigation of the one-dimensional Hubbard model in the following chapters. They were obtained by Lieb and Wu in [298]. The wave function (3.91)-(3.94) is not in Lieb and Wu's original paper, but in the form presented here first appears in the literature in Woynarovich's article [481]. The SO(4) highest weight properties of the Bethe ansatz states were proven in [125]. The norm formula (3.120) was proposed in [175]. Its proof is one of the interesting open problems for the one-dimensional Hubbard model.

The reader who is interested in the derivations of the results presented in this chapter may continue with the appendices. In appendix 3.A we recall some basic facts about the formalism of the second quantization. The remaining appendices contain rigorous derivations of the Lieb-Wu equations, the Bethe ansatz wave functions, and the SO(4) highest weight properties in the general N-particle case. We also discuss the limiting cases of large and small coupling and the continuum limit in appendix 3.G. Alternative accounts of the derivation of the Lieb-Wu equations can be found in [95, 423].

Appendices to Chapter 3

3.A Scalar products and projection operators

Here we recall a number of textbook formulae for switching from second to first quantization and back that are needed in Chapter 3.

We shall use the shorthand notation

$$\mathbb{Z}_L = \{1, \ldots, L\}. \tag{3.A.1}$$

For the values $\uparrow$, $\downarrow$ of the spin variables we identify $\uparrow$ with 1 and $\downarrow$ with 2. It is then natural to define $\uparrow < \downarrow$. The coordinates x_j and spin variables a_j of electrons on a ring of L sites take values in $\mathbb{Z}_L$ and $\mathbb{Z}_2$, respectively. The positions and spins of N electrons are determined by row vectors $\mathbf{x} \in \mathbb{Z}_L^N$ and $\mathbf{a} \in \mathbb{Z}_2^N$.

Let us recall the construction of the Fock space. The Fock space is built by the action of fermionic creation operators $c_{x,a}^\dagger$ on a Fock vacuum $|0\rangle$. The Fock vacuum is annihilated by any annihilation operator,

$$c_{x,a}|0\rangle = 0, \tag{3.A.2}$$

$x \in \mathbb{Z}_L$, $a \in \mathbb{Z}_2$. In order to be able to consider expectation values of operators, we need a dual Fock vacuum $\langle 0|$ that satisfies $\langle 0|c_{x,a}^\dagger = 0$ and

$$\langle 0|0\rangle = 1. \tag{3.A.3}$$

Using (3.A.2), (3.A.3) and the canonical anticommutation relations between the Fermi operators, we can calculate expectation values $\langle 0|A|0\rangle$ of arbitrary operators A that are linear combinations of products of Fermi operators. For instance,

$$\langle 0|c_{x,a}c_{y,b}^\dagger|0\rangle = \langle 0|(\delta_{x,y}\delta_{a,b} - c_{y,b}^\dagger c_{x,a})|0\rangle = \delta_{x,y}\delta_{a,b}. \tag{3.A.4}$$

We see that $\langle x, a| = \langle 0|c_{x,a}$ is dual to $|x, a\rangle = c_{x,a}^\dagger|0\rangle$. More generally,

$$\langle \mathbf{x}, \mathbf{a}| = \langle 0|c_{x_1,a_1} \ldots c_{x_N,a_N} \tag{3.A.5}$$

is dual to $|\mathbf{x}, \mathbf{a}\rangle$.

Clearly, the scalar product $\langle \mathbf{x}, \mathbf{a}|\mathbf{y}, \mathbf{b}\rangle$ of two Wannier states vanishes, if the number of electrons in both states is different. In other words, two Wannier states with different

73

numbers of particles are orthogonal to each other. If the number of particles in two Wannier states $|\mathbf{x}, \mathbf{a}\rangle$, $|\mathbf{y}, \mathbf{b}\rangle$ is the same, say N, their scalar product is

$$\langle \mathbf{x}, \mathbf{a} | \mathbf{y}, \mathbf{b} \rangle = \det A \,, \tag{3.A.6}$$

where A is an $N \times N$-matrix with elements

$$A_{jk} = \delta_{x_j, y_k} \delta_{a_j, b_k} \,. \tag{3.A.7}$$

Equation (3.A.6) can be proven by induction over N: For $N = 1$ equation (3.A.4) applies and the formula is true. Assume it is true for some positive integer $N - 1$. Then

$$\langle \mathbf{x}, \mathbf{a} | \mathbf{y}, \mathbf{b} \rangle = \langle 0 | c_{x_1, a_1} \cdots c_{x_N, a_N} c^\dagger_{y_N, b_N} \cdots c^\dagger_{y_1, b_1} | 0 \rangle$$

$$= \langle 0 | c_{x_1, a_1} \cdots c_{x_{N-1}, a_{N-1}} c^\dagger_{y_{N-1}, b_{N-1}} \cdots c^\dagger_{y_1, b_1} | 0 \rangle \delta_{x_N, y_N} \delta_{a_N, b_N}$$

$$- \langle 0 | c_{x_1, a_1} \cdots c_{x_{N-1}, a_{N-1}} c^\dagger_{y_N, b_N} c^\dagger_{y_{N-2}, b_{N-2}} \cdots c^\dagger_{y_1, b_1} | 0 \rangle \delta_{x_N, y_{N-1}} \delta_{a_N, b_{N-1}}$$

$$+ \cdots$$

$$+ (-1)^{N-1} \langle 0 | c_{x_1, a_1} \cdots c_{x_{N-1}, a_{N-1}} c^\dagger_{y_N, b_N} \cdots c^\dagger_{y_2, b_2} | 0 \rangle \delta_{x_N, y_1} \delta_{a_N, b_1} \,. \tag{3.A.8}$$

Here we used the elementary anticommutators to move c_{x_N, a_N} to the very right. The right hand side of equation (3.A.8) is precisely the Laplace expansion of $\det A$ with respect to the last row. Hence, by hypothesis, (3.A.6) is true for every positive N.

It is sometimes useful to consider pairs $(x_j, a_j) \in \mathbb{Z}_L \times \mathbb{Z}_2$ as the 'coordinates of an electron' (see [283]). Such pairs have a natural ordering '<' defined by $(x_j, a_j) < (x_k, a_k)$, if $x_j \leq x_k$ and $a_j < a_k$ for $x_j = x_k$.

If $(x_j, a_j) = (x_k, a_k)$ for some $j \neq k$, then there are two identical creation operators in the expression $c^\dagger_{x_N, a_N} \cdots c^\dagger_{x_1, a_1} | 0 \rangle$ defining the Wannier state $|\mathbf{x}, \mathbf{a}\rangle$, and $|\mathbf{x}, \mathbf{a}\rangle = 0$. On the other hand, if $(x_j, a_j) \neq (x_k, a_k)$ for all $j \neq k$, then the Wannier state $|\mathbf{x}, \mathbf{a}\rangle$ is non-zero, and there is a permutation $P \in \mathfrak{S}^N$ such that $|\mathbf{x}, \mathbf{a}\rangle = \text{sign}(P) |\mathbf{x}P, \mathbf{a}P\rangle$ and $(x_{P(j)}, a_{P(j)}) < (x_{P(j+1)}, a_{P(j+1)})$ for $j = 1, \ldots L - 1$. Therefore we can describe the Wannier basis of the N-particle subspace of the space of states $\mathcal{H}^{(L)}$ of electrons on an L-site ring as

$$\mathcal{B}_N = \left\{ |\mathbf{x}, \mathbf{a}\rangle \in \mathcal{H}^{(L)} \big| (1, \uparrow) \leq (x_1, a_1) < \cdots < (x_N, a_N) \leq (L, \downarrow) \right\}. \tag{3.A.9}$$

If we further define $\mathcal{B}_0 = \{|0\rangle\}$, we can decompose the basis $\mathcal{B}$ of Wannier states (2.5) into

$$\mathcal{B} = \bigcup_{N=0}^{2L} \mathcal{B}_N \,. \tag{3.A.10}$$

For the projection operators P_N onto the N-particle subspace spanned by $\mathcal{B}_N$ we obtain the expression

$$
\begin{aligned}
P_N &= \sum_{(1,\uparrow)\leq(x_1,a_1)<\cdots<(x_N,a_N)\leq(L,\downarrow)} |\mathbf{x},\mathbf{a}\rangle\langle\mathbf{x},\mathbf{a}| \\
&= \frac{1}{N!} \sum_{P\in\mathfrak{S}^N} \sum_{(1,\uparrow)\leq(x_{P(1)},a_{P(1)})<\cdots<(x_{P(N)},a_{P(N)})\leq(L,\downarrow)} |\mathbf{x}P,\mathbf{a}P\rangle\langle\mathbf{x}P,\mathbf{a}P| \\
&= \frac{1}{N!} \sum_{P\in\mathfrak{S}^N} \sum_{(1,\uparrow)\leq(x_{P(1)},a_{P(1)})<\cdots<(x_{P(N)},a_{P(N)})\leq(L,\downarrow)} |\mathbf{x},\mathbf{a}\rangle\langle\mathbf{x},\mathbf{a}| \\
&= \frac{1}{N!} \sum_{\mathbf{x}\in\mathbb{Z}_L^N} \sum_{\mathbf{a}\in\mathbb{Z}_2^N} |\mathbf{x},\mathbf{a}\rangle\langle\mathbf{x},\mathbf{a}|\,. \qquad (3.A.11)
\end{aligned}
$$

Here we used $|\mathbf{x}P,\mathbf{a}P\rangle\langle\mathbf{x}P,\mathbf{a}P| = |\mathbf{x},\mathbf{a}\rangle\langle\mathbf{x},\mathbf{a}|$ for all $P\in\mathfrak{S}^N$ in the third equation and $|\mathbf{x},\mathbf{a}\rangle = 0$ for $(x_j,a_j) = (x_k,a_k)$, $j\neq k$, in the fourth equation. It follows from (3.A.11) that

$$
\mathrm{id} = |0\rangle\langle 0| + \sum_{N=1}^{2L} \frac{1}{N!} \sum_{\mathbf{x}\in\mathbb{Z}_L^N} \sum_{\mathbf{a}\in\mathbb{Z}_2^N} |\mathbf{x},\mathbf{a}\rangle\langle\mathbf{x},\mathbf{a}| \qquad (3.A.12)
$$

is a partition of the identity operator into the projectors onto the N-particle subspaces. Furthermore, applying (3.A.12) to an N-particle state $|\psi\rangle$ we obtain the identity

$$
|\psi\rangle = \frac{1}{N!} \sum_{\mathbf{x}\in\mathbb{Z}_L^N} \sum_{\mathbf{a}\in\mathbb{Z}_2^N} \psi(\mathbf{x};\mathbf{a})|\mathbf{x},\mathbf{a}\rangle\,, \qquad (3.A.13)
$$

where $\psi(\mathbf{x};\mathbf{a}) = \langle\mathbf{x},\mathbf{a}|\psi\rangle$ is the N-particle wave function. Conversely, if the N-particle wave function $\psi(\mathbf{x};\mathbf{a})$ is totally antisymmetric and if we take (3.A.13) as a definition of an N-particle state $|\psi\rangle$, then

$$
\begin{aligned}
\langle\mathbf{x},\mathbf{a}|\psi\rangle &= \frac{1}{N!} \sum_{\mathbf{y}\in\mathbb{Z}_L^N} \sum_{\mathbf{b}\in\mathbb{Z}_2^N} \psi(\mathbf{y};\mathbf{b})\langle\mathbf{x},\mathbf{a}|\mathbf{y},\mathbf{b}\rangle \\
&= \frac{1}{N!} \sum_{P\in\mathfrak{S}^N} \sum_{\mathbf{y}\in\mathbb{Z}_L^N} \sum_{\mathbf{b}\in\mathbb{Z}_2^N} \mathrm{sign}(P)\psi(\mathbf{y};\mathbf{b})\delta_{x_1,y_{P(1)}}\delta_{a_1,b_{P(1)}}\cdots\delta_{x_N,y_{P(N)}}\delta_{a_N,b_{P(N)}} \\
&= \frac{1}{N!} \sum_{P\in\mathfrak{S}^N} \psi(\mathbf{x};\mathbf{a}) = \psi(\mathbf{x};\mathbf{a})\,. \qquad (3.A.14)
\end{aligned}
$$

Here we used the antisymmetry, $\mathrm{sign}(P)\psi(\mathbf{y},\mathbf{b}) = \psi(\mathbf{y}P,\mathbf{b}P)$, in the third equation.

If $\psi(\mathbf{x};\mathbf{a})$ is a totally antisymmetric eigenfunction of the cyclic N-particle Hamiltonian $H_N^{(L)}$, equation (3.10), with eigenvalue E, then the corresponding N-particle state $|\psi\rangle$, equation (3.A.13), is an eigenstate of the Hubbard Hamiltonian (2.22), for we have

$$H|\psi\rangle = \frac{1}{N!} \sum_{\mathbf{y} \in \mathbb{Z}_L^N} \sum_{\mathbf{b} \in \mathbb{Z}_2^N} \langle \mathbf{y}, \mathbf{b} | \psi \rangle H | \mathbf{y}, \mathbf{b} \rangle$$

$$= \frac{1}{(N!)^2} \sum_{\mathbf{x}, \mathbf{y} \in \mathbb{Z}_L^N} \sum_{\mathbf{a}, \mathbf{b} \in \mathbb{Z}_2^N} \langle \mathbf{x}, \mathbf{a} | H | \mathbf{y}, \mathbf{b} \rangle \langle \mathbf{y}, \mathbf{b} | \psi \rangle | \mathbf{x}, \mathbf{a} \rangle$$

$$= \frac{1}{N!} \sum_{\mathbf{x} \in \mathbb{Z}_L^N} \sum_{\mathbf{a} \in \mathbb{Z}_2^N} \langle \mathbf{x}, \mathbf{a} | H | \psi \rangle | \mathbf{x}, \mathbf{a} \rangle$$

$$= \frac{1}{N!} \sum_{\mathbf{x} \in \mathbb{Z}_L^N} \sum_{\mathbf{a} \in \mathbb{Z}_2^N} (H_N^{(L)} \psi)(\mathbf{x}; \mathbf{a}) | \mathbf{x}, \mathbf{a} \rangle = E | \psi \rangle. \tag{3.A.15}$$

Here we used (3.A.14) in the first equation, (3.A.12) in the second and third equation and (3.8) in the fourth equation.

For the scalar product of two N-particle states $|\psi\rangle$ and $|\varphi\rangle$ we obtain

$$\langle \psi | \varphi \rangle = \frac{1}{(N!)^2} \sum_{\mathbf{x}, \mathbf{y} \in \mathbb{Z}_L^N} \sum_{\mathbf{a}, \mathbf{b} \in \mathbb{Z}_2^N} \overline{\psi(\mathbf{x}; \mathbf{a})} \varphi(\mathbf{y}; \mathbf{b}) \langle \mathbf{x}, \mathbf{a} | \mathbf{y}, \mathbf{b} \rangle$$

$$= \frac{1}{(N!)^2} \sum_{P \in \mathfrak{S}^N} \sum_{\mathbf{x}, \mathbf{y} \in \mathbb{Z}_L^N} \sum_{\mathbf{a}, \mathbf{b} \in \mathbb{Z}_2^N} \mathrm{sign}(P) \delta_{x_1, y_{P(1)}} \delta_{a_1, b_{P(1)}} \cdots \delta_{x_N, y_{P(N)}} \delta_{a_N, b_{P(N)}} \overline{\psi(\mathbf{x}; \mathbf{a})} \varphi(\mathbf{y}; \mathbf{b})$$

$$= \frac{1}{(N!)^2} \sum_{P \in \mathfrak{S}^N} \sum_{\mathbf{y} \in \mathbb{Z}_L^N} \sum_{\mathbf{b} \in \mathbb{Z}_2^N} \mathrm{sign}(P) \overline{\psi(\mathbf{y} P; \mathbf{b} P)} \varphi(\mathbf{y}; \mathbf{b})$$

$$= \frac{1}{N!} \sum_{\mathbf{x} \in \mathbb{Z}_L^N} \sum_{\mathbf{a} \in \mathbb{Z}_2^N} \overline{\psi(\mathbf{x}; \mathbf{a})} \varphi(\mathbf{x}; \mathbf{a}). \tag{3.A.16}$$

We used (3.A.13) in the first equation, (3.A.6) in the second equation and the fact that N-particle wave functions are totally antisymmetric in the fourth equation. Note that the right hand side of (3.A.16) is also equal to the scalar product (3.11), whence its definition.

As a corollary to equation (3.A.16) we obtain the following formula for the square of the norm of an N-particle state,

$$\|\psi\|^2 = \langle \psi | \psi \rangle = \frac{1}{N!} \sum_{\mathbf{x} \in \mathbb{Z}_L^N} \sum_{\mathbf{a} \in \mathbb{Z}_2^N} |\psi(\mathbf{x}; \mathbf{a})|^2. \tag{3.A.17}$$

3.B Derivation of Bethe ansatz wave functions and Lieb-Wu equations

This appendix contains a detailed derivation of the Bethe ansatz wave functions and of the Lieb-Wu equations in the general N-particle case. Since the derivation is slightly lengthy, let us outline what we are going to do.

(i) In the next section we try to motivate the choice (3.B.19) of the Bethe ansatz wave function and explain some of its elementary properties.

(ii) In Section 3.B.2 the wave function (3.B.19) is inserted into the Schrödinger equation (3.16) which results in a single equation (3.B.22) for the amplitudes $A(\mathbf{k}P|\mathbf{a}Q)$. This equation is then interpreted in the language of an auxiliary spin system and is shown to be equivalent to the recursion relation (3.B.31) over the symmetric group $\mathfrak{S}^N$.

(iii) In Section 3.B.3 (with some details postponed to appendix 3.C) we show that the recursion has a unique solution independent of the realization of a permutation as a product of transpositions of nearest neighbours. This brings us into first contact with the important Yang-Baxter equation (3.B.35).

(iv) In Section 3.B.4 we impose the periodic boundary conditions (3.17) onto the Bethe ansatz wave function (3.B.19) with amplitudes defined by (3.B.31). This reduces the solution of the N-particle problem to that of an auxiliary eigenvalue problem for an inhomogeneous spin system related to the XXX Heisenberg chain.

(v) In Section 3.B.5 we solve the spin problem by an algebraic technique called the algebraic Bethe ansatz (we shall have to say more about it in Chapter 12). The eigenvalue (3.B.85) of the spin problem depends on the quasi momenta $\{k_j\}$ in the Bethe ansatz wave function. They appear as inhomogeneities in the spin problem and have to be chosen in accordance with the periodic boundary conditions. This determines the Lieb-Wu equations.

3.B.1 The Bethe ansatz wave function

In the following we construct joint *antisymmetric* solutions of the Schrödinger equation (3.16) and the equation (3.22) for the shift operator, i.e., we require the solutions to satisfy

$$\psi(\mathbf{x}R; \mathbf{a}R) = \mathrm{sign}(R)\psi(\mathbf{x}; \mathbf{a}) \tag{3.B.1}$$

for all $R \in \mathfrak{S}^N$. Here we resort to the notation for the symmetric group introduced in Section 3.3.1. Its full power and usefulness will become apparent below. Equation (3.B.1) encodes the fact that electrons are fermions which ought to satisfy the Pauli principle. The antisymmetry of the wave function is essential for the Bethe ansatz to work. The so-called bosonic Hubbard model which has the same 'first quantized' Hamiltonian (3.15) but totally symmetric wave functions can not be solved by Bethe ansatz (see [83]).

The Bethe ansatz starts with a clever guess (an ansatz) for the form of the so-called Bethe ansatz wave function. It is a multiparametric function which has the essential properties of the solution. The actual calculation then determines the parameters (the quantum numbers) in the Bethe ansatz wave function.

The form of the Bethe ansatz wave function can not be rigorously derived, but will be fully justified only a posteriori, at the end of the calculation. It can only be motivated, and the physical intuition behind it can be explained. This is the aim of the following considerations which are based on three ideas: locality of the Schrödinger equation, existence of 'higher conserved quantities' and antisymmetry of the wave function.

Let us first see what can be learned from locality and antisymmetry. The Schrödinger equation (3.16) has to be satisfied for all $\mathbf{x} \in \mathbb{Z}^N$. Because of the antisymmetry (3.B.1), however, it suffices to consider it in one of the the the elementary domains

$$\overline{D}_Q = \left\{ \mathbf{x} \in \mathbb{Z}^N \mid x_{Q(1)} \leq \cdots \leq x_{Q(N)} \right\}, \tag{3.B.2}$$

$Q \in \mathfrak{S}^N$, say in $\overline{D}_{\mathrm{id}}$. In particular, the Schrödinger equation must be satisfied off the boundaries of $\overline{D}_Q$, i.e., in the sets

$$D_Q = \left\{ \mathbf{x} \in \mathbb{Z}^N \mid x_{Q(1)} < \cdots < x_{Q(N)} \right\}. \tag{3.B.3}$$

But for $\mathbf{x} \in D_Q$ no two electrons occupy the same lattice site, and the local interaction does not matter. We have

$$(H_N - E)\psi(\mathbf{x}; \mathbf{a}) = -\left\{ \sum_{j=1}^{N} (\Delta_j^+ + \Delta_j^-) + E \right\} \psi(\mathbf{x}; \mathbf{a}). \tag{3.B.4}$$

This is an equation for non-interacting electrons, since $[\Delta_j^{\pm}, \Delta_k^{\pm}] = 0$. The corresponding one-particle problem $\Delta^+ f(n) = f(n+1) = \alpha f(n)$, $n \in \mathbb{Z}$, has the solution $f(n) = \alpha^n f(0) = A\mathrm{e}^{\mathrm{i}kn}$ ($f(0)$, α, A, k complex numbers), which is unique up to normalization. Out of the one-particle solutions we can construct solutions $f(\mathbf{x}|\mathbf{k}) = A\mathrm{e}^{\mathrm{i}\langle \mathbf{k}, \mathbf{x} \rangle}$ of (3.B.4). We find that

$$(H_N - E)f(\mathbf{x}|\mathbf{k}) = 0, \tag{3.B.5}$$

if and only if

$$E = -2 \sum_{j=1}^{N} \cos(k_j). \tag{3.B.6}$$

The N-particle shift operator $\hat{U}_N$ commutes with the N-particle Hamiltonian, $[H_N, \hat{U}_N] = 0$. Therefore it is possible to construct a system of common eigenfunctions of H_N and $\hat{U}_N$. The functions $f(\mathbf{x}|\mathbf{k})$ satisfy

$$(\hat{U}_N - \omega)f(\mathbf{x}|\mathbf{k}) = 0 \tag{3.B.7}$$

for all $\mathbf{x} \in \mathbb{Z}^N$, if and only if

$$\omega = \mathrm{e}^{\mathrm{i}(k_1 + \cdots + k_N)}. \tag{3.B.8}$$

We conclude from the above considerations that the translationally invariant eigenfunctions of the Hubbard Hamiltonian must be linear combinations of the functions $f(\mathbf{x}|\mathbf{k})$ with degenerate eigenvalues E and ω, (3.B.6) and (3.B.8).

There is a trivial degeneracy due to the invariance under permutations of the individual k_j: the common solution $f(\mathbf{x}|\mathbf{k})$ of (3.B.4) and (3.B.7) is degenerate with $f(\mathbf{x}|\mathbf{k}P)$ for all $P \in \mathfrak{S}^N$. In general, for a fixed vector $\mathbf{k}$, we have no further degeneracy (E is also invariant under a change of signs of the individual k_j, but ω is not). Still, more degenerate solutions may exist with independent vectors $\mathbf{k}'$ satisfying

$$\mathrm{e}^{\mathrm{i}(k_1+\cdots+k_N)} = \mathrm{e}^{\mathrm{i}(k_1'+\cdots+k_N')}, \qquad \sum_{j=1}^{N} \cos(k_j) = \sum_{j=1}^{N} \cos(k_j'). \qquad (3.\mathrm{B}.9)$$

It is now an interesting observation that the equations (3.B.9) restrict the possible choices of $\mathbf{k}$. They can be interpreted as constraints on $\mathbf{k}$. Let us consider the case $N = 2$. In this case (3.B.9) implies (a) $\{k_1, k_2\} = \{k_1', k_2'\}$ mod 2π, or (b) $k_1 + k_2 = k_1' + k_2' = \pi$ mod 2π. Case (b) corresponds to $E = 0$ and has been discussed in Section 3.2. In the generic situation $E \neq 0$ case (a) applies, and $f(\mathbf{x}|\mathbf{k})$ and $f(\mathbf{x}|\mathbf{k}')$ are non-degenerate, unless $\{k_1, k_2\} = \{k_1', k_2'\}$ mod 2π.

Imagine there were more mutually commuting quantities I_n, having $f(\mathbf{x}|\mathbf{k})$ as a common eigenfunction and commuting with $\hat{U}$ and H. There would be more constraints than just the two equations (3.B.9). If the number of constraints would equal the number of particles, this could render $f(\mathbf{x}|\mathbf{k})$ and $f(\mathbf{x}|\mathbf{k}')$ to be non-degenerate as common eigenfunctions of *all* of the I_n, $\hat{U}$ and H for generic choice of $\mathbf{k}$ and $\mathbf{k}' \neq \mathbf{k}P$.

For the Hubbard model, higher conserved quantities actually exist that have the above properties (see Chapter 12). This motivates the ansatz

$$\psi_Q(\mathbf{x}|\mathbf{k}) = \sum_{P\in\mathfrak{S}^N} a(P, Q)\mathrm{e}^{\mathrm{i}\langle \mathbf{k}P, \mathbf{x}\rangle} = \sum_{P\in\mathfrak{S}^N} \tilde{a}(P, Q)\mathrm{e}^{\mathrm{i}\langle \mathbf{k}P, \mathbf{x}Q\rangle} \qquad (3.\mathrm{B}.10)$$

for a joint solution of the Schrödinger equation (3.16) and the equation for the shift operator (3.B.7) in the sector D_Q. Note that $\tilde{a}(P, Q) = a(PQ^{-1}, Q)$. The ansatz (3.B.10) is called Bethe ansatz or Bethe-Yang hypothesis and was introduced by C. N. Yang in his seminal paper [493] on the electron gas with delta-function interaction. Let us emphasize that the peculiar feature of (3.B.10) is the occurrence of only *one* wave vector $\mathbf{k}$.

There is the following physical picture behind (3.B.10), which was stressed by Zamolodchikov and Zamolodchikov [502]. Assume for a while the components of $\mathbf{k}$ be real. Then (3.B.10) describes a scattering state of N particles. For a classical system of N interacting particles passing from a sector Q to a sector Q' means to change the order of the particles. Thus, the particles have scattered each other. Assuming the same wave vector $\mathbf{k}$ in every sector Q means to assume that the *individual* k_j are conserved under scattering of the particles. The particles merely exchange their momenta. Mere exchange of momenta is characteristic of two-particle scattering. This means we assume the N-particle scattering process to be a sequence of two-particle scattering processes.

Generally, there is no conservation of the set of individual momenta $\{k_j\}$ in N-particle scattering. Conservation of the set of individual k_j is typical for integrable systems. For a beautiful classical example see [330].

So far the ansatz (3.B.10) says nothing about the internal degrees of freedom of the particles and nothing about symmetry. Let us try to include these features: the Hubbard model is a model of electrons. We are thus seeking for solutions $\psi(\mathbf{x}; \mathbf{a}|\mathbf{k})$ of the Schrödinger equation (3.16) which satisfy Pauli's principle (3.B.1). The dependence of our ansatz wave function (3.B.10) on spin $\mathbf{a}$ and wave vector $\mathbf{k}$ has to be incorporated into the coefficients, $\tilde{a}(P, Q) = \tilde{a}(P, Q, \mathbf{k}, \mathbf{a})$.

Let $\mathbf{x} \in D_Q, \mathbf{y} = \mathbf{x}R, \mathbf{b} = \mathbf{a}R, R \in \mathfrak{S}^N$. Then $x_{Q(j)} = y_{R^{-1}Q(j)}$ and $\mathbf{y} \in D_{R^{-1}Q}$. The wave function (3.B.10) in the sector $R^{-1}Q$ is

$$\psi_{R^{-1}Q}(\mathbf{y}; \mathbf{b}|\mathbf{k}) = \sum_{P \in \mathfrak{S}^N} \tilde{a}(P, R^{-1}Q, \mathbf{k}, \mathbf{b}) e^{i\langle \mathbf{k}P, \mathbf{y}R^{-1}Q \rangle}$$

$$= \sum_{P \in \mathfrak{S}^N} \tilde{a}(P, R^{-1}Q, \mathbf{k}, \mathbf{a}R) e^{i\langle \mathbf{k}P, \mathbf{x}Q \rangle}. \qquad (3.B.11)$$

The Pauli principle (3.B.1) requires this to be equal to

$$\text{sign}(R)\psi_Q(\mathbf{x}; \mathbf{a}|\mathbf{k}) = \sum_{P \in \mathfrak{S}^N} \text{sign}(R)\tilde{a}(P, Q, \mathbf{k}, \mathbf{a}) e^{i\langle \mathbf{k}P, \mathbf{x}Q \rangle}, \qquad (3.B.12)$$

for arbitrary $\mathbf{x} \in D_Q$ and arbitrary $\mathbf{k}$. Hence, the Pauli principle is satisfied if and only if

$$\tilde{a}(P, R^{-1}Q, \mathbf{k}, \mathbf{a}R) = \text{sign}(R)\tilde{a}(P, Q, \mathbf{k}, \mathbf{a}). \qquad (3.B.13)$$

This equation fixes $\tilde{a}$ in all sectors D_Q, once it is known in a specific sector. In order to simplify further calculations we make the ansatz

$$\tilde{a}(P, Q, \mathbf{k}, \mathbf{a}) = \text{sign}(PQ)A(\mathbf{k}P|\mathbf{a}Q), \qquad (3.B.14)$$

which clearly satisfies (3.B.13) and acquires further motivation by the results of Section 3.2 about the two-particle problem.

So far we have constructed an ansatz wave function with the following properties. It is defined inside the sectors D_Q, where it satisfies the Schrödinger equation (3.16) and the equation of the shift operator (3.B.7) (with the same values of E and ω in every sector). It is totally antisymmetric under simultaneous permutations of $\mathbf{x}$ and $\mathbf{a}$, equation (3.B.1). Our specification of $\tilde{a}$ in equation (3.B.14) also forces antisymmetry under exchange of the components k_j of the wave vector $\mathbf{k}$. For, let $\mathbf{x} \in D_Q, R \in \mathfrak{S}^N$. Then, using (3.B.10)

$$\psi(\mathbf{x}; \mathbf{a}|\mathbf{k}R) = \sum_{P \in \mathfrak{S}^N} \text{sign}(PQ)A(\mathbf{k}RP|\mathbf{a}Q) e^{i\langle \mathbf{k}RP, \mathbf{x}Q \rangle} = \text{sign}(R)\psi(\mathbf{x}; \mathbf{a}|\mathbf{k}), \qquad (3.B.15)$$

where the second equality follows from $R\mathfrak{S}^N = \mathfrak{S}^N$ and $\text{sign}(R^{-1}) = \text{sign}(R)$. We infer from (3.B.15) that $\psi(\mathbf{x}; \mathbf{a}|\mathbf{k})$ vanishes identically (or is singular), if two of the k_j are the same. We may therefore assume below, that the k_j be mutually distinct.

Let us now try to extend our ansatz to the boundaries of the sectors $\overline{D}_Q$. First we have to make sure that (3.B.10) and (3.B.14) define $\psi(\mathbf{x}; \mathbf{a}|\mathbf{k})$ uniquely on the boundaries. $\mathbf{x}$ is on the boundary of $\overline{D}_Q$, if there is an n, such that $x_{Q(n)} = x_{Q(n+1)}$. $\mathbf{x}$ is then invariant under permutation of $x_{Q(n)}$ and $x_{Q(n+1)}$,

$$\mathbf{x}Q = \mathbf{x}Q\Pi_{n,n+1}, \qquad (3.B.16)$$

and also belongs to the boundary of $\overline{D}_{Q\Pi_{n,n+1}}$. Let us write $\Pi = \Pi_{n,n+1}$ for short. Then uniqueness of the wave function at the joint boundary of $\overline{D}_Q$ and $\overline{D}_{Q\Pi}$ requires

$$
\begin{aligned}
0 &= \sum_{P \in \mathfrak{S}^N} \operatorname{sign}(PQ) e^{i\langle kP, xQ \rangle} \big(A(kP|aQ) + A(kP|aQ\Pi) \big) \\
&= \sum_{P \in \mathfrak{A}^N} \Big\{ \operatorname{sign}(PQ) e^{i\langle kP, xQ \rangle} \big(A(kP|aQ) + A(kP|aQ\Pi) \big) \\
&\qquad + \operatorname{sign}(P\Pi Q) e^{i\langle kP\Pi, xQ \rangle} \big(A(kP\Pi|aQ) + A(kP\Pi|aQ\Pi) \big) \Big\} \\
&= \sum_{P \in \mathfrak{A}^N} \operatorname{sign}(PQ) e^{i\langle kP, xQ \rangle} \big(A(kP|aQ) + A(kP|aQ\Pi) \\
&\qquad - A(kP\Pi|aQ) - A(kP\Pi|aQ\Pi) \big) \,.
\end{aligned}
\tag{3.B.17}
$$

Thus, a condition sufficient for uniqueness is

$$
A(kP|aQ) + A(kP|aQ\Pi) = A(kP\Pi|aQ) + A(kP\Pi|aQ\Pi) \,.
\tag{3.B.18}
$$

In the first line of (3.B.17) we used (3.B.16) and $\operatorname{sign}(\Pi) = -1$. In the second line we used $\mathfrak{S}^N = \mathfrak{A}^N \cup \mathfrak{A}^N \Pi$, and in the third line $\langle kR, xR \rangle = \langle k, x \rangle$ and $\Pi^2 = \operatorname{id}$. Recall that Π is any transposition of nearest neighbours.

Provided that (3.B.18) is satisfied, our wave function is now uniquely defined as a function on $\mathbb{Z}^N$ by the equation

$$
\psi(x; a|k) = \sum_{P \in \mathfrak{S}^N} \operatorname{sign}(PQ) A(kP|aQ) e^{i\langle kP, xQ \rangle} \,.
\tag{3.B.19}
$$

For every given x the permutation Q on the right hand side has to be chosen such that $x \in \overline{D}_Q$. $\psi(x; a|k)$ solves the Schrödinger equation (3.16) for every $x \in D_Q$, $Q \in \mathfrak{S}^N$. In the next section we shall stipulate $\psi(x; a|k)$ to satisfy the Schrödinger equation at the boundaries of the sectors $\overline{D}_Q$. This will provide conditions on the amplitudes $A(kP|aQ)$.

3.B.2 Equations for the amplitudes

Let us start with the case of three or more electrons sitting at the same site. Then $x_{Q(n-1)} = x_{Q(n)} = x_{Q(n+1)}$ at the boundary of some sector $\overline{D}_Q$, and, say, $a_{Q(n)} = a_{Q(n+1)}$ (recall that a_j takes only two values, $\uparrow$, $\downarrow$). It follows that $\psi(x; a|k) = 0$, and thus

$$
\begin{aligned}
(H_N - E)\psi(x; a|k) &= -\sum_{j=1}^N (\Delta_j^+ + \Delta_j^-)\psi(x; a|k) \\
&= -\sum_{j=1}^N \left(\Delta_{Q(j)}^+ + \Delta_{Q(j)}^- \right) \psi(x; a|k) \\
&= -\left(\Delta_{Q(n)}^+ + \Delta_{Q(n)}^- + \Delta_{Q(n+1)}^+ + \Delta_{Q(n+1)}^- \right) \psi(x; a|k) \\
&= -\Big(\psi(x + e^{Q(n)}; a|k) + \psi(x - e^{Q(n)}; a|k) \\
&\qquad + \psi(x + e^{Q(n+1)}; a|k) + \psi(x - e^{Q(n+1)}; a|k) \Big) = 0 \,.
\end{aligned}
\tag{3.B.20}
$$

Here we used the fact that $\psi(\mathbf{x}; \mathbf{a}|\mathbf{k})$ is antisymmetric in $x_{Q(n)}$ and $x_{Q(n+1)}$, which follows from $a_{Q(n)} = a_{Q(n+1)}$. In the fifth row of (3.B.20) this antisymmetry implies $\psi(\mathbf{x} \pm \mathbf{e}^{Q(n+1)}; \mathbf{a}|\mathbf{k}) = -\psi(\mathbf{x} \pm \mathbf{e}^{Q(n)}; \mathbf{a}|\mathbf{k})$.

We have seen that the Schrödinger equation (3.16) is satisfied, if three or more electrons are sitting at the same site of the lattice. We are thus left with the problem of pairs of electrons occupying the same lattice site. Let us assume we have precisely one such pair, $x_{Q(n)} = x_{Q(n+1)}$, and all other x_j are mutually distinct. It follows that

$$(H_N - E)\psi(\mathbf{x}; \mathbf{a}|\mathbf{k})$$

$$= -\big(\Delta^+_{Q(n)} + \Delta^-_{Q(n)} + \Delta^+_{Q(n+1)} + \Delta^-_{Q(n+1)}\big)\psi(\mathbf{x}; \mathbf{a}|\mathbf{k})$$

$$\quad - \sum_{\substack{j=1 \\ j \neq n, n+1}}^{N} \big(\Delta^+_{Q(j)} + \Delta^-_{Q(j)}\big)\psi(\mathbf{x}; \mathbf{a}|\mathbf{k}) + (4u - E)\psi(\mathbf{x}; \mathbf{a}|\mathbf{k})$$

$$= -\big(\psi(\mathbf{x} + \mathbf{e}^{Q(n)}; \mathbf{a}|\mathbf{k}) + \psi(\mathbf{x} - \mathbf{e}^{Q(n)}; \mathbf{a}|\mathbf{k}) + \psi(\mathbf{x} + \mathbf{e}^{Q(n+1)}; \mathbf{a}|\mathbf{k})$$

$$\quad + \psi(\mathbf{x} - \mathbf{e}^{Q(n+1)}; \mathbf{a}|\mathbf{k})\big)$$

$$\quad + \sum_{P \in \mathfrak{S}^N} \text{sign}(PQ)A(\mathbf{k}P|\mathbf{a}Q)\big(4u + 2\cos(k_{P(n)}) + 2\cos(k_{P(n+1)})\big)\,\mathrm{e}^{\mathrm{i}\langle \mathbf{k}P, \mathbf{x}Q\rangle}$$

$$= \sum_{P \in \mathfrak{S}^N} \Big\{\text{sign}(PQ\Pi)A(\mathbf{k}P|\mathbf{a}Q\Pi)\big(-\mathrm{e}^{\mathrm{i}\langle \mathbf{k}P, (\mathbf{x}+\mathbf{e}^{Q(n)})Q\Pi\rangle} - \mathrm{e}^{\mathrm{i}\langle \mathbf{k}P, (\mathbf{x}-\mathbf{e}^{Q(n+1)})Q\Pi\rangle}\big)$$

$$\quad + \text{sign}(PQ)A(\mathbf{k}P|\mathbf{a}Q)\big(-\mathrm{e}^{\mathrm{i}\langle \mathbf{k}P, (\mathbf{x}-\mathbf{e}^{Q(n)})Q\rangle} - \mathrm{e}^{\mathrm{i}\langle \mathbf{k}P, (\mathbf{x}+\mathbf{e}^{Q(n+1)})Q\rangle}$$

$$\quad + \big[4u + 2\cos(k_{P(n)}) + 2\cos(k_{P(n+1)})\big]\,\mathrm{e}^{\mathrm{i}\langle \mathbf{k}P, \mathbf{x}Q\rangle}\big)\Big\}$$

$$= \sum_{P \in \mathfrak{S}^N} \text{sign}(PQ)\mathrm{e}^{\mathrm{i}\langle \mathbf{k}P, \mathbf{x}Q\rangle}\Big\{A(\mathbf{k}P|\mathbf{a}Q\Pi)\big(\mathrm{e}^{\mathrm{i}k_{P(n+1)}} + \mathrm{e}^{-\mathrm{i}k_{P(n)}}\big)$$

$$\quad + A(\mathbf{k}P|\mathbf{a}Q)\big(\mathrm{e}^{\mathrm{i}k_{P(n)}} + \mathrm{e}^{-\mathrm{i}k_{P(n+1)}} + 4u\big)\Big\}$$

$$= \sum_{P \in \mathfrak{A}^N} \text{sign}(PQ)\mathrm{e}^{\mathrm{i}\langle \mathbf{k}P, \mathbf{x}Q\rangle}\Big\{A(\mathbf{k}P|\mathbf{a}Q\Pi)\big(\mathrm{e}^{\mathrm{i}k_{P(n+1)}} + \mathrm{e}^{-\mathrm{i}k_{P(n)}}\big)$$

$$\quad - A(\mathbf{k}P\Pi|\mathbf{a}Q\Pi)\big(\mathrm{e}^{\mathrm{i}k_{P(n)}} + \mathrm{e}^{-\mathrm{i}k_{P(n+1)}}\big)$$

$$\quad + A(\mathbf{k}P|\mathbf{a}Q)\big(\mathrm{e}^{\mathrm{i}k_{P(n)}} + \mathrm{e}^{-\mathrm{i}k_{P(n+1)}} + 4u\big)$$

$$\quad - A(\mathbf{k}P\Pi|\mathbf{a}Q)\big(\mathrm{e}^{\mathrm{i}k_{P(n+1)}} + \mathrm{e}^{-\mathrm{i}k_{P(n)}} + 4u\big)\Big\}$$

$$= \sum_{P \in \mathfrak{A}^N} \text{sign}(PQ)\mathrm{e}^{\mathrm{i}\langle \mathbf{k}P, \mathbf{x}Q\rangle}$$

$$\Big\{\big(\cos(k_{P(n)}) + \cos(k_{P(n+1)})\big)\big(A(\mathbf{k}P|\mathbf{a}Q\Pi) - A(\mathbf{k}P\Pi|\mathbf{a}Q\Pi)$$

$$\quad + A(\mathbf{k}P|\mathbf{a}Q) - A(\mathbf{k}P\Pi|\mathbf{a}Q)\big)$$

$$\quad + \mathrm{i}\big(\sin(k_{P(n+1)}) - \sin(k_{P(n)})\big)\big(A(\mathbf{k}P|\mathbf{a}Q\Pi) + A(\mathbf{k}P\Pi|\mathbf{a}Q\Pi)$$

$$\quad - A(\mathbf{k}P|\mathbf{a}Q) - A(\mathbf{k}P\Pi|\mathbf{a}Q)\big)$$

$$\quad + 4u\big(A(\mathbf{k}P|\mathbf{a}Q) - A(\mathbf{k}P\Pi|\mathbf{a}Q)\big)\Big\}$$

$$= \sum_{P \in \mathfrak{A}^N} \text{sign}(PQ) e^{i \langle kP, xQ \rangle} \{ 4u \big(A(\mathbf{k}P|\mathbf{a}Q) - A(\mathbf{k}P\Pi|\mathbf{a}Q) \big)$$

$$+ i \big(\sin(k_{P(n)}) - \sin(k_{P(n+1)}) \big) \big(A(\mathbf{k}P|\mathbf{a}Q) + A(\mathbf{k}P\Pi|\mathbf{a}Q) $$

$$- A(\mathbf{k}P|\mathbf{a}Q\Pi) - A(\mathbf{k}P\Pi|\mathbf{a}Q\Pi) \big) \} \, . \tag{3.B.21}$$

Here we used $\mathbf{x} + \mathbf{e}^{Q(n)}, \mathbf{x} - \mathbf{e}^{Q(n+1)} \in \overline{D}_{Q\Pi}$ and $\mathbf{x} - \mathbf{e}^{Q(n)}, \mathbf{x} + \mathbf{e}^{Q(n+1)} \in \overline{D}_Q$ in the third equation and $(\mathbf{x} + \mathbf{e}^{Q(n)})Q\Pi = \mathbf{x}Q + \mathbf{e}^{n+1}$, $(\mathbf{x} - \mathbf{e}^{Q(n+1)})Q\Pi = \mathbf{x}Q - \mathbf{e}^n$ in the fourth equation. In the seventh equation we used the uniqueness condition (3.B.18).

From the right hand side of (3.B.21) we can read off a sufficient condition for the Schrödinger equation to be satisfied,

$$\big(\sin(k_{P(n)}) - \sin(k_{P(n+1)}) \big) \big(A(\mathbf{k}P|\mathbf{a}Q) - A(\mathbf{k}P|\mathbf{a}Q\Pi) \big) - 4iu \, A(\mathbf{k}P|\mathbf{a}Q)$$

$$= - \big(\sin(k_{P(n)}) - \sin(k_{P(n+1)}) \big) \big(A(\mathbf{k}P\Pi|\mathbf{a}Q) - A(\mathbf{k}P\Pi|\mathbf{a}Q\Pi) \big) - 4iu \, A(\mathbf{k}P\Pi|\mathbf{a}Q) \, . \tag{3.B.22}$$

If a second pair of electrons occupies another lattice site, we get the same equation (3.B.22). Thus, (3.B.18) and (3.B.22) are sufficient for $\psi(\mathbf{x}; \mathbf{a}|\mathbf{k})$ to be uniquely defined and to satisfy the Schrödinger equation for all $\mathbf{x} \in \mathbb{Z}^N$.

Can we construct a set of amplitudes $A(\mathbf{k}P|\mathbf{a}Q)$ that satisfies both equations, (3.B.18) and (3.B.22)? In order to study this question we reformulate equations (3.B.18) and (3.B.22) in a more convenient language. For fixed $\mathbf{k}P$ the amplitude $A(\mathbf{k}P|\mathbf{a}Q)$ is a function of the N spin variables a_j, which take on values $a_j = \uparrow, \downarrow$ each. We may therefore interpret $A(\mathbf{k}P|\mathbf{a}Q)$ as representing the coordinates of a 2^N-dimensional vector. The corresponding auxiliary vector space may be understood as the 2^N-dimensional space of states of an N-site spin-$\frac{1}{2}$ chain. Let us introduce the canonical basis of this vector space. It is constructed from tensor products of basis vectors $\mathbf{e}_\uparrow = \binom{1}{0}$ and $\mathbf{e}_\downarrow = \binom{0}{1}$. A basis vector may be written as $|\mathbf{a}\rangle = \mathbf{e}_{a_1} \otimes \cdots \otimes \mathbf{e}_{a_N}$. These basis vectors are mutually orthonormal under the canonical hermitian scalar product,

$$\langle \mathbf{a}|\mathbf{b} \rangle = \langle \mathbf{e}_{a_1}, \mathbf{e}_{b_1} \rangle \dots \langle \mathbf{e}_{a_N}, \mathbf{e}_{b_N} \rangle = \delta_{b_1}^{a_1} \dots \delta_{b_N}^{a_N} \, . \tag{3.B.23}$$

There is a natural action of the symmetric group on basis vectors in our spin chain Hilbert space,

$$Q|\mathbf{a}\rangle = |\mathbf{a}Q^{-1}\rangle \, , \tag{3.B.24}$$

which defines a representation of the symmetric group,

$$P(Q|\mathbf{a}\rangle) = P|\mathbf{a}Q^{-1}\rangle = |(\mathbf{a}Q^{-1})P^{-1}\rangle = |\mathbf{a}(PQ)^{-1}\rangle = (PQ)|\mathbf{a}\rangle \, . \tag{3.B.25}$$

This representation is unitary: we have

$$\langle \mathbf{a}|Q^+|\mathbf{b}\rangle = \langle \mathbf{a}Q^{-1}|\mathbf{b}\rangle = \delta_{b_1}^{a_{Q^{-1}(1)}} \dots \delta_{b_N}^{a_{Q^{-1}(N)}}$$

$$= \delta_{b_{Q(1)}}^{a_1} \dots \delta_{b_{Q(N)}}^{a_N} = \langle \mathbf{a}|\mathbf{b}Q\rangle = \langle \mathbf{a}|Q^{-1}|\mathbf{b}\rangle \, , \tag{3.B.26}$$

and thus $Q^+ = Q^{-1}$.

Let us define

$$|\mathbf{k}P\rangle = \sum_{b_1,\ldots,b_N=\uparrow,\downarrow} A(\mathbf{k}P|\mathbf{b})|\mathbf{b}\rangle . \tag{3.B.27}$$

Then it follows from (3.B.23) that

$$A(\mathbf{k}P|\mathbf{a}Q) = \langle \mathbf{a}Q|\mathbf{k}P\rangle . \tag{3.B.28}$$

Using (3.B.28), (3.B.24), and (3.B.26), we can rewrite the uniqueness condition (3.B.18) and the condition (3.B.22) arising from the Schrödinger equation in the language of the spin problem. The uniqueness condition turns into

$$(1 + \Pi)|\mathbf{k}P\Pi\rangle = (1 + \Pi)|\mathbf{k}P\rangle . \tag{3.B.29}$$

Similarly, equation (3.B.22) is equivalent to

$$\left[(s_{P(n)} - s_{P(n+1)})(1 - \Pi) - 4iu\right]|\mathbf{k}P\rangle = \left[-(s_{P(n)} - s_{P(n+1)})(1 - \Pi) - 4iu\right]|\mathbf{k}P\Pi\rangle . \tag{3.B.30}$$

Here we have introduced the short hand notation $s_n = \sin(k_n)$ which in the following will prove to be convenient. Equation (3.B.30) can be easily solved for $|\mathbf{k}P\Pi\rangle$. Since $\Pi^2 = 1$, multiplication by $\left[-(s_{P(n)} - s_{P(n+1)})(1 + \Pi) - 4iu\right]$ from the left gives

$$|\mathbf{k}P\Pi\rangle = Y_{n,n+1}(s_{P(n)} - s_{P(n+1)})|\mathbf{k}P\rangle , \tag{3.B.31}$$

where we introduced the operators

$$Y_{jk}(\lambda) = \frac{2iu + \lambda\,\Pi_{jk}}{2iu + \lambda} . \tag{3.B.32}$$

Instead of (3.B.18) and (3.B.22) we can now deal with equations (3.B.29) and (3.B.31). This innocent looking reformulation of the original equations has strong immediate consequences. It turns the remaining calculations into a beautiful piece of algebra based on the properties of the so-called Y-operators (3.B.32) first introduced by C. N. Yang in [493].

The first implication that can be drawn from (3.B.29) and (3.B.31) is that (3.B.18) is automatically satisfied for any set of amplitudes solving (3.B.22). This can be seen as follows. We multiply equation (3.B.31), which is equivalent to (3.B.22), by $1 + \Pi$ from the left. Then, since $(1 + \Pi)\Pi = (1 + \Pi)$, (3.B.31) turns into (3.B.29), which is equivalent to (3.B.18). Hence, equation (3.B.22) is a sufficient condition for the Bethe wave function $\psi(\mathbf{x}; \mathbf{a}|\mathbf{k})$, equation (3.B.19), to be uniquely defined *and* to satisfy the Schrödinger equation.

3.B.3 Consistency

Our original problem of finding joint solutions of (3.B.18) and (3.B.22) has now reduced to the problem of finding solutions of (3.B.22). At the same time the recursive structure of (3.B.31) allows us to construct these solutions. We see from (3.B.22) that the Y-operators (3.B.32) induce the action of the nearest neighbour transpositions Π on the vectors $\mathbf{k}$

that label the states $|\mathbf{k}\rangle$ of the auxiliary spin problem. Since the symmetric group $\mathfrak{S}^N$ is generated by the transpositions of nearest neighbours, all spin states $|\mathbf{k}P\rangle$, $P \in \mathfrak{S}^N$, can be constructed from $|\mathbf{k}\rangle$ by repeated use of (3.B.31). Equivalently, (3.B.31) connects every state $|\mathbf{k}P\rangle$, $P \in \mathfrak{S}^N$, with any other state $|\mathbf{k}\tilde{P}\rangle$, $\tilde{P} \in \mathfrak{S}^N$. It follows that all amplitudes $A(\mathbf{k}P|\mathbf{a}) = \langle \mathbf{a}|\mathbf{k}P\rangle$ in the sector $\overline{D}_{\mathrm{id}}$ can be generated out of $A(\mathbf{k}|\mathbf{a})$. The amplitudes in the other sectors are simply $A(\mathbf{k}P|\mathbf{a}Q) = \langle \mathbf{a}Q|\mathbf{k}P\rangle = \langle \mathbf{a}|Q|\mathbf{k}P\rangle$.

The remaining problem we have to face is that of the uniqueness of the amplitudes constructed by iterative use of (3.B.31). This problem is known as the consistency problem. It is due to the fact that a representation of a permutation $P \in \mathfrak{S}^N$ as a product of nearest neighbour transpositions is not unique.

Let us consider the example $N = 3$. Then we have, for instance, the identity

$$\Pi_{12}\Pi_{23}\Pi_{12} - \Pi_{23}\Pi_{12}\Pi_{23} - \Pi_{13} . \tag{3.B.33}$$

There are thus two different ways to create $A(\mathbf{k}\Pi_{13}|\mathbf{a})$ out of $A(\mathbf{k}|\mathbf{a})$ by application of Y-operators, equation (3.B.31),

$$
\begin{aligned}
A(\mathbf{k}\Pi_{13}|\mathbf{a}) &= A(\mathbf{k}\Pi_{12}\Pi_{23}\Pi_{12}|\mathbf{a}) = A(\mathbf{k}\Pi_{23}\Pi_{12}\Pi_{23}|\mathbf{a}) \\
&= \langle \mathbf{a}|Y_{12}(s_2 - s_3)Y_{23}(s_1 - s_3)Y_{12}(s_1 - s_2)|\mathbf{k}\rangle \\
&= \langle \mathbf{a}|Y_{23}(s_1 - s_2)Y_{12}(s_1 - s_3)Y_{23}(s_2 - s_3)|\mathbf{k}\rangle .
\end{aligned}
\tag{3.B.34}
$$

For this equation to be true for arbitrary states $|\mathbf{a}\rangle$ and $|\mathbf{k}\rangle$ the triple products of Y-operators on both sides of the third equation (3.B.34) must agree. It is crucial that this is indeed the case. The Y-operators $Y_{jk}(\lambda)$ satisfy the so-called Yang-Baxter equation,

$$Y_{jk}(\lambda)Y_{kl}(\lambda + \mu)Y_{jk}(\mu) = Y_{kl}(\mu)Y_{jk}(\lambda + \mu)Y_{kl}(\lambda) , \tag{3.B.35}$$

which can be proven rather easily by direct calculation (see appendix 3.C). It further follows from the definition (3.B.32) of $Y_{jk}(\lambda)$ that

$$Y_{jk}^{-1}(\lambda) = Y_{jk}(-\lambda) . \tag{3.B.36}$$

Using equations (3.B.35) and (3.B.36) we show in appendix 3.C that all amplitudes $A(\mathbf{k}P|\mathbf{a}Q)$ are uniquely defined by (3.B.31) once the amplitude $A(\mathbf{k}|\mathbf{a})$ is given. This is a situation typical for a scattering problem, where the amplitude of the incoming wave is arbitrary and fixes the amplitudes of the scattered and the reflected waves.

3.B.4 Periodic boundary conditions

So far we have solved the Schrödinger equation (3.16) regarded as a mere difference equation, not as an eigenvalue problem. In this subsection we shall impose the periodic boundary conditions (3.17) on the Bethe ansatz wave functions. This will result in an eigenvalue problem for the amplitudes $A(\mathbf{k}P|\mathbf{a}Q)$, which can be interpreted as an auxiliary spin problem. The auxiliary spin problem is solved below in Section 3.B.5 by means of the algebraic Bethe ansatz.

Let $\mathbf{x} \in \{1, \ldots, L\}^N$. The condition $(\Delta_j^- - \Delta_{j,L}^-)\psi(\mathbf{x}, \mathbf{a}|\mathbf{k}) = 0$ is trivially satisfied for $x_j = 2, \ldots, L$. If $x_j = 1$, then there is an element $Q \in \mathfrak{S}^N$, such that $\mathbf{x} \in \overline{D}_Q$ and $x_j = x_{Q(1)}$. In this case the periodic boundary conditions are non-trivial,

$$\psi(\mathbf{x} - (1 - L)\mathbf{e}^{Q(1)}; \mathbf{a}|\mathbf{k}) = \psi(\mathbf{x} - \mathbf{e}^{Q(1)}; \mathbf{a}|\mathbf{k}) . \tag{3.B.37}$$

Similarly, the condition $(\Delta_j^+ - \Delta_{j,L}^+)\psi(\mathbf{x}, \mathbf{a}|\mathbf{k}) = 0$ is trivial for $x_j = 1, \ldots, L - 1$. If $x_j = L$, then there is an element $Q \in \mathfrak{S}^N$, such that $\mathbf{x} \in \overline{D}_Q$ and $x_j = x_{Q(N)}$. We obtain the non-trivial condition

$$\psi(\mathbf{x} + (1 - L)\mathbf{e}^{Q(N)}; \mathbf{a}|\mathbf{k}) = \psi(\mathbf{x} + \mathbf{e}^{Q(N)}; \mathbf{a}|\mathbf{k}) . \tag{3.B.38}$$

Let us discuss equation (3.B.37). Let $\mathbf{y} = \mathbf{x} - \mathbf{e}^{Q(1)}$ and $\mathbf{z} = \mathbf{y} + L\mathbf{e}^{Q(1)}$. Then $\mathbf{y} \in \overline{D}_Q$, $z_{Q(1)} = y_{Q(1)} + L$ and $z_{Q(j)} = y_{Q(j)}$, $j = 2, \ldots, N$. It follows that $z_{Q(2)} \leq \cdots \leq z_{Q(N)} \leq z_{Q(1)}$, which may be equivalently stated as $z_{QU_N(1)} \leq \cdots \leq z_{QU_N(N)}$. Here

$$U_N = \Pi_{12} \ldots \Pi_{N-1,N} \tag{3.B.39}$$

is the generator of the cyclic subgroup of order N of the symmetric group. We conclude that

$$\psi(\mathbf{y} + L\mathbf{e}^{Q(1)}; \mathbf{a}|\mathbf{k}) = \psi(\mathbf{z}; \mathbf{a}|\mathbf{k})$$
$$= \sum_{P \in \mathfrak{S}^N} \text{sign}(PQU_N) A(\mathbf{k}P|\mathbf{a}QU_N) \mathbf{e}^{i\langle \mathbf{k}P, (\mathbf{y} + L\mathbf{e}^{Q(1)})QU_N \rangle}$$
$$= \sum_{P \in \mathfrak{S}^N} \text{sign}(PQ) A(\mathbf{k}PU_N|\mathbf{a}QU_N) \mathbf{e}^{ik_{P(1)}L} \mathbf{e}^{i\langle \mathbf{k}P, \mathbf{y}Q \rangle} . \tag{3.B.40}$$

Comparing (3.B.40) with the expression for $\psi(\mathbf{y}; \mathbf{a}|\mathbf{k})$, we see that the condition

$$A(\mathbf{k}PU_N|\mathbf{a}QU_N) = \mathbf{e}^{-ik_{P(1)}L} A(\mathbf{k}P|\mathbf{a}Q) \tag{3.B.41}$$

is sufficient for (3.B.37) to be satisfied.

A very similar reasoning can be applied to equation (3.B.38). Let $\mathbf{y} = \mathbf{x} + \mathbf{e}^{Q(N)}$ and $\mathbf{z} = \mathbf{y} - L\mathbf{e}^{Q(N)}$. Then $\mathbf{y} \in \overline{D}_Q$, $z_{Q(N)} = y_{Q(N)} - L$ and $z_{Q(j)} = y_{Q(j)}$ for $j = 1, \ldots, N - 1$. It follows that $z_{Q(N)} \leq z_{Q(1)} \leq \cdots \leq z_{Q(N-1)}$, which is equivalent to $z_{QU_N^{-1}(1)} \leq \cdots \leq z_{QU_N^{-1}(N)}$. We thus obtain

$$\psi(\mathbf{y} - L\mathbf{e}^{Q(N)}; \mathbf{a}|\mathbf{k}) = \psi(\mathbf{z}; \mathbf{a}|\mathbf{k})$$
$$= \sum_{P \in \mathfrak{S}^N} \text{sign}(PQU_N^{-1}) A(\mathbf{k}P|\mathbf{a}QU_N^{-1}) \mathbf{e}^{i\langle \mathbf{k}P, (\mathbf{y} - L\mathbf{e}^{Q(N)})QU_N^{-1} \rangle}$$
$$= \sum_{P \in \mathfrak{S}^N} \text{sign}(PQ) A(\mathbf{k}PU_N^{-1}|\mathbf{a}QU_N^{-1}) \mathbf{e}^{-ik_{P(N)}L} \mathbf{e}^{i\langle \mathbf{k}P, \mathbf{y}Q \rangle} . \tag{3.B.42}$$

Hence, a sufficient condition for (3.B.38) to hold is

$$A(\mathbf{k}P|\mathbf{a}Q) = \mathbf{e}^{-ik_{P(N)}L} A(\mathbf{k}PU_N^{-1}|\mathbf{a}QU_N^{-1}) . \tag{3.B.43}$$

Let us set $\tilde{P} = PU_N^{-1}$, $\tilde{Q} = QU_N^{-1}$. Then (3.B.43) turns into

$$A(\mathbf{k}\tilde{P}U_N|\mathbf{a}\tilde{Q}U_N) = e^{-ik_{\tilde{P}(1)}L}A(\mathbf{k}\tilde{P}|\mathbf{a}\tilde{Q}), \qquad (3.B.44)$$

which is equivalent to (3.B.41). We are thus left with a single condition, namely (3.B.41), that suffices to guarantee periodicity of the wave functions (3.B.19).

Now (3.B.31) allows us to transform (3.B.41) into an eigenvalue problem. First of all we obtain

$$
\begin{aligned}
A(\mathbf{k}PU_N|\mathbf{a}QU_N) &= \langle \mathbf{a}QU_N|\mathbf{k}PU_N\rangle = \langle \mathbf{a}Q|U_N|\mathbf{k}PU_{N-1}\Pi_{N-1,N}\rangle \\
&= \langle \mathbf{a}Q|U_N Y_{N-1,N}(s_{PU_{N-1}(N-1)} - s_{PU_{N-1}(N)})|\mathbf{k}PU_{N-1}\rangle \\
&= \langle \mathbf{a}Q|U_N Y_{N-1,N}(s_{P(1)} - s_{P(N)})|\mathbf{k}PU_{N-1}\rangle \\
&= \langle \mathbf{a}Q|U_N Y_{N-1,N}(s_{P(1)} - s_{P(N)})\ldots Y_{1,2}(s_{P(1)} - s_{P(2)})|\mathbf{k}P\rangle \\
&= \langle \mathbf{a}Q|U_{N-1}\Pi_{N-1,N} Y_{N-1,N}(s_{P(1)} - s_{P(N)})\ldots Y_{1,2}(s_{P(1)} - s_{P(2)})|\mathbf{k}P\rangle \\
&= \langle \mathbf{a}Q|\Pi_{1,N} Y_{1,N}(s_{P(1)} - s_{P(N)})\Pi_{1,N-1} Y_{1,N-1}(s_{P(1)} - s_{P(N-1)})\cdot \\
&\quad \ldots \cdot \Pi_{1,2} Y_{1,2}(s_{P(1)} - s_{P(2)})|\mathbf{k}P\rangle \\
&= e^{-ik_{P(1)}L}\langle \mathbf{a}Q|\mathbf{k}P\rangle .
\end{aligned}
\qquad (3.B.45)
$$

This suggests to define the operator

$$X_{jk}(\lambda) = \Pi_{jk} Y_{jk}(\lambda). \qquad (3.B.46)$$

It then follows from the last two lines of equation (3.B.45) that (3.B.41) is equivalent to the eigenvalue problem

$$X_{1,N}(s_{P(1)} - s_{P(N)})\ldots X_{1,2}(s_{P(1)} - s_{P(2)})|\mathbf{k}P\rangle = e^{-ik_{P(1)}L}|\mathbf{k}P\rangle , \qquad (3.B.47)$$

which has to be solved for all $P \in \mathfrak{S}^N$.

Remark. Applying (3.B.31) to the eigenvalue problem (3.B.47) it is not difficult to see that (3.B.47) is equivalent to

$$
\begin{aligned}
X_{j,j-1}(s_{P(j)} - s_{P(j-1)})&\ldots X_{j,1}(s_{P(j)} - s_{P(1)})X_{j,N}(s_{P(j)} - s_{P(N)}) \\
&\ldots \cdot X_{j,j+1}(s_{P(j)} - s_{P(j+1)})|\mathbf{k}P\rangle = e^{-ik_{P(j)}L}|\mathbf{k}P\rangle
\end{aligned}
\qquad (3.B.48)
$$

for $j = 1,\ldots,N$, where we introduced the conventions $X_{1,0}(\lambda) = X_{1,N}(\lambda)$ and $X_{N,N+1}(\lambda) = X_{N,1}(\lambda)$.

Although the mutual compatibility of equations (3.B.48) for fixed P and $j = 1,\ldots,N$ is clear by construction, the reader may wonder what is the reason behind it. As before it is again the Yang-Baxter equation. The Yang-Baxter equation for the operators $X_{jk}(\lambda)$ is obtained from (3.B.35) by multiplication with $\Pi_{jk}\Pi_{jl}\Pi_{kl} = \Pi_{kl}\Pi_{jl}\Pi_{jk}$ from the right,

$$X_{jk}(\lambda)X_{jl}(\lambda+\mu)X_{kl}(\mu) = X_{kl}(\mu)X_{jl}(\lambda+\mu)X_{jk}(\lambda). \qquad (3.B.49)$$

Instead of applying (3.B.49) and the inversion relation $X_{jk}(\lambda)X_{jk}(-\lambda) = \mathrm{id}$ directly to (3.B.48), which is one possible way of proving the compatibility of the equations (3.B.48) for $j = 1, \ldots, N$, we shall postpone a direct proof until the next section where we derive a more appropriate formulation of (3.B.48).

3.B.5 Algebraic solution of the spin problem

We shall now identify the eigenvalue problem (3.B.48) as an eigenvalue problem for the transfer matrix of an inhomogeneous XXX spin chain and then solve it by means of the algebraic Bethe ansatz. For more information about the general method the reader is referred to Chapter 12. Here we only give a brief but self-contained account of the aspects needed to solve our concrete problem.

Let us first express the transpositions Π_{jk} in terms of spin operators,

$$\Pi_{jk} = \tfrac{1}{2}(\mathrm{id} + \sigma_j^\alpha \sigma_k^\alpha). \tag{3.B.50}$$

Here we introduced the usual embedding of Pauli matrices into the space of endomorphisms on the space of states,

$$\sigma_j^\alpha = I_2^{\otimes(j-1)} \otimes \sigma^\alpha \otimes I_2^{\otimes(N-j)}, \tag{3.B.51}$$

$j = 1, \ldots, N$. In our notation I_2 is the 2×2 unit matrix.

Next, we shall need extensions of our spin chain Hilbert space by one or two sites, respectively. We will indicate this by supplying superscripts (a) or (ab) to the operators and call these additional sites auxiliary spaces. Then we can define

$$T(\lambda|\mathbf{k}P) = X_{aN}^{(a)}(\lambda - iu - s_{P(N)}) \ldots X_{a1}^{(a)}(\lambda - iu - s_{P(1)}), \tag{3.B.52}$$

with the subscript a referring to the auxiliary space. It follows that

$$\begin{aligned}
T(s_{P(1)} + iu|\mathbf{k}P) &= X_{aN}^{(a)}(s_{P(1)} - s_{P(N)}) \ldots X_{a2}^{(a)}(s_{P(1)} - s_{P(2)}) X_{a1}^{(a)}(0) \\
&= \Pi_{a1} X_{1N}^{(a)}(s_{P(1)} - s_{P(N)}) \ldots X_{12}^{(a)}(s_{P(1)} - s_{P(2)}) \\
&= \tfrac{1}{2}(I_2 \otimes \mathrm{id} + \sigma^\alpha \otimes \sigma_1^\alpha)(I_2 \otimes (X_{1N}(s_{P(1)} - s_{P(N)}) \ldots X_{12}(s_{P(1)} - s_{P(2)}))).
\end{aligned} \tag{3.B.53}$$

Here we used the fact that $X_{a1}^{(a)}(0) = \Pi_{a1}$ in the second equation. Taking the trace in auxiliary space we obtain

$$\mathrm{tr}\big(T(s_{P(1)} + iu|\mathbf{k}P)\big) = X_{1N}(s_{P(1)} - s_{P(N)}) \ldots X_{12}(s_{P(1)} - s_{P(2)}). \tag{3.B.54}$$

With the abbreviation

$$t(\lambda|\mathbf{k}P) = \mathrm{tr}(T(\lambda|\mathbf{k}P)) \tag{3.B.55}$$

equation (3.B.47) is an eigenvalue problem for the operator $t(s_{P(1)} + iu|\mathbf{k}P)$. The operator $t(\lambda|\mathbf{k}P)$ may be interpreted as the transfer matrix of an inhomogeneous XXX spin chain, $T(\lambda|\mathbf{k}P)$ as the corresponding monodromy matrix (see Chapter 12).

Let us now identify the standard objects of the algebraic Bethe ansatz, namely the R-matrix and the L-matrix. For this purpose we shall write our operators as matrices with respect to the auxiliary spaces. By appropriately shifting the spectral parameters and multiplying by Π_{ab} from the left the Yang-Baxter equation (3.B.49) takes the form

$$Y_{ab}^{(ab)}(\lambda - \mu)X_{an}^{(ab)}(\lambda - iu)X_{bn}^{(ab)}(\mu - iu) = X_{an}^{(ab)}(\mu - iu)X_{bn}^{(ab)}(\lambda - iu)Y_{ab}^{(ab)}(\lambda - \mu).$$
(3.B.56)

Equation (3.B.56) can be understood as matrix equation in auxiliary space when we use the usual conventions for tensor products of matrices,

$$Y_{ab}^{(ab)}(\lambda) = \check{R}(\lambda) \otimes I_2^{\otimes N} = \check{R}(\lambda),$$
(3.B.57a)

$$X_{an}^{(ab)}(\lambda - iu) = \frac{\lambda + iu(\sigma^\alpha \otimes I_2) \otimes \sigma_n^\alpha}{\lambda + iu} = \frac{\lambda + iu(\sigma^\alpha \sigma_n^\alpha \otimes I_2)}{\lambda + iu} = L_n(\lambda) \otimes I_2,$$
(3.B.57b)

$$X_{bn}^{(ab)}(\mu - iu) = \frac{\mu + iu(I_2 \otimes \sigma^\alpha) \otimes \sigma_n^\alpha}{\mu + iu} = \frac{\mu + iu(I_2 \otimes \sigma^\alpha \sigma_n^\alpha)}{\mu + iu} = I_2 \otimes L_n(\mu).$$
(3.B.57c)

The matrices $\check{R}(\lambda)$ and $L_n(\lambda)$ can be read off from these equations,

$$\check{R}(\lambda) = \frac{4iu + \lambda(I_2 \otimes I_2 + \sigma^\alpha \otimes \sigma^\alpha)}{4iu + 2\lambda} = \begin{pmatrix} 1 & 0 & 0 & 0 \\ 0 & b(\lambda) & c(\lambda) & 0 \\ 0 & c(\lambda) & b(\lambda) & 0 \\ 0 & 0 & 0 & 1 \end{pmatrix},$$
(3.B.58)

where

$$b(\lambda) = \frac{2iu}{\lambda + 2iu}, \qquad c(\lambda) = \frac{\lambda}{\lambda + 2iu},$$
(3.B.59)

and

$$L_n(\lambda) = \frac{\lambda + iu\, \sigma^\alpha \sigma_n^\alpha}{\lambda + iu} = c(2\lambda) + b(2\lambda)\sigma^\alpha \sigma_n^\alpha$$

$$= \begin{pmatrix} c(2\lambda) + b(2\lambda)\sigma_n^z & 2b(2\lambda)\sigma_n^- \\ 2b(2\lambda)\sigma_n^+ & c(2\lambda) - b(2\lambda)\sigma_n^z \end{pmatrix}.$$
(3.B.60)

$\check{R}(\lambda)$ is called the R-matrix, $L_n(\lambda)$ the L-matrix at site n.

Inserting (3.B.57) into (3.B.56) we obtain

$$\check{R}(\lambda - \mu)\big(L_n(\lambda) \otimes L_n(\mu)\big) = \big(L_n(\mu) \otimes L_n(\lambda)\big)\check{R}(\lambda - \mu).$$
(3.B.61)

Furthermore, the monodromy matrix (3.B.52) can be written as

$$T(\lambda|\mathbf{k}P) = L_N(\lambda - s_{P(N)}) \cdots L_1(\lambda - s_{P(1)}),$$
(3.B.62)

and, using the fact that the entries of L-matrices with different site indices mutually commute, we can iterate (3.B.61) to obtain

$$\check{R}(\lambda - \mu)\big(T(\lambda|\mathbf{k}P) \otimes T(\mu|\mathbf{k}P)\big) = \big(T(\mu|\mathbf{k}P) \otimes T(\lambda|\mathbf{k}P)\big)\check{R}(\lambda - \mu).$$
(3.B.63)

This is a set of quadratic commutation relation between the matrix elements of the monodromy matrix $T(\lambda|\mathbf{k}P)$. We will later say (see Chapter 12) that $T(\lambda|\mathbf{k}P)$ is a representation of the Yang-Baxter algebra with R-matrix $\check{R}(\lambda)$. (3.B.63) will be the central tool in the following calculations.

Remark. Multiplying (3.B.63) by $\check{R}^{-1}(\lambda - \mu)$ from the right and taking the trace in the tensor product of auxiliary spaces, we obtain

$$[t(\lambda|\mathbf{k}P), t(\mu|\mathbf{k}P)] = 0. \tag{3.B.64}$$

$t(\lambda|\mathbf{k}P)$ is a commutating family of transfer matrices. This observation is the starting point for a simple proof of the compatibility of the equations (3.B.48): Let us slightly generalize the definition (3.B.52),

$$T^{(j)}(\lambda|\mathbf{k}P) = X^{(a)}_{aj-1}(\lambda - iu - s_{P(j-1)}) \dots X^{(a)}_{a1}(\lambda - iu - s_{P(1)})$$
$$\times X^{(a)}_{aN}(\lambda - iu - s_{P(n)}) \dots X^{(a)}_{aj}(\lambda - iu - s_{P(j)}), \tag{3.B.65}$$

for $j = 1, \dots, N$. Repeating the steps in (3.B.53) and (3.B.54) we conclude that

$$\mathrm{tr}\left(T^{(j)}(s_{P(j)} + iu|\mathbf{k}P)\right) = X_{j,j-1}(s_{P(j)} - s_{P(j-1)}) \dots X_{j1}(s_{P(j)} - s_{P(1)})$$
$$\times X_{jN}(s_{P(j)} - s_{P(N)}) \dots X_{j,j+1}(s_{P(j)} - s_{P(j+1)}). \tag{3.B.66}$$

On the other hand,

$$T^{(j)}(\lambda|\mathbf{k}P) = L_{j-1}(\lambda - s_{P(j-1)}) \dots L_1(\lambda - s_{P(1)}) L_N(\lambda - s_{P(N)}) \dots L_j(\lambda - s_{P(j)}) \;. \tag{3.B.67}$$

Taking the trace of (3.B.67) and using the cyclic invariance of the trace for a product of matrices with mutually commuting entries we arrive at the conclusion that

$$\mathrm{tr}\left(T^{(j)}(s_{P(j)} + iu|\mathbf{k}P)\right) = \mathrm{tr}\left(T(s_{P(j)} + iu|\mathbf{k}P)\right) = t(s_{P(j)} + iu|\mathbf{k}P). \tag{3.B.68}$$

By (3.B.68) and (3.B.66) we identify the products of operators on the left hand side of (3.B.48) with $t(s_{P(j)} + iu|\mathbf{k}P)$. Then the compatibility of the equations (3.B.48) follows from (3.B.64).

We are now prepared to solve (3.B.47) by means of the algebraic Bethe ansatz. We shall write

$$T(\lambda|\mathbf{k}P) = \begin{pmatrix} A(\lambda) & B(\lambda) \\ C(\lambda) & D(\lambda) \end{pmatrix}. \tag{3.B.69}$$

Spelling out (3.B.63) yields a set of 16 quadratic equations for A, B, C, D. In particular,

$$B(\lambda)B(\mu) = B(\mu)B(\lambda), \tag{3.B.70}$$

$$A(\lambda)B(\mu) = \frac{B(\mu)A(\lambda)}{c(\mu - \lambda)} - \frac{b(\mu - \lambda)B(\lambda)A(\mu)}{c(\mu - \lambda)}, \tag{3.B.71}$$

$$D(\lambda)B(\mu) = \frac{B(\mu)D(\lambda)}{c(\lambda - \mu)} - \frac{b(\lambda - \mu)B(\lambda)D(\mu)}{c(\lambda - \mu)}. \tag{3.B.72}$$

Eigenstates of $t(\lambda|\mathbf{k}P)$ can be constructed by acting with products of operators $B(\lambda)$ on a reference state $|0\rangle = |\uparrow \ldots \uparrow\rangle = \binom{1}{0}^{\otimes N}$. The L-matrices act on this state as

$$L_n(\lambda)|0\rangle = \begin{pmatrix} c(2\lambda) + b(2\lambda) & * \\ 0 & c(2\lambda) - b(2\lambda) \end{pmatrix}|0\rangle = \begin{pmatrix} 1 & * \\ 0 & \frac{\lambda - iu}{\lambda + iu} \end{pmatrix}|0\rangle. \tag{3.B.73}$$

Let us denote the eigenvalues of $A(\lambda)$ and $D(\lambda)$ on $|0\rangle$ by $a(\lambda)$ and $d(\lambda)$, respectively. Then it follows from the definition (3.B.62) and from (3.B.73) that

$$a(\lambda) = 1, \quad d(\lambda) = \prod_{j=1}^{N} \frac{\lambda - s_{P(j)} - iu}{\lambda - s_{P(j)} + iu}. \tag{3.B.74}$$

Let us consider the ansatz

$$|\mathbf{k}P\rangle = |\mathbf{k}P, \boldsymbol{\lambda}\rangle = B(\lambda_1) \ldots B(\lambda_M)|0\rangle, \tag{3.B.75}$$

where $\boldsymbol{\lambda} = (\lambda_1, \ldots, \lambda_M)$, $M \leq N/2$. Because of (3.B.70), this state is symmetric in the λ_j, $|\mathbf{k}P, \boldsymbol{\lambda}Q\rangle = |\mathbf{k}P, \boldsymbol{\lambda}\rangle$ for all $Q \in \mathfrak{S}^M$. Applying $A(\lambda)$ and $D(\lambda)$ to the state $|\mathbf{k}P, \boldsymbol{\lambda}\rangle$ we obtain expressions of the following form,

$$A(\lambda)|\mathbf{k}P, \boldsymbol{\lambda}\rangle = \prod_{j=1}^{M} \frac{1}{c(\lambda_j - \lambda)}|\mathbf{k}P, \boldsymbol{\lambda}\rangle + B(\lambda) \sum_{j=1}^{M} F_j(\lambda) \prod_{\substack{k=1 \\ k \neq j}}^{M} B(\lambda_k)|0\rangle, \tag{3.B.76}$$

$$D(\lambda)|\mathbf{k}P, \boldsymbol{\lambda}\rangle = d(\lambda) \prod_{j=1}^{M} \frac{1}{c(\lambda - \lambda_j)}|\mathbf{k}P, \boldsymbol{\lambda}\rangle + B(\lambda) \sum_{j=1}^{M} \tilde{F}_j(\lambda) \prod_{\substack{k=1 \\ k \neq j}}^{M} B(\lambda_k)|0\rangle. \tag{3.B.77}$$

The functions $F_j(\lambda)$ and $\tilde{F}_j(\lambda)$ will be determined below. Equations (3.B.76) and (3.B.77) follow from (3.B.70)-(3.B.72) and (3.B.74). Consider, for instance, (3.B.71). When moving A past B two terms are generated, the first term on the right-hand side of (3.B.71), where A and B keep their arguments, and the second term, where the arguments are interchanged. When moving A through a product of B's, there is precisely one term where all B's keep their arguments. It is resulting from repeated use of the first term on the right-hand side of (3.B.71) and gives the first term on the right-hand side of (3.B.76). All other possibilities are exhausted by replacing one of the arguments in the product of B's, say λ_j, by λ, and we therefore have the second term on the right-hand side of (3.B.76).

Let us calculate $F_1(\lambda)$. When moving A through the product of B's, we first apply the second term on the right hand side of (3.B.71) and then repeatedly apply the first term. This gives *all* terms proportional to $B(\lambda) \prod_{k=2}^{M} B(\lambda_k)|0\rangle$. Thus,

$$F_1(\lambda) = -\frac{b(\lambda_1 - \lambda)}{c(\lambda_1 - \lambda)} \prod_{k=2}^{M} \frac{1}{c(\lambda_k - \lambda_1)}. \tag{3.B.78}$$

The general term $F_j(\lambda)$ is obtained by first moving $B(\lambda_j)$ to the left in the product of B's, using (3.B.70), and then using the same reasoning as for $F_1(\lambda)$,

$$F_j(\lambda) = -\frac{b(\lambda_j - \lambda)}{c(\lambda_j - \lambda)} \prod_{\substack{k=1 \\ k \neq j}}^{M} \frac{1}{c(\lambda_k - \lambda_j)}. \tag{3.B.79}$$

Similarly,

$$\tilde{F}_j(\lambda) = -\frac{b(\lambda - \lambda_j)d(\lambda_j)}{c(\lambda - \lambda_j)} \prod_{\substack{k=1 \\ k \neq j}}^{M} \frac{1}{c(\lambda_j - \lambda_k)}. \tag{3.B.80}$$

Now $F_j(\lambda) + \tilde{F}_j(\lambda) = 0$ for $j = 1, \ldots, M$ is a sufficient (and necessary) condition for $|\mathbf{k}P, \boldsymbol{\lambda}\rangle$ being an eigenvector of $A(\lambda) + D(\lambda)$. Let us write this condition more explicitly,

$$F_j(\lambda) + \tilde{F}_j(\lambda) = \frac{2iu}{\lambda - \lambda_j} \left[\prod_{\substack{k=1 \\ k \neq j}}^{M} \frac{1}{c(\lambda_k - \lambda_j)} - d(\lambda_j) \prod_{\substack{k=1 \\ k \neq j}}^{M} \frac{1}{c(\lambda_j - \lambda_k)} \right] = 0. \tag{3.B.81}$$

Here we have used the definitions of $b(\lambda)$ and $c(\lambda)$, equation (3.B.59). Equation (3.B.81) has to hold for all $\lambda \in \mathbb{C}$. This is only possible, if the term in square brackets vanishes. Hence, (3.B.81) is equivalent to

$$d(\lambda_j) = \prod_{\substack{k=1 \\ k \neq j}}^{M} \frac{c(\lambda_j - \lambda_k)}{c(\lambda_k - \lambda_j)} = \prod_{\substack{k=1 \\ k \neq j}}^{M} \frac{\lambda_j - \lambda_k - 2iu}{\lambda_j - \lambda_k + 2iu}. \tag{3.B.82}$$

Inserting here the expression (3.B.74) for $d(\lambda)$ we arrive at

$$\prod_{l=1}^{N} \frac{\lambda_j - \sin k_l - iu}{\lambda_j - \sin k_l + iu} = \prod_{\substack{k=1 \\ k \neq j}}^{M} \frac{\lambda_j - \lambda_k - 2iu}{\lambda_j - \lambda_k + 2iu}, \quad j = 1, \ldots, M. \tag{3.B.83}$$

This is a set of Bethe equations for the so-called spin rapidities λ_j. If (3.B.83) is satisfied, then $|\mathbf{k}P, \boldsymbol{\lambda}\rangle$ is an eigenvector of $A(\lambda) + D(\lambda) = t(\lambda|\mathbf{k}P)$ with eigenvalue (see (3.B.76), (3.B.77))

$$\Lambda(\lambda) = \prod_{j=1}^{M} \frac{1}{c(\lambda_j - \lambda)} + d(\lambda) \prod_{j=1}^{M} \frac{1}{c(\lambda - \lambda_j)}. \tag{3.B.84}$$

The reader will have recognized that (3.B.83) is the 'spin part' (3.96) of the Lieb-Wu equations.

We have now solved the eigenvalue problem (3.B.47) for the transfer matrix $t(s_{P(1)} + iu|\mathbf{k}P)$. $|\mathbf{k}P, \boldsymbol{\lambda}\rangle$ is an eigenvector of $t(s_{P(1)} + iu|\mathbf{k}P)$, if (3.B.83) holds. The corresponding eigenvalue that follows from (3.B.84) by insertion of (3.B.59) and (3.B.74) is

$$\Lambda(s_{P(1)} + iu) = \prod_{j=1}^{M} \frac{\lambda_j - s_{P(1)} + iu}{\lambda_j - s_{P(1)} - iu}. \tag{3.B.85}$$

Solving equation (3.B.47) means to find eigenvectors and eigenvalues, such that $\Lambda(s_{P(1)} + iu) = e^{-ik_{P(1)}L}$, which leads to

$$e^{ik_{P(1)}L} = \prod_{j=1}^{M} \frac{\lambda_j - \sin(k_{P(1)}) - iu}{\lambda_j - \sin(k_{P(1)}) + iu}. \tag{3.B.86}$$

Since this equation must be fulfilled for all $P \in \mathfrak{S}^N$, we find the set of equations

$$e^{ik_l L} = \prod_{j=1}^{M} \frac{\lambda_j - \sin k_l - iu}{\lambda_j - \sin k_l + iu}, \qquad l = 1, \ldots, N, \tag{3.B.87}$$

which together with (3.B.83) determines the parameters of the Bethe wave functions (3.B.19) and is easily identified as the 'charge part' (3.95) of the Lieb-Wu equations.

It follows from rather general considerations that the spin rapidities λ_j have to be mutually distinct ('Pauli principle for bosons', see [270]).

3.B.6 Summary

Let us present a brief summary: for every solution $\{\{k_l\}_{l=1}^{N}, \{\lambda_j\}_{j=1}^{M}\}$ of the Lieb-Wu equations

$$e^{ik_l L} = \prod_{j=1}^{M} \frac{\lambda_j - \sin k_l - iu}{\lambda_j - \sin k_l + iu}, \qquad l = 1, \ldots, N, \tag{3.B.88}$$

$$\prod_{l=1}^{N} \frac{\lambda_j - \sin k_l - iu}{\lambda_j - \sin k_l + iu} = \prod_{\substack{k=1 \\ k \neq j}}^{M} \frac{\lambda_j - \lambda_k - 2iu}{\lambda_j - \lambda_k + 2iu}, \qquad j = 1, \ldots, M, \tag{3.B.89}$$

where the sets $\{k_l\}_{l=1}^{N}$ and $\{\lambda_j\}_{j=1}^{M}$ consist of mutually distinct complex numbers, the wave function

$$\psi(\mathbf{x}; \mathbf{a}|\mathbf{k}; \boldsymbol{\lambda}) = \sum_{P \in \mathfrak{S}^N} \text{sign}(PQ)\langle \mathbf{a}Q|kP, \boldsymbol{\lambda}\rangle \, e^{i(kP, \mathbf{x}Q)} \tag{3.B.90}$$

solves the Schrödinger equation (3.16) and satisfies the periodic boundary conditions (3.17). It is therefore an eigenfunction of the cyclic N-particle Hamiltonian (3.10) with eigenvalue

$$E = -2 \sum_{j=1}^{N} \cos(k_j). \tag{3.B.91}$$

It is also an eigenfunction of the shift operator with eigenvalue

$$\omega = e^{i(k_1 + \cdots + k_N)}. \tag{3.B.92}$$

Recalling our definition (2.54) of the momentum operator we conclude that the eigenfunctions (3.B.90) carry lattice momentum

$$P = (k_1 + \cdots + k_N) \bmod 2\pi. \tag{3.B.93}$$

The eigenfunctions (3.B.90) have the following symmetry properties,

$$\psi(\mathbf{x}R; \mathbf{a}R|\mathbf{k}; \lambda) = \text{sign}(R)\psi(\mathbf{x}; \mathbf{a}|\mathbf{k}; \lambda), \quad R \in \mathfrak{S}^N, \tag{3.B.94}$$

$$\psi(\mathbf{x}; \mathbf{a}|\mathbf{k}R; \lambda) = \text{sign}(R)\psi(\mathbf{x}; \mathbf{a}|\mathbf{k}; \lambda), \quad R \in \mathfrak{S}^N, \tag{3.B.95}$$

$$\psi(\mathbf{x}; \mathbf{a}|\mathbf{k}; \lambda R) = \psi(\mathbf{x}; \mathbf{a}|\mathbf{k}; \lambda), \quad R \in \mathfrak{S}^M. \tag{3.B.96}$$

They are antisymmetric under simultaneous exchange of position and spin variables, they are antisymmetric under exchange of momenta k_l, and they are symmetric under exchange of spin rapidities λ_j.

3.C Some technical details

This appendix contains some technical details which were left out in the derivation of the Bethe ansatz wave function in appendix 3.B.

3.C.1 Yang-Baxter equation

We show that $Y_{jk}(\lambda)$ satisfies the Yang-Baxter equation (3.B.35). Because of the homogeneity of the Yang-Baxter equation we may replace $Y_{jk}(\lambda)$ by $\tilde{Y}_{jk}(\lambda) = 1 + \lambda\Pi_{jk}$. For simplicity we may further set $j = 1, k = 2, l = 3$ in (3.B.35). Then

$$\begin{aligned}
(1 + \lambda\Pi_{12})&(1 + (\lambda + \mu)\Pi_{23})(1 + \mu\Pi_{12}) - (1 + \mu\Pi_{23})(1 + (\lambda + \mu)\Pi_{12})(1 + \lambda\Pi_{23}) \\
&= \mu(\lambda + \mu)(\Pi_{23}\Pi_{12} - \Pi_{23}\Pi_{12}) + \lambda(\lambda + \mu)(\Pi_{12}\Pi_{23} - \Pi_{12}\Pi_{23}) \\
&\quad + \lambda\mu(\Pi_{12}\Pi_{12} - \Pi_{23}\Pi_{23}) + \lambda\mu(\lambda + \mu)(\Pi_{12}\Pi_{23}\Pi_{12} - \Pi_{23}\Pi_{12}\Pi_{23}) \\
&= 0,
\end{aligned} \tag{3.C.1}$$

where we have used (3.B.33) and the identity $\Pi_{jk}\Pi_{jk} = 1$. Equation (3.C.1) is equivalent to (3.B.35).

3.C.2 The consistency problem

In order to show the consistency of the equations (3.B.31) we have to recall an abstract definition of the symmetric group. The symmetric group $\mathfrak{S}^N$ is generated by the identity id and the transpositions of nearest neighbours $\Pi_{n,n+1}$, $n = 1, \ldots, N - 1$, modulo the relations

$$\Pi_{n,n+1}\Pi_{n+1,n+2}\Pi_{n,n+1} = \Pi_{n+1,n+2}\Pi_{n,n+1}\Pi_{n+1,n+2}, \tag{3.C.2a}$$

$$\Pi_{n,n+1}\Pi_{m,m+1} = \Pi_{m,m+1}\Pi_{n,n+1} \quad \text{for } |n - m| > 1, \tag{3.C.2b}$$

$$\Pi_{n,n+1}\Pi_{n,n+1} = \text{id}. \tag{3.C.2c}$$

Equation (3.C.2a) is called the braid relation. Note that (3.C.2b) is applicable only for $N > 3$.

Due to (3.C.2a)–(3.C.2c) a representation of a permutation as a product of transpositions of nearest neighbours is not unique. Therefore the use of equations (3.C.2a)–(3.C.2c) in (3.B.31) imposes certain consistency conditions on the Y-operators.

Let us consider the braid relation. Suppose $P \in \mathfrak{S}^N$ contains a string of the form $\Pi_{n,n+1}\Pi_{n+1,n+2}\Pi_{n,n+1}$, i.e., $P = Q\Pi_{n,n+1}\Pi_{n+1,n+2}\Pi_{n,n+1}R$ with $Q, R \in \mathfrak{S}^N$. Then, on the one hand,

$$
\begin{aligned}
|\mathbf{k}P\rangle &= \ldots |\mathbf{k}Q\Pi_{n,n+1}\Pi_{n+1,n+2}\Pi_{n,n+1}\rangle \\
&= \ldots Y_{n,n+1}(s_{Q\Pi_{n,n+1}\Pi_{n+1,n+2}(n)} - s_{Q\Pi_{n,n+1}\Pi_{n+1,n+2}(n+1)}) \\
&\quad \times Y_{n+1,n+2}(s_{Q\Pi_{n,n+1}(n+1)} - s_{Q\Pi_{n,n+1}(n+2)}) \cdot Y_{n,n+1}(s_{Q(n)} - s_{Q(n+1)})|\mathbf{k}Q\rangle \\
&= \ldots Y_{n,n+1}(s_{Q(n+1)} - s_{Q(n+2)})Y_{n+1,n+2}(s_{Q(n)} - s_{Q(n+2)}) \\
&\quad \times Y_{n,n+1}(s_{Q(n)} - s_{Q(n+1)})|\mathbf{k}Q\rangle\,,
\end{aligned}
\tag{3.C.3}
$$

where the dots denote the string of Y-operators which induces the action of the permutation R. On the other hand,

$$
\begin{aligned}
|\mathbf{k}P\rangle &= \ldots |\mathbf{k}Q\Pi_{n+1,n+2}\Pi_{n,n+1}\Pi_{n+1,n+2}\rangle \\
&= \ldots Y_{n+1,n+2}(s_{Q\Pi_{n+1,n+2}\Pi_{n,n+1}(n+1)} - s_{Q\Pi_{n+1,n+2}\Pi_{n,n+1}(n+2)}) \\
&\quad \times Y_{n,n+1}(s_{Q\Pi_{n+1,n+2}(n)} - s_{Q\Pi_{n+1,n+2}(n+1)}) \cdot Y_{n+1,n+2}(s_{Q(n+1)} - s_{Q(n+2)})|\mathbf{k}Q\rangle \\
&= \ldots Y_{n+1,n+2}(s_{Q(n)} - s_{Q(n+1)})Y_{n,n+1}(s_{Q(n)} - s_{Q(n+2)}) \\
&\quad \times Y_{n+1,n+2}(s_{Q(n+1)} - s_{Q(n+2)})|\mathbf{k}Q\rangle\,.
\end{aligned}
\tag{3.C.4}
$$

Now for the Bethe ansatz to be consistent the expressions for $|\mathbf{k}P\rangle$ in equation (3.C.3) and (3.C.4) have to agree for arbitrary $\mathbf{k}$. This is the case, if and only if

$$
\begin{aligned}
&Y_{n,n+1}(s_{Q(n+1)} - s_{Q(n+2)})Y_{n+1,n+2}(s_{Q(n)} - s_{Q(n+2)})Y_{n,n+1}(s_{Q(n)} - s_{Q(n+1)}) \\
&= Y_{n+1,n+2}(s_{Q(n)} - s_{Q(n+1)})Y_{n,n+1}(s_{Q(n)} - s_{Q(n+2)})Y_{n+1,n+2}(s_{Q(n+1)} - s_{Q(n+2)})\,,
\end{aligned}
\tag{3.C.5}
$$

which is nothing but the Yang-Baxter equation (3.B.35) proven in appendix A.

Similarly, equation (3.C.2b) leads to the consistency condition

$$
\begin{aligned}
&Y_{n,n+1}(s_{Q(n)} - s_{Q(n+1)})Y_{m,m+1}(s_{Q(m)} - s_{Q(m+1)}) \\
&= Y_{m,m+1}(s_{Q(m)} - s_{Q(m+1)})Y_{n,n+1}(s_{Q(n)} - s_{Q(n+1)}) \quad \text{for } |n - m| > 1\,.
\end{aligned}
\tag{3.C.6}
$$

This condition is trivial.

From the last equation, (3.C.2c), we obtain

$$
Y_{n,n+1}(s_{Q(n)} - s_{Q(n+1)})Y_{n,n+1}(s_{Q(n+1)} - s_{Q(n)}) = 1\,,
\tag{3.C.7}
$$

which is always true as a consequence of equation (3.B.36).

3.D Highest weight property of the Bethe ansatz states with respect to total spin

We give a proof of the highest weight theorem presented in Section 3.4. The proof is based on the fact that the action of the spin operators $S^{\pm}$ and S^z on the Bethe ansatz states reduces to an action of associated spin operators on the eigenstates $|\mathbf{k}P, \lambda\rangle$ of the inhomogeneous transfer matrix $t(\lambda|\mathbf{k}P)$. The highest weight theorem then follows from the su(2) invariance of $t(\lambda|\mathbf{k}P)$.

3.D.1 Spin operators in fermionic and in spin chain representation

Let us introduce some more notation for spin chains. When dealing with spin chains it is often useful to switch from the local basis $\{\sigma^x, \sigma^y, \sigma^z, I_2\}$ to the canonical gl(2) basis $\{e_1^1, e_1^2, e_2^1, e_2^2\}$, where

$$e_1^1 = \begin{pmatrix} 1 & 0 \\ 0 & 0 \end{pmatrix}, \ e_1^2 = \begin{pmatrix} 0 & 1 \\ 0 & 0 \end{pmatrix}, \ e_2^1 = \begin{pmatrix} 0 & 0 \\ 1 & 0 \end{pmatrix}, \ e_2^2 = \begin{pmatrix} 0 & 0 \\ 0 & 1 \end{pmatrix}. \tag{3.D.1}$$

These matrices act as

$$e_a^b \mathbf{e}_c = \delta_c^b \mathbf{e}_a \tag{3.D.2}$$

on the unit vectors $\mathbf{e}_\uparrow = \begin{pmatrix} 1 \\ 0 \end{pmatrix}$, $\mathbf{e}_\downarrow = \begin{pmatrix} 0 \\ 1 \end{pmatrix}$ and obey the multiplication rule

$$e_a^b e_c^d = \delta_c^b e_a^d . \tag{3.D.3}$$

Every operator $A \in \text{End}(\mathbb{C}^2)$ can be represented as $A = A_b^a e_a^b$, where

$$A_b^a = A_d^c \, \mathbf{e}^a e_c^d \mathbf{e}_b = \mathbf{e}^a A \mathbf{e}_b = \langle \mathbf{e}_a, A\mathbf{e}_b \rangle . \tag{3.D.4}$$

With every operator $A \in \text{End}(\mathbb{C}^2)$ we can associate an operator

$$A_n^s = I_2^{\otimes(n-1)} \otimes A \otimes I_2^{\otimes(N-n)} , \tag{3.D.5}$$

acting on the space of states of an N-site spin-$\frac{1}{2}$ chain, and the completely symmetrized one-site operator

$$A^s = \sum_{n=1}^N A_n^s . \tag{3.D.6}$$

The operator A_n^s has matrix elements

$$\langle \mathbf{a}|A_n^s|\mathbf{b}\rangle = \delta_{b_1}^{a_1} \ldots \delta_{b_{n-1}}^{a_{n-1}} A_{b_n}^{a_n} \delta_{b_{n+1}}^{a_{n+1}} \ldots \delta_{b_N}^{a_N} . \tag{3.D.7}$$

Similarly, we can associate a fermionic operator with A. Define 2×2-matrices S_n by their matrix elements

$$S_{n \, a}^{\ b} = c_{n,a}^\dagger c_{n,b} . \tag{3.D.8}$$

Then

$$A_n^f = \operatorname{tr}(A S_n) \tag{3.D.9}$$

induces an action of A on the nth site of an electronic lattice model, and we may define the completely symmetric one-site operator

$$A^f = \sum_{n=1}^{N} A_n^f . \tag{3.D.10}$$

The spin operators S^α, $\alpha = x, y, z$, and the particle number operator $\hat{N}$ are of this form.

3.D.2 Action of spin operators on Bethe ansatz states

Let us calculate the action of A^f on an N-particle state $|\psi\rangle$. Using (3.A.13) and (3.A.11) we obtain

$$
\begin{aligned}
A^f |\psi\rangle &= \frac{1}{N!} \sum_{\mathbf{y} \in \mathbb{Z}_L^N} \sum_{\mathbf{b} \in \mathbb{Z}_2^N} \psi(\mathbf{y}; \mathbf{b}) \, A^f |\mathbf{y}, \mathbf{b}\rangle \\
&= \frac{1}{(N!)^2} \sum_{\mathbf{x}, \mathbf{y} \in \mathbb{Z}_L^N} \sum_{\mathbf{a}, \mathbf{b} \in \mathbb{Z}_2^N} |\mathbf{x}, \mathbf{a}\rangle \langle \mathbf{x}, \mathbf{a}| A^f |\mathbf{y}, \mathbf{b}\rangle \langle \mathbf{y}, \mathbf{b}|\psi\rangle \\
&= \frac{1}{N!} \sum_{\mathbf{x} \in \mathbb{Z}_L^N} \sum_{\mathbf{a} \in \mathbb{Z}_2^N} \langle \mathbf{x}, \mathbf{a}| A^f |\psi\rangle |\mathbf{x}, \mathbf{a}\rangle .
\end{aligned}
\tag{3.D.11}
$$

The matrix element $\langle \mathbf{x}, \mathbf{a}| A^f |\psi\rangle$ is easily calculated, observing that the commutator of A^f and the Fermi operator c_{x_n,a_n} is

$$[c_{x_n,a_n}, A^f] = [c_{x_n,a_n}, c^\dagger_{x_n,a} c_{x_n,b}] A_b^a = \{c_{x_n,a_n}, c^\dagger_{x_n,a}\} c_{x_n,b} A_b^a = A_{b_n}^{a_n} c_{x_n,b_n} . \tag{3.D.12}$$

It follows that

$$
\begin{aligned}
\langle \mathbf{x}, \mathbf{a}| A^f |\psi\rangle &= \sum_{n=1}^{N} A_{b_n}^{a_n} \langle 0| c_{x_1,a_1} \cdots c_{x_{n-1},a_{n-1}} c_{x_n,b_n} c_{x_{n+1},a_{n+1}} \cdots c_{x_N,a_N} |\psi\rangle \\
&= \sum_{n=1}^{N} \delta_{b_1}^{a_1} \cdots \delta_{b_{n-1}}^{a_{n-1}} A_{b_n}^{a_n} \delta_{b_{n+1}}^{a_{n+1}} \cdots \delta_{b_N}^{a_N} \langle 0| c_{x_1,b_1} \cdots c_{x_N,b_N} |\psi\rangle \\
&= \sum_{\mathbf{b} \in \mathbb{Z}_2^N} \langle \mathbf{a}| A^s |\mathbf{b}\rangle \psi(\mathbf{x}; \mathbf{b}) .
\end{aligned}
\tag{3.D.13}
$$

Let us apply this formula to the Bethe ansatz eigenstate $|\psi_{\mathbf{k},\lambda}\rangle$. We obtain

$$
\begin{aligned}
\langle \mathbf{x}, \mathbf{a}| A^f |\psi_{\mathbf{k},\lambda}\rangle &= \sum_{\mathbf{b} \in \mathbb{Z}_2^N} \sum_{P \in \mathfrak{S}^N} \operatorname{sign}(PQ) \langle \mathbf{a}| A^s |\mathbf{b}\rangle \langle \mathbf{b}Q| k P, \lambda\rangle e^{i\langle kP, xQ\rangle} \\
&= \sum_{P \in \mathfrak{S}^N} \operatorname{sign}(PQ) \langle \mathbf{a}| A^s Q| kP, \lambda\rangle e^{i\langle kP, xQ\rangle} \\
&= \sum_{P \in \mathfrak{S}^N} \operatorname{sign}(PQ) \langle \mathbf{a}Q| A^s| kP, \lambda\rangle e^{i\langle kP, xQ\rangle} .
\end{aligned}
\tag{3.D.14}
$$

Here we have used $\langle \mathbf{b}Q| = \langle \mathbf{b}|Q$ (see Section 3.B.2) and $\sum_{\mathbf{b}\in\mathbb{Z}_2^N} |\mathbf{b}\rangle\langle \mathbf{b}| = \mathrm{id}$ in the second equation and $[A^s, Q] = 0$ in the third equation. Equations (3.D.13) and (3.D.14) show how A^f induces the action of A^s on the spin part of the Bethe ansatz wave functions.

Let us consider the simplest example. For $A = I_2$ we have $A^f = \sum_{n=1}^N c_{n,a}^\dagger c_{n,a} = \hat{N}$ and $A^s = N \cdot \mathrm{id}$. Thus, (3.D.13) and (3.D.14) imply that

$$\hat{N}|\psi_{\mathbf{k},\boldsymbol{\lambda}}\rangle = N|\psi_{\mathbf{k},\boldsymbol{\lambda}}\rangle \,, \tag{3.D.15}$$

which shows nothing but the fact that the Bethe ansatz states are N-particle states. In the next section we shall consider the action of spin operators on Bethe ansatz states.

3.D.3 su(2) invariance of the spin problem

In the previous section we have shown that the action of Fermi operators A^f of the form (3.D.10) induces an action of associated spin operators A^s, equation (3.D.6), on the spin states $|\mathbf{k}P, \boldsymbol{\lambda}\rangle$ that determine the amplitudes $\langle \mathbf{a}Q|\mathbf{k}P, \boldsymbol{\lambda}\rangle$ in the Bethe ansatz wave functions. Here we consider the cases $A = \sigma^\alpha$, $\alpha = x, y, z$, i.e., $A^f = 2S^\alpha$. The corresponding spin operators on the spin chain space of states are

$$\Sigma^\alpha = \sum_{n=1}^N \sigma_n^\alpha \,. \tag{3.D.16}$$

These operators generate a representation of the Lie algebra su(2),

$$[\Sigma^\alpha, \Sigma^\beta] = 2\mathrm{i}\varepsilon^{\alpha\beta\gamma} \Sigma^\gamma \,. \tag{3.D.17}$$

The ladder operators in this representation are

$$\Sigma^\pm = \Sigma^x \pm \mathrm{i}\Sigma^y \,. \tag{3.D.18}$$

They correspond to $A = \sigma^\pm$, $A^f = 2S^\pm$.

We are now going to prove that the eigenstates $|\mathbf{k}P, \boldsymbol{\lambda}\rangle$ of the transfer matrix $t(\lambda|\mathbf{k}P)$ (see equation (3.B.55)) are su(2) highest weight states with highest weight $N - 2M$. This means that

$$\Sigma^+|\mathbf{k}P, \boldsymbol{\lambda}\rangle = 0 \tag{3.D.19}$$

and

$$\Sigma^z|\mathbf{k}P, \boldsymbol{\lambda}\rangle = (N - 2M)|\mathbf{k}P, \boldsymbol{\lambda}\rangle \,, \tag{3.D.20}$$

if the Bethe ansatz equations (3.B.83) are satisfied. We shall further show that the transfer matrix $t(\lambda|\mathbf{k}P)$ commutes with the su(2) generators,

$$[\Sigma^\pm, t(\lambda|\mathbf{k}P)] = [\Sigma^z, t(\lambda|\mathbf{k}P)] = 0 \,. \tag{3.D.21}$$

Equations (3.D.19), (3.D.20) and (3.D.21) follow from the 'su(2) invariance' of the monodromy matrix $T(\lambda|kP)$, which can be stated as follows:

$$[T(\lambda|kP), \sigma^{\alpha} + \Sigma^{\alpha}] = 0, \qquad (3.D.22)$$

for $\alpha = x, y, z$. Here the σ^{α} are 2×2-matrices in auxiliary space, and the Σ^{α} are spin operators in the spin chain space of states, i.e., they act in the same space as the matrix elements of $T(\lambda|kP)$.

We first prove the basic formula (3.D.22) and show afterwards how (3.D.19)–(3.D.21) follow. Equation (3.D.22) is a consequence of a similar statement for the L-operators (see (3.B.60)), $L_n(\lambda) = c(2\lambda) + b(2\lambda)\sigma^{\alpha}\sigma_n^{\alpha}$. On the one hand we have

$$[L_n(\lambda), \sigma_n^{\alpha}] = -b(2\lambda)\sigma^{\beta}[\sigma_n^{\alpha}, \sigma_n^{\beta}] = -b(2\lambda)2i\varepsilon^{\alpha\beta\gamma}\sigma^{\beta}\sigma_n^{\gamma}, \qquad (3.D.23)$$

on the other hand

$$[L_n(\lambda), \sigma^{\alpha}] = -b(2\lambda)\sigma_n^{\beta}[\sigma^{\alpha}, \sigma^{\beta}] = -b(2\lambda)2i\varepsilon^{\alpha\beta\gamma}\sigma_n^{\beta}\sigma^{\gamma} = b(2\lambda)2i\varepsilon^{\alpha\beta\gamma}\sigma^{\beta}\sigma_n^{\gamma}. \quad (3.D.24)$$

Thus, adding the latter two equations,

$$[L_n(\lambda), \sigma^{\alpha} + \sigma_n^{\alpha}] = 0. \qquad (3.D.25)$$

This result is easily lifted to the level of the monodromy matrix,

$$
\begin{aligned}
[T(\lambda|kP), \Sigma^{\alpha}] &= \sum_{n=1}^{N} [L_N(\lambda) \ldots L_1(\lambda), \sigma_n^{\alpha}] \\
&= \sum_{n=1}^{N} L_N(\lambda) \ldots L_{n+1}(\lambda)[L_n(\lambda), \sigma_n^{\alpha}]L_{n-1}(\lambda) \ldots L_1(\lambda) \\
&= -\sum_{n=1}^{N} L_N(\lambda) \ldots L_{n+1}(\lambda)[L_n(\lambda), \sigma^{\alpha}]L_{n-1}(\lambda) \ldots L_1(\lambda) \\
&= -[T(\lambda|kP), \sigma^{\alpha}], \qquad (3.D.26)
\end{aligned}
$$

and equation (3.D.22) is proven. In the above calculation we used $[L_n(\lambda), \sigma_m^{\alpha}] = 0$ for $m \neq n$ in the second equation, (3.D.25) in the third equation, and the Leibniz rule in the last equation.

Let us now reformulate the invariance equation (3.D.22) for the monodromy matrix. We multiply (3.D.22) by a numerical 2×2 matrix A and take the trace in auxiliary space. Then

$$\left[\Sigma^{\alpha}, \text{tr}\{A\, T(\lambda|\mathbf{k})\}\right] = \text{tr}\{[\sigma^{\alpha}, A]T(\lambda|\mathbf{k})\}. \qquad (3.D.27)$$

Setting $A = I_2$ we obtain (3.D.21). Setting $A = \sigma^{-}$ and $\alpha = +$ or $\alpha = z$, respectively, we obtain

$$[\Sigma^{+}, B(\lambda)] = A(\lambda) - D(\lambda), \qquad (3.D.28a)$$

$$[\Sigma^{z}, B(\lambda)] = -2B(\lambda) \qquad (3.D.28b)$$

which will be needed in the proof of (3.D.19) and (3.D.20). More generally we may define $J^{\alpha}(\lambda) = \text{tr}\{\sigma^{\alpha}T(\lambda|\mathbf{k})\}$. These linear combinations of monodromy matrix elements obviously satisfy

$$[\Sigma^\alpha, J^\beta(\lambda)] = 2i\varepsilon^{\alpha\beta\gamma} J^\gamma(\lambda), \tag{3.D.29}$$

which means they transform like a vector representation of su(2).

Let us proceed with the proof of (3.D.19) and (3.D.20). Equation (3.D.20) is easier to prove than (3.D.19). It follows from (3.D.28b) and from $\Sigma^z|0\rangle = N|0\rangle$,

$$\begin{aligned}
\Sigma^z|\mathbf{k}P, \boldsymbol{\lambda}\rangle &= \Sigma^z B(\lambda_1)\dots B(\lambda_M)|0\rangle \\
&= B(\lambda_1)\dots B(\lambda_M)(\Sigma^z - 2M)|0\rangle \\
&= (N - 2M)|\mathbf{k}P, \boldsymbol{\lambda}\rangle.
\end{aligned} \tag{3.D.30}$$

Note that this equation is valid, even if the Bethe ansatz equations (3.B.83) are not satisfied.

By way of contrast, the Bethe ansatz equations are needed to prove (3.D.19). First of all, $\Sigma^+|0\rangle = 0$. Hence, (3.D.28a) implies that

$$\begin{aligned}
\Sigma^+|\mathbf{k}P, \boldsymbol{\lambda}\rangle &= \Sigma^+ B(\lambda_1)\dots B(\lambda_M)|0\rangle \\
&= \sum_{n=1}^{M} B(\lambda_1)\dots B(\lambda_{n-1})\big(A(\lambda_n) - D(\lambda_n)\big)B(\lambda_{n+1})\dots B(\lambda_M)|0\rangle.
\end{aligned} \tag{3.D.31}$$

For the evaluation of the expression on the right hand side of this equation we use the commutation relations (3.B.71), (3.B.72) between $B(\lambda)$ and $A(\mu)$, $D(\mu)$. Our arguments are similar to the arguments used in the derivation of the second level Bethe ansatz equations (3.B.83) in appendix 3.B.5: When commuting $A(\lambda_n)$ and $D(\lambda_n)$ successively to the right in equation (3.D.31), two terms are generated in every step, one term, in which A and D keep their arguments, and another term in which A and D interchange their arguments with B. Therefore, taking into account that $A(\lambda)|0\rangle = |0\rangle$ and $D(\lambda)|0\rangle = d(\lambda)|0\rangle$ (see appendix 3.B.5), the right-hand side of (3.D.31) must be equal to a sum over products of $M-1$ operators $B(\lambda)$ acting on the ferromagnetic state,

$$\Sigma^+|\mathbf{k}P, \boldsymbol{\lambda}\rangle = \sum_{n=1}^{M} \big(G_n - \tilde{G}_n\, d(\lambda_n)\big) \prod_{\substack{m=1 \\ m\neq n}}^{M} B(\lambda_m)|0\rangle. \tag{3.D.32}$$

Let us consider the coefficient of the first term in this sum. Contributions to this term can only stem from the first term,

$$\big(A(\lambda_1) - D(\lambda_1)\big)B(\lambda_2)\dots B(\lambda_M)|0\rangle, \tag{3.D.33}$$

in the sum on the right-hand side of (3.D.31), since all other terms contain $B(\lambda_1)$. When moving the operators $A(\lambda_1)$ and $D(\lambda_1)$ to the right, they have to keep their arguments in every step in order to produce a contribution to the first term on the right-hand side of (3.D.32). Thus, it follows from (3.B.71), (3.B.72) that

$$G_1 - \tilde{G}_1\, d(\lambda_1) = \prod_{m=2}^{M} \frac{1}{c(\lambda_m - \lambda_1)} - d(\lambda_1) \prod_{m=2}^{M} \frac{1}{c(\lambda_1 - \lambda_m)}. \tag{3.D.34}$$

Like in the derivation of the Bethe ansatz equations we can use the symmetry of the Bethe ansatz states $|\mathbf{k}P, \boldsymbol{\lambda}\rangle$ with respect to permutations of the λ_j to conclude that all the coefficients are of the same form. Thus,

$$G_n - \tilde{G}_n \, d(\lambda_n) = \prod_{\substack{m=1 \\ m \neq n}}^{M} \frac{1}{c(\lambda_n - \lambda_m)} \left[\prod_{\substack{m=1 \\ m \neq n}}^{M} \frac{c(\lambda_n - \lambda_m)}{c(\lambda_m - \lambda_n)} - d(\lambda_n) \right], \qquad (3.\text{D}.35)$$

for $n = 1, \ldots, M$. Obviously, these coefficients vanish if the Bethe ansatz equations (3.B.83) are satisfied, and our proof of (3.D.19) is complete. Alternatively, (3.D.19) can be proven by induction over M, which we suggest as an exercise to the reader.

Let us discuss the implications of (3.D.19)–(3.D.21). The spin states $|\mathbf{k}P, \boldsymbol{\lambda}\rangle$ are highest weight states with respect to the total spin, which is conserved by the transfer matrix $t(\lambda|\mathbf{k}P)$. Since the value $N - 2M$ of the total spin must be positive, we must have $M \leq N/2$. Furthermore, because of (3.D.21), all states $(\Sigma^-)^n|\mathbf{k}P, \boldsymbol{\lambda}\rangle$, $n = 1, \ldots, N - 2M$, are eigenstates of $t(\lambda|\mathbf{k}P)$ with the common eigenvalue $\Lambda(\lambda)$ (see (3.B.84)).

Because of (3.D.11) and (3.D.14) similar statements are true for the Bethe ansatz eigenstates of the Hubbard Hamiltonian. First of all, we must have $M \leq N/2$. Equations (3.D.19), (3.D.20) in conjunction with (3.D.11) and (3.D.14) imply that the Bethe ansatz eigenstates of the Hubbard Hamiltonian are highest weight with respect to the spin representation S^α, $\alpha = x, y, z$, of su(2), i.e., we have

$$S^+|\psi_{\mathbf{k},\boldsymbol{\lambda}}\rangle = 0, \qquad (3.\text{D}.36)$$

$$S^z|\psi_{\mathbf{k},\boldsymbol{\lambda}}\rangle = \tfrac{1}{2}(N - 2M)|\psi_{\mathbf{k},\boldsymbol{\lambda}}\rangle. \qquad (3.\text{D}.37)$$

Therefore, the states

$$(S^-)^n|\psi_{\mathbf{k},\boldsymbol{\lambda}}\rangle = \frac{1}{N!} \sum_{\mathbf{x} \in \mathbb{Z}_L^N} \sum_{\mathbf{a} \in \mathbb{Z}_2^N} \sum_{P \in \mathfrak{S}^N} \mathrm{sign}(PQ)\langle \mathbf{a}Q|(\Sigma^-)^n|\mathbf{k}P, \boldsymbol{\lambda}\rangle e^{i\langle \mathbf{k}P, \mathbf{x}Q\rangle}|\mathbf{x}, \mathbf{a}\rangle \qquad (3.\text{D}.38)$$

for $n = 0, \ldots, N - 2M$ form a degenerate multiplet of eigenstates of the Hubbard Hamiltonian. The multiplet has dimension $N - 2M + 1$. This result together with a similar result for the η-representation of su(2) will play an important role, when we count the number of Bethe ansatz states in Chapter 4.

3.E Explicit expressions for the amplitudes in the Bethe ansatz wave functions

The amplitudes

$$A(\mathbf{k}P|\mathbf{a}Q) = \langle \mathbf{a}Q|\mathbf{k}P, \boldsymbol{\lambda}\rangle \qquad (3.\text{E}.1)$$

in the Bethe ansatz wave functions carry the information about the spin configuration of the state. In appendix 3.B we expressed these amplitudes through the action of spin wave creation operators $B(\lambda)$ on the ferromagnetic state $|0\rangle = \mathbf{e}_\uparrow^{\otimes N}$,

$$\langle \mathbf{a}Q|\mathbf{k}P, \boldsymbol{\lambda}\rangle = \langle \mathbf{a}Q|B(\lambda_1) \ldots B(\lambda_M)|0\rangle. \qquad (3.\text{E}.2)$$

This representation is very convenient for many purposes. It is used, for instance, in appendix 3.D to prove the highest weight property of the Bethe ansatz states with respect to total spin. Sometimes however, we wish to have a more explicit expression. This will be derived below.

Let us define the states

$$|\{y_j\}_{j=1}^M\rangle = \prod_{j=1}^M \sigma_{y_j}^- |0\rangle \,. \tag{3.E.3}$$

Obviously, the set

$$\left\{ |\{y_j\}_{j=1}^M\rangle \in (\mathbb{C}^2)^{\otimes N} \,\middle|\, 1 \le y_1 < \cdots < y_M \le N \right\} \tag{3.E.4}$$

is a basis of the subspace with fixed value $\frac{1}{2}(N - 2M)$ of the z-component of the total spin of the spin chain space of states. As we saw in appendix 3.D the eigenstates $|\mathbf{k}P, \boldsymbol{\lambda}\rangle$ of the transfer matrix $t(\lambda|\mathbf{k}P)$ are in this subspace. Hence, we get the expansion

$$|\mathbf{k}P, \boldsymbol{\lambda}\rangle = \sum_{1 \le y_1 < \cdots < y_M \le N} \Phi(\mathbf{y}|\mathbf{k}P, \boldsymbol{\lambda})|\{y_j\}_{j=1}^M\rangle \,, \tag{3.E.5}$$

where $\mathbf{y} = (y_1, \ldots, y_M)$. The coefficients $\Phi(\mathbf{y}|\mathbf{k}P, \boldsymbol{\lambda})$ are called the coordinate Bethe ansatz wave functions of the inhomogeneous XXX spin chain. Their homogeneous limit, $k_j = 0$, was obtained by H. Bethe in his famous article [60]. Here we explain, how the wave functions $\Phi(\mathbf{y}|\mathbf{k}P, \boldsymbol{\lambda})$ are obtained within the scheme of the algebraic Bethe ansatz. We shall show that

$$\Phi(\mathbf{y}|\mathbf{k}P, \boldsymbol{\lambda}) = \sum_{R \in \mathfrak{S}^M} A(\lambda R) \prod_{\ell=1}^M \left(\frac{2iu}{\lambda_{R(\ell)} - s_{P(y_\ell)} + iu} \prod_{j=1}^{y_\ell - 1} \frac{\lambda_{R(\ell)} - s_{P(j)} - iu}{\lambda_{R(\ell)} - s_{P(j)} + iu} \right) , \tag{3.E.6}$$

where

$$A(\boldsymbol{\lambda}) = \prod_{1 \le m < n \le M} \frac{\lambda_m - \lambda_n - 2iu}{\lambda_m - \lambda_n} \,. \tag{3.E.7}$$

In our proof we shall resort to a formidable formula for the 'iteration of the co-multiplication' which was obtained in [214, 215]. In order to explain this formula it is helpful to consider a simple special case first. Let us construct an abstract 'two-site model' by dividing the product of L-operators in the definition (3.B.62) of the monodromy matrix into two parts,

$$T_1(\lambda|\mathbf{k}P) = L_n(\lambda - s_{P(n)}) \ldots L_1(\lambda - s_{P(1)}) \,, \tag{3.E.8}$$

$$T_2(\lambda|\mathbf{k}P) = L_N(\lambda - s_{P(N)}) \ldots L_{n+1}(\lambda - s_{P(n+1)}) \,. \tag{3.E.9}$$

Then,

$$T(\lambda|\mathbf{k}P) = T_2(\lambda|\mathbf{k}P)T_1(\lambda|\mathbf{k}P) \,. \tag{3.E.10}$$

Both, T_1 and T_2 are 2×2-matrices in auxiliary space,

$$T_\alpha(\lambda|\mathbf{k}) = \begin{pmatrix} A_\alpha(\lambda) & B_\alpha(\lambda) \\ C_\alpha(\lambda) & D_\alpha(\lambda) \end{pmatrix}, \tag{3.E.11}$$

$\alpha = 1, 2$. By construction, the matrix elements of T_1 commute with the matrix elements of T_2, and both matrices satisfy the same commutation relations as T.

$$\check{R}(\lambda - \mu)\big(T_\alpha(\lambda|\mathbf{k}P) \otimes T_\alpha(\mu|\mathbf{k}P)\big) = \big(T_\alpha(\mu|\mathbf{k}P) \otimes T_\alpha(\lambda|\mathbf{k}P)\big)\check{R}(\lambda - \mu), \tag{3.E.12}$$

$\alpha = 1, 2$. The matrix elements of T can be represented in terms of the matrix elements of T_1 and T_2. In particular,

$$B(\lambda) = A_2(\lambda)B_1(\lambda) + B_2(\lambda)D_1(\lambda). \tag{3.E.13}$$

We conclude that

$$|\mathbf{k}P, \lambda\rangle = \prod_{j=1}^{M} \big(A_2(\lambda_j)B_1(\lambda_j) + B_2(\lambda_j)D_1(\lambda_j)\big)|0\rangle. \tag{3.E.14}$$

Now we know from the form of the L-operators (see (3.B.60)) that

$$A_\alpha(\lambda)|0\rangle = |0\rangle, \quad D_\alpha(\lambda)|0\rangle = d_\alpha(\lambda)|0\rangle \tag{3.E.15}$$

with

$$d_1(\lambda) = \prod_{j=1}^{n} \frac{\lambda - s_{P(j)} - iu}{\lambda - s_{P(j)} + iu}, \quad d_2(\lambda) = \prod_{j=n+1}^{N} \frac{\lambda - s_{P(j)} - iu}{\lambda - s_{P(j)} + iu}. \tag{3.E.16}$$

We further know the commutation relations between $A_\alpha(\lambda)$, $D_\alpha(\lambda)$ and $B_\alpha(\lambda)$, $\alpha = 1, 2$. They follow from (3.E.12) and are of the same form as (3.B.71), (3.B.72). Using these commutation relations and the equations (3.E.15), (3.E.16), we can expand the product on the right hand side of (3.E.14). We obtain

$$|\mathbf{k}P, \lambda\rangle = \sum_{(S_1, S_2) \in p_2\left(\{\lambda_j\}_{j=1}^{M}\right)} \left[\prod_{\lambda_{m_1}^{(1)} \in S_1} \prod_{\lambda_{m_2}^{(2)} \in S_2} \frac{d_1(\lambda_{m_2}^{(2)})}{c(\lambda_{m_2}^{(2)} - \lambda_{m_1}^{(1)})} B_1(\lambda_{m_1}^{(1)})B_2(\lambda_{m_2}^{(2)}) \right]|0\rangle. \tag{3.E.17}$$

Here $p_2\big(\{\lambda_j\}_{j=1}^{M}\big)$ is the set of all ordered pairs (S_1, S_2) of subsets $S_1, S_2 \subset \{\lambda_j\}_{j=1}^{M}$, such that $S_1 \cup S_2 = \{\lambda_j\}_{j=1}^{M}$ and $S_1 \cap S_2 = \emptyset$. For instance, if $M = 2$,

$$p_2(\{\lambda_1, \lambda_2\}) = \Big\{ (\{\lambda_1, \lambda_2\}, \emptyset), (\{\lambda_1\}, \{\lambda_2\}), (\{\lambda_2\}, \{\lambda_1\}), (\emptyset, \{\lambda_1, \lambda_2\}) \Big\}. \tag{3.E.18}$$

If one of the subsets is empty, we replace the corresponding factors in the products by 1. Equation (3.E.17) can be proven by induction over M. We recommend the reader to verify it for $M = 1$ and $M = 2$.

We may now iterate the above procedure, for example, by dividing T_2 into two factors. After k steps we arrive at

$$T(\lambda|\mathbf{k}P) = T_k(\lambda|\mathbf{k}P)\ldots T_1(\lambda|\mathbf{k}P)\,, \tag{3.E.19}$$

where each $T_\alpha(\lambda|\mathbf{k}P)$ consists of a string of L-operators as in (3.E.8) and (3.E.9). The matrix elements of T_α commute with the matrix elements of T_β, if $\alpha \neq \beta$, and like for the two-site model we have

$$T_\alpha(\lambda|\mathbf{k}) = \begin{pmatrix} A_\alpha(\lambda) & B_\alpha(\lambda) \\ C_\alpha(\lambda) & D_\alpha(\lambda) \end{pmatrix}\,, \tag{3.E.20}$$

$$A_\alpha(\lambda)|0\rangle = |0\rangle\,, \qquad D_\alpha(\lambda)|0\rangle = d_\alpha(\lambda)|0\rangle\,, \tag{3.E.21}$$

$\alpha = 1, \ldots, k$, where the $d_\alpha(\lambda)$ are given by similar expressions as in (3.E.16). Within this notation the k-site generalization of (3.E.17) is

$$|\mathbf{k}P, \lambda\rangle = \sum_{(S_1,\ldots,S_k)\in p_k\left(\{\lambda_j\}_{j=1}^M\right)} \left[\prod_{1\leq\alpha<\beta\leq k} \prod_{\lambda_{m_\alpha}^{(\alpha)}\in S_\alpha} \prod_{\lambda_{m_\beta}^{(\beta)}\in S_\beta} \frac{d_\alpha(\lambda_{m_\beta}^{(\beta)})}{c(\lambda_{m_\beta}^{(\beta)} - \lambda_{m_\alpha}^{(\alpha)})}\right]$$

$$\times \left[\prod_{\alpha=1}^{k} \prod_{\lambda_{m_\alpha}^{(\alpha)}\in S_\alpha} B_\alpha(\lambda_{m_\alpha}^{(\alpha)})\right]|0\rangle\,. \tag{3.E.22}$$

The set $p_k\left(\{\lambda_j\}_{j=1}^M\right)$ is the set of all ordered k-tuples $(S_1, \ldots, S_k)$ of subsets $S_\alpha \subset \{\lambda_j\}_{j=1}^M$, $\alpha = 1, \ldots, k$, such that $\bigcup_{\alpha=1}^{k} S_\alpha = \{\lambda_j\}_{j=1}^M$ and $S_\alpha \cap S_\beta = \emptyset$ for $\alpha \neq \beta$. If one of the sets S_α is empty, then, by definition, the corresponding product is equal to 1. Equation (3.E.22) was obtained in [215]. Note that it is valid in the more general context of the so-called generalized model (see e.g. [270]). It can be proven by induction over k.

We now consider the particular case $k = N$, where N is the length of the spin chain. Then each factor $T_\alpha(\lambda|\mathbf{k}P), \alpha = 1, \ldots, N$, must be identified with an individual L-operator $L_\alpha(\lambda - s_{P(\alpha)})$. From the definition (3.B.60) of the L-operators we obtain

$$d_\alpha(\lambda) = c(2(\lambda - s_{P(\alpha)})) - b(2(\lambda - s_{P(\alpha)})) = \frac{\lambda - s_{P(\alpha)} - iu}{\lambda - s_{P(\alpha)} + iu} \tag{3.E.23}$$

and

$$B_\alpha(\lambda) = 2b(2(\lambda - s_{P(\alpha)}))\sigma_\alpha^- = \frac{2iu\sigma_\alpha^-}{\lambda - s_{P(\alpha)} + iu}\,. \tag{3.E.24}$$

Thus, $B_\alpha(\lambda) \sim \sigma_\alpha^-$, and therefore

$$B_\alpha(\lambda_1)B_\alpha(\lambda_2) = 0\,. \tag{3.E.25}$$

The latter equation leads to several simplifications in (3.E.22). If any of the sets S_α, $\alpha = 1, \ldots, N$, contains more than one element, then, because of (3.E.25), the corresponding product $\prod_{\lambda_{m_\alpha}^{(\alpha)}\in S_\alpha} B_\alpha(\lambda_{m_\alpha}^{(\alpha)})$ on the right-hand side of (3.E.22) is zero. Let us consider the general non-vanishing term in the sum on the right-hand side of (3.E.22). For this term M of the sets S_α, say S_{α_j}, $j = 1, \ldots, M$, are non-empty. They contain precisely one element each, which must be one of the λ_j. Hence, there is a permutation $R \in \mathfrak{S}^M$, such that $S_{\alpha_j} = \{\lambda_{R(j)}\}$. The remaining sets S_α are empty. All N-tuples $(S_1, \ldots, S_N)$ that lead to non vanishing term in the sum in (3.E.22) are thus uniquely characterized by M integers α_j with $1 \le \alpha_1 < \cdots < \alpha_M \le N$ and a permutation $R \in \mathfrak{S}^M$. We may therefore transform the sum in (3.E.22) into a sum over the α_j and the permutations R,

$$
|kP, \lambda\rangle = \sum_{(S_1,\ldots,S_N)\in p_N\left(\{\lambda_j\}_{j=1}^M\right)} \left[\prod_{1\le\alpha<\beta\le N} \prod_{\lambda_{m_\alpha}^{(\alpha)}\in S_\alpha} \prod_{\lambda_{m_\beta}^{(\beta)}\in S_\beta} \frac{1}{c(\lambda_{m_\beta}^{(\beta)} - \lambda_{m_\alpha}^{(\alpha)})} \right]
$$

$$
\times \left[\prod_{1\le\alpha<\beta\le N} \prod_{\lambda_{m_\beta}^{(\beta)}\in S_\beta} d_\alpha(\lambda_{m_\beta}^{(\beta)}) \right] \cdot \left[\prod_{\alpha=1}^N \prod_{\lambda_{m_\alpha}^{(\alpha)}\in S_\alpha} B_\alpha(\lambda_{m_\alpha}^{(\alpha)}) \right] |0\rangle
$$

$$
= \sum_{1\le\alpha_1<\cdots<\alpha_M\le N} \sum_{R\in\mathfrak{S}^M} \left[\prod_{1\le m<n\le M} \frac{1}{c(\lambda_{R(n)} - \lambda_{R(m)})} \right]
$$

$$
\times \left[\prod_{n=1}^M \prod_{\alpha=1}^{\alpha_n-1} d_\alpha(\lambda_{R(n)}) \right] \cdot \left[\prod_{n=1}^M B_{\alpha_n}(\lambda_{R(n)}) \right] |0\rangle . \tag{3.E.26}
$$

Here we have taken into account that the products over empty sets contribute by definition a factor of 1. Let us finally insert (3.E.23), (3.E.24) and the explicit form of $c(\lambda)$, equation (3.B.59). We end up with

$$
|kP, \lambda\rangle = \sum_{1\le\alpha_1<\cdots<\alpha_M\le N} \sum_{R\in\mathfrak{S}^M} \left[\prod_{1\le m<n\le M} \frac{\lambda_{R(m)} - \lambda_{R(n)} - iu}{\lambda_{R(m)} - \lambda_{R(n)}} \right]
$$

$$
\times \left[\prod_{n=1}^M \frac{2iu}{\lambda_{R(n)} - s_{P(\alpha_n)} + iu} \prod_{\alpha=1}^{\alpha_n-1} \frac{\lambda_{R(n)} - s_{P(\alpha)} - iu}{\lambda_{R(n)} - s_{P(\alpha)} + iu} \right] \cdot \left[\prod_{n=1}^M \sigma_{\alpha_n}^- \right] |0\rangle , \tag{3.E.27}
$$

which after a suitable redefinition of variables turns into the desired result, i.e., into (3.E.6), (3.E.7).

3.F Lowest weight theorem for the η-pairing symmetry

This appendix contains a proof of equation (3.101): we show that for any even lattice length L all Bethe ansatz states $|\psi_{k,\lambda}\rangle$, equation (3.89), are annihilated by the operator η^-. We present a simplified version of the argument originally given in [125].

The operator η^- reduces the particle number by two (see equation (2.87)). The result of acting with η^- on any N-particle state $|\psi_N\rangle$ is therefore an $(N-2)$-particle state. The corresponding $(N-2)$-particle wave function is (compare appendix 3.A)

$$(\eta^- \psi_N)(x_1, \ldots, x_{N-2}; a_1, \ldots, a_{N-2})$$

$$= \langle (x_1, \ldots, x_{N-2}), (a_1, \ldots, a_{N-2}) | \eta^- | \psi_N \rangle$$

$$= \sum_{x=1}^{L} (-1)^x \langle 0 | c_{x_1, a_1} \cdots c_{x_{N-2}, a_{N-2}} c_{x,\uparrow} c_{x,\downarrow} | \psi_N \rangle$$

$$= \sum_{x=1}^{L} (-1)^x \psi_N(x_1, \ldots, x_{N-2}, x, x; a_1, \ldots, a_{N-2}, \uparrow, \downarrow)$$

$$= \sum_{a_{N-1}, a_N = \uparrow, \downarrow} \delta_{a_{N-1}, \uparrow} \delta_{a_N, \downarrow} \sum_{x_{N-1}, x_N = 1}^{L} (-1)^{x_N} \delta_{x_{N-1}, x_N} \psi_N(\mathbf{x}; \mathbf{a}), \qquad (3.F.1)$$

where we have inserted the definition (2.80b) of η^- in the third line. Replacing $\psi_N(\mathbf{x}; \mathbf{a})$ with any Bethe ansatz wave function $\psi(\mathbf{x}; \mathbf{a} | \mathbf{k}; \boldsymbol{\lambda})$ we see that our assertion $\eta^- | \psi_{\mathbf{k}, \boldsymbol{\lambda}} \rangle = 0$ is equivalent to the following

Lemma 1. *For even length of the lattice L every Bethe ansatz wave function $\psi(\mathbf{x}; \mathbf{a} | \mathbf{k}; \boldsymbol{\lambda})$, determined by equations (3.91)–(3.96) with $2M \leq N \leq L$, satisfies the identity*

$$\sum_{x_{N-1}, x_N = 1}^{L} (-1)^{x_N} \delta_{x_{N-1}, x_N} \psi(\mathbf{x}; \mathbf{a} | \mathbf{k}; \boldsymbol{\lambda}) = 0. \qquad (3.F.2)$$

Remark. Let us emphasize that we consider only finite charge momenta and finite spin rapidities, i.e., k_j, λ_ℓ are supposed to take values in $\mathbb{C}$ and not on the Riemann sphere. The corresponding solutions of the Lieb-Wu equations with $2M \leq N \leq L$ will be termed *regular* as in [125].

The proof of lemma 1 will be divided into two essential steps. In step one the summation over x_{N-1} and x_N is carried out. Using the evenness of L and the fact that $\mathbf{k}$ and $\boldsymbol{\lambda}$ satisfy the Lieb-Wu equations we shall see that the validity of the lemma depends on a simple identity for the spin part $\langle \mathbf{a} Q | \mathbf{k} P, \boldsymbol{\lambda} \rangle$ of the Bethe ansatz wave function (3.91)–(3.94). The second essential step in the proof is then to establish this identity.

Step 1. We denote the sum on the left-hand side of (3.F.2) by S. The first problem when we calculate S, and for this purpose sum expressions involving the Bethe ansatz wave functions

$$\psi(\mathbf{x}; \mathbf{a} | \mathbf{k}; \boldsymbol{\lambda}) = \sum_{P \in \mathfrak{S}^N} \text{sign}(PQ) \langle \mathbf{a} Q | \mathbf{k} P, \boldsymbol{\lambda} \rangle \, e^{i\langle \mathbf{k} P, \mathbf{x} Q \rangle} \qquad (3.F.3)$$

over $x = x_{N-1} = x_N$, is the dependence of the permutation Q on $\mathbf{x}$. We have to make this dependence explicit. For this purpose we fix $x_1, \ldots, x_{N-2}$. Then a permutation $Q_1 \in \mathfrak{S}^N$ exists, such that $Q_1(1) = N - 1$, $Q_1(2) = N$ and $x_{Q_1(3)} \leq \cdots \leq x_{Q_1(N)}$. When x runs from 1 to L, then Q runs through $N - 1$ different permutations characterized by x lying between two successive values $x_{Q_1(j)}$ and $x_{Q_1(j+1)}$, $j = 3, \ldots, N - 1$, or lying below $x_{Q_1(3)}$ or above

$x_{Q_1(N)}$, respectively. We will call these permutations Q_ℓ. They can be simply expressed in terms of the generators $U_\ell = \Pi_{12}\Pi_{23}\ldots\Pi_{\ell-1,\ell}$ of the cyclic subgroups of order ℓ of $\mathfrak{S}^N$. We have

$$Q_\ell = Q_1 U_{\ell+1}^2, \quad \ell = 2, \ldots, N - 1. \tag{3.F.4}$$

Using these specific permutations we can now rewrite the left-hand side of (3.F.2) as

$$S = \sum_{\ell=1}^{N-1} \sum_{x \in I_\ell} \sum_{P \in \mathfrak{S}^N} \text{sign}(PQ_\ell)\langle \mathbf{a}Q_\ell|\mathbf{k}P\rangle(-1)^x e^{i\langle \mathbf{k}P|\mathbf{x}Q_\ell\rangle}\Big|_{x_{N-1}=x_N=x} \tag{3.F.5}$$

$$= \sum_{\ell=1}^{N-1} \sum_{P \in \mathfrak{S}^N} \text{sign}(PQ_1)\langle \mathbf{a}Q_1 U_{\ell+1}^2|\mathbf{k}P U_{\ell+1}^2\rangle \sum_{x \in I_\ell}(-1)^x e^{i\langle \mathbf{k}P|\mathbf{x}Q_1\rangle}\Big|_{r_{N-1}=r_N=r}.$$

In the second equation we took into account that $\text{sign}(U_\ell^2) = 1$ and we shifted the summation over P by $U_{\ell+1}^2$. We suppressed the argument λ in the spin part of the wave function. As a further convenient shorthand notation we introduced the sets

$$I_\ell = \begin{cases} \{x_{Q_1(\ell+1)}, \ldots, x_{Q_1(\ell+2)} - 1\} & \text{for } x_{Q_1(\ell+1)} < x_{Q_1(\ell+2)} \\ \emptyset & \text{else} \end{cases}. \tag{3.F.6}$$

This definition works for $\ell = 2, \ldots, N - 2$, but not at the boundaries $\ell = 1, N - 1$, where we define $I_1 = \{1, \ldots, x_{Q_1(3)} - 1\}$ and $I_{N-1} = \{x_{Q_1(N)}, \ldots, L\}$ instead and employ again the convention that sets I_1 of 'zero length' ($x_{Q_1(3)} = 1$) are empty. The number of elements in I_ℓ will be denoted by $|I_\ell|$.

Having rewritten S in the form (3.F.5) we can perform the summations over x by means of the geometric sum formula. The only exception we have to treat separately is when $k_{P(1)} + k_{P(2)} = \pi$ and thus $-\exp\big(i(k_{P(1)} + k_{P(2)})\big) = 1$. In this case we obtain

$$S = \sum_{\ell=1}^{N-1} \sum_{P \in \mathfrak{S}^N} \text{sign}(PQ_1)|I_\ell|e^{i\sum_{j=3}^N k_{P(j)}x_{Q_1(j)}}\langle \mathbf{a}Q_1 U_{\ell+1}^2|\mathbf{k}P U_{\ell+1}^2\rangle$$

$$= \sum_{\ell=1}^{N-1} \sum_{P \in \mathfrak{A}^N} \text{sign}(PQ_1)|I_\ell|e^{i\sum_{j=3}^N k_{P(j)}x_{Q_1(j)}}$$
$$\big(\langle \mathbf{a}Q_1 U_{\ell+1}^2|\mathbf{k}P U_{\ell+1}^2\rangle - \langle \mathbf{a}Q_1 U_{\ell+1}^2|\mathbf{k}P \Pi_{12} U_{\ell+1}^2\rangle\big) \tag{3.F.7}$$

which vanishes, because the terms in the brackets on the right-hand side cancel each other: $|\mathbf{k}P\Pi_{12}U_\ell^2\rangle = |\mathbf{k}PU_\ell^2\Pi_{\ell-1,\ell}\rangle = Y_{\ell-1,\ell}(s_{P(1)} - s_{P(2)})|\mathbf{k}PU_\ell^2\rangle = |\mathbf{k}PU_\ell^2\rangle$. Here we used equation (3.B.31), again the abbreviation $s_j = \sin(k_j)$ introduced in appendix 3.B, and the fact that $Y_{\ell-1,\ell}(0) = 1$.

Now we have proven lemma 1 for the special case $k_{P(1)} + k_{P(2)} = \pi$. For $k_{P(1)} + k_{P(2)} \neq \pi$ we use the geometric sum formula to carry out the summation over x in equation

(3.F.5). We obtain

$$\sum_{x\in I_\ell}(-1)^x e^{i\langle kP|xQ_1\rangle}\Big|_{x_{N-1}=x_N=x}$$

$$=\frac{e^{i\sum_{j=3}^N k_{P(j)}x_{Q_1(j)}}}{1+e^{i(k_{P(1)}+k_{P(2)})}}\left[\left(-e^{i(k_{P(1)}+k_{P(2)})}\right)^{x_{Q_1(\ell+1)}}-\left(-e^{i(k_{P(1)}+k_{P(2)})}\right)^{x_{Q_1(\ell+2)}}\right]\qquad(3.F.8)$$

for $\ell=2,\dots,N-2$. For $\ell=1$ and $\ell=N-1$ the terms in the square brackets have to be modified. For $\ell=1$ the first term must be replaced by $-e^{i(k_{P(1)}+k_{P(2)})}$, while for $\ell=N-1$ we have to replace the second term by $(-e^{i(k_{P(1)}+k_{P(2)})})^{L+1}$. It is exactly these two boundary contributions that cancel each other for even L by means of equation (3.B.45), since

$$\langle \mathbf{a}Q_1 U_N^2|\mathbf{k}PU_N^2\rangle(-e^{i(k_{P(1)}+k_{P(2)})})^{L+1}=-(-1)^L\langle\mathbf{a}Q_1|\mathbf{k}P\rangle e^{i(k_{P(1)}+k_{P(2)})}.\qquad(3.F.9)$$

Note that equation (3.B.45) which we used twice to obtain the right-hand side is only valid if the Lieb-Wu equations are satisfied.

Inserting now equation (3.F.8) and the two single remaining boundary terms into equation (3.F.5) we obtain after appropriately shifting the summation indices

$$S=\sum_{\ell=3}^N(-1)^{x_{Q_1(\ell)}}\sum_{P\in\mathfrak{S}^N}\text{sign}(PQ_1)e^{i\left[(k_{P(1)}+k_{P(2)}+k_{P(\ell)})x_{Q_1(\ell)}+\sum_{\substack{j=3\\j\neq\ell}}^N k_{P(j)}x_{Q_1(j)}\right]}$$

$$\times\frac{\langle\mathbf{a}Q_1 U_\ell^2|\mathbf{k}PU_\ell^2\rangle-\langle\mathbf{a}Q_1 U_{\ell-1}^2|\mathbf{k}PU_{\ell-1}^2\rangle}{1+e^{i(k_{P(1)}+k_{P(2)})}}.\qquad(3.F.10)$$

This is almost the final expression we wished to derive in step one of our proof. To finish step one we introduce one more little piece of notation. By $\mathfrak{S}_\ell^3$, $\ell=3,\dots,N$, we denote the subgroup of $\mathfrak{S}^N$ of all bijective maps of the set $\{1,2,\ell\}$ onto itself. Then, using once more the translational invariance of the sum over P, we can rewrite S as

$$S=\sum_{\ell=3}^N(-1)^{x_{Q_1(\ell)}}\frac{1}{6}\sum_{P\in\mathfrak{S}^N}\text{sign}(PQ_1)e^{i\left[(k_{P(1)}+k_{P(2)}+k_{P(\ell)})x_{Q_1(\ell)}+\sum_{\substack{j=3\\j\neq\ell}}^N k_{P(j)}x_{Q_1(j)}\right]}$$

$$\times\sum_{\Gamma\in\mathfrak{S}_\ell^3}\text{sign}(\Gamma)\frac{\langle\mathbf{a}Q_1 U_\ell^2|\mathbf{k}P\Gamma U_\ell^2\rangle-\langle\mathbf{a}Q_1 U_{\ell-1}^2|\mathbf{k}P\Gamma U_{\ell-1}^2\rangle}{1+e^{i(k_{P\Gamma(1)}+k_{P\Gamma(2)})}}.\qquad(3.F.11)$$

Step 2. We shall show that the spin part of the Bethe ansatz wave function satisfies the relation

$$\sum_{\Gamma\in\mathfrak{S}_\ell^3}\text{sign}(\Gamma)\frac{\langle\mathbf{a}Q_1 U_\ell^2|\mathbf{k}P\Gamma U_\ell^2\rangle-\langle\mathbf{a}Q_1 U_{\ell-1}^2|\mathbf{k}P\Gamma U_{\ell-1}^2\rangle}{1+e^{i(k_{P\Gamma(1)}+k_{P\Gamma(2)})}}=0\qquad(3.F.12)$$

for all $P\in\mathfrak{S}^N$ and for $\ell=3,\dots,N$. Then lemma 1 will follow from (3.F.11).

Let us denote the left-hand side of (3.F.12) by C. Observing that

$$|\mathbf{k}P\Gamma\Pi_{12}U_\ell^2\rangle=|\mathbf{k}P\Gamma U_\ell^2\Pi_{\ell-1,\ell}\rangle=Y_{\ell-1,\ell}(s_{P\Gamma(1)}-s_{P\Gamma(2)})|\mathbf{k}P\Gamma U_\ell^2\rangle\qquad(3.F.13)$$

and using the explicit expression (3.B.32) for the Y-operator and the coset decomposition $\mathfrak{S}_\ell^3 = \mathfrak{A}_\ell^3 \cup \mathfrak{A}_\ell^3 \Pi_{12}$ we obtain

$$C = \sum_{\Gamma \in \mathfrak{A}_\ell^3} \text{sign}(\Gamma) \frac{s_{P\Gamma(1)} - s_{P\Gamma(2)}}{1 + e^{i(k_{P\Gamma(1)} + k_{P\Gamma(2)})}} \left[\frac{\langle \mathbf{a} Q_1 U_\ell^2 | (1 - \Pi_{\ell-1,\ell}) | \mathbf{k} P \Gamma U_\ell^2 \rangle}{2iu + s_{P\Gamma(1)} - s_{P\Gamma(2)}} \right.$$

$$\left. - \frac{\langle \mathbf{a} Q_1 U_{\ell-1}^2 | (1 - \Pi_{\ell-2,\ell-1}) | \mathbf{k} P \Gamma U_{\ell-1}^2 \rangle}{2iu + s_{P\Gamma(1)} - s_{P\Gamma(2)}} \right]. \tag{3.F.14}$$

Next we introduce the shorthand notation

$$B_\ell(P\Gamma) = \frac{\langle \mathbf{a} Q_1 U_\ell^2 | (1 - \Pi_{\ell-1,\ell}) | \mathbf{k} P \Gamma U_\ell^2 \rangle}{2iu + s_{P\Gamma(1)} - s_{P\Gamma(2)}}. \tag{3.F.15}$$

We shall show below that

$$B_\ell(P\Gamma) - B_{\ell-1}(P\Gamma) = B_\ell(P) - B_{\ell-1}(P) \tag{3.F.16}$$

for all $\Gamma \in \mathfrak{S}_\ell^3$. Assuming this for the moment to be true we can complete the proof of (3.F.12):

$$C = (B_\ell(P) - B_{\ell-1}(P)) \sum_{\Gamma \in \mathfrak{A}_\ell^3} \text{sign}(\Gamma) \frac{s_{P\Gamma(1)} - s_{P\Gamma(2)}}{1 + e^{i(k_{P\Gamma(1)} + k_{P\Gamma(2)})}}$$

$$= \frac{i}{2}(B_\ell(P) - B_{\ell-1}(P)) \sum_{\Gamma \in \mathfrak{A}_\ell^3} \text{sign}(\Gamma) \left(e^{-i(k_{P\Gamma(1)})} - e^{-i(k_{P\Gamma(2)})} \right) = 0. \tag{3.F.17}$$

We have thus reduced the proof of lemma 1 to the verification of the invariance equation (3.F.16), which is a straightforward but slightly cumbersome matter and can be done by resorting to the explicit form (3.91)–(3.94) of the spin part of the Bethe ansatz wave function. More precisely, we shall only need the explicit form of the functions $F_\mathbf{k}(\lambda, y)$, equation (3.93). The amplitudes $A(\lambda)$, equation (3.94), do not depend on the vector $\mathbf{k}$. In the following we only need that they satisfy the recurrence relation

$$A(\lambda R \Pi_{m,m+1}) = \frac{\lambda_{P(m)} - \lambda_{P(m+1)} + 2iu}{\lambda_{P(m)} - \lambda_{P(m+1)} - 2iu} A(\lambda R) \tag{3.F.18}$$

for all $R \in \mathfrak{S}^M$.

The matrix $\frac{1}{2}(1 - \Pi_{\ell-1,\ell})$ is the antisymmetrization operator in the indices $\ell - 1$ and ℓ. Hence, $B_\ell(P)$ is zero for $a_{Q_1 U_\ell^2(\ell-1)} = a_{Q_1(1)} = a_{N-1} = a_{Q_1 U_\ell^2(\ell)} = a_{Q_1(2)} = a_N$ and changes sign if we interchange a_{N-1} and a_N. We may therefore choose $a_{N-1} = \downarrow$ and $a_N = \uparrow$ for the remaining part of the proof. The sequence of 'down spin coordinates' $y_m^{(\ell)}$, $m = 1, \ldots, M$, is the same for $\langle \mathbf{a} Q_1 U_\ell^2 |$ and $\langle \mathbf{a} Q_1 U_\ell^2 \Pi_{\ell-1,\ell} | = \langle \mathbf{a} Q_1 U_\ell^2 | \Pi_{\ell-1,\ell}$ with precisely one exception, say $y_{m_\ell}^{(\ell)}$, which is $\ell - 1$ for $\langle \mathbf{a} Q_1 U_\ell^2 |$ and ℓ for $\langle \mathbf{a} Q_1 U_\ell^2 | \Pi_{\ell-1,\ell}$. Using this in the explicit form (3.91)–(3.94) of the spin part of the Bethe ansatz wave function results in

$$B_\ell(P) = 2iu \sum_{R \in \mathfrak{S}^M} A(\lambda R) \Big[\prod_{m=1,2,\ell} \frac{1}{\lambda_{R(m_\ell)} - s_{P(m)} + iu} \Big] \Big[\prod_{n=3}^{\ell-1} \frac{\lambda_{R(m_\ell)} - s_{P(n)} - iu}{\lambda_{R(m_\ell)} - s_{P(n)} + iu} \Big]$$

$$\times \Big[\prod_{\substack{m=1 \\ m \neq m_\ell}}^{M} F_{\mathbf{k}PU_\ell^2}(\lambda_{R(m)}, y_m^{(\ell)}) \Big] (\lambda_{R(m_\ell)} - s_{P(\ell)} - iu). \tag{3.F.19}$$

We have now rewritten $B_\ell(P)$ in such a way that the first three factors under the sum on the right hand side are manifestly invariant under permutations of $k_{P(1)}$, $k_{P(2)}$ and $k_{P(\ell)}$. The last factor, $\lambda_{R(m_\ell)} - s_{P(\ell)} - iu$, is clearly not invariant. How about the factors $F_{\mathbf{k}PU_\ell^2}(\lambda_{R(m)}, y_m^{(\ell)})$? The answer depends on the value of $y_m^{(\ell)}$: it is easy to see that

$$F_{\mathbf{k}PU_\ell^2}(\lambda, y) = F_{\mathbf{k}P\Gamma U_\ell^2}(\lambda, y) = F_{\mathbf{k}PU_{\ell-1}^2}(\lambda, y) \tag{3.F.20}$$

for all $\Gamma \in \mathfrak{S}_\ell^3$ unless $y = \ell - 2, \ell - 1, \ell$. This fact will become important below.

Next we replace ℓ by $\ell - 1$ in (3.F.19) and subtract the result from (3.F.19). We distinguish two cases by looking at the relevant pieces of the sequences of spins $a_{Q_{\ell-1}}(n)$ and $a_{Q_{\ell-2}}(n)$ (see table 3.F.1). Case (i): $a_{Q_1(\ell)} = \uparrow$ (called the up-spin case in [125]). We see from the table that $m_{\ell-1} = m_\ell$ and $y_m^{(\ell-1)} = y_m^{(\ell)} \neq \ell - 2, \ell - 1, \ell$ for $m \neq m_\ell$. Using (3.F.19) and (3.F.20) it follows that

$$B_\ell(P) - B_{\ell-1}(P) = 2iu \sum_{R \in \mathfrak{S}^M} A(\lambda R) \Big[\prod_{m=1,2,\ell} \frac{1}{\lambda_{R(m_\ell)} - s_{P(m)} + iu} \Big]$$

$$\times \Big[\prod_{n=3}^{\ell-1} \frac{\lambda_{R(m_\ell)} - s_{P(n)} - iu}{\lambda_{R(m_\ell)} - s_{P(n)} + iu} \Big] \Big[\prod_{\substack{m=1 \\ m \neq m_\ell}}^{M} F_{\mathbf{k}PU_\ell^2}(\lambda_{R(m)}, y_m^{(\ell)}) \Big] (-2iu) \tag{3.F.21}$$

with $y_m^{(\ell)} \neq \ell - 2, \ell - 1, \ell$. Taking into account (3.F.20) we see that this expression is indeed invariant under all permutations of $k_{P(1)}$, $k_{P(2)}$ and $k_{P(\ell)}$.

Case (ii): $a_{Q_1(\ell)} = \downarrow$ (down-spin case of [125]). We infer from table 3.F.1 that $m_{\ell-1} = m_\ell - 1$, $y_{m_\ell}^{(\ell-1)} = \ell$, $y_{m_{\ell-1}}^{(\ell)} = \ell - 2$, and $y_m^{(\ell-1)} = y_m^{(\ell)} \neq \ell - 2, \ell - 1, \ell$ for $m \neq m_\ell, m_\ell - 1$.

Table 3.F.1. *The two different cases in the evaluation of* $B_\ell(P) - B_{\ell-1}(P)$

$\uparrow$-spin case	$n = \ell - 2$	$n = \ell - 1$	$n = \ell$
$a_{Q_{\ell-2}(n)}$	$\downarrow$	$\uparrow$	$\uparrow$
$a_{Q_{\ell-1}(n)}$	$\uparrow$	$\downarrow$	$\uparrow$
$\downarrow$-spin case	$n = \ell - 2$	$n = \ell - 1$	$n = \ell$
$a_{Q_{\ell-2}(n)}$	$\downarrow$	$\uparrow$	$\downarrow$
$a_{Q_{\ell-1}(n)}$	$\downarrow$	$\downarrow$	$\uparrow$

Using again (3.F.19) and (3.F.20) we obtain this time

$$
B_\ell(P) - B_{\ell-1}(P) = 2iu \sum_{R \in \mathfrak{S}^M} A(\lambda R) \Big[\prod_{\substack{m=1 \\ m \neq m_\ell - 1, m_\ell}}^{M} F_{\mathbf{k}PU_\ell^2}(\lambda_{R(m)}, y_m^{(\ell)}) \Big]
$$

$$
\times \Big\{ F_{\mathbf{k}PU_\ell^2}(\lambda_{R(m_\ell-1)}, \ell-2) \Big[\prod_{m=1,2,\ell} \frac{1}{\lambda_{R(m_\ell)} - s_{P(m)} + iu} \Big] \Big[\prod_{n=3}^{\ell-1} \frac{\lambda_{R(m_\ell)} - s_{P(n)} - iu}{\lambda_{R(m_\ell)} - s_{P(n)} + iu} \Big]
$$

$$
\times (\lambda_{R(m_\ell)} - s_{P(\ell)} - iu) - \Big[\prod_{m=1,2,\ell} \frac{1}{\lambda_{R(m_\ell-1)} - s_{P(m)} + iu} \Big] \Big[\prod_{n=3}^{\ell-1} \frac{\lambda_{R(m_\ell-1)} - s_{P(n)} - iu}{\lambda_{R(m_\ell-1)} - s_{P(n)} + iu} \Big]
$$

$$
\times (\lambda_{R(m_\ell-1)} - s_{P(\ell)} + iu) F_{\mathbf{k}PU_{\ell-1}^2}(\lambda_{R(m_\ell)}, \ell) \Big\}. \tag{3.F.22}
$$

Here we insert the explicit expressions for $F_{\mathbf{k}PU_\ell^2}(\lambda, \ell-2)$ and $F_{\mathbf{k}PU_{\ell-1}^2}(\lambda, \ell)$ and pull out the common factors. We arrive at

$$
B_\ell(P) - B_{\ell-1}(P) = (2iu)^2 \sum_{R \in \mathfrak{S}^M} A(\lambda R) \Big[\prod_{\substack{m=1 \\ m \neq m_\ell - 1, m_\ell}}^{M} F_{\mathbf{k}PU_\ell^2}(\lambda_{R(m)}, y_m^{(\ell)}) \Big]
$$

$$
\times \Big[\prod_{n=3}^{\ell-1} \frac{\lambda_{R(m_\ell-1)} - s_{P(n)} - iu}{\lambda_{R(m_\ell-1)} - s_{P(n)} + iu} \cdot \frac{\lambda_{R(m_\ell)} - s_{P(n)} - iu}{\lambda_{R(m_\ell)} - s_{P(n)} + iu} \Big]
$$

$$
\times \Big[\prod_{m=1,2,\ell} \frac{1}{\lambda_{R(m_\ell-1)} - s_{P(m)} + iu} \cdot \frac{1}{\lambda_{R(m_\ell)} - s_{P(m)} + iu} \Big]
$$

$$
\times \Big\{ (\lambda_{R(m_\ell-1)} - s_{P(1)} + iu)(\lambda_{R(m_\ell-1)} - s_{P(2)} + iu)(\lambda_{R(m_\ell)} - s_{P(\ell)} - iu)
$$

$$
- (\lambda_{R(m_\ell-1)} - s_{P(\ell)} + iu)(\lambda_{R(m_\ell)} - s_{P(1)} - iu)(\lambda_{R(m_\ell)} - s_{P(2)} - iu) \Big\}. \tag{3.F.23}
$$

The only factor which does not have the required symmetry under permutations of $k_{P(1)}$ $k_{P(2)}$ and $k_{P(\ell)}$ is the factor in curly brackets. We rewrite it using the trivial identities

$$
\lambda_{R(m_\ell)} - s_{P(\ell)} - iu = \lambda_{R(m_\ell)} - \lambda_{R(m_\ell-1)} - 2iu + \lambda_{R(m_\ell-1)} - s_{P(\ell)} + iu \,,
$$

$$
\lambda_{R(m_\ell-1)} - s_{P(\ell)} + iu = -(\lambda_{R(m_\ell)} - \lambda_{R(m_\ell-1)} - 2iu) + \lambda_{R(m_\ell)} - s_{P(\ell)} - iu
$$

as

$$
\{\dots\} = (\lambda_{R(m_\ell)} - \lambda_{R(m_\ell-1)} - 2iu)\big[(\lambda_{R(m_\ell-1)} - s_{P(1)} + iu)(\lambda_{R(m_\ell-1)} - s_{P(2)} + iu)
$$

$$
+ (\lambda_{R(m_\ell)} - s_{P(1)} - iu)(\lambda_{R(m_\ell)} - s_{P(2)} - iu) \big]
$$

$$
+ (\lambda_{R(m_\ell-1)} - s_{P(1)} + iu)(\lambda_{R(m_\ell-1)} - s_{P(2)} + iu)(\lambda_{R(m_\ell-1)} - s_{P(\ell)} + iu)
$$

$$
- (\lambda_{R(m_\ell)} - s_{P(1)} - iu)(\lambda_{R(m_\ell)} - s_{P(2)} - iu)(\lambda_{R(m_\ell)} - s_{P(\ell)} - iu) \,. \tag{3.F.24}
$$

The second and the third term on the right-hand side have now the required symmetry, but the first term has not. We observe, however, that the three factors in square brackets on the right-hand side of equation (3.F.23) are not only symmetric in $k_{P(1)}$, $k_{P(2)}$ and $k_{P(\ell)}$ but also in $\lambda_{R(m_\ell-1)}$ and $\lambda_{R(m_\ell)}$, while $A(\lambda R \Pi_{m_\ell-1,m_\ell})$ is given by equation (3.F.18). We can thus use our usual trick of decomposing the symmetric group as $\mathfrak{S}^M = \mathfrak{A}^M \cup \mathfrak{A}^M \Pi_{m_\ell-1,m_\ell}$ and reducing the sum in (3.F.23) to a sum over $\mathfrak{A}^M$. This way the first term on the right-hand side of equation (3.F.24) gets reduced to

$$-4iu(\lambda_{R(m_\ell)} - \lambda_{R(m_\ell-1)})(\lambda_{R(m_\ell)} - \lambda_{R(m_\ell-1)} - 2iu).$$

Since this expression is independent of $\mathbf{k}$ at all, we have finally succeeded in showing that $B_\ell(P) - B_{\ell-1}(P)$ is symmetric in $k_{P(1)}$, $k_{P(2)}$ and $k_{P(\ell)}$, and the proof of lemma 1 is complete.   □

3.G Limiting cases of the Bethe ansatz solution

The Bethe ansatz solution of the Hubbard model has several interesting limiting cases. In appendices 2.A and 2.B we considered the strong coupling limits and the continuum limit of the Hamiltonian. Here we discuss the same limits within the context of the Bethe ansatz solution. The strong coupling limits will lead us to the Bethe ansatz solutions of the t-0 model and of the isotropic Heisenberg chain. We will also see how the Bethe ansatz can be used to obtain the spectrum of the t-J Hamiltonian (2.A.33). In addition we shall briefly touch upon the limit of weak coupling in which we shall 'discover' the so-called Gaudin model [156].

3.G.1 Strong coupling limits

Formally the limit $u \to \infty$ is easily performed in the Bethe ansatz wave function and in the Lieb-Wu equations. We just have to observe that the scale of the spin rapidities is at our disposal. Hence we may define $u\Lambda_\ell = \lambda_\ell$ and then take the limit in (3.95) and (3.96). We immediately obtain the following equations,

$$e^{ik_j L} = \prod_{\ell=1}^{M} \frac{\Lambda_\ell - i}{\Lambda_\ell + i}, \quad j = 1, \ldots, N, \tag{3.G.1}$$

$$\left(\frac{\Lambda_\ell - i}{\Lambda_\ell + i}\right)^N = \prod_{\substack{m=1 \\ m \neq \ell}}^{M} \frac{\Lambda_\ell - \Lambda_m - 2i}{\Lambda_\ell - \Lambda_m + 2i}, \quad \ell = 1, \ldots, M. \tag{3.G.2}$$

We see that the equations for the charge momenta and spin rapidities decouple. The functions $\sin(k_j)$ which appeared as inhomogeneities in the auxiliary spin problem have all vanished from the equations. Equations (3.G.2) are the Bethe ansatz equations of the XXX Heisenberg chain [60] (see Chapter 12.1.7). The expression on the right-hand side of (3.G.1) is the eigenvalue of the shift operator for the spin chain. We therefore write

$$e^{iP_s} = \prod_{\ell=1}^{M} \frac{\Lambda_\ell - i}{\Lambda_\ell + i}, \tag{3.G.3}$$

where P_s denotes the corresponding momentum eigenvalue. It may take values $P_s = m2\pi/N$, $m = 0, 1, \ldots, N - 1$ as can, for instance, be seen by multiplying (3.G.2) over all $\ell = 1, \ldots, M$. It follows that equations (3.G.1) have the solutions

$$k_j = \frac{n_j 2\pi}{L} + \frac{P_s}{L}, \quad j = 1, \ldots, N, \tag{3.G.4}$$

$n_j \in \{0, \ldots, L - 1\}$. These are the quantum numbers of a system of free spinless fermions under twisted boundary conditions with twist angle $P_s \in [0, 2\pi]$. They determine the eigenvalues of the Hubbard Hamiltonian (2.22) (which is more suitable for taking the limit $u \to \infty$ than (2.31)) as

$$E = -2 \sum_{j=1}^{N} \cos(k_j). \tag{3.G.5}$$

For the Bethe ansatz wave function (3.91) we achieve a similar decoupling. After rescaling $u\Lambda_\ell = \lambda_\ell$ in (3.92) and taking the limit $u \to \infty$ it turns into

$$\psi_\infty(\mathbf{x}; \mathbf{a} | \mathbf{k}; \boldsymbol{\lambda}) = \langle \mathbf{a} Q | \boldsymbol{\lambda} \rangle \sum_{P \in \mathfrak{S}^N} \text{sign}(PQ) e^{i\langle \mathbf{k}P, \mathbf{x}Q \rangle} = \langle \mathbf{a} Q | \boldsymbol{\lambda} \rangle \det\left(e^{ik_m x_n}\right), \tag{3.G.6}$$

where the amplitudes $\langle \mathbf{a} Q | \boldsymbol{\lambda} \rangle$ are now independent of $\mathbf{k}$ and are of the form of the Bethe ansatz wave functions of the XXX Heisenberg chain,

$$\langle \mathbf{a} Q | \boldsymbol{\lambda} \rangle = C \sum_{P \in \mathfrak{S}^M} \left[\prod_{1 \leq m < n \leq M} \frac{\Lambda_{P(m)} - \Lambda_{P(n)} - 2i}{\Lambda_{P(m)} - \Lambda_{P(n)}} \right] \prod_{\ell=1}^{M} \left(\frac{\Lambda_{P(\ell)} - i}{\Lambda_{P(\ell)} + i} \right)^{y_\ell} \tag{3.G.7}$$

with $C = (2i)^M \prod_{\ell=1}^{M} (\Lambda_\ell - i)^{-1}$. The y_ℓ are coordinates of down spins on electrons as described in Section 3.3.2.

What is the interpretation of these results? In order to answer this question let us first count the number of independent wave functions (3.G.6). For each fixed number N of electrons there are 2^N linearly independent wave functions (3.G.7) of the spin system[1] and $\binom{L}{N}$ inequivalent ways of choosing k_j in accordance with (3.G.4), giving a total number of $\sum_{N=0}^{L} \binom{L}{N} 2^N = 3^L$ wave functions of the form (3.G.6). 3^L is the dimension of the space of states of the various t-J Hamiltonians of appendix 2.A which is the space of electronic states with double occupancy of sites excluded. This observation fits together well with the fact that, because of the Slater determinant factor, the wave functions ψ_∞ in equation (3.G.6) vanish if any two coordinates, x_j, x_k, $j \neq k$, coincide.

In fact, the wave functions ψ_∞ give a set of eigenfunctions of the t-0 Hamiltonian (2.A.40) which becomes complete when supplemented with the su(2) symmetry of the spin system. This can be verified by using the right-hand side of (3.G.6) as an ansatz wave function in the eigenvalue problem of the t-0 Hamiltonian. One can proceed in much the same way as in the nested Bethe ansatz calculation for the Hubbard model in appendix 3.B.

---

[1] To obtain this number we also have to take into account the su(2) symmetry of the spin system (see Section 3.D).

It turns out that the 'spin problem' of diagonalizing the operator $t(s_{P(1)} - iu|\mathbf{k}P) = X_{1,n}(s_{P(1)} - s_{P(N)}) \ldots X_{1,2}(s_{P(1)} - s_{P(2)})$ on the left hand side of (3.B.47) gets replaced with the eigenvalue problem for the shift operator $\hat{U} = \Pi_{1,2} \ldots \Pi_{N-1,N}$ which is obtained from $t(s_{P(1)} - iu|\mathbf{k}P)$ for $u \to \infty$.[2] The shift operator $\hat{U}$ is diagonal in the basis of Bethe ansatz states of the XXX Hamiltonian. Hence, we obtain the Bethe ansatz solution (3.G.1), (3.G.2), (3.G.6) of the t-0 model.

Of course, the shift operator is highly degenerate and is diagonal in other bases. A particularly convenient choice is the basis of eigenstates of the XX chain (see e.g. Section 12.2.1) which is equivalent to a system of free spinless fermions. In this basis the spin part $\langle \mathbf{a} Q | \boldsymbol{\lambda} \rangle$ of the wave function ψ_∞ also assumes a determinantal form and the set (3.G.2) of Bethe ansatz equations for the spin problem gets replaced by

$$e^{iq_\ell N} = (-1)^{M+1}, \quad \ell = 1, \ldots, M \tag{3.G.8}$$

(see [179, 219]). In these variables the momentum of the spin system entering the right-hand side of (3.G.1) through (3.G.3) is $P_s = \sum_{\ell=1}^{M} q_\ell$. The use of the XX chain eigenfunctions instead of the more complicated XXX chain eigenfunctions made is possible to study dynamical correlations of the t-0 model in [219].

However, if one is interested in the large u limit of the Hubbard model rather than in the highly degenerate t-0 model it is preferable to stick with the XXX chain wavefunctions for the spin part of ψ_∞, because ψ_∞ then includes the $u \to \infty$ limit of the ground state of the Hubbard model (which becomes unique for infinitesimal positive $1/u$). Ogata and Shiba identified this limit state and used it to study the ground state one-point Green functions and spin-spin correlation function of the Hubbard model in the strong coupling limit numerically [344] (see also [345]). They showed [344] that the limit state gives the ground-state energy correctly to the order $1/u$ (rather than 1).

It is, of course, possible to expand the whole Bethe ansatz solution of the Hubbard model systematically in $1/u$. To give an example of how this works let us calculate the leading $1/u$ corrections to the energy eigenvalues. They turn out to be fairly simple.

We assume we are given a solution of the Lieb-Wu equations (3.95), (3.96) having an asymptotic expansion of the form

$$k_j^{(LW)} = k_j + \frac{\Delta k_j}{u} + \mathcal{O}(1/u^2), \quad j = 1, \ldots, N, \tag{3.G.9}$$

$$\lambda_\ell^{(LW)} = u\Lambda_\ell + \Delta\Lambda_\ell + \mathcal{O}(1/u), \quad \ell = 1, \ldots, M. \tag{3.G.10}$$

Inserting this solution into the Lieb-Wu equations and comparing order by order in $1/u$ we find that the k_j and Λ_ℓ satisfy equations (3.G.1) and (3.G.2). The first order corrections Δk_j

---

[2] This observation may lead us to the interpretation of $t(s_{P(1)} - iu|\mathbf{k}P)$ as a shift operator for the inhomogeneous system. The precise sense in which this is true can be seen in appendix 12.B.

and $\Delta\Lambda_\ell$ satisfy a set of linear equations with coefficients depending on k_j and Λ_ℓ,

$$\Delta k_j L = 2 \sum_{\ell=1}^{M} \frac{\Delta\Lambda_\ell - \sin(k_j)}{\Lambda_\ell^2 + 1}, \quad j = 1, \ldots, N, \tag{3.G.11}$$

$$\sum_{j=1}^{N} \frac{\Delta\Lambda_\ell - \sin(k_j)}{\Lambda_\ell^2 + 1} = 2 \sum_{m=1}^{M} \frac{\Delta\Lambda_\ell - \Delta\Lambda_m}{(\Lambda_\ell - \Lambda_m)^2 + 4}, \quad \ell = 1, \ldots, M. \tag{3.G.12}$$

Using (3.G.12) we can eliminate the variables $\Delta\Lambda_\ell$ from (3.G.11). We obtain

$$\Delta k_j = \frac{e_s}{L} \sum_{n=1}^{N} (\sin(k_j) - \sin(k_n)), \tag{3.G.13}$$

where

$$e_s = -\frac{2}{N} \sum_{\ell=1}^{M} \frac{1}{\Lambda_\ell^2 + 1}, \tag{3.G.14}$$

which the reader may know as the energy per lattice site of the XXX Heisenberg chain for antiferromagnetic exchange coupling $J = 1$. Using (3.G.13) and (3.G.9) we obtain the energy eigenvalues of the Hubbard Hamiltonian (2.22) up to the order $1/u$,

$$E = -2 \sum_{j=1}^{N} \cos(k_j) + \frac{2e_s}{u} \left[\frac{N}{L} \sum_{j=1}^{N} \sin^2(k_j) - \frac{1}{L} \left(\sum_{j=1}^{N} \sin(k_j) \right)^2 \right]. \tag{3.G.15}$$

These are also the energy eigenvalues of the t-J Hamiltonian (2.A.33) up to the order $1/u$. At half-filling $N = L$ the k_j are uniquely fixed (for given P_s) by equation (3.G.4) and the above expression for the energy simplifies to

$$\frac{E}{L} = \frac{e_s}{u} = -\frac{2}{uL} \sum_{\ell=1}^{M} \frac{1}{\Lambda_\ell^2 + 1}. \tag{3.G.16}$$

Comparing with the results of appendix 2.A, in particular with the expression (2.A.37), we see that we have indeed solved the isotropic Heisenberg chain. Its energy eigenvalues are given by (3.G.16), where the spin rapidities satisfy the Bethe ansatz equations (3.G.2).

3.G.2 Continuum limit and Bethe ansatz solution of the model of electrons with delta-function interaction

In appendix 2.B we showed that the Hubbard model can be viewed as a lattice regularization of the continuum model of electrons with mutual delta-function interaction. The Heisenberg equation of motion for the field operators of this model is called the non-linear Schrödinger equation. The corresponding Hamiltonian is

$$H_{NLS} = \int_0^\ell dx \left[\left(\partial_x \Psi_a^\dagger(x) \right) \partial_x \Psi_a(x) \pm 2c\, \Psi_a^\dagger(x) \Psi_b^\dagger(x) \Psi_b(x) \Psi_a(x) \right]. \tag{3.G.17}$$

In appendix 2.B we obtained this Hamiltonian from the Hubbard Hamiltonian by an appropriate rescaling of the fields and variables. We introduced a lattice constant a_0 and considered it to be small. Discrete space variables n were rescaled as $x = a_0 n$ and momenta on the lattice k as $q = k/a_0$. The total length ℓ of the continuum model is therefore related to the number of lattice sites L as $\ell = a_0 L$. We further found out that we have to rescale the Hubbard coupling u as $c = u/a_0$. Let us also introduce rescaled spin rapidities $\mu = \lambda/a_0$, insert the new variables into the Lieb-Wu equations (3.95), (3.96) and perform the limit $a_0 \to 0$. We obtain

$$
e^{iq_j \ell} = \prod_{\ell=1}^{M} \frac{\mu_\ell - q_j - ic}{\mu_\ell - q_j + ic}, \qquad j = 1, \ldots, N, \tag{3.G.18}
$$

$$
\prod_{j=1}^{N} \frac{\mu_\ell - q_j - ic}{\mu_\ell - q_j + ic} = \prod_{\substack{m=1 \\ m \neq \ell}}^{M} \frac{\mu_\ell - \mu_m - 2ic}{\mu_\ell - \mu_m + 2ic}, \qquad \ell = 1, \ldots, M. \tag{3.G.19}
$$

The solutions q_j of these equations determine the energy eigenvalues of the Hamiltonian (3.G.17) which are obtained by using the expression (3.97) for the energy of the Hubbard model in (2.B.19) and taking the limit $a_0 \to 0$:

$$
E = \sum_{j=1}^{N} q_j^2. \tag{3.G.20}
$$

With no further effort one also obtains the eigenfunctions of the NLS Hamiltonian from those of the Hubbard Hamiltonian (see (3.91)–(3.94)) by first rescaling and then taking the limit of vanishing lattice constant. The only structural change in the formulae is that the functions $\sin k_j$ get replaced by q_j. In particular, the form of the wavefunctions in the continuum ψ^c remains the same as on the lattice,

$$
\psi^c(\mathbf{x}; \mathbf{a} | \mathbf{q}; \boldsymbol{\mu}) = \sum_{P \in \mathfrak{S}^N} \text{sign}(PQ) \langle \mathbf{a}Q | \mathbf{q}P, \boldsymbol{\mu} \rangle \, e^{i\langle \mathbf{q}P, \mathbf{x}Q \rangle}, \tag{3.G.21}
$$

with $\mathbf{q} = (q_1, \ldots, q_N)$ and $\boldsymbol{\mu} = (\mu_1, \ldots, \mu_M)$. The spin dependent amplitudes $\langle \mathbf{a}Q | \mathbf{q}P, \boldsymbol{\mu} \rangle$ are now

$$
\langle \mathbf{a}Q | \mathbf{q}P, \boldsymbol{\mu} \rangle = \sum_{R \in \mathfrak{S}^M} A(\boldsymbol{\mu}R) \prod_{\ell=1}^{M} F_{\mathbf{q}P}(\mu_{R(\ell)}; y_\ell), \tag{3.G.22}
$$

where $F_{\mathbf{q}}(\mu; y)$ is

$$
F_{\mathbf{q}}(\mu; y) = \frac{2ic}{\mu - q_y + ic} \prod_{j=1}^{y-1} \frac{\mu - q_j - ic}{\mu - q_j + ic}, \tag{3.G.23}
$$

and where the amplitudes $A(\boldsymbol{\mu})$ of the auxiliary spin model wavefunctions are given by

$$
A(\boldsymbol{\mu}) = \prod_{1 \leq m < n \leq M} \frac{\mu_m - \mu_n - 2ic}{\mu_m - \mu_n}. \tag{3.G.24}
$$

The variables y_ℓ are again coordinates of down spins on electrons. Their values depend on **a** and Q.

The above results were first obtained by M. Gaudin [154] and C. N. Yang [493] by directly applying the ideas of the nested Bethe ansatz as described in appendix 3.B to the Hamiltonian (3.G.17). Here we obtained them as an extra bonus out of the Bethe ansatz solution of the Hubbard model. Starting from the Bethe ansatz equations (3.G.18), (3.G.19) and the expression (3.G.20) we could develop the theory of the one-dimensional electron gas with delta-function interaction in much the same way as we develop the theory of the one-dimensional Hubbard model in the following chapters of this book.

Let us conclude our digression on the NLS Hamiltonian with the 'derivation' of a conjecture for the norm of the eigenfunctions which cannot be found in the literature. We first observe (see (2.B.5)) that the continuum analogue of the Wannier states is $|\mathbf{x}, \mathbf{a}\rangle_c = \Psi^\dagger_{a_N}(x_N) \ldots \Psi^\dagger_{a_1}(x_1)|0\rangle = \lim_{a_0 \to 0} a_0^{-\frac{N}{2}} |\mathbf{x}/a_0, \mathbf{a}\rangle$. On the other hand, the lattice constant a_0 plays the role of the volume element dx_j when turning from summation to integration in the continuum limit (see (2.B.7)). This means that the eigenstates in the continuum, $|\psi^c_{\mathbf{q},\mu}\rangle$, are obtained as

$$|\psi^c_{\mathbf{q},\mu}\rangle = \lim_{a_0 \to 0} a_0^{\frac{N}{2}} |\Psi_{a_0\mathbf{q},a_0\mu}\rangle = \frac{1}{N!} \int_0^\ell dx_1 \cdots \int_0^\ell dx_N \sum_{\mathbf{a} \in \mathbb{Z}_2^N} \psi^c(\mathbf{x}; \mathbf{a}|\mathbf{q}; \mu)|\mathbf{x}, \mathbf{a}\rangle_c . \quad (3.G.25)$$

Hence, we can calculate the squared norm $\|\psi^c_{\mathbf{q},\mu}\|^2 = \langle \psi^c_{\mathbf{q},\mu}|\psi^c_{\mathbf{q},\mu}\rangle$ of the eigenstates in the continuum limit as

$$\|\psi^c_{\mathbf{q},\mu}\|^2 = \lim_{a_0 \to 0} a_0^N \langle \Psi_{a_0\mathbf{q},a_0\mu}|\Psi_{a_0\mathbf{q},a_0\mu}\rangle . \quad (3.G.26)$$

This limit can be easily performed in (3.120),

$$\|\psi^c_{\mathbf{q},\mu}\|^2 = |2c|^M \left|\det \begin{pmatrix} \mathcal{S}^c_{qq} & \mathcal{S}^c_{q\mu} \\ \mathcal{S}^c_{\mu q} & \mathcal{S}^c_{\mu\mu} \end{pmatrix}\right| \prod_{1 \le j < k \le M} \left[1 + \frac{4c^2}{(\mu_j - \mu_k)^2}\right]. \quad (3.G.27)$$

The determinant on the right hand side of this equation is the determinant of an $(N + M) \times (N + M)$-matrix consisting of four blocks with matrix elements

$$\left(\mathcal{S}^c_{qq}\right)_{mn} = \delta_{m,n}\left[\ell + \sum_{l=1}^M \frac{2c}{c^2 + (\mu_l - q_n)^2}\right], \quad m, n = 1, \ldots, N, \quad (3.G.28)$$

$$\left(\mathcal{S}^c_{q\mu}\right)_{mn} = \left(\mathcal{S}^c_{\mu q}\right)_{nm} = -\frac{2c}{c^2 + (\mu_m - q_n)^2}, \quad m = 1, \ldots, M, \quad n = 1, \ldots, N, \quad (3.G.29)$$

$$\left(\mathcal{S}^c_{\mu\mu}\right)_{mn} = \delta_{m,n}\left[\sum_{j=1}^N \frac{2c}{c^2 + (\mu_n - q_j)^2} - \sum_{l=1}^M \frac{4c}{(2c)^2 + (\mu_n - \mu_l)^2}\right]$$
$$+ \frac{4c}{(2c)^2 + (\mu_m - \mu_n)^2}, \quad m, n = 1, \ldots, M. \quad (3.G.30)$$

The determinant in (3.G.27) may again be interpreted as the Hessian determinant of a properly defined action $S^c(\mathbf{q}; \boldsymbol{\mu})$ which generates the Bethe ansatz equations (3.G.18) and (3.G.19) and is obtained from $S(\mathbf{k}; \boldsymbol{\lambda})$, equation (3.114), in the limit

$$S^c(\mathbf{q}; \boldsymbol{\mu}) = \lim_{a_0 \to 0} \frac{S(a_0\mathbf{q}; a_0\boldsymbol{\mu}) - NL}{a_0}. \tag{3.G.31}$$

We leave it as an exercise to the reader to write this action explicitly.

3.G.3 Weak coupling limit

For $u \to 0$ the Hubbard Hamiltonian (2.31) turns into the tight-binding Hamiltonian H_0 (equation (2.12) with $t = 1$) which describes free electrons on the lattice (see (2.20)). What happens to the Lieb-Wu equations (3.95), (3.96) and to the Bethe ansatz wave function (3.91) in this limit?

The charge momenta k_j stay finite in the free electron limit. It follows that the function $F_\mathbf{k}(\lambda; y)$, equation (3.93), vanishes and therefore also the spin part $\langle \mathbf{a}Q|\mathbf{k}P, \boldsymbol{\lambda} \rangle$ of the Bethe ansatz wave function. In order to obtain a non-vanishing weak-coupling limit we change the normalization and first multiply the Bethe ansatz wave function by $(2iu)^{-M}$. Then it follows that

$$\lim_{u \to 0} \frac{\langle \mathbf{a}Q|\mathbf{k}P, \boldsymbol{\lambda} \rangle}{(2iu)^M} = \sum_{R \in \mathfrak{S}^M} \prod_{\ell=1}^{M} \frac{1}{\lambda_{R(\ell)} - \sin(k_{P(y_\ell)})}. \tag{3.G.32}$$

The functions on the right-hand side are the coordinate Bethe ansatz wave functions of the (rational) inhomogeneous Gaudin model [156]. Performing the same limit in the Lieb-Wu equations we obtain

$$e^{ik_j L} = 1, \quad j = 1, \dots, N, \tag{3.G.33a}$$

$$\sum_{j=1}^{N} \frac{1}{\lambda_\ell - \sin(k_j)} = \sum_{\substack{m=1 \\ m \neq \ell}}^{M} \frac{2}{\lambda_\ell - \lambda_m}, \quad \ell = 1, \dots, M. \tag{3.G.33b}$$

Equation (3.G.33a) determines the spectrum to be free fermionic. Equations (3.G.33b), on the other hand, are the Bethe ansatz equations of the inhomogeneous Gaudin model. Here they determine the spin part (3.G.32) of the Bethe ansatz wave functions of the Hubbard model for vanishing coupling $u \to 0$.

Our above results indicate that the $u \to 0$ limit of the Bethe ansatz solution of the Hubbard model is rather non-trivial. The Bethe ansatz equations (3.G.33) decouple, but the wave functions do not turn into Slater determinants, as one might have expected for free electrons. They rather involve the eigenfunctions of an anisotropic Gaudin model as the spin parts of wave functions which are still of Bethe ansatz form. This appears less puzzling if one considers the large degeneracy of a system of free fermions which makes it possible to construct more than one basis of eigenfunctions.

Further aspects of the limit $u \to 0$ that were discussed in the literature and that will be taken up in later chapters are the behaviour of the ground-state energy [113, 326, 431] (see Chapter 6) and of the elementary excitations [142].

4

String hypothesis

Eigenstates of the Hubbard Hamiltonian are described in terms of the solutions of the Lieb-Wu equations (3.95), (3.96)

$$e^{ik_j L} = \prod_{\ell=1}^{M} \frac{\lambda_\ell - \sin k_j - iu}{\lambda_\ell - \sin k_j + iu}, \quad j = 1, \ldots, N,$$

$$\prod_{j=1}^{N} \frac{\lambda_\ell - \sin k_j - iu}{\lambda_\ell - \sin k_j + iu} = \prod_{\substack{m=1 \\ m \neq \ell}}^{M} \frac{\lambda_\ell - \lambda_m - 2iu}{\lambda_\ell - \lambda_m + 2iu}, \quad \ell = 1, \ldots, M. \tag{4.1}$$

In the following we will concentrate on *regular* solutions of these equations as defined in Section 3.3.2. Physical quantities like energy and momentum are expressed directly in terms of the roots of the Lieb-Wu equations, which we also call *spectral parameters*. These roots are in general *complex* numbers. The problem we are faced with now is that there is no simple analytical or numerical method for solving a large number of coupled, nonlinear, algebraic equations. As long as the number of roots is small, say four or five, one may determine all solutions by numerical means (see for example Ref. [95]). However, we are ultimately interested in the thermodynamic limit and therefore need another way of analyzing the Lieb-Wu equations. It turns out that for very large lengths L of the lattice most of the roots of (4.1) arrange themselves in regular patterns in the complex plane. These patterns are called 'strings'. We will describe shortly how to find these patterns. If one makes the assumption that all roots of (4.1) form strings one can turn (4.1) into a set of equations involving only the real parts of the roots. This procedure is known as the 'string hypothesis' and was first introduced for the case of the spin-$\frac{1}{2}$ Heisenberg model by H. Bethe in [60], see also [433]. Finally, the resulting equations are turned into a set of coupled integral equations, which can be solved analytically in various limits. This approach is the standard way of analyzing Bethe ansatz equations in integrable models and is believed to give exact results in the thermodynamic limit.

For finite systems one may easily establish that the string hypothesis is not strictly correct. While most of the roots indeed arrange themselves into strings, there sometimes are significant deviations from the ideal patterns and there also are solutions that cannot be described in the framework of the string hypothesis. An instructive illustration of these issues for the case of the spin-$\frac{1}{2}$ Heisenberg chain can be found in Refs. [28,29,46,79,94, 124,151,208,228,328,343]. A crucial point is that as far as the isotropic Heisenberg chain and the Hubbard model are concerned, the string hypothesis appears to give exact results for thermodynamic quantities like the (free) energy per site, dispersions of excited states in the infinite volume, or dressed scattering matrices as long as we consider only situations in which we have finite densities of Bethe roots in the infinite volume limit. However, even in these cases the string hypothesis is not free of problems as was shown for the Heisenberg chain in Refs. [30,483,484]. In the thermodynamic limit at finite temperatures the validity of the string hypothesis was demonstrated by A. M. Tsvelik and P. B. Wiegmann in [460].

Violations of the string hypothesis can be quite important if we are interested in the corrections to the energies in a large but finite volume. These finite-size corrections are very important as it is possible to determine the large-distance asymptotics of correlation functions from them (see Chapter 8). In integrable models like the integrable spin-S Takhtajan-Babujian model [31,444], where the zero-temperature ground state involves string solutions, it is necessary to take into account (small) deviations from the string hypothesis in the finite volume in order to obtain the correct finite-size corrections to energy levels [16,145,146]. For the Hubbard model this particular type of complication is absent.

4.1 String configurations

Let us now discuss how to find the particular patterns into which the roots of the Lieb-Wu equations arrange themselves. The basic idea is very simple: take L large and then make one of the spectral parameters complex. The left-hand side of the equation for that particular spectral parameter will then be exponentially large and this implies that we must be very close to a pole in one of the factors on the right-hand side of the equations.

4.1.1 k-Λ strings

Let us look at a specific example, namely $N = 2$, $M = 1$, corresponding to a state with one spin-up and one spin-down electron. We denote the corresponding spectral parameters by k_1, k_2 and Λ'. Let us now take

$$k_1 = q - i\xi \,, \tag{4.2}$$

where q, ξ are real and $\xi > 0$. The Lieb-Wu equations are

$$e^{ik_1 L} = \frac{\Lambda' - \sin(k_1) - iu}{\Lambda' - \sin(k_1) + iu} \,, \tag{4.3}$$

$$e^{ik_2 L} = \frac{\Lambda' - \sin(k_2) - iu}{\Lambda' - \sin(k_2) + iu} \,, \tag{4.4}$$

$$1 = \prod_{j=1}^{2} \frac{\Lambda' - \sin(k_j) - iu}{\Lambda' - \sin(k_j) + iu} \,. \tag{4.5}$$

As $\xi > 0$ the left hand side of equation (4.3) is exponentially large ($\propto \exp(\xi L)$) for large L. The only way to fulfil (4.3) is for the r.h.s. to be exponentially close to a pole, i.e. k_1

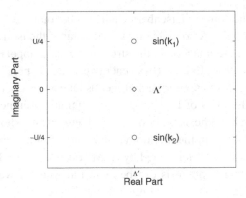

Fig. 4.1. A simple k-Λ string.

must be related to Λ' by

$$\Lambda' = \sin(k_1) - iu + \mathcal{O}(e^{-\xi L}). \tag{4.6}$$

However, this implies that the first factor on the right hand side of (4.5) is exponentially large, which in turn forces the second factor to be very close to zero. This fixes k_2 to be

$$\sin(k_2) = \Lambda' - iu + \mathcal{O}(e^{-\xi L}). \tag{4.7}$$

We may replace the Lieb-Wu equations (4.3) and (4.4) with exponential accuracy by (4.6) and (4.7). The remaining Lieb-Wu equation (4.5) then determines Λ'. Putting everything together (and taking care of the fact that $\xi > 0$) we have found a 'k-Λ string' solution of the form

$$k_1 = \pi - \arcsin(\Lambda' + iu) + \mathcal{O}(e^{-\delta L}),$$
$$k_2 = \pi - \arcsin(\Lambda' - iu) + \mathcal{O}(e^{-\delta L}),$$
$$\Lambda' \quad \text{real}, \tag{4.8}$$

where δ is some constant and the branch of $\arcsin(x)$ is fixed as $-\pi/2 \leq \mathrm{Re}(\arcsin(x)) \leq \pi/2$. Why do we call this pattern a *string*? In order to see this, let us consider the quantities $\sin(k_{1,2})$ rather than $k_{1,2}$. As is shown in figure 4.1, $\sin(k_{1,2})$ and Λ' are arranged like pearls on a string in the complex plane. The above analysis is straightforwardly generalized: one finds that $2m$ k's can combine with m Λ's to form a 'k-Λ string of length $2m$', which has the form ($u > 0$)

$$k^1 = \pi - \arcsin(\Lambda'^m + miu),$$
$$k^2 = \arcsin(\Lambda'^m + (m - 2)iu),$$
$$k^3 = \pi - k^2,$$
$$\vdots \tag{4.9}$$
$$k^{2m-2} = \arcsin(\Lambda'^m - (m - 2)iu),$$
$$k^{2m-1} = \pi - k^{2m-2},$$
$$k^{2m} = \pi - \arcsin(\Lambda'^m - miu),$$

and

$$\Lambda'^{m,j} = \Lambda'^m + (m - 2j + 1)iu\,, \quad j = 1, \ldots m. \tag{4.10}$$

Here m denotes the 'length' of the string and j counts the λ's involved in a given string. Λ'^m is the real center of the k-Λ string.

4.1.2 A composition principle

The simple calculation above has shown us that there exist particular solutions of the Lieb-Wu equations that are very well approximated by strings. It is not difficult to see that such string solutions may be combined to form other solutions of the Lieb-Wu equations. In other words, strings are the basic building blocks that make up general solutions.

Let us illustrate this 'composition principle' for the case $L = 6$, $N = 5$, $M = 2$ and $u = 1.25$, i.e. two spin-down and three spin-up electrons on a six-site lattice. The roots of the Lieb-Wu equations can be easily obtained by a numerical root-finding algorithm such as Newton's method. One solution is

$$\Lambda'_1 = 1.4230\,, \quad k_{1,2} = 2.3616 \pm 1.3310\,i\,,$$
$$\sin(k_{1,2}) = 1.42385 \pm 1.25142\,i\,,$$
$$\Lambda'_2 = -0.0484\,, \quad k_{3,4} = 3.1665 \pm 1.0499\,i\,,$$
$$\sin(k_{3,4}) = -0.0400 \pm 1.2534\,i,$$
$$k_5 = -0.5842. \tag{4.11}$$

We may consider this solution to be composed of two k-Λ strings of length two (as is easily verified by comparison with (4.13)) and one real k, namely k_5.

A second solution is

$$\Lambda_1 = 0.5402\,, \quad k_{1,2} = 2.8085 \pm 1.0870\,i\,,$$
$$\sin(k_{1,2}) = 0.5399 \pm 1.2419\,i\,,$$
$$k_3 = -1.7305\,, \quad k_4 = 0.7330\,, \quad k_5 = 1.6637\,,$$
$$\Lambda_2 = -0.7253. \tag{4.12}$$

It is composed of one k-Λ string of length two, one real Λ (a 'one-string' of Λ's) and three real k's, namely k_3, k_4, k_5.

As a given solution of the Lieb-Wu equations may 'contain' several k-Λ-strings (our first example above contains two) we introduce an index α to distinguish between them and denote a k-Λ string of length m by

$$k_\alpha^1 = \pi - \arcsin(\Lambda'^m_\alpha + miu),$$
$$k_\alpha^2 = \arcsin(\Lambda'^m_\alpha + (m-2)iu),$$
$$k_\alpha^3 = \pi - k_\alpha^2,$$

$$\vdots \tag{4.13}$$

$$k_\alpha^{2m-2} = \arcsin(\Lambda'^m_\alpha - (m-2)iu),$$
$$k_\alpha^{2m-1} = \pi - k_\alpha^{2m-2},$$
$$k_\alpha^{2m} = \pi - \arcsin(\Lambda'^m_\alpha - miu),$$

and

$$\Lambda'^{m,j}_\alpha = \Lambda'^m_\alpha + (m-2j+1)iu\,, \quad j = 1, \ldots m. \tag{4.14}$$

4.1.3 Λ strings

It is clear from our construction that (4.13) are the only types of strings involving k's. However, there are also strings involving only Λ's. In order to see this, let us consider a situation where there are N real k_j's, where N is very large. The Lieb-Wu equations for the M Λ's, where we assume that $M \ll N$, are

$$\prod_{j=1}^N \frac{\Lambda_l - \sin(k_j) - iu}{\Lambda_l - \sin(k_j) + iu} = \prod_{\substack{m=1 \\ m\neq l}}^M \frac{\Lambda_l - \Lambda_m - 2ui}{\Lambda_l - \Lambda_m + 2ui}\,. \tag{4.15}$$

If we take $\mathrm{Im}(\Lambda_1) > 0$, then the left-hand side of the equation for Λ_1 tends to zero exponentially fast in N. As $M \ll N$ this forces one of the factors on the right-hand side of the equation to be very small. Without loss of generality we may take

$$\Lambda_2 = \Lambda_1 - 2ui + \mathcal{O}(e^{-\gamma N})\,, \tag{4.16}$$

where γ is some constant. Repeating these arguments for Λ_2 and so on we find that it is possible for m Λ's to form a 'Λ string', that is a configuration of roots that is symmetric around the real axis and has a spacing of $2iu$ between consecutive roots. We will denote such strings by

$$\Lambda^{m,j}_\alpha = \Lambda^m_\alpha + (m-2j+1)iu + \mathcal{O}(e^{-\gamma N}). \tag{4.17}$$

Here α distinguishes the strings of the same length m, and $j = 1, \ldots, m$ counts the λ's involved in the αth Λ string of length m. Λ^m_α is the real center of the string. The solutions (4.17) and (4.13), (4.14) were first found by M. Takahashi in [435]. The fact that most solutions of the Lieb-Wu equations are well-described in terms of strings has been established for small lattices by solving the Lieb-Wu equations numerically, see e.g. [95].

4.2 String solutions as bound states

It is clear from the form of the Bethe ansatz wave function (3.91)–(3.94) that solutions of the Lieb-Wu equations with only real k_j's and λ_ℓ's correspond to eigenstates of the Hubbard Hamiltonian that are superpositions of plane waves. By their very construction string solutions involve complex solutions of the Lieb-Wu equations. What then is the physical nature of the eigenstates of the Hubbard Hamiltonian described by string solutions? The answer is that they describe *bound states* of electrons in the sense that the many-body wave function exhibits an exponential decay with respect to the difference of the coordinates of electrons involved in a string. In the remainder of this section we illustrate this point by considering simple examples.

4.2.1 k-Λ strings

Let us consider a two electron wave function where one of the electrons has spin up and one spin down respectively ($N = 2$, $M = 1$). The Lieb-Wu equations are given by

$$e^{ik_j L} = \frac{\Lambda - \sin(k_j) - iu}{\Lambda - \sin(k_j) + iu} \qquad j = 1, 2\,,$$

$$1 = \prod_{j=1}^{2} \frac{\Lambda - \sin(k_j) - iu}{\Lambda - \sin(k_j) + iu}\,. \tag{4.18}$$

We concentrate on solutions to (4.18) which in the limit $L \to \infty$ take the form of a k-Λ string[1]

$$k_1 = \pi - \arcsin(\Lambda + iu)\,,$$

$$k_2 = \pi - \arcsin(\Lambda - iu)\,. \tag{4.19}$$

For finite but large L there are deviations from (4.19), which are evaluated in detail in Appendix 4.A. There it is shown that the solution of (4.18) for finite L can be expressed as

$$k_1 = q - i\xi = k_2^*\,, \quad \xi > 0\,,$$

$$\Lambda = \sin(q)\cosh(\xi)\,, \tag{4.20}$$

where $q = \frac{m\pi}{L}$, $m = -L + 1, \dots, L$ and ξ is subject to the equation

$$\sinh(\xi) = -\frac{u}{\cos(q)} \frac{\sinh(\xi L)}{\cosh(\xi L) - (-1)^m}\,. \tag{4.21}$$

For $L \gg \max(\frac{2}{u}, 1)$ equations (4.20)–(4.21) imply that

$$\sin(k_1) = \Lambda + iu + i\epsilon = (\sin(k_2))^*\,, \tag{4.22}$$

where ϵ is exponentially small in L

$$\epsilon = \mathcal{O}(e^{-\xi L})\,. \tag{4.23}$$

---

[1] For brevity we denote the center of the k-Λ string by Λ rather than Λ'.

The corresponding unnormalized wave function is obtained from the general formulae (3.91)–(3.94).

$$\psi(x_1, x_2; \downarrow, \uparrow \mid k_1, k_2; \Lambda) = \theta_H(x_2 - x_1) \left[\frac{2iu\, e^{ik_1 x_1 + ik_2 x_2}}{\Lambda - \sin(k_1) + iu} - \frac{2iu\, e^{ik_2 x_1 + ik_1 x_2}}{\Lambda - \sin(k_2) + iu} \right]$$

$$+ \theta_H(x_1 - x_2) \left[\frac{2iu\, e^{ik_1 x_1 + ik_2 x_2}}{\Lambda - \sin(k_1) + iu} \frac{\Lambda - \sin(k_2) - iu}{\Lambda - \sin(k_2) + iu} \right.$$

$$\left. - \frac{2iu\, e^{ik_2 x_1 + ik_1 x_2}}{\Lambda - \sin(k_2) + iu} \frac{\Lambda - \sin(k_1) - iu}{\Lambda - \sin(k_1) + iu} \right], \tag{4.24}$$

where $\theta_H(x)$ is the Heaviside function. Using (4.22) and (4.23) we can evaluate the wave function explicitly

$$\psi(x_1, x_2; \downarrow, \uparrow \mid k_1, k_2; \Lambda) = \frac{-2u}{\epsilon} \left[\theta_H(x_2 - x_1)\, e^{ik_1 x_1 + ik_2 x_2} \right.$$

$$\left. + \theta_H(x_1 - x_2)\, e^{ik_2 x_1 + ik_1 x_2} \right] + \cdots \tag{4.25}$$

Taking the limit $L \to \infty$ while keeping $|x_1 - x_2|$ fixed, we may drop the subleading terms indicated by the dots in (4.25) and arrive at the simple result

$$\psi(x_1, x_2; \downarrow, \uparrow \mid k_1, k_2; \Lambda) = A\, e^{iq(x_1 + x_2)}\, e^{-\xi |x_1 - x_2|}, \tag{4.26}$$

where A is a normalization constant. Equation (4.26) shows that the k-Λ string indeed describes a bound state of two electrons. It is shown in Section 7.9 that the energy of the bound state (4.26) is *higher* than the energy of a scattering state of two electrons. This sounds very strange! However, in our previous discussion of the wave-function (4.26) in Section 3.2.4 we showed that the bound-state property is a *lattice effect* and does not survive a continuum limit.

Longer k-Λ strings also describe bound states as may be shown by an analogous calculation, see e.g. the Appendix of [123].

4.2.2 Λ strings

Let us consider two down-spin electrons forming a Λ string and $N - 2$ up-spin electrons, i.e. a solution to the Lieb-Wu equations (4.1) with

$$\Lambda_1 = \Lambda_2^* = \lambda - iu + i\epsilon, \ \lambda \in \mathbb{R},$$

$$k_j \in \mathbb{R}, \ j = 1, \dots, N, \tag{4.27}$$

where $|\epsilon| \ll 1$. Let us denote the positions of the down spins *among the electrons* by y_1 and $y_2 > y_1$ as illustrated in figure 4.2. What we want to show is that if we take N to infinity while keeping the 'distance' $d = y_2 - y_1$ fixed, then the wave function corresponding to (4.27) decays exponentially in d. In other words, the two down spins form a bound state (in the sense defined above) on the lattice formed by the electrons.

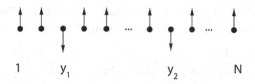

Fig. 4.2. The two down spins on the lattice formed by the N electrons.

The wave function associated with the roots (4.27) in the sector Q defined by $1 \leq x_{Q(1)} \leq x_{Q(2)} \leq \cdots \leq x_{Q(N)}$ is of the form

$$\psi(\mathbf{x}; \mathbf{a} | \mathbf{k}; \Lambda) = \sum_{P \in \mathfrak{S}_N} \text{sign}(PQ) \langle \mathbf{a}Q | \mathbf{k}P, \Lambda \rangle \, e^{i \sum_{j=1}^{N} k_{P_j} x_{Q_j}}, \tag{4.28}$$

where we have defined vectors

$$\mathbf{x} = (x_1, \ldots, x_N), \quad \mathbf{k} = (k_1, \ldots, k_N),$$
$$\mathbf{a} = (\downarrow, \downarrow, \uparrow, \ldots, \uparrow), \quad \Lambda = (\Lambda_1, \Lambda_2), \tag{4.29}$$

and the amplitudes $\langle \mathbf{a}Q | \mathbf{k}P, \Lambda \rangle$ are given by

$$\langle \mathbf{a}Q | \mathbf{k}P, \Lambda \rangle = A(\Lambda_1, \Lambda_2) \Big\{ F_{\mathbf{k}P}(\Lambda_1; y_1) F_{\mathbf{k}P}(\Lambda_2; y_2)$$
$$+ \frac{A(\Lambda_2, \Lambda_1)}{A(\Lambda_1, \Lambda_2)} F_{\mathbf{k}P}(\Lambda_2; y_1) F_{\mathbf{k}P}(\Lambda_1; y_2) \Big\}. \tag{4.30}$$

The terms on the r.h.s. of (4.30) are evaluated as follows. Combining the definition (3.94) of $A(\Lambda)$ with the Lieb-Wu equations (3.96) we obtain

$$\frac{A(\Lambda_2, \Lambda_1)}{A(\Lambda_1, \Lambda_2)} = \frac{\Lambda_2 - \Lambda_1 - 2iu}{\Lambda_2 - \Lambda_1 + 2iu} = \prod_{j=1}^{N} \frac{\Lambda_2 - \sin(k_j) - iu}{\Lambda_2 - \sin(k_j) + iu}. \tag{4.31}$$

The $F_{\mathbf{k}P}$-functions are defined by (3.93) and with the help of the Lieb-Wu equations (3.95) their products in (4.30) can be cast in the form

$$F_{\mathbf{k}P}(\Lambda_1; y_1) F_{\mathbf{k}P}(\Lambda_2; y_2) = \exp\left(i \sum_{j=1}^{y_1-1} k_{P(j)} L \right) \prod_{l=y_1}^{y_2-1} \frac{\Lambda_2 - \sin(k_{P(l)}) - iu}{\Lambda_2 - \sin(k_{P(l)}) + iu}$$
$$\times \frac{(2iu)^2}{[\Lambda_1 - \sin(k_{P(y_1)}) + iu][\Lambda_2 - \sin(k_{P(y_2)}) + iu]}, \tag{4.32}$$

$$F_{\mathbf{k}P}(\Lambda_2; y_1) F_{\mathbf{k}P}(\Lambda_1; y_2) = \exp\left(i \sum_{j=1}^{y_2-1} k_{P(j)} L \right) \prod_{l=y_1}^{y_2-1} \frac{\Lambda_2 - \sin(k_{P(l)}) + iu}{\Lambda_2 - \sin(k_{P(l)}) - iu}$$
$$\times \frac{(2iu)^2}{[\Lambda_2 - \sin(k_{P(y_1)}) + iu][\Lambda_1 - \sin(k_{P(y_2)}) + iu]}, \tag{4.33}$$

We want to show that the wave function (4.28) decays exponentially with respect to the difference of the coordinates y_1 and y_2. The only nontrivial y dependence is in the amplitudes

$\langle \mathbf{a}Q | \mathbf{k}P, \Lambda \rangle$ and it suffices to prove that they decay exponentially. Given that the momenta k_j are real by assumption we may easily establish the following inequality

$$\left| \frac{\Lambda_2 - \sin(k_j) - iu}{\Lambda_2 - \sin(k_j) + iu} \right| = \left| \frac{\lambda - \sin(k_j) - i\epsilon}{\lambda - \sin(k_j) + 2iu - i\epsilon} \right| < 1. \tag{4.34}$$

The inequality (4.34) ensures that (4.32) decays exponentially for $y_2 \gg y_1$. This implies that the first term on the r.h.s. of (4.30) exhibits the desired exponential decay. On the other hand, the second term on the r.h.s. of (4.30) vanishes in the limit $N \to \infty$ if we keep $y_2 - y_1$ fixed. This is established by multiplying (4.33) by (4.31) and then using the inequality (4.34).

4.3 Takahashi's equations

The string hypothesis for the Hubbard model was formulated by M. Takahashi in 1972 [435]. It postulates that all regular[2] solutions $\{k_j\}$, $\{\lambda_l\}$ of the Lieb-Wu equations (4.1) are composed of the three different classes of strings discussed above.

  (i) A single real momentum k_j.
 (ii) m Λ's combining into a Λ string (4.17). This includes the case $m = 1$, which is just a single Λ_α.
(iii) $2m$ k's and m Λ's combining into a k-Λ string (4.13), (4.14).

For large lattices ($L \gg 1$) and a large number of electrons ($N \gg 1$), almost all strings are close to ideal, i.e. the imaginary parts of the k's and λ's are almost equally spaced. The string hypothesis assumes that almost all solutions of the Lieb-Wu equations (4.1) are approximately given by (4.13), (4.14), (4.17) with exponentially small corrections of order $\mathcal{O}(\exp(-\delta L))$, where δ is real and positive and depends on the specific string under consideration.

Using the string hypothesis, the task of solving the Lieb-Wu equations can be significantly simplified. Let us consider solutions of the Lieb-Wu equations for some particular, fixed values of N and M. Within the framework of the string hypothesis, every solution can be represented in terms of a particular *configuration* of strings: it contains M_n Λ-strings and M'_n k-Λ strings of length n ($n = 1, 2, \ldots$) and $\mathcal{M}_e$ single k_j's (i.e. k's not associated with k-Λ strings). We call $\mathcal{M}_e$, M_n, M'_n *occupation numbers* of the string configuration under consideration. The occupation numbers satisfy the 'sum rules'

$$M = \sum_{n=1}^{\infty} n(M_n + M'_n)$$

$$N = \mathcal{M}_e + \sum_{n=1}^{\infty} 2n M'_n . \tag{4.35}$$

---

[2] We recall that regular solutions are defined by the property that all spectral parameters are finite, i.e. $|k_j| < \infty$, $\lambda_l < \infty$.

Applying this prescription to the Lieb-Wu equations (4.1) and then taking the logarithm of the resulting equations, we arrive at the following form of Bethe ansatz equations for the real 'centers' of the strings, which we call *Takahashi's equations*

$$k_j L = 2\pi I_j - \sum_{n=1}^{\infty} \sum_{\alpha=1}^{M_n} \theta \left(\frac{\sin k_j - \Lambda_\alpha^n}{nu} \right) - \sum_{n=1}^{\infty} \sum_{\alpha=1}^{M_n'} \theta \left(\frac{\sin k_j - \Lambda_\alpha'^n}{nu} \right), \quad (4.36)$$

$$\sum_{j=1}^{N-2M'} \theta \left(\frac{\Lambda_\alpha^n - \sin k_j}{nu} \right) = 2\pi J_\alpha^n + \sum_{m=1}^{\infty} \sum_{\beta=1}^{M_m} \Theta_{nm} \left(\frac{\Lambda_\alpha^n - \Lambda_\beta^m}{u} \right), \quad (4.37)$$

$$2L \operatorname{Re} \left[\arcsin(\Lambda_\alpha'^n + niu) \right] = 2\pi J_\alpha'^n + \sum_{j=1}^{N-2M'} \theta \left(\frac{\Lambda_\alpha'^n - \sin k_j}{nu} \right)$$

$$+ \sum_{m=1}^{\infty} \sum_{\beta=1}^{M_m'} \Theta_{nm} \left(\frac{\Lambda_\alpha'^n - \Lambda_\beta'^m}{u} \right). \quad (4.38)$$

Here we assumed the length of the lattice L to be even. The functions θ and Θ_{nm} in (4.36)–(4.38) are defined as $\theta(x) = 2\arctan(x)$, and

$$\Theta_{nm}(x) = \begin{cases} \theta \left(\frac{x}{|n-m|} \right) + 2\theta \left(\frac{x}{|n-m|+2} \right) + \cdots + 2\theta \left(\frac{x}{n+m-2} \right) + \theta \left(\frac{x}{n+m} \right), & \text{if } n \neq m \\ 2\theta \left(\frac{x}{2} \right) + 2\theta \left(\frac{x}{4} \right) + \cdots + 2\theta \left(\frac{x}{2n-2} \right) + \theta \left(\frac{x}{2n} \right), & \text{if } n = m. \end{cases}$$

$$(4.39)$$

I_j, J_α^n, and $J_\alpha'^n$ are integer or half-odd integer numbers that arise due to the multivaluedness of the logarithm. We have

$$I_j \text{ is } \begin{cases} \text{integer} & \text{if } \sum_m (M_m + M_m') \text{ is even} \\ \text{half-odd integer} & \text{if } \sum_m (M_m + M_m') \text{ is odd,} \end{cases} \quad (4.40)$$

$$J_\alpha^n \text{ is } \begin{cases} \text{integer} & \text{if } N - M_n \text{ is odd} \\ \text{half-odd integer} & \text{if } N - M_n \text{ is even,} \end{cases} \quad (4.41)$$

$$J_\alpha'^n \text{ is } \begin{cases} \text{integer} & \text{if } L - N + M_n' \text{ is odd} \\ \text{half-odd integer} & \text{if } L - N + M_n' \text{ is even.} \end{cases} \quad (4.42)$$

The integers M_n and M_m' are the numbers of Λ strings of length n, and k-Λ strings of length m in a specific solution of the system (4.36)–(4.38). Finally, M' is the total number of Λ's involved in k-Λ strings

$$M' = \sum_{n=1}^{\infty} n M_n'. \quad (4.43)$$

The integer (half-odd integer) numbers in (4.36)–(4.38) have ranges

$$-\frac{L}{2} < I_j \le \frac{L}{2}, \tag{4.44}$$

$$|J^n_\alpha| \le \frac{1}{2}\left(N - 2M' - \sum_{m=1}^{\infty} t_{nm} M_m - 1\right), \tag{4.45}$$

$$|J'^n_\alpha| \le \frac{1}{2}\left(L - N + 2M' - \sum_{m=1}^{\infty} t_{nm} M'_m - 1\right), \tag{4.46}$$

where

$$t_{mn} = 2\min(m,n) - \delta_{mn}. \tag{4.47}$$

The ranges (4.45)–(4.46) may be obtained by following Yang and Yang's procedure [496]. One defines so-called 'counting-functions', e.g. for Λ-strings of length n we define

$$Lz_n(\Lambda) = \sum_{j=1}^{N-2M'} \theta\left(\frac{\Lambda - \sin k_j}{nu}\right) - \sum_{m=1}^{\infty}\sum_{\beta=1}^{M_m} \Theta_{nm}\left(\frac{\Lambda - \Lambda^m_\beta}{u}\right). \tag{4.48}$$

By its definition the counting function evaluated on a root of Takahashi's equations fulfils

$$z_n(\Lambda^n_a) = \frac{2\pi J^n_\alpha}{L}. \tag{4.49}$$

The counting functions will play an important role later on, see Chapter 5. A crucial property of the counting functions is that they are monotonically increasing functions of their arguments. For our present purposes we note that by virtue of the monotonicity property we have

$$J^n_\alpha < \frac{L}{2\pi}\lim_{\Lambda\to\infty} z_n(\Lambda). \tag{4.50}$$

The limit $\Lambda \to \infty$ is easily evaluated, which leads to the result (4.45) once we make use of our knowledge of whether the J^n_α are integers or half-odd integers. The range for J'^n_α is obtained in an analogous way. Finally, the range of the I_j's follows from the observation that Takahashi's equations for k_j are invariant under shifting k_j by 2π.

In terms of the parameters of the ideal strings the total energy and momentum (3.97) are expressed as

$$P = \left[\sum_{j=1}^{N-2M'} k_j - \sum_{n=1}^{\infty}\sum_{\alpha=1}^{M'_n}\left(2\,\mathrm{Re}\arcsin\left(\Lambda'^n_\alpha + niu\right) - (n+1)\pi\right)\right] \mod 2\pi, \tag{4.51}$$

$$E = -2\sum_{j=1}^{N-2M'}\cos(k_j) + 4\sum_{n=1}^{\infty}\sum_{\alpha=1}^{M'_n}\mathrm{Re}\sqrt{1 - \left(\Lambda'^n_\alpha + niu\right)^2} + u(L - 2N). \tag{4.52}$$

Equations (4.36)–(4.46) can be used to study all excitations of the Hubbard model in the thermodynamic limit (see Chapter 7). They are the basis for the derivation of Takahashi's integral equations [435], which determine the thermodynamics of the Hubbard model (see Chapter 5). These and other applications of (4.36)–(4.46) are based on the following assumptions.

(i) Any set of non-repeating (half odd) integers I_j, J_α^n, $J_\alpha'^n$ subject to the constraints (4.44)–(4.46) specifies precisely one regular solution $\{k_j\}$, $\{\Lambda_\alpha^n\}$, $\{\Lambda_\alpha'^n\}$ of equations (4.36)–(4.38).

(ii) The solutions $\{k_j\}$, $\{\Lambda_\alpha^n\}$, $\{\Lambda_\alpha'^n\}$ of (4.36)–(4.38) specified by a set of non-repeating (half odd) integers I_j, J_α^n, $J_\alpha'^n$ subject to (4.44)–(4.46) are in one-to-one correspondence to solutions of the Lieb-Wu equations (4.1).

(iii) For large L and N almost all solutions $\{k_j\}$, $\{\lambda_l\}$ of the Lieb-Wu equations (4.1) are exponentially close to the corresponding solution $\{k_j\}$, $\{\Lambda_\alpha^n\}$, $\{\Lambda_\alpha'^n\}$ of Takahashi's equations, which means that the strings contained in $\{k_j\}$, $\{\lambda_l\}$ are well approximated by the ideal strings determined by $\{k_j\}$, $\{\Lambda_\alpha^n\}$, $\{\Lambda_\alpha'^n\}$.

How well the correspondence between the roots of the Lieb-Wu equations and Takahashi's equations works for finite L is an interesting question. Some aspects of this issue are addressed in Appendix 4.A. For small lattices $L \leq 6$ and fixed u this question has been investigated numerically in Ref. [95] and good agreement has been found.

4.4 Completeness of the Bethe ansatz

As a first application of the string hypothesis, let us now address the issue whether the Bethe ansatz actually gives a complete set of eigenstates of the Hubbard Hamiltonian. On a lattice of L sites there are 4^L eigenstates. Using the string hypothesis together with the SO(4) symmetry and the highest-weight theorem it was shown in [123] that one can construct 4^L linearly independent eigenstates from the solutions of the Lieb-Wu equations: the Bethe ansatz gives a *complete* set of eigenstates. The proof of completeness of the Bethe ansatz given in [123] is based on assumptions (i) and (ii) above. Note that assumption (ii) does *not* mean that the classification of the solutions of the Lieb-Wu equations (4.1) into strings is actually given by (4.36)–(4.46). There may be a *redistribution* between different kinds of solutions. This phenomenon was observed in a number of Bethe ansatz solvable models and was carefully studied for several simple examples in Refs. [60, 123, 124]. It was found that the redistribution did in no case affect the total number of solutions of the Bethe ansatz equations. In Appendix 4.A we discuss deviations from the string hypothesis for some particular examples.

Using (i) and (ii) above, the proof of completeness reduces to a combinatorial problem based on equations (4.44)–(4.46) [123]. From (4.44)–(4.46) we may read off the numbers of allowed values of the (half odd) integers I_j, J_α^n, $J_\alpha'^n$ in a given configuration $\{M_n\}$, $\{M_n'\}$ of strings. These are

(i) L for a free k_j (not involved in a k-Λ string),

(ii) $N - 2M' - \sum_{m=1}^{\infty} t_{nm} M_m$ for a Λ string of length n,

(iii) $L - N + 2M' - \sum_{m=1}^{\infty} t_{nm} M'_m$ for a k-Λ string of length n.

The total number of ways to select the I_j, J_α^n, J'^n_α (recall that they are assumed to be non-repeating) for a given configuration $\{M_n\}$, $\{M'_n\}$ is simply given as a product over binomial coefficients[3]

$$n(\{M_n\}, \{M'_n\}) = C_{N-2M'}^L \prod_{n=1}^{\infty} C_{M_n}^{N-2M'-\sum_{m=1}^{\infty} t_{nm} M_m}$$

$$\times \prod_{n=1}^{\infty} C_{M'_n}^{L-N+2M'-\sum_{m=1}^{\infty} t_{nm} M'_m}. \tag{4.53}$$

Hence, the number of regular Bethe ansatz states for given numbers N of electrons and M of down spins is

$$n_{\text{reg}}(M, N) = \sum_{\{M_n\}, \{M'_n\}} n(\{M_n\}, \{M'_n\}), \tag{4.54}$$

where the summation is over all configurations of strings which satisfy the obvious constraints $N - 2M' \geq 0$ and $M = \sum_{m=1}^{\infty} m(M_m + M'_m)$.

As has been shown in Appendices 3.D and 3.F all regular Bethe ansatz states are lowest weight states with respect to the SO(4) symmetry. This implies that each regular Bethe ansatz state with N electrons and M down spins gives rise to a SO(4) multiplet of eigenstates of the Hamiltonian of dimension

$$\dim_{M,N} = (N - 2M + 1)(L - N + 1). \tag{4.55}$$

Hence the total number of states in the SO(4) extended Bethe ansatz is

$$n_{\text{tot}}(L) = \sum_{M,N} n_{\text{reg}}(M, N) \dim_{M,N} = \sum_{M,N} n_{\text{reg}}(M, N)(N - 2M + 1)(L - N + 1), \tag{4.56}$$

where the sum is over all M, N with $0 \leq 2M \leq N \leq L$. The sums (4.54) and (4.56) are evaluated in Appendix 4.B following the original calculation [123]. It turns out that

$$n_{\text{tot}}(L) = 4^L, \tag{4.57}$$

which is the dimension of the Hilbert space of the Hubbard model on an L-site chain.

The essential ingredients of the proof of completeness can be summarized as follows:

(i) Impose periodic boundary conditions.

(ii) Construct an explicit expression for the wave functions. We use the form (3.91)–(3.94).

(iii) Introduce the concept of *regular* Bethe ansatz states (see Section 3.3.2). This eliminates infinite k's and λ's, whose multiplicities are not under control.

(iv) Prove the lowest weight theorem (3.101), (3.105) for regular Bethe ansatz states. This establishes that each regular Bethe ansatz state gives rise to a SO(4) multiplet of eigenstates of the Hamiltonian of dimension (4.55).

---

[3] For a definiton see Eqn. (4.B.2).

(v) Enumerate all regular Bethe ansatz states for fixed numbers of electrons N and down spins M by means of Takahashi's integers (4.44)–(4.46).

(vi) Calculate the total number of states by taking into account the SO(4) multiplet associated with each regular Bethe ansatz state.

4.5 Higher-level Bethe ansatz

The string hypothesis is a statement on the allowed patterns of Bethe ansatz roots, i.e. the distributions of roots in the complex plane compatible with the Bethe ansatz equations. As we have seen above, the patterns are derived by considering small numbers of roots on large lattices. In the Hubbard model this corresponds to very low densities of electrons. On the other hand, we are mainly interested in the physical properties of the Hubbard model at a finite electron density in the thermodynamic limit. It is not a priori clear that the string hypothesis will provide a good description of this limit. The direct application of the string hypothesis in the analogous case of the spin-$\frac{1}{2}$ Heisenberg XXZ chain was first criticized by C. Destri and J.H. Lowenstein in [98].

A priori there are two reasons for the potential failure of the string hypothesis. Let us revisit the reasoning leading to the form (4.17) for ideal Λ-strings. We used the fact that the l.h.s. of the Lieb-Wu equations (4.15) goes to zero exponentially in the electron number N if Λ_l has a positive imaginary part. This implies that the r.h.s. of the Lieb-Wu equations (4.15) must vanish in the thermodynamic limit

$$N, L \to \infty, \quad \frac{N}{L} \text{ fixed.} \tag{4.58}$$

at finite density. As long as the number of down spins M remains finite we conclude that one of the numerators on the r.h.s. converges to 0. This chain of reasoning may break down if: (i) for some reason the imaginary part of the complex root tends to zero in the thermodynamic limit such that the l.h.s. does not necessarily converge to 0, or (ii) we are considering a state with a system size dependent number of roots such that the r.h.s. of the Lieb-Wu equation involves an increasing number of factors, i.e.

$$M, L \to \infty, \quad \frac{M}{L} \text{ fixed.} \tag{4.59}$$

In case (ii) the r.h.s. of the Lieb-Wu equations (4.15) can be a product of finite factors but nevertheless vanish in the limit (4.59). This case is of particular importance for studying low-lying excitations above the ground state at a fixed density. Detailed studies in the spin-$\frac{1}{2}$ XXZ chain [30, 471, 483, 484], the XYZ model [259] and the Hubbard model [258] have shown that although the string hypothesis does not hold per se, its application still yields exact results for e.g. the spectrum of low-lying excitations.

Appendices to Chapter 4

4.A On deviations from the string hypothesis

In this appendix we discuss in more detail the relation between the roots of the Lieb-Wu equations and the solutions to Takahashi's equations. We will show that in general the correspondence between them unsurprisingly is not perfect. For simplicity we concentrate on the particular example of a k-Λ string solution $N = 2$, $M = 1$ on a lattice of L sites, which we already encountered in Section 4.1. It is convenient to rewrite equations (4.3), (4.4) by

  (i) multiplying (4.3) and (4.4), using (4.5) to replace the right-hand side of the resulting equation by 1 and then taking the logarithm, and

  (ii) solving (4.3) and (4.4) for $\sin k_j - \Lambda$.

The result is

$$(k_1 + k_2) \bmod 2\pi = \frac{2\pi}{L} m, \quad m = -\left[\frac{L-1}{2}\right], \dots, \left[\frac{L}{2}\right], \tag{4.A.1}$$

$$\sin k_j - \Lambda = u \, \mathrm{ctg}\left(\frac{k_j L}{2}\right), \quad j = 1, 2. \tag{4.A.2}$$

It is easy to see from (4.3)–(4.5) that k_2 actually must be the complex conjugate of k_1 for the particular solution we are interested in. We therefore may set

$$k_1 = q - \mathrm{i}\xi, \qquad k_2 = q + \mathrm{i}\xi, \tag{4.A.3}$$

with real q and real, positive ξ. It then follows from (4.A.2) that

$$\sin(q + \mathrm{i}\xi) = \Lambda + u \frac{\sin(qL) - \mathrm{i}\,\mathrm{sh}(\xi L)}{\mathrm{ch}(\xi L) - \cos(qL)}, \tag{4.A.4}$$

or, if we separate real and imaginary parts of this equation,

$$\sin(q)\mathrm{ch}(\xi) = \Lambda + u \frac{\sin(qL)}{\mathrm{ch}(\xi L) - \cos(qL)}, \tag{4.A.5}$$

$$\cos(q)\mathrm{sh}(\xi) = -u \frac{\mathrm{sh}(\xi L)}{\mathrm{ch}(\xi L) - \cos(qL)}. \tag{4.A.6}$$

The real part q of $k_{1,2}$ is readily determined by using equation (4.A.1),

$$q = m\frac{\pi}{L}, \quad m = -L+1, \ldots, L. \tag{4.A.7}$$

Given q we may then use (4.A.6) to determine the imaginary part ξ as a function of $q = m\frac{\pi}{L}$, and finally we can use (4.A.5) to obtain Λ. As $qL = m\pi$ we have $\sin(qL) = 0$, $\cos(qL) = (-1)^m$, and as a result (4.A.5) and (4.A.6) decouple

$$\Lambda = \sin(q)\mathrm{ch}(\xi), \tag{4.A.8}$$

$$\mathrm{sh}(\xi) = -\frac{u}{\cos(q)}\frac{\mathrm{sh}(\xi L)}{\mathrm{ch}(\xi L) - (-1)^m} \equiv f(\xi, m). \tag{4.A.9}$$

For later convenience we define a function $g(\xi)$ by

$$g(\xi) = \frac{\mathrm{sh}(\xi L)}{\mathrm{ch}(\xi L) - (-1)^m} = \begin{cases} \tanh\left(\frac{\xi L}{2}\right), & \text{for } m \text{ odd} \\[2mm] \coth\left(\frac{\xi L}{2}\right), & \text{for } m \text{ even}. \end{cases} \tag{4.A.10}$$

At this point we are left with a nonlinear equation (4.A.9) for ξ, which cannot be solved in a closed analytical form. As usual in such cases it is very instructive to depict the equation graphically, which is done in figure 4.A.1. Depending on the value of m there is either exactly one solution of (4.A.9) or no solution at all. The latter is easily understood by noting that due to the fact that $g(\xi) > 0$ for all $\xi > 0$, equation (4.A.9) can have solutions for positive ξ only if

$$\begin{aligned} \tfrac{\pi}{2} < |q| \leq \pi & \quad \text{for } u > 0, \\ |q| < \tfrac{\pi}{2} & \quad \text{for } u < 0. \end{aligned} \tag{4.A.11}$$

In order to keep things simple, let us concentrate on the repulsive case $u > 0$. It follows from (4.A.10) that (4.A.9) *always* has exactly one solution ξ_m for every even m which satisfies (4.A.11). A fact that can be gleaned from figure 4.A.1 is that in order for a solution with

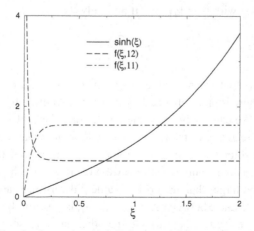

Fig. 4.A.1. Sketch of equation (4.A.9) for $u = 0.25$, $L = 20$ and $m = 11, 12$.

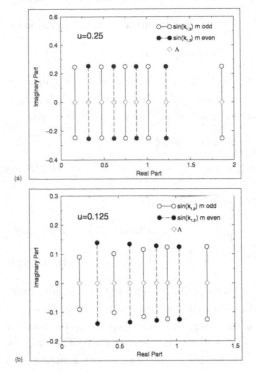

Fig. 4.A.2. Length 2 k-Λ string solutions for a $L = 20$ site lattice and (a) $u = 0.25$ (b) $u = 0.125$. We see that there is precisely one solution for each value of m.

odd m to exist, the slope of $f(\xi, m)$ at $\xi = 0$ must be greater than the slope of $\sinh(\xi)$, i.e.

$$-\frac{uL}{2\cos(q)} > 1. \tag{4.A.12}$$

This is satisfied for all q with $\frac{\pi}{2} < |q| < \pi$ if and only if

$$L > \frac{2}{u}. \tag{4.A.13}$$

What we have found is that whenever the length of the lattice L is sufficiently large and (4.A.13) is satisfied, there is precisely one k-Λ two string solution for every m that satisfies (4.A.7) and (4.A.11). We can easily count these solutions: for odd L there are L solutions and for even L there are $L - 1$ of them. In order to illustrate what these solutions actually look like, we plot all solutions with $\Lambda > 0$ for $L = 20$ and two different values of u in figure 4.A.2. In order to obtain simple patterns we plot $\sin(k_{1,2})$ rather than $k_{1,2}$. As long as uL is sufficiently much larger than the critical value 2 determined above, the 'spacing' of the strings, that is the distance between Λ and $\sin(k_{1,2})$, is essentially constant and equal to u. This is examplified in figure 4.A.2 (a). On the other hand, as soon as uL approaches 2 there are considerable deviations in the spacings as can be seen in figure 4.A.2(b). There are two branches, corresponding to even and odd m respectively. The spacing of the even (odd) m branch increases (decreases) as Λ becomes small.

Let us compare these results to what we would predict on the basis of Takahashi's equations (4.36)–(4.38) and the allowed values of (half-odd) integers (4.44)–(4.46). The only nonzero occupation number for the solution we are considering is $M'_1 = 1$. Hence the allowed range of integers is $|J'^1| \leq \frac{1}{2}(L - t_{11} - 1) = \frac{1}{2}(L - 2)$, which means that there are $L - 1$ possible values of J'^1. This agrees with what we just found since L was assumed to be even in (4.44)–(4.46). Let us furthermore investigate how well the actual solutions of the Lieb-Wu equations are reproduced by the roots of Takahashi's equations. The latter are

$$2L \, \text{Re} \left[\arcsin(\Lambda'^1_\alpha + iu) \right] = 2\pi \left[-\frac{L}{2} + \alpha \right], \quad \alpha = 1, \ldots L - 1, \tag{4.A.14}$$

and by construction we have $\sin(k_\alpha^{1,2}) = \Lambda'^1_\alpha \pm iu$. Solving (4.A.14) for $L = 20$ we find that the solutions of Takahashi's equations for $u = 0.25$ are very close to the actual solutions of the Lieb-Wu equations: the relative errors are at most 1.4% in the imaginary parts and much less than a percent in the real parts. For smaller u the situation gets worse as is apparent from figure (4.A.2)(b): the imaginary parts of $\sin(k_{1,2})$ obtained by solving the Lieb-Wu equations deviate very significantly from $\pm iu$. On the other hand the differences between the real centers of the strings Λ and the Λ'^1_α obtained from (4.A.14) are still very small: the relative errors are less than a percent.

One obvious question is what happens when we decrease u below the 'critical' value $2/L$. Now the inequality (4.A.12) is violated for one or more odd m and as a result there are fewer than $L - 1$ k-Λ string solutions (for even L) with $N = 2$ and $M = 1$ to the Lieb-Wu equations. The disappearance of these solutions as u is decreased from an initially large $u > 2/L$ can be visualized as follows. At first the spacing $\sin(k_2) - \sin(k_1)$ of the odd-m solutions decreases below the ideal value $2u$. This effect is particularly pronounced for the solutions with small real parts of $\sin(k_{1,2})$ (and thus Λ) and can be seen in figure 4.A.2(b). As u approaches $2/L$ from above, the spacings of the two solutions with $\pm\Lambda$ closest to zero turn to zero: the strings 'collapse'. For values of u smaller than $2/L$, there are two solutions with real $k_{1,2}$ *in place of the collapsed k-Λ strings*. More details of this process can be found in Refs. [123] and [95].

On the other hand the number of solutions of Takahashi's equations (4.36)–(4.38) is by construction independent of u. Hence, if u becomes too small, the solutions of the Lieb-Wu equations are in general quite different from the solutions of Takahashi's equations. At first sight this looks like a very serious problem. However, we are really interested in taking L very large while keeping u fixed. In this limit the condition (4.A.12) is always satisfied. Of course this does not guarantee that the string hypothesis is valid for *all* solutions of the Lieb-Wu equations.

4.B Details about the enumeration of eigenstates

In this appendix we evaluate the total number of states obtained by combining the SO(4) symmetry with the Bethe ansatz. Our presentation closely parallels the original paper [123].

We recall that the numbers of allowed values for the (half-odd) integers I_j, J_α^n, $J_\alpha^{\prime n}$ corresponding to each of the fundamental strings are

(i) L for a 'single' k_i

(ii) $N - 2M' - \sum_{m=1}^\infty t_{nm} M_m$ for a Λ-string of length n

(iii) $L - N + 2M' - \sum_{m=1}^\infty t_{nm} M'_m$ for a k-Λ-string of length n.

As before we denote the number of single k_i's by $\mathcal{M}_e$. The number of ways to choose the (half-odd) integers I_j, J_α^n, $J_\alpha^{\prime n}$ in a solution with occupation numbers $\mathcal{M}_e$, M_m and M'_m is therefore given by[1]

$$n(\mathcal{M}_e, \{M_m\}, \{M'_m\}) = C_{\mathcal{M}_e}^L \prod_{n=1}^\infty C_{M_n}^{N - 2M' - \sum_{m=1}^\infty t_{nm} M_m}$$

$$\times \prod_{n=1}^\infty C_{M'_n}^{L - N + 2M' - \sum_{m=1}^\infty t_{nm} M'_m}. \qquad (4.B.1)$$

Here C_m^n are binomial coefficients

$$C_m^n = \frac{n!}{m!(n-m)!}. \qquad (4.B.2)$$

The total number of regular solutions of the Lieb-Wu equations for given numbers of electrons N and electrons with spin down M is now obtained by summing $n(\mathcal{M}_e, \{M_m\}, \{M'_m\})$ over all the occupation numbers $\mathcal{M}_e$, M_m and M'_m, subject to the constraints (4.44)–(4.46) that specify the allowed ranges of (half-odd) integers. In what follows it will be convenient to introduce the number of up spins[2]

$$N_\uparrow = N - M. \qquad (4.B.3)$$

Every solution to the Lieb-Wu equations (4.1) gives rise to a regular Bethe ansatz state, which comes with an entire SO(4) multiplet of eigenstates of the Hamiltonian. The dimension $\dim_{M,N}$ of this multiplet is given by

$$\dim_{M,N_\uparrow} = (N_\uparrow - M + 1)(L - N_\uparrow - M + 1). \qquad (4.B.4)$$

The number of eigenstates that are obtained by combining the Bethe ansatz *and* the $SO(4)$ symmetry is therefore given by[3]

$$\# \text{ (eigenstates)} =$$

$$\sum_{\substack{M \geq 0 \\ N_\uparrow - M \geq 0 \\ N_\uparrow + M \leq L}} \sum_{\substack{N_\uparrow \geq 0}} \left[\sum_{\substack{\mathcal{M}_e = 0 \\ N_\uparrow + M = \mathcal{M}_e + 2\sum_{m=1}^\infty m M'_m}}^\infty \sum_{\substack{M_m = 0 \\ N_\uparrow - M = \mathcal{M}_e - 2\sum_{m=1}^\infty m M_m}}^\infty \sum_{\substack{M'_m = 0}}^\infty n(\mathcal{M}_e, \{M_m\}, \{M'_m\}) \right] \dim_{M,N_\uparrow}. \qquad (4.B.5)$$

---

[1] We recall that the (half-odd) integers are assumed to be non-repeating.

[2] We recall that the number of down spins is denoted by M throughout.

[3] The constraints in the summations in brackets arise from the 'sum rules' (4.35) $M = \sum_{m=1}^\infty m[M_m + M'_m]$ and $N = \mathcal{M}_e + 2\sum_{m=1}^\infty m M'_m$.

Table 4.B.1. $L = 2$. n *denotes the number of regular Bethe ansatz states*
of a given type. There are a total number of $16 = 4^2$ *eigenstates of the*
hamiltonian

$\mathcal{M}_e$	M_1	M_1'	M	$N_\uparrow$	n	$\dim_{M,N_\uparrow}$	#(states)
0	0	0	0	0	1	3	3
1	0	0	0	1	2	4	8
2	0	0	0	2	1	3	3
2	1	0	1	1	1	1	1
0	0	1	1	1	1	1	1
							16

In order to have a complete set this total number must be equal to 4^L. The counting of the eigenstates obtained from the $SO(4)$ extended nested Bethe ansatz has thus been reduced to a purely algebraic problem.

4.B.1 Simple examples: 2 and 4 site lattices

Before turning to the general case let us see how things work for the 2-site and 4-site models. The 2-site model ($L = 2$) was discussed in detail in Ref. [125], where a complete set of $4^2 = 16$ eigenstates of the Hamiltonian was constructed explicitly. In table 4.B.1 we show how these eigenstates are classified in terms of the SO(4) extended Bethe ansatz.

The wave functions and energies of all 16 eigenstates of the $L = 2$ site system are given in Ref. [125]. The total number of 16 states splits into 2 singlets, 2 triplets and 2 quadruplets of $SO(4)$. The ground state is the singlet with $M_1 = 1$ for the case $u > 0$ and the singlet with $M_1' = 1$ for the case $u < 0$. In both cases it has energy $E_0 = -\sqrt{4u^2 + 16}$. The counting for the 4-site model ($L = 4$) is presented in table 4.B.2, where we show how the total number of $4^4 = 256$ is obtained. We note that the total number of regular Bethe ansatz states is only 60.

An explicit enumeration of a complete set of eigenstates for the $L = 6$ site system has been carried out in Ref. [95].

4.B.2 Counting eigenstates

In this subsection we will evaluate the right hand side of (4.B.5) explicitly for general even L and show that it equals 4^L. This will establish completeness. We will split this proof into two steps as follows. In the first step we will prove Lemmas 2 and 3. In the second step we will then use these lemmas to perform the summation in (4.B.5). To keep the notation

Table 4.B.2. *L = 4. There are 60 regular Bethe ansatz states, which, when weighted with the correct SO(4) multiplicities, give a total of* $256 = 4^4$ *eigenstates of the Hamiltonian. n denotes the number of regular Bethe states.*

$\mathcal{M}_e$	M_1	M_2	M_1'	M_2'	M	$N_\uparrow$	n	$\dim_{M,N_\uparrow}$	#(states)
0	0	0	0	0	0	0	1	5	5
1	0	0	0	0	0	1	4	8	32
2	0	0	0	0	0	2	6	9	54
3	0	0	0	0	0	3	4	8	32
4	0	0	0	0	0	4	1	5	5
2	1	0	0	0	1	1	6	3	18
0	0	0	1	0	1	1	3	3	9
3	1	0	0	0	1	2	8	4	32
1	0	0	1	0	1	2	8	4	32
4	2	0	0	0	2	2	1	1	1
4	0	1	0	0	2	2	1	1	1
2	1	0	1	0	2	2	6	1	6
0	0	0	2	0	2	2	1	1	1
0	0	0	0	1	2	2	1	1	1
4	1	0	0	0	1	3	3	3	9
2	0	0	1	0	1	3	6	3	18
							60		256

simple we define

$$P_n = \mathcal{N} - \sum_{m=1}^{\infty} t_{nm} M_m , \tag{4.B.6}$$

$$n(\{M_m\}) = \prod_{n=1}^{\infty} C_{M_n}^{P_n} , \tag{4.B.7}$$

where t_{nm} is given by (4.47).

Lemma 2. *The following identity holds:*

$$\sum_{\substack{M_1,M_2,\ldots=0 \\ \sum_{m=1}^{\infty} m M_m = \mathcal{M}}}^{\infty} n(\{M_m\}) = C_{\mathcal{M}}^{\mathcal{N}} - C_{\mathcal{M}-1}^{\mathcal{N}} . \tag{4.B.8}$$

Proof. We first solve for $M_1 = \mathcal{M} - \sum_{m=2}^{\infty} m M_m$ and substitute this back into the left-hand side of (4.B.8). The quantities P_n reduce to

$$P_n = \mathcal{N} - 2\mathcal{M} + M_n + 2 \sum_{m=n+1}^{\infty} (m - n) M_m . \tag{4.B.9}$$

Next we consider the summation over M_2. Although the summand on the left-hand side of (4.B.8) has the form of an infinite product, only two of the factors contain M_2. Singling these out, we find that the summation over M_2 is as follows

$$\Omega_2 = \sum_{M_2=0}^{\infty} C_{M_2}^{\mathcal{N}-2\mathcal{M}+M_2+2\sum_{m=3}^{\infty}(m-2)M_m} \, C_{\mathcal{M}-\sum_{m=2}^{\infty} mM_m}^{\mathcal{N}-\mathcal{M}+\sum_{m=3}^{\infty}(m-2)M_m} \, . \tag{4.B.10}$$

In order to carry out the sum over M_2 we will make use of the identity

$$\sum_{\alpha=0}^{\infty} C_\alpha^{B+\alpha} \, x^\alpha = (1-x)^{-1-B} \, , \tag{4.B.11}$$

which is easily proved by mathematical induction. As a simple consequence we have

$$(1-x^2)^{-1-\omega}(1+x)^\eta \, |_{x^A} = \sum_{\alpha=0}^{\infty} C_\alpha^{\omega+\alpha} \, C_{A-2\alpha}^\eta \, , \tag{4.B.12}$$

where the notation $|_{x^A}$ denotes the coefficient of the power x^A. The right-hand side of (4.B.10) is of the same form as (4.B.12), so that

$$\Omega_2 = (1-x^2)^{-1-[\mathcal{N}-2\mathcal{M}+2(M_3+2M_4+\cdots)]}(1+x)^{\mathcal{N}-\mathcal{M}+(M_3+2M_4+\cdots)} \, |_{x^{\mathcal{M}-3M_3-\cdots}}$$

$$= (1+x)^{\mathcal{N}-\mathcal{M}}(1-x^2)^{-\mathcal{N}+2\mathcal{M}-1} \prod_{n=3}^{\infty} (\mathcal{Z}_n^{(0)})^{M_n} \, |_{x^{\mathcal{M}}}$$

$$= \frac{1}{2\pi i} \oint \frac{dx}{x^{\mathcal{M}+1}} (1+x)^{\mathcal{N}-\mathcal{M}}(1-x^2)^{-\mathcal{N}+2\mathcal{M}-1} \prod_{n=3}^{\infty} (\mathcal{Z}_n^{(0)})^{M_n} \, , \tag{4.B.13}$$

where we have introduced the notation

$$\mathcal{Z}_n^{(0)} = \frac{x^n}{(1-x)^{2(n-2)}(1+x)^{n-2}} \, . \tag{4.B.14}$$

In the last line of (4.B.13) we extracted the coefficient of $x^{\mathcal{M}}$ by performing a contour integral around the origin $x = 0$. After carrying out the M_2 summation, equation (4.B.8) reads

$$\sum_{\substack{M_1,M_2,\ldots=0 \\ \sum_{m=1}^{\infty} mM_m=\mathcal{M}}}^{\infty} n(\{M_m\}) = \frac{1}{2\pi i} \oint \frac{dx}{x^{\mathcal{M}+1}} A(x) \, , \tag{4.B.15}$$

where

$$A(x) = (1+x)^{\mathcal{N}-\mathcal{M}}(1-x^2)^{-1-\mathcal{N}+2\mathcal{M}}$$

$$\times \sum_{M_3,M_4,\ldots=0}^{\infty} \prod_{n=3}^{\infty} C_{M_n}^{\mathcal{N}-2\mathcal{M}+M_n+2\sum_{m=n+1}^{\infty}(m-n)M_m} \prod_{l=3}^{\infty} (\mathcal{Z}_l^{(0)})^{M_l} \, . \tag{4.B.16}$$

In the next step we use (4.B.12) to carry out the summation over M_3

$$\Omega_3 = \sum_{M_3=0}^{\infty} C_{M_3}^{\mathcal{N}-2\mathcal{M}+M_3+2\sum_{n=4}^{\infty}(n-3)M_n} \, (\mathcal{Z}_3^{(0)})^{M_3}$$

$$= \left(1 - \mathcal{Z}_3^{(0)}\right)^{-1-\mathcal{N}+2\mathcal{M}-2\sum_{n=4}^{\infty}(n-3)M_n} . \tag{4.B.17}$$

The full expression for $A(x)$ is then reduced to

$$A(x) = (1+x)^{\mathcal{N}-\mathcal{M}}(1-x^2)^{-1-\mathcal{N}+2\mathcal{M}} \left(1 - \mathcal{Z}_3^{(0)}\right)^{-1-\mathcal{N}+2\mathcal{M}}$$

$$\times \sum_{M_4,M_5,\dots=0}^{\infty} \prod_{n=4}^{\infty} C_{M_n}^{\mathcal{N}-2\mathcal{M}+M_n+2\sum_{m=n+1}^{\infty}(m-n)M_m} \prod_{l=4}^{\infty} (\mathcal{Z}_l^{(1)})^{M_l} , \tag{4.B.18}$$

where $\mathcal{Z}_n^{(1)}$ are defined through

$$\mathcal{Z}_n^{(1)} = \mathcal{Z}_n^{(0)} \left(1 - \mathcal{Z}_3^{(0)}\right)^{-2(n-3)} . \tag{4.B.19}$$

The summations over all remaining M_n, $n \geq 4$ are precisely of the same structure as the M_3 summation and are carried out analogously. It is easy to see that this results in

$$A(x) = (1+x)^{\mathcal{N}-\mathcal{M}} F(x)^{-1-\mathcal{N}+2\mathcal{M}} , \tag{4.B.20}$$

$$F(x) = (1-x^2) \prod_{m=3}^{\infty} \left(1 - \mathcal{Z}_m^{(m-3)}\right) . \tag{4.B.21}$$

The quantities $\mathcal{Z}_n^{(m)}$ are defined recursively by

$$\mathcal{Z}_n^{(m)} = \mathcal{Z}_n^{(m-1)} \left(1 - \mathcal{Z}_{m+2}^{(m-1)}\right)^{-2(n-m-2)} . \tag{4.B.22}$$

Our task is now to find a closed expression for $F(x)$ by exploiting this relation. Defining functions $U_m(x)$ such that

$$U_2 = x^{-2} , \quad U_m = \frac{1}{\mathcal{Z}_m^{(m-3)}} , \quad m \geq 3 , \tag{4.B.23}$$

we can express $F(x)$ as

$$F(x) = \prod_{m=2}^{\infty} \left(1 - \frac{1}{U_m}\right) . \tag{4.B.24}$$

We will now show that the functions $U_m(x)$ satisfy the following recursion relation, which we denote by RR $\mathtt{I_p}$

$$\mathtt{RR\ I_p}: \quad (U_{p+3} - 1)^2 = U_{p+4}U_{p+2} , \quad p \geq 0 . \tag{4.B.25}$$

Together with the initial conditions

$$U_2 = x^{-2}, \quad U_3 = \frac{(1-x)^2(1+x)}{x^3} \qquad (4.B.26)$$

these relations determine all $U_m(x)$ and thereby $F(x)$. In order to prove (4.B.25), it is useful to establish a second recursion relation RR II_p, which involves certain other $\mathcal{Z}$'s

$$\text{RR II}_\text{p}: \qquad \frac{\mathcal{Z}_{n+1}^{(p)}}{\mathcal{Z}_n^{(p)}} = \frac{U_{p+2}}{U_{p+3}}, \qquad p \geq 0, \ n \geq p+3. \qquad (4.B.27)$$

Let us now prove both RR I_p and RR II_p by mathematical induction. We start with $\mathcal{Z}_n^{(0)}$ as defined in (4.B.14), and U_2 and U_3 as given in (4.B.26). One easily checks that RR $\text{II}_{p=0}$ is valid. Using (4.B.22) to construct $U_4 = 1/\mathcal{Z}_4^{(1)}$ and then making use of RR $\text{II}_{p=0}$, one obtains RR $\text{I}_{p=0}$. This establishes the validity of both RR $\text{I}_{p=0}$ and RR $\text{II}_{p=0}$, which is the starting point for the proof by induction.

Let us now assume that RR I_p and RR II_p hold for a given p. Using this induction assumption and the definition of $\mathcal{Z}_n^{(m)}$ (4.B.22) we establish the validity of RR II_{p+1}. Finally, RR I_{p+1} follows from RR II_{p+1} and (4.B.22). This completes the induction step.

One may easily verify that the expressions

$$U_j = \left(\frac{a(x)^{j+1} - a(x)^{-j-1}}{a(x) - a(x)^{-1}} \right)^2 \qquad (4.B.28)$$

with

$$a(x) = \frac{1}{2} \left(\sqrt{\frac{1-3x}{x}} + \sqrt{\frac{1+x}{x}} \right) \qquad (4.B.29)$$

satisfy the recursion relations (4.B.25) and the initial conditions (4.B.26). The function $F^2(x)$ is then expressed as a convergent product

$$F^2(x) = \prod_{m=2}^\infty \frac{(U_m - 1)^2}{U_m^2} = \prod_{m=2}^\infty \frac{U_{m+1} U_{m-1}}{U_m^2} = \lim_{l \to \infty} \frac{U_1}{U_2} \frac{U_{l+1}}{U_l} = a(x)^2 x(x+1), \quad (4.B.30)$$

where we have defined $U_1(x) = \frac{x+1}{x}$, in accord with (4.B.25), and where we used (4.B.25) in the second equality. Combining (4.B.8), (4.B.15), (4.B.16), (4.B.21), (4.B.24), (4.B.29) and (4.B.30) we arrive at the following representation for the number of regular Bethe ansatz states with $\mathcal{M}$ overturned spins

$$\sum_{\substack{M_m=0 \\ \mathcal{M}=\sum_{m=1}^{\infty} m M_m}}^{\infty} n(\{M_m\})$$

$$= \frac{1}{2\pi i} \oint \frac{dx}{x^{\mathcal{M}+1}} (1+x)^{\mathcal{N}-\mathcal{M}} \left(\frac{(1+x) + \sqrt{(1+x)(1-3x)}}{2} \right)^{2\mathcal{M}-\mathcal{N}-1}$$

$$= \frac{1}{2\pi i} \oint \frac{2\,dy}{y^{\mathcal{M}+1}} \left(2(1+y) \right)^{\mathcal{N}-\mathcal{M}} \left(1+y+\sqrt{1-y^2} \right)^{2\mathcal{M}-\mathcal{N}-1}, \qquad (4.\text{B}.31)$$

where the contour is a small circle around the origin and we used the substitution $y = \frac{2x}{1-x}$. Substituting $y^{-1} = \cosh\phi$, the integral reduces to $I_+ - I_-$, where

$$I_\pm = \frac{1}{2\pi} \int_0^{2\pi} d\varphi \, e^{-\phi(\mathcal{N}-\mathcal{M}\mp 1)} (1+e^\phi)^{\mathcal{N}-1} = C_{\mathcal{N}-\mathcal{M}\mp 1}^{\mathcal{N}-1}. \qquad (4.\text{B}.32)$$

Here we have used $\phi = \Lambda - i\varphi$, with $\Lambda \to \infty$. This finally establishes the result (4.B.8) since

$$\sum_{\substack{M_m=0 \\ \mathcal{M}=\sum_{m=1}^{\infty} m M_m}}^{\infty} n(\{M_m\}) = C_{\mathcal{N}-\mathcal{M}-1}^{\mathcal{N}-1} - C_{\mathcal{N}-\mathcal{M}+1}^{\mathcal{N}-1} \equiv C_{\mathcal{M}}^{\mathcal{N}} - C_{\mathcal{M}-1}^{\mathcal{N}}. \qquad (4.\text{B}.33)$$

Here the last equality is easily verified by a direct calculation. This concludes the proof of Lemma 2. $\qquad\square$

Lemma 3.

$$\sum_{\mathcal{M}=0}^{[\mathcal{N}/2]} \left[C_{\mathcal{M}}^{\mathcal{N}} - C_{\mathcal{M}-1}^{\mathcal{N}} \right] [\mathcal{N} - 2\mathcal{M} + 1] = 2^{\mathcal{N}}. \qquad (4.\text{B}.34)$$

Proof. We rewrite the left hand side of (4.B.34) as

$$\sum_{\mathcal{M}=0}^{[\mathcal{N}/2]} C_{\mathcal{M}}^{\mathcal{N}} (\mathcal{N} - 2\mathcal{M} + 1) - \sum_{\mathcal{M}=0}^{[\mathcal{N}/2]-1} C_{\mathcal{M}}^{\mathcal{N}} (\mathcal{N} - 2\mathcal{M} - 1)$$

$$= 2 \sum_{\mathcal{M}=0}^{[\mathcal{N}/2]-1} C_{\mathcal{M}}^{\mathcal{N}} + C_{[\frac{\mathcal{N}}{2}]}^{\mathcal{N}} \left(1 + \mathcal{N} - 2\left[\frac{\mathcal{N}}{2}\right] \right)$$

$$= \sum_{\mathcal{M}=0}^{[\mathcal{N}/2]-1} C_{\mathcal{M}}^{\mathcal{N}} + C_{[\frac{\mathcal{N}}{2}]}^{\mathcal{N}} \left(1 + \mathcal{N} - 2\left[\frac{\mathcal{N}}{2}\right] \right) + \sum_{\mathcal{N}-[\mathcal{N}/2-1]}^{\mathcal{N}} C_{\mathcal{M}}^{\mathcal{N}}$$

$$= \sum_{\mathcal{M}=0}^{\mathcal{N}} C_{\mathcal{M}}^{\mathcal{N}} = 2^{\mathcal{N}}. \qquad (4.\text{B}.35)$$

$\qquad\square$

Lemmas 2 and 3 have a natural interpretation in the context of the spin-$\frac{1}{2}$ Heisenberg XXX model [433]. Equation (4.B.8) gives the total number of regular Bethe ansatz states (defined by $\mathcal{M} \leq [\mathcal{N}/2]$) with $\mathcal{M}$ overturned spins in the XXX model on a lattice of length $\mathcal{N}$. The second formula shows that the total number of states obtained by combining the regular Bethe ansatz with the $SU(2)$ symmetry equals $2^{\mathcal{N}}$, which is the dimension of the Hilbert space of the XXX model. These relations thus establish the completeness of the $SU(2)$ extended Bethe ansatz for the XXX model. The fact that identities that have their origin in the XXX model play a role here should not come as a surprise. Indeed, our method of solution of the Hubbard model is the *nested* Bethe ansatz. The solutions to the Bethe ansatz are specified by two sets $\{k_j\}$ and $\{\Lambda_\alpha\}$ of spectral parameters. The k_j's are momenta associated with charge degrees of freedom, whereas the Λ_α's, which describe the 'nesting' of the Bethe ansatz, are rapidities of spin excitations of the type encountered in the Heisenberg XXX model. This should make clear that the second stage of the nested Bethe ansatz for the Hubbard model is really a spin-problem, which is very similar to the Bethe ansatz analysis of the Heisenberg XXX model. This fact can be seen directly in our construction of the wave functions in Appendix 3.B. Our two-step procedure for performing the summation is natural from the point of view of the nesting: in the first step we sum over the spin degrees of freedom, and in the second step we then sum over the charge degrees of freedom as well.

The total number of states obtained from the SO(4) extended Bethe ansatz for the Hubbard model is given by (4.B.1) and (4.B.5). The summations over the multiplicities M_m and over the difference $N_\uparrow - M$ in the summation (4.B.5) can be carried out by using Lemmas 2 and 3 respectively, if we substitute $\mathcal{M} \to \frac{1}{2}(\mathcal{M}_e - N_\uparrow + M)$ and $\mathcal{N} \to \mathcal{M}_e$. (Under these summations the total number of electrons N is kept fixed.) The summation that remains after this 'spin summation' is

$$
\# \text{ (eigenstates)} = \sum_{N=0}^{L} (L - N + 1)
$$

$$
\times \sum_{\substack{\mathcal{M}_e=0 \\ N=\mathcal{M}_e+2\sum_{m=1}^{\infty} m M_m'}}^{N} \sum_{M_m'=0}^{\infty} 2^{\mathcal{M}_e} \, C_{\mathcal{M}_e}^{L} \prod_{n=1}^{\infty} C_{M_n'}^{L-N_e+\sum_{m=1}^{\infty}(2m-t_{nm})M_m'} .
\tag{4.B.36}
$$

Next we carry out the sums over all M_n''s, using a similar kind of 'summation device' to that employed in the proof of Lemma 2. As a consequence of (4.B.11) we have

$$
(1 - x^2)^{-1-B}(1 + 2x)^L = \sum_{M_1'=0}^{\infty} \sum_{p=0}^{L} C_{M_1'}^{B+M_1'} \, C_p^L \, 2^p \, x^{2M_1'+p} ,
\tag{4.B.37}
$$

and therefore

$$
\frac{1}{2\pi i} \oint \frac{dx}{x^{\gamma+1}} (1 - x^2)^{-1-B}(1 + 2x)^L = \sum_{M_1'=0}^{\infty} C_{M_1'}^{B+M_1'} \, C_{\gamma-2M_1'}^{L} \, 2^{\gamma-2M_1'} .
\tag{4.B.38}
$$

The integration is along a small contour around zero. Defining $E = L - N$, $\gamma = N - 2\sum_{m=2}^{\infty} m M'_m$ and $B = E + 2\sum_{m=2}^{\infty}(m-1)M'_m$, the r.h.s. of (4.B.38) becomes the summation over M'_1 in (4.B.36), if we solve the constraint in the sum in (4.B.36) for $\mathcal{M}_e = N - 2\sum_{m=1}^{\infty} m M'_m$. Using (4.B.11) in (4.B.36) we then obtain the following expression for the number of eigenstates:

$$\# \,(\text{eigenstates}) = \frac{1}{2\pi i} \oint \frac{dx}{x^{L+1}(1-x^2)}(1+2x)^L \sum_{E=0}^{L}(E+1)\frac{x^E}{(1-x^2)^E}F(x), \quad (4.B.39)$$

where

$$F(x) = \sum_{\substack{M'_m=0 \\ m\geq 2}}^{\infty} \prod_{n=2}^{\infty} C_{M'_n}^{E+M'_n+2\sum_{m=n+1}^{\infty}(m-n)M'_m} \prod_{m=2}^{\infty} \left(\mathcal{Z}_m^{(0)}\right)^{M'_m}, \quad (4.B.40)$$

and

$$\mathcal{Z}_m^{(0)} = x^{2m}(1-x^2)^{2(1-m)}. \quad (4.B.41)$$

The summations over M'_2, M'_3, ... have precisely the form of the l.h.s. of (4.B.11) and can thus be performed easily. The result is

$$F(x) = \prod_{m=2}^{\infty}(1 - \mathcal{Z}_m^{(m-2)})^{-1-E}, \quad (4.B.42)$$

where

$$\mathcal{Z}_m^{(p)} = \frac{\mathcal{Z}_m^{(p-1)}}{(1 - \mathcal{Z}_{p+1}^{(p-1)})^{2(m-p-1)}}. \quad (4.B.43)$$

It can be shown along the lines of the proof of Lemma 2, that the quantities $U_m = \frac{1}{\mathcal{Z}_m^{(m-2)}}$ obey the recursion relation

$$(U_{p+2} - 1)^2 = U_{p+3}U_{p+1}, \quad p \geq 0 \quad (4.B.44)$$

with initial conditions

$$U_1 = x^{-2}, \qquad U_2 = \frac{(1-x^2)^2}{x^4}. \quad (4.B.45)$$

Equation (4.B.27) is replaced by

$$\frac{\mathcal{Z}_{n+1}^{(p)}}{\mathcal{Z}_n^{(p)}} = \frac{U_{p+1}}{U_{p+2}}, \quad p \geq 0, \ n \geq p+2. \quad (4.B.46)$$

Equation (4.B.39) now can be written as

$$\# \,(\text{eigenstates}) = \frac{1}{2\pi i} \oint \frac{dx}{x^{L+1}}(1+2x)^L \sum_{E=0}^{L}(E+1)x^E[f(x)]^{-E-1}, \quad (4.B.47)$$

where

$$f(x) = \prod_{l=1}^{\infty} (1 - U_l^{-1}).$$ (4.B.48)

The solution of the recursion relation (4.B.44) is again of the form (4.B.28), i.e., $U_j = \left(\frac{a(x)^{j+1} - a(x)^{-j-1}}{a(x) - a(x)^{-1}} \right)^2$, where now

$$a(x) = \frac{1}{2x} + \sqrt{\frac{1}{4x^2} - 1}$$ (4.B.49)

as a consequence of the new initial conditions (4.B.45). Insertion of the resulting expression for U_l into (4.B.48) leads to the following result for the function $f(x)$:

$$2 f(x) = 1 + \sqrt{1 - 4x^2} = 2x \, a(x).$$ (4.B.50)

Equation (4.B.47) can now be rewritten as

$$\# \text{ (eigenstates)} = \sum_{E=0}^{L} (E + 1) \, I(E),$$ (4.B.51)

where

$$I(E) = \frac{1}{2\pi i} \oint d\left(-\frac{1}{x}\right) \left(\frac{1}{x} + 2\right)^{L} [a(x)]^{-E-1}.$$ (4.B.52)

The contour integration can be worked out just as we did in the proof of Lemma 2. Defining $\alpha = \Lambda - i\varphi$ with $\Lambda \gg 1$ and substituting $x = \frac{1}{e^{\alpha} + e^{-\alpha}}$ we obtain

$$I(E) = I_+(E) - I_-(E),$$ (4.B.53)

where

$$I_\pm(E) = \frac{1}{2\pi} \int_0^{2\pi} d\varphi \, e^{\pm\alpha} \, e^{-(1+E)\alpha} \left(e^{\frac{\alpha}{2}} + e^{-\frac{\alpha}{2}}\right)^{2L}.$$ (4.B.54)

Expanding

$$\left(e^{\frac{\alpha}{2}} + e^{-\frac{\alpha}{2}}\right)^{2L} = \sum_{p=0}^{2L} C_p^{2L} \, e^{\alpha(L-p)},$$ (4.B.55)

and then using

$$\frac{1}{2\pi} \int_0^{2\pi} d\varphi \, e^{\pm i(n\varphi)} = \delta_{n,0},$$ (4.B.56)

in the resulting expression, we find that

$$I_+(E) = C_{L-E}^{2L}, \qquad I_-(E) = C_{L-E-2}^{2L}.$$ (4.B.57)

Using these results in (4.B.53) and then in (4.B.51) we are left with only a single summation

$$\# \text{ (eigenstates)} = \sum_{E=0}^{L}(E+1)\left[C_{L-E}^{2L} - C_{L-E-2}^{2L}\right]. \tag{4.B.58}$$

This summation can be performed the same way as (4.B.35) and we finally obtain the desired result

$$\# \text{ (eigenstates)} = 4^{L}. \tag{4.B.59}$$

This concludes our two-step evaluation of the sum (4.B.5).

Using the results derived above, we can obtain a closed expression for the number of regular Bethe ansatz states for given numbers M and $N_\uparrow$ of spin-down and spin-up electrons:

$$\sum_{\substack{\mathcal{M}_e=0 \\ N=\mathcal{M}_e+2\sum mM'_m \\ M=\sum m(M_m+M'_m)}}^{\infty} \sum_{M_m=0}^{\infty} \sum_{M'_m=0}^{\infty} n(\mathcal{M}_e, \{M_m\}, \{M'_m\})$$

$$= C_{N_\uparrow}^{L}\left[C_M^L + C_{M-2}^L\right] - \left[C_{N_\uparrow+1}^L + C_{N_\uparrow-1}^L\right]C_{M-1}^L. \tag{4.B.60}$$

This formula is the close analogue of the result (4.B.8) for the **XXX** Heisenberg model.   □

5

Thermodynamics in the Yang-Yang approach

Let us now turn to the determination of thermodynamic quantities and the zero-temperature excitation spectrum in the thermodynamic limit. A convenient way to construct the spectrum was pioneered by C. N. Yang and C. P. Yang for the case of the delta-function Bose gas [496]. The starting point are the Bethe Ansatz equations in the finite volume. They are used to derive a set of coupled, nonlinear integral equations called thermodynamic Bethe Ansatz (TBA) equations, which describe the thermodynamics of the model at finite temperatures. The quantities entering these equations have a natural interpretation in terms of dressed energies of elementary excitations. Yang and Yang's formalism is a natural generalization of the thermodynamics of the free Fermi gas to interacting systems. In order to elucidate this point we briefly review the calculations for the thermodynamics of noninteracting electrons.

5.1 A point of reference: noninteracting electrons

Let us consider a tight-binding model of noninteracting electrons, described by the Hamiltonian

$$\mathcal{H} = -\sum_{j=1}^{L}\sum_{\sigma} c_{j,\sigma}^{\dagger} c_{j+1,\sigma} + \text{h.c.} - \mu \widehat{N} - 2BS^z , \tag{5.1}$$

The spectrum is easily determined by means of Fourier transformation. We define the electron annihilation operator in momentum space by

$$\tilde{c}_{k_l,\sigma} = \sum_{j=1}^{L} \exp(-ijk_l)\, c_{j,\sigma} , \tag{5.2}$$

where the momenta are quantized according to

$$k_j = \frac{2\pi n_j}{L} , \quad n_j = -\frac{L}{2}, \dots, \frac{L}{2} - 1. \tag{5.3}$$

In momentum space the Hamiltonian reads

$$\mathcal{H} = \frac{1}{L}\sum_{l,\sigma} [-2\cos(k_l) - \mu_\sigma]\, \tilde{n}_\sigma(k_l) , \tag{5.4}$$

149

where $\tilde{n}_\sigma(k_l) = \tilde{c}^\dagger_{k_l,\sigma} \tilde{c}_{k_l,\sigma}$ and where we have defined chemical potentials for spin-up and spin-down electrons by

$$\mu_\uparrow = \mu + B , \quad \mu_\downarrow = \mu - B . \tag{5.5}$$

The quantity

$$\varepsilon_0^{(\sigma)}(k) = -2\cos(k) - \mu_\sigma ,$$

is the bare energy of spin-σ electrons. In the thermodynamic limit the Hamiltonian becomes

$$\mathcal{H} = \int_{-\pi}^{\pi} \frac{dk}{2\pi} \sum_\sigma \varepsilon_0^{(\sigma)}(k) \, \tilde{n}_\sigma(k) , \tag{5.6}$$

where the Fermi creation and annihilation operators now fulfil the anticommutation relations

$$\{\tilde{c}_{k,\sigma}, \tilde{c}^\dagger_{k',\sigma'}\} = 2\pi \, \delta(k - k') \, \delta_{\sigma,\sigma'} . \tag{5.7}$$

The momentum density of states for electrons with spin σ is constant

$$\rho_L^{(\sigma)}(k_l) = \frac{1}{L(k_{l+1} - k_l)} = \frac{1}{2\pi}. \tag{5.8}$$

Let us now turn to the description of thermodynamic properties. We will construct the thermodynamic limit starting from a finite but large volume L. We then keep the densities of up and down spins constant, while taking $L \to \infty$

$$\frac{N_\sigma}{L} = n_\sigma . \tag{5.9}$$

Let us consider a state characterized by the N_σ momenta $k_1^\sigma < k_2^\sigma < \ldots < k_{N_\sigma}^\sigma$. As the momentum density of states is constant, there is a flat distribution of L 'vacancies' for the N_σ momenta. A single particle state with given momentum k_a and spin σ can be either occupied, in which case we speak of a 'particle' with spin σ and momentum k_a being present, or unoccupied, in which case we speak of a 'hole' with quantum numbers k_a and σ. Let us introduce densities for particles $\rho_{L,p}^{(\sigma)}(k)$ and holes $\rho_{L,h}^{(\sigma)}(k)$

# of particles with spin σ and momentum in $[k, k + \Delta k] = \rho_{L,p}^{(\sigma)}(k) \, \Delta k$,

# of holes with spin σ and momentum in $[k, k + \Delta k] = \rho_{L,h}^{(\sigma)}(k) \, \Delta k$. $\tag{5.10}$

By construction we have

$$\rho_L^{(\sigma)}(k) = \rho_{L,p}^{(\sigma)}(k) + \rho_{L,h}^{(\sigma)}(k) = \frac{1}{2\pi}. \tag{5.11}$$

In the finite volume the energy per site of an eigenstate of $\mathcal{H}$ with N_σ electrons with spin σ and corresponding momenta $\{k_a^\sigma\}$ is given by

$$
\begin{aligned}
e_L &= \frac{1}{L} \sum_{\sigma=\uparrow,\downarrow} \sum_{a=1}^{N_\sigma} [-2\cos(k_a^\sigma) - \mu_\sigma] \\
&= \sum_a \sum_\sigma [-2\cos(k_a^\sigma) - \mu_\sigma] \frac{k_{a+1}^\sigma - k_a^\sigma}{L[k_{a+1}^\sigma - k_a^\sigma]} \\
&= \sum_a \sum_\sigma [-2\cos(k_a^\sigma) - \mu_\sigma](k_{a+1}^\sigma - k_a^\sigma) \, \rho_{L,p}^{(\sigma)}(k_a) .
\end{aligned}
\tag{5.12}
$$

We note that the last line of (5.12) is of the form of a discretized integral. Now we take the infinite volume limit for fixed densities n_σ. In a finite volume the momentum densities of states are defined on a discrete set of momenta. In the thermodynamic limit they turn into functions of the continuous momentum

$$
\rho_\alpha^{(\sigma)}(k) = \lim_{L\to\infty} \rho_{L,\alpha}^{(\sigma)}(k_a) , \quad \alpha = p, h.
\tag{5.13}
$$

Employing this prescription and turning sums into integrals, we obtain the following expression for the energy per site of an eigenstate of $\mathcal{H}$ described by a particle density $\rho_p^{(\sigma)}(k)$ in the thermodynamic limit

$$
e = \int_{-\pi}^{\pi} dk \sum_\sigma [-2\cos(k) - \mu_\sigma] \, \rho_p^{(\sigma)}(k) .
\tag{5.14}
$$

Let us now describe the entropy, which is defined as the logarithm of the available states. The number of states with spin σ and momentum in the interval $[k, k + \Delta k]$ is equal to the number of possibilities for distributing $L \Delta k \, \rho_{L,p}^{(\sigma)}(k)$ particles among $L \Delta k \, \rho_L^{(\sigma)}(k)$ vacancies, which is equal to

$$
\exp(\Delta S) = \prod_\sigma \frac{[L\Delta k \, \rho_L^{(\sigma)}(k)]!}{[L\Delta k \, \rho_{L,p}^{(\sigma)}(k)]! \, [L\Delta k \, \rho_{L,h}^{(\sigma)}(k)]!} .
\tag{5.15}
$$

For large L we are dealing with factorials of large numbers, which can be approximated by using Stirling's formula. The $\mathcal{O}(L)$ contribution to the entropy is then given by

$$
\Delta S = L\Delta k \sum_\sigma \rho_L^{(\sigma)}(k) \ln[\rho_L^{(\sigma)}(k)] - \rho_{L,p}^{(\sigma)}(k) \ln[\rho_{L,p}^{(\sigma)}(k)] - \rho_{L,h}^{(\sigma)}(k) \ln[\rho_{L,h}^{(\sigma)}(k)].
\tag{5.16}
$$

In the thermodynamic limit the entropy per site becomes

$$
s = \int_{-\pi}^{\pi} dk \sum_\sigma \left(\rho^{(\sigma)}(k) \ln[\rho^{(\sigma)}(k)] - \rho_p^{(\sigma)}(k) \ln[\rho_p^{(\sigma)}(k)] - \rho_h^{(\sigma)}(k) \ln[\rho_h^{(\sigma)}(k)] \right).
\tag{5.17}
$$

Combining equations (5.17) and (5.14) we may express the Gibbs free energy per site as

$$f = \int_{-\pi}^{\pi} dk \sum_{\sigma} [-2\cos(k) - \mu_\sigma] \, \rho_p^{(\sigma)}(k)$$

$$-T \int_{-\pi}^{\pi} dk \sum_{\sigma} \left\{ \mathcal{L}\left[\rho^{(\sigma)}(k)\right] - \mathcal{L}\left[\rho_p^{(\sigma)}(k)\right] - \mathcal{L}\left[\rho_h^{(\sigma)}(k)\right]\right\}, \qquad (5.18)$$

where we have introduced the shorthand notation

$$\mathcal{L}[f(x)] = f(x) \, \ln[f(x)] . \qquad (5.19)$$

We now regard the Gibbs free energy to be a functional of the particle and hole densities $\rho_{p,h}^{(\sigma)}(k)$, which are still subject to the constraint (5.11). In the state of thermodynamic equilibrium the free energy must be stationary with respect to variations of the particle/hole densities

$$0 = \delta f = \int_{-\pi}^{\pi} dk \left[\sum_{\sigma} \frac{\delta f}{\delta \rho_p^{(\sigma)}(k)} \delta\rho_p^{(\sigma)}(k) + \frac{\delta f}{\delta \rho_h^{(\sigma)}(k)} \delta\rho_h^{(\sigma)}(k) \right]. \qquad (5.20)$$

The constraint (5.11) tells us that the variations of $\rho_h^{(\sigma)}(k)$ and $\rho_p^{(\sigma)}(k)$ are not independent, but rather

$$\delta\rho_h^{(\sigma)}(k) = -\delta\rho_p^{(\sigma)}(k). \qquad (5.21)$$

Evaluating the functional derivatives in (5.20) and then using (5.11) and (5.21) we obtain an equation for the *ratio* of hole and particle densities

$$\xi^{(\sigma)}(k) = \frac{\rho_h^{(\sigma)}(k)}{\rho_p^{(\sigma)}(k)},$$

$$T \ln\left[\xi^{(\sigma)}(k)\right] = -2\cos(k) - \mu_\sigma = \varepsilon_0^{(\sigma)}(k). \qquad (5.22)$$

The particle and hole densities in thermal equilibrium are expressed in terms of $\xi^{(\sigma)}(k)$ as

$$\rho_p^{(\sigma)}(k) = \frac{1}{2\pi[1 + \xi^{(\sigma)}(k)]} = \frac{1}{2\pi[1 + \exp\left(\varepsilon_0^{(\sigma)}(k)/T\right)]},$$

$$\rho_h^{(\sigma)}(k) = \frac{\xi^{(\sigma)}(k)}{2\pi[1 + \xi^{(\sigma)}(k)]} = \frac{\exp\left(\varepsilon_0^{(\sigma)}(k)/T\right)}{2\pi[1 + \exp\left(\varepsilon_0^{(\sigma)}(k)/T\right)]}. \qquad (5.23)$$

In (5.23) we recognize the well-known expressions for the Fermi-Dirac momentum occupation numbers $\bar{n}^{(\sigma)}(k) = 2\pi\rho_p^{(\sigma)}(k)$ for noninteracting particles and holes with a dispersion $\varepsilon^{(\sigma)}(k)$, see e.g. Chapter 8 of Ref. [200]. Inserting (5.23) back into the equation for the Gibbs free energy we arrive at

$$f = -T \int_{-\pi}^{\pi} \frac{dk}{2\pi} \sum_{\sigma} \ln\left[1 + \exp(-\varepsilon_0^{(\sigma)}(k)/T)\right]. \qquad (5.24)$$

Equation (5.24) is the well known result for the free energy of a Fermi gas, where the particles have an energy $\varepsilon_0^{(\sigma)}(k)$.

5.2 Thermodynamic Bethe Ansatz (TBA) equations

Let us now try to generalize the steps we went through for noninteracting electrons to the case of the Hubbard model. We shall follow Takahashi's derivation of the TBA equations for the repulsive Hubbard model [435]. The analogous calculations for the attractive case can be found in [287]. Let us recall that we are working with the following form of the Hubbard Hamiltonian

$$
\mathcal{H} = -\sum_{j=1}^{L}\sum_{\sigma} c_{j,\sigma}^{\dagger}c_{j+1,\sigma} + c_{j+1,\sigma}^{\dagger}c_{j,\sigma} + 4u \sum_{j=1}^{L}(n_{j,\uparrow} - \tfrac{1}{2})(n_{j,\downarrow} - \tfrac{1}{2})
$$
$$
-\mu \sum_{j=1}^{L}\left(n_{j,\uparrow} + n_{j,\downarrow}\right) - B \sum_{j=1}^{L}\left(n_{j,\uparrow} - n_{j,\downarrow}\right). \tag{5.25}
$$

There are three essential steps to our derivation.

- In Step 1 we show how to describe a *general* eigenstate of the Hamiltonian in the thermodynamic limit

$$
L \to \infty, \qquad N_{\sigma} \to \infty, \qquad \frac{N_{\sigma}}{L} = \text{fixed}, \qquad \sigma = \uparrow, \downarrow. \tag{5.26}
$$

This is done by turning Takahashi's equations (4.36)–(4.38) into a system of coupled integral equations for root densities of particles and holes. The integral equations will allow us to express the root densities of holes in terms of those of particles. The analogous relation for noninteracting fermions is given by equation (5.11).
- In Step 2 we express the entropy and the Gibbs free energy in terms of the root distribution functions.
- Finally, in Step 3 we minimize the Gibbs free energy with respect to the root densities and obtain a set of coupled, nonlinear integral equations that describe the state of thermodynamic equilibrium.

Steps 2 and 3 are analogous, albeit significantly more complicated, to our derivation for noninteracting electrons in Section 5.1.

Step 1: Our starting point are Takahashi's equations (4.36)–(4.38) and expressions for energy (4.52) and momentum (4.51) for large but finite L. The key fact we will use is that in the framework of the string hypothesis there is a one-to-one correspondence between sets of (half-odd) integers and sets of spectral parameters, i.e.

$$
\{I_j, J_{\alpha}^n, J_{\beta}'^m\} \longleftrightarrow \{k_j, \Lambda_{\alpha}^n, \Lambda_{\beta}'^m\}. \tag{5.27}
$$

Every permitted set $\{I_j, J_{\alpha}^n, J_{\beta}'^m\}$ *uniquely* specifies a solution $\{k_j, \Lambda_{\alpha}^n, \Lambda_{\beta}'^m\}$ of Takahashi's equations and thus an eigenfunction of the Hamiltonian. In what follows it will be useful to adopt the following picture. The integers I_j have a certain allowed range, in which we

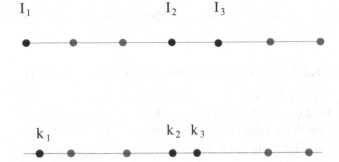

Fig. 5.1. Distributions of (half-odd) integers I_j and corresponding distribution of spectral parameters k_j for a particular solution of Takahashi's equations.

choose a particular distribution when we specify the set (5.27). Due to (5.27) there is a corresponding pattern of spectral parameters k_j. This can be visualized as shown in figure 5.1. Analogous pictures hold for the (half-odd) integers J_α^n ($J_\alpha'^n$) and the corresponding spectral parameters Λ_α^n ($\Lambda_\alpha'^n$). In what follows it will be very useful to view the distributions of spectral parameters in terms of *particles* and *holes*.

A very important property of Takahashi's equations (4.36)–(4.38) is that as we approach the thermodynamic limit $L \to \infty$, N/L and M/L fixed (finite densities of electrons and spin down electrons), the roots of (4.36)–(4.38) become dense

$$k_{j+1} - k_j = \mathcal{O}(L^{-1}), \quad \Lambda_{\alpha+1}^n - \Lambda_\alpha^n = \mathcal{O}(L^{-1}), \quad \Lambda_{\alpha+1}'^n - \Lambda_\alpha'^n = \mathcal{O}(L^{-1}). \quad (5.28)$$

We now define so-called *counting functions* y, z_n, z_n' as follows

$$Ly(k) = kL + \sum_{n=1}^{\infty}\sum_{\alpha=1}^{M_n} \theta\left(\frac{\sin k - \Lambda_\alpha^n}{nu}\right) + \sum_{n=1}^{\infty}\sum_{\alpha=1}^{M_n'} \theta\left(\frac{\sin k - \Lambda_\alpha'^n}{nu}\right), \quad (5.29)$$

$$Lz_n(\Lambda) = \sum_{j=1}^{N-2M'} \theta\left(\frac{\Lambda - \sin k_j}{nu}\right) - \sum_{m=1}^{\infty}\sum_{\beta=1}^{M_m} \Theta_{nm}\left(\frac{\Lambda - \Lambda_\beta^m}{u}\right), \quad (5.30)$$

$$Lz_n'(\Lambda') = L[\arcsin(\Lambda' + niu) + \arcsin(\Lambda' - niu)]$$
$$- \sum_{j=1}^{N-2M'} \theta\left(\frac{\Lambda' - \sin k_j}{nu}\right) - \sum_{m=1}^{\infty}\sum_{\beta=1}^{M_m'} \Theta_{nm}\left(\frac{\Lambda' - \Lambda_\beta'^m}{u}\right). \quad (5.31)$$

By definition the counting functions satisfy the following equations when evaluated for a given solution of the Takahashi equations

$$y(k_j) = \frac{2\pi I_j}{L}, \quad z_n'(\Lambda_\alpha'^n) = \frac{2\pi J_\alpha'^n}{L}, \quad z_n(\Lambda_\alpha^n) = \frac{2\pi J_\alpha^n}{L}. \quad (5.32)$$

The counting functions have the important property that they are monotonically increasing functions of their arguments. Let us see what the counting functions look like for a particular example. We choose $L = 6$, $u = 1.25$ and $N = 3$. The solution of the Bethe Ansatz

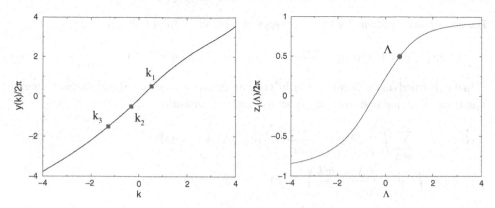

Fig. 5.2. Counting functions $y(k)/2\pi$, $z_1(\Lambda)/2\pi$ and spectral parameters k_1, k_2, k_3, Λ of the solution (5.33) of the Lieb-Wu equations.

equations we consider has $M = 1$ and is given by

$$k_1 = 0.54266222, \quad k_2 = -0.31583772, \quad k_3 = -1.2740220,$$
$$\Lambda_1 = 0.587983. \tag{5.33}$$

In figure 5.2 we plot the counting functions $y(k)/2\pi$ and $z_1(\Lambda)/2\pi$ for our example and also indicate the spectral parameters k_j and Λ. We see that the counting functions are indeed monotonically increasing and that $y(k_1) = \pi$, $y(k_2) = -\pi$, $y(k_3) = -3\pi$ and $z_1(\Lambda) = \pi$.

In the next step we define *root densities*, which are related to the counting functions as follows. By definition the counting functions 'enumerate' the Bethe Ansatz roots e.g.

$$L[y(k_j) - y(k_n)] = 2\pi(I_j - I_n). \tag{5.34}$$

For a given solution of Takahashi's equations (4.36)–(4.38) some of the (half-odd) integers between I_j and I_n will be 'occupied' i.e. there will be a corresponding root k, whereas others will be omitted. We describe the corresponding k-values in terms of a root density $\rho^p(k)$ for 'particles' and a density $\rho^h(k)$ for 'holes'. These root densities for particles and holes are the analogs of the particle and hole densities of state for noninteracting electrons. In a very large system we then have by definition (here the property (5.28) is essential)

$$L\rho^p(k)\,dk = \text{number of particles in } dk,$$
$$L\rho^h(k)\,dk = \text{number of holes in } dk. \tag{5.35}$$

Combining equations (5.34) and (5.35) it is then clear that in the thermodynamic limit we have

$$2\pi[\rho^p(k) + \rho^h(k)] = \frac{dy(k)}{dk}. \tag{5.36}$$

The analogous equations for the other types of roots in (4.36)–(4.38) are

$$2\pi \left[\sigma_n^P(\Lambda) + \sigma_n^h(\Lambda)\right] = \frac{dz_n(\Lambda)}{d\Lambda}, \qquad 2\pi \left[\sigma_n'^P(\Lambda) + \sigma_n'^h(\Lambda)\right] = \frac{dz_n'(\Lambda)}{d\Lambda}. \quad (5.37)$$

In the thermodynamic limit Takahashi's equations can now be turned into coupled integral equations involving both counting functions and root densities

$$y(k) = k + \sum_{n=1}^{\infty} \int_{-\infty}^{\infty} d\Lambda \, \theta \left(\frac{\sin k - \Lambda}{nu}\right) \left[\sigma_n'^P(\Lambda) + \sigma_n^P(\Lambda)\right], \quad (5.38)$$

$$z_n(\Lambda) = \int_{-\pi}^{\pi} dk \, \theta \left(\frac{\Lambda - \sin k}{nu}\right) \rho^P(k)$$

$$- \sum_{m=1}^{\infty} \int_{-\infty}^{\infty} d\Lambda' \, \Theta_{nm} \left(\frac{\Lambda - \Lambda'}{u}\right) \sigma_m^P(\Lambda'), \quad (5.39)$$

$$z_n'(\Lambda) = \arcsin(\Lambda + niu) + \arcsin(\Lambda - niu)$$

$$- \int_{-\pi}^{\pi} dk \, \theta \left(\frac{\Lambda - \sin k}{nu}\right) \rho^P(k) - \sum_{m=1}^{\infty} \int_{-\infty}^{\infty} d\Lambda' \, \Theta_{nm} \left(\frac{\Lambda - \Lambda'}{u}\right) \sigma_m'^P(\Lambda').$$

$$(5.40)$$

By differentiating (5.38)–(5.39) we obtain a set of equations that contains only the root densities for particles and holes

$$\rho^P(k) + \rho^h(k) = \frac{1}{2\pi} + \cos k \sum_{n=1}^{\infty} \int_{-\infty}^{\infty} d\Lambda \, a_n(\Lambda - \sin k) \left[\sigma_n'^P(\Lambda) + \sigma_n^P(\Lambda)\right],$$

$$\sigma_n^h(\Lambda) = -\sum_{m=1}^{\infty} A_{nm} * \sigma_m^P \Big|_{\Lambda} + \int_{-\pi}^{\pi} dk \, a_n(\sin k - \Lambda) \, \rho^P(k),$$

$$\sigma_n'^h(\Lambda) = \frac{1}{\pi} \mathrm{Re} \frac{1}{\sqrt{1 - (\Lambda - inu)^2}} - \sum_{m=1}^{\infty} A_{nm} * \sigma_m'^P \Big|_{\Lambda}$$

$$- \int_{-\pi}^{\pi} dk \, a_n(\sin k - \Lambda) \, \rho^P(k). \quad (5.41)$$

Here we have introduced the shorthand notation

$$a_n(x) = \frac{1}{2\pi} \frac{2nu}{(nu)^2 + x^2}, \quad (5.42)$$

and A_{nm} is an integral operator acting on a function f as

$$A_{nm} * f \Big|_{x} = \delta_{nm} f(x) + \int_{-\infty}^{\infty} \frac{dy}{2\pi} \frac{d}{dx} \Theta_{nm} \left(\frac{x - y}{u}\right) f(y). \quad (5.43)$$

Equations (5.41) can be used to express the densities of holes in terms of densities of particles. They are the analog of equation (5.11) for noninteracting electrons.

Step 2: In order to obtain an expression for the Gibbs free energy we need to determine the entropy. A general state is characterized by its distribution of Bethe Ansatz roots. In the thermodynamic limit, states can be described in terms of the root densities of particles and holes. We therefore can view the entropy as a functional of the root densities. For a given set of particle and hole densities, the contribution to the entropy of, e.g., states with k lying in the interval $[k, k + \Delta k]$ can be determined as follows:

The number of vacancies for k in the interval $[k, k + \Delta k]$ is $L(\rho^p(k) + \rho^h(k))\Delta k$. Of these vacancies $L\rho^p(k)\Delta k$ are occupied. The total number of states with k lying in the interval $[k, k + \Delta k]$ is

$$\frac{(L[\rho^p(k) + \rho^h(k)]\Delta k)!}{(L\rho^p(k)\Delta k)!(L\rho^h(k)\Delta k)!}, \qquad (5.44)$$

where '!' denotes the factorial. The corresponding contribution dS to the entropy is the logarithm of this number of states, which becomes large as we approach the thermodynamic limit. Hence we can use Stirling's formula to approximate the factorials in (5.44). In this way we obtain the following expression for the total entropy per site of the Hubbard model

$$
\begin{aligned}
s = \int_{-\pi}^{\pi} dk \, & \left\{ \mathcal{L}\left[\rho^p(k) + \rho^h(k)\right] - \mathcal{L}[\rho^p(k)] - \mathcal{L}[\rho^h(k)]\right\} \\
& + \sum_{n=1}^{\infty} \int_{-\infty}^{\infty} d\Lambda \left\{ \mathcal{L}\left[\sigma_n'^p(\Lambda) + \sigma_n'^h(\Lambda)\right] - \mathcal{L}\left[\sigma_n'^p(\Lambda)\right] - \mathcal{L}\left[\sigma_n'^h(\Lambda)\right]\right\} \\
& + \sum_{n=1}^{\infty} \int_{-\infty}^{\infty} d\Lambda \left\{ \mathcal{L}\left[\sigma_n^p(\Lambda) + \sigma_n^h(\Lambda)\right] - \mathcal{L}\left[\sigma_n^p(\Lambda)\right] - \mathcal{L}\left[\sigma_n^h(\Lambda)\right]\right\}, \qquad (5.45)
\end{aligned}
$$

where $\mathcal{L}[f(x)]$ was defined in (5.19). The Gibbs free energy per site is

$$
\begin{aligned}
f(\mu, B, T) & = e - \mu n_c - 2Bm - Ts \\
& = \int_{-\pi}^{\pi} dk \, [-2\cos k - \mu - 2u - B]\, \rho^p(k) \\
& \quad + \sum_{n=1}^{\infty} \int_{-\infty}^{\infty} d\Lambda \left[4\mathrm{Re}\sqrt{1 - (\Lambda - inu)^2} - n(2\mu + 4u)\right] \sigma_n'^p(\Lambda) \\
& \quad + \sum_{n=1}^{\infty} \int_{-\infty}^{\infty} d\Lambda \, 2nB\, \sigma_n^p(\Lambda) - Ts + u. \qquad (5.46)
\end{aligned}
$$

Here μ is a chemical potential, B is a magnetic field, T is the temperature, n_c is the particle density and m the magnetization per site. We consider only positive values of magnetic field $B \geq 0$ and negative values of chemical potential $\mu \leq 0$. All other parameter regions can be obtained from the one we consider by employing the discrete symmetries of the Hamiltonian (see Section 2.2.4). Due to the symmetry of the Hamiltonian (in zero field) (2.31) under interchange of up and down spins we have (2.63)

$$f(\mu, -B, T) = f(\mu, B, T). \qquad (5.47)$$

Similarly, if we employ the particle-hole transformation

$$c_{j,\uparrow}^{\dagger} \leftrightarrow c_{j,\downarrow}(-1)^{j}, \quad c_{j,\downarrow}^{\dagger} \leftrightarrow c_{j,\uparrow}(-1)^{j} \tag{5.48}$$

we may derive the equality (2.64)

$$f(-\mu, B, T) = f(\mu, B, T) + 2\mu. \tag{5.49}$$

Step 3: The state of thermodynamic equilibrium minimizes the Gibbs free energy f (5.46). As f is a functional of the root densities, the state of thermodynamic equilibrium must be stationary with respect to variations in a maximal set of *independent* root densities. As the hole densities are expressed in terms of the particle densities by (5.41), this yields the requirement

$$
\begin{aligned}
0 = \delta f = &\int_{-\pi}^{\pi} dk \left[\frac{\delta f}{\delta \rho^{p}(k)} \delta \rho^{p}(k) + \frac{\delta f}{\delta \rho^{h}(k)} \delta \rho^{h}(k) \right] \\
&+ \sum_{n=1}^{\infty} \int_{-\infty}^{\infty} d\Lambda \left[\frac{\delta f}{\delta \sigma_{n}'^{p}(\Lambda)} \delta \sigma_{n}'^{p}(\Lambda) + \frac{\delta f}{\delta \sigma_{n}'^{h}(\Lambda)} \delta \sigma_{n}'^{h}(\Lambda) \right] \\
&+ \sum_{n=1}^{\infty} \int_{-\infty}^{\infty} d\Lambda \left[\frac{\delta f}{\delta \sigma_{n}^{p}(\Lambda)} \delta \sigma_{n}^{p}(\Lambda) + \frac{\delta f}{\delta \sigma_{n}^{h}(\Lambda)} \delta \sigma_{n}^{h}(\Lambda) \right],
\end{aligned}
\tag{5.50}
$$

where we have to take (5.41) into account as constraint equations. We first evaluate the functional derivatives in (5.50) by using the explicit representation (5.46), (5.45) for the free energy f. For example, we have

$$\frac{\delta f}{\delta \rho^{p}(k)} = -2\cos k - \mu - 2u - B - T \ln \left[1 + \frac{\rho^{h}(k)}{\rho^{p}(k)} \right]. \tag{5.51}$$

The constraint equations (5.41) are then used to express all hole densities as well as their variations in terms of particle densities and their variations. For example,

$$\delta \rho^{h}(k) = -\delta \rho^{p}(k) + \cos(k) \int_{-\infty}^{\infty} d\Lambda \sum_{n=1}^{\infty} a_{n}(\Lambda - \sin k) \left[\delta \sigma_{n}'^{p}(\Lambda) + \delta \sigma_{n}^{p}(\Lambda) \right]. \tag{5.52}$$

Finally we require the coefficients of the independent variations $\delta \rho^{p}(k)$, $\delta \sigma_{n}^{p}(\Lambda)$ and $\delta \sigma_{n}'^{p}(\Lambda)$ to vanish. In this way one obtains a set of equations for the ratios

$$
\begin{aligned}
\zeta(k) &= \rho^{h}(k)/\rho^{p}(k), \\
\eta_{n}(\Lambda) &= \sigma_{n}^{h}(\Lambda)/\sigma_{n}^{p}(\Lambda), \\
\eta_{n}'(\Lambda) &= \sigma_{n}'^{h}(\Lambda)/\sigma_{n}'^{p}(\Lambda),
\end{aligned}
\tag{5.53}
$$

$$\ln \zeta(k) = \frac{-2\cos k - \mu - 2u - B}{T}$$

$$+ \sum_{n=1}^{\infty} \int_{-\infty}^{\infty} d\Lambda \, a_n(\sin k - \Lambda) \ln\left(1 + \frac{1}{\eta_n'(\Lambda)}\right)$$

$$- \sum_{n=1}^{\infty} \int_{-\infty}^{\infty} d\Lambda \, a_n(\sin k - \Lambda) \ln\left(1 + \frac{1}{\eta_n(\Lambda)}\right). \tag{5.54}$$

$$\ln\left(1 + \eta_n(\Lambda)\right) = -\int_{-\pi}^{\pi} dk \, \cos(k) \, a_n(\sin k - \Lambda) \ln\left(1 + \frac{1}{\zeta(k)}\right)$$

$$+ \frac{2nB}{T} + \sum_{m=1}^{\infty} A_{nm} * \ln\left(1 + \frac{1}{\eta_m}\right)\bigg|_{\Lambda}. \tag{5.55}$$

$$\ln\left(1 + \eta_n'(\Lambda)\right) + \int_{-\pi}^{\pi} dk \, \cos(k) \, a_n(\sin k - \Lambda) \ln\left(1 + \frac{1}{\zeta(k)}\right)$$

$$= \frac{4\mathrm{Re}\sqrt{1 - (\Lambda - inu)^2} - 2n\mu - 4nu}{T} + \sum_{m=1}^{\infty} A_{nm} * \ln\left(1 + \frac{1}{\eta_m'}\right)\bigg|_{\Lambda}. \tag{5.56}$$

The equations (5.54)–(5.56) are called 'Thermodynamic Bethe Ansatz equations' or TBA equations. We note that (5.41) together with (5.54)–(5.56) completely determine the densities of holes *and* particles in the state of thermal equilibrium. The analog of the TBA equations for noninteracting electrons is (5.22).

The Gibbs free energy per site is expressed in terms of solutions of (5.54)–(5.56) as

$$f = -T \int_{-\pi}^{\pi} \frac{dk}{2\pi} \ln\left(1 + \frac{1}{\zeta(k)}\right) + u$$

$$- T \sum_{n=1}^{\infty} \int_{-\infty}^{\infty} \frac{d\Lambda}{\pi} \ln\left(1 + \frac{1}{\eta_n'(\Lambda)}\right) \mathrm{Re}\frac{1}{\sqrt{1 - (\Lambda - inu)^2}}. \tag{5.57}$$

As we will see it is very useful (in particular for taking the zero temperature limit) to recast (5.54)–(5.56) in a different form. We first note that the inverse of the integral operator A_{mn} is given by

$$A_{kn}^{-1} * f\bigg|_x = \delta_{k,n} f(x) - (\delta_{k-1,n} + \delta_{k+1,n}) \int_{-\infty}^{\infty} dy \, s(x - y) \, f(y), \tag{5.58}$$

where

$$s(x) = \frac{1}{4u \cosh(\pi x / 2u)} = \frac{1}{2\pi} \int_{-\infty}^{\infty} d\omega \, \frac{\exp(-i\omega x)}{2\cosh(\omega u)}. \tag{5.59}$$

By acting with A^{-1} on (5.55) and (5.56) we obtain the following 'tridiagonal form' of the TBA equations

$$\ln \eta_1(\Lambda) = s * \ln(1 + \eta_2)\Big|_\Lambda - \int_{-\pi}^{\pi} dk \, \cos(k) \, s(\Lambda - \sin k) \, \ln\left(1 + \frac{1}{\zeta(k)}\right) ,$$

$$\ln \eta_n(\Lambda) = s * \ln\left([1 + \eta_{n-1}][1 + \eta_{n+1}]\right)\Big|_\Lambda , \quad n = 2, 3, \ldots \tag{5.60}$$

$$\ln \eta_1'(\Lambda) = s * \ln(1 + \eta_2')\Big|_\Lambda - \int_{-\pi}^{\pi} dk \, \cos(k) \, s(\Lambda - \sin k) \, \ln(1 + \zeta(k)) ,$$

$$\tag{5.61}$$

$$\ln \eta_n'(\Lambda) = s * \ln\left([1 + \eta_{n-1}'][1 + \eta_{n+1}']\right)\Big|_\Lambda , \quad n = 2, 3, \ldots \tag{5.62}$$

Here s is an integral operator with kernel (5.59), so that e.g.

$$s * \ln(1 + \eta_2)\Big|_\Lambda = \int_{-\infty}^{\infty} d\lambda \, s(\Lambda - \lambda) \, \ln[1 + \eta_2(\lambda)]. \tag{5.63}$$

Equations (5.60),(5.62) need to supplemented by the following 'boundary conditions', which are obtained directly by taking n to infinity in (5.54)–(5.56)

$$\lim_{n \to \infty} \frac{\ln \eta_n}{n} = \frac{2B}{T} , \quad \lim_{n \to \infty} \frac{\ln \eta_n'}{n} = \frac{-2\mu}{T} . \tag{5.64}$$

Equation (5.54) can be recast in the form

$$\ln \zeta(k) = -\frac{2 \cos k}{T} - \frac{1}{T} \int_{-\infty}^{\infty} dy \, s(\sin k - y) \left[4 \mathrm{Re} \sqrt{1 - (y - iu)^2} \right]$$

$$+ \int_{-\infty}^{\infty} dy \, s(\sin k - y) \, \ln\left(\frac{1 + \eta_1'(y)}{1 + \eta_1(y)}\right) . \tag{5.65}$$

Equations (5.60), (5.62) and (5.65) are derived by using the identities given in Chapter 17.1.

Now we are in a position to define *dressed energies* by

$$\kappa(k) = T \ln(\zeta(k)) , \tag{5.66}$$

$$\varepsilon_n(\Lambda) = T \ln(\eta_n(\Lambda)) , \tag{5.67}$$

$$\varepsilon_n'(\Lambda) = T \ln(\eta_n'(\Lambda)) . \tag{5.68}$$

As was first shown by C. N. Yang and C. P. Yang for the delta-function Bose gas [496], the quantities defined in this way describe the dressed energies of elementary excitations in the zero temperature limit. We will discuss this point in detail in Chapter 7. We note that for noninteracting electrons we had $\xi^{(\sigma)}(k) = \varepsilon_0^{(\sigma)}(k)$, i.e. the ratio of hole and particle densities of states is equal to the exponential of the 'bare' (or single-particle) energy divided by temperature. In the Hubbard model the dressed energies are very different from the bare ones as a result of the electron-electron interactions.

5.3 Thermodynamics

The expression for the Gibbs free energy density (5.57) can be simplified to [436]

$$f = e_0 - \mu - u - T \int_{-\pi}^{\pi} dk \, \rho_0(k) \ln(1 + \zeta(k))$$

$$-T \int_{-\infty}^{\infty} d\Lambda \, \sigma_0(\Lambda) \ln(1 + \eta_1(\Lambda)), \tag{5.69}$$

where

$$\rho_0(k) = \frac{1}{2\pi} + \cos(k) \int_{-\infty}^{\infty} \frac{d\omega}{2\pi} \frac{J_0(\omega) \cos(\omega \sin(k))}{1 + \exp(2u|\omega|)},$$

$$\sigma_0(\Lambda) = \int_{-\infty}^{\infty} \frac{d\omega}{2\pi} \frac{J_0(\omega) \exp(-i\omega\Lambda)}{2 \cosh(u\omega)},$$

$$e_0 = -u - 4 \int_{0}^{\infty} \frac{d\omega}{\omega} \frac{J_0(\omega) J_1(\omega)}{1 + \exp(\omega 2u)} \tag{5.70}$$

and J_0 and J_1 are Bessel functions.

We note that ρ_0, σ_0 and e_0 are the root density for real k's, the root density for real Λ's and the ground state energy per site for the half filled repulsive Hubbard model, respectively [298] (see Eqn. (6.B.14)). Since the occurrence of quantities related to the half filled Hubbard model in (5.69) may be surprising, we would like to emphasize that (5.69), (5.70) hold for all negative values of the chemical potential μ, i.e. for all particle densities between zero and one.

The representation (5.69) is convenient as it shows that the Gibbs free energy is determined by the dressed energies for real k's and real Λ's only. In order to derive (5.69) the following identities are useful

$$\int_{-\pi}^{\pi} dk \, a_n(\sin k - \Lambda) \rho_0(k) = \frac{1}{\pi} \text{Re} \frac{1}{\sqrt{1 - (\Lambda - inu)^2}},$$

$$A_{n1} * \sigma_0 \bigg|_{\Lambda} = \int_{-\infty}^{\infty} \frac{d\omega}{2\pi} J_0(\omega) e^{-nu|\omega|} e^{-i\omega\Lambda} = \int_{-\pi}^{\pi} \frac{dk}{2\pi} a_n(\Lambda - \sin k)$$

$$= \int_{-\pi}^{\pi} dk \, a_n(\sin k - \Lambda) \rho_0(k). \tag{5.71}$$

At very low temperatures $T \ll B$ it is possible to determine the Gibbs free energy by using an expansion of the TBA equations (5.54)–(5.56) for small T [436]. The TBA equations essentially reduce to only two coupled equations for ζ and η_1 in this limit. For generic values of B and arbitrary temperatures one needs to resort to a numerical solution of (5.54)–(5.56). In order to do so, one needs to truncate the infinite towers of equations for Λ and k-Λ strings at some finite value of their respective lengths. In [240, 469] such a truncated set of equations was solved by iteration. The integrals were discretized by using approximately 50 (100) points for the k (Λ) integrations. The results of these computations are compared to the results of the Quantum Transfer Matrix approach in chapter 13.

5.4 Infinite temperature limit

Let us consider the TBA equations in the limit

$$T \to \infty, \quad \frac{B}{T}, \frac{\mu}{T} \quad \text{fixed.} \tag{5.72}$$

In this limit we may neglect the driving terms in the TBA equations (5.54)–(5.56), which implies that $\zeta(k)$, $\eta_n(\Lambda)$, $\eta'_n(\Lambda)$ are *constants*,

$$\zeta(k) \longrightarrow \zeta,$$
$$\eta_n(\Lambda) \longrightarrow \eta_n,$$
$$\eta'_n(\Lambda) \longrightarrow \eta'_n. \tag{5.73}$$

The TBA equations in their tridiagonal form (5.60)–(5.65) then turn into simple recursion relations

$$2\ln \eta_n = \ln(1 + \eta_{n-1}) + \ln(1 + \eta_{n+1}), \quad \lim_{n\to\infty} \frac{\ln \eta_n}{n} = \frac{2B}{T},$$

$$2\ln \eta'_n = \ln(1 + \eta'_{n-1}) + \ln(1 + \eta'_{n+1}), \quad \lim_{n\to\infty} \frac{\ln \eta'_n}{n} = \frac{-2\mu}{T},$$

$$2\ln \zeta = \ln\left[\frac{1 + \eta'_1}{1 + \eta_1}\right]. \tag{5.74}$$

Here we have defined $\eta_0 = 0$ and $\eta'_0 = 0$. The solution of (5.74) is

$$\eta_n = \left[\frac{\sinh\left(\frac{(n+1)B}{T}\right)}{\sinh\left(\frac{B}{T}\right)}\right]^2 - 1,$$

$$\eta'_n = \left[\frac{\sinh\left(\frac{(n+1)\mu}{T}\right)}{\sinh\left(\frac{\mu}{T}\right)}\right]^2 - 1,$$

$$\zeta = \frac{\cosh\left(\frac{\mu}{T}\right)}{\cosh\left(\frac{B}{T}\right)}. \tag{5.75}$$

The free energy per site can be calculated from (5.69)

$$f \simeq -T \ln\left[1 + 2\exp\left(\frac{\mu}{T}\right)\cosh\left(\frac{B}{T}\right) + \exp\left(\frac{2\mu}{T}\right)\right]. \tag{5.76}$$

It follows that the entropy per site s in the limit $T \to \infty$ is

$$s = \ln(4). \tag{5.77}$$

This is what one expects for a system with four degrees of freedom per site.[1]

---

[1] The partition function on a system with L sites is $Z(T) = \text{tr}[\exp(-H/T)]$ and $Z(T \to \infty) = \text{tr}\, 1 = 4^L$.

5.5 Zero temperature limit

In the limit $T \to 0$ the TBA equations simplify in an essential way. The key to these simplifications is that many of the dressed energies $\varepsilon_n(\Lambda)$, $\varepsilon'_m(\Lambda)$ are positive for all values of Λ and as a consequence their contributions to the right hand sides of the TBA equations (5.54)–(5.56) vanish. For example, by equation (5.60) we have

$$\varepsilon_n(\Lambda) = T \, \ln \eta_n(\Lambda) > 0 \quad n = 2, 3, \dots \tag{5.78}$$

Together with the 'boundary conditions' (5.64) this implies that for any $B > 0$

$$\lim_{T \to 0} \ln \left(1 + \frac{1}{\eta_n(\Lambda)} \right) = 0 \,, \quad n \geq 2. \tag{5.79}$$

Similarly (5.62) implies that

$$\varepsilon'_n(\Lambda) > 0 \quad n = 2, 3, \dots \tag{5.80}$$

and in conjunction with (5.64) we arrive at

$$\lim_{T \to 0} \ln \left(1 + \frac{1}{\eta'_n(\Lambda)} \right) = 0 \,, \quad n \geq 2. \tag{5.81}$$

Finally, as is shown in Appendix 5.A, we have

$$\varepsilon'_1(\Lambda) > 0 \,; \quad \lim_{T \to 0} \ln \left(1 + \frac{1}{\eta'_1(\Lambda)} \right) = 0 \,. \tag{5.82}$$

5.5.1 Dressed energies

Using (5.79)–(5.82) in the TBA equations (5.54)–(5.56) we arrive at the following set of equations that describe the zero temperature limit

$$\kappa(k) = -2\cos k - \mu - 2u - B + \int_{-\infty}^{\infty} d\Lambda \, a_1(\sin k - \Lambda)\varepsilon_1^-(\Lambda), \tag{5.83}$$

$$\varepsilon_1(\Lambda) = 2B + \int_{-\pi}^{\pi} dk \, \cos(k) \, a_1(\sin k - \Lambda) \, \kappa^-(k)$$
$$- \int_{-\infty}^{\infty} d\Lambda' \, a_2(\Lambda - \Lambda') \, \varepsilon_1^-(\Lambda') \,, \tag{5.84}$$

$$\varepsilon_n(\Lambda) = 2nB + \int_{-\pi}^{\pi} dk \, \cos(k) \, a_n(\sin k - \Lambda)\kappa^-(k) - A_{n1} * \varepsilon_1^- \Big|_\Lambda \,, \tag{5.85}$$

$$\varepsilon'_n(\Lambda) = 4\mathrm{Re}\sqrt{1 - (\Lambda - inu)^2} - 2n\mu - 4nu$$
$$+ \int_{-\pi}^{\pi} dk \, \cos(k) \, a_n(\sin k - \Lambda) \, \kappa^-(k) \,. \tag{5.86}$$

Here the quantities κ^- and ε_1^- are defined by

$$f(x) = f^+(x) + f^-(x), \tag{5.87}$$

$$f^-(x) = \begin{cases} f(x) & \text{if } f(x) < 0 \\ 0 & \text{if } f(x) \geq 0 \end{cases} ; \quad f^+(x) = \begin{cases} 0 & \text{if } f(x) < 0 \\ f(x) & \text{if } f(x) \geq 0 \end{cases}. \tag{5.88}$$

It is shown in Appendix 5.B that $\kappa(k)$ and $\varepsilon_1(\Lambda)$ are symmetric functions of their arguments and that they increase monotonously with $k > 0$ and $\Lambda > 0$ respectively. This means that $\kappa(k)$ is negative in an interval $[-Q, Q]$ and positive outside this interval

$$\kappa(\pm Q) = 0, \qquad \kappa(k) \begin{cases} < 0 & \text{for } |k| < Q \\ > 0 & \text{for } |k| > Q. \end{cases} \tag{5.89}$$

Similarly we have for $\varepsilon_1(\Lambda)$

$$\varepsilon_1(\pm A) = 0, \qquad \varepsilon_1(\Lambda) \begin{cases} < 0 & \text{for } |\Lambda| < A \\ > 0 & \text{for } |\Lambda| > A. \end{cases} \tag{5.90}$$

5.5.2 Root densities

The equations (5.41) for the root densities simplify as well in the $T \to 0$ limit. By exploiting the positivity of the dressed energies we obtain

$$\lim_{T \to 0} \frac{\sigma_n^P(\Lambda)}{\sigma_n^h(\Lambda)} = \lim_{T \to 0} \exp\left(-\frac{\varepsilon_n(\Lambda)}{T}\right) = 0, \ n \geq 2,$$

$$\lim_{T \to 0} \frac{\sigma'^P_n(\Lambda)}{\sigma'^h_n(\Lambda)} = \lim_{T \to 0} \exp\left(-\frac{\varepsilon'_n(\Lambda)}{T}\right) = 0, \ n \geq 1,$$

$$\lim_{T \to 0} \frac{\sigma_1^P(\Lambda)}{\sigma_1^h(\Lambda)} = \lim_{T \to 0} \exp\left(-\frac{\varepsilon_1(\Lambda)}{T}\right) = 0 \text{ if } |\Lambda| > A,$$

$$\lim_{T \to 0} \frac{\rho^P(k)}{\rho^h(k)} = \lim_{T \to 0} \exp\left(-\frac{\kappa(k)}{T}\right) = 0 \text{ if } |k| > Q. \tag{5.91}$$

Assuming that the root densities are smooth, bounded functions,[2] this implies that

$$\sigma_n^P(\Lambda) = 0, \ n \geq 2,$$
$$\sigma'^P_n(\Lambda) = 0, \ n \geq 1,$$
$$\sigma_1^P(\Lambda) = 0, \ \text{if } |\Lambda| > A,$$
$$\rho^P(k) = 0, \ \text{if } |k| > Q. \tag{5.92}$$

---

[2] It can be shown that this is indeed the case for small temperatures.

Making use of these simplifications we can cast the integral equations for the root densities in the form

$$\rho^P(k) = \theta_H(Q - |k|)\left[\frac{1}{2\pi} + \cos k \int_{-A}^{A} d\Lambda\, a_1(\sin k - \Lambda)\, \sigma_1^P(\Lambda)\right],$$

$$\sigma_1^P(\Lambda) = \theta_H(A - |\Lambda|)\left[-\int_{-A}^{A} d\Lambda'\, a_2(\Lambda - \Lambda')\sigma_1^P(\Lambda')\right.$$

$$\left. + \int_{-Q}^{Q} dk\, a_1(\sin k - \Lambda)\, \rho^P(k)\right], \tag{5.93}$$

$$\rho^h(k) = \theta_H(|k| - Q)\left[\frac{1}{2\pi} + \cos k \int_{A}^{A} d\Lambda\, a_1(\sin k - \Lambda)\, \sigma_1^P(\Lambda)\right],$$

$$\sigma_1^h(\Lambda) = \theta_H(|\Lambda| - A)\left[-\int_{-A}^{A} d\Lambda'\, a_2(\Lambda - \Lambda')\sigma_1^P(\Lambda')\right.$$

$$\left. + \int_{-Q}^{Q} dk\, a_1(\sin k - \Lambda)\, \rho^P(k)\right],$$

$$\sigma_n^h(\Lambda) = -A_{n1} * \sigma_1^P\Big|_\Lambda + \int_{-Q}^{Q} dk\, a_n(\sin k - \Lambda)\, \rho^P(k)\,,\, n \geq 2\,,$$

$$\sigma_n'^h(\Lambda) = \frac{1}{\pi}\text{Re}\frac{1}{\sqrt{1 - (\Lambda - inu)^2}} - \int_{-Q}^{Q} dk\, a_n(\sin k - \Lambda)\, \rho^P(k), \tag{5.94}$$

where $\theta_H(x)$ is the Heaviside step function. We note that at zero temperature the functional forms of $\rho^P(k)$ and $\rho^h(k)$ and similarly of $\sigma_1^P(\Lambda)$ and $\sigma_1^h(\Lambda)$ are identical. Hence it is convenient to define quantities

$$\rho(k) = \rho^P(k) + \rho^h(k)\,,$$

$$\sigma_1(\Lambda) = \sigma_1^P(\Lambda) + \sigma_1^h(\Lambda)\,. \tag{5.95}$$

Then $\rho(k)$ describes the root density of particles if $|k| \leq Q$ and the root density of holes if $|k| \geq Q$ respectively. In terms of $\rho(k)$ and $\sigma_1(\Lambda)$ the integral equations (5.93) and (5.94) take a more compact form

$$\sigma_1(\Lambda) = \int_{-Q}^{Q} dk\, a_1(\sin k - \Lambda)\, \rho(k) - \int_{-A}^{A} d\Lambda'\, a_2(\Lambda - \Lambda')\sigma_1(\Lambda'),$$

$$\rho(k) = \frac{1}{2\pi} + \cos k \int_{-A}^{A} d\Lambda\, a_1(\sin k - \Lambda)\, \sigma_1(\Lambda). \tag{5.96}$$

As we will see later on, equations (5.96) describe the distribution of roots in the zero temperature ground state of the Hubbard model. They were first obtained by E. H. Lieb and F. Y. Wu in [298].

5.5.3 Dressed momenta

The total momentum (4.51) can be rewritten by using (4.36)-(4.38) in the following useful manner

$$
P = \frac{2\pi}{L} \left(\sum_{j=1}^{N-2M'} I_j + \sum_{m=1}^{\infty} \sum_{\beta=1}^{M_m} J_\beta^m - \sum_{n=1}^{\infty} \sum_{\alpha=1}^{M_n'} J_\alpha'^n \right) + \pi \sum_{n=1}^{\infty} \sum_{\alpha=1}^{M_n'} (n+1)
$$

$$
= \sum_{j=1}^{N-2M'} y(k_j) + \sum_{m=1}^{\infty} \sum_{\beta=1}^{M_m} z_m(\Lambda_\beta^m) - \sum_{n=1}^{\infty} \sum_{\alpha=1}^{M_n'} z_n'(\Lambda_\alpha'^n) + \pi \sum_{n=1}^{\infty} \sum_{\alpha=1}^{M_n'} (n+1).
$$

$$(5.97)$$

Using this expression for the total momentum we now can identify the dressed momenta of various types of 'excitations'. We find that an additional real k with $|k| > Q$ ('particle') or hole in the sea of k's ($|k| < Q$) carry momentum $\pm p(k)$ respectively, where

$$
p(k) = y(k) = 2\pi \int_0^k dk' \, \rho(k) \, . \tag{5.98}
$$

Similarly, the dressed momentum of an additional root Λ or a hole in the sea of Λ_α^1's with spectral parameter Λ is $\pm p_1(\Lambda)$, where

$$
p_1(\Lambda) = z_1(\Lambda) = -2\pi \int_\Lambda^\infty d\Lambda' \, \sigma_1(\Lambda') + z_1(\infty). \tag{5.99}
$$

The momentum associated with adding a Λ-string of length 2 or larger is

$$
p_n(\Lambda) = z_n(\Lambda) = -2\pi \int_\Lambda^\infty d\Lambda' \sigma_n^h(\Lambda') + z_n(\infty) \, , \quad n \geq 2. \tag{5.100}
$$

Finally, a k-Λ string of length n has dressed momentum

$$
p_n'(\Lambda) = -z_n'(\Lambda) + \pi(n+1)
$$

$$
= 2\pi \int_\Lambda^\infty d\Lambda' \, \sigma_n'^h(\Lambda') - z_n'(\infty) + \pi(n+1). \tag{5.101}
$$

5.5.4 Zero temperature limit in zero magnetic field

If we take the limit of vanishing magnetic field, equations (5.83)–(5.86) simplify further. Taking $B \to 0$ corresponds to $A \to \infty$. In the absence of a magnetic field, the magnetization must be zero [293] as otherwise the spin rotational SU(2) symmetry would be broken, which is forbidden in one dimension by the Mermin-Wagner theorem. For $A = \infty$ we have

$$
\frac{N_\downarrow + N_\uparrow}{L} = \int_{-Q}^{Q} dk \, \rho^P(k) \, ,
$$

$$
\frac{N_\downarrow}{L} = \int_{-A}^{A} d\lambda \, \sigma_1(\lambda) = \frac{1}{2} \int_{-Q}^{Q} dk \, \rho^P(k) \, , \tag{5.102}
$$

where the last equality is obtained by inserting the integral equation (5.93). We thus find that for $A = \infty$, the numbers of up and down spins are the same $N_\downarrow = N_\uparrow$ and the magnetization is indeed zero.

Using Fourier transformation, equations (5.83) and (5.85) can be simplified

$$\kappa(k) = -2\cos k - \mu - 2u + \int_{-Q}^{Q} dk' \cos k' \, R(\sin k' - \sin k) \, \kappa(k') \,,$$

$$\varepsilon_1(\Lambda) = \int_{-Q}^{Q} dk \frac{\cos k}{4u} \frac{1}{\cosh \frac{\pi}{2u}(\Lambda - \sin k)} \kappa(k) \,,$$

$$\varepsilon_n(\Lambda) = 0 \quad n = 2, 3, \dots \tag{5.103}$$

where

$$R(x) = \int_{-\infty}^{\infty} \frac{d\omega}{2\pi} \frac{\exp(i\omega x)}{1 + \exp(2u|\omega|)} \,. \tag{5.104}$$

The vanishing of the dressed energies of Λ-strings of lengths greater than one i.e. $\varepsilon_n(\Lambda) = 0$ for $n \geq 2$ is due to the absence of a magnetic field. For finite magnetic fields all $\varepsilon_n(\Lambda)$ will be nontrivial functions. The equations for the root densities simplify to

$$\rho(k) = \frac{1}{2\pi} + \int_{-Q}^{Q} dk' \cos k \, R(\sin k' - \sin k) \, \rho(k') \,,$$

$$\sigma_1(\Lambda) = \int_{-Q}^{Q} dk \frac{1}{4u} \frac{1}{\cosh \frac{\pi}{2u}(\Lambda - \sin k)} \rho(k) \,,$$

$$\sigma_n^p(\Lambda) = 0 \quad n = 2, 3, \dots \qquad \sigma_m^h(\Lambda) = 0 = \sigma_m'^p(\Lambda) \quad m = 1, 2, \dots$$

$$\sigma_n'^h(\Lambda) = \frac{1}{\pi} \mathrm{Re} \frac{1}{\sqrt{1 - (\Lambda - inu)^2}} - \int_{-Q}^{Q} dk \, a_n(\sin k - \Lambda) \, \rho(k) \,. \tag{5.105}$$

Equations (5.105) describe the ground state of the repulsive Hubbard model at zero temperature and zero magnetic field. There is one Fermi sea of k's (charge degrees of freedom) with Fermi rapidity $\pm Q$ and a second Fermi sea of Λ^1's, which are filled on the entire real axis.

The expressions (5.99)–(5.101) for the dressed momenta can be simplified as well

$$p_1(\Lambda) = \pi \frac{N}{2L} - 2 \int_{-Q}^{Q} dk \, \arctan \left[\exp \left(-\frac{\pi}{2u}(\Lambda - \sin k) \right) \right] \rho(k), \tag{5.106}$$

$$p_n'(\Lambda) = -2\mathrm{Re} \arcsin(\Lambda - inu)$$
$$+ \int_{-Q}^{Q} dk \, 2 \arctan \left(\frac{\Lambda - \sin k}{nu} \right) \rho(k) + \pi(n + 1). \tag{5.107}$$

The result (5.106) was first obtained by Coll [88].

Appendices to Chapter 5

5.A Zero temperature limit for $\varepsilon_1'(\Lambda)$

In order to establish that $\varepsilon_1'(\Lambda) \geq 0$ we rewrite Eqn. (5.61) as

$$
\ln \eta_1'(\Lambda) = s * \ln(1 + \eta_2') \Big|_\Lambda
$$
$$
- \int_0^{\pi/2} dk \, \cos(k) \, s(\Lambda - \sin k) \ln \left[\frac{1 + \zeta(k)}{1 + \zeta(\pi - k)} \right]
$$
$$
- \int_{-\pi/2}^0 dk \, \cos(k) \, s(\Lambda - \sin k) \ln \left[\frac{1 + \zeta(k)}{1 + \zeta(-\pi - k)} \right]. \qquad (5.A.1)
$$

The first term on the right hand side of (5.A.1) is positive. Using (5.54) we can straightforwardly establish the following identities

$$
\ln \zeta(\pi - k) - \ln \zeta(k) = \frac{4 \cos k}{T} \geq 0 \quad \text{for } 0 \leq k \leq \pi/2,
$$
$$
\ln \zeta(-\pi - k) - \ln \zeta(k) = \frac{4 \cos k}{T} \geq 0 \quad \text{for } -\pi/2 \leq k \leq 0. \qquad (5.A.2)
$$

These equations imply that the second and third terms on the right-hand side of (5.A.1) are positive and therefore

$$
\varepsilon_1'(\Lambda) = T \ln \eta_1'(\Lambda) \geq 0. \qquad (5.A.3)
$$

5.B Properties of the integral equations at $T = 0$

In this appendix we show that the solutions $\kappa(k)$ and $\varepsilon_1(\Lambda)$ of the integral equations (5.83) and (5.84) are symmetric functions of their arguments and that they increase monotonously with $k > 0$ and $\Lambda > 0$ respectively. Our discussion follows the appendix of [435]. The uniqueness and monotonicity properties of the solutions to the integral equations for the root densities was proved by E. H. Lieb and F. Y. Wu in [300].

We first rewrite the integral equation for $\varepsilon_1(\Lambda)$ by way of Fourier transformation as

$$\varepsilon_1(\Lambda) = B + \int_{-\infty}^{\infty} d\Lambda' \, R(\Lambda - \Lambda') \, \varepsilon_1^+(\Lambda')$$

$$+ \int_{-\pi}^{\pi} dk \, \cos(k) \, s(\Lambda - \sin k) \, \kappa^-(k) , \tag{5.B.1}$$

where $R(\Lambda)$ and $s(\Lambda)$ are defined in (5.104) and (5.59) respectively. The resulting set (5.B.1), (5.83) of coupled integral equations is then solved by a double iteration

$$\kappa(k) = \lim_{n \to \infty} \kappa^{(n)}(k) , \quad \varepsilon_1(\Lambda) = \lim_{n \to \infty} \varepsilon_1^{(n)} ,$$

$$\varepsilon_1^{(n)}(\Lambda) = \lim_{m \to \infty} \varepsilon_1^{(n,m)}(\Lambda) ,$$

$$\kappa^{(1)}(k) = -2\cos(k) - \mu - 2u - B ,$$

$$\varepsilon_1^{(n,1)}(\Lambda) = 2B ,$$

$$\varepsilon_1^{(n,m+1)}(\Lambda) = B + \int_{-\infty}^{\infty} d\Lambda' \, R(\Lambda - \Lambda') \, \varepsilon_1^{(n,m)+}(\Lambda')$$

$$+ \int_{-\pi}^{\pi} dk \, \cos(k) \, s(\Lambda - \sin k) \, \kappa^{(n)-}(k) , \tag{5.B.2}$$

$$\kappa^{(n+1)}(k) = \kappa^{(1)}(k) + \int_{-\infty}^{\infty} d\Lambda \, a_1(\sin k - \Lambda) \, \varepsilon_1^{(n)-}(\Lambda) . \tag{5.B.3}$$

It is straightforward to show by mathematical induction that the functions constructed in this iterative manner are symmetric

$$\kappa^{(n)}(-k) = \kappa^{(n)}(k) ,$$

$$\varepsilon_1^{(n)}(-\Lambda) = \varepsilon_1^{(n)}(\Lambda) . \tag{5.B.4}$$

The existence of the limits

$$\kappa(k) = \lim_{n \to \infty} \kappa^{(n)}(k) , \quad \varepsilon_1(\Lambda) = \lim_{n \to \infty} \varepsilon_1^{(n)} , \tag{5.B.5}$$

is established by the following three Lemmas.

Lemma 4.

$$0 \geq \int_{-\pi}^{\pi} dk \, \cos(k) \, s(\Lambda - \sin k) \, \kappa^{(n)-}(k)$$

$$\geq \int_{-\pi}^{\pi} dk \, \cos(k) \, s(\Lambda - \sin k) \, \kappa^{(n)}(k) \tag{5.B.6}$$

$$\geq -2 \int_{-\pi}^{\pi} dk \, \cos^2(k) \, s(\Lambda - \sin k) . \tag{5.B.7}$$

Proof. Let us begin with the first inequality in Lemma 4. It is easy to see that

$$\int_{-\pi}^{\pi} dk \ \cos(k) \ s(\Lambda - \sin k) \ \kappa^{(n)-}(k)$$

$$= \int_{0}^{\pi/2} dk \ \cos(k) \ s(\Lambda - \sin k) \left[\kappa^{(n)-}(k) - \kappa^{(n)-}(\pi - k) \right]$$

$$+ \int_{-\pi/2}^{0} dk \ \cos(k) \ s(\Lambda - \sin k) \left[\kappa^{(n)-}(k) - \kappa^{(n)-}(-\pi - k) \right]. \qquad (5.B.8)$$

On the other hand, one can derive directly from the definition of $\kappa^{(n)}$ (5.B.3) that

$$\kappa^{(n)}(\pi - k) - \kappa^{(n)}(k) - 4\cos(k) \geq 0 \quad \text{for } k \in [0, \pi/2],$$
$$\kappa^{(n)}(-\pi - k) - \kappa^{(n)}(k) = 4\cos(k) \geq 0 \quad \text{for } k \in [-\pi/2, 0]. \qquad (5.B.9)$$

Equations (5.B.9) in turn imply the following inequalities for $\kappa^{(n)\alpha}$, $\alpha = \pm$

$$\kappa^{(n)\alpha}(\pi - k) \geq \kappa^{(n)\alpha}(k), \quad k \in [0, \pi/2]$$
$$\kappa^{(n)\alpha}(-\pi - k) \geq \kappa^{(n)\alpha}(k), \quad k \in [-\pi/2, 0]. \qquad (5.B.10)$$

As a consequence of (5.B.10) both terms on the right hand side of (5.B.8) are negative, which proves the first inequality in Lemma 4. A completely analogous calculation yields

$$\int_{-\pi}^{\pi} dk \ \cos(k) \ s(\Lambda - \sin k) \ \kappa^{(n)+}(k) \leq 0, \qquad (5.B.11)$$

which establishes the second inequality in Lemma 4. The final inequality in Lemma 4 is obtained by substituting the expression (5.B.3) for $\kappa^{(n)}$ into equation (5.B.6) and then using Eqn. 17.1.   □

Lemma 5. *The limit* $\varepsilon_1^{(n)}(\Lambda) = \lim\limits_{m \to \infty} \varepsilon_1^{(n,m)}(\Lambda)$ *exists.*

Proof. By mathematical induction we can prove that

$$\varepsilon_1^{(n,m+1)}(\Lambda) \leq \varepsilon_1^{(n,m)}(\Lambda). \qquad (5.B.12)$$

Induction start: Using that $\varepsilon_1^{(n,1)}(\Lambda) = \varepsilon_1^{(n,1)+}(\Lambda) = 2B$ we have

$$\varepsilon_1^{(n,2)}(\Lambda) - \varepsilon_1^{(n,1)}(\Lambda) = \int_{-\pi}^{\pi} dk \ \cos(k) \ s(\Lambda - \sin k) \ \kappa^{(n)-}(k), \qquad (5.B.13)$$

which is negative by Lemma 4.

Induction step: Assuming (5.B.12) to hold for $m \leq k$, we obtain

$$\varepsilon_1^{(n,k+1)}(\Lambda) - \varepsilon_1^{(n,k)}(\Lambda) = \int_{-\infty}^{\infty} d\Lambda' \ R(\Lambda - \Lambda') \left[\varepsilon_1^{(n,k)+}(\Lambda') - \varepsilon_1^{(n,k-1)+}(\Lambda') \right] \leq 0,$$

because the integrand is always positive. This completes the induction step and establishes that $\varepsilon_1^{(n,m)}(\Lambda)$ decreases under iteration in m.

On the other hand we can obtain a lower bound by using Lemma 4

$$\varepsilon_1^{(n,m)}(\Lambda) \geq -2 \int_{-\pi}^{\pi} dk \, \cos^2 k \, s(\Lambda - \sin k) \, . \tag{5.B.14}$$

This completes the proof. □

Lemma 6. *The following inequalities hold*

$$\varepsilon_1^{(n,m)}(\Lambda) \geq \varepsilon_1^{(n+1,m)}(\Lambda) \, , \tag{5.B.15}$$

$$\kappa^{(n)}(k) \geq \kappa^{(n+1)}(k) \, , \tag{5.B.16}$$

$$\varepsilon_1^{(n)}(\Lambda) \geq \varepsilon_1^{(n+1)}(\Lambda) \, . \tag{5.B.17}$$

Proof. All three inequalities are proved by mathematical induction in n.
Induction start: It is obvious from the definition (5.B.3) that $\kappa^{(1)}(k) \geq \kappa^{(2)}(k)$. Using this fact, we can show by mathematical induction in m that $\varepsilon_1^{(1,m)}(\Lambda) \geq \varepsilon_1^{(2,m)}(\Lambda)$: the case $m = 1$ is obvious, $m = 2$ is equivalent to proving that

$$\int_{-\pi}^{\pi} dk \, \cos(k) \, s(\Lambda - \sin(k))[\kappa^{(1)-}(k) - \kappa^{(2)-}(k)] \geq 0. \tag{5.B.18}$$

The integral (5.B.18) is rewritten as

$$\int_{0}^{\pi/2} dk \, \cos(k) \, [s(\Lambda - \sin(k)) + s(\Lambda + \sin(k))] \, \{f_1(k) - f_2(k)\} \, , \tag{5.B.19}$$

where

$$f_n(k) = \kappa^{(n)-}(k) - \kappa^{(n)-}(\pi - k) \, . \tag{5.B.20}$$

By equation (5.B.9) we have

$$f_n(k) = \kappa^{(n)-}(k) - [\kappa^{(n)}(k) + 4\cos(k)]^- $$
$$= \begin{cases} \kappa^{(n)-}(k) & \text{if } (\kappa^{(n)}(k) + 4\cos(k)) \geq 0 \, , \\ -4\cos(k) & \text{if } (\kappa^{(n)}(k) + 4\cos(k)) \leq 0 \, . \end{cases} \tag{5.B.21}$$

Using the fact that $\kappa^{(1)}(k) \geq \kappa^{(2)}(k)$ we obtain

$$f_1(k) - f_2(k) = \begin{cases} 0 & \text{if } \kappa^{(1)}(k) \leq -4\cos(k) \, , \\ \kappa^{(1)-}(k) - \kappa^{(2)-}(k) & \text{if } \kappa^{(2)}(k) \geq -4\cos(k) \, , \\ \kappa^{(1)-}(k) + 4\cos(k) & \text{if } \kappa^{(1)}(k) \geq -4\cos(k) \geq \kappa^{(2)}(k), \end{cases} \tag{5.B.22}$$

which establishes that $f_1(k) - f_2(k) \geq 0$ for $k \in [0, \pi/2]$. This completes the proof that $\varepsilon_1^{(1,m)}(\Lambda) \geq \varepsilon_1^{(2,m)}(\Lambda)$. The induction over m is carried out by assuming that $\varepsilon_1^{(1,m-1)}(\Lambda) \geq$

$\varepsilon_1^{(2,m-1)}(\Lambda)$ and calculating

$$\varepsilon_1^{(1,m)}(\Lambda) - \varepsilon_1^{(2,m)}(\Lambda) = \int_{-\pi}^{\pi} dk \, \cos(k) \, s(\Lambda - \sin(k))[\kappa^{(1)-}(k) - \kappa^{(2)-}(k)]$$

$$+ \int_{-\infty}^{\infty} d\Lambda' \, R(\Lambda' - \Lambda)[\varepsilon_1^{(1,m-1)+}(\Lambda') - \varepsilon_1^{(2,m-1)+}(\Lambda')].$$

$$(5.B.23)$$

The first term on the r.h.s. of (5.B.23) is greater than zero by (5.B.18) and the second term is positive as the integrand is always positive by virtue of the induction assumption. Taking the limit $m \to \infty$, which exists by Lemma 5, we obtain that $\varepsilon_1^{(1)}(\Lambda) \geq \varepsilon_1^{(2)}(\Lambda)$. This establishes the validity of the induction start $n = 1$.

Induction step: The inequality $\varepsilon_1^{(n-1)}(\Lambda) \geq \varepsilon_1^{(n)}(\Lambda)$ implies $\kappa^{(n)}(k) \geq \kappa^{(n+1)}(k)$ courtesy of the definition (5.B.3). The proof that $\varepsilon_1^{(n,m)}(\Lambda) \geq \varepsilon_1^{(n+1,m)}(\Lambda)$ is accomplished by mathematical induction in m in complete analogy to the $n = 1$ case. Finally taking the limit $m \to \infty$ gives $\varepsilon_1^{(n)}(\Lambda) \geq \varepsilon_1^{(n+1)}(\Lambda)$. This completes the proof of Lemma 6. □

Lemma 6 shows that both $\kappa^{(n)}(k)$ and $\varepsilon_1^{(n)}(\Lambda)$ decrease under iteration. On the other hand, they can be bounded from below by using the inequality (5.B.14). This shows that the limit $n \to \infty$ exists.

Finally, the monotonicity properties of the dressed energies are established by the following two Lemmas.

Lemma 7. *(a)* $\varepsilon_1^{(n,m)}(\Lambda)$ *is a monotonously increasing function (MIF) of* Λ *for* $\Lambda \in [0, \infty]$. *(b)* $\kappa^{(n)}(k)$ *is a MIF of* k *for* $k \in [0, \pi/2]$.

Proof. We rewrite the recursion relation for $\varepsilon_1^{(n,m+1)}(\Lambda)$ as

$$\varepsilon_1^{(n,m+1)}(\Lambda) = B + \int_{-\infty}^{\infty} d\Lambda' \, R(\Lambda - \Lambda') \, \varepsilon_1^{(n,m)+}(\Lambda')$$

$$+ \int_{-\infty}^{\infty} d\Lambda' \, s(\Lambda - \Lambda') \, \varphi^{(n)}(\Lambda'), \qquad (5.B.24)$$

where

$$\varphi^{(n)}(\Lambda) = \int_{-\pi}^{\pi} dk \, \cos(k) \, \delta(\Lambda - \sin k) \, \kappa^{(n)-}(k)$$

$$= \begin{cases} 0 & \text{if } |\Lambda| > 1 \\ -4\sqrt{1 - \Lambda^2} & \text{if } |\Lambda| < 1 \quad \text{and} \quad \kappa^{(n)}(z) < -4\sqrt{1 - \Lambda^2} \\ \kappa^{(n)-}(z) & \text{if } |\Lambda| < 1 \quad \text{and} \quad \kappa^{(n)}(z) > -4\sqrt{1 - \Lambda^2}, \quad (5.B.25) \end{cases}$$

where $z = \arcsin(\Lambda)$. The function $\varphi^{(n)}(\Lambda)$ is a symmetric, continuous function of Λ and increases monotonously with $\Lambda > 0$ if $\kappa^{(n)}(k)$ is a MIF of k in the interval $[0, \pi/2]$. We are now in a position to prove Lemma 7 by mathematical induction.

Induction start: $\varepsilon_1^{(1,1)}(\Lambda)$ is a MIF of Λ. $\kappa^{(1)}(k)$ is a MIF in the interval $[0, \pi/2]$, implying that $\varphi^{(1)}(\Lambda)$ is a MIF of Λ. Mathematical induction in m establishes that $\varepsilon_1^{(1,m)}(\Lambda)$ defined by the recursion relation (5.B.24) is a MIF of Λ and taking the limit $m \to \infty$ we obtain that $\varepsilon_1^{(1)}(\Lambda)$ has the same property. Application of the recursion relation (5.B.3) shows that $\kappa^{(2)}(k)$ is a MIF of k in $[0, \pi/2]$.

Induction step: Assuming that $\kappa^{(n)}(k)$ is a MIF of k in the interval $[0, \pi/2]$ it follows that $\varphi^{(n)}(\Lambda)$ defined by (5.B.25) is a MIF of Λ. This fact allows us to show by mathematical induction in m that $\varepsilon_1^{(n,m)}(\Lambda)$ as defined in (5.B.24) and hence also $\varepsilon_1^{(n)}(\Lambda)$ are MIF of Λ. Application of the recursion relation (5.B.3) then establishes that $\kappa^{(n+1)}(k)$ is a MIF of k in $[0, \pi/2]$.   □

Lemma 8. *$\kappa^{(n)}(k)$ is a MIF of k for $k \in [\pi/2, \pi]$.*

Proof. We rewrite the equation (5.B.2) for $\varepsilon^{(n)}(\Lambda)$ by replacing $\kappa^{(n)-}(k)$ by $\kappa^{(n)}(k) - \kappa^{(n)+}(k)$ in the integrand of the second integral and then utilizing the recursion (5.B.3) for $\kappa^{(n)}(k)$ together with the 'symmetric integration Lemma' (17.1). This yields

$$\varepsilon_1^{(n)}(\Lambda) = B + \int_{-\infty}^{\infty} d\Lambda' \, R(\Lambda - \Lambda') \, \varepsilon_1^{(n)+}(\Lambda')$$

$$+ \int_{-\pi}^{\pi} dk \, \cos(k) \, s(\Lambda - \sin k) \, [-2\cos(k) - \kappa^{(n)+}(k)]. \qquad (5.B.26)$$

Substituting this into (5.B.3) and using (17.11) and (17.12) to simplify the integrals we have

$$\kappa^{(n+1)}(k) = \left\{ \kappa^{(1)}(k) + B - 2\int_{-\pi}^{\pi} dk' \, \cos^2(k') \, R(\sin(k) - \sin(k')) \right\}$$

$$- \int_{-\infty}^{\infty} d\Lambda \, s(\Lambda - \sin(k)) \, \varepsilon_1^{(n)+}(\Lambda)$$

$$- \int_{-\pi}^{\pi} dk' \, \cos(k') \, R(\sin(k) - \sin(k')) \, \kappa^{(n)+}(k'). \qquad (5.B.27)$$

We are now in a position to prove the Lemma by mathematical induction.

Induction start: it follows from the definition (5.B.2) that $\kappa^{(1)}(k)$ is a MIF of k in the interval $[\pi/2, \pi]$.

Induction step: Assuming that $\kappa^{(n)}(k)$ is a MIF of k in the interval $[\pi/2, \pi]$ we will prove that $\kappa^{(n+1)}(k)$ has the same property.

We have already shown that $\varepsilon_1^{(n)}$ is a symmetric function of Λ and increases monotonously for $\Lambda > 0$. This implies that the second term on the r.h.s. of (5.B.27) is a MIF of k in the interval $[\pi/2, \pi]$. The third term on the r.h.s. of (5.B.27) is rewritten as

$$\int_0^{\pi/2} dk' \, \cos(k') \left[R(\sin(k) - \sin(k')) + R(\sin(k) + \sin(k')) \right]$$
$$\times \left\{ \kappa^{(n)+}(\pi - k') - \kappa^{(n)+}(k') \right\}. \tag{5.B.28}$$

The term in braces is a monotonously decreasing function of k' by virtue of the induction assumption and Lemma 7. This in turn implies that the integral is a MIF of k in the interval $[\pi/2, \pi]$. Finally, the first term on the r.h.s. is equal to

$$-2\cos(k) - \mu - 2u - 2 \int_{-\infty}^{\infty} \frac{d\omega}{\omega} \frac{J_1(\omega) \exp(-i\omega \sin(k))}{1 + \exp(2u|\omega|)} \equiv \kappa_0(k) - \mu. $$
$$\tag{5.B.29}$$

Here $\kappa_0(k)$ is the dressed energy for a half filled band in zero magnetic field, see (6.B.8). Using the series representation (6.B.9) we can easily see that $\kappa_0(k)$ is a MIF of k in the interval $[\pi/2, \pi]$. This completes the proof of Lemma 8. $\qquad\square$

6

Ground state properties in the thermodynamic limit

In this chapter we study the ground state properties in the thermodynamic limit. In particular we determine the ground state phase diagram and calculate density, magnetization, magnetic susceptibility and compressibility as functions of the chemical potential and the magnetic field. This is done by analyzing the integral equations that determine the distributions of Bethe Ansatz roots in the ground state. This method was introduced for the case of the spin-$\frac{1}{2}$ Heisenberg model by Hulthén [207]. Our discussion is based on Takahashi's work [435, 436].

6.1 A point of reference: noninteracting electrons

Before tackling the general case let us consider noninteracting electrons ($u = 0$). Here zero temperature properties can be determined in a simple way by going to momentum space.[1] The energy for an electron with spin σ and momentum k is

$$\varepsilon_\sigma(k) = -2\cos(k) - \mu_\sigma, \qquad (6.1)$$

where $\mu_\uparrow = \mu + B$ and $\mu_\downarrow = \mu - B$. In the ground state the band for spin σ is filled in the interval $[-k_{F,\sigma}, k_{F,\sigma}]$

$$k_{F,\sigma} = \begin{cases} 0 & \text{if } \mu_\sigma < -2 \\ \arccos(-\mu_\sigma/2) & \text{if } -2 \leq \mu_\sigma \leq 2 \\ \pi & \text{if } \mu_\sigma > 2. \end{cases} \qquad (6.2)$$

The density and magnetization per site are given by

$$n_c = \sum_{\sigma=\uparrow,\downarrow} \int_{-k_{F,\sigma}}^{k_{F,\sigma}} dk \, \rho_\sigma(k) = \frac{1}{\pi} \sum_\sigma k_{F,\sigma},$$

$$m = \frac{1}{2} \left[\int_{-k_{F,\uparrow}}^{k_{F,\uparrow}} dk \, \rho_\uparrow(k) - \int_{-k_{F,\downarrow}}^{k_{F,\downarrow}} dk \, \rho_\downarrow(k) \right] = \frac{1}{2\pi} [k_{F,\uparrow} - k_{F,\downarrow}], \qquad (6.3)$$

---

[1] Of course the limit $u \to 0$ can also be studied by starting with the Bethe Ansatz solution, but the analysis is more complicated [142].

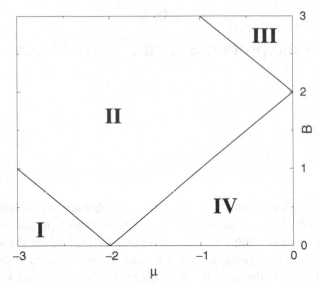

Fig. 6.1. Ground state phase diagram as a function of chemical potential μ and magnetic field B for noninteracting electrons ($u = 0$).

where we have used that the densities of up and down spins are constant, cf (5.8)

$$\rho_\sigma(k) = \frac{1}{2\pi}, \qquad \sigma = \uparrow, \downarrow. \tag{6.4}$$

The ground state free energy per site is

$$f = e - \mu\, n_c - 2B\, m = \sum_\sigma \int_{-k_{F,\sigma}}^{k_{F,\sigma}} dk\, \rho_\sigma(k)\, \varepsilon_\sigma(k)$$

$$= -\frac{1}{\pi} \sum_\sigma 2\sin(k_{F,\sigma}) + \mu_\sigma k_{F,\sigma}. \tag{6.5}$$

If we consider a less than half filled band ($\mu < 0$) we arrive at the phase diagram for noninteracting electrons shown in figure 6.1. Using the picture of the ground state in terms of two bands for spin up and spin down electrons respectively we obtain four different phases:

- Phase I: $k_{F,\sigma} = 0$: Empty Lattice.
  Both bands are empty, the ground state is the empty lattice. The density and magnetization are zero. The chemical potential must be sufficiently negative $\mu \leq -2 - B$ in order for the empty lattice to be the ground state.
- Phase II: $k_{F,\downarrow} = 0, 0 < k_{F,\uparrow} < \pi$: Partially filled, spin polarized band.
  The spin down band is empty, the spin up band is partially filled. The magnetization is equal to half the density, which varies between 0 and 1. We have

$$n_c = 2m = \frac{k_{F,\uparrow}}{\pi} = \frac{1}{\pi} \arccos\left(-\frac{\mu + B}{2}\right). \tag{6.6}$$

The charge susceptibility is

$$\chi_c(\mu, B) = \frac{\partial n_c}{\partial \mu} = \frac{1}{\pi\sqrt{4 - (\mu + B)^2}}. \tag{6.7}$$

It diverges for $B \to \pm 2 - \mu$, i.e. as we approach the boundaries with phases I and III.

- Phase III: $k_{F,\downarrow} = 0, k_{F,\uparrow} = \pi$: Half filled, spin polarized band.

The spin down band is empty, the spin up band is entirely filled. In this phase the density is 1 and the magnetisation is $\frac{1}{2}$.

- Phase IV: $0 < k_{F,\sigma} < \pi$: Partially filled and magnetized band.

Both bands are partially filled. The density varies between 0 and 1 and the magnetisation between 0 and $\frac{1}{2}$

$$n_c = \frac{1}{\pi}\left[\arccos\left(-\frac{\mu + B}{2}\right) + \arccos\left(-\frac{\mu - B}{2}\right)\right],$$

$$m = \frac{1}{2\pi}\left[\arccos\left(-\frac{\mu + B}{2}\right) - \arccos\left(-\frac{\mu - B}{2}\right)\right]. \tag{6.8}$$

The spin and charge susceptibilities are given by

$$\chi_c(\mu, B) = \frac{1}{\pi}\left[\frac{1}{\sqrt{4 - (\mu + B)^2}} + \frac{1}{\sqrt{4 - (\mu - B)^2}}\right],$$

$$\chi_s(\mu, B) = \frac{\partial m}{\partial B} = \frac{1}{2}\chi_c(\mu, B). \tag{6.9}$$

The susceptibilities diverge at the boundary with Phase II, signalling a quantum phase transition. The critical exponent associated with this transition is $\frac{1}{2}$.

6.2 Defining equations

The $T \to 0$ limit of the thermodynamic equations can be used to characterize the ground state of the system. The only dressed energies that can be negative are $\kappa(k)$ and $\varepsilon_1(\Lambda)$. From the definitions (5.68) and (5.53) it follows that if a dressed energy is positive for all values of spectral parameter, then the corresponding root density of particles is identically zero. This means that the ground state can only contain real k's and real Λ's. The integral equations for the dressed energies are

$$\kappa(k) = -2\cos k - \mu - 2u - B + \int_{-A}^{A} d\Lambda\, a_1(\sin k - \Lambda)\, \varepsilon_1(\Lambda),$$

$$\varepsilon_1(\Lambda) = 2B + \int_{-Q}^{Q} dk\, \cos(k)\, a_1(\sin k - \Lambda)\kappa(k)$$

$$- \int_{-A}^{A} d\Lambda'\, a_2(\Lambda - \Lambda')\, \varepsilon_1(\Lambda'). \tag{6.10}$$

Here the integration boundaries $\pm Q$ and $\pm A$ are the points at which the dressed energies switch sign, so that they are determined as functions of the chemical potential and the magnetic fields via the conditions

$$\kappa(\pm Q) = 0, \qquad \varepsilon_1(\pm A) = 0. \tag{6.11}$$

If $\kappa(k)$ $(\varepsilon_1(\Lambda))$ does not switch sign, then obviously $Q = 0$ or $Q = \pi$ $(A = \infty$ or $A = 0)$. In general, equations (6.10) cannot be solved analytically, but it is quite straightforward to solve them numerically with high accuracy. How to do this is outlined in Appendix 6.A.

The integral equations describing the root densities of the ground state are (5.95) and were first obtained in [298]

$$\rho(k) = \frac{1}{2\pi} + \cos k \int_{-A}^{A} d\Lambda \, a_1(\sin k - \Lambda) \, \sigma_1(\Lambda), \tag{6.12}$$

$$\sigma_1(\Lambda) = \int_{-Q}^{Q} dk \, a_1(\Lambda - \sin k) \, \rho(k) - \int_{-A}^{A} d\Lambda' \, a_2(\Lambda - \Lambda') \sigma_1(\Lambda'). \tag{6.13}$$

The integrated densities yield the total number of electrons per site and the number of down spin electrons per site respectively

$$\int_{-Q}^{Q} dk \, \rho(k) = \frac{N}{L}, \qquad \int_{-A}^{A} d\Lambda \, \sigma_1(\Lambda) = \frac{M_1}{L} = \frac{N_{\downarrow}}{L}. \tag{6.14}$$

Hence the particle density n_c and the magnetization per site m are

$$n_c = \frac{N}{L} = \int_{-Q}^{Q} dk \, \rho(k),$$

$$m = \frac{\langle S^z \rangle}{L} = \frac{N - 2M}{2L} = \frac{1}{2} \left[\int_{-Q}^{Q} dk \, \rho(k) - 2 \int_{-A}^{A} d\Lambda \sigma_1(\Lambda) \right]. \tag{6.15}$$

The spin and charge susceptibilities are defined as

$$\chi_s(\mu, B) = \frac{\partial m}{\partial B}, \qquad \chi_c(\mu, B) = \frac{\partial n_c}{\partial \mu}. \tag{6.16}$$

We note that in the parameter region we are working in we always have $0 \le m \le n_c/2$. Last but not least the ground state free energy per site is given by (5.46), (5.57)

$$f = e - \mu n_c - 2Bm = \int_{-Q}^{Q} dk \, (-2\cos k - \mu - 2u - B) \, \rho(k)$$

$$+ 2B \int_{-A}^{A} d\Lambda \, \sigma_1(\Lambda) + u$$

$$= \int_{-Q}^{Q} \frac{dk}{2\pi} \kappa(k) + u. \tag{6.17}$$

6.3 Ground state phase diagram

Let us now discuss the ground state phase diagram for the interacting case $u > 0$. This can be done either as a function of the chemical potential μ and the magnetic field B, or as a function of the particle density n_c and the magnetization per site m, which are related to the root densities by (6.14) and (6.15). It is important to note that in the grand canonical ensemble the density is a function of *both* the chemical potential and the magnetic field. The same is true for the magnetization. This means that keeping the chemical potential fixed and varying the magnetic field will change the density.

The different phases are most easily identified by considering the integration boundaries Q and A as control parameters. We first note that $Q = 0$ implies $A = 0$ and corresponds to a completely empty band, i.e. $N = 0$. On the other hand, one can easily see by using the identity (6.12) and (17.1) that $Q = \pi$ implies that the band is half filled, i.e. there is one electron per site

$$\frac{N}{L} = \int_{-\pi}^{\pi} dk \, \rho(k) = 1. \tag{6.18}$$

It is also useful to consider the two limiting cases $A = 0$ and $A = \infty$. In the latter case we find from (6.14) and (6.12)–(6.13)

$$\frac{N_\downarrow}{L} = \int_{-\infty}^{\infty} d\Lambda \sigma_1(\Lambda) = \frac{1}{2} \int_{-Q}^{Q} dk \, \rho(k) = \frac{N}{2L}, \tag{6.19}$$

i.e. the magnetization of the ground state is zero. The first equality in (6.19) follows directly from (6.14) and the second equality is obtained by integrating (6.13) between $-\infty$ and ∞. On the other hand, for $A = 0$ the ground state is completely magnetized as $N_\downarrow = 0$ by (6.14). These considerations allow us to distinguish between the following five phases at zero temperature, which are shown in figures 6.4 and 6.5 in the canonical and grand canonical ensemble respectively.

- Phase I: $Q = 0$, $A = 0$: Empty band.
  This region corresponds to an empty band, i.e. zero density of electrons $n_c = m = 0$. As the ground state is the empty lattice, the dressed energies (6.10) must be always positive. This yields the condition

  $$\mu \le \mu_0(B) = -2 - 2u - B. \tag{6.20}$$

  In other words, the chemical potential must be sufficiently negative in order for the ground state to be given by the empty lattice. This is intuitively obvious.
- Phase II: $0 < Q < \pi$, $A = 0$: Partially filled, spin polarized band.
  This phase corresponds to electron densities between zero (empty band) and one (half filled band) $0 < n_c < 1$ and completely polarized spins $m = n_c/2$. The integral equations for the dressed energies simplify to

  $$\kappa(k) = -2\cos k - \mu - 2u - B, \quad \kappa(\pm Q) = 0,$$
  $$\varepsilon_1(\Lambda) = 2B + \int_{-Q}^{Q} dk \, \cos(k) \, a_1(\Lambda - \sin k) \, \kappa(k) \ge 0. \tag{6.21}$$

  Let us now determine the ranges of B and μ that correspond to Phase II. The condition $\kappa(\pm Q) = 0$ fixes Q as a function of B and μ

  $$\cos Q = -\frac{1}{2}(\mu + B + 2u). \tag{6.22}$$

As we have by definition $Q < \pi$, the magnetic field must be smaller than an upper critical field B_u

$$B \le B_u = 2 - \mu - 2u. \tag{6.23}$$

For $B > B_u$ we have $Q = \pi$ and the band is half filled. The requirement $\varepsilon_1(\Lambda) \ge 0$ implies that for fixed μ the magnetic field B must be larger than a critical value B_c, which is given implicitly as the solution of the equation

$$B \ge B_c = \frac{2u}{\pi} \int_0^Q dk \, \cos k \, \frac{\cos k - \cos Q}{u^2 + (\sin k)^2}. \tag{6.24}$$

For $u \to \infty$ this equation can be used to express B_c in a simple way as a function of the electron density

$$B_c = \frac{1}{u} \left[n_c - \frac{\sin 2\pi n_c}{2\pi} \right] + \mathcal{O}(1/u^3). \tag{6.25}$$

The root density for real k's is constant as the integral on the r.h.s. of (6.12) vanishes

$$\rho(k) = \frac{1}{2\pi}. \tag{6.26}$$

This implies that the density is given by

$$n_c = \frac{1}{\pi} \arccos \left(1 - \frac{\mu - \mu_0(B)}{2} \right). \tag{6.27}$$

- Phase III: $Q = \pi$, $A = 0$: Half filled, spin polarized band.
  This region corresponds to a half filled band $n_c = 1$ and completely polarized spins $m = 1/2$. The integral equations for the dressed energies can be solved explicitly

$$\kappa(k) = -2 \cos k - \mu - 2u - B,$$
$$\varepsilon_1(\Lambda) = 2B - 4\mathrm{Re}\sqrt{1 - (\Lambda - iu)^2} + 4u. \tag{6.28}$$

The requirements $\kappa(k) \le 0$ and $\varepsilon_1(\Lambda) \ge 0$ imply the conditions

$$B \ge B_0 = 2\sqrt{1 + u^2} - 2u,$$
$$\mu \ge 2 - 2u - B. \tag{6.29}$$

- Phase IV: $0 < Q < \pi$, $0 < A \le \infty$: Partially filled and magnetized band.
  This phase corresponds to $0 < n_c < 1$, $0 \le m < n_c/2$. In general the integral equations can only be solved numerically in this region. However, simplifications occur for $B = 0$ that for example allow to make analytic progress in the limits of small densities $n_c \approx 0$ and densities close to half-filling $n_c \approx 1$. These results are summarized in Sections 6.B.3 and 6.B.2 of Appendix 6.B.
- Phase V: $Q = \pi$, $0 < A \le \infty$: Half filled, partially magnetized band.

This phase corresponds to $n_c = 1$, $0 \le m < n_c/2$. Equations (6.10) simplify to

$$\kappa(k) = -2\,\cos k - \mu - 2u - B + \int_{-A}^{A} d\Lambda\, a_1(\sin k - \Lambda)\,\varepsilon_1(\Lambda),$$

$$\varepsilon_1(\Lambda) = 2B - 4\mathrm{Re}\sqrt{1 - (\Lambda - iu)^2} + 4u$$

$$- \int_{-A}^{A} d\Lambda'\, a_2(\Lambda - \Lambda')\,\varepsilon_1(\Lambda'). \tag{6.30}$$

By construction we have $\kappa(k) \le 0$ for all $k \in [-\pi, \pi]$. In the interior of Phase V $\kappa(k)$ is strictly negative for all values of k and in particular we have $\kappa(\pm\pi) < 0$. The boundary between regions IV and V is determined by the condition $\kappa(\pm\pi) = 0$.

For zero magnetic field $B = 0$, which corresponds to $A = \infty$, the integral equations (6.30) can be solved explicitly by Fourier transformation as is shown in Appendix 6.B.1. The chemical potential μ_- separating phases IV and V at $B = 0$ can then be obtained by setting $\kappa(\pi) = 0$ in the explicit expression (6.B.8) for $\kappa(k)$. We find

$$\mu_-(u) = 2 - 2u - 2\int_0^\infty \frac{d\omega}{\omega}\, \frac{J_1(\omega)e^{-\omega u}}{\cosh(\omega u)}, \tag{6.31}$$

where J_1 is a Bessel function.[2]

We emphasize that throughout Region V the density is equal to 1, but the chemical potential varies. This implies that μ is not an invertible function of n_c. Physically this is very interesting: as soon as the band is half filled (for fixed B), it is no longer possible to force additional electrons into the ground state by increasing the chemical potential by a small amount. The only way this can happen is if all eigenstates of the Hamiltonian with one additional electron are separated in energy from the ground state by a finite gap. This in turn implies that in phase V the Hubbard model describes an insulator for all $u > 0$. The insulating state is of a rather unconventional nature: it is a *Mott insulator* [158,331,332]. Lieb and Wu's demonstration that the half filled Hubbard model is an insulator was based on proving that the chemical potential has a discontinuity. Denoting the ground state energy for the Hubbard model with N electrons by $E(N, u)$, we have

$$\mu_-(u) = E(L, u) - E(L - 1, u), \tag{6.32}$$

where L is the length of the lattice. On the other hand, the chemical potential necessary to force one extra electron into the half filled ground state is by definition

$$\mu_+(u) = E(L + 1, u) - E(L, u). \tag{6.33}$$

In order to calculate $\mu_+(u)$ we employ the Shiba transformation (2.61), which tells us that $E(L + 1, u) = E(L - 1, u)$ and hence

$$\mu_+(u) = -\mu_-(u). \tag{6.34}$$

---

[2] The term $-2u$ is not present in Lieb and Wu's expression in their 1968 paper [298] because they consider a Hamiltonian $H = -\sum_{j,\sigma} c_{j,\sigma}^\dagger c_{j+1,\sigma} + c_{j+1,\sigma}^\dagger c_{j,\sigma} + 4u \sum_j n_{j,\uparrow} n_{j,\downarrow}$, whereas our discussion is for a Hamiltonian $H = -\sum_{j,\sigma} c_{j,\sigma}^\dagger c_{j+1,\sigma} + c_{j+1,\sigma}^\dagger c_{j,\sigma} + u \sum_j (2n_{j,\uparrow} - 1)(2n_{j,\downarrow} - 1)$, which amounts to a shift in the chemical poential by $-2u$.

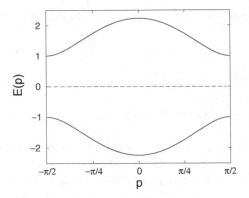

Fig. 6.2. Band insulator.

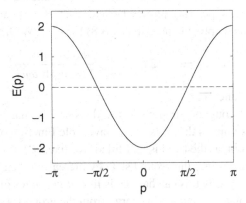

Fig. 6.3. Tight-binding band of the half filled Hubbard model at $u = 0$.

The jump in the chemical potential at half filling is

$$\mu_+ - \mu_- = -4 + 4u + 4 \int_0^\infty \frac{d\omega}{\omega} \frac{J_1(\omega)e^{-\omega u}}{\cosh(\omega u)}. \tag{6.35}$$

The result (6.35) was first obtained by Lieb and Wu [298, 300].

In order to see better why we are dealing with an unusual insulating state, let us compare it to the generic case of a band insulator. In the latter the valence band is full and separated from the conduction band by a band gap. An example of a band insulator is the tight-binding Hamiltonian

$$H = -t \sum_{j=1}^{L} \left[c_{j,\sigma}^\dagger c_{j+1,\sigma} + c_{j+1,\sigma}^\dagger c_{j,\sigma} \right] + V \sum_{j=1}^{L} (-1)^j n_j, \tag{6.36}$$

which describes electrons moving in a potential that alternates from site to site. The spectrum of (6.36) is easily determined by Fourier transformation and is shown in figure 6.2 for $V = 1$

$$E(p) = \pm\sqrt{V^2 + 4t^2 \cos^2(p)}. \tag{6.37}$$

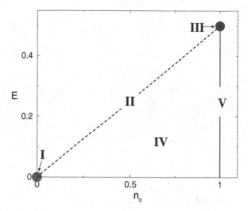

Fig. 6.4. Ground state phase diagram as a function of electron density n_c and magnetization m.

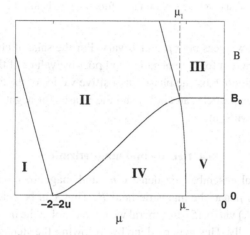

Fig. 6.5. Ground state phase diagram as a function of chemical potential μ and magnetic field B. The special values shown are $B_0 = 2\sqrt{1 + u^2} - 2u$, $\mu_1 = 2 - 2\sqrt{1 + u^2}$ and $\mu_-(u)$ is defined in (6.31).

We see that adding an extra electron to the conduction band or making a hole in the valence band costs an energy larger than the band gap $2V$. The situation in the half filled Hubbard model is quite different. If we neglect the interactions (i.e. set $u = 0$) we obtain the simple cosine band shown in figure 6.3, which implies that the ground state is *metallic*. This shows that the insulating nature of the half filled ground state in the repulsive Hubbard model is driven entirely by electron-electron interactions! The phase transition from the metallic phase at $u = 0$ to the Mott insulating state at $u > 0$ is an example of a *Mott transition* [158, 332].

We are now in the position to exhibit the full ground state phase diagram of the repulsive Hubbard model. In figures 6.4 and 6.5 we display the phase diagram as a function of particle density n_c and magnetization m and as a function of the chemical potential μ and the magnetic field B respectively. It is clear from these plots that n_c and m are not invertible functions of the chemical potential and the magnetic field. For example, throughout region

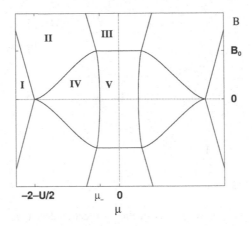

Fig. 6.6. Complete ground state phase diagram as a function of chemical potential μ and magnetic field B.

III n_c and m are fixed, whereas both μ and B vary. For the sake of completeness we show the complete phase diagram for both negative and positive values of the chemical potential and magnetic field in figure 6.6. The phases at positive values of the chemical potential are obtained by exchanging particles and holes and the phases for negative magnetic fields by exchanging up and down spins.

6.4 Density and magnetization

In the grand canonical ensemble the density n_c and magnetization m are functions of the chemical potential μ and the magnetic field B. They can be calculated from the root densities $\rho(k)$ and $\sigma_1(\Lambda)$ via (6.15). In general we need to solve the integral equations for the root densities numerically. This may be done by following the steps outlined in Appendix 6.A. There are several limiting cases in which analytical solutions are possible. These are discussed in Appendices 6.B.1, 6.B.2, 6.B.3 and 6.C.

6.4.1 Fixed B

Let us first consider the density as function of the chemical potential for fixed magnetic field B.

Zero magnetic field For $B = 0$ we obtain the behaviour shown in figure 6.7. For $\mu < \mu_0(0) = -2 - 2u$ the density is zero. In the vicinity of the boundary between Phases I and IV the density in Phase IV increases in a universal square root fashion (6.B.32) as is shown in Appendix 6.B.3

$$n_c \simeq \frac{1}{\pi}\sqrt{\mu - \mu_0(0)}. \tag{6.38}$$

The density then increases monotonically with μ until it reaches 1 at $\mu_-(u) < 0$ (6.31) (boundary between phases IV and V). Increasing μ further does not change n_c, it stays

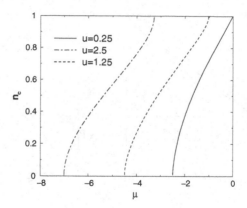

Fig. 6.7. Density as function of the chemical potential for $B = 0$ and $u = 0.25, 1.25, 2.5$. Once n_c reaches 1 (half-filling), it stays constant in the range of μ shown ($\mu < 0$).

fixed at 1 throughout the Mott insulating phase V. In the vicinity of half-filling, the density behaves like

$$n_c(\mu, B = 0) \simeq 1 - c_1\sqrt{\mu_-(u) - \mu}, \tag{6.39}$$

where the coefficient c_1 is calculated in Appendix 6.B.2 (see equation (6.B.26)). This is an example of the *commensurate-incommensurate* phase transition [111, 193, 363, 381]. The ground state at $\mu > \mu_-$ is half filled and commensurate with the symmetries of the underlying lattice. For $\mu < \mu_-$ the ground state is partially filled and the long-distance asymptotics of correlation functions exhibit oscillating behaviour with a characteristic wave number related to the band filling.

Finite magnetic field The behaviour of $n_c(\mu)$ in a strong magnetic field, which leads to a complete polarization of the ground state, has been discussed in Section (6.3). In phase II we have (6.27)

$$n_c = \frac{1}{\pi}\arccos\left(1 - \frac{\mu - \mu_0(B)}{2}\right), \tag{6.40}$$

where we recall that $\mu_0(B) = -2 - 2u - B$. At low densities, i.e. close to the boundary between phases II and I, this again yields a square root behaviour

$$n_c \simeq \frac{1}{\pi}\sqrt{\mu - \mu_0(B)} + \mathcal{O}\left([\mu - \mu_0(B)]^{\frac{3}{2}}\right). \tag{6.41}$$

In figure 6.8 we plot the density as function of the chemical potential for $u = 1.25$ and three different values of the magnetic field B. We see that the density has a cusp at the values of μ, where the transition between phases II and IV occurs for the given values of B. For fields above the critical field B_0 phase IV no longer exists and there is a transition from phase II to phase III instead. We have $B_0 \approx 0.702$ for $u = 1.25$.

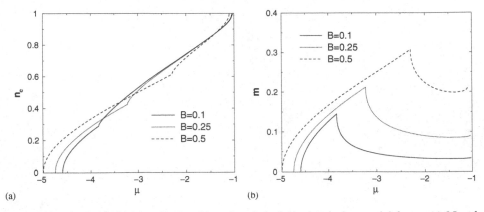

Fig. 6.8. Density (a) and Magnetization (b) as function of the chemical potential for $u = 1.25$ and several values of B.

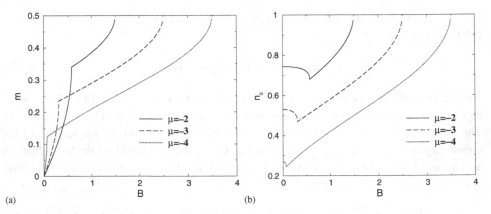

Fig. 6.9. (a) Magnetization and (b) Density as function of the applied field for three different values of chemical potential μ and $u = 1.25$.

6.4.2 Fixed μ

Let us now determine the magnetization m and the density n_c as functions of the applied field B for fixed values of the chemical potential μ. Solving the relevant integral equations numerically yields the magnetization curves shown in figure 6.9(a). At small fields m increases linearly with B

$$m(B)\big|_{\mu=\text{const}} \propto B. \tag{6.42}$$

The constant of proportionality is the magnetic susceptibility in zero field and can be obtained by taking the limit $B \to 0$ in equation (6.76) below.

At the critical field B_c corresponding to the transition between phases II and IV there is a cusp in the magnetization. For large fields B we eventually cross the phase boundary between phase II and III, which occurs at $B_u = 2 - \mu - 2u$. In phase III we have $n_c = 2m = 1$. Approaching the phase boundary from within phase II, both n_c and m exhibit a characteristic

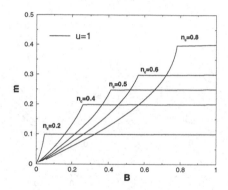

Fig. 6.10. Magnetization as function of the applied field at fixed particle density n_c for $u = 1$.

square root behaviour as can be seen from (6.40)

$$n_c = 2m = \frac{1}{\pi} \arccos \left(\frac{B_u - B}{2} - 1 \right)$$

$$\simeq 1 - \frac{1}{\pi} \sqrt{B_u - B} + \mathcal{O} \left([B_u - B]^{\frac{3}{2}} \right). \qquad (6.43)$$

6.4.3 Fixed n_c

We also may consider the magnetization as a function of the magnetic field while keeping the particle density n_c fixed. In figure 6.10 we show the magnetization curves for five different values of n_c and $u = 1$. As a function of B the magnetization increases from zero and reaches its saturation value $\frac{n_c}{2}$ at the critical field B_c defined in (6.24). For larger fields $B > B_c$ the magnetization remains constant.

The magnetization at half-filling was first calculated by Takahashi in [435,436]. Magnetic properties in the less than half filled band were first calculated by Shiba [396].

6.5 Spin and charge velocities

In the Fermi gas the Fermi velocity v_F is defined as the group velocity of elementary excitations. It is obtained by taking the derivative of the quasiparticle energy with respect to the momentum at the Fermi surface. In the Hubbard model the ground state in phase IV can be thought of as being described by two partially filled 'Fermi seas' of spectral parameters k_j and Λ_α^1 respectively and it is possible to define two characteristic velocities

$$v_c = \left. \frac{\partial \kappa(p)}{\partial p} \right|_{p=p(Q)} = \left. \frac{\kappa'(k)}{p'(k)} \right|_{k=Q}, \qquad (6.44)$$

$$v_s = \left. \frac{\partial \varepsilon_1(p_1)}{\partial p_1} \right|_{p_1=p_1(A)} = \left. \frac{\varepsilon_1'(\Lambda)}{p_1'(\Lambda)} \right|_{\Lambda=A}. \qquad (6.45)$$

Here $'$ denotes the derivative with respect to the argument and $p(k)$ and $p_1(\Lambda)$ are the contributions to the total momentum associated with adding a particle with spectral parameters

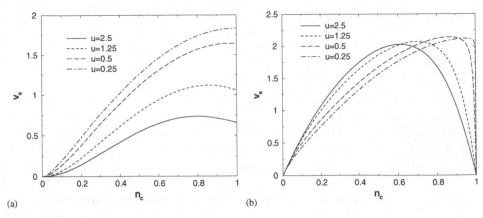

Fig. 6.11. (a) Spin velocity and (b) charge velocity as functions of the density n_c for several values of u in zero magnetic field.

k and Λ to the respective Fermi seas. Their explicit expressions have been given before in (5.98), (5.99). The labelling of the velocities is chosen to reflect the fact that $v_{c,s}$ are the group velocities of charge and spin excitations respectively. This fact is proved in Chapter 7. In order to determine the charge and spin velocities we first need to solve the following set of coupled integral equations for the derivatives of the dressed energies, which are obtained from (6.10) by taking derivatives and exploiting the fact that $\kappa(\pm Q) = \varepsilon_1(\pm A) = 0$:

$$\kappa'(k) = 2\sin(k) + \cos(k) \int_{-A}^{A} d\Lambda\, a_1(\Lambda - \sin k)\, \varepsilon_1'(\Lambda),$$

$$\varepsilon_1'(\Lambda) = \int_{-Q}^{Q} dk\, a_1(\Lambda - \sin k)\, \kappa'(k) - \int_{-A}^{A} d\Lambda'\, a_2(\Lambda - \Lambda')\, \varepsilon_1'(\Lambda'). \tag{6.46}$$

The denominators of (6.44),(6.45) are proportional to the root densities at $\pm Q$, $\pm A$ as can be seen from (5.98), (5.99).

$$p'(Q) = 2\pi\rho(Q), \quad p_1'(A) = 2\pi\sigma_1(A). \tag{6.47}$$

In figure 6.11 we plot $v_{c,s}$ as functions of the density of electrons n_c for several values of u and zero magnetic field $B = 0$. This parameter regime corresponds to phase IV. For $n_c \to 1$ the charge velocity goes to zero, signalling the occurrence of a quantum phase transition to the Mott insulating phase V at $n_c = 1$, see figure 6.4. For $n_c \to 0$ both velocities approach zero, corresponding to a situation where the respective 'Fermi seas' become very shallow. At zero density $n_c = 0$ we have a transition to phase I. The spin and charge velocities were analyzed by H. J. Schulz in his review article [383].

6.6 Susceptibilities

The susceptibilities are obtained by taking derivatives of the magnetization m with respect to the magnetic field and of the density n_c with respect to the chemical potential. In phases I, II and III this is quite simple to do. However, in phases IV and V m and n_c are themselves given in terms of solutions of integral equations and taking derivatives is not an entirely trivial procedure. Of course one could simply implement the derivatives by taking finite differences, e.g.

$$\chi_c\big|_{B=\text{const}} \approx \frac{n_c(\mu + \Delta\mu) - n_c(\mu)}{\Delta\mu}, \tag{6.48}$$

but depending on the properties of $n_c(\mu)$ this can give rather inaccurate results. In order to obtain good numerical accuracy it is better to express the susceptibilities in terms of functions that fulfil linear integral equations as suggested by F. Woynarovich and K. Penc in [492]. Here we follow a slightly different path.

6.6.1 Phases I and III

In phases I (empty band) and Phase III (half filled, spin polarized band) the charge and the magnetic susceptibility are both zero.

6.6.2 Phase II

In phase II (spin-polarized partially filled band) the density is given by (6.40) and the magnetization takes its maximal value $m = n_c/2$. The susceptibilities are

$$\chi_c(\mu, B) = 2\chi_s(\mu, B) = \frac{1}{\pi\sqrt{4 - (\mu + 2u + B)^2}}. \tag{6.49}$$

6.6.3 Phase IV: matrix notations

In order to simplify the necessary manipulations of coupled integral equations we introduce a unifying matrix notation. We introduce vectors of integration variables $x_{c,s}$ and integration boundaries $X_{c,s}$ by

$$(x_c, x_s) = (k, \Lambda), \quad (X_c, X_s) = (Q, A). \tag{6.50}$$

Next we introduce a vector notation for the root densities and dressed energies as well as the corresponding driving terms in the integral equations

$$\begin{aligned}
r_c(x_c) &= \rho(k), & r_c^{(0)}(x_c) &= \tfrac{1}{2\pi}, \\
r_s(x_s) &= \sigma(\Lambda), & r_s^{(0)}(x_s) &= 0, \\
e_c(x_c) &= \kappa(k), & e_c^{(0)}(x_c) &= -2\cos(k) - \mu - 2u - B, \\
e_s(x_s) &= \varepsilon_1(\Lambda), & e_s^{(0)}(x_s) &= 2B.
\end{aligned} \tag{6.51}$$

Finally, we define a matrix integral operator with kernels

$$\begin{aligned}
K_{cc}(x_c, y_c) &= 0, \\
K_{sc}(x_c, x_s) &= a_1(\sin(x_c) - x_s), \\
K_{cs}(x_c, x_s) &= \cos(x_c)a_1(\sin(x_c) - x_s), \\
K_{ss}(x_s, y_s) &= -a_2(x_s - y_s).
\end{aligned} \tag{6.52}$$

In terms of the notations introduced above, the integral equations for the root densities and dressed energies can be written in simple forms

$$r_a(x_a) = r_a^{(0)}(x_a) + \sum_b K_{ab} * r_b\big|_{x_a}, \tag{6.53}$$

$$e_a(x_a) = e_a^{(0)}(x_a) + \sum_b K_{ab}^T * e_b\big|_{x_a}. \tag{6.54}$$

Here $\hat{K}^T$ is the transpose of the matrix integral operator $\hat{K}$ and $*$ as usual denotes convolution. The susceptibilities are given by

$$\chi_s(\mu, B) = \frac{\partial m}{\partial B}\bigg|_{\mu=\mathrm{const}} = \sum_{a=c,s} \frac{\partial m}{\partial X_a} \frac{\partial X_a}{\partial B},$$

$$\chi_c(\mu, B) = \frac{\partial n_c}{\partial \mu}\bigg|_{B=\mathrm{const}} = \sum_{a=c,s} \frac{\partial n_c}{\partial X_a} \frac{\partial X_a}{\partial \mu}. \tag{6.55}$$

In what follows it is convenient to express the magnetization per site in terms of the particle densities for electrons n_c and spin down electrons n_s ($2m = n_c - 2n_s$), where

$$n_a = \int_{-X_a}^{X_a} dx_a \, r_a(x_a). \tag{6.56}$$

The calculation of $\chi_{c,s}$ is naturally split into two parts: we first determine the derivatives of the $n_{c,s}$ with respect to the integration boundaries $X_{c,s}$ and then the derivatives of $X_{c,s}$ with respect to the magnetic field and chemical potential.

(i) The derivatives of the electron density n_c and the down-spin density n_s with respect to the integration boundaries $X_{c,s}$ are

$$\frac{\partial n_a}{\partial X_b} = 2r_b(X_b) Z_{ab}, \tag{6.57}$$

$$Z = \begin{pmatrix} \xi_{cc}(Q) & \xi_{cs}(A) \\ \xi_{sc}(Q) & \xi_{ss}(A) \end{pmatrix}, \tag{6.58}$$

where ξ is the so-called dressed charge matrix, defined in terms of the coupled integral equations

$$\xi_{ab}(x_b) = \delta_{ab} + \sum_d \int_{-X_d}^{X_d} dx_d \, \xi_{ad}(x_d) K_{db}(x_d, x_b). \tag{6.59}$$

The formal solution of (6.59) is

$$\xi_{ab}(x_b) = \int_{-X_a}^{X_a} dx_a \, (1 - \hat{K})_{ab}^{-1}(x_a, x_b). \tag{6.60}$$

Equation (6.57) is proved as follows: taking the derivative of (6.56) we have

$$\frac{\partial n_a}{\partial X_b} = 2\delta_{ab} r_a(X_a) + \int_{-X_a}^{X_a} dx_a \, \frac{\partial r_a(x_a)}{\partial X_b}. \tag{6.61}$$

The derivatives of the root densities fulfil the following integral equation, which is obtained from (6.53)

$$\frac{\partial r_a(x_a)}{\partial X_b} = r_{ab}^{(0)}(x_a) + \sum_d \int_{-X_d}^{X_d} dx_d\, K_{ad}(x_a, x_d)\, \frac{\partial r_d(x_d)}{\partial X_b}, \tag{6.62}$$

$$r_{ab}^{(0)}(x_a) = [K_{ab}(x_a, X_b) + K_{ab}(x_a, -X_b)]\, r_b(X_b). \tag{6.63}$$

Equation (6.62) is formally solved by

$$\frac{\partial r_a(x_a)}{\partial X_b} = \sum_d \int_{-X_d}^{X_d} dx_d\, (1 - \hat{K})_{ad}^{-1}(x_a, x_d)\, r_{db}^{(0)}(x_d). \tag{6.64}$$

Inserting (6.64) into (6.61) we obtain

$$\frac{\partial n_a}{\partial X_b} = r_b(X_b) \int_{-X_a}^{X_a} dx_a \left[(1 - \hat{K})_{ab}^{-1}(x_a, X_b) + (1 - \hat{K})_{ab}^{-1}(x_a, -X_b)\right]. \tag{6.65}$$

Finally we use (6.60) to obtain (6.57).

(ii) Next we determine the derivatives of the integration boundaries X_a with respect to μ and B. To simplify notation we introduce

$$(\mu_c, \mu_s) = (\mu, B). \tag{6.66}$$

The integration boundaries $\{X_a\}$ are fixed in terms of the chemical potentials $\{\mu_a\}$ by the condition (6.11), which in matrix notations reads

$$e_a(X_a) = 0. \tag{6.67}$$

Taking the derivative of (6.67) with respect to μ_b and using the integral equation (6.54) for $e_a(x_a)$ we have

$$0 = \frac{\partial e_a(X_a)}{\partial \mu_b}$$

$$= \frac{\partial e_a^{(0)}(X_a)}{\partial \mu_b} + \sum_d \int_{-X_d}^{X_d} dx_d \left[K_{ad}^T(X_a, x_d) \frac{\partial e_d(x_d)}{\partial \mu_b} + e_d(x_d) \frac{\partial K_{ad}^T(X_a, x_d)}{\partial \mu_b} \right]. \tag{6.68}$$

The r.h.s. is simplified by first using the identity

$$\frac{\partial e_a^{(0)}(X_a)}{\partial \mu_b} = \frac{\partial e_a^{(0)}(x_a)}{\partial x_a}\bigg|_{x_a = X_a} \frac{\partial X_a}{\partial \mu_b} + \frac{\partial e_a^{(0)}(x_a)}{\partial \mu_b}\bigg|_{x_a = X_a}, \tag{6.69}$$

and then the following integral equation for the derivatives of the dressed energies for generic arguments with respect to the chemical potentials

$$\frac{\partial e_a(x_a)}{\partial \mu_b} = \frac{\partial e_a^{(0)}(x_a)}{\partial \mu_b} + \sum_d \int_{-X_d}^{X_d} dx_d\, K_{ad}^T(x_a, x_d) \frac{\partial e_d(x_d)}{\partial \mu_b}. \tag{6.70}$$

We obtain

$$0 = \frac{\partial X_a}{\partial \mu_b} \left[\frac{\partial e_a^{(0)}(x_a)}{\partial x_a} + \sum_d \int_{-X_d}^{X_d} dx_d \, e_d(x_d) \frac{\partial K_{ad}^T(x_a, x_d)}{\partial x_a} \right] \bigg|_{x_a = X_a} + \frac{\partial e_a(x)}{\partial \mu_b} \bigg|_{x = X_a}. \tag{6.71}$$

The term in brackets on the r.h.s. of (6.71) is nothing but

$$\frac{\partial e_a(x_a)}{\partial x_a} \bigg|_{x_a = X_a}, \tag{6.72}$$

and the solution of (6.70) is expressed in terms of the dressed charge matrix as

$$\frac{\partial e_a(x_a)}{\partial \mu_b} \bigg|_{x_a = X_a} = \sum_d (1 - \hat{K}^T)_{ad}^{-1} * \frac{\partial e_d^{(0)}}{\partial \mu_b} \bigg|_{X_a} = \sum_d Z_{da} \frac{\partial e_d^{(0)}(x_d)}{\partial \mu_b}. \tag{6.73}$$

Putting everything together we arrive the desired set of equations determining $\frac{\partial X_a}{\partial \mu_b}$

$$\sum_d Z_{da} \frac{\partial e_d^{(0)}(x_d)}{\partial \mu_b} + \frac{\partial e_a(x_a)}{\partial x_a} \bigg|_{x_a = X_a} \frac{\partial X_a}{\partial \mu_b} = 0. \tag{6.74}$$

The explicit form of (6.74) is

$$\frac{\partial Q}{\partial \mu} = \frac{Z_{cc}}{\kappa'(Q)}, \quad \frac{\partial Q}{\partial B} = \frac{Z_{cc} - 2Z_{sc}}{\kappa'(Q)}, \quad \frac{\partial A}{\partial \mu} = \frac{Z_{cs}}{\varepsilon'(A)}, \quad \frac{\partial A}{\partial B} = \frac{Z_{cs} - 2Z_{ss}}{\varepsilon'(A)}. \tag{6.75}$$

Using (6.74) and (6.57) in (6.55) and then recalling the definitions (6.44), (6.45) of the spin and charge velocities, the susceptibilities can be expressed as

$$\chi_s(\mu, B) = \frac{(Z_{cs} - 2Z_{ss})^2}{2\pi v_s} + \frac{(Z_{cc} - 2Z_{sc})^2}{2\pi v_c}, \tag{6.76}$$

$$\chi_c(\mu, B) = \frac{Z_{cc}^2}{\pi v_c} + \frac{Z_{cs}^2}{\pi v_s}. \tag{6.77}$$

It is sometimes convenient to consider the magnetic susceptibility at fixed density n_c rather than at fixed chemical potential. Similarly one may want to know χ_c at a fixed magnetization rather than for fixed magnetic field. These quantities can be expressed in terms of the dressed charge matrix and the spin and charge velocities as well. It is shown in Section 8.4 that

$$\chi_c(\mu, m) = \frac{4}{\pi} \frac{(\det Z)^2}{v_c(Z_{cs} - 2Z_{ss})^2 + v_s(Z_{cc} - 2Z_{sc})^2}, \tag{6.78}$$

$$\chi_s(n_c, B) = \frac{(\det Z)^2}{\pi} \frac{1}{v_c Z_{cs}^2 + v_s Z_{cc}^2}. \tag{6.79}$$

6.6.4 Phase V

It is shown in Section 8.4 that the magnetic susceptibility in Phase V (half filled, partially magnetized band) can be expressed as

$$\chi_s(B, n_c = 1) = \frac{\xi_s^2(A)}{\pi v_s},$$ (6.80)

where $\xi_s(A)$ is the dressed charge for the half filled band, defined via the integral equation (8.55)

$$\xi_s(\Lambda) = 1 - \int_{-A}^{A} d\Lambda' \, a_2(\Lambda - \Lambda') \, \xi_s(\Lambda').$$ (6.81)

The magnetic susceptibility at half-filling was first calculated by M. Takahashi in [429,430]. The magnetic susceptibility below half-filling in zero field was studied by H. Shiba in [396], G. V. Uimin and S. V. Fomichev [463] and by T. Usuki, N. Kawakami and A. Okiji in [468]. The spin and charge susceptibilities for finite magnetic field were determined by J. M. P. Carmelo, P. Horsch and A. A. Ovchinnikov in [78].

6.7 Ground state energy

In general the ground state free energy (6.17) is a complicated function of μ and B (or n_c and m) and has to be calculated from the solutions of the integral equations (6.10). However, there are several limiting cases in which analytic results are available. At half-filling (we set $\mu = 0$) and zero field $B = 0$ the integral equations for the dressed energies (6.10) can be solved by Fourier transformation as is shown in Appendix 6.B.1. Using these results one derives the following integral representation for the ground state energy, which was first obtained by E. Lieb and F. Y. Wu [298]

$$f(u, \mu = 0, B = 0) = -u - 4 \int_0^\infty \frac{d\omega}{\omega} \frac{J_0(\omega)J_1(\omega)}{1 + \exp(2u\omega)} \equiv e(1).$$ (6.82)

As a function of u the free energy per site has branch points at $u = \pm i/n, n = 1, 2, \ldots$ [431]. Hence $u = 0$ is an accumulation point of branch points and perturbative expansions around the noninteracting theory $u = 0$ have zero radius of convergence.

The large-u expansion of (6.82) was obtained by M. Takahashi in [431]

$$f(u, \mu = 0, B = 0) = -u - \ln(2)\frac{1}{u}$$
$$+ \sum_{n=2}^{\infty} (-1)^{n-1} \left[\frac{(2n-1)!!}{2n!!} \right]^2 \frac{\zeta(2n-1)}{2n-1} \left[1 - \frac{1}{2^{2n-2}} \right] \left(\frac{1}{u} \right)^{2n-1},$$ (6.83)

and converges as long as $u > 1$. Here $\zeta(z)$ is the Riemann zeta function and $(2n-1)!! = (2n-1)(2n-3) \cdots 3 \, 1$. The $\mathcal{O}(u^{-3})$ term was used in [437] to determine the next-nearest neighbour spin-spin correlation function in the spin-$\frac{1}{2}$ Heisenberg model.

The small-u expansion of (6.82) was derived in [113] and is an asymptotic expansion with zero radius of convergence

$$f(u, \mu = 0, B = 0) = -\frac{4}{\pi} - \frac{7\zeta(3)}{\pi^3} u^2$$
$$- \sum_{n=2}^{\infty} \frac{\zeta(2n+1)}{\pi^{2n+1}} \frac{(2^{2n+1} - 1)(2n - 1)\left[(2n - 3)!!\right]^3}{(2n - 2)!! \, 2^{2n-2}} u^{2n}.$$

(6.84)

In an almost half filled band $n_c \approx 1$ and $B = 0$ it is possible to derive an expansion of the ground state energy in powers of $\delta = 1 - n_c$ [376]. Some details are given in Appendix 6.B.2. The results are

$$f = e(n_c) - \mu n_c,$$

$$e(n_c) = e(1) - \mu_-(u)\delta + \frac{a_1^2 \alpha_1}{3} \delta^3 + \mathcal{O}(\delta^4),$$

$$n_c(\mu) \approx 1 - (a_1^2 \alpha_1)^{-\frac{1}{2}} \sqrt{\mu_-(u) - \mu},$$

(6.85)

where a_1 and α_1 are constants given in (6.B.21) and (6.B.24) respectively and where $\mu_-(u)$ is given by (6.31).

Similarly, in the low density limit $n_c \ll 1$ the ground state energy can be calculated in an expansion in powers of n_c. We discuss this limit in Appendix 6.B.3. The results are

$$e(n_c) = u - (2 + 2u)n_c + \frac{\pi^2}{3} n_c^3 + \mathcal{O}(n_c^4),$$

$$n_c(\mu) \approx \frac{1}{\pi} \sqrt{\mu - \mu_0(0)},$$

(6.86)

where $\mu_0(0)$ is given by (6.20). In figure 6.12 we plot the ground state energy as a function of the chemical potential for $B = 0$ and different values of u.

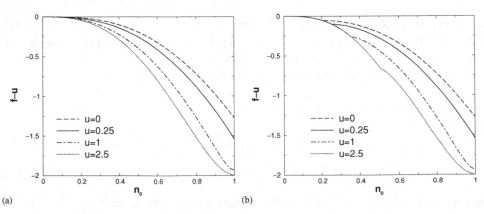

Fig. 6.12. Ground-state free energy (shifted by u) as a function of electron density for several values of u; (a) in zero magnetic field (b) for $B = 0.2$.

Appendices to Chapter 6

6.A Numerical solution of integral equations

Here we discuss how to solve the coupled integral equations (6.10) numerically. Let us rewrite them in a shorthand notation as

$$\kappa(k) = -2\cos k - \mu - 2u - B + K^T_{cs} * \varepsilon_1\big|_k,$$
$$\varepsilon_1(\Lambda) = 2B + K^T_{sc} * \kappa\big|_\Lambda + K^T_{ss} * \varepsilon_1\big|_\Lambda. \tag{6.A.1}$$

Here K^T_{ab} are integral operators defined by (6.10), for example

$$K^T_{cs} * \varepsilon_1\big|_k = \int_{-A}^{A} \frac{d\Lambda}{\pi} \frac{U/4}{(U/4)^2 + (\sin k - \Lambda)^2} \varepsilon_1(\Lambda). \tag{6.A.2}$$

The main problem we are facing in solving (6.10) is that the integration boundaries $\pm Q$ and $\pm A$ are functions of the magnetic field B and the chemical potential μ, and are fixed by the conditions

$$\kappa(\pm Q) = 0, \qquad \varepsilon_1(\pm A) = 0. \tag{6.A.3}$$

We are therefore dealing with a 'chicken vs egg' problem: in order to solve the integral equations (6.A.1), we first need to know Q and A for given μ and B. On the other hand, Q and A are determined as functions of μ and B by the conditions (6.A.3), which involve the solution of the integral equation (6.A.1). A crude way of solving this problem would be to proceed as follows.

1. Fix B and μ.
2. Choose values for Q and A.
3. Solve (6.A.1) numerically.
4. Check equations (6.A.3). If they are fulfilled within some error we are done. If not go back to step 2.

We will follow a different path, see e.g. [492]. The key observation is that μ and B enter the linear integral equations (6.A.1) *linearly*. We start by choosing the integration boundaries Q and A. We then determine $\mu(Q, A)$ and $B(Q, A)$ from the solutions of a set of three pairs of coupled linear integral equations. These are constructed as follows

$$\bar{\kappa}(k) = -2\cos k - 2u + K_{cs}^T * \bar{\varepsilon}_1\big|_k,$$

$$\bar{\varepsilon}_1(\Lambda) = K_{sc}^T * \bar{\kappa}\big|_\Lambda + K_{ss}^T * \bar{\varepsilon}_1\big|_\Lambda,$$

$$\xi_{ac}(k) = \delta_{ac} + \sum_{d=c,s} \xi_{ad} * K_{dc}\big|_k,$$

$$\xi_{as}(\Lambda) = \delta_{as} + \sum_{d=c,s} \xi_{ad} * K_{ds}\big|_\Lambda. \tag{6.A.4}$$

Here ξ_{ab} are the elements of the dressed charge matrix (6.59) and $\bar{\kappa}$ and $\bar{\varepsilon}$ fulfil the integral equations for the dressed energies for zero magnetic field and zero chemical potential. By construction we have

$$\kappa(k) = \bar{\kappa}(k) - (\mu + B)\xi_{cc}(k) + 2B\xi_{sc}(k),$$

$$\varepsilon_1(\Lambda) = \bar{\varepsilon}_1(\Lambda) - (\mu + B)\xi_{cs}(\Lambda) + 2B\xi_{ss}(\Lambda). \tag{6.A.5}$$

We now solve the three sets of coupled integral equations numerically for fixed values of Q and A. We then determine the corresponding values of μ and B from the conditions (6.A.3), which in terms of the quantities $Z_{ca} = \xi_{ac}(Q)$, $Z_{sa} = \xi_{as}(A)$, $\bar{\kappa}(Q)$, $\bar{\varepsilon}_1(A)$ read

$$B = \frac{Z_{cc}\bar{\varepsilon}_1(A) - Z_{cs}\bar{\kappa}(Q)}{2(Z_{cs}Z_{sc} - Z_{ss}Z_{cc})},$$

$$\mu = \frac{\bar{\varepsilon}_1(A)(2Z_{sc} - Z_{cc}) - \bar{\kappa}(Q)(2Z_{ss} - Z_{cs})}{2(Z_{cs}Z_{sc} - Z_{ss}Z_{cc})}. \tag{6.A.6}$$

What remains to be done is to solve (6.A.4) numerically for fixed values of Q and A. This is easily done by using for example the Numerical Recipes routine `fred2` [364]. We note that in order to use this routine, the coupled integral equations have to be written in a 'block-diagonal' form as we will now explain for the example of coupled equations for two quantities $X(k)$ and $Y(\Lambda)$

$$X(k) = X_0(k) + G_{cc} * X\big|_k + G_{cs} * Y\big|_k,$$

$$Y(\Lambda) = Y_0(\Lambda) + G_{sc} * X\big|_\Lambda + G_{ss} * Y\big|_\Lambda, \tag{6.A.7}$$

where G_{ab} are some integral operators. The integrals are computed by means of a Gaussian quadrature

$$\int_{-Q}^{Q} dk\, f(k) \longrightarrow \sum_{j=1}^{N} u_j\, f(k_j),$$

$$\int_{-A}^{A} d\Lambda\, g(\Lambda) \longrightarrow \sum_{k=1}^{M} v_k\, g(\Lambda_k). \tag{6.A.8}$$

Here $\{u_j\}$ and $\{v_k\}$ are the weights of the quadrature rule, while the N points k_j and M

points Λ_k are the abscissas. We now define vectors

$$
f_j = \begin{cases} X(k_j) & \text{if } j \le N, \\ Y(\Lambda_{j-N}) & \text{if } j > N, \end{cases} \tag{6.A.9}
$$

$$
g_j = \begin{cases} X_0(k_j) & \text{if } j \le N, \\ Y_0(\Lambda_{j-N}) & \text{if } j > N, \end{cases} \tag{6.A.10}
$$

$$
w_j = \begin{cases} u_j & \text{if } j \le N, \\ v_{j-N} & \text{if } j > N, \end{cases} \tag{6.A.11}
$$

as well as a block-diagonal matrix

$$
M_{jl} = \begin{cases} G_{cc}(k_j, k_l) & \text{if } j \le N, \, l \le N, \\ G_{cs}(k_j, \Lambda_l) & \text{if } j \le N, \, l > N \\ G_{sc}(\Lambda_j, k_l) & \text{if } j > N, \, l \le N, \\ G_{ss}(\Lambda_j, \Lambda_l) & \text{if } j > N, \, l > N. \end{cases} \tag{6.A.12}
$$

The coupled integral equations can now be written in the following discretized form

$$
f_j = g_j + \sum_{l=1}^{N+M} w_l \, M_{jl} \, f_l. \tag{6.A.13}
$$

Equation (6.A.13) is now of the same form as (18.1.4) of [364] and can be solved by the routine `fred2`.

6.B Ground state properties in zero magnetic field

In absence of a magnetic field the integral equations for the dressed energies and root densities are simpler, because the Λ-integration extends over the entire real axis.[1] This allows us to simplify all integral equations by Fourier transformation. For example, from the second equation of (6.10) for $B = 0$, $A = \infty$ we obtain

$$
\begin{aligned}
\tilde{\varepsilon}_1(\omega) &= \int_{-\infty}^{\infty} d\Lambda \, \exp(i\omega\Lambda) \, \varepsilon_1(\Lambda) \\
&= \int_{-Q}^{Q} dk \, \kappa(k) \, \cos(k) \, \exp\left[-u|\omega| + i\omega\sin(k)\right] - \tilde{\varepsilon}_1(\omega) \, \exp\left[-2u|\omega|\right].
\end{aligned} \tag{6.B.1}
$$

Collecting terms and Fourier transforming back we arrive at

$$
\varepsilon_1(\Lambda) = \int_{-Q}^{Q} dk \, \frac{\cos k}{4u \cosh\frac{\pi}{2u}(\Lambda - \sin k)} \kappa(k). \tag{6.B.2}
$$

---

[1] We recall that we have shown in the beginning of Section 6.3 that the integration boundary A approaches infinity as we take the magnetic field to zero.

Finally, inserting (6.B.2) into the right hand side of the equation (6.10) for $\kappa(k)$ we obtain

$$\kappa(k) = -2\cos k - \mu - 2u + \int_{-Q}^{Q} dk'\cos k'\ R(\sin k' - \sin k)\,\kappa(k'),\qquad (6.B.3)$$

where

$$R(x) = \int_{-\infty}^{\infty} \frac{d\omega}{2\pi}\frac{\exp(i\omega x)}{1 + \exp(2u|\omega|)}.\qquad (6.B.4)$$

The integral equations for root densities can be simplified similarly with the result

$$\rho(k) = \frac{1}{2\pi} + \int_{-Q}^{Q} dk'\cos k\ R(\sin k' - \sin k)\,\rho(k'),\qquad |k| \le Q,$$

$$\sigma_1(\Lambda) = \int_{-Q}^{Q} dk\,\frac{1}{4u\cosh\frac{\pi}{2u}(\Lambda - \sin k)}\rho(k).\qquad (6.B.5)$$

6.B.1 Half filled band

For $Q = \pi$, i.e. a half filled band, the integral equations (6.B.3), (6.B.5), can be solved by Fourier series techniques. However, a simpler way to obtain $\kappa(k)$ and $\varepsilon_1(\Lambda)$ in this case is to take (6.30) as a starting point. Setting $A = \infty$ in (6.30) we can calculate $\varepsilon_1(\Lambda)$ by Fourier transformation. Using that

$$\int_{-\pi}^{\pi} \frac{dk}{\pi}\,2\cos^2(k)\,\frac{u}{u^2 + (\sin k - \Lambda)^2} = 4\mathrm{Re}\sqrt{1 - (\Lambda - iu)^2} - 4u,\qquad (6.B.6)$$

we obtain for the Fourier transform

$$\tilde{\varepsilon}_1(\omega) = -\int_{-\pi}^{\pi} dk\,\frac{\cos^2(k)\,\exp(i\omega\sin k)}{\cosh(u\omega)}$$

$$= -\frac{2\pi J_1(\omega)}{\omega\cosh(u\omega)}.\qquad (6.B.7)$$

Here the second equality is obtained via integration by parts and $J_1(x)$ is a Bessel function. Using the result (6.B.7) in (6.30) one arrives at the following integral representations

$$\kappa(k) = -2\cos k - \mu - 2u - 2\int_0^{\infty} \frac{d\omega}{\omega}\,\frac{J_1(\omega)\cos(\omega\sin k)e^{-\omega u}}{\cosh(\omega u)} \equiv \kappa_0(k) - \mu,$$

$$\varepsilon_1(\Lambda) = -2\int_0^{\infty} \frac{d\omega}{\omega}\,\frac{J_1(\omega)\,\cos(\omega\Lambda)}{\cosh(\omega u)} \equiv \varepsilon_0(\Lambda).\qquad (6.B.8)$$

In the limits of large and small values of u it is possible to derive expansions for $\kappa_0(k)$. For small values of u the following series representation is useful [324]

$$\kappa_0(k) = -\frac{4}{u}\sum_{n=0}^{\infty} \frac{K_1(u_n)}{u_n}\cosh(u_n\sin(k)),\qquad k \in (\frac{\pi}{2}, \pi].\qquad (6.B.9)$$

where $u_n = (n + \frac{1}{2})\frac{\pi}{u}$ and K_1 is a modified Bessel function. For small $u \ll 1$ we have

$u_n \gg 1$ and thus can use the asymptotics of the modified Bessel function to obtain

$$
\begin{aligned}
\kappa_0(k) \approx\; & -\frac{8}{\pi}\sqrt{u}\,\exp\left(-\frac{\pi}{2u}\right)\cosh\left(\frac{\pi}{2u}\sin(k)\right)\left[1+\frac{3u}{4\pi}-\frac{15u^2}{32\pi^2}+\cdots\right] \\
& -\frac{8}{\pi}\sqrt{\frac{u}{27}}\,\exp\left(-\frac{3\pi}{2u}\right)\cosh\left(\frac{3\pi}{2u}\sin(k)\right)\left[1+\frac{u}{4\pi}-\frac{5u^2}{96\pi^2}+\cdots\right] \\
& +\cdots
\end{aligned}
\tag{6.B.10}
$$

This expansion converges very rapidly as long as k is sufficiently far away from $\pi/2$

$$
1-\sin(k)\gg u. \tag{6.B.11}
$$

The behaviour for k in the interval $[0,\pi/2)$ is obtained from (6.B.9) by using the relation $\kappa_0(\pi-k)=\kappa_0(k)+4\cos(k)$. For large u one easily derives the following $1/u$-expansion

$$
\begin{aligned}
\kappa_0(k) =\; & -2u-2\cos(k)-\frac{\ln(2)}{u}+\frac{3\zeta(3)}{64u^3}[1+4\sin^2(k)] \\
& -\frac{15\zeta(5)}{2048u^5}[1+12\sin^2(k)+8\sin^4(k)]+\mathcal{O}(u^{-7}),
\end{aligned}
\tag{6.B.12}
$$

where $\zeta(x)$ is the Riemann zeta function. The maximum of $\kappa_0(k)$ is taken at $k=\pi$, where we have [431]

$$
\kappa_0(\pi)=2-2u-\frac{\ln(2)}{u}+\sum_{n=2}^{\infty}(-1)^n\frac{(2n-3)!!}{(2n)!!}\frac{\zeta(2n-1)}{2^{2n-3}}\left[1-\frac{1}{2^{2n-2}}\right]u^{-2n+1}. \tag{6.B.13}
$$

The integral equations for the root densities can be solved in a way completely analogous to the dressed energies with the result

$$
\begin{aligned}
\rho(k) &= \frac{1}{2\pi}+\cos(k)\int_{-\infty}^{\infty}\frac{d\omega}{2\pi}\frac{J_0(\omega)\cos(\omega\sin(k))}{1+\exp(2u|\omega|)}\equiv\rho_0(k), \\
\sigma_1(\Lambda) &= \int_{-\infty}^{\infty}\frac{d\omega}{2\pi}\frac{J_0(\omega)}{2\cosh(u\omega)}\exp(-i\omega\Lambda)\equiv\sigma_0(\Lambda).
\end{aligned}
\tag{6.B.14}
$$

Equations (6.B.14) were first obtained by E.H. Lieb and F.Y. Wu in [298].

6.B.2 The almost half filled band

For n_c slightly less than 1 it is possible to solve the integral equations for the dressed energies and the root densities by an iterative procedure [376]. The following symmetry properties for $\rho(k)$ and $\kappa(k)$ are easily derived from the integral equations (6.B.5) and (6.B.3) respectively

$$
\rho(-k)=\rho(k),\quad \rho(\pi-k)=-\rho(k)+\frac{1}{\pi}, \tag{6.B.15}
$$
$$
\kappa(-k)=\kappa(k),\quad \kappa(\pi-k)=\kappa(k)+4\cos(k). \tag{6.B.16}
$$

Using these properties the integral equations for $\rho(k)$ and $\kappa(k)$ can be rewritten in a form suitable for a solution by iteration. We will concentrate on $\rho(k)$ but note that $\kappa(k)$ can be dealt with in an analogous fashion. Using the symmetry properties (6.B.15) we rewrite (6.B.5) as

$$\rho(k) = \rho_0(k) - \cos(k) \int_0^{\pi-Q} dk' \, \check{\mathbf{R}}(k, k') \, \rho(\pi - k'),$$

$$\check{\mathbf{R}}(k, k') = R(\sin(k) - \sin(k')) + R(\sin(k) + \sin(k')), \qquad (6.B.17)$$

where $\rho_0(k)$ is the root density at half filling and $B = 0$ (6.B.14) and $R(x)$ is the kernel defined in (6.B.4). The advantage of (6.B.17) is that $\pi - Q$ is small, which permits a solution by iteration. Rather than using $\pi - Q$ as expansion parameter we will use the deviation of the density from one

$$\delta \equiv 1 - n_c = 1 - \int_{-Q}^{Q} dk \, \rho(k) = 2 \int_0^{\pi-Q} dk \, \rho(\pi - k). \qquad (6.B.18)$$

Inserting the expansions

$$\pi - Q = \sum_{n=1}^{\infty} a_n \delta^n, \quad \rho(k) = \sum_{n=0}^{\infty} \rho_n(k) \delta^n, \qquad (6.B.19)$$

into (6.B.17) one can determine the coefficients a_n and the functions $\rho_n(k)$ iteratively. The integrals on the right hand sides of both (6.B.17) and (6.B.19) are taken by Taylor expanding around $k' = 0$. For example, we have

$$\check{\mathbf{R}}(k, k') \, \rho_n(\pi - k') = \sum_{m=0}^{\infty} \frac{d^m}{dx^m}\bigg|_{x=0} \left[\check{\mathbf{R}}(k, x) \, \rho_n(\pi - x) \right] \frac{k'^m}{m}!$$

$$\equiv \sum_{m=0}^{\infty} \alpha_{nm}(k) \, k'^m. \qquad (6.B.20)$$

The first few terms of this expansion are [376]

$$a_1 = \frac{1}{2\rho_0(\pi)}, \qquad\qquad \rho_1(k) = -\cos(k) \, R(\sin k),$$

$$a_2 = -2R(0)a_1^2, \qquad\qquad \rho_2(k) = 0,$$

$$a_3 = 4a_1^3[R(0)]^2 - \frac{a_1^4}{3}\rho_0''(\pi), \qquad \rho_3(k) = -\frac{a_1^2}{6} R''(\sin k) \cos(k).$$

$$(6.B.21)$$

We can use these results to obtain an expansion of the ground state free energy per site (6.17)

$$f = u - (\mu + 2u)n_c - 2 \int_{-\pi}^{\pi} dk \, \rho(k) \, \cos(k) - 4 \int_0^{\pi-Q} dk \, \rho(\pi - k) \, \cos(k)$$

$$= e(n_c) - \mu n_c, \qquad (6.B.22)$$

where $e(n_c)$ is the ground state energy per site in the canonical ensemble. Using (6.B.21) we can determine the first few terms in the expansion of $e(n_c)$ around $n_c = 1$

$$e(n_c) = e(1) - \mu_-(u)\,\delta + \frac{a_1^2\alpha_1}{3}\,\delta^3 + \mathcal{O}(\delta^4),$$ (6.B.23)

where $\mu_-(u)$ is the chemical potential at the transition to half-filling defined in (6.31) and

$$\alpha_1 = 1 - 2\int_0^\infty d\omega\,\frac{\omega J_1(\omega)}{1 + \exp(2u\omega)}.$$ (6.B.24)

The chemical potential can now be obtained as

$$\mu = \frac{\partial e(n_c)}{\partial n_c} = \mu_-(u) - a_1^2\alpha_1\,\delta^2 + \mathcal{O}(\delta^3).$$ (6.B.25)

Inverting this relation we arrive at

$$n_c(\mu) \approx 1 - (a_1^2\alpha_1)^{-\frac{1}{2}}\sqrt{\mu_-(u) - \mu}.$$ (6.B.26)

6.B.3 *Low density*

At low densities $n_c \ll 1$ we can again solve (6.B.5) by iteration, using n_c as the small parameter. The integral equation for $\rho(k)$ can be written as

$$\rho(k) = \frac{1}{2\pi} + \cos(k)\int_0^Q dk'\,\check{R}(k, k')\,\rho(k'),$$ (6.B.27)

which makes it clear that the calculation is completely analogous to the one we just did for n_c close to 1. Expanding

$$Q = \sum_{j=1}^\infty \check{a}_j\,(n_c)^j, \quad \rho(k) = \frac{1}{2\pi} + \sum_{j=1}^\infty \check{\rho}_j(k)\,(n_c)^j,$$ (6.B.28)

we obtain

$$\begin{aligned}
&\check{a}_1 = \pi, && \check{\rho}_1(k) = \cos(k)\,R(\sin k), \\
&\check{a}_2 = -2\pi^2 R(0), && \check{\rho}_2(k) = 0, \\
&\check{a}_3 = 4\pi^3[R(0)]^2, && \check{\rho}_3(k) = \frac{\pi^2}{6}R''(\sin k)\cos(k).
\end{aligned}$$ (6.B.29)

These results can be obtained from (6.B.21) by setting $\rho_0(k) = \frac{1}{2\pi}$ and $\check{\rho}_n(k) = \rho_n(\pi - k)$. The ground state energy in the canonical ensemble can now be calculated

$$e(n_c) = u + \int_{-Q}^Q dk(-2\cos(k) - 2u)\,\rho(k)$$

$$= u - (2 + 2u)n_c + \frac{\pi^2}{3}n_c^3 + \mathcal{O}(n_c^4).$$ (6.B.30)

The chemical potential as a function of the density is given by

$$\mu = \frac{\partial e(n_c)}{\partial n_c} = -(2 + 2u) + \pi^2 n_c^2 + \mathcal{O}(n_c^3).$$ (6.B.31)

Inverting this relation we have

$$n_c(\mu) \approx \frac{1}{\pi}\sqrt{\mu - \mu_0(0)}, \tag{6.B.32}$$

where $\mu_0(B) = -2 - 2u - B$ has been defined in (6.20).

6.C Small magnetic fields at half filling: application of the Wiener-Hopf method

Let us consider the half filled case in a weak magnetic field $B \to 0$ for fixed u.[2] The integral equations for $\kappa(k)$ and $\varepsilon_1(\Lambda)$ are given by (6.30) and the integration boundary A is very large. As $\kappa(k)$ is obtained by integrating $\varepsilon_1(\Lambda)$ let us concentrate on the integral equation for the latter and rewrite it in the following form

$$\varepsilon_1(\Lambda) = \varepsilon_1^{(0)}(\Lambda) - \int_{-A}^{A} \frac{d\Lambda'}{2\pi} \frac{4u}{(2u)^2 + (\Lambda - \Lambda')^2} \varepsilon_1(\Lambda'),$$

$$\varepsilon_1^{(0)}(\Lambda) = 2B - 4\mathrm{Re}\sqrt{1 - (\Lambda - iu)^2} + 4u. \tag{6.C.1}$$

Fourier transforming (6.C.1) we arrive at

$$\widetilde{\varepsilon_1}(\omega) = 2\pi B\delta(\omega) - \frac{2\pi J_1(\omega)}{\omega \cosh(u\omega)}$$

$$+ \frac{1}{1 + \exp(2u|\omega|)} \int_A^{\infty} + \int_{-\infty}^{-A} d\Lambda' \exp(i\omega\Lambda') \varepsilon_1(\Lambda'). \tag{6.C.2}$$

Fourier transforming back and using that $\varepsilon_1(-\Lambda) = \varepsilon_1(\Lambda)$ we obtain the following integral equation for the function $y(\Lambda) \equiv \varepsilon_1(\Lambda + A)$

$$y(\Lambda) = g_0(\Lambda) + \int_0^{\infty} d\Lambda' \left[R(\Lambda - \Lambda') + R(\Lambda + \Lambda' + 2A) \right] y(\Lambda'),$$

$$g_0(\Lambda) = B - \int_{-\infty}^{\infty} \frac{d\omega}{\omega} \frac{J_1(\omega)}{\cosh(u\omega)} \exp(i\omega[\Lambda + A]), \tag{6.C.3}$$

where $R(x)$ is given by (5.104). Equation (6.C.3) can be solved in terms of an expansion [495]

$$y(\Lambda) = \sum_{n=0}^{\infty} y_n(\Lambda), \tag{6.C.4}$$

where $y_n(\Lambda)$ are defined as solutions of the Wiener-Hopf equations

$$y_n(\Lambda) = g_n(\Lambda) + \int_0^{\infty} d\Lambda' R(\Lambda - \Lambda') y_n(\Lambda'),$$

$$g_n(\Lambda) = \int_0^{\infty} d\Lambda' R(\Lambda + \Lambda' + 2A) y_{n-1}(\Lambda'), \ n \geq 1. \tag{6.C.5}$$

---

[2] The case $u \ll B \ll 1$ is quite different, see for example [492].

Equations (6.C.5) can now be solved iteratively: if we know $y_{n-1}(\Lambda)$ we can determine $y_n(\Lambda)$. The usefulness of this procedure is based on the fact that the driving terms $g_n(\Lambda)$ and therefore the solutions $y_n(\Lambda)$ become 'smaller' (in precisely which sense will become clear later) as n increases because A is large. Our discussion follows the general procedure set out in [495] and its implementation for the Hubbard model in [492].

6.C.1 General structure

Let us assume that we know the function $y_{n-1}(\Lambda)$ and hence the 'driving term' $g_n(\Lambda)$. We define functions $\tilde{y}_n^{\pm}(\omega)$ which are analytic and the upper and lower half-planes respectively

$$\tilde{y}_n^{+}(\omega) = \int_0^{\infty} d\Lambda \, \exp(i\omega\Lambda) \, y_n(\Lambda),$$

$$\tilde{y}_n^{-}(\omega) = \int_{-\infty}^0 d\Lambda \, \exp(i\omega\Lambda) \, y_n(\Lambda). \tag{6.C.6}$$

In terms of these functions we can express the Fourier transform of the equation (6.C.5)

$$\frac{1}{1 + \exp(-2u|\omega|)} \, \tilde{y}_n^{+}(\omega) + \tilde{y}_n^{-}(\omega) = \tilde{g}_n(\omega), \tag{6.C.7}$$

where $\tilde{g}_n(\omega)$ is the Fourier transform of $g_n(\Lambda)$. The key idea is now to split (6.C.7) into the sum of two pieces that are analytic in the upper and lower half-planes respectively. In order to achieve this goal we will employ the factorisation

$$1 + \exp(-2u|\omega|) = G^{+}(\omega) \, G^{-}(\omega),$$

$$G^{+}(\omega) = G^{-}(-\omega) = \frac{\sqrt{2\pi}}{\Gamma(\frac{1}{2} - i\frac{u\omega}{\pi})} \left(-i\frac{u\omega}{\pi}\right)^{-i\frac{u\omega}{\pi}} \exp\left(i\frac{u\omega}{\pi}\right). \tag{6.C.8}$$

The functions $G^{\pm}(\omega)$ are analytic in the upper and lower half-planes respectively and are normalised such that

$$\lim_{|\omega| \to \infty} G^{\pm}(\omega) = 1. \tag{6.C.9}$$

Some special values of the functions $G^{\pm}$ are

$$G^{-}(0) = \sqrt{2}, \qquad G^{-}\left(-\frac{i\pi}{2u}\right) = \sqrt{\frac{\pi}{e}}. \tag{6.C.10}$$

Using (6.C.8) in (6.C.7) we arrive at

$$\frac{\tilde{y}_n^{+}(\omega)}{G^{+}(\omega)} + G^{-}(\omega) \, \tilde{y}_n^{-}(\omega) = G^{-}(\omega) \, \tilde{g}_n(\omega). \tag{6.C.11}$$

In the next step we decompose the right-hand side of (6.C.11) into a sum of two functions

$Q^{\pm}(\omega)$ that are analytic in the upper and lower half-planes respectively

$$G^-(\omega)\, \tilde{g}_n(\omega) \equiv Q_n^+(\omega) + Q_n^-(\omega). \tag{6.C.12}$$

This implies that

$$\tilde{y}_n{}^+(\omega) = G^+(\omega)Q_n^+(\omega),$$

$$\tilde{y}_n{}^-(\omega) = \frac{Q_n^-(\omega)}{G^-(\omega)}. \tag{6.C.13}$$

6.C.2 Solution of the equation for $y_0(\Lambda)$

We have

$$\tilde{g}_0(\omega) = 2\pi\, B\delta(\omega) - \frac{2\pi\, J_1(\omega)\, \exp(-i\omega A)}{\omega\, \cosh(u\omega)}. \tag{6.C.14}$$

Next we have to determine the decomposition into $Q_0^{\pm}(\omega)$. The δ-function piece is easily done

$$2\pi\, \delta(\omega) = i\left(\frac{1}{\omega + i\varepsilon} - \frac{1}{\omega - i\varepsilon}\right). \tag{6.C.15}$$

The second term in (6.C.14) is a meromorphic function of ω with simple poles at the points

$$\omega_n = i\frac{\pi}{2u}(2n + 1),\ n \in Z. \tag{6.C.16}$$

We note that there is no pole at $\omega = 0$. The decomposition of the factor $1/\cosh(u\omega)$ giving rise to these poles into functions $\chi^{\pm}$ analytic in the upper and lower half-plane respectively is

$$\frac{1}{\cosh(u\omega)} = \chi^+(\omega) + \chi^-(\omega),$$

$$\chi^-(\omega) = \frac{1}{\cosh(u\omega)} - \frac{i}{u}\sum_{n=0}^{\infty}(-1)^n\frac{1}{\omega + \omega_n},$$

$$\chi^+(\omega) = \frac{i}{u}\sum_{n=0}^{\infty}(-1)^n\frac{1}{\omega + \omega_n}. \tag{6.C.17}$$

Using (6.C.17) we can write $f^-(\omega)/\cosh(u\omega)$ for any function $f^-(z)$ that is analytic and bounded in the lower half-plane as the sum of two functions $F^{\pm}(\omega)$ analytic in the upper/lower half-plane

$$\frac{f^-(\omega)}{\cosh(u\omega)} = F^+(\omega) + F^-(\omega),$$

$$F^+(\omega) = \frac{i}{u}\sum_{n=0}^{\infty}(-1)^n\frac{f^-(-\omega_n)}{\omega + \omega_n}. \tag{6.C.18}$$

Applying (6.C.18) to (6.C.12) and (6.C.14)

$$Q_0^+(\omega) = \frac{iBG^-(0)}{\omega + i\varepsilon}$$
$$- 4i \sum_{n=0}^{\infty}(-1)^n \frac{G^-(-i\frac{\pi}{2u}(2n+1))I_1(\frac{\pi}{2u}(2n+1))}{(2n+1)(\omega + i\frac{\pi}{2u}(2n+1))} \exp\left(-\frac{\pi(2n+1)A}{2u}\right),$$

$$(6.C.19)$$

where $I_1(z)$ is a Bessel function. We note that it is essential that $A > 1$ as otherwise the expansion in (6.C.19) does not converge. The function $Q_0^-(\omega)$ can be determined in an analogous way

$$Q_0^-(\omega) = \frac{iBG^-(0)}{\omega - i\varepsilon} - \frac{2\pi J_1(\omega) \exp(-i\omega A)G^-(\omega)}{\omega \cosh(u\omega)}$$
$$+ 4i \sum_{n=0}^{\infty}(-1)^n \frac{G^-(-i\frac{\pi}{2u}(2n+1))I_1(\frac{\pi}{2u}(2n+1))}{(2n+1)(\omega + i\frac{\pi}{2u}(2n+1))} \exp\left(-\frac{\pi(2n+1)A}{2u}\right).$$

$$(6.C.20)$$

The functions $\tilde{y}_0^{\pm}(\omega)$ are now easily obtained by using (6.C.13), e.g.

$$\tilde{y}_0^+(\omega) = G^+(\omega)\left[\frac{iBG^-(0)}{\omega + i\varepsilon} - 4i\frac{G^-(-i\frac{\pi}{2u})I_1(\frac{\pi}{2u})}{\omega + i\frac{\pi}{2u}} \exp\left(-\frac{\pi A}{2u}\right) + \ldots\right]. \quad (6.C.21)$$

Equation (6.C.19) can be used to determine the integration boundary A as a function of B as follows. By definition we have $y(0) = \varepsilon_1(A) = 0$, so that

$$0 = y(0) = \lim_{\omega \to \infty} -i\omega \, \tilde{y}^+(\omega). \quad (6.C.22)$$

In the leading approximation we replace $\tilde{y}^+(\omega)$ by $\tilde{y}_0^+(\omega)$ in (6.C.22). This results in the following equation

$$B = 4 \sum_{n=0}^{\infty}(-1)^n \frac{G^-(-i\frac{\pi}{2u}(2n+1))I_1(\frac{\pi}{2u}(2n+1))}{(2n+1)G^-(0)} \exp\left(-\frac{\pi(2n+1)A}{2u}\right).$$

$$(6.C.23)$$

Here we have used that $\lim_{\omega \to \infty} G^{\pm}(\omega) = 1$. We may now solve (6.C.23) for the dependence of the integration boundary A on the magnetic field

$$A \simeq \frac{2u}{\pi} \ln(B_c/B) + \mathcal{O}(B), \quad (6.C.24)$$

$$B_c = \frac{4G^-(-i\frac{\pi}{2u})I_1(\frac{\pi}{2u})}{G^-(0)} = \sqrt{\frac{2\pi}{e}} \, 2I_1\left(\frac{\pi}{2u}\right). \quad (6.C.25)$$

The subleading contributions in (6.C.24) are not important as they turn out to be much smaller than the leading contributions due to $y_n(\Lambda)$ with $n = 1, 2, \ldots$.

6.C.3 Equation for $y_1(\Lambda)$

The Fourier transform of the driving term of the equation (6.C.5) for $n = 1$ is

$$\tilde{g}_1(\omega) = \frac{\exp(-i2A\omega)\,\tilde{y}_0{}^+(-\omega)}{1 + \exp(2u|\omega|)},$$

$$= \left[1 - \frac{1}{G^+(\omega)G^-(\omega)}\right]\exp(-i2A\omega)\,\tilde{y}_0{}^+(-\omega). \qquad (6.C.26)$$

The function $Q_1^+(\omega)$ defined in (6.C.12) is then given by

$$Q_1^+(\omega) = \frac{1}{2\pi i}\int_{-\infty}^{\infty} dx\,\frac{1}{x - \omega - i\varepsilon}\,G^-(x)\,\tilde{g}_1(x). \qquad (6.C.27)$$

The integrand has a branch cut along the negative imaginary axis and by deforming the contour of integration we can rewrite (6.C.27) as an integral around the branch cut

$$Q_1^+(\omega) = \frac{1}{2\pi i}\int_0^{\infty} dx\,\frac{\tilde{y}_0{}^+(ix)}{x - i\omega}\,\exp(-2Ax)\left[\frac{1}{G^+(-ix - \varepsilon)} - \frac{1}{G^+(-ix + \varepsilon)}\right]$$

$$= \frac{2}{(2\pi)^{3/2}}\int_0^{\infty} dx\,\frac{\tilde{y}_0{}^+(ix)}{x - i\omega}\,\exp(-2Ax)\,\exp\left(\frac{ux}{\pi}\left[\ln\left(\frac{ux}{\pi}\right) - 1\right]\right)$$

$$\times \operatorname{Im}\left[\Gamma\left(\frac{1}{2} - \frac{u(x - i\varepsilon)}{\pi}\right)\exp(iux)\right]. \qquad (6.C.28)$$

We note that the integrand in (6.C.28) is regular at the positions of the poles of the Gamma-function in the limit $\varepsilon \to 0$. As $A \gg 1$ the integral in (6.C.28) can be approximated by expanding the terms other than $\exp(-2Ax)$ in the integrand in a power series around $x = 0$. As long as ω is not too small this gives

$$\tilde{y}_1^+(\omega) = G^+(\omega)\frac{Bu}{\pi\sqrt{2A}}\frac{1}{-i\omega} + \mathcal{O}(A^{-2}), \qquad (6.C.29)$$

$$y_1(0) = \lim_{\omega \to \infty} -i\omega\,\tilde{y}_1^+(\omega) \approx \frac{Bu}{\pi\sqrt{2A}}. \qquad (6.C.30)$$

Equation (6.C.30) permits us to determine the leading correction to the expression (6.C.24) for the integration boundary. Equation (6.C.22) is approximated by

$$0 = y_0(0) + y_1(0), \qquad (6.C.31)$$

which gives

$$A \simeq \frac{2u}{\pi}\ln(B_c/B) - \frac{u}{2\pi}\frac{1}{\ln(B_c/B)}. \qquad (6.C.32)$$

6.C.4 Dressed energies

Using the above results it is possible to obtain approximate expressions for the dressed energies $\kappa(k)$ and $\varepsilon_1(\Lambda)$. Let us start with the integral equation for $\varepsilon_1(\Lambda)$ (6.C.1). The following identity is easily proved using (17.14)

$$\int_{-A}^{A} d\Lambda \, a_1(\Lambda - \sin k) \, \varepsilon_1(\Lambda) = - \int_{A}^{\infty} d\Lambda \, [a_1(\Lambda - \sin k) + a_1(\Lambda + \sin k)] \, \varepsilon_1(\Lambda)$$
$$+ \int_{-\infty}^{\infty} d\Lambda \, a_1(\Lambda - \sin k) \, \varepsilon_1^{(0)}(\Lambda)$$
$$- \int_{-A}^{A} d\Lambda \, a_3(\Lambda - \sin k) \, \varepsilon_1(\Lambda). \qquad (6.C.33)$$

Iterating (6.C.33) and using

$$\sum_{n=0}^{\infty} (-1)^n a_{2n+1}(x) = \frac{1}{4u \cosh(\pi x / 2u)} \equiv s(x), \qquad (6.C.34)$$

we obtain

$$\int_{-A}^{A} d\Lambda \, a_1(\Lambda - \sin k) \, \varepsilon_1(\Lambda) = - \int_{A}^{\infty} d\Lambda \, [s(\Lambda - \sin k) + s(\Lambda + \sin k)] \, \varepsilon_1(\Lambda)$$
$$+ \int_{-\infty}^{\infty} d\Lambda \, s(\Lambda - \sin k) \, \varepsilon_1^{(0)}(\Lambda). \qquad (6.C.35)$$

Inserting (6.C.35) into the integral equation (6.30) for $\kappa(k)$ we find

$$\kappa(k) = -2 \cos k - \mu - 2u - 2 \int_0^{\infty} \frac{d\omega}{\omega} \frac{J_1(\omega) \exp(-u|\omega|)}{\cosh(u\omega)} \cos(\omega \sin k)$$
$$- \int_0^{\infty} d\Lambda \, y(\Lambda) \, [s(\Lambda + A - \sin k) + s(\Lambda + A + \sin k)]. \qquad (6.C.36)$$

Expanding $s(\Lambda + A \pm \sin k)$ in a geometric series in $\exp(-\frac{\pi}{2u}[\Lambda + A \pm \sin(k)])$ this can be rewritten as

$$\kappa(k) = -2 \cos k - \mu - 2u - 2 \int_0^{\infty} \frac{d\omega}{\omega} \frac{J_1(\omega) \exp(-u|\omega|)}{\cosh(u\omega)} \cos(\omega \sin k)$$
$$- \frac{1}{u} \sum_{n=0}^{\infty} (-1)^n \, \tilde{y}^+(i\alpha_n) \exp(-\alpha_n A) \, \cosh[\alpha_n \sin(k)], \qquad (6.C.37)$$

where

$$\alpha_n = \frac{\pi}{2u}(2n + 1). \qquad (6.C.38)$$

The general expression for the ground state energy per site is given by (6.17).

$$f(u, \mu, B) \simeq -\mu - u - 4 \int_0^{\infty} \frac{d\omega}{\omega} \frac{J_0(\omega) J_1(\omega)}{1 + \exp(2u\omega)}$$
$$- \frac{1}{u} \sum_{n=0}^{\infty} (-1)^n \, \tilde{y}^+(i\alpha_n) \, \exp(-\alpha_n A) \, I_0(\alpha_n). \qquad (6.C.39)$$

We can use our results for $\tilde{y}_{1,2}(\omega)$ and A (6.C.32) to evaluate the leading contributions to the ground state free energy. We find

$$f(u, \mu, B) \simeq -\mu - u - 4 \int_0^\infty \frac{d\omega}{\omega} \frac{J_0(\omega)J_1(\omega)}{1 + \exp(2u\omega)}$$

$$-\frac{1}{2\pi} \frac{I_0(\pi/2u)}{I_1(\pi/2u)} B^2 \left[1 + \frac{1}{2\ln(B_c/B)} \right]. \tag{6.C.40}$$

We recall that we are in a regime where B is very small and where the chemical potential varies only in the interval $[\mu_-, 0]$, where μ_- is defined in (6.31). The magnetization is given by

$$m = -\frac{1}{2} \frac{\partial f(u, \mu, B)}{\partial B} \approx \frac{1}{2\pi} \frac{I_0(\pi/2u)}{I_1(\pi/2u)} B \left[1 + \frac{1}{2\ln(B_c/B)} \right]. \tag{6.C.41}$$

Thus we arrive at the following expression for the ground state energy per site

$$f(u, \mu, B) \simeq f(u, \mu, 0) - 2\pi \frac{I_1(\pi/2u)}{I_0(\pi/2u)} m^2 \left[1 - \frac{1}{2\ln(m_c/m)} \right], \tag{6.C.42}$$

where $m_c = B_c \frac{I_0(\pi/2u)}{2\pi I_1(\pi/2u)}$. There is a simple interpretation for the expressions (6.C.41) and (6.C.42): the spin velocity at half-filling ($\mu > \mu_-$) and zero field ($B = 0$) is given by

$$v_s = \lim_{\lambda \to \infty} \frac{\frac{d\varepsilon_1(\lambda)}{d\lambda}}{\frac{dp_1(\lambda)}{d\lambda}} = 2 \frac{I_1(\pi/2u)}{I_0(\pi/2u)}, \tag{6.C.43}$$

where the dressed energy $\varepsilon_1(\lambda)$ is given by (6.B.8) and the dressed momentum $p_1(\lambda)$ is calculated from (5.99) and (6.B.14). Using the expression for v_s we can write the magnetization as

$$m \approx \frac{B}{v_s\pi} \left[1 + \frac{1}{2\ln(B_c/B)} \right]. \tag{6.C.44}$$

The logarithmic correction signals the presence of a marginally irrelevant operator (interaction of spin currents) in the spin sector; the situation is completely analogous to the Heisenberg spin-1/2 chain [10, 61, 189, 341] in a weak field [14, 305] or the supersymmetric t-J model [377]. The easiest way to see this is to construct the continuum limit in the regime $U/t \ll 1$, see e.g. Refs. [7, 168]. This is done in Section 10.3.

7

Excited states at zero temperature

In this chapter we determine the spectrum of excited states. We establish that in contrast to the case of noninteracting electrons $u = 0$, excitations for $u \neq 0$ cannot be described in terms of electronic degrees of freedom, that is (dressed) electrons and holes. Instead, low-lying excited states involve collective modes of the spin and charge degrees of freedom respectively. This phenomenon is known as *spin-charge separation*.

In Sections 7.2–7.4 we consider the half-filled band (Phase V), where excitations involving the charge degrees of freedom have a gap, while the spin excitations remain gapless. In Section 7.6 we study the effects of a magnetic field on the excitation spectrum over the half-filled ground state (Phase V). In Section 7.7 we discuss the less than half-filled band, where both spin and charge degrees of freedom are gapless (Phase IV). Finally, in Section 7.9 we briefly touch upon the excitation spectrum over the empty ground state (Phase I).

As we have seen, in Phases IV and V the ground state is described by a filled Fermi sea of Bethe ansatz roots k_j and a second filled Fermi sea of roots Λ_α^1. On general grounds we expect low-lying excitations to be described by distributions of roots that are close to the root distribution of the ground state. An essential point to note is that the ground state energy scales like the system size L for large L, whereas the excited state we are considering differ in energy from the ground state only by $\mathcal{O}(1)$.

Low-lying excited states are Bethe ansatz states with given, small numbers of holes in the distributions of k's and Λ's and given, small numbers of 'extra' roots k_j, Λ_β^n and $\Lambda_\gamma'^n$. Here small means that we consider a ground state with N electrons and M down spins and excited states with $N + n$ electrons and $M + m$ down spins such that

$$\lim_{L \to \infty} \frac{n}{L} = 0, \quad \lim_{L \to \infty} \frac{m}{L} = 0. \tag{7.1}$$

The main idea behind our construction of excited states is to make full use of the results of the zero-temperature limit of the TBA equations. In particular, as we have seen, the TBA analysis tells us which dressed energies and dressed momenta are non-vanishing at $T = 0$. This then tells us how small changes of the root distribution of the ground state will affect the energy and momentum.

209

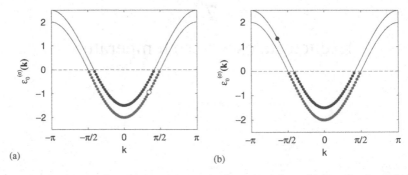

Fig. 7.1. Examples of (a) spin-up hole and (b) spin-down particle excitations for the tight binding model of noninteracting electrons in a magnetic field.

7.1 A point of reference: noninteracting electrons

Before turning to the excitation spectrum for the Hubbard model let us recall briefly the situation in the absence of interactions, i.e. $u = 0$. In momentum space the Hamiltonian can be written as (5.6)

$$\mathcal{H} = \int_{-\pi}^{\pi} \frac{dk}{2\pi} \sum_{\sigma=\uparrow,\downarrow} \varepsilon_0^{(\sigma)}(k)\, \tilde{n}_\sigma(k), \tag{7.2}$$

where $\varepsilon_0^{(\uparrow)}(k) = -2\cos(k) - \mu - B$ and $\varepsilon_0^{(\downarrow)}(k) = -2\cos(k) - \mu + B$ are the bare energies for electrons with spin up and down respectively and $\tilde{n}_\sigma(k)$ is the number operator of electrons with spin σ and momentum k. The ground state is obtained by filling all negative energy states, see Section 6.1. In the general case this corresponds to filling two Fermi seas with Fermi momenta $k_{F,\sigma}$ given by (6.2). For definiteness we consider the case where $0 < k_{F,\sigma} < \pi$ in what follows. It is easy to see that

$$[\tilde{n}_\sigma(k), \mathcal{H}] = 0, \tag{7.3}$$

which implies that the number of electrons with given spin and momentum is a conserved quantity. This in turn allows us to construct excited eigenstates of the Hamiltonian by adding electrons with given momentum and spin to the ground state or alternatively removing them from the Fermi sea.

7.1.1 'Single-particle' excitations

The simplest excitations over the ground state are obtained by adding ('particle' excitation) or removing ('hole' excitation) one electron. Particle excitations have charge $-e$, spin σ and a dispersion

$$E_p^{(\sigma)}(P) = \varepsilon_0^{(\sigma)}(P), \qquad |P| > k_{F,\sigma}. \tag{7.4}$$

Hole excitations have charge e, spin $-\sigma$ and a dispersion

$$E_h^{(\sigma)}(P) = -\varepsilon_0^{(\sigma)}(P), \qquad |P| < k_{F,\sigma}. \tag{7.5}$$

We show examples of such particle and hole excitations in figure 7.1.

7.1.2 'Particle-hole' excitations

Multiparticle excited states are scattering states of particle and hole excitations. The energy and momentum of a state with N_p particles with momenta $\{p_j\}$ and spins $\{\sigma_j\}$ and N_h holes with momenta $\{\bar{p}_k\}$ and spins $\{\bar{\sigma}_k\}$ are

$$E = \sum_{j=1}^{N_p} \varepsilon_0^{(\sigma_j)}(p_j) - \sum_{k=1}^{N_h} \varepsilon_0^{(\bar{\sigma}_k)}(\bar{p}_k),$$

$$P = \sum_{j=1}^{N_p} p_j - \sum_{k=1}^{N_h} \bar{p}_k, \quad |\bar{p}_k| < k_{F,\bar{\sigma}_k}, \quad |p_j| > k_{F,\sigma_k}. \tag{7.6}$$

7.2 Zero magnetic field and half-filled band

In this section we determine the spectrum of low-lying excitation in the half-filled repulsive Hubbard model in the absence of a magnetic field. We will see that any low-lying state can be thought of as a scattering state of an even number of *elementary* excitations [120, 121].

7.2.1 Elementary excitations

There are two different types of elementary excitations:

- Gapped, spinless excitations carrying charge $\mp e$. They are sometimes called *antiholons* and *holons* respectively. The dressed energy $\mathcal{E}_{\bar{h},h}(k)$ and dressed momentum $\mathcal{P}_{\bar{h},h}(k)$ of these excitations are given in (7.8). They transform under the spin-$\frac{1}{2}$ respresentation of the η-pairing SU(2) symmetry and hence the $(\frac{1}{2}, 0)$ representation of $SU(2) \times SU(2)$.
- Gapless, charge-neutral excitations carrying spin $\pm\frac{1}{2}$. Such excitations are called *spinons*. Their dressed energy $\mathcal{E}_{\bar{s},s}(\Lambda)$ and dressed momentum $\mathcal{P}_{\bar{s},s}(\Lambda)$ are given in (7.8). They transform under the spin-1/2 respresentation of the spin SU(2) symmetry and hence the $(0, \frac{1}{2})$ respresentation of $SU(2) \times SU(2)$.

It is important to distinguish these elementary excitations from 'physical' excitations, which are the *permitted combinations* of elementary excitations. In other words, not any combination of elementary excitations is allowed, but only those consistent with the *selection rules* (4.44).

We introduce the following terminology: we call the set $\{\mathcal{M}_e, M_n, M_n' | n = 1 \ldots \infty\}$ of the numbers of real k's, Λ-strings of length n and k-Λ-strings of length n *occupation numbers* of the corresponding excitation. This is in contrast to our usage of the term *quantum numbers*, which is reserved for the eigenvalues of energy, momentum, S^z, $\mathbf{S}^2 = \vec{S} \cdot \vec{S}$, η^z and $\eta^2 = \vec{\eta} \cdot \vec{\eta}$. In what follows we will establish the following.

Classification of Excitations

All low-lying excited states of the half-filled repulsive Hubbard model are scattering states of an even number of elementary excitations. The excitation energies and momenta are sums over the energies and momenta of the constituent elementary excitations

$$E = \sum_{\{\alpha_j\}} \mathcal{E}_{\alpha_j}, \quad P = \sum_{\{\alpha_j\}} \mathcal{P}_{\alpha_j}, \tag{7.7}$$

where α_j is an index labelling the type of elementary excitation (holon h, antiholon $\bar{h}$, $S^z = \frac{1}{2}$ spinon s, $S^z = -\frac{1}{2}$ spinon $\bar{s}$). It is shown below that

$$\mathcal{E}_s(\Lambda) = \mathcal{E}_{\bar{s}}(\Lambda) = 2 \int_0^\infty \frac{d\omega}{\omega} \frac{J_1(\omega) \cos(\omega\Lambda)}{\cosh(\omega u)},$$

$$\mathcal{P}_s(\Lambda) = \mathcal{P}_{\bar{s}}(\Lambda) = \frac{\pi}{2} - \int_0^\infty \frac{d\omega}{\omega} \frac{J_0(\omega) \sin(\omega\Lambda)}{\cosh(u\omega)},$$

$$\mathcal{E}_h(k) = \mathcal{E}_{\bar{h}}(k) = 2\cos k + 2u + 2 \int_0^\infty \frac{d\omega}{\omega} \frac{J_1(\omega)\cos(\omega \sin k)e^{-\omega u}}{\cosh(\omega u)},$$

$$\mathcal{P}_h(k) = \mathcal{P}_{\bar{h}}(k) + \pi = \frac{\pi}{2} - k - 2 \int_0^\infty \frac{d\omega}{\omega} \frac{J_0(\omega) \sin(\omega \sin(k))}{1 + \exp(2u|\omega|)}.$$

$$\tag{7.8}$$

The energies of the elementary excitations are plotted as functions of their momenta in figures 7.2 and 7.3. The antiholon energy has a minimum at $\mathcal{P}_{\bar{h}} = \frac{\pi}{2}$ and the holon energy at $\mathcal{P}_h = -\frac{\pi}{2}$. Both holon and antiholon momenta cover the entire Brillouin zone. The momentum of the spinon varies in the interval $[0, \pi]$ and covers only half the Brillouin zone. The low-energy modes occur at $\mathcal{P}_s = 0$ and $\mathcal{P}_s = \pi$.

The classification of excitations was first proposed in [120, 121]. An important consequence of this classification is that at half filling spin-charge separation *on the level of the quantum numbers of elementary excitations* holds not only at low energies, but extends to any finite energy in the thermodynamic limit. On the other hand we will see that at finite energies the scattering matrix between holons and spinons becomes nontrivial, which establishes the presence of interactions between them.

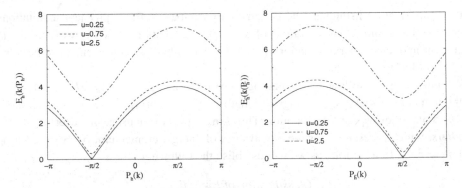

Fig. 7.2. Energies of the elementary holon and antiholon excitations as functions of their momenta for $u = 0.25$, $u = 0.75$ and $u = 2.5$. The gap increases with u and is very small for $u = 0.25$.

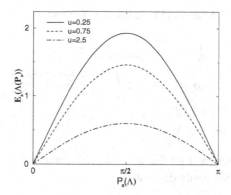

Fig. 7.3. Energy of the elementary spinon excitation as function of its momentum for $u = 0.25$, $u = 0.75$ and $u = 2.5$. For any $u > 0$ there are gapless modes at $\mathcal{P}_s = 0, \pi$.

Given that all elementary excitations carry definite $SU(2) \times SU(2)$ quantum numbers, all excited states can be characterized by $SU(2) \times SU(2)$ quantum numbers as well. More precisely, as all excited states contain an even number of elementary excitations, they are classified by $SO(4) = SU(2) \times SU(2)/\mathbb{Z}_2$ quantum numbers. For example, in the two-particle sector we can consider scattering states of two spinons, two (anti)holons or one spinon and one (anti)holon. Our classification of excitations stated above then tells us that there are the following excitations:

$$
\begin{aligned}
\text{two spinons} \quad & (0, \tfrac{1}{2}) \otimes (0, \tfrac{1}{2}) = (0, 1) \oplus (0, 0), \\
\text{two (anti)holons} \quad & (\tfrac{1}{2}, 0) \otimes (\tfrac{1}{2}, 0) = (1, 0) \oplus (0, 0), \\
\text{spinon} - \text{(anti)holon} \quad & (0, \tfrac{1}{2}) \otimes (\tfrac{1}{2}, 0) = (\tfrac{1}{2}, \tfrac{1}{2}).
\end{aligned}
\tag{7.9}
$$

In Section 7.2.4 we explicitly construct the highest-weight states of all these multiplets from the Bethe ansatz.

Before turning to the explicit construction of low-lying excited states over the ground state and the proof of our classification of excitations, we first derive explicit expressions for the dressed energies and momenta for the half-filled band in the absence of a magnetic field. As we have seen above, at half filling the chemical potential is not fixed but can vary between 0 and μ_-. For definiteness we choose $\mu = 0$ in what follows.

The integration boundaries in the integral equations for the dressed energies in this case are $Q = \pi$ and $A = \infty$. The integral equations for the dressed energies (5.103), root densities (5.105) and momenta (5.98)–(5.101) at $T = 0$ can be simplified further. For the dressed energies the easiest way of seeing this is the following. Our starting point are the coupled integral equations (5.83)–(5.86). Inserting (5.83) into (5.85) for $n = 1$ we obtain a linear integral equation involving only $\varepsilon_1(\Lambda)$, which can be solved by Fourier transformation. Inserting the result of this calculation into (5.83) we obtain $\kappa(k)$. All other dressed energies can be calculated once $\varepsilon_1(\Lambda)$ and $\kappa(k)$ are known. We find

$$\varepsilon_1(\Lambda) = -2 \int_0^\infty \frac{d\omega}{\omega} \frac{J_1(\omega) \cos(\omega\Lambda)}{\cosh(\omega u)} ,$$

$$\kappa(k) = -2\cos k - 2u - 2 \int_0^\infty \frac{d\omega}{\omega} \frac{J_1(\omega)\cos(\omega \sin k)e^{-\omega u}}{\cosh(\omega u)}$$

$$\varepsilon_n(\Lambda) = 0 , \quad n \geq 2 ,$$

$$\varepsilon_n'(\Lambda) = 0 , \quad n \geq 1. \tag{7.10}$$

The integral equations for the root densities simplify in a similar manner. The following integral representations are easily derived

$$\rho(k) = \frac{1}{2\pi} + \cos(k) \int_{-\infty}^\infty \frac{d\omega}{2\pi} \frac{J_0(\omega)}{1 + \exp(2u|\omega|)} \exp(-i\omega \sin(k)) \equiv \rho_0(k),$$

$$\sigma_1(\Lambda) = \int_{-\infty}^\infty \frac{d\omega}{2\pi} \frac{J_0(\omega)}{2\cosh(u\omega)} \exp(-i\omega\Lambda) \equiv \sigma_0(\Lambda). \tag{7.11}$$

Finally we can determine the dressed momenta (5.98)–(5.101) from (7.11) using Fourier transformation

$$p(k) = k + 2 \int_0^\infty \frac{d\omega}{\omega} \frac{J_0(\omega)}{1 + \exp(2u|\omega|)} \sin(\omega \sin(k)) ,$$

$$p_1(\Lambda) = \int_0^\infty \frac{d\omega}{\omega} \frac{J_0(\omega)}{\cosh(u\omega)} \sin(\omega\Lambda) ,$$

$$p_n'(\Lambda) = \pi(n+1) , \tag{7.12}$$

where we have used (17.15) to obtain the last identity. In figure 7.4 we plot the results (7.10) and (7.12) for $\kappa(k)$ and $p(k)$ for three different values of u. We see that $p(k)$ is a monotonically increasing function of k and varies in the interval $[-\pi, \pi]$. The dressed energy $\kappa(k)$ is always strictly negative, but $\kappa(\pm\pi)$ is very close to zero for small values of u.

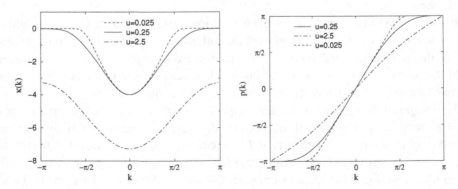

Fig. 7.4. Dressed energy $\kappa(k)$ and momentum $p(k)$ as functions of the spectral parameter k for three different values of u.

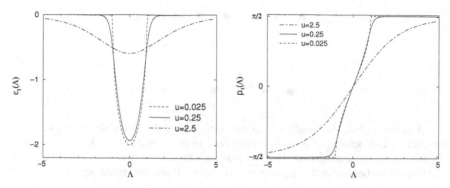

Fig. 7.5. Dressed energy $\varepsilon(\Lambda)$ and momentum $p_1(\Lambda)$ as functions of the spectral parameter Λ for three different values of u.

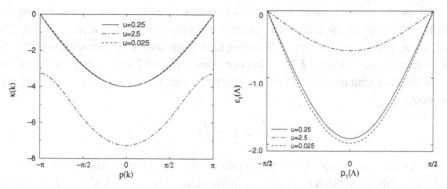

Fig. 7.6. Dressed energies $\kappa(k)$ and $\varepsilon_1(\Lambda)$ as functions of the dressed momenta $p(k)$ and $p_1(\Lambda)$ respectively, for three different values of u.

In figure 7.5 we plot the results (7.10) and (7.12) for $\varepsilon_1(\Lambda)$ and $p_1(\Lambda)$ on the interval $\Lambda \in [-5, 5]$ for three different values of u. We see that the dressed momentum $p_1(\Lambda)$ is a monotonically increasing function of Λ and varies in the interval $[-\frac{\pi}{2}, \frac{\pi}{2}]$. The dressed energy $\varepsilon_1(\Lambda)$ tends to zero for $\Lambda \to \pm\infty$. By inverting the functions $p(k)$ and $p_1(\Lambda)$ we can eliminate the auxiliary parameters k and Λ and plot the dressed energies as functions of the respective dressed momenta. This is done in figure 7.6. We note that the curves for $u = 0.25$ and $u = 0.025$ are very close to one another.

From our knowledge of the dressed energies and dressed momenta we can infer several important properties. Firstly, we see that most of the dressed energies and momenta are identically zero. This tells us that adding the corresponding type of Bethe ansatz root to the ground state distribution does not change the total energy and momentum. In other words, these roots merely control degeneracies of the spectrum. The only nontrivial nonvanishing dressed energies and momenta are the ones corresponding to the roots that make up the two Fermi seas of the ground state. This tells us that making a hole at position k_h in the distribution of k's costs energy $-\kappa(k_h)$ and adds $-p(k_h)$ to the total momentum. Similarly, making a

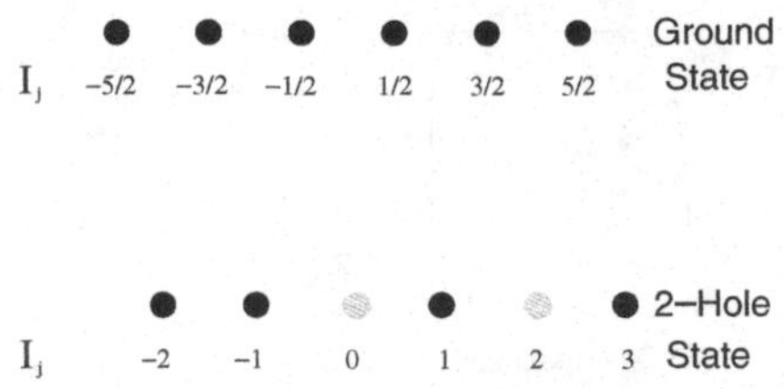

● ● ● ● ● ● Ground
I_j –5/2 –3/2 –1/2 1/2 3/2 5/2 State

● ● ◦ ● ◦ ● 2–Hole
I_j –2 –1 0 1 2 3 State

Fig. 7.7. Example of the extra contribution to the momentum: in the ground state the I_j's are half-odd integers and are distributed symmetrically around zero. In the excited state we have two holes, which contribute to the momentum according to their positions. In addition there is an extra contribution of π to the momentum because in the excited state the integers I_j are distributed asymmetrically around zero.

hole at position Λ_h in the distribution of Λ^1's costs energy $-\varepsilon_1(\Lambda_h)$ and adds $-p_1(\Lambda_h)$ to the total momentum. A special case are k-Λ strings. They have zero dressed energy and therefore do not represent 'dynamical' excitations but contribute a constant $\pi(n+1)$ to the momentum. There is one subtlety concerning the momentum. If the occupation numbers of a given state are such that the I_j's are integers, there is an additional, constant contribution of π to the momentum. This is because the vacancies for integer I_j's are by definition *asymmetric*

$$\frac{L}{2} < I_j \leq \frac{L}{2} \,. \tag{7.13}$$

If all vacancies were occupied, the momentum would be equal to π. This can be most easily seen by using the expression (5.97) for the momentum. When we calculate the difference of momenta between a given excited state and the ground state, we have to take into account this constant contribution in addition to the contribution of the holes in the distribution of I_j's. This subtlety is illustrated in figure 7.7.

An important point to keep in mind here is that of course not any combination of holes in the ground state distribution and any set of additional roots corresponding to k-Λ strings and longer Λ strings is permitted, but only those that fulfil the selection rules (4.44). As an example we shall consider low-lying excited states containing only two elementary excitations in Section 7.2.4.

7.2.2 Holon and spinon band widths

From our knowledge of the dressed energies and momenta we may infer the 'band widths' of the elementary holon and spinon excitations. It is clear from Fig.7.4 that the holon 'band width' is simply

$$W_h = \kappa(\pi) - \kappa(0) = 4 \,, \tag{7.14}$$

Reinstituting the hopping amplitude we find that the holon 'band width' is independent of U and equal to $4t$.

The spinon 'band width' is given by

$$W_s = -\varepsilon_1(0) = 2 \int_0^\infty \frac{d\omega}{\omega} \frac{J_1(\omega)}{\cosh(u\omega)}. \tag{7.15}$$

By analogy with the Heisenberg spin-$\frac{1}{2}$ chain we may associate the spinon 'band width' with an effective Heisenberg exchange constant J_{eff} by

$$J_{\text{eff}} = \frac{2}{\pi} W_s. \tag{7.16}$$

In the large-u limit the integral (7.15) is easily evaluated and reinsitituting the hopping t we obtain

$$J_{\text{eff}} \simeq \frac{4t^2}{U}. \tag{7.17}$$

This agrees with the result of the strong coupling expansion of Appendix 2.A (see Eqn. (2.A.37)).

Charge gap at half filling Excitations involving the charge degrees of freedom are gapped at half filling. The 'charge gap' Δ is determined from the minimum of the holon energy $\mathcal{E}_h(k)$ (7.8)

$$\Delta = \min_k \mathcal{E}_h(k) = \mathcal{E}_h(\pm\pi) = -\kappa(\pm\pi)$$
$$= -2 + 2u + 2 \int_0^\infty \frac{d\omega}{\omega} \frac{J_1(\omega)\, e^{-\omega u}}{\cosh(\omega u)}. \tag{7.18}$$

Small and large u expansion of (7.18) are given by (6.B.10) and (6.B.12). The charge gap (7.18) is plotted as a function of u in figure 7.8. We recall that the charge gap is measured in units of the hopping matrix element t and $u = U/4t$.

7.2.3 *Spin and charge velocities at half filling*

The spin and charge velocities at half filling can be inferred from the expressions for the dressed energies (7.10) and momenta (7.12). The spinon dispersion becomes soft around

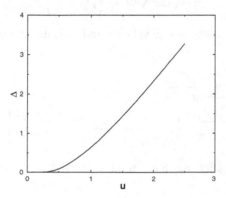

Fig. 7.8. Charge gap at half filling, Δ as a function of u.

$\mathcal{P}_s = 0, \pi$, which corresponds to $\Lambda \to \pm\infty$. At small momentum the spinon energy is simply proportional to the spinon momentum

$$\mathcal{E}_s(\Lambda) \simeq v_s \, \mathcal{P}_s(\Lambda) \,, \qquad \Lambda \to \infty. \tag{7.19}$$

Hence the spin velocity may be defined as

$$v_s = \lim_{\Lambda \to \infty} \frac{\partial \mathcal{E}_s(\Lambda)}{\partial \Lambda} \left[\frac{\partial \mathcal{P}_s(\Lambda)}{d\Lambda} \right]^{-1} . \tag{7.20}$$

Using (7.10) and (7.12) we obtain integral representations for the derivatives. The leading contribution to the integrals for large Λ comes from the pole of the integrands closest to the real axis. This gives the following result for the spin velocity[1]

$$v_s = 2 \frac{I_1 \left(\frac{\pi}{2u} \right)}{I_0 \left(\frac{\pi}{2u} \right)} \,, \tag{7.21}$$

where $I_n(z)$ are Bessel functions. We note that

$$v_s \xrightarrow[u \to 0]{} 2 \,,$$

$$v_s \xrightarrow[u \gg 1]{} \frac{\pi}{2u} \,. \tag{7.22}$$

In the strong coupling limit the effective Heisenberg exchange integral is $J_{\text{eff}} = \frac{4t^2}{U}$ and hence

$$v_s(u \gg 1) = \frac{\pi a_0}{2} J_{\text{eff}} \,, \tag{7.23}$$

where we have reinstalled proper units. The result (7.23) is equal to the spin velocity in the spin-$\frac{1}{2}$ Heisenberg chain (see e.g. [132, 132]) as expected on the basis of the strong-coupling expansion in Appendix 2.A.

Let us now turn to the charge velocity. The minima of the holon dispersion are at $k = \pm\pi$, which corresponds to $\mathcal{P}_h = -\frac{\pi}{2} \bmod 2\pi$. In the vicinity of the minimum we have a massive relativistic dispersion

$$\mathcal{E}_h^2(\mathcal{P}_h) \approx \Delta^2 + v_c^2 \left(\mathcal{P}_h + \frac{\pi}{2} \right)^2 \,, \tag{7.24}$$

where v_c, by definition, is the charge velocity and Δ is the charge gap (7.18). Hence we may calculate v_c from

$$v_c^2 = \lim_{\mathcal{P}_h \to -\frac{\pi}{2}} \frac{\mathcal{E}_h^2(\mathcal{P}_h) - \Delta^2}{(\mathcal{P}_h + \frac{\pi}{2})^2}$$

$$= \Delta \frac{\partial^2 \mathcal{E}_h(\mathcal{P}_h)}{\partial \mathcal{P}_h^2} \bigg|_{\mathcal{P}_h = -\frac{\pi}{2}} = \Delta \frac{\partial^2 \mathcal{E}_h(k)}{\partial k^2} \left[\frac{\partial \mathcal{P}_h(k)}{\partial k} \right]^{-2} \bigg|_{k = \pm\pi} \,, \tag{7.25}$$

where in the last equality we have used that $\frac{\partial \mathcal{E}_h(k)}{\partial k} \big|_{k = \pm\pi} = 0$. Using the integral

---

[1] In order to restore proper units the dimensionless velocities have to be multiplied by ta_0.

representations for $\mathcal{E}_h$ and $\mathcal{P}_h$ we obtain the following result for the charge velocity at half filling

$$v_c = 2\frac{\left[-1 + u + 2\int_0^\infty \frac{d\omega}{\omega} \frac{J_1(\omega)}{1+\exp(2u\omega)}\right]^{\frac{1}{2}} \left[1 - 2\int_0^\infty d\omega\, \omega \frac{J_1(\omega)}{1+\exp(2u\omega)}\right]^{\frac{1}{2}}}{\left[1 - 2\int_0^\infty d\omega \frac{J_0(\omega)}{1+\exp(2u\omega)}\right]}. \quad (7.26)$$

We note the following limits of weak and strong coupling

$$v_c \xrightarrow[u\to 0]{} 2\,,$$

$$v_c \xrightarrow[u\gg 1]{} 2\sqrt{u}\,. \quad (7.27)$$

7.2.4 Two-particle sector

In this subsection we construct all excited states that contain only two elementary excitations. These can be either (anti)holons or spinons.

Charge Triplet Excitation   This is a gapped excitation with occupation numbers $M_1 = \frac{L}{2} - 1$ and $\mathcal{M}_e = L - 2$. This results in quantum numbers $S^z = 0$, $\eta^z = -1$. As we are constructing a Bethe ansatz state we can use the highest-weight theorem to infer the quantum numbers $\mathbf{S}^2 = 0$ and $\eta^2 = 2$. The resulting $SO(4)$ representation is thus $(1, 0)$. The allowed ranges of the integers I_j and half-odd integers J_α^1 are

$$|J_\alpha^1| \le \frac{L}{4} - 1\,, \quad -\frac{L}{2} < I_j \le \frac{L}{2}\,. \quad (7.28)$$

Hence there are $\frac{L}{2} - 1$ vacancies for the $\frac{L}{2} - 1$ Λ_α^1's and L vacancies for the $L - 2$ k's. As a result we are left with two *holes* in the distribution of k's. Let us denote their positions by k_1 and k_2 respectively. The two holes are the only dynamical objects and carry an energy and momentum equal to

$$E_{CT}(k_1, k_2) = -\kappa(k_1) - \kappa(k_2) = \mathcal{E}_h(k_1) + \mathcal{E}_h(k_2)\,,$$

$$P_{CT}(k_1, k_2) = -p(k_1) - p(k_2) + \pi = \mathcal{P}_h(k_1) + \mathcal{P}_h(k_2)\,, \quad (7.29)$$

where $\kappa(k)$ and $p(k)$ are given by (7.10) and (7.12) respectively and $\mathcal{E}_h, \mathcal{P}_h$ by (7.8). The extra contribution of π to the momentum arises because the I_j's are integers (see the discussion above figure 7.7).[2] The charge triplet excitation has a minimal gap of 2Δ, where Δ is given by (7.18).

The full SO(4) multiplet is obtained by acting repeatedly with $\eta^\dagger$ on the lowest-weight state we just constructed. One subtlety to keep in mind is that this action shifts the total momentum by π, see (2.88). As a result the momentum of the $\eta^z = 0$ state of the charge-triplet is $-p(k_1) - p(k_2)$.

Charge Singlet Excitation This excitation has occupation numbers $M_1 = \frac{L}{2} - 1$, $M_1' = 1$ and $\mathcal{M}_e = L - 2$. The corresponding quantum numbers are $S^z = 0$, $\eta^z = 0$ and by the

---

[2] We assume throughout this chapter that $L = 2 \times$ odd integer.

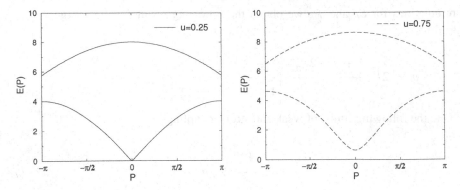

Fig. 7.9. Upper and lower boundaries of the scattering continuum for the charge singlet excitation for $u = 0.25$ (solid line) and $u = 0.75$ (dashed line).

highest-weight theorem $\mathbf{S}^2 = 0$ and $\eta^2 = 0$. We are therefore dealing with a singlet representation $(0, 0)$ of $SO(4)$. The ranges of the half-odd integers I_j, half-odd integers J_α^1 and integer J'^1_1 are

$$-\frac{L}{2} < I_j \le \frac{L}{2} , \quad |J_\alpha^1| \le \frac{L}{4} - 1 , \quad J'^1_1 = 0, \tag{7.30}$$

which implies that we again have two holes at positions $k_{1,2}$ in the distribution of k's. In addition we now have a single k-Λ string at a position Λ', which however carries zero dressed energy as $\varepsilon'_1(\Lambda') \equiv 0$ by (7.10), and only contributes a constant $p'_1(\Lambda') = 2\pi$ to the momentum as follows from (7.12). Therefore energy and momentum of the charge singlet excitation are given by

$$E_{CS}(k_1, k_2) = -\kappa(k_1) - \kappa(k_2) = \mathcal{E}_h(k_1) + \mathcal{E}_{\bar{h}}(k_2) ,$$
$$P_{CS}(k_1, k_2) = (-p(k_1) - p(k_2)) \mod (2\pi) = \mathcal{P}_h(k_1) + \mathcal{P}_{\bar{h}}(k_2). \tag{7.31}$$

The charge singlet excitation is degenerate with the $\eta^z = 0$ state of the charge triplet.

One can easily check that there are no other excitations involving only two holes in the Fermi sea of k's and no holes in the Fermi sea of Λ^1's. The charge triplet excitation as well as other excited states involving k-Λ strings were first studied by F. Woynarovich [481,482].

Spin Triplet Excitation This gapless excitation has occupation numbers $M_1 = \frac{L}{2} - 1$, and $\mathcal{M}_e = L$ and quantum numbers $S^z = 1, \eta^z = 0, \mathbf{S}^2 = 2, \eta^2 = 0$. The corresponding $SO(4)$ representation is $(0, 1)$. The ranges of the integers I_j and half-odd integers J_α^1 are

$$-\frac{L}{2} < I_j \le \frac{L}{2} , \quad |J_\alpha^1| \le \frac{L}{4} , \tag{7.32}$$

which implies that we have two holes in the distribution of Λ's and no holes in the sea of k's. The resulting energy and momentum are

$$E_{ST}(\Lambda_1, \Lambda_2) = -\varepsilon_1(\Lambda_1) - \varepsilon_1(\Lambda_2) = \mathcal{E}_s(\Lambda_1) + \mathcal{E}_s(\Lambda_2) ,$$
$$P_{ST}(\Lambda_1, \Lambda_2) = -p_1(\Lambda_1) - p_1(\Lambda_2) + \pi = \mathcal{P}_s(\Lambda_1) + \mathcal{P}_s(\Lambda_2). \tag{7.33}$$

The extra contribution of π to the momentum arises because the I_j's are integers (see the discussion above).

The spin triplet excitation in the half-filled Hubbard model is very similar to the spin triplet Heisenberg spin-$\frac{1}{2}$ antiferromagnet [132, 333]. This is precisely what one expects on the basis of the mapping of the half-filled Hubbard model on the Heisenberg model discussed in Appendix 2.A. The single-spinon dispersion was first constructed by A. A. Ovchinnikov [352] and reexamined by T.C. Choy and W. Young in [84]. A detailed analysis of spin excitations was carried out by F. Woynarovich in [486] and by means of functional relations by A. Klümper, A. Schadschneider and J. Zittartz [258]. The spin triplet excitation can be probed by inelastic Neutron Scattering Experiments.

Spin Singlet Excitation There is exactly one other excitation involving two holes in the Λ sea and no holes in the k sea. Its occupation numbers are $\mathcal{M}_e = L$, $M_1 = \frac{L}{2} - 2$ and $M_2 = 1$ and the corresponding ranges of the integers I_j, J_α^1 and J_1^2 are

$$-\frac{L}{2} < I_j \le \frac{L}{2}, \quad |J_\alpha^1| \le \frac{L-2}{4}, \quad J^2 = 0. \tag{7.34}$$

The quantum numbers are $S^z = 0$, $\eta^z = 0$, $\mathbf{S}^2 = 0$, $\eta^2 = 0$ so that we are dealing with another $SO(4)$ singlet $(0, 0)$. Energy and momentum of the spin singlet excitation are degenerate with those of the spin triplet

$$E_{SS}(\Lambda_1, \Lambda_2) = -\varepsilon_1(\Lambda_1) - \varepsilon_1(\Lambda_2) = \mathcal{E}_s(\Lambda_1) + \mathcal{E}_{\bar{s}}(\Lambda_2),$$
$$P_{SS}(\Lambda_1, \Lambda_2) = -p_1(\Lambda_1) - p_1(\Lambda_2) + \pi = \mathcal{P}_s(\Lambda_1) + \mathcal{P}_{\bar{s}}(\Lambda_2). \tag{7.35}$$

The extra contribution of π to the momentum arises because the I_j's are integers (see the discussion above). The spin singlet excitation was first constructed by F. Woynarovich in [485].

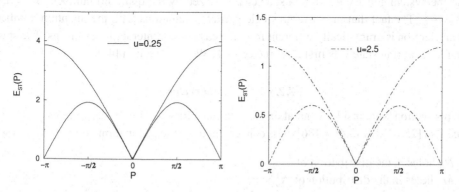

Fig. 7.10. Upper and lower boundaries of the scattering continuum for the spin triplet excitation for $u = 0.25$ and $u = 2.5$.

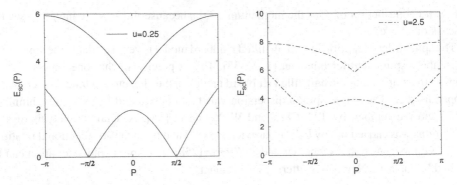

Fig. 7.11. Upper and lower boundaries of the scattering continuum for the spin-charge scattering state for $u = 0.25$ and $u = 2.5$.

Spin-Charge scattering states The last type of two-particle excitation is gapped and has the quantum numbers of a physical electron or hole. It involves one hole in each of the k and Λ seas. Its occupation numbers are $\mathcal{M}_e = L - 1$, $M_1 = \frac{L}{2} - 1$ and it has the quantum numbers of a hole (a missing electron) $S^z = \frac{1}{2}, \eta^z = -\frac{1}{2}, \mathbf{S}^2 = \frac{3}{4}, \eta^2 = \frac{3}{4}$. This corresponds to the $(\frac{1}{2}, \frac{1}{2})$ respresentation of $SO(4)$. The ranges of the integers I_j and J_α^1 are

$$-\frac{L}{2} < I_j \le \frac{L}{2} , \quad |J_\alpha^1| \le \frac{L - 2}{4} . \tag{7.36}$$

The energy and momentum of the spin-charge scattering states are given by

$$E_{\mathrm{SC}}(\Lambda, k) = -\varepsilon_1(\Lambda) - \kappa(k) = \mathcal{E}_s(\Lambda) + \mathcal{E}_h(k) ,$$
$$P_{\mathrm{SC}}(\Lambda, k) = -p_1(\Lambda) - p(k) + \pi = \mathcal{P}_s(\Lambda) + \mathcal{P}_h(k). \tag{7.37}$$

The extra contribution of π to the momentum arises because the I_j's are integers (see the discussion above). The spin-charge scattering state has a minimal gap of Δ (7.18). This type of excitation is of particular importance for photoemission experiments like Angle Resolved Photoemission Spectroscopy (ARPES) [71, 91], where an electron is removed and the system gets excited into a state characterized by the quantum numbers $S^z = \pm\frac{1}{2}$, $\eta^z = -\frac{1}{2}$. The fact that there are no single-particle excitations with the quantum numbers of an electron is dramatically different from the case of noninteracting electrons. The spin-charge scattering state was first constructed by F. Woynarovich in [485].

7.2.5 2N particle sector

In this section we consider general excited states, following the discussion given in [121] (see also [258, 481, 482, 485, 486]). It is convenient to fix the occupation numbers as follows:

- N_h holes in the distribution of k_j's;
- M_h holes in the distribution of Λ_α^1's;
- M_n spectral parameters Λ_β^n where $n \ge 2$;
- M_n' spectral parameters $\Lambda_\beta'^n$.

The total numbers of electrons N and of real spectral parameters Λ_α^1 of such an excitation are

$$N = L - N_h + 2M' ,$$

$$M_1 = \frac{1}{2}(L - N_h - M_h) - \sum_{m=2}^{\infty} M_m , \qquad (7.38)$$

where $M' = \sum_{m=1}^{\infty} m M'_m$. The $SO(4)$ quantum numbers are given by

$$\eta^z = \frac{N - L}{2} , \qquad S^z = \frac{N - 2M}{2} = \frac{M_h}{2} - \sum_{m\geq 2}(m - 1)M_m . \qquad (7.39)$$

The numbers of holons (n_h), antiholons ($n_{\bar{h}}$), $S^z = \frac{1}{2}$ spinons (n_s) and $S^z = -\frac{1}{2}$ spinons ($n_{\bar{s}}$) are calculated from the relations $N_h = n_h + n_{\bar{h}}$, $M_h - n_s + n_s$, $2\eta^z = n_{\bar{h}} - n_h$ and $2S^z = (n_s - n_{\bar{s}})$. The result is

$$n_h = N_h - M' , \qquad n_{\bar{h}} = M' ,$$

$$n_s = M_h - \sum_{m\geq 2}(m - 1)M_m , \qquad n_{\bar{s}} = \sum_{m\geq 2}(m - 1)M_m . \qquad (7.40)$$

The total energy is simply given by the sums of the dressed energies of the holes in the distributions of k's and Λ^1's

$$E = -\sum_{j=1}^{N_h} \kappa(k_j^h) - \sum_{\alpha=1}^{M_h} \varepsilon_1(\Lambda_\alpha^h) . \qquad (7.41)$$

By the definition (7.8) of the energies of the elementary excitations this is equal to the sum of the energies of holons, antiholons and spinons

$$E = \sum_{j=1}^{n_h} \mathcal{E}_h(k_j^h) + \sum_{j=n_h+1}^{N_h} \mathcal{E}_{\bar{h}}(k_j^h) + \sum_{\alpha=1}^{n_s} \mathcal{E}_s(\Lambda_\alpha^h) + \sum_{\alpha=n_s+1}^{M_h} \mathcal{E}_{\bar{s}}(\Lambda_\alpha^h) , \qquad (7.42)$$

where we have chosen the spectral parameters $k_1^h, \ldots, k_{n_h}^h$ and $k_{n_h+1}^h, \ldots, k_{N_h}^h$ to correspond to holons and antiholons respectively and similarly $\Lambda_1^h, \ldots, k_{n_s}^h$ and $\Lambda_{n_s+1}^h, \ldots, \Lambda_{M_h}^h$ to correspond spinons and antispinons.

From the definition (5.97) and the discussion at the beginning of Section (7.2) it follows that the momentum of the excited state is given by

$$P = -\sum_{j=1}^{N_h} p(k_j^h) - \sum_{\alpha=1}^{M_h} p_1(\Lambda_\alpha^h) + \pi \sum_{n=1}^{\infty}(n + 1)M'_n + \pi[1 + \sum_{m=1}^{\infty} M_m + M'_m] , \qquad (7.43)$$

where the last term is designed to give an extra contribution of π if the I_j's are integers. The *raison d' être* for this contribution has been discussed above figure 7.7. We want to show that (7.43) is equal to the sum of the momenta of the constituent elementary excitations

$$P = \sum_{j=1}^{n_h} \mathcal{P}_h(k_j^h) + \sum_{j=n_h+1}^{N_h} \mathcal{P}_{\bar{h}}(k_j^h) + \sum_{\alpha=1}^{n_s} \mathcal{P}_s(\Lambda_\alpha^h) + \sum_{\alpha=n_s+1}^{M_h} \mathcal{P}_{\bar{s}}(\Lambda_\alpha^h) , \qquad (7.44)$$

where $\mathcal{P}_\alpha$ are defined in (7.8). By definition of the $\mathcal{P}_\alpha$ we have

$$P = -\sum_{j=1}^{N_h} p(k_j^h) - \sum_{\alpha=1}^{M_h} p_1(\Lambda_\alpha^h) + \frac{\pi}{2}\left[n_h - n_{\bar{h}} + n_s + n_{\bar{s}}\right]. \tag{7.45}$$

Hence we want to show that

$$\frac{\pi}{2}\left[n_h - n_{\bar{h}} + n_s + n_{\bar{s}}\right] = \pi\left[1 + \sum_{n=1}^{\infty}(n+2)M'_n + M_n\right] \mod(2\pi). \tag{7.46}$$

Using (7.40) a straightforward calculation shows that the left hand side of (7.46) is equal to

$$\frac{\pi L}{2} - \pi\sum_{n=1}^{\infty}[nM'_n + M_n]. \tag{7.47}$$

For $L = 2 \times$ odd integer, the case we have been discussing, this is the same as the right hand side of (7.46). This proves that (7.8) is indeed the correct expression for the momenta of the elementary excitations in the sense that in order for the total momentum to be additive, the constant (spectral parameter independent) contributions must be chosen exactly as in (7.8).

In order to prove our classification of elementary excitations we still have to show that the total number $N(N_h, M_h, S^z, \eta^z)$ of excited SO(4) lowest-weight[3] states with N_h holes in the distribution of k's and M_h holes in the distribution of Λ^1's with fixed quantum numbers S^z and η^z is predicted correctly.

According to our classification, this number is simply equal to the number of lowest-weight scattering states of N_h holons and M_h spinons with the given values for $S^z \geq 0$ and $\eta^z \leq 0$. Therefore $N(N_h, M_h, S^z, \eta^z)$ can be expressed as a product

$$N(N_h, \eta^z)\, N(M_h, S^z), \tag{7.48}$$

where $N(N_h, \eta^z)$ is the number of lowest-weight states with a given η^z in the N_h-fold tensor product of spin-$\frac{1}{2}$ representations of the η-pairing $SU(2)$ and similarly $N(M_h, S^z)$ is the number of highest-weight states with a given S^z in the M_h-fold tensor product of spin-$\frac{1}{2}$ representations of the spin-$SU(2)$. From representation theory of $SU(2)$ we can infer that

$$N(N_h, \eta^z) = C^{N_h}_{\frac{N_h}{2}+\eta^z} - C^{N_h}_{\frac{N_h}{2}+\eta^z-1},$$

$$N(M_h, S^z) = C^{M_h}_{\frac{M_h}{2}-S^z} - C^{M_h}_{\frac{M_h}{2}-S^z-1}, \tag{7.49}$$

where C^a_b are binomial coefficients. Let us now reproduce this result directly from the Bethe ansatz. The different SO(4) lowest-weight states correspond to different distributions of the (half-odd) integers J^n_α ($n \geq 2$), J'^n_α of the 'non-dynamical' spectral parameters. Using

---

[3] Recall that 'SO(4) lowest weight state' means that the state is a lowest weight state with respect to the η-pairing SU(2) and a highest-weight state with respect to the spin-SU(2) symmetry algebra.

(4.44)–(4.46) one easily establishes the following ranges for the (half-odd) integers

$$|J_\alpha^n| \le \frac{1}{2}\left[M_h - \sum_{m=2}^{\infty}(t_{nm} - 2)M_m - 1 \right], \quad n \ge 2,$$

$$|J'^n_\alpha| \le \frac{1}{2}\left[N_h - \sum_{m=1}^{\infty} t_{nm}M'_m - 1 \right]. \tag{7.50}$$

This means for example that there are $N_h - \sum_{m=1}^{\infty} t_{nm}M'_m$ vacancies for the (half-odd) integers J'^n_α and accordingly

$$C_{M'_n}^{N_h - \sum_{m=1}^{\infty} t_{nm}M'_m} \tag{7.51}$$

different ways of distributing them. In this way we obtain the following result for the total number of states with given occupation numbers and values of η^z and S^z

$$N(N_h, M_h, S^z, \eta^z) = \left[\sum_{\substack{M'_1, M'_2, \dots \\ \sum_{m=1}^{\infty} mM'_m = \frac{N_h}{2} + \eta^z}} \prod_{n=1}^{\infty} C_{M'_n}^{N_h - \sum_{m=1}^{\infty} t_{mn}M'_m} \right]$$

$$\times \left[\sum_{\substack{M_2, M_3, \dots \\ \sum_{m=2}^{\infty}(m-1)M_m = \frac{M_h}{2} - S^z}} \prod_{n=2}^{\infty} C_{M_n}^{M_h - \sum_{m=2}^{\infty}(t_{mn} - 2)M_m} \right]. \tag{7.52}$$

These sums are precisely of the same structure as the ones we encountered in establishing the completeness of the Bethe ansatz in Chapter 4.B. Indeed, the first factor in (7.52) is of the same form as (4.B.8). The second factor can be brought to the same form by a simple relabelling of the summation variables $M_m = \tilde{M}_{m-1}$ for $m \ge 2$ and using that $t_{n+1\,m+1} = t_{nm} + 2$. Using (4.B.8) in both factors we obtain

$$N(N_h, M_h, S^z, \eta^z) = N(N_h, \eta^z)\, N(M_h, S^z), \tag{7.53}$$

where $N(N_h, \eta^z)$ and $N(M_h, S^z)$ are given by (7.49). The number of excited states obtained directly from the Bethe ansatz is thus indeed identical to the one obtained from our classification of general excited states.

7.3 Root-density formalism

There are several ways of constructing the ground state and low-lying excited states in a Bethe ansatz solvable model. In this section we dicuss a method widely used in the literature. We choose to call it the 'root-density formalism' as the fundamental objects are distribution functions for Bethe ansatz roots in the thermodynamic limit. Historically the first to use this method was Hulthén [207] in his study of the spin-1/2 Heisenberg antiferromagnetic chain. For the Hubbard model the method was used in many works on the zero temperature excitation spectrum [88, 352, 481, 482, 486]. The root density formalism is useful for determining scattering phase shifts of elementary excitations, see Section 7.4. Our discussion is based on F. Woynarovich's work [481, 482, 486].

Let us recall the notation $\mathcal{M}_e = N - 2M'$ for the number of k's that are not associated with k-Λ strings. For definiteness we consider the case where the lattice length is $L = 2 \times$ odd integer.

7.3.1 The half-filled ground state

The ground state has occupation numbers $\mathcal{M}_e = L$, $M_1 = L/2$ and the Bethe roots fulfil the equations

$$k_j L = 2\pi I_j - \sum_{\alpha=1}^{L/2} \theta\left(\frac{\sin k_j - \Lambda_\alpha^1}{u}\right) ,$$

$$\sum_{j=1}^{L} \theta\left(\frac{\Lambda_\alpha^1 - \sin k_j}{u}\right) = 2\pi J_\alpha + \sum_{\beta=1}^{L/2} \theta\left(\frac{\Lambda_\alpha^1 - \Lambda_\beta^1}{2u}\right) . \tag{7.54}$$

We now assume that for large system sizes L the distributions of k_j's and Λ_α's become dense, just like we did when we derived the thermodynamic equations (5.28)

$$k_{j+1} - k_j = \mathcal{O}(L^{-1}), \quad \Lambda_{\alpha+1}^1 - \Lambda_\alpha^1 = \mathcal{O}(L^{-1}) . \tag{7.55}$$

Subtracting (7.54) for consecutive roots we obtain for the first equation in (7.54)

$$L(k_{j+1} - k_j) = 2\pi - \sum_{\alpha=1}^{L/2} \frac{\cos(k_j)2u}{u^2 + (\sin(k_j) - \Lambda_\alpha^1)^2}(k_{j+1} - k_j) + O(L^{-1}) , \tag{7.56}$$

where we have used that $I_{j+1} - I_j = 1$ and (7.55) to Taylor-expand. In the next step we define root densities for finite, large L by

$$\rho^{(L)}(k_j) = \frac{1}{L(k_{j+1} - k_j)} , \quad \sigma_1^{(L)}(\Lambda_\alpha^1) = \frac{1}{L(\Lambda_{\alpha+1}^1 - \Lambda_\alpha^1)} . \tag{7.57}$$

The limit $L \to \infty$ of (7.57) exists due to (7.55) and we denote the limiting densities by

$$\rho(k) = \lim_{L\to\infty} \rho^{(L)}(k) , \quad \sigma_1(\Lambda) = \lim_{L\to\infty} \sigma_1^{(L)}(\Lambda) . \tag{7.58}$$

In terms of the root densities (7.57) we can rewrite (7.56) and the analogous equation for the roots Λ_α as

$$\rho^{(L)}(k_j) = \frac{1}{2\pi} + \frac{\cos k_j}{2\pi} \frac{1}{L} \sum_{\alpha=1}^{L/2} \frac{2u}{u^2 + (\sin k_j - \Lambda_\alpha^1)^2} , \tag{7.59}$$

$$\sigma^{(L)}(\Lambda_\alpha^1) = \frac{1}{2\pi L} \sum_{j=1}^{L} \frac{2u}{u^2 + (\sin k_j - \Lambda_\alpha^1)^2}$$

$$- \frac{1}{2\pi L} \sum_{\beta=1}^{L/2} \frac{4u}{4u^2 + (\Lambda_\beta^1 - \Lambda_\alpha^1)^2} . \tag{7.60}$$

In the next step we take the limit $L \to \infty$. Let us multiply and divide the last term in (7.59) by $\Lambda_{\alpha+1} - \Lambda_\alpha$

$$\rho^{(L)}(k_j) = \frac{1}{2\pi} + \frac{\cos k_j}{2\pi} \sum_{\alpha=1}^{L/2} (\Lambda_{\alpha+1}^1 - \Lambda_\alpha^1) \frac{2u}{u^2 + (\sin k_j - \Lambda_\alpha^1)^2} \sigma_1^{(L)}(\Lambda_\alpha^1). \quad (7.61)$$

If we now take $L \to \infty$ the last term simply turns into an integral. Carrying out the analogous calculation for (7.60) we arrive at the following set of coupled integral equations

$$\rho(k) = \frac{1}{2\pi} + \frac{\cos k}{2\pi} \int_{-A}^{A} d\Lambda \, \frac{2u}{u^2 + (\sin k - \Lambda)^2} \sigma_1(\Lambda), \quad (7.62)$$

$$\sigma_1(\Lambda) = \int_{-Q}^{Q} \frac{dk}{2\pi} \frac{2u}{u^2 + (\Lambda - \sin k)^2} \rho(k)$$

$$- \int_{-A}^{A} \frac{d\Lambda'}{2\pi} \frac{4u}{(2u)^2 + (\Lambda - \Lambda')^2} \sigma_1(\Lambda'). \quad (7.63)$$

The integration boundaries Q and A are fixed by the requirements that we are at half filling and the magnetization is zero

$$\int_{-Q}^{Q} dk \, \rho(k) = \frac{M_e}{L} = 1,$$

$$\int_{-A}^{A} d\Lambda \, \sigma_1(\Lambda) = \frac{M_1}{L} = \frac{1}{2}, \quad (7.64)$$

which yield $Q = \pi$ and $A = \infty$. Equations (7.63) are now easily solved by Fourier transformation with the results (7.11):

$$\rho_0(k) = \frac{1}{2\pi} + \cos(k) \int_{-\infty}^{\infty} \frac{d\omega}{2\pi} \frac{J_0(\omega)}{1 + \exp(2u|\omega|)} \exp(-i\omega \sin(k)),$$

$$\sigma_0(\Lambda) = \int_{-\infty}^{\infty} \frac{d\omega}{2\pi} \frac{J_0(\omega)}{2\cosh(u\omega)} \exp(-i\omega\Lambda). \quad (7.65)$$

We recall that the subscript '0' indicates that we are dealing with the root densities of the half-filled ground state. The derivation of equations (7.65) a priori is exact up to order $\mathcal{O}(L^{-1})$. In the thermodynamic limit the ground state energy per site is given by

$$e_{GS} = \lim_{L \to \infty} \frac{1}{L} \sum_{j=1}^{L} [-2\cos k_j - 2u] + u$$

$$= - \lim_{L \to \infty} 2 \sum_{j=1}^{L} (k_{j+1} - k_j) \cos k_j \, \rho^{(L)}(k_j) - u$$

$$= -2 \int_{-\pi}^{\pi} dk \, \cos(k) \rho(k) - u. \quad (7.66)$$

Inserting the integral representation (7.11) for $\rho(k)$ into (7.66) we reproduce the Lieb-Wu result (6.82).

Equations (7.63) can also be obtained directly from the equations for the counting functions (5.29)–(5.31). For the ground state we have

$$y(k) = k + \frac{1}{L} \sum_{\alpha=1}^{M_1} \theta\left(\frac{\sin k - \Lambda_\alpha^1}{u}\right), \tag{7.67}$$

$$z_1(\Lambda) = \frac{1}{L} \sum_{j=1}^{M_e} \theta\left(\frac{\Lambda - \sin k_j}{u}\right) - \frac{1}{L} \sum_{\alpha=1}^{M_1} \theta\left(\frac{\Lambda - \Lambda_\alpha^1}{2u}\right). \tag{7.68}$$

Taking derivatives with respect to k and Λ respectively and using the relations (5.36), (5.37), we obtain

$$\rho^p(k) + \rho^h(k) = \frac{1}{2\pi} + \frac{\cos k}{L} \sum_{\alpha=1}^{M_1} a_1(\sin k - \Lambda_\alpha^1), \tag{7.69}$$

$$\sigma_1^p(\Lambda) + \sigma_1^h(\Lambda) = \frac{1}{L} \sum_{j=1}^{M_e} a_1(\Lambda - \sin k_j) - \frac{1}{L} \sum_{\alpha=1}^{M_1} a_2(\Lambda - \Lambda_\alpha^1). \tag{7.70}$$

In the ground state all vacancies are filled, so that $\rho^h(k) = 0$ and $\sigma_1^h(k) = 0$. Finally, turning the sums into integrals we arrive at (7.63).

7.3.2 General excited states

Let us now consider a general excited state allowed by the selection rules (4.44) with N_h holes in the distribution of k's, M_h holes in the distribution of Λ_α^1's, M_l roots Λ_α^l ($l \geq 2$) and M_n' roots $\Lambda_\alpha'^n$. The equations for the counting functions $y(k)$ and $z_1(\Lambda)$ now read

$$y(k) = k + \frac{1}{L} \sum_{\alpha=1}^{M_1} \theta\left(\frac{\sin k - \Lambda_\alpha^1}{u}\right) + \frac{1}{L} \sum_{l=2}^{\infty} \sum_{\beta=1}^{M_l} \theta\left(\frac{\sin k - \Lambda_\beta^l}{lu}\right) \tag{7.71}$$

$$+ \frac{1}{L} \sum_{l=1}^{\infty} \sum_{\beta=1}^{M_l'} \theta\left(\frac{\sin k - \Lambda_\beta'^l}{lu}\right),$$

$$z_1(\Lambda) = \frac{1}{L} \sum_{j=1}^{M_e} \theta\left(\frac{\Lambda - \sin k_j}{u}\right) - \frac{1}{L} \sum_{\alpha=1}^{M_1} \theta\left(\frac{\Lambda - \Lambda_\alpha^1}{2u}\right)$$

$$- \frac{1}{L} \sum_{l=2}^{\infty} \sum_{\beta=1}^{M_l} \Theta_{1l}\left(\frac{\Lambda - \Lambda_\beta^l}{u}\right). \tag{7.72}$$

Taking again derivatives with respect to k and Λ and turning the sums over k_j and Λ_α^1 into integrals we arrive at

$$\rho^p(k) + \rho^h(k) = \frac{1}{2\pi} + \cos(k) \int_{-\infty}^{\infty} d\Lambda \, a_1(\sin k - \Lambda) \sigma_1^p(\Lambda) \tag{7.73}$$

$$+ \frac{\cos k}{L} \sum_{l=2}^{\infty} \sum_{\beta=1}^{M_l} a_l(\sin k - \Lambda_\beta^l) + \frac{\cos k}{L} \sum_{l=1}^{\infty} \sum_{\beta=1}^{M_l'} a_l(\sin k - \Lambda_\beta'^l),$$

$$\sigma_1^p(\Lambda) + \sigma_1^h(\Lambda) = \int_{-\pi}^{\pi} dk \, a_1(\Lambda - \sin k) \, \rho^p(k) - \int_{-\infty}^{\infty} d\Lambda' \, a_2(\Lambda - \Lambda') \, \sigma_1^p(\Lambda')$$

$$- \frac{1}{L} \sum_{l=2}^{\infty} \sum_{\beta=1}^{M_l} A_{1l}(\Lambda - \Lambda_\beta^l) . \qquad (7.74)$$

The point is that we know what the root densities for holes are. By construction there are N_h holes in the k-sea with corresponding rapidities k_n^h and similarly there are M_h holes in the Λ-sea at positions Λ_n^h. Hence, to leading order in L^{-1}, we have

$$\rho^h(k) = \frac{1}{L} \sum_{n=1}^{N_h} \delta(k - k_n^h) ,$$

$$\sigma_1^h(\Lambda) = \frac{1}{L} \sum_{m=1}^{M_h} \delta(\Lambda - \Lambda_m^h) . \qquad (7.75)$$

Apart from the integral equations describing the two Fermi seas we have considered so far, we still have Takahashi's equations determining the spectral parameters Λ_β^m for $m \geq 2$ and Λ'^m_β. For example, we have

$$2\pi J'^m_\beta = 2L \, \text{Re} \left(\arcsin \left[\Lambda'^m_\beta + imu \right] \right) - \sum_{j=1}^{M_e} \theta \left(\frac{\Lambda'^m_\beta - \sin k_j}{mu} \right)$$

$$- \sum_{l=1}^{\infty} \sum_{\gamma=1}^{M'_l} \Theta_{ml} \left(\frac{\Lambda'^m_\beta - \Lambda'^l_\gamma}{u} \right) . \qquad (7.76)$$

Here we can again turn the sum over the k_j's into an integral. Putting everything together and dropping in the index 'p' for the particle root densities we arrive at the following set of equations characterizing a general low-lying excitation over the half-filled ground state

$$\rho(k) = \frac{1}{2\pi} + \cos(k) \int_{-\infty}^{\infty} d\Lambda \, a_1(\sin k - \Lambda) \, \sigma_1(\Lambda) - \frac{1}{L} \sum_{n=1}^{N_h} \delta(k - k_n^h)$$

$$+ \frac{\cos k}{L} \sum_{l=2}^{\infty} \sum_{\beta=1}^{M_l} a_l(\sin k - \Lambda_\beta^l) + \frac{\cos k}{L} \sum_{l=1}^{\infty} \sum_{\beta=1}^{M'_l} a_l(\sin k - \Lambda'^l_\beta) , \qquad (7.77)$$

$$\sigma_1(\Lambda) = \int_{-\pi}^{\pi} dk \, a_1(\Lambda - \sin k) \, \rho(k) - \int_{-\infty}^{\infty} d\Lambda' \, a_2(\Lambda - \Lambda') \, \sigma_1(\Lambda')$$

$$- \frac{1}{L} \sum_{l=2}^{\infty} \sum_{\beta=1}^{M_l} A_{1l}(\Lambda - \Lambda_\beta^l) - \frac{1}{L} \sum_{m=1}^{M_h} \delta(\Lambda - \Lambda_m^h) , \qquad (7.78)$$

$$\frac{2\pi J'^m_\beta}{L} = 2\text{Re}\left(\arcsin\left[\Lambda'^m_\beta + imu\right]\right) - \int_{-\pi}^{\pi} dk\ \theta\left(\frac{\Lambda'^m_\beta - \sin k}{mu}\right)\rho(k)$$

$$- \frac{1}{L}\sum_{l=1}^{\infty}\sum_{\gamma=1}^{M'_l}\Theta_{ml}\left(\frac{\Lambda'^m_\beta - \Lambda'^l_\gamma}{u}\right),\tag{7.79}$$

$$\frac{2\pi J^n_\alpha}{L} = \int_{-\pi}^{\pi} dk\ \theta\left(\frac{\Lambda^n_\alpha - \sin k}{nu}\right)\rho(k) - \int_{-\infty}^{\infty} d\Lambda\ \Theta_{n1}\left(\frac{\Lambda^n_\alpha - \Lambda}{u}\right)\sigma_1(\Lambda)$$

$$- \frac{1}{L}\sum_{m=2}^{\infty}\sum_{\beta=1}^{M_m}\Theta_{nm}\left(\frac{\Lambda^n_\alpha - \Lambda^m_\beta}{u}\right).\tag{7.80}$$

What about the equations for the holes? All they do is fix the relation between the integer or half-odd integer numbers characterizing the positions of the holes and the corresponding spectral parameters k^h_n and Λ^h_m. As long as we are content with parametrizing the positions of the holes by the spectral parameters we do not have to solve these equations.

So far our discussion has been very general. In order to see how to utilize equations (7.78) and (7.80) let us reconsider all excitations involving two elementary excitations. Apart from rederiving our previous results for the energies and momenta of these excited states we will obtain results that will be used to determine the two-particle scattering matrix in Section 7.4.

7.3.3 Charge singlet excitation

We start with the charge singlet excitation, which we recall is characterized by occupation numbers $\mathcal{M}_e = L - 2$, $M'_1 = 1$ and $M_1 = \frac{L}{2} - 1$. Both the I_j's and the J^1_α's are half-odd integers whereas the single J'^1_1 is an integer with

$$-\frac{L}{2} < I_j \le \frac{L}{2}, \qquad |J^1_\alpha| \le \frac{L}{4} - 1, \qquad J'^1_1 = 0.\tag{7.81}$$

There are 2 holes in the k-sea whereas the Λ^1 sea is completely filled as there are exactly $\frac{L}{2} - 1$ vacancies. We denote the half-odd integers parametrizing the holes in the k-sea by $I^h_{1,2}$ and the corresponding spectral parameters by $k^h_{1,2}$. The equations for the root densities (7.77), (7.78), (7.80) become

$$\rho(k) = \frac{1}{2\pi} + \cos(k)\int_{-\infty}^{\infty} d\Lambda\ a_1(\sin k - \Lambda)\,\sigma_1(\Lambda) - \frac{1}{L}\sum_{n=1}^{2}\delta(k - k^h_n)$$

$$+ \frac{\cos k}{L}a_1(\sin k - \Lambda'^1_1),\tag{7.82}$$

$$\sigma_1(\Lambda) = \int_{-\pi}^{\pi} dk\ a_1(\Lambda - \sin k)\,\rho(k) - \int_{-\infty}^{\infty} d\Lambda'\ a_2(\Lambda - \Lambda')\,\sigma_1(\Lambda')$$

$$0 = 2\text{Re}\left(\arcsin\left[\Lambda'^1_1 + iu\right]\right) - \int_{-\pi}^{\pi} dk\ \theta\left(\frac{\Lambda'^1_1 - \sin k}{u}\right)\rho(k).\tag{7.83}$$

As the integral equations are linear, we may employ the decomposition[4]

$$\rho(k) = \rho_0(k) + \frac{1}{L}\rho_{CS}(k),$$

$$\sigma_1(\Lambda) = \sigma_0(\Lambda) + \frac{1}{L}\sigma_{CS}(\Lambda). \tag{7.84}$$

It is easy to see that $\rho_0(k)$ and $\sigma_0(\Lambda)$ fulfil precisely the same integral equations as the root densities of the half-filled ground state (7.63) and are thus given by (7.11). The densities $\rho_{CS}(k)$ and $\sigma_{CS}(\Lambda)$ satisfy the following coupled integral equations

$$\rho_{CS}(k) = \cos(k) \int_{-\infty}^{\infty} d\Lambda \, a_1(\sin k - \Lambda) \, \sigma_{CS}(\Lambda) - \sum_{n=1}^{2} \delta(k - k_n^h)$$

$$+ \cos(k) \, a_1(\sin k - \Lambda_1'), \tag{7.85}$$

$$\sigma_{CS}(\Lambda) = \int_{-\pi}^{\pi} dk \, a_1(\Lambda - \sin k) \, \rho_{CS}(k) - \int_{-\infty}^{\infty} d\Lambda' \, a_2(\Lambda - \Lambda') \, \sigma_{CS}(\Lambda'), \tag{7.86}$$

These equations are solved by Fourier transformation as follows. We insert (7.85) into (7.86) and observe that due to the 'symmetric integration Lemma' (17.1) several terms vanish. We then Fourier transform and obtain

$$\sigma_{CS}(\Lambda) = -\sum_{j=1}^{2} \int_{-\infty}^{\infty} \frac{d\omega}{2\pi} \frac{\exp(i\omega[\sin k_j^h - \Lambda])}{2\cosh(u\omega)} = -\sum_{j=1}^{2} s(\Lambda - \sin k_j^h), \tag{7.87}$$

where $s(x) = \frac{1}{4u\cosh(\pi x/2u)}$ has been defined in (5.59). Using (7.87) in (7.85) we then can determine ρ_{CS}

$$\rho_{CS}(k) = \cos(k) \, a_1(\Lambda_1' - \sin k) - \sum_{j=1}^{2} \left[\delta(k - k_j^h) + \cos(k) \, R(\sin(k) - \sin(k_j^h)) \right], \tag{7.88}$$

where $R(x)$ is given by (5.104). Finally, we can solve the equation for Λ_1'. Using the integral representation (7.65) for $\rho_0(k)$, the 'symmetric integration Lemma' (17.1) and then (17.16), we find

$$\int_{-\pi}^{\pi} dk \, \theta \left(\frac{\Lambda_1' - \sin k}{u} \right) \rho_0(k) = 2\text{Re} \left(\arcsin \left[\Lambda_1' + iu \right] \right), \tag{7.89}$$

which in conjunction with (7.83) implies that

$$0 = \int_{-\pi}^{\pi} dk \, \theta \left(\frac{\Lambda_1' - \sin k}{u} \right) \rho_{CS}(k). \tag{7.90}$$

---

[4] We note that equations (7.84) define *all four* quantities on the r.h.s.

Inserting the explicit expression (7.88) for $\rho_{CS}(k)$ into this equation and using (17.1) we obtain

$$\theta\left(\frac{\Lambda'^1_1 - \sin k^h_1}{u}\right) + \theta\left(\frac{\Lambda'^1_1 - \sin k^h_2}{u}\right) = 0 \,. \tag{7.91}$$

This fixes Λ'^1_1 in terms of the positions k^h_j of the two holes

$$\Lambda'^1_1 = \frac{\sin(k^h_1) + \sin(k^h_2)}{2} \,. \tag{7.92}$$

We are now in a position to calculate the energy and momentum of the charge singlet excitation. From the general expression (4.52) for the energy we obtain

$$E_{CS} = \lim_{L\to\infty} [E - L e_{GS}]$$
$$= \int_{-\pi}^{\pi} dk\, \rho_{CS}(k)\,(-2\cos(k) - 2u) + 4\mathrm{Re}\sqrt{1 - (\Lambda'^1_1 - iu)^2} - 4u. \tag{7.93}$$

Using the explicit form (7.88) for $\rho_{CS}(k)$ together with the identities (17.1) and (17.13) we find

$$E_{CS} = \sum_{j=1}^{2}\left[2\cos(k^h_j) + 2u + 2\int_{-\infty}^{\infty} \frac{d\omega}{\omega}\, \frac{J_1(\omega)\exp(i\omega\sin(k^h_j))}{1 + \exp(2u|\omega|)}\right], \tag{7.94}$$

which agrees with our previous result (7.31). The momentum is calculated exactly as in Section 7.2. We start with equation (5.97), which for the charge singlet excitation reads

$$P_{CS} = \frac{2\pi}{L}\left(\sum_{j=1}^{L-2} I_j + \sum_{\beta=1}^{M_1} J^1_\beta - J'^1_1\right). \tag{7.95}$$

As both the I_j's and the J^1_α's are half-odd integers and $J'^1_1 = 0$ this becomes

$$P_{CS} = -\frac{2\pi}{L}\sum_{j=1}^{2} I^h_j = -\sum_{j=1}^{2} y(k^h_j)\,, \tag{7.96}$$

where the counting function $y(k)$ is given by

$$y(k) = k + \frac{1}{L}\sum_{\alpha=1}^{M_1}\theta\left(\frac{\sin k - \Lambda^1_\alpha}{u}\right) + \frac{1}{L}\theta\left(\frac{\sin k - \Lambda'^1_1}{u}\right)$$
$$= k + \int_{-\infty}^{\infty} d\Lambda\, \theta\left(\frac{\sin k - \Lambda}{u}\right)\sigma_0(\Lambda) + \mathcal{O}(L^{-1}). \tag{7.97}$$

Here $\sigma_0(\Lambda)$ is given by (7.65). Carrying out the remaining integral we obtain

$$P_{CS} = -\sum_{j=1}^{2}\left[k^h_j + \int_{-\infty}^{\infty} \frac{d\omega}{i\omega}\, \frac{J_0(\omega)\exp(i\omega\,\sin k^h_j)}{1 + \exp(2u|\omega|)}\right]. \tag{7.98}$$

This agrees with (7.31).

7.3.4 Charge triplet excitation

The charge triplet excitation has occupation numbers $M_1 = \frac{L}{2} - 1$ and $\mathcal{M}_e = L - 2$ and can be treated in the same way as the charge singlet. There are again two holes in the distribution of k's and no holes in the distribution of Λ_α^1's. Specifying our general expression for the root densities (7.78) to the case at hand and using a decomposition of the form

$$\rho(k) = \rho_0(k) + \frac{1}{L}\rho_{CT}(k),$$

$$\sigma_1(\Lambda) = \sigma_0(\Lambda) + \frac{1}{L}\sigma_{CT}(\Lambda). \tag{7.99}$$

we obtain a set of coupled equations for $\rho_{CT}(k)$ and σ_{CT}, that can be solved in complete analogy with the charge singlet case. We find

$$\sigma_{CT}(\Lambda) = -\sum_{j=1}^{2}\int_{-\infty}^{\infty}\frac{d\omega}{2\pi}\frac{\exp(i\omega[\sin k_j^h - \Lambda])}{2\cosh(u\omega)} = -\sum_{j=1}^{2}s(\Lambda - \sin k_j^h), \tag{7.100}$$

$$\rho_{CT}(k) = -\sum_{j=1}^{2}\left[\delta(k - k_j^h) + \cos(k)\,R(\sin(k) - \sin(k_j^h))\right]. \tag{7.101}$$

It is easily verified that the energy and momentum of the charge triplet excitation are equal to our previous result (7.29).

7.3.5 Spin singlet excitation

The spin singlet excitation has occupation numbers $\mathcal{M}_e = L$, $M_1 = \frac{L}{2} - 2$ and $M_2 = 1$. The integer corresponding to the Λ-string of length 2 is equal to zero, $J_1^2 = 0$, by virtue of (4.44). There are two holes with corresponding spectral parameters $\Lambda_{1,2}^h$ in the distribution of Λ_α^1's and no holes in the distribution of k's. The equations determining the root densities are given by (7.77), (7.78) and (7.80)

$$\rho(k) = \frac{1}{2\pi} + \cos(k)\int_{-\infty}^{\infty}d\Lambda a_1(\sin k - \Lambda)\,\sigma_1(\Lambda) + \frac{\cos k}{L}a_2(\sin k - \Lambda_1^2) \tag{7.102}$$

$$\sigma_1(\Lambda) = \int_{-\pi}^{\pi}dk\,a_1(\Lambda - \sin k)\,\rho(k) - \int_{-\infty}^{\infty}d\Lambda'\,a_2(\Lambda - \Lambda')\,\sigma_1(\Lambda')$$

$$-\frac{1}{L}A_{12}(\Lambda - \Lambda_1^2) - \frac{1}{L}\sum_{m=1}^{2}\delta(\Lambda - \Lambda_m^h), \tag{7.103}$$

$$0 = L\int_{-\pi}^{\pi}dk\,\theta\left(\frac{\Lambda_1^2 - \sin k}{2u}\right)\rho(k) - L\int_{-\infty}^{\infty}d\Lambda\,\Theta_{21}\left(\frac{\Lambda_1^2 - \Lambda}{u}\right)\sigma_1(\Lambda). \tag{7.104}$$

We introduce a decomposition

$$\rho(k) = \rho_0(k) + \frac{1}{L}\rho_{SS}(k)\,,$$

$$\sigma_1(\Lambda) = \sigma_0(\Lambda) + \frac{1}{L}\sigma_{SS}(\Lambda)\,,\qquad(7.105)$$

and then solve the equations for $\rho_{SS}(k)$ and $\sigma_{SS}(\Lambda)$ by Fourier techniques. Using (17.12) we find

$$\sigma_{SS}(\Lambda) = -\sum_{j=1}^{2}[\delta(\Lambda - \Lambda_j^h) - R(\Lambda - \Lambda_j^h)] - a_1(\Lambda - \Lambda_1^2)\,,\qquad(7.106)$$

$$\rho_{SS}(k) = -\cos(k)\sum_{j=1}^{2}s(\Lambda_j^h - \sin k)\,,\qquad(7.107)$$

where $R(x)$ and $s(x)$ are defined in (5.104) and (5.59) respectively. Finally we need to solve the equation for Λ_1^2. Noting that by virtue of (17.15)

$$\int_{-\infty}^{\infty} d\Lambda\, \Theta_{21}\left(\frac{x - \Lambda}{u}\right)\sigma_0(\Lambda) = 2\mathrm{Re}\,(\arcsin[x + 2iu])\,,\qquad(7.108)$$

and then using (7.89) we see that the $\mathcal{O}(L)$ part of equation (7.104) is satisfied for any Λ_1^2. The $\mathcal{O}(1)$ part gives the condition

$$\sum_{j=1}^{2}\theta\left(\frac{\Lambda_j^h - \Lambda_1^2}{u}\right) = 0\,,\qquad(7.109)$$

which has the solution

$$\Lambda_1^2 = \frac{\Lambda_1^h + \Lambda_2^h}{2}\,.\qquad(7.110)$$

The energy of the spin singlet excitation is

$$E_{SS} = \int_{-\pi}^{\pi} dk\, \rho_{SS}(k)\,(-2\cos(k) - 2u)$$

$$= \sum_{j=1}^{2}\int_{-\infty}^{\infty}\frac{d\omega}{\omega}\,\frac{J_1(\omega)\,\exp(-i\omega\Lambda_j^h)}{\cosh(u\omega)}\,.\qquad(7.111)$$

Finally, the momentum is found to be

$$P_{SS} = \pi - \sum_{j=1}^{2}\int_{-\infty}^{\infty}\frac{d\omega}{i\omega}\,\frac{J_0(\omega)\,\exp(i\omega\Lambda_j^h)}{2\cosh(u\omega)}\,.\qquad(7.112)$$

7.3.6 Spin triplet excitation

The calculations for the spin triplet excitation are very similar to those we just did for the spin singlet. We find

$$\sigma_{ST}(\Lambda) = -\sum_{j=1}^{2}[\delta(\Lambda - \Lambda_j^h) - R(\Lambda - \Lambda_j^h)] \,,$$

$$\rho_{ST}(k) = -\cos(k)\sum_{j=1}^{2} s(\Lambda_j^h - \sin k) \,. \qquad (7.113)$$

Energy and momentum are the same as for the spin singlet.

7.3.7 Spin-charge scattering state

The spin-charge scattering state has occupation numbers $\mathcal{M}_e = L - 1$, $M_1 = \frac{L}{2} - 1$ and thus one hole in the k-sea and one hole on the Λ^1-sea. We denote the corresponding spectral parameters by k^h and Λ^h. From (7.77), (7.78) we have the following equations for the root densities

$$\rho(k) = \frac{1}{2\pi} + \cos(k) \int_{-\infty}^{\infty} d\Lambda\, a_1(\sin k - \Lambda)\, \sigma_1(\Lambda) - \frac{1}{L}\delta(k - k^h) \,,$$

$$\sigma_1(\Lambda) = \int_{-\pi}^{\pi} dk\, a_1(\Lambda - \sin k)\, \rho(k) - \int_{-\infty}^{\infty} d\Lambda'\, a_2(\Lambda - \Lambda')\, \sigma_1(\Lambda')$$
$$- \frac{1}{L}\delta(\Lambda - \Lambda^h) \,. \qquad (7.114)$$

Using the decomposition

$$\rho(k) = \rho_0(k) + \frac{1}{L}\rho_{SC}(k) \,,$$

$$\sigma_1(\Lambda) = \sigma_0(\Lambda) + \frac{1}{L}\sigma_{SC}(\Lambda) \,, \qquad (7.115)$$

we obtain (using (17.11))

$$\rho_{SC}(k) = -\cos(k)\, s(\Lambda^h - \sin k) - [\delta(k - k^h) + \cos(k)\, R(\sin(k) - \sin(k^h))] \,,$$

$$\sigma_{SC}(\Lambda) = -[\delta(\Lambda - \Lambda^h) - R(\Lambda - \Lambda^h)] - s(\Lambda - \sin k^h) \,. \qquad (7.116)$$

The energy of the spin-charge scattering state is then given by

$$E_{SC} = \int_{-\pi}^{\pi} dk\, \rho_{CS}(k)\, (-2\cos(k) - 2u)$$

$$= \int_{-\infty}^{\infty} \frac{d\omega}{\omega}\, \frac{J_1(\omega)\, \exp(-i\omega\Lambda^h)}{\cosh(u\omega)}$$

$$+ 2\cos(k^h) + 2u + 2\int_{-\infty}^{\infty} \frac{d\omega}{\omega}\, \frac{J_1(\omega)\exp(i\omega \sin(k^h))}{1 + \exp(2u|\omega|)} \,, \qquad (7.117)$$

which agrees with (7.37). The momentum is given by

$$P_{SC} = \pi - \int_{-\infty}^{\infty} \frac{d\omega}{i\omega} \frac{J_0(\omega) \exp(i\omega\Lambda_j^h)}{2\cosh(u\omega)} - k^h - \int_{-\infty}^{\infty} \frac{d\omega}{i\omega} \frac{J_0(\omega) \exp(i\omega \sin k^h)}{1 + \exp(2u|\omega|)},$$

$$(7.118)$$

which agrees with (7.37).

7.4 Scattering matrix

So far we have determined the dispersions and the $SU(2) \times SU(2)$ quantum numbers of the elementary excitations in the half-filled Hubbard chain. In this section we go one step further and determine the two-particle scattering matrix (S-matrix), which gives us information on how elementary excitations interact in a given excited state. The S-matrix also determines the low-temperature thermodynamics in relativistic integrable theories [505]. The scattering of elementary excitations in the half-filled Hubbard chain is of a very special kind: it is *factorizable* [52, 322, 339, 493, 503]. The scattering process of $2N$ particles is described as follows: at $t = -\infty$ $2N$ particles with momenta $p_1 > p_2 > \ldots > p_{2N}$ are arranged in our one-dimensional space such that $x_1 \ll x_2 \ll \ldots \ll x_{2N}$, i.e. the fastest particle is the leftmost one and the slowest the rightmost one. In the interaction region the particles collide *two at a time*. The set of momenta and $SU(2) \otimes SU(2)$ quantum numbers are conserved in each collision: the scattering is completely elastic. Two-particle scattering processes can be represented graphically as in figure 7.12(a). The corresponding S-matrix element is denoted by

$$S_{\alpha_1\alpha_2}^{\beta_1\beta_2}(p_1, p_2).$$

$$(7.119)$$

After $N(2N - 1)$ pair collisions the particles are arranged along the spatial direction in order of increasing momenta, i.e. the fastest particle is now the rightmost one. In the final state of scattering at $t = \infty$ we have $x_1 \gg x_2 \gg \ldots \gg x_{2N}$. A four-particle scattering process is depicted in figure 7.12(b). The factorization of the scattering implies that the N-particle S-matrix is expressed as a product over two-particle S-matrices

$$S^{(N)}(p_1, \ldots, p_N) = \prod_{j=2}^{N} \prod_{k=1}^{j-1} S(p_k, p_j).$$

$$(7.120)$$

The same scattering process can be represented by different diagrams as is shown in figure 7.13 for the example of three-particle scattering. As they describe the same process, the corresponding products of two-particle S-matrices must be the same. This condition is equivalent to a Yang-Baxter equation for the two-particle S-matrices

$$S_{\alpha_1\alpha_2}^{\gamma_1\gamma_2}(p_1, p_2) S_{\gamma_1\alpha_3}^{\beta_1\gamma_3}(p_1, p_3) S_{\gamma_2\gamma_3}^{\beta_2\beta_3}(p_2, p_3) = S_{\alpha_2\alpha_3}^{\gamma_2\gamma_3}(p_2, p_3) S_{\alpha_1\gamma_3}^{\gamma_1\beta_3}(p_1, p_3) S_{\gamma_1\gamma_2}^{\beta_1\beta_2}(p_1, p_2).$$

$$(7.121)$$

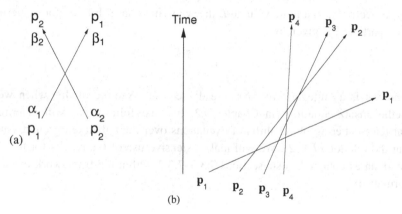

Fig. 7.12. Two- and four-particle scattering processes in a theory with factorizable scattering.

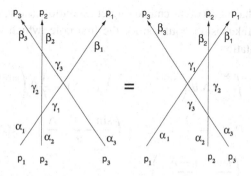

Fig. 7.13. Yang-Baxter equation for factorizable three-particle scattering.

In quantum mechanical scattering theory the S-matrix can be extracted from the asymptotics of the wave-function of the scattering state [283]. The boundary conditions of the quantum mechanical problem are free. This is in contrast to the periodic boundary conditions imposed in the Bethe ansatz solution, which complicates the problem of determining phase shifts. In Ref. [266] V. E. Korepin developed a general method for calculating the exact two-particle S-matrix directly from the Bethe ansatz equations. It is based on the fact that the Bethe ansatz equations have an interpretation in terms of scattering of particles. The same holds true for the Bethe ansatz equations describing low-lying excitations over the true ground state, and using this interpretation it is possible to calculate the phase shifts for excited states. Korepin's method was subsequently applied to many integrable models. In recent years these methods have been generalized to accommodate open boundary conditions [104, 116, 185]. It also has been shown by direct calculation that the multi-particle scattering factorizes [103].

An approach equivalent to Korepin's was suggested by N. Andrei and C. Destri in [19]. It is based on the relation of the scattering phase shift to the quantization condition for the momentum in a finite volume. Consider the scattering of two particles with momenta p_1 and

p_2 respectively. In a finite volume L the scattering phase shift δ^{12} for scattering of particle 1 on particle 2 is given by

$$p_1 + \frac{\delta^{12}}{L} = \frac{2\pi n}{L} \,, \tag{7.122}$$

where n is an integer. We have already seen an example of this when we derived the Bethe ansatz equations in Chapter 3.2. In what follows we will determine the two-particle scattering phase shifts of excitations over the half-filled ground state by exploiting the relation (7.122). We will make extensive use of the results for the respective excited states obtained in subsections 7.3.3–7.3.7 within the framework of the root density formalism.

7.4.1 Charge sector

Let us begin with the phase shift of the charge singlet excitation (see subsection 7.3.3 above). Our starting point is Takahashi's equation for the first hole, which we take to describe a holon elementary excitation

$$\frac{2\pi I_1^h}{L} = k_1^h + \int_{-\infty}^{\infty} d\Lambda\, \theta\left(\frac{\sin(k_1^h) - \Lambda}{u}\right) \sigma_1(\Lambda) + \frac{1}{L}\theta\left(\frac{\sin(k_1^h) - \Lambda'_1^1}{u}\right)$$

$$= \frac{\pi}{2} - \mathcal{P}_h(k_1^h) + \frac{1}{L}\int_{-\infty}^{\infty} d\Lambda\, \theta\left(\frac{\sin(k_1^h) - \Lambda}{u}\right) \sigma_{\mathrm{CS}}(\Lambda)$$

$$+ \frac{1}{L}\theta\left(\frac{\sin(k_1^h) - \Lambda'_1^1}{u}\right)\,, \tag{7.123}$$

where $\mathcal{P}_h(k_1^h) = \frac{\pi}{2} - p(k_1^h)$ is the dressed momentum of the holon (7.8), (7.12) and where we have used (7.84) and (7.11). We see that (7.123) is precisely of the form (7.122)! This observation allows us to determine the charge-singlet phase shift $\delta_{\mathrm{CS}}^{12}$. We have to be a bit careful, because I_1^h is actually a half-odd integer number, which gives an extra contribution of π to the phase shift. Using the expressions (7.87) for σ_{CS} and (7.92) for Λ'_1^1 and the integral (17.18) we obtain the following result

$$\delta_{\mathrm{CS}}^{12} = -\theta\left(k_{12}\right) + i \ln\left[\frac{\Gamma\left(\frac{1}{2} + i\,\frac{k_{12}}{2}\right)\Gamma\left(1 - i\,\frac{k_{12}}{2}\right)}{\Gamma\left(\frac{1}{2} - i\,\frac{k_{12}}{2}\right)\Gamma\left(1 + i\,\frac{k_{12}}{2}\right)}\right]\,, \tag{7.124}$$

where

$$k_{12} = \frac{\sin(k_1^h) - \sin(k_2^h)}{2u} \geq 0\,. \tag{7.125}$$

The condition $k_{12} > 0$ is necessary for scattering to occur. The analogous calculation for the charge triplet is easily done using the explicit results for the root densities (7.100) and (7.101). I_1^h is now an integer, so that there is no extra contribution to the phase shift. We find the following result for the charge triplet phase shift

$$\delta_{CT}^{12} = \pi + i \ln \left[\frac{\Gamma\left(\frac{1}{2} + i\,\frac{k_{12}}{2}\right) \Gamma\left(1 - i\,\frac{k_{12}}{2}\right)}{\Gamma\left(\frac{1}{2} - i\,\frac{k_{12}}{2}\right) \Gamma\left(1 + i\,\frac{k_{12}}{2}\right)} \right]. \tag{7.126}$$

Having determined the phase shifts we are now in a position to write down the two-particle scattering matrices. The scattering phases $\exp(i\delta_{CS}^{12})$ and $\exp(i\delta_{CT}^{12})$ are the eigenvalues of the two-particle scattering matrix in the η-pairing singlet and triplet sectors respectively. In total there are four two-particle states in the charge sector: holon-holon, antiholon-antiholon and two holon-antiholon states. We introduce bases for the initial (in) and final (out) states of the scattering process in the usual way

$$|k_1, k_2\rangle_{\alpha_1\alpha_2}^{\text{in}}, \quad |k_1, k_2\rangle_{\alpha_1\alpha_2}^{\text{out}}, \quad k_1 > k_2, \quad \alpha_j \in \{h, \bar{h}\}. \tag{7.127}$$

Here $\alpha = h, \bar{h}$ is an η-pairing SU(2) index corresponding to holon and antiholon respectively. The 'in-states' ('out-states') correspond to an ordering of the particles by decreasing (increasing) momentum. This means that for 'in-states' the left-most particle has the largest momentum, the second particle from the left has the second largest momentum and so on. The S-matrix elements are defined by

$$|k_1, k_2\rangle_{\alpha_1\alpha_2}^{\text{in}} = \check{S}_{\alpha_1\alpha_2}^{\beta_1\beta_2}(k_1, k_2)|k_1, k_2\rangle_{\beta_1\beta_2}^{\text{out}}. \tag{7.128}$$

Scattering in holon-holon and antiholon-antiholon states is diagonal by construction and the corresponding S-matrix elements are

$$\check{S}_{\alpha\alpha}^{\alpha\alpha}(k_1, k_2) = \exp(i\delta_{CT}^{12}) = -\frac{\Gamma\left(\frac{1}{2} - i\,\frac{k_{12}}{2}\right) \Gamma\left(1 + i\,\frac{k_{12}}{2}\right)}{\Gamma\left(\frac{1}{2} + i\,\frac{k_{12}}{2}\right) \Gamma\left(1 - i\,\frac{k_{12}}{2}\right)}. \tag{7.129}$$

Let us now consider scattering of a holon and an antiholon. The two eigenstates of the S-matrix are the charge-triplet and charge-singlet states

$$|k_1, k_2\rangle_t^{\text{in}} = \exp(i\delta_{CT}^{12})|k_1, k_2\rangle_t^{\text{out}},$$

$$|k_1, k_2\rangle_s^{\text{in}} = \exp(i\delta_{CS}^{12})|k_1, k_2\rangle_s^{\text{out}} \tag{7.130}$$

where

$$|k_1, k_2\rangle_{t,s}^{\text{in/out}} = \frac{1}{\sqrt{2}} \left[|k_1, k_2\rangle_{h\bar{h}}^{\text{in/out}} \pm |k_1, k_2\rangle_{\bar{h}h}^{\text{in/out}} \right]. \tag{7.131}$$

Using the definition (7.128) and the fact that the physics is unchanged if we interchange holons and antiholons (charge conjugation symmetry)

$$\check{S}_{\alpha_1\alpha_2}^{\beta_1\beta_2}(k_1, k_2) = \check{S}_{\bar{\alpha}_1\bar{\alpha}_2}^{\bar{\beta}_1\bar{\beta}_2}(k_1, k_2), \tag{7.132}$$

we can extract the S-matrix elements from (7.130)–(7.131)

$$\check{S}_{h\bar{h}}^{h\bar{h}}(k_1, k_2) = \frac{1}{2} \left[\exp(i\,\delta_{CT}^{12}) + \exp(i\,\delta_{CS}^{12}) \right],$$

$$\check{S}_{h\bar{h}}^{\bar{h}h}(k_1, k_2) = \frac{1}{2} \left[\exp(i\,\delta_{CT}^{12}) - \exp(i\,\delta_{CS}^{12}) \right]. \tag{7.133}$$

Using the explicit expressions (7.124) and (7.127) for the phase shifts we can write the above results in the following compact form

$$\check{S}^{\beta_1\beta_2}_{\alpha_1\alpha_2}(k_1,k_2) = -\frac{\Gamma\left(\frac{1}{2} - i\frac{k_{12}}{2}\right)\Gamma\left(1 + i\frac{k_{12}}{2}\right)}{\Gamma\left(\frac{1}{2} + i\frac{k_{12}}{2}\right)\Gamma\left(1 - i\frac{k_{12}}{2}\right)}$$

$$\times \left[\delta_{\alpha_1\beta_1}\delta_{\alpha_2\beta_2}\frac{k_{12}}{k_{12} - i} - \delta_{\alpha_1\beta_2}\delta_{\alpha_2\beta_1}\frac{i}{k_{12} - i}\right], \qquad (7.134)$$

where $\alpha_j, \beta_j = h, \bar{h}$. It is easily checked that the expression (7.134) is invariant under general SU(2) rotations U

$$\check{S}(k_1, k_2) = U^\dagger \, \check{S}(k_1, k_2) \, U. \qquad (7.135)$$

The S-matrix (7.134) was first derived from the factorization conditions [237, 503] by B. Berg *et al.* in [54, 55].

7.4.2 Spin sector

Let us now turn to scattering in the spin sector. Let us first determine the phase shift for the spin singlet excitation (see subsection 7.3.5 above). Takahashi's equation for the first hole in the distribution of the Λ^1_α's reads

$$\frac{2\pi J^h_1}{L} = \int_{-\pi}^{\pi} dk \, \theta\left(\frac{\Lambda^h_1 - \sin(k)}{u}\right) \rho(k) - \int_{-\infty}^{\infty} d\Lambda \, \theta\left(\frac{\Lambda^h_1 - \Lambda}{2u}\right) \sigma(\Lambda)$$

$$- \frac{1}{L}\Theta_{12}\left(\frac{\Lambda^h_1 - \Lambda^2_1}{u}\right), \qquad (7.136)$$

where Λ^2_1 is the spectral parameter corresponding to the Λ-string of length two. Using the results of subsection 7.3.5 we can rewrite (7.136) as

$$\frac{2\pi J^h_1}{L} = \frac{\pi}{2} - \mathcal{P}_s(\Lambda^h_1) + \frac{1}{L}\int_{-\pi}^{\pi} dk \, \theta\left(\frac{\Lambda^h_1 - \sin(k)}{u}\right) \rho_{SS}(k)$$

$$- \frac{1}{L}\int_{-\infty}^{\infty} d\Lambda \, \theta\left(\frac{\Lambda^h_1 - \Lambda}{2u}\right) \sigma_{SS}(\Lambda) - \frac{1}{L}\Theta_{12}\left(\frac{\Lambda^h_1 - \Lambda^2_1}{u}\right), \qquad (7.137)$$

where $\rho_{SS}(k)$ and $\sigma_{SS}(\Lambda)$ are given by (7.107) and $\mathcal{P}_s(\Lambda)$ is the dressed momentum of a spinon (7.8). Comparing (7.137) to (7.122) we can identify the phase-shift for the spin singlet excitation as

$$\delta^{12}_{SS} = -L\mathcal{P}_s(\Lambda^h_1) - 2\pi J^h_1. \qquad (7.138)$$

Using (7.107) and (7.110) and carrying out the remaining integrals we obtain

$$\delta^{12}_{SS} = \pi + \theta(\Lambda_{12}) + i \, \ln\left[\frac{\Gamma\left(\frac{1}{2} - i\frac{\Lambda_{12}}{2}\right)\Gamma\left(1 + i\frac{\Lambda_{12}}{2}\right)}{\Gamma\left(\frac{1}{2} + i\frac{\Lambda_{12}}{2}\right)\Gamma\left(1 - i\frac{\Lambda_{12}}{2}\right)}\right], \qquad (7.139)$$

where

$$\Lambda_{12} = \frac{\Lambda_1^h - \Lambda_2^h}{2u} > 0 . \tag{7.140}$$

The analogous calculation for the phase shift of the spin triplet excitation gives

$$\delta_{ST}^{12} = i \ln \left[\frac{\Gamma\left(\frac{1}{2} - i \frac{\Lambda_{12}}{2}\right) \Gamma\left(1 + i \frac{\Lambda_{12}}{2}\right)}{\Gamma\left(\frac{1}{2} + i \frac{\Lambda_{12}}{2}\right) \Gamma\left(1 - i \frac{\Lambda_{12}}{2}\right)} \right], \tag{7.141}$$

where the extra π arises because $J_{1,2}^h$ are half-odd integers for the spin triplet excitation. We now can write down the two particle S-matrix in the spin sector by repeating the steps we went through for holon-antiholon scattering. Introducing a spin index $\alpha = s, \bar{s}$ for spinons with $S^z = \pm \frac{1}{2}$ we can cast the result in the following form

$$\hat{S}_{\alpha_1\alpha_2}^{\beta_1\beta_2}(\Lambda_1, \Lambda_2) = \frac{\Gamma\left(\frac{1}{2} + i \frac{\Lambda_{12}}{2}\right) \Gamma\left(1 - i \frac{\Lambda_{12}}{2}\right)}{\Gamma\left(\frac{1}{2} - i \frac{\Lambda_{12}}{2}\right) \Gamma\left(1 + i \frac{\Lambda_{12}}{2}\right)}$$
$$\times \left[\delta_{\alpha_1\beta_1} \delta_{\alpha_2\beta_2} \frac{\Lambda_{12}}{\Lambda_{12} + i} + \delta_{\alpha_1\beta_2} \delta_{\alpha_2\beta_1} \frac{i}{\Lambda_{12} + i} \right], \tag{7.142}$$

where $\alpha_j, \beta_j = s, \bar{s}$. The S-matrix (7.142) is again SU(2) invariant as is required by the spin-rotational symmetry. As a function of the 'uniformising' spectral parameter, the S-matrix (7.142) is the same as in the spin-$\frac{1}{2}$ Heisenberg model [132] and in the Kondo model [18, 20, 460].

7.4.3 Scattering of spin and charge

Finally let us consider scattering of spinons and (anti)holons. The phase-shifts can be extracted from the spin-charge scattering state dicsussed in subsection 7.3.7. The equation for the hole in the distribution of the k_j's is

$$\frac{2\pi I^h}{L} = k^h + \int_{-\infty}^{\infty} d\Lambda \, \theta \left(\frac{\sin(k^h) - \Lambda}{u} \right) \sigma_1(\Lambda) , \tag{7.143}$$

where $\sigma_1(\Lambda)$ is given by (7.115), (7.116). Comparing (7.143) to the quantization condition (7.122) we see that the phase shift is given by

$$\delta_{SC}^{12} = -L\mathcal{P}_h(k^h) - 2\pi I^h , \tag{7.144}$$

where the dressed momentum $\mathcal{P}_h(k^h)$ of the holon is given by (7.8). Inserting (7.143) into (7.144) and then using (7.115) and (7.116) we obtain the following result for the phase shift

$$\delta_{SC}^{12} = \pi - \int_{-\infty}^{\infty} \frac{d\omega}{i\omega} \frac{\exp\left(-i\omega[\sin(k^h) - \Lambda^h]\right)}{2\cosh(u\omega)}$$
$$= 2 \arctan\left[\exp\left(\frac{\pi}{2u} [\sin(k^h) - \Lambda^h] \right) \right] + \frac{\pi}{2} . \tag{7.145}$$

For scattering to occur we need

$$\sin(k^h) \geq \Lambda^h \,. \tag{7.146}$$

By construction the S-matrix describing scattering of a spinon on a holon is diagonal (the SO(4) quantum numbers cannot change during the scattering process). We can express it in the form

$$
\begin{aligned}
\tilde{S}^{\beta_1\beta_2}_{\alpha_1\alpha_2}(k, \Lambda) &= i\frac{1 + i\exp\left(\frac{\pi}{2u}[\sin(k) - \Lambda]\right)}{1 - i\exp\left(\frac{\pi}{2u}[\sin(k) - \Lambda]\right)}\delta_{\alpha_1\beta_1}\delta_{\alpha_2\beta_2}\\
&= -\frac{1 + i\tanh\left(\frac{\pi}{4u}[\sin(k) - \Lambda]\right)}{1 - i\tanh\left(\frac{\pi}{4u}[\sin(k) - \Lambda]\right)}\delta_{\alpha_1\beta_1}\delta_{\alpha_2\beta_2}\,.
\end{aligned} \tag{7.147}
$$

where $\alpha_1, \beta_1 \in \{h, \bar{h}\}$ and $\alpha_2, \beta_2 \in \{s, \bar{s}\}$.

The S-matrix for the half-filled Hubbard model in zero magnetic field was determined in [120, 121] and for a less than half-filled band in [18].

7.5 'Physical' Bethe ansatz equations

Let us return to the general excited state discussed in 7.3.2: there are N_h and M_h holes in the distributions of k's and Λ^1_α's respectively, M_l roots Λ^l_α ($l \geq 2$) and M'_n roots Λ'^n_α. Using the formalism developed above, it is possible to derive a set of equations for the spectral parameters of the holes and extra roots. These equations are the analog of Takahashi's equations, which we recall describe excited states over the reference state of the Bethe ansatz, the empty lattice. The equations we will derive now describe excited states over the true ground state of the half-filled Hubbard model. For this reason they are known as 'physical Bethe ansatz equations'. Our discussion follows Ref. [490].

The root densities describing a general excitation over the half-filled ground state are given in terms of the integral equations (7.78). The solution of these equations is

$$
\begin{aligned}
\rho(k) &= \rho_0(k) + \frac{1}{L}\rho'(k)\,,\\
\sigma_1(\Lambda) &= \sigma_0(\Lambda) + \frac{1}{L}\sigma'_1(\Lambda)\,,
\end{aligned} \tag{7.148}
$$

where $\rho_0(k)$ and $\sigma_0(\Lambda)$ are the root densities describing the half-filled ground state (7.11) and

$$
\begin{aligned}
\rho'(k) = &-\sum_{j=1}^{N_h}\left[\delta(k - k^h_j) + \cos(k)R(\sin(k) - \sin(k^h_j))\right]\\
&- \cos(k)\sum_{m=1}^{M_h}s(\sin(k) - \Lambda^h_m) + \cos(k)\sum_{l=1}^{\infty}\sum_{\beta=1}^{M'_l}a_l(\sin(k) - \Lambda'^l_\beta)\,,
\end{aligned} \tag{7.149}
$$

$$\sigma_1'(\Lambda) = -\sum_{j=1}^{N_h} s(\Lambda - \sin(k_j^h)) - \sum_{m=1}^{M_h} [\delta(\Lambda - \Lambda_m^h) - R(\Lambda - \Lambda_m^h)]$$

$$-\sum_{l=2}^{\infty}\sum_{\beta=1}^{M_l} a_{l-1}(\Lambda - \Lambda_\beta^l). \tag{7.150}$$

Given the root densities (7.149) and (7.150) describing the excited state, we may write down the equations fixing the positions of the holes k_j^h ($j = 1, \dots, N_h$), Λ_m^h ($m = 1, \dots, M_h$) and the 'extra' roots Λ_α^n ($l \geq 2, n = 1, \dots, M_n$), $\Lambda_\beta'^m$ ($m = 1, \dots, M_m'$). The equations for the spectral parameters k_j^h describing holons and antiholons are

$$2\pi I_j^h = L\left[\frac{\pi}{2} - \mathcal{P}_h(k_j^h)\right] - \sum_{l=1}^{N_h} \Phi(\sin(k_j^h) - \sin(k_l^h))$$

$$+ \sum_{m=1}^{M_h} \Psi(\sin(k_j^h) - \Lambda_m^h) + \sum_{m=1}^{\infty}\sum_{\gamma=1}^{M_m'} \theta\left(\frac{\sin(k_j^h) - \Lambda_\gamma'^m}{mu}\right), \tag{7.151}$$

where $\mathcal{P}_h(k)$ is the momentum of a holon with spectral parameter k (7.8) and $\Phi(x)$ and $\Psi(x)$ are closely related to the holon-holon (7.126) and holon-spinon (7.145) scattering phase shifts respectively

$$\Phi(x) = i \ln\left[\frac{\Gamma\left(\frac{1}{2} + i\frac{x}{4u}\right)\Gamma\left(1 - i\frac{x}{4u}\right)}{\Gamma\left(\frac{1}{2} - i\frac{x}{4u}\right)\Gamma\left(1 + i\frac{x}{4u}\right)}\right],$$

$$\Psi(x) = \frac{\pi}{2} - 2\arctan\left[\exp\left(\frac{\pi x}{2u}\right)\right]. \tag{7.152}$$

The equations for spectral parameters Λ_m^h describing the spinon excitations are found to be

$$2\pi J_m^h = L\left[\frac{\pi}{2} - \mathcal{P}_s(\Lambda_m^h)\right] + \sum_{n=1}^{M_h} \Phi(\Lambda_m^h - \Lambda_n^h)$$

$$+ \sum_{j=1}^{N_h} \Psi(\Lambda_m^h - \sin(k_j^h)) - \sum_{l=2}^{\infty}\sum_{\beta=1}^{M_l} \theta\left(\frac{\Lambda_m^h - \Lambda_\beta^l}{(l-1)u}\right). \tag{7.153}$$

Here $\mathcal{P}_s(\Lambda)$ is the spinon momentum (7.8). We note that $-\Phi(\Lambda_m^h - \Lambda_n^h)$ coincides up to a constant with the spinon-spinon triplet phase shift (7.139). Last but not least the equations fixing the 'extra' roots are obtained from equations (7.99) and (7.80). We find

$$2\pi J_\alpha^n = \sum_{n=1}^{M_h} \theta\left(\frac{\Lambda_\alpha^n - \Lambda_m^h}{(n-1)u}\right) - \sum_{l=2}^{\infty}\sum_{\beta=1}^{M_l} \Theta_{n-1l-1}\left(\frac{\Lambda_\alpha^n - \Lambda_\beta^l}{u}\right), \tag{7.154}$$

$$2\pi J_\beta'^m = \sum_{j=1}^{N_h} \theta\left(\frac{\Lambda_\beta'^m - \sin(k_j^h)}{mu}\right) - \sum_{l=1}^{\infty}\sum_{\gamma=1}^{M_l'} \Theta_{ml}\left(\frac{\Lambda_\beta'^m - \Lambda_\gamma'^l}{u}\right). \tag{7.155}$$

It is straightforward to check that the physical Bethe ansatz equations (7.151)–(7.155) reproduce all results we obtained before in the two-particle sector.

7.6 Finite magnetic field and half-filled band

Let us now consider the excitation spectrum at half filling in the presence of a magnetic field. We set the chemical potential equal to zero, $\mu = 0$ and constrain our discussion to Phase V, i.e. values of B such that

$$B < B_0 = 2\sqrt{1 + u^2} - 2u \, . \tag{7.156}$$

The integral equations (5.83)–(5.85) for the dressed energies can be simplified to

$$\varepsilon_1(\Lambda) = 2B + 4u - 4\mathrm{Re}\sqrt{1 - (\Lambda - iu)^2} - \int_{-A}^{A} d\Lambda' \, a_2(\Lambda - \Lambda') \, \varepsilon_1(\Lambda') \, ,$$

$$\kappa(k) = -2\cos(k) - 2u - B + \int_{-A}^{A} d\Lambda \, a_1(\sin(k) - \Lambda) \, \varepsilon_1(\Lambda) \, ,$$

$$\varepsilon_n(\Lambda) = 2nB + 4nu - 4\mathrm{Re}\sqrt{1 - (\Lambda - inu)^2}$$

$$- \int_{-A}^{A} d\Lambda' \left[a_{n-1}(\Lambda - \Lambda') + a_{n+1}(\Lambda - \Lambda') \right] \varepsilon_1(\Lambda') \, , \quad n \geq 2 \, ,$$

$$\varepsilon_n'(\Lambda) = 0 \, , \quad n \geq 1. \tag{7.157}$$

The dressed energies $\varepsilon_n(\Lambda)$ are all positive for $n \geq 2$ and simple lower bounds are derived in Appendix 7.B. Similarly we obtain the following set of integral equations for the root densities from (5.93) and (5.94) by using (17.17)[5]

$$\sigma_1(\Lambda) = \frac{1}{\pi} \mathrm{Re} \frac{1}{\sqrt{1 - (\Lambda - iu)^2}} - \int_{-A}^{A} d\Lambda' \, a_2(\Lambda - \Lambda') \, \sigma_1(\Lambda') \, ,$$

$$\rho(k) = \frac{1}{2\pi} + \cos(k) \int_{-A}^{A} d\Lambda \, a_1(\sin(k) - \Lambda) \, \sigma_1(\Lambda) \, ,$$

$$\sigma_n^h(\Lambda) = \frac{1}{\pi} \mathrm{Re} \frac{1}{\sqrt{1 - (\Lambda - inu)^2}}$$

$$- \int_{-A}^{A} d\Lambda' \left[a_{n-1}(\Lambda - \Lambda') + a_{n+1}(\Lambda - \Lambda') \right] \sigma_1(\Lambda') \, , \quad n \geq 2 \, ,$$

$$\sigma_n'^h(\Lambda) = 0 \, , \quad n \geq 1. \tag{7.158}$$

The dressed momenta can again be calculated from the root densities by (5.98)–(5.101). The dispersions $\kappa(p)$ and $\varepsilon_1(p_1)$ are shown in figure 7.14 for $u = 0.25$ and several values of the applied magnetic field.

In a weak field, the dispersions are very close to the zero field results shown in figure 7.6. Increasing the value B leads to relatively small changes in $\kappa(p)$, whereas $\varepsilon_1(p_1)$ is pushed to higher energies and covers a larger momentum range. For $B = 1.5$, which is close to the saturation field for $u = 0.25$, $B_0 \approx 1.56155$, the momentum range in which $\varepsilon_1(p_1) \leq 0$ has almost shrunk to zero. In figure 7.15 we plot $\varepsilon_2(p_2)$ for $u = 0.25$ and several values of the applied magnetic field. We see that increasing the field pushes the dispersion up to higher

---

[5] We still use the convention (5.95).

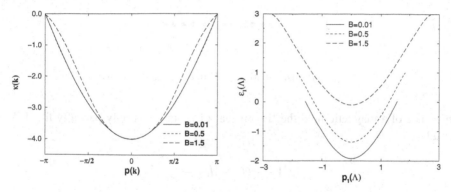

Fig. 7.14. Dressed energies $\kappa(k)$ and $\varepsilon_1(\Lambda)$ as functions of the dressed momenta $p(k)$ and $p_1(\Lambda)$ respectively, for $u = 0.25$ and three different values of B. The curves for $B = 0.01$ and $B = 0.5$ are indistinguishable in the plot.

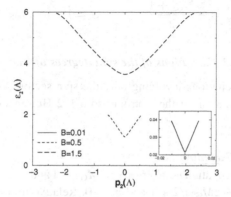

Fig. 7.15. Dressed energy $\varepsilon_2(\Lambda)$ as function of the dressed momentum $p_2(\Lambda)$ for three different values of B and $u = 0.25$. The inset is an enlargement of the dispersion for $B = 0.01$. As the magnetic field B decreases, the dispersion collapses to zero energy and momentum.

energies while increasing the momentum range covered. In the limit $B \to 0$, the dispersion collapses to zero. The dispersions $\varepsilon_n(p_n)$ of longer Λ-strings look similar.

In the following discussion of the ground state and excitations we concentrate on regular Bethe states. As usual further states are obtained by acting with the SO(4) raising and lowering operators. The magnetic field breaks the spin rotational symmetry and as a result acting with the spin lowering operator S^- on a regular Bethe state with energy E (and S^z eigenvalue different from zero) gives an eigenstate of the Hamiltonian with energy $E + 2B$.

7.6.1 Ground state

The ground state is characterized by the occupation numbers $\mathcal{M}_e = L$, $M_1 = M_{GS} < L/2$. We furthermore take M_{GS} to be an odd integer. As a result the I_j's are half-odd integers and the J_α^1's are integers. The quantum numbers of the ground state are

$$S^z = \frac{L - 2M_{GS}}{2}, \qquad \eta^z = 0. \tag{7.159}$$

$$\mathbf{J}_\alpha^1 \quad \circ\circ\circ\circ \; \bullet\bullet\bullet\bullet\bullet\bullet\bullet\bullet\bullet\bullet\bullet \; \circ\circ\circ\circ$$

$$-(M_{\mathrm{GS}}-1)/2 \qquad\qquad\qquad (M_{\mathrm{GS}}-1)/2$$

Fig. 7.16. Distribution of J_α^1 in the ground state.

In presence of a magnetic field the 'Fermi sea' of roots Λ_α^1 is only partially filled. More precisely

$$|J_\alpha^1| \le \frac{1}{2}(L - M_{\mathrm{GS}} - 1), \tag{7.160}$$

so that we have $L - M_{\mathrm{GS}}$ vacancies for M_{GS} roots and thus $L - 2M_{\mathrm{GS}}$ holes. The ground state is obtained by filling a 'Fermi sea of integers J_α^1', see figure 7.16

$$J_\alpha^1 = -\frac{M_{\mathrm{GS}}+1}{2} + \alpha, \quad \alpha = 1, \dots, M_{\mathrm{GS}}. \tag{7.161}$$

7.6.2 Excitations of the spin degrees of freedom

Let us first consider excitations involving only the spin sector. As we will see these are quite similar to the excitation of the isotropic spin-1/2 Heisenberg chain in a magnetic field [235, 333, 447].

$\delta S^z = 1$ 'Two-Spinon' Excitation

This excitation has occupation numbers $M_1 = M_{\mathrm{GS}} - 1$ and $\mathcal{M}_e = L$. Hence it has quantum numbers $S^z = (L - 2M_{\mathrm{GS}})/2 + 1$ and $\eta^z = 0$. Relative to the ground state we have $\delta S^z = 1$. The ranges of the integers I_j and half-odd integers J_α^1 are

$$-\frac{L}{2} < I_j \le \frac{L}{2}, \quad |J_\alpha^1| \le \frac{1}{2}[L - M_{\mathrm{GS}}]. \tag{7.162}$$

Now there are $L - M_{\mathrm{GS}} + 1$ vacancies for the $M_{\mathrm{GS}} - 1$ Λ_α^1's, so that compared to the ground state there are two additional holes. We first consider the case, where the half-odd integers $J_{1,2}^h$ associated with these holes fulfil

$$-\frac{M_{\mathrm{GS}}}{2} \le J_{1,2}^h \le \frac{M_{\mathrm{GS}}}{2}. \tag{7.163}$$

This distribution is shown schematically in figure 7.17. This means that we have two holes in the distribution of spectral parameters Λ_α^1 and concomitantly are dealing with a two-parametric excitation. Given that $\delta S^z = 1$ we may associate quantum numbers $S^z = 1/2$ and $\eta^z = 0$ with each hole. The energy and momentum are

$$E_{\mathrm{hh}}^+(\Lambda_1, \Lambda_2) = -\varepsilon_1(\Lambda_1) - \varepsilon_1(\Lambda_2),$$
$$P_{\mathrm{hh}}^+(\Lambda_1, \Lambda_2) = -p_1(\Lambda_1) - p_1(\Lambda_2) + \pi. \tag{7.164}$$

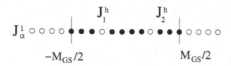

Fig. 7.17. Distribution of J_α^1 in the $\delta S^z = 1$ two-spinon excitation.

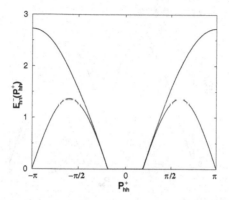

Fig. 7.18. Upper and lower boundaries of the $S^z = 1$ two-spinon continuum for $u = 0.25$ and $B = 0.5$.

Here the superscript indicates that we are considering a $\delta S^z = 1$ excitation and the extra contribution of π to the momentum arises because the I_j's are integers (see the discussion above figure 7.7). In particular for weak fields this excitation is very similar to the spin-triplet excitation[6] and in the limit $B \to 0$ goes over in the latter. On the basis of this similarity we will call the two elementary excitations making up the $\delta S^z = 1$ state 'spinons'. In figure 7.18(a) we show the lower and upper boundaries of the 'two-hole' scattering continuum for $u = 0.25$ and $B = 0.5$. We see that the excitation only exists in part of the Brillouin zone: there are no states at small momentum. As B increases the continuum occupies a smaller and smaller part of the Brillouin zone until it altogether disappears at the saturation field $B_0 = 2\sqrt{1 + u^2} - 2u$, where the ground state becomes fully spin polarized.

It is by no means necessary to choose the distribution of J_α^1 in the way we have done, i.e. such that we obtain two holes. This is contrast to the zero field case, where we necessarily end up with two holes. An equally valid distribution of half-odd integers J_α^1 is shown in figure 7.19. Now we have one hole at J^h and one particle at J^p, with corresponding spectral parameters Λ_h and Λ_p. By construction $J^{p,h}$ have ranges

$$|J^p| > \frac{M_{GS} - 2}{2} > |J^h|. \tag{7.165}$$

---

[6] More precisely to its highest weight state.

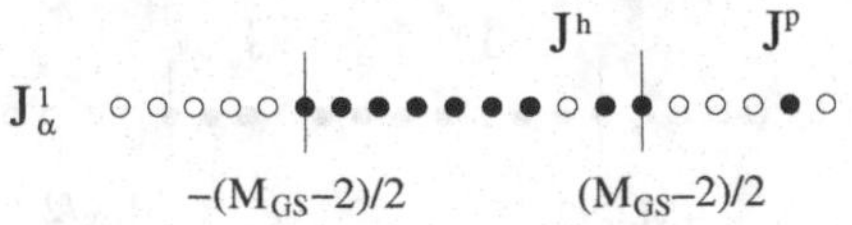

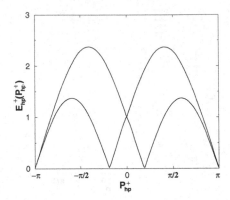

Fig. 7.19. Distribution of J_α^1 for a 'particle-hole' $\delta S^z = 1$ state.

Fig. 7.20. Upper and lower boundaries of the $S^z = 1$ 'hole-particle' continuum for $u = 0.25$ and $B = 0.5$.

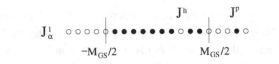

Fig. 7.21. Distribution of J_α^1 for the 'particle-hole' $\delta S^z = 1$ state interpreted as a special case of a three-hole one-particle excitation.

The energy and momentum are

$$E_{\mathrm{hp}}^+(\Lambda_h, \Lambda_p) = -\varepsilon_1(\Lambda_h) + \varepsilon_1(\Lambda_p)\,,$$
$$P_{\mathrm{hp}}^+(\Lambda_h, \Lambda_p) = -p_1(\Lambda_h) + p_1(\Lambda_p) + \pi. \qquad (7.166)$$

The upper and lower boundaries of this 'hole-particle' continuum are shown in figure 7.20. It is clear from its construction that this excitation disappears in both limits $B \to 0$ and $B \to B_0$.

By construction the particle-hole excitation is two-parametric. However, it is perhaps more natural to consider it as a special case of a four-parametric excitation. In figure 7.21 we have replotted the distribution of the J_α^1's by choosing the 'Fermi integer' to be $\frac{M_{\mathrm{GS}}}{2}$ rather than $\frac{M_{\mathrm{GS}}-2}{2}$. In this way of looking at things the particle-hole state corresponds to the special limit of a three-hole one-particle excitation, where two of the holes are located at the two 'Fermi points', i.e. their associated spectral parameters are $\pm A$. Now we may

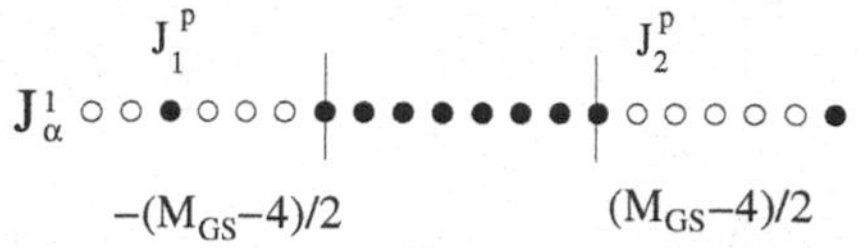

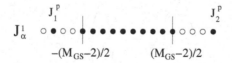

Fig. 7.22. Distribution of J_α^1 in the $\delta S^z = 1$ 'particle-particle' excitation.

$$J_\alpha^1 \quad \circ \bullet \circ \circ \circ \bullet \bullet \bullet \bullet \bullet \bullet \bullet \bullet \bullet \bullet \bullet \circ \circ \circ \bullet$$

$$-(M_{GS}-2)/2 \qquad (M_{GS}-2)/2$$

Fig. 7.23. Distribution of J_α^1 in the $\delta S^z = -1$ 'particle-particle' excitation.

associate quantum numbers $S^z = \pm 1/2$ with holes and particles respectively and consider them as two species of spinons.

A third configuration of the J_α^1's that leads to a two-parametric excitation is to have two particles, charaterized by the half-odd integers $J_{1,2}^P$ with $|J_{1,2}^P| > (M_{GS} - 4)/2$ and corresponding spectral parameters $\Lambda_{1,2}$. The distribution of J_α^1 is shown schematically in figure 7.22. Energy and momentum of this particle-particle excitation are

$$E_{pp}^+(\Lambda_1, \Lambda_2) = \varepsilon_1(\Lambda_1) + \varepsilon_1(\Lambda_2) \,,$$
$$P_{pp}^+(\Lambda_2, \Lambda_2) = p_1(\Lambda_1) + p_1(\Lambda_2) + \pi. \tag{7.167}$$

The particle-particle state may alternatively be considered as a special case of a (four-hole two-particle) six-spinon excitation, where four holes are located at the 'Fermi points' $\pm A$.

$\delta S^z = -1$ 'Two-Spinon' Excitation

Let us now consider states with occupation numbers $M_1 = M_{GS} + 1$ and $\mathcal{M}_e = L$. The quantum numbers are $S^z = (L - 2M_{GS})/2 - 1$ and $\eta^z = 0$. Relative to the ground state we have $\delta S^z = -1$. The ranges of the integers I_j and half-odd integers J_α^1 are

$$-\frac{L}{2} < I_j \le \frac{L}{2} \,, \quad |J_\alpha^1| \le \frac{1}{2}[L - M_{GS} - 2] \,. \tag{7.168}$$

Now there are $L - M_{GS} - 1$ vacancies for the $M_{GS} + 1$ Λ_α^1's, so that compared to the ground state there are two additional particles. Energy and momentum of this particle-particle excitation are

$$E_{pp}^-(\Lambda_1, \Lambda_2) = \varepsilon_1(\Lambda_1) + \varepsilon_1(\Lambda_2) \,,$$
$$P_{pp}^-(\Lambda_2, \Lambda_2) = p_1(\Lambda_1) + p_1(\Lambda_2) + \pi. \tag{7.169}$$

We note that these are the same as for the $\delta S^z = 1$ particle-particle excitation. However, whereas we may interpret the latter as a special case of a six-spinon excitation, the $\delta S^z = -1$

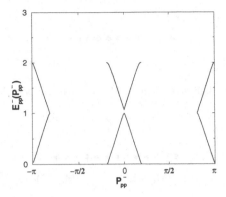

Fig. 7.24. Upper and lower boundaries of the $S^z = -1$ 'particle-particle' continuum for $u = 0.25$ and $B = 0.5$.

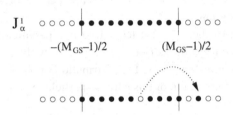

Fig. 7.25. A 'particle-hole' excitation of the spin degrees of freedom.

state involves only two spinons. The boundaries of the $\delta S^z = -1$ two-spinon continuum are shown in figure 7.24.

$\delta S^z = 0$ *'Particle-Hole' Excitation*

Given that the Fermi sea of Λ_α^1 is only partially filled, we may construct a 'particle-hole' excitation by the process described in figure 7.25. By construction this state has the same quantum numbers as the ground state, but describes a two-parametric excitation of the spin degrees of freedom. Denoting the spectral parameters corresponding to the 'particle' and 'hole' as Λ_p and Λ_h respectively, we have

$$E_{hp}^0(\Lambda_h, \Lambda_p) = -\varepsilon_1(\Lambda_h) + \varepsilon_1(\Lambda_p),$$
$$P_{hp}^0(\Lambda_h, \Lambda_p) = -p_1(\Lambda_h) + p_1(\Lambda_p). \tag{7.170}$$

We note that this excitation exists only in a nonzero magnetic field and disappears in the limit $B \to 0$. In figure 7.26 we plot the boundaries of the continuum of states belonging to the particle-hole excitation. We note that the only difference to the $\delta S^z = 1$ particle-hole excitation is a shift of momentum by π. We expect the particle-hole excitation to dominate the 'longitudinal' spin response (i.e. the zz-component of the dynamical structure factor) at low energies of the half-filled Hubbard model in a sufficiently strong magnetic field.

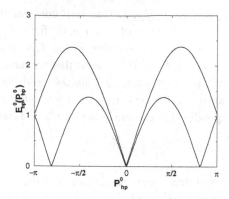

Fig. 7.26. Upper and lower boundaries of the $S^z = 0$ 'hole-particle' continuum for $u = 0.25$ and $B = 0.5$.

$\delta S^z = -1$ *'Magnon' Excitation*

Another state with $\delta S^z = -1$ is obtained by the choice of occupation numbers $\mathcal{M}_e = L$, $M_1 = M_{GS} - 1$ and $M_2 = 1$. Now both the I_j's and J_α^1's are half-odd integers and J_1^2 is an integer. We find

$$|J_\alpha^1| \le \frac{1}{2}[L - M_{GS} - 2] , \quad |J_1^2| \le \frac{1}{2}[L - 2M_{GS} - 2] . \tag{7.171}$$

As compared to the ground state there is one less vacancy for the J_α^1, but as there is one less root the number of holes is the same as in the ground state. Hence we choose

$$J_\alpha^1 = -\frac{M_{GS}}{2} + \alpha , \quad \alpha = 1, \ldots, M_{GS} - 1, \tag{7.172}$$

and do not associate any excitation with the Fermi sea of the Λ_α^1's. This implies that the excitation described by the above occupation numbers is *one-parametric* with energy and momentum given by

$$E_{Mag}(\Lambda) = \varepsilon_2(\Lambda) ,$$
$$P_{Mag}(\Lambda) = p_2(\Lambda). \tag{7.173}$$

We note that this excitation is gapped (see Appendix 7.B)

$$E_{Mag}(\Lambda) \ge \varepsilon_2(0) \ge 2B . \tag{7.174}$$

The dispersion for the magnon excitation for $u = 0.25$ and three values of B is shown in figure 7.15. An interesting question is why the magnon excitation is stable and does not decay into two $S^z = -1/2$ spinons. It turns out that the magnon dispersion is just very slightly higher in energy than the $\delta S^z = -1$ two-spinon continuum, which is shown in figure 7.24.

$\delta S^z = 0$ *'Two-Spinon-Magnon' Excitation* Let us now consider the analog of the spin-singlet excitation (7.35) in the presence of a magnetic field. The occupation numbers are $\mathcal{M}_e = L$, $M_1 = M_{GS} - 2$, $M_2 = 1$, resulting in the I_j's being integers and $|J_\alpha^1| \leq \frac{1}{2}(L - M_{GS} - 1)$, $|J_\alpha^2| \leq \frac{1}{2}(L - 2M_{GS})$. This means that we have two extra holes in the distribution of J_α^1's and the excitation is *three-parametric*, as there is a finite range for the integer J_1^2 and concomitantly also for the corresponding rapidity Λ of the 2-string. Let us denote the integers corresponding to the extra holes by $J_{1,2}^h$ and consider only the case

$$|J_{1,2}^h| \leq \frac{M_{GS} - 1}{2}. \tag{7.175}$$

Then the dynamical degrees of freedom are parametrized by the spectral parameters $\Lambda_{1,2}^h$ of the holes in the distribution of the J_α^1's and Λ of the extra 2-string. Energy and momentum are

$$E_{2sM}(\Lambda_1^h, \Lambda_2^h, \Lambda) = -\varepsilon_1(\Lambda_1^h) - \varepsilon_1(\Lambda_2^h) + \varepsilon_2(\Lambda),$$
$$P_{2sM}(\Lambda_1^h, \Lambda_2^h, \Lambda) = -p_1(\Lambda_1^h) - p_1(\Lambda_2^h) + p_2(\Lambda) + \pi. \tag{7.176}$$

When the magnetic field becomes small, the range of $p_2(\Lambda)$ as well as the energy $\varepsilon_2(\Lambda)$ collapses to zero and we recover the spin-singlet excitation discussed before, see equation (7.35). The cases where instead of having two holes in the distribution of J_α^1 we have a particle and a hole or two particles can be treated in the same way as for the $\delta S^z = 1$ two-spinon excitation.

7.6.3 Excitations involving the charge sector

Charge Singlet Excitation This excitation has occupation numbers $M_1 = M_{GS} - 1, M_1' = 1$ and $\mathcal{M}_e = L - 2$. The corresponding quantum numbers are $S^z = (L - 2M_{GS})/2$, $\eta^z = 0$, i.e. the same as for the ground state. The ranges of the half-odd integers I_j, half-odd integers J_α^1 and integer $J_1'^1$ are

$$-\frac{L}{2} < I_j \leq \frac{L}{2}, \quad |J_\alpha^1| \leq \frac{1}{2}[L - M_{GS} - 2], \quad J_1'^1 = 0, \tag{7.177}$$

which implies that we have two holes at positions $k_{1,2}^h$ in the distribution of k's. In the distribution of Λ^1's there is one less vacancy than in the ground state, but also one less root. Hence no degrees of freedom are excited here and the distribution of J_α^1 remains symmetric around zero. We also have a single k-Λ string at a position Λ', which however carries zero dressed energy as $\varepsilon_1'(\Lambda') \equiv 0$ by (7.10), and only contributes a constant $p_1'(\Lambda') = 2\pi$ to the momentum as follows from (7.12). Therefore energy and momentum of the charge singlet excitation are given by

$$E_{CS}(k_1^h, k_2^h) = -\kappa(k_1^h) - \kappa(k_2^h),$$
$$P_{CS}(k_1^h, k_2^h) = -p(k_1^h) - p(k_2^h) \mod (2\pi). \tag{7.178}$$

This excitation is very similar to the charge-singlet in zero magnetic field, see figure (7.9).

Spin-Charge Scattering State Finally, let us consider the spin-charge scattering state with occupation numbers $\mathcal{M}_e = L - 1$, $M_1 = M_{GS} - 1$. The differences in quantum numbers compared to the ground state are $\delta S^z = \frac{1}{2}$ and $\eta^z = -\frac{1}{2}$ so that the excited state corresponds to the removal of one electron from the ground state. There are L vacancies for the integers I_j so there we have one hole at position k^h in the distribution of the k_j's. We have the same number of vacancies for the integers J_α^1 as in the ground state but one less root, which implies we have one additional hole in the distribution of the J_α^1. Let us denote the corresponding integer by J^h and restrict our attention to the case $J^h \leq \frac{M_{GS}-1}{2}$. Then energy and momentum are given by

$$E_{sc}(k^h, \Lambda^h) = -\varepsilon_1(\Lambda^h) - \kappa(k^h)\,,$$
$$P_{sc}(k^h, \Lambda^h) = -p_1(\Lambda^h) - p(k^h) + \pi. \tag{7.179}$$

By construction this excitation reduces to the spin-charge scattering state considered previously when the magnetic field goes to zero. The case where we have a particle-like excitation in the distribution of J_α^1 instead of a hole can be treated in complete analogy with the two-spinon excitation considered above.

7.7 Zero magnetic field and less than half-filled band

Let us now consider a less than half-filled band $0 < n_c < 1$ and zero magnetic field. Here the ground state is metallic and in contrast to the half-filled case there is no single-particle gap, i.e. there exist gapless excitations with the quantum numbers of an electron. For definiteness let us take the total number of electrons $N_{GS} = 2 \times$ odd. The occupation numbers of the ground state are $M_1 = \frac{\mathcal{M}_e}{2} = \frac{N_{GS}}{2}$, which implies that I_j are half-odd integers and J_α^1 are integers. In the ground state all vacancies are filled symmetrically around zero

$$I_j = j - \frac{N_{GS} + 1}{2}\,, \qquad j = 1, \ldots, N_{GS}\,,$$
$$J_\alpha^1 = \alpha - \frac{N_{GS} + 2}{4}\,, \qquad \alpha = 1, \ldots, \frac{N_{GS}}{2}\,. \tag{7.180}$$

The distributions of the corresponding spectral parameters in the thermodynamic limit are described by (6.B.5). In order to determine the excitation spectrum we will again make use of the dressed energies and momenta obtained from the solution of the TBA equations.

There are three different kinds of *elementary* excitations.

- The first type of elementary excitation is gapless, carries no spin and has charge $\mp e$. It corresponds to adding a particle to or making a hole in the distribution of k's. Such excitations are sometimes called *antiholons* and *holons* respectively[7] and have dressed energy $\pm\kappa(k)$ and dressed momentum $\pm p(k)$.

---

[7] We will follow this nomenclature, although it is somewhat imprecise, as we have already used the same terminology for the elementary excitations in the charge sector at half filling, which are different in nature.

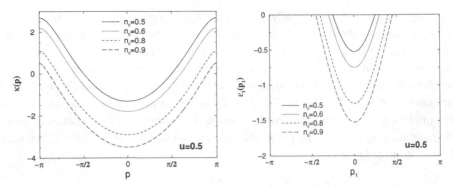

Fig. 7.27. Dressed energies $\kappa(p)$ and $\varepsilon_1(p_1)$ as functions of the dressed momenta for $u = 0.5$ and several densities.

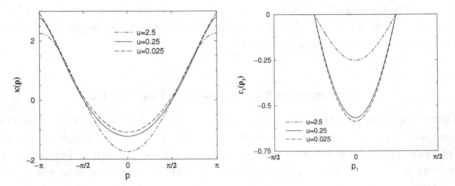

Fig. 7.28. Dressed energies $\kappa(p)$ and $\varepsilon_1(p_1)$ as functions of the dressed momenta for $n_c = 0.5$ and several values of u.

- The second type of elementary excitation is gapless and carries spin-$\frac{1}{2}$ but no charge. It corresponds to a hole in the distribution of Λ^1's. Such excitations are called *spinons*. They have dressed energy $-\varepsilon_1(\Lambda)$ and dressed momentum $-p_1(\Lambda)$.
- There is an infinite number of different types of gapped excitations that carry charge but no spin. They correspond to adding a k-Λ string of length n to the ground state. Their dressed energies are $\varepsilon'_n(\Lambda)$, their dressed momenta are $p'_n(\Lambda)$.

In figures 7.27 and 7.28 we plot the dressed energies $\kappa(p(k))$ and $\varepsilon_1(p_1(\Lambda))$ as functions of the dressed momenta for several values of n_c and u.

The resulting dispersions can be interpreted in terms of a band picture for spinons and antiholons. The antiholon band covers the entire Brillouin zone whereas the spinon band only covers the interval $[-\frac{\pi n_c}{2}, \frac{\pi n_c}{2}]$, where n_c is the density. In the ground state the spinon band is filled completely whereas the antiholon band is filled in the interval $[-\pi n_c, \pi n_c]$.

This can be seen by recalling that the total momentum can be expressed as

$$P = \frac{2\pi}{L} \left(\sum_{j=1}^{N-2M'} I_j + \sum_{m=1}^{\infty} \sum_{\beta=1}^{M_m} J_\beta^m - \sum_{n=1}^{\infty} \sum_{\alpha=1}^{M_n'} J_\alpha'^n \right) + \pi \sum_{n=1}^{\infty} \sum_{\alpha=1}^{M_n'} (n+1). \tag{7.181}$$

The ground state has zero momentum and corresponds to symmetric distributions of I_j and J_α^1 (7.180). Making use of the monotonicity properties of $p(k)$ and $p_1(\Lambda)$ we conclude that the ground state indeed corresponds to filling all spinon states with $p_1(\Lambda) \in [-\frac{\pi n_c}{2}, \frac{\pi n_c}{2}]$ and antiholon states with $p(k) \in [-\pi n_c, \pi n_c]$. It is apparent from figure 7.28 that the bandwidth of the antiholon band does not depend on u. It is easy to see using the integral equation for $\kappa(k)$ that the bandwidth is actually equal to $\kappa(\pi) - \kappa(0) = 4$ (or $4t$ if we restore the hopping matrix element t). On the other hand the bandwidth of the spinon band depends sensitively on u. This should be the case as in the large-U limit the effective Heisenberg exchange is $4t^2/U$.

Let us emphasize that so far we have talked about *elementary* excitations in the repulsive Hubbard model below half filling. Like in the half-filled case it is important to distinguish these from 'physical' excitations, which are the permitted combinations of elementary excitations. As in the half-filled case there are no excitations involving only one spinon or one (anti)holon. The simplest gapless excitations involve a pair of elementary excitations and we will consider them next. Unlike in the half-filled case there exist single-particle excitations below half filling. These involve k-Λ strings and all have gaps. We will consider examples of such excitations below.

7.7.1 Charge-neutral excitations

Example 1 Particle-hole excitation.

This is a two-parametric gapless physical excitation with spin and charge zero, i.e. its quantum numbers as well as its occupation numbers are the same as the ones of the ground state. It is obtained by removing a spectral parameter k_h with $|k_h| < Q$ from the ground-state distribution of k's and adding a spectral parameter k_p with $|k_p| > Q$. Its energy and momentum are

$$E_{ph} = \kappa(k_p) - \kappa(k_h),$$
$$P_{ph} = p(k_p) - p(k_h). \tag{7.182}$$

This excitation is allowed by the selection rules (4.44) as in the ground state only the half-odd integers $|I_j| \leq (N_{GS} - 1)/2$ are occupied and thus the possibility of removing a root corresponding to $|I_h| \leq (N_{GS} - 1)/2$ and adding a root corresponding to $|I_p| > (N_{GS} - 1)/2$ exists. The particle-hole excitation was first studied by means of the root density approach by C.F. Coll in [88]. In figure 7.29 we show the particle-hole spectrum for densities $n_c = 0.6$ and $n_c = 0.8$. As we approach half filling the phase-space for particles shrinks to zero and it is clear from figure 7.29 that as the density increases the total area of the particle-hole continuum diminishes.

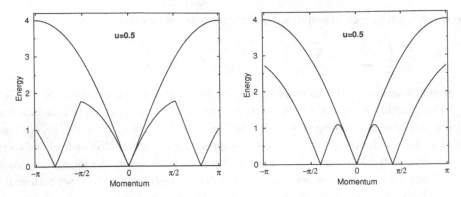

Fig. 7.29. Particle-hole excitation for $u = 0.5$ and densities $n_c = 0.6$ and $n_c = 0.8$. Shown are the lower and upper boundaries of the continuum.

The low-energy modes occur at momenta 0 and $\pm 2\pi n_c$ mod 2π. This can be easily understood in terms of the 'band' picture discussed above.

Example 2a Spin triplet excitation.

Let us consider an excitation involving the spin degrees of freedom next. If we change the number of down spins by one while keeping the number of electrons fixed we obtain an excitation with spin 1. Recalling that in the ground state we have N_{GS} electrons out of which $M_1 = N_{GS}/2$ have spin down, the excited state will have occupation numbers $\mathcal{M}_e = N_{GS}$ and $M_1 = N_{GS}/2 - 1$. The selection rules (4.44) then read

$$-\frac{L}{2} < I_j \le \frac{L}{2}\,, \qquad |J_\alpha^1| \le \frac{N_{GS}}{4}\,. \tag{7.183}$$

The first condition is irrelevant as we are below half filling, but the second one tells us that there are two more vacancies than there are roots. In other words, flipping one spin leads to *two* holes in the distribution of Λ^1's. There is one more subtlety we have to take care of: changing the number of down spins by one, while keeping the number of electrons fixed leads to a shift of all I_j in (4.36) by either $\frac{1}{2}$ or $-\frac{1}{2}$. In other words the I_j's are now integers whereas they were half-odd integers for the ground state. The consequence of this shift is a constant contribution of $\pm \pi \frac{N_{GS}}{L}$ to the momentum of the excited state as can be seen from (5.97). This leads to two 'branches' of the same excitation. The possible distributions for the I_j's in the spin triplet excitation are shown in figure 7.30.

Taking this into account we obtain a gapless two-spinon scattering state with energy and momentum

$$E_{\text{trip}} = -\varepsilon_1(\Lambda_1^h) - \varepsilon_1(\Lambda_2^h)\,,$$
$$P_{\text{trip}} = -p_1(\Lambda_1^h) - p_1(\Lambda_2^h) \pm \pi n_c\,. \tag{7.184}$$

In figures 7.31 and 7.32 we show the spin-triplet spectrum for densities $n_c = 0.6$ and $n_c = 0.8$. In figure 7.31 we employ an extended zone scheme, which shows that the general

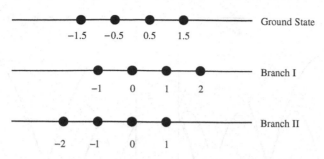

Fig. 7.30. Distribution of I_j in the ground state and for the two branches of the spin triplet excitation.

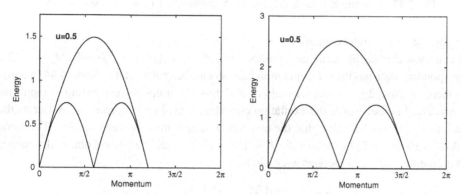

Fig. 7.31. Spin-triplet excitation for $u = 0.5$ and densities $n_c = 0.6$ and $n_c = 0.8$. Shown are the lower and upper boundaries of the continuum for the positive branch. The negative branch is obtained by reversing the sign of the momentum. Note that we have not folded back to the first Brillouin zone.

form of the continuum stays the same as the density changes. In figure 7.32 we fold back to the first Brillouin zone. The zero energy modes in the spin-triplet spectrum occur at wave numbers 0, $\pm \pi n_c$ and $\pm 2\pi n_c \bmod (2\pi)$.

In the Hubbard model the spin-triplet excitations were first studied by A. A. Ovchinnikov [352] and by C. F. Coll [88]. The situation encountered here is similar to the spin-$\frac{1}{2}$ Heisenberg chain [132] in the sense that the spin-triplet excitation is a scattering continuum of two spin-$\frac{1}{2}$ objects. Furthermore there is a spin-singlet excitation, which is precisely degenerate with the triplet (see Example 2b below). This fits nicely into a picture based on spin-$\frac{1}{2}$ objects: scattering states of two spinons give precisely one spin 1 and one spin 0 multiplet $\frac{1}{2} \otimes \frac{1}{2} = 1 \oplus 0$. Finally, when we approach half filling, the spin-triplet continuum goes over into the $S = 1$ two-spinon scattering continuum of the half-filled Hubbard model discussed previously in subsection 7.2.4.

On the other hand there are differences as well: in the less than half-filled Hubbard model the Fermi momentum is generally incommensurate, which leads to incommensurabilities in the spin excitations (see figure 7.31) as is apparent from figures 7.31–7.32.

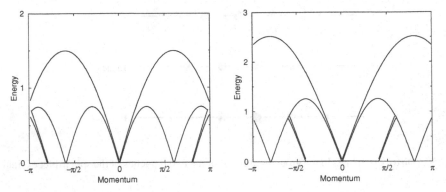

Fig. 7.32. Spin-triplet excitation for $u = 0.5$ and densities $n_c = 0.6$ and $n_c = 0.8$.

Example 2b Spin singlet excitation.

Let us now choose the occupation numbers as $\mathcal{M}_e = N_{GS}$, $M_1 = \frac{N_{GS}}{2} - 2$, $M_2 = 1$. The corresponding state has the same quantum numbers as the ground state. From (4.44) we find that there are $N_{GS}/2$ vacancies for real Λ's and thus two holes corresponding to rapidities Λ_1^h and Λ_2^h. In other words the excitation considered involves two spinons. As far as the 2-string is concerned we find that the associated integer must be zero $J_1^2 = 0$. The same shift as in Example 2a occurs for the I_j's. Using (5.103) and (5.97) we obtain the energy and momentum of the associated excitation

$$E_{\text{sing}} = -\varepsilon_1(\Lambda_1^h) - \varepsilon_1(\Lambda_2^h) \,,$$
$$P_{\text{sing}} = -p_1(\Lambda_1^h) - p_1(\Lambda_2^h) \pm \pi n_c \,. \tag{7.185}$$

We see that the spin singlet is precisely degenerate with the spin triplet considered above. This indicates a symmetry beyond simple spin rotational invariance: it is a consequence of the $sl(2)$ Yangian symmetry of the Hamiltonian in zero magnetic field in the infinite volume (see Chapter 14).

7.7.2 Charged excitations

Example 3 'Antiholon-Spinon' excitation.

Let us now consider an excitation with the quantum numbers of an electron, i.e. spin $\pm \frac{1}{2}$ and charge $-e$ with respect to the ground state. We choose $\mathcal{M}_e = N_{GS} + 1$, $M_1 = \frac{N_{GS}}{2}$. It follows from (4.40), (4.41) and (4.45) that I_j are half-odd integers and J_α^1 are half-odd integers with range

$$|J_\alpha^1| \leq \frac{N_{GS}}{4}. \tag{7.186}$$

Compared to the ground state there is one extra spectral parameter k and therefore the excitation involves one antiholon. Equation (7.186) implies that there are $\frac{N_{GS}}{2} + 1$ vacancies

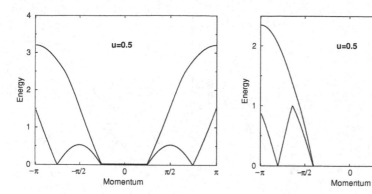

Fig. 7.33. Antiholon-spinon excitation for $u = 0.5$ and densities $n_c = 0.5$ and $n_c = 0.8$. Shown are the lower and upper boundaries of the continuum

and thus one hole in the distribution of the spectral parameters Λ_α^1. Let us denote the spectral parameter of this hole, which corresponds to a spinon, by Λ^h. Energy and momentum of the antiholon-spinon excitation are thus

$$E_{\bar{h}s} = \kappa(k) - \varepsilon_1(\Lambda^h)\,,$$
$$P_{\bar{h}s} = p(k) - p_1(\Lambda^h)\,. \tag{7.187}$$

Soft modes occur at momenta $\pm\frac{\pi n_c}{2}$, and $\pm\frac{3\pi n_c}{2}\bmod(2\pi)$. The excitation we have constructed has spin $S^z = \frac{1}{2}$ and is a highest-weight state of the spin-SU(2) symmetry. The $S^z = -\frac{1}{2}$ excitation is obtained by acting with the spin lowering operator S^- and is degenerate with the $S^z = \frac{1}{2}$ state. By construction the support of this excitation disappears as we approach half filling as the range in momentum space for the k-particle vanishes. The antiholon-spinon excitation was analyzed by H. J. Schulz in [384] by means of a numerical solution of the Bethe ansatz equations on a lattice with 40 sites.

Example 4 'Holon-Spinon' excitation.

Similarly we can construct an excitation where one electron is removed from the ground state. We choose $\mathcal{M}_e = N_{GS} - 1$, $M_1 = \frac{N_{GS}}{2} - 1$. It follows from (4.40), (4.41) and (4.45) that I_j are integers and J_α^1 are integers with range

$$|J_\alpha^1| \le \frac{N_{GS} - 2}{4}\,. \tag{7.188}$$

Compared to the ground state there is one less spectral parameter k and therefore the excitation involves one holon with corresponding spectral parameter k^h. Like for the spin-triplet the I_j's are integers so that there is a contribution $\pm\frac{N_{GS}}{L}$ to the total momentum. Equation (7.188) implies that there are $\frac{N_{GS}}{2}$ vacancies and thus one hole in the distribution of the spectral parameters Λ_α^1. Let us denote the spectral parameter of this hole, which

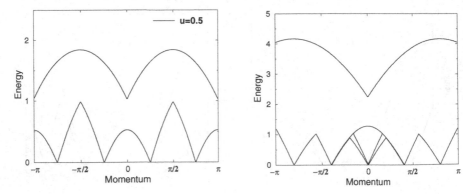

Fig. 7.34. Holon-spinon excitation for $u = 0.5$ and densities $n_c = 0.5$ and $n_c = 0.8$. Shown are the lower and upper boundaries of the continuum.

corresponds to a spinon, by Λ^h. Energy and momentum of the holon-spinon excitations are

$$E_{hs} = -\kappa(k^h) - \varepsilon_1(\Lambda^h) \,,$$
$$P_{hs} = -p(k^h) - p_1(\Lambda^h) \pm \pi n_c \,. \tag{7.189}$$

There are soft modes at momenta $\pm\frac{\pi n_c}{2}$, $\pm\frac{3\pi n_c}{2} \bmod (2\pi)$ and $\pm\frac{5\pi n_c}{2} \bmod (2\pi)$. We note that the 'hole' in the continuum around zero momentum is a result of folding back to the first Brillouin zone. At half filling this excitation reduces to the spin-charge scattering state of section 7.2. The holon-spinon excitation was analyzed by H. J. Schulz in [384] by means of a numerical solution of the Bethe ansatz equations on a lattice with 40 sites.

Example 5 k-Λ string of length 2.

    Last but not least let us consider the simplest excitation involving a k-Λ-string. One possibility is to choose the occupation numbers as $\mathcal{M}_e = N_{GS} - 2$, $M_1 = N_{GS}/2 - 1$, $M'_1 = 1$. This excitation has the same quantum numbers as the ground state. There are $\frac{N_{GS}}{2} - 1$ vacancies for an identical number of half-odd integers J^1_α and hence no holes in the distribution of Λ^1_α. We also keep the distribution of I_j fixed in such a way that $I_{j+1} - I_j = 1$. It is easily checked that this excitation is allowed by (4.44) and that its energy and momentum are

$$E_{k-\Lambda} = \varepsilon'_1(\Lambda) \,, \qquad P_{k-\Lambda} = p'_1(\Lambda) \,, \tag{7.190}$$

where $\Lambda \in (-\infty, \infty)$. In figure 7.35 the dispersion of a k-Λ string of length 2 is shown for $u = 0.125$ and $u = 0.5$ and several values of density n_c. We see that the range of momenta collapses to zero as we approach half filling. At the same time the dressed energy approaches zero. This is in agreement with the results for a half-filled band [121], where both the dressed energy and the range of momentum are identically zero.

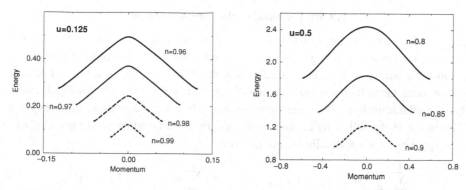

Fig. 7.35. Dispersion of a k-Λ string excitation of length 2 for several values of u and density n_c.

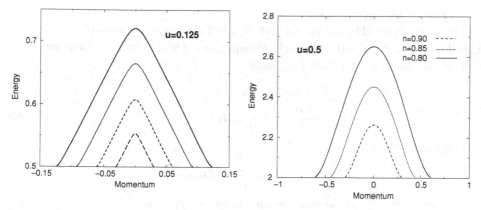

Fig. 7.36. Dispersion of a k-Λ string excitation of length 2 for several values of u and density n_c, where the contribution $-2\mu - 4u$ has been subtracted.

In order to further exhibit this collapse we subtract the offset $-2\mu - 4u$. The resulting curves are displayed in figure 7.36.

The 2-string excitation has the same spin and η^z quantum numbers as for example particle-hole excitations (Example 1). However, the 2-string dispersion does not overlap with the particle-hole continuum shown in figure 7.29 as it occurs at higher energies. Hence a decay of the 2-string into a particle-hole pair is not possible for kinematic reasons.

7.8 Finite magnetic field and less than half-filled band

The excitation spectrum below half filling in the presence of a magnetic field can be determined in an analogous way. Now both k-Λ strings and Λ-strings of arbitrary length become dynamical objects. Excitations involving such strings are however gapped for the reasons stated in Sections 7.6 and 7.7.

Some excited states have been considered by J. M. P. Carmelo *et al.* in [77].

7.9 Empty band in the infinite volume

The ground state of the empty band is given by the reference state of the Bethe ansatz. Therefore, excitations in a large, finite volume are obtained by solving the Bethe ansatz equations for one, two, three, etc. particles. The result is a very complicated set of many-electron states. Let us look at some examples. We stress that a part of the following discussion is based on the string hypothesis and therefore should be taken *cum grano salis* as explained in Chapter 4. In the following discussion we will drop the constant contribution of uL to the energy, i.e. discuss a Hamiltonian of the form

$$H = -\sum_{j=1}^{L} \sum_{a=\uparrow,\downarrow} (c_{j,a}^{\dagger} c_{j+1,a} + c_{j+1,a}^{\dagger} c_{j,a}) + 4u \sum_{j=1}^{L} n_{j\uparrow} n_{j\downarrow} - (2u + \mu)\hat{N}. \qquad (7.191)$$

Example 1 One-electron state.

This state is obtained by setting $N = \mathcal{M}_e = 1$ in Takahashi's equations (4.36)–(4.38) or equivalently $N = 1$, $M = 0$ in the Lieb-Wu equations (3.95), (3.96). By construction the electron has spin up and for finite L we have

$$E = -2\cos(k) - 2u - \mu, \quad P = k,$$
$$k = \frac{2\pi I_1}{L}, \quad -\frac{L}{2} < I_1 \le \frac{L}{2}. \qquad (7.192)$$

In the infinite-volume limit

$$L \to \infty, \quad \frac{I_1}{L} = \text{fixed}, \qquad (7.193)$$

one recovers a simple cosine-band. As discussed in Appendices 3.D and 3.F the one-electron state (let us denote it by $|k\rangle$) is a highest-weight state of the spin SU(2) and a lowest weight state of the η-pairing SU(2) symmetry of the Hubbard model. As a result descendant states are obtained by acting with symmetry operators. For example a down-spin electron excitation is obtained by acting with the spin-lowering operator, i.e. $S^-|k\rangle$. It has the same energy and momentum as $|k\rangle$. We must remember that $\mu \neq 0$ so that the Hamiltonian does not commute with η, but instead

$$[\eta^+, H] = 2\mu \, \eta^+. \qquad (7.194)$$

This implies that states of the form $(\eta^+)^n |k\rangle$ with $n \le L - 1$ have an energy $E - 2n\mu$.

Example 2 Two-electron spin triplet state.

Taking $\mathcal{M}_e = 2$ in Takahashi's equations and all other occupation numbers zero (or equivalently $N = 2$, $M = 0$ in the Lieb-Wu equations), we obtain a two-electron state with $S^z = 1$ and

$$E = -2\cos(k_1) - 2\cos(k_2) - 4u - 2\mu, \quad P = k_1 + k_2, \qquad (7.195)$$
$$k_j = \frac{2\pi I_j}{L}, \quad j = 1, 2, \quad -\frac{L}{2} < I_1 \neq I_2 \le \frac{L}{2}. \qquad (7.196)$$

In the infinite volume limit this turns into a two-particle scattering continuum of states with

$$E = -4 \cos\left(\frac{P}{2}\right) \cos\left(\frac{x}{2}\right) - 4u - 2\mu \,, \tag{7.197}$$

where both the total (P) and relative (x) momentum vary in the interval $[-\pi, \pi]$. It is again possible to construct further states by acting with the $SO(4)$ ladder operators.

Example 3 Two-electron spin singlet state.

The occupation numbers $\mathcal{M}_e = 2$, $M_1 = 1$ in Takahashi's equations give a state with $S^z = 0$. From the highest-weight theorem it follows that $\mathbf{S}^2 = 0$ so that we are dealing with a spin singlet. The expression for energy and momentum are the same as for the spin triplet (7.195), whereas the Bethe ansatz equations now read

$$\exp(ik_j L) = \frac{\Lambda - \sin(k_j) - iu}{\Lambda - \sin(k_j) + iu} \,, \quad j = 1, 2 \tag{7.198}$$

$$1 = \left[\frac{\Lambda - \sin(k_1) - iu}{\Lambda - \sin(k_1) + iu}\right]\left[\frac{\Lambda - \sin(k_2) - iu}{\Lambda - \sin(k_2) + iu}\right]. \tag{7.199}$$

Equation (7.199) is easily solved

$$\Lambda = \frac{1}{2}\left[\sin(k_1) + \sin(k_2)\right], \tag{7.200}$$

and inserting this back into (7.198) we obtain two coupled polynomial equations in $\exp(ik_j)$ of order $L + 1$

$$\exp(ik_1 L) = \exp(-ik_2 L) = \frac{\sin(k_1) - \sin(k_2) + 2iu}{\sin(k_1) - \sin(k_2) - 2iu} . \tag{7.201}$$

For fixed L these equations can only be solved numerically. However, simplifications occur in the infinite-volume limit. This is most easily seen by taking the logarithm of (7.201)

$$k_1 = -\frac{i}{L} \ln\left[\frac{\sin(k_1) - \sin(k_2) + 2iu}{\sin(k_1) - \sin(k_2) - 2iu}\right] + \frac{2\pi n_1}{L} . \tag{7.202}$$

A similar equation holds for k_2. In the infinite-volume limit $L \to \infty$, $\frac{n_j}{L} =$ fixed the first term in (7.202) drops out and we obtain the same result (7.197) for the energy as for the two-electron spin triplet state. Thus in the infinite-volume limit the singlet and triplet states are degenerate. This degeneracy, which is related to the Yangian symmetry (see Chapter 15), is broken in the finite volume.

Example 4 k-Λ-string of length 2.

Finally let us consider the case $M'_1 = 1$ and all other occupation numbers zero. The resulting excitation is a spin singlet two-electron state. Energy and momentum are

$$E = 4 \, \mathrm{Re}\sqrt{1 - (\Lambda' - iu)^2} - 4u - 2\mu \,,$$

$$P = -2 \, \mathrm{Re} \arcsin(\Lambda' - iu) \,, \tag{7.203}$$

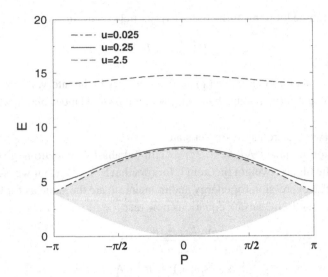

Fig. 7.37. Dispersion of a k-Λ string of length 2 for different values of u in the infinite-volume limit compared to the two-electron scattering continuum (shaded area). The chemical potential has been fixed as $\mu = -2 - 2u$.

where Λ' is a solution of

$$\exp\left(-2iL \, \mathrm{Re} \, \arcsin(\Lambda' - iu)\right) = \exp(iPL) = 1 \ . \tag{7.204}$$

In the infinite-volume limit we obtain the dispersion shown in figure 7.37 where we have fixed the chemical potential to lie at the boundary of the empty-band phase $\mu = -2 - 2u$. We see that the k-Λ string dispersion is always above the two-electron scattering continuum. For small u it sits just on top of the continuum whereas for large u it occurs at a very high energy. This is somewhat counter-intuitive as the k-Λ string describes a bound state of two electrons in the sense that its wave function decays exponentially with respect to the relative distance between the two electrons. As explained in Section 3.2.4 the fact that bound states lie above the scattering continua in energy is a lattice effect.

Appendices to Chapter 7

7.A Relating root-density and dressed-energy formalisms

In this appendix we discuss a method that relates the Yang-Yang approach of Sections 7.2 and 7.7 directly to the root-density approach we used in Section 7.3. This is particularly useful for the less than half-filled band, where the excitation energies are given implicitly in terms of the solutions of coupled integral equations that have to be solved numerically. As the integral equations for the root densities and the dressed energies are quite different, it is not immediately obvious that the expressions for the excitation energies obtained in the two approaches are equivalent. The method discussed below permits one to prove this equivalence. For definiteness we will concentrate on the particular example of the charge-singlet exitation at half filling in zero magnetic field, the method we use is however completely general. The integral equations for the root densities are (7.85), (7.86)

$$\rho_{CS}(k) = \cos(k) \int_{-\infty}^{\infty} d\Lambda \, a_1(\sin k - \Lambda) \, \sigma_{CS}(\Lambda) - \sum_{n=1}^{2} \delta(k - k_n^h)$$

$$+ \cos(k) \, a_1(\sin k - \Lambda'_1^1),$$

$$\sigma_{CS}(\Lambda) = \int_{-\pi}^{\pi} dk \, a_1(\Lambda - \sin k) \, \rho_{CS}(k) - \int_{-\infty}^{\infty} d\Lambda' \, a_2(\Lambda - \Lambda') \, \sigma_{CS}(\Lambda') \, .$$

$$\text{(7.A.1)}$$

We rewrite these equations using the matrix notation introduced in Section 6.6.3:

$$r_{CS,a}(x_a) = r_{CS,a}^{(0)}(x_a) + \sum_b \int_{-X_b}^{X_b} dx'_b \, K_{ab}(x_a, x'_b) \, r_{CS,b}(x'_b) \, . \qquad \text{(7.A.2)}$$

The solution of (7.A.2) is given by

$$r_{CS,a}(x_a) = \sum_b \int_{-X_b}^{X_b} dx'_b \, (1 - K)_{ab}^{-1}(x_a, x'_b) \, r_{CS,b}^{(0)}(x'_b)$$

$$= \sum_b \sum_{n=0}^{\infty} \int_{-X_b}^{X_b} dx'_b \, K_{ab}^n(x_a, x'_b) \, r_{CS,b}^{(0)}(x'_b) \, , \qquad \text{(7.A.3)}$$

265

where $K^n(x, y)$ is the n-fold convolution of the matrix kernel K_{ab}

$$
K_{ab}^{n+1}(x_a, x_b') = \sum_{c_1,\dots,c_n} \int_{-X_{c_1}}^{X_{c_1}} dx_{c_1}^{(1)} \dots \int_{-X_{c_n}}^{X_{c_n}} dx_{c_n}^{(n)} \, K_{ac_1}(x_a, x_{c_1}^{(1)})
$$
$$
\times K_{c_1 c_2}(x_{c_1}^{(1)}, x_{c_2}^{(2)}) \dots K_{c_n b}(x_{c_n}^{(n)}, x_b') . \tag{7.A.4}
$$

In the matrix notation the integral equations for the dressed energies take the form

$$
e_a(x_a) = e_a^{(0)}(x_a) + \sum_b \int_{-X_b}^{X_b} dx_b' \, K_{ab}^T(x_a, x_b') \, e_b(x_b') , \tag{7.A.5}
$$

and are formally solved by

$$
e_a(x_a) = \int_{-X_b}^{X_b} dx_b' \, (1 - K^T)_{ab}^{-1}(x_a, x_b') \, e_b^{(0)}(x_b')
$$
$$
= \sum_{n=0}^{\infty} \sum_b \int_{-X_b}^{X_b} dx_b' \, (K^T)_{ab}^n(x_a, x_b') \, e_b^{(0)}(x_b')
$$
$$
= \sum_{n=0}^{\infty} \sum_b \int_{-X_b}^{X_b} dx_b' \, e_b^{(0)}(x_b') \, K_{ba}^n(x_b', x_a) , \tag{7.A.6}
$$

where the last line is readily proved by induction over n. The energy of the charge-singlet excitation in the framework of the root density formalism is given by equation (7.93), which can be written as

$$
E_{\text{CS}} = 4\text{Re}\sqrt{1 - (\Lambda_1'^1 - iu)^2} - 4u + \sum_a \int_{-X_a}^{X_a} dx_a \, e_a^{(0)}(x_a) \, r_{\text{CS},a}(x_a)
$$
$$
= 4\text{Re}\sqrt{1 - (\Lambda_1'^1 - iu)^2} - 4u + \sum_a \int_{-X_a}^{X_a} dx_a \, e_a(x_a) \, r_{\text{CS},a}^{(0)}(x_a) , \tag{7.A.7}
$$

where the second line follows from (7.A.3) and (7.A.6). Using the explicit form for the driving terms $r_{\text{CS},a}^{(0)}(x_a)$ we obtain

$$
E_{\text{CS}} = \left[4\text{Re}\sqrt{1 - (\Lambda_1'^1 - iu)^2} - 4u \right] - \sum_{j=1}^{2} \kappa(k_j^h)
$$
$$
+ \int_{-\pi}^{\pi} dk \, \kappa(k) \, \cos(k) \, a_1(\sin k - \Lambda_1^1). \tag{7.A.8}
$$

Finally, using the integral equation (5.103) for $\kappa(k)$ and the 'symmetric integration Lemma' (17.1) one can rewrite the last term in (7.A.8) as

$$
- \int_{-\pi}^{\pi} dk \, 2\cos^2(k) \, a_1(\sin k - \Lambda_1^1), \tag{7.A.9}
$$

which precisely cancels the first term in (7.A.8) by virtue of (17.13). This proves that the energy calculated in the root density formalism is the same as the one calculated using dressed energies. The crucial point is that we did not need an explicit solution of the integral equations in our derivation.

7.B Lower bounds for $\varepsilon_n(0)$, $n \geq 2$ at half filling in a finite magnetic field

In this appendix we derive lower bounds for the dressed energies $\varepsilon_n(\Lambda)$ for $n \geq 2$ at half filling and in presence of a nonzero magnetic field B. Our starting point are equations (7.157)

$$\varepsilon_n(\Lambda) = 2nB + 4nu - 4\mathrm{Re}\sqrt{1 - (\Lambda - inu)^2}$$

$$- \int_{-A}^{A} d\Lambda' \left[a_{n-1}(\Lambda - \Lambda') + a_{n+1}(\Lambda - \Lambda') \right] \varepsilon_1(\Lambda') . \tag{7.B.1}$$

The functions $\varepsilon_n(\Lambda)$ are monotonically increasing for positive Λ and symmetric around $\Lambda = 0$ and hence

$$\varepsilon_n(\Lambda) \geq \varepsilon_n(0) . \tag{7.B.2}$$

On the other hand, $\varepsilon_1(\Lambda) \geq 0$ for all $|\Lambda| > A$, which implies that

$$\varepsilon_n(0) \geq 2nB + 4nu - 4\mathrm{Re}\sqrt{1 + n^2u^2} - \int_{-\infty}^{\infty} d\Lambda\, a_{n-1}(\Lambda)\, \varepsilon_1(\Lambda)$$

$$- \int_{-A}^{A} d\Lambda a_{n+1}(\Lambda)\, \varepsilon_1(\Lambda) . \tag{7.B.3}$$

Using the integral equation (7.157) for $\varepsilon_1(\Lambda)$ we may express the first integral on the r.h.s. as

$$\int_{-\infty}^{\infty} d\Lambda\, a_{n-1}(\Lambda)\, \varepsilon_1(\Lambda) = 2B + 4nu - 4\mathrm{Re}\sqrt{1 + n^2u^2}$$

$$- \int_{-A}^{A} d\Lambda\, a_{n+1}(\Lambda)\, \varepsilon_1(\Lambda) . \tag{7.B.4}$$

Combining equations (7.B.3) and (7.B.4) we obtain a simple lower bound for the dressed energies of Λ-strings of length $n \geq 2$

$$\varepsilon_n(0) \geq 2(n - 1)B , \quad n \geq 2 . \tag{7.B.5}$$

8

Finite size corrections at zero temperature

In this chapter we want to refine the analysis of the ground state and low-lying excitations of the Hubbard model in the phases with gapless modes, i.e. phases II, IV and V discussed in Chapters 6, 7, by taking account of corrections which are important when considering Hubbard chains of finite length L. For the generic case, i.e. away from half-filling in a magnetic field, the finite-size corrections to the spectrum of the Hubbard model have been calculated by F. Woynarovich [487]. These results are the basis for our discussion in the following Chapter 9 of the asymptotic behaviour of correlation functions within the conformal approach [6,51,62,75] and thereby will allow us to make contact with Haldane's Luttinger liquid approach for the description of one-dimensional strongly correlated electron systems [189–192]

8.1 Generic case – the repulsive Hubbard model in a magnetic field

To investigate how the thermodynamic limit is approached we have to take into account finite-size corrections in our previous derivation of integral equations from Takahashi's equations. This analysis has to be performed separately for each of the phases with gapless excitations identified before. From a technical point of view the most complex situation is found in phase IV – the partially filled, partially magnetized band with two massless modes. The finite-size scaling behaviour in the phases with a single gapless mode can be studied using the same techniques and we will point out the differences to the 'generic' case studied in this section later in this chapter.

Since we are interested in the ground state and low-lying excitations we restrict ourselves to solutions of the Bethe Ansatz equations which have been identified as the most important ones for the low-energy sector of the repulsive Hubbard model at less than half-filling in a magnetic field (phase IV) before, namely those described in terms of finite densities of real roots k and λ only.[1]

8.1.1 Finite-size corrections to Takahashi's equations

From Takahashi's equations (4.36)–(4.38) we have obtained the defining equations (5.29), (5.30) for the counting functions $y(k)$ and $z_1(\lambda)$. These functions are well-defined objects

---

[1] This implies that we are considering states with $N = \mathcal{M}_e$ electrons and $M = M_1$ overturned spins in the following.

for any system size. Extending the definition (5.36), (5.37) of root densities to finite systems we can define finite size densities of the Bethe Ansatz roots (the notation $u^{(L)}$ for finite size quantities is used to avoid confusion with the corresponding quantities in the thermodynamic limit)

$$2\pi \left[\rho^{(L)}(k) + \rho^{h,(L)}(k)\right] = \frac{\partial y^{(L)}(k)}{\partial k}, \quad 2\pi \left[\sigma_1^{(L)}(\lambda) + \sigma_1^{h,(L)}(k)\right] = \frac{\partial z_1^{(L)}(\lambda)}{\partial \lambda}. \quad (8.1)$$

On the level of the Bethe Ansatz integral equations finite-size corrections for the densities (i.e. corrections to equations (5.93) of sub-leading order in $1/L$) are taken into account using the method introduced by de Vega and Woynarovich [470] (see also Refs. [63,64,145,216, 488]). It has been applied to the repulsive Hubbard model at half-filling by Woynarovich and Eckle [489] and away from half-filling by Woynarovich [487]. The finite-size corrections to the spectrum of the attractive Hubbard model have been studied in Ref. [65].

This method is based on application of the Euler-McLaurin summation formula to the discrete equations above (f' denotes the derivative of f)

$$\frac{1}{L} \sum_{n=n_1}^{n_2} f\left(\frac{n}{L}\right) = \int_{n_-/L}^{n_+/L} dx\, f(x) + \frac{1}{24L^2} \left\{ f'\left(\frac{n_-}{L}\right) - f'\left(\frac{n_+}{L}\right) \right\} + \cdots \quad (8.2)$$

where we have introduced $n_- = n_1 - \frac{1}{2}, n_+ = n_2 + \frac{1}{2}$.

To use this formula we restrict ourselves to a subset of Bethe Ansatz states which contains the zero temperature ground state and an important class of low-lying excitations: consider solutions to Takahashi's equations (4.36)–(4.38) for a set of consecutive quantum numbers

$$I_j = I_- + j - \frac{1}{2}, \quad j = 1, \ldots, N,$$

$$J_\alpha^1 = J_-^1 + \alpha - \frac{1}{2}, \quad \alpha = 1, \ldots, M. \quad (8.3)$$

For such states the finite size densities (8.1) are given by the following equations

$$\rho^{(L)}(k) = \frac{1}{2\pi} + \frac{\cos k}{L} \sum_{\alpha=1}^{M} a_1(\sin k - \lambda_\alpha), \quad (8.4)$$

$$\sigma_1^{(L)}(\lambda) = \frac{1}{L} \sum_{j=1}^{N} a_1(\lambda - \sin k_j) - \frac{1}{L} \sum_{\beta=1}^{M} a_2(\lambda - \lambda_\beta). \quad (8.5)$$

With $I_+ = I_- + N$ and $J_+^1 = J_-^1 + M$ we define the numbers $Q_\pm$ and $A_\pm$ for this state by

$$y^{(L)}(Q_\pm) = 2\pi I_\pm/L, \quad z_1^{(L)}(A_\pm) = 2\pi J_\pm^1/L. \quad (8.6)$$

Now the Euler-McLaurin formula (8.2) is directly applicable to equations (8.4), (8.5) resulting in a system of two coupled linear integral equations for $r^{(L)} = (r_c^{(L)}, r_s^{(L)}) \equiv (\rho^{(L)}, \sigma_1^{(L)})$ of the form (we denote by $\int_x$ integration over the intervals $[X_-, X_+]$ where

$X_{c,\pm} = Q_\pm$ and $X_{s,\pm} = A_\pm$, respectively)

$$x_c(k) = x_c^{(0)}(k) + \cos k \int_s d\lambda' \, a_1(\sin k - \lambda') \, x_s(\lambda') ,$$

$$x_s(\lambda) = x_s^{(0)}(\lambda) + \int_c dk' \, a_1(\lambda - \sin k') \, x_c(k') \tag{8.7}$$

$$- \int_s d\lambda' \, a_2(\lambda - \lambda') \, x_s(\lambda')$$

or, symbolically $(X = \{Q_\pm, A_\pm\})$

$$x(k, \lambda) = x^{(0)}(k, \lambda) + \mathbb{K}(k, \lambda; k', \lambda'|X) \otimes x(k', \lambda') \tag{8.8}$$

where $\mathbb{K}$ is the integral operator with kernel

$$K(k, \lambda; k', \lambda') = \begin{pmatrix} 0 & \cos k \, a_1(\sin k - \lambda') \\ a_1(\lambda - \sin k') & -a_2(\lambda - \lambda') \end{pmatrix}. \tag{8.9}$$

Compared to the integral equations (6.12), (6.13) for $r = (\rho, \sigma_1)$ obtained within the root density formalism in the limit $L \to \infty$ the only difference are the driving terms containing additional terms of order $(1/L^2)$. Using (8.1) to eliminate the counting functions the driving terms read

$$r_c^{(0)} = \rho^{(0)}(k) = \frac{1}{2\pi} + \frac{\cos k}{24L^2} \left\{ \frac{a_1'(\sin k - A_+)}{r_s^{(L)}(A_+)} - \frac{a_1'(\sin k - A_-)}{r_s^{(L)}(A_-)} \right\} + o\left(\frac{1}{L^2}\right), \tag{8.10}$$

$$r_s^{(0)} = \sigma_1^{(0)}(\lambda) = \frac{1}{24L^2} \left\{ \cos Q_+ \frac{a_1'(\lambda - \sin Q_+)}{r_c^{(L)}(Q_+)} - \cos Q_- \frac{a_1'(\lambda - \sin Q_-)}{r_c^{(L)}(Q_-)} \right.$$

$$\left. - \frac{a_2'(\lambda - A_+)}{r_s^{(L)}(A_+)} + \frac{a_2'(\lambda - A_-)}{r_s^{(L)}(A_-)} \right\} + o\left(\frac{1}{L^2}\right). \tag{8.11}$$

These expressions are implicit functions of X. Note further, that the solution $r^{(L)}$ of (8.7) at these integration boundaries enter the expressions for $r_c^{(0)}$ and $r_s^{(0)}$. From the linearity of the integral equations (8.7) it is clear that to order $1/L^2$ their solution for the finite-size density can be written as $(i \in \{c, s\})$

$$r_i^{(L)}(x) = r_i^{(\infty)}(x) + \frac{1}{24L^2} \sum_{k \in \{c,s\}} \left\{ \frac{f_{ik}^{(+)}(x)}{r_k^{(L)}(X_{k,+})} + \frac{f_{ik}^{(-)}(x)}{r_k^{(L)}(X_{k,-})} \right\} \tag{8.12}$$

with $r^{(\infty)}$ and $f_{\cdot k}^{(\pm)}$ being solutions of the integral equation (8.7) with driving terms replaced by

$$r^{(\infty)} : \quad x_c^{(0)}(k) = \frac{1}{2\pi}, \quad x_s^{(0)}(\lambda) = 0$$

$$f_{\cdot c}^{(\pm)} : \quad x_c^{(0)}(k) = 0, \quad x_s^{(0)}(\lambda) = \pm \cos Q_\pm a_1'(\lambda - \sin Q_\pm) \tag{8.13}$$

$$f_{\cdot s}^{(\pm)} : \quad x_c^{(0)}(k) = \pm \cos k \, \frac{a_1'(\sin k - A_\pm)}{2\pi}, \quad x_s^{(0)}(\lambda) = \mp a_2'(\lambda - A_\pm).$$

By construction the driving terms for the functions $f^{(\pm)}$ can be written as

$$\mp \frac{\partial}{\partial x'_k} K_{ik}(x;x') \bigg|_{x'=X_\pm} . \tag{8.14}$$

8.1.2 Finite-size corrections to the energy

To compute the finite-size corrections to the energies we have to perform similar manipulations with an appropriate expression for the energy of the Bethe Ansatz state corresponding to our choice (8.3) of quantum numbers. The expression used in Chapter 6 to study the ground state properties in the $L \to \infty$ limit relied on the thermodynamic approach and was based on the dressed energies $e(x) = (\kappa(k), \varepsilon_1(\lambda))$. These were given in terms of the integral equations (6.10) similar to (8.8), namely

$$e(k,\lambda) = e^{(0)}(k,\lambda) + \mathbb{K}^\top(k,\lambda;k',\lambda'|X) \otimes e(k',\lambda') . \tag{8.15}$$

The driving terms are the bare energies $e^{(0)}$

$$e^{(0)} = (e_c^{(0)}(k), e_s^{(0)}(\lambda)) = (-2\cos k - \mu - 2u - B, 2B) . \tag{8.16}$$

The kernel of the integral operator $\mathbb{K}^\top$ is the transpose of (8.9)

$$K^\top(k,\lambda;k',\lambda') = \begin{pmatrix} 0 & a_1(\sin k - \lambda') \\ \cos k'\, a_1(\lambda - \sin k') & -a_2(\lambda - \lambda') \end{pmatrix} . \tag{8.17}$$

The method outlined above for the densities, i.e. taking into account finite size corrections to the integral equations (8.15), is not applicable within this approach. Instead we apply the Euler-McLaurin formula (8.2) to the expression equation (3.97) for the energy of the Bethe Ansatz state within the Lieb-Wu approach and obtain

$$\begin{aligned} E &= Lu + \sum_{j=1}^N e_c^{(0)}(k_j) + \sum_{\alpha=1}^M e_s^{(0)}(\lambda_\alpha) \\ &= Lu + L \sum_{i\in\{c,s\}} \int_{J_i} dx\, e_i^{(0)}(x) r_i^{(L)}(x) \\ &\quad - \frac{1}{24L} \sum_{i\in\{c,s\}} \left\{ \frac{e_i^{(0)\prime}(X_{i,+})}{r_i^{(L)}(X_{i,+})} - \frac{e_i^{(0)\prime}(X_{i,-})}{r_i^{(L)}(X_{i,-})} \right\} + o\left(\frac{1}{L}\right) . \end{aligned} \tag{8.18}$$

Using (8.12) this becomes

$$
E = L\,u + L \sum_{i \in \{c,s\}} \int_{J_i} dx\, e_i^{(0)}(x) r_i^{(\infty)}(x)
$$

$$
- \frac{1}{24L} \sum_{i \in \{c,s\}} \left(\frac{e_i^{(0)'}(X_{i,+})}{r_i^{(L)}(X_{i,+})} + \sum_{k \in \{c,s\}} \int_{J_i} dx\, e_i^{(0)}(x) \frac{f_{ik}^{(+)}(x)}{r_k(X_{k,+})} \right.
$$

$$
\left. - \frac{e_i^{(0)'}(X_{i,-})}{r_i^{(L)}(X_{i,-})} + \sum_{k \in \{c,s\}} \int_{J_i} dx\, e_i^{(0)}(x) \frac{f_{ik}^{(-)}(x)}{r_k(X_{k,-})} \right). \tag{8.19}
$$

The first two terms in this expression $L\,u + L\,e_\infty(X)$ with

$$
e_\infty(X) = \sum_{i \in \{c,s\}} \int_{J_i} dx\, e_i^{(0)}(x) r_i^{(\infty)}(x) \tag{8.20}
$$

are simply L times the energy density for a state with given values of $Q_\pm$ and $A_\pm$ in the infinite system. Since we are mostly interested in the finite size scaling properties of the ground state and low-lying excitations we should expand (8.19) around the Bethe Ansatz state which minimizes these terms for given values of the chemical potential μ and magnetic field B. This is just the zero temperature ground state of the Hubbard model as determined in Chapter 6. This state is characterized by the vanishing of the dressed energies at the Fermi points $k = \pm Q$, $\lambda = \pm A$ (6.11). Using the formal solutions of the integral equations for the dressed energy and the functions $f_{ik}^{(\pm)}$ we find

$$
\pm e_i^{(0)'}(X_{i,\pm}) + \sum_{k \in \{c,s\}} \int_{J_i} dx\, e_k^{(0)}(x) f_{ki}^{(\pm)}(x)
$$

$$
= \pm e_i^{(0)'}(X_{i,\pm}) \mp \sum_{k,\ell \in \{c,s\}} \int_{J_i} dx\, e_k^{(0)}(x) \left[\left(\frac{1}{1 - \mathbb{K}} \right)_{k\ell} \times K_{e\ell}' \right](x; X_\pm)
$$

$$
= \pm e_i^{(0)'}(X_{i,\pm}) \mp \sum_{k \in \{c,s\}} \int_{J_i} dx\, e_k(x) K_{ki}'\left(x; \left[\left(\frac{1}{1 - K} \right)_{k\ell} \times K_{ei} \right] X_\pm \right) = \pm e_i'(X_{i,\pm}).
$$

$$
\tag{8.21}
$$

(K′ denotes the derivative with respect to the second argument of K.) With $X_\pm = \pm X_0$ and the resulting symmetry of the functions e_i and r_i in this case (8.19) can be rewritten as:

$$
E = L\,u + L e_\infty(X) - \frac{1}{12L} \sum_{i \in \{c,s\}} \frac{e_i'(X_{0,i})}{r_i^{(L)}(X_{0,i})}. \tag{8.22}
$$

To the order in the system size L considered here we can replace the densities $r_i^{(L)}$ in the last term by their values in the thermodynamic limit. Hence, we identify the expressions for the Fermi velocities v_i of the low-lying charge and magnetic excitations over the ground state (6.44), (6.45)

$$\left.\frac{e_i'(x)}{r_i^{(\infty)}(x)}\right|_{x=X_{0,i}} = 2\pi\, v_i \tag{8.23}$$

and obtain

$$E = L\,u + L e_\infty(X) - \frac{\pi}{6L}\,(v_c + v_s)\ . \tag{8.24}$$

The last term is the leading finite-size correction to the ground state energy of the one-dimensional repulsive Hubbard model in the generic case. As will be discussed later this result can be used to justify the field theoretical description of this model in terms of two free bosons for any non negative value of the coupling constant u.

To proceed we have to expand $e_\infty(X)$ up to second order in the variation of the Fermi points X. The linear terms in $\Delta X = (X \mp X_0)$ vanish as a consequence of $e_i(\pm X_{0,i}) = 0$. Variation to second order in ΔX gives (note that $e_{GS} = e_\infty(X_0)$ is the ground state energy per site of the system)

$$e_\infty(X) = e_{GS} + \pi \sum_{i\in\{c,s\}} v_i$$

$$\times \left(\left\{ r_i^{(\infty)}(X_{0,i})\left(X_{i,+} - X_{0,i}\right)\right\}^2 + \left\{ r_i^{(\infty)}(X_{0,i})\left(X_{i,-} + X_{0,i}\right)\right\}^2 \right). \tag{8.25}$$

8.1.3 The dressed charge matrix

To complete our analysis of the finite-size spectrum of the repulsive Hubbard chain in a magnetic field we have to express the variations ΔX in terms of the deviation of the particle number N and magnetization M and the asymmetry of the distribution of quantum numbers $2D_i = n_+^i + n_-^i$ from their ground state values. This is possible with the counting functions and their relation to the densities $r_i^{(L)}$ (8.1):

$$\frac{N_i}{L} = \int_i dx\, r_i^{(L)}(x)\ ,$$

$$2\frac{D_c}{L} = \left\{ \int_{-\pi}^{Q_-} - \int_{Q_+}^{\pi} \right\} dk\, r_c^{(L)}(k) - \frac{1}{\pi}\int_s d\lambda\, \theta\,(\lambda/u)\, r_s^{(L)}(\lambda),$$

$$2\frac{D_s}{L} = \left\{ \int_{-\infty}^{A_-} - \int_{A_+}^{\infty} \right\} dk\, r_s^{(L)}(k). \tag{8.26}$$

Since the variations ΔX enter equation (8.25) quadratically and we have to compute the finite size correction to the energy density to order $1/L$ it is sufficient to replace the densities in these expressions by their leading terms $r_i^{(\infty)}$. We now consider the total densities defined in (8.26) to be the independent variables determining the boundaries of integration $X_{j,\pm}$.

Taking derivatives of the first of these equations w.r.t. $n_k = N_k/L$ we obtain

$$
\delta_{ik} = \sum_j \left(\left\{ r_i^{(\infty)}(X_{i,+})\delta_{ij} + \int_i \mathrm{d}x \, \frac{\partial r_i^{(\infty)}}{\partial X_{j,+}} \right\} \frac{\partial X_{j,+}}{\partial n_k} \right.
$$
$$
\left. + \left\{ -r_i^{(\infty)}(X_{i,-})\delta_{ij} + \int_i \mathrm{d}x \, \frac{\partial r_i^{(\infty)}}{\partial X_{j,-}} \right\} \frac{\partial X_{j,-}}{\partial n_k} \right)
$$
$$
= \pm 2 \sum_j \left\{ \delta_{ij} + \int_i \mathrm{d}x \, g_{ij}(x) \right\} r_j^{(\infty)}(X_{0,j}) \frac{\partial X_{j,\pm}}{\partial n_k} \tag{8.27}
$$

where $g_{i.}$ are the solutions of (8.8) with driving term $K_{i.}(k, \lambda; Q, A)$ and we have again used the symmetries arising from the fact that $X_\pm = \pm X_0$ for the ground state. Equation (8.27) can be simplified further: using the formal solution of the integral equation for g_{ij} in terms of a von Neumann series, i.e.

$$
g_{ij}(x) = K_{ij}(x; X_0) + \sum_k \int_k \mathrm{d}x' \, K_{ik}(x; x') K_{kj}(x'; X_0) + \cdots \tag{8.28}
$$

one finds that $\delta_{ij} + \int_i \mathrm{d}x \, g_{ij}(x) = Z_{ij}$ with

$$
Z = \begin{pmatrix} \xi_{cc}(Q) & \xi_{cs}(A) \\ \xi_{sc}(Q) & \xi_{ss}(A) \end{pmatrix}, \tag{8.29}
$$

where ξ is the so-called dressed charge matrix, defined in terms of the integral equation (**1** is the 2×2 unit matrix)

$$
\xi = \mathbf{1} + \xi \otimes \mathbb{K}. \tag{8.30}
$$

In components these equations read

$$
\xi_{cc}(k) = 1 + \int_s \mathrm{d}\lambda' \, \xi_{cs}(\lambda') a_1(\lambda' - \sin k),
$$
$$
\xi_{cs}(\lambda) = \int_c \mathrm{d}k' \, \cos k' \xi_{cc}(k') a_1(\sin k' - \lambda) - \int_s \mathrm{d}\lambda' \, \xi_{cs}(\lambda') a_2(\lambda' - \lambda),
$$
$$
\xi_{sc}(k) = \int_s \mathrm{d}\lambda' \, \xi_{ss}(\lambda') a_1(\lambda' - \sin k), \tag{8.31}
$$
$$
\xi_{ss}(\lambda) = 1 + \int_c \mathrm{d}k' \, \cos k' \xi_{sc}(k') a_1(\sin k' - \lambda) - \int_s \mathrm{d}\lambda' \, \xi_{ss}(\lambda') a_2(\lambda' - \lambda).
$$

Hence, equation (8.27) gives the identity

$$
\delta_{ik} = \pm 2 Z_{ij} \, r_j^{(\infty)}(X_{0,j}) \frac{\partial X_{j,\pm}}{\partial n_k}. \tag{8.32}
$$

Similarly, we proceed with the second pair of equations (8.26) for $d_k = D_k/L$ to obtain $(Y = (\pi, \infty))$:

$$\delta_{ik} = \sum_j \left\{ \delta_{ij} + \left(\int_{-Y_i}^{-X_{0,i}} - \int_{X_{0,i}}^{Y_i} \right) dx\ g_{ij}(x) \right.$$
$$\left. - \frac{\delta_{ic}}{\pi} \left(\delta_{js}\theta(A/u) + \int_s dx\ \theta(\lambda/u) g_{sj}(x) \right) \right\} r_j^{(\infty)}(X_{0,j}) \frac{\partial X_{j,\pm}}{\partial d_k}.$$

To express the remaining integrals in this equation through the dressed charge matrix we consider the derivatives of ξ_{ij} with respect to its argument x_j. Comparison of the resulting integral equations for $z_{ij} = \partial \xi_{ij}/\partial x_i$ with the ones for the functions g_{ij} leads to the following identities:

$$z_{lj}(x) = \sum_k Z_{lk} \left(g_{jk}(x) - g_{jk}(-x) \right). \tag{8.33}$$

Integrating these equations from $X_{0,i}$ to Y_i gives

$$\xi_{ij}(Y_i) - Z_{ij} = \sum_k Z_{ik} \left(\int_{-Y_i}^{-X_{0,i}} - \int_{X_{0,i}}^{Y_i} \right) dx\ g_{jk}(x). \tag{8.34}$$

Now, using

$$\xi_{ij}(Y_i) = \delta_{ij} + \delta_{jc} \int_s d\lambda\ \xi_{is}(\lambda) a_1(\lambda)$$

$$= \delta_{ij} + \frac{\delta_{jc}}{\pi} \left(Z_{is}\theta(A/u) + \sum_k Z_{ik} \int_s d\lambda\ \theta(\lambda/u) g_{sk}(\lambda) \right)$$

we obtain

$$\delta_{ik} = \left(Z^\top \right)_{ij}^{-1} r_j^{(\infty)}(X_{0,j}) \frac{\partial X_{j,\pm}}{\partial d_k}. \tag{8.35}$$

With equations (8.32) and (8.35) we have obtained the Jacobian for the transformation from the variations ΔX to the deviation ΔN_i of the numbers of particles and overturned spins from their ground state values and the total currents D_i in the state. This leads to the main result of this section, namely the expression for the finite size correction to the energy of low lying excited states with given changes of particle numbers and/or currents:

$$\Delta E(\Delta \mathbf{N}, \mathbf{D}) = L\ (e_\infty(X) - e_{GS})$$
$$= \frac{2\pi}{L} \left[\frac{1}{4} \Delta \mathbf{N}^\top \left(Z^\top \right)^{-1} V Z^{-1} \Delta \mathbf{N} + \mathbf{D}^\top Z V Z^\top \mathbf{D} + \sum_{k \in \{c,s\}} v_k \left(N_k^+ + N_k^- \right) \right]$$
$$+ o\left(\frac{1}{L} \right). \tag{8.36}$$

Here $V = \text{diag}(v_c, v_s)$ is a 2×2 matrix with the Fermi velocities (8.23) on the diagonal. $N_i^\pm$ are non negative integers enumerating the number of particle-hole pairs in the vicinity of the Fermi points – these contributions follow from a simple extension of the arguments used above when holes are considered in the distribution (8.3) of quantum numbers near the points $k = \pm Q$ and $\lambda = \pm A$.

Simple counting gives an analogous expression for the momentum of the excited state: denoting the Fermi momenta (of electrons rather than the objects considered in the Bethe Ansatz) by $k_{F,\sigma}$

$$k_{F,\uparrow(\downarrow)} = \frac{1}{2}(\pi n_c \pm 2\pi m) \tag{8.37}$$

the momentum can be written as

$$\Delta P(\Delta \mathbf{N}, \mathbf{D}) = \frac{2\pi}{L}\left(\Delta \mathbf{N}^\top \cdot \mathbf{D} + \sum_{k\in\{c,s\}}\left(N_k^+ - N_k^-\right)\right) + 2D_c k_{F,\uparrow} + 2(D_c + D_s)k_{F,\downarrow} \tag{8.38}$$

As a consequence of the the constraints on the parity of the Bethe Ansatz quantum numbers n^c and n^s which appeared in the derivation of (3.109) there are similar conditions on the parities of the numbers $\Delta \mathbf{N}$ and $\mathbf{D}$ characterizing the excited states in (8.36) and (8.38): In the thermodynamic limit the ground state is the unique state with $\Delta N_c = \Delta N_s = D_c = D_s = 0$. For excited states, the vector $\Delta \mathbf{N}$ has integer components denoting the change in the number of electrons and down spins with respect to this ground state. Due to the constraints mentioned above the numbers D_c and D_s are integer or half-odd integer depending on the parities of ΔN_c and ΔN_s:

$$D_c = \frac{\Delta N_c + \Delta N_s}{2} \bmod 1 \,, \quad D_s = \frac{\Delta N_c}{2} \bmod 1 \,. \tag{8.39}$$

8.2 Special cases

The expressions derived in the previous section show that the complete spectrum of low-lying states in the Hubbard model can be parametrized in terms of just four numbers, namely the elements of the dressed charge matrix (8.29) (apart from the Fermi velocities v_c and v_s which are known from the discussion of the thermodynamic limit already). Given the parameters of the system, i.e. the interaction strength u and the chemical potential and magnetic field the entries of Z are easily computed numerically.

In the present section we want to discuss a few cases where the expressions derived above simplify significantly and even allow for an analytical calculation of some of the elements of Z.

8.2.1 Zero magnetic field

For vanishing magnetic field the ground state of the one-dimensional Hubbard model has zero magnetization and is a spin-$SU(2)$ singlet. As noted in the discussion of ground state properties (see Section 6.B) this leads to the fact that the distribution of spin rapidities λ

covers the real axis ($A = \infty$) which in turn allows to simplify the Bethe Ansatz integral equations for the root densities and dressed energies by Fourier transformation. Similarly, the coupled equations (8.31) for the dressed charge matrix reduce to a simple scalar one in this case [487].

Integrating out ξ_{cs} from the first two of the integral equations (8.31) we are left with a scalar equation for $\xi_{cc} \equiv \xi$ (the kernel $R(x)$ is defined in (5.104)):

$$\xi(k) = 1 + \int_{-Q}^{Q} dk' \, \cos k' \, R(\sin k - \sin k') \, \xi(k'). \tag{8.40}$$

The resulting driving term in the integral equation for ξ_{cs} vanishes as $1/A^2$, this leads to $Z_{cs} = 0$ for the corresponding entry in the dressed charge matrix. Proceeding analogously for the second pair of integral equations for the dressed charge matrix, one obtains $\xi_{sc}(k) = \xi(k)/2$. The remaining element of ξ is determined by the integral equation

$$\xi_{ss}(x) \sim 1 + o(A^{-2}) + \int_{-A}^{A} dx' \, a_2(x - x') \, \xi_{ss}(x'), \quad A \to \infty. \tag{8.41}$$

$Z_{ss} = \lim_{A \to \infty} \xi_{ss}(A)$ can be obtained using the Wiener-Hopf method giving the following expression for the dressed charge matrix in a vanishing magnetic field:

$$Z = \begin{pmatrix} \xi(Q) & 0 \\ \frac{1}{2}\xi(Q) & \frac{1}{2}\sqrt{2} \end{pmatrix}. \tag{8.42}$$

Further results can be obtained in the limit of strong and weak coupling, respectively. Rewriting the integral equation for ξ as

$$\xi(z) = 1 + \int_{-z_0}^{z_0} dz' \, \tilde{R}(z - z') \, \xi(z'), \tag{8.43}$$

$$\tilde{R}(z) = \frac{1}{2\pi} \int_0^{\infty} d\omega \, \frac{e^{-\omega}}{\cosh \omega} \cos \omega z,$$

where $z = \sin k/u$, it becomes clear that the entries of the matrix Z in (8.42) depend on $z_0 = \sin Q/u$ only. Now equation (8.43) can be solved in the limiting cases $u \gg \sin Q$ and $0 < u \ll \sin Q$:

For large $u/\sin Q$ the equation for ξ can be solved by iteration with the result

$$\xi(z_0) = 1 + \frac{\sin Q}{\pi u} \ln 2. \tag{8.44}$$

For any fixed coupling strength $\sin Q$ becomes small for small particle density ($Q \to 0$) and near half-filling ($Q \to \pi$).

Using the expansions from Section 6.B.3 we can write down the dressed charge at fixed u for small densities

$$\xi(z_0) = 1 + \frac{\ln 2}{u} n_c, \quad n_c \ll \min(1, u). \tag{8.45}$$

Similarly, close to half filling $n_c = 1 - \delta$ (δ is the doping) we can use the expansion (6.B.19) and obtain:

$$\xi(z_0) = 1 + \frac{\ln 2}{2\pi u \rho_0(\pi)} \delta, \qquad \delta \ll u \rho_0(\pi). \qquad (8.46)$$

We recall that from equation (6.B.14)

$$2\pi \rho_0(\pi) = 1 - \int_0^\infty d\omega \, \frac{e^{-u\omega}}{\cosh(u\omega)} J_0(\omega)$$

$$= 1 - 2 \sum_{k=1}^\infty \frac{(-1)^{k+1}}{\sqrt{1 + (2uk)^2}}. \qquad (8.47)$$

Note that, due to the essential singularity associated with the Mott transition at half filling the expansion (8.46) is useful for sufficiently large values of u only: for $u = 0.1$ we find $u\rho_0(\pi) \approx 1.5 \cdot 10^{-8}$.

For $u \gg 1$ the integral equations simplify and we obtain ξ for arbitrary values of the density explicitly

$$\xi(z_0) = 1 + \frac{\ln 2}{\pi u} \sin \pi n_c, \qquad u \gg 1. \qquad (8.48)$$

In Appendix 8.A we use a perturbative scheme based on the Wiener–Hopf method to solve equation (8.43) for small $u / \sin Q$ and obtain

$$\xi(z_0) = \sqrt{2} \left(1 - \frac{u}{2\pi \sin Q} \right). \qquad (8.49)$$

This value of $\xi(z_0)$ also governs the low-energy spectrum for small u close to half-filling (see the discussion above).

In summary we have shown that the dressed charge matrix for the one-dimensional Hubbard model below half-filling in zero magnetic field is of the form (8.42) for any value of the repulsive interaction $u \geq 0$. The remaining parameter $\xi(Q)$ varies in the interval

$$1 \leq \xi(z_0) \leq \sqrt{2}. \qquad (8.50)$$

A numerical solution of the Bethe Ansatz integral equations for different values of $\sin Q/u$ gives the lines of constant $\xi(Q)$ in the n_c-u-plane (see figure 8.1). We note that the finite size spectrum depends on the Fermi velocities v_c and v_s in addition to $\xi(Q)$.

8.2.2 Partially filled spin-polarized band

In phase II all electrons have spin up and excitations with magnetization less than $n_c/2$ have a gap. This implies that the low-lying states are some of the plane-wave eigenstates of the tight-binding Hamiltonian (2.12). The dispersion of the gapless charge excitations as found in Section 6.3 is just the bare energy $e_c^{(0)}(k)$ from (8.16) and the complete excitation spectrum is that of free lattice electrons with Fermi momentum $k_{F\uparrow} = \pi n_c$ and Fermi velocity $v_\uparrow = 2 \sin k_{F\uparrow}$.

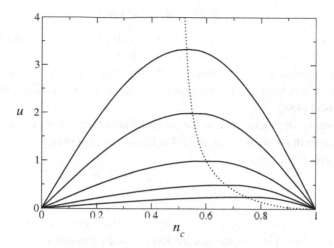

Fig. 8.1. Contours of constant dressed charge $\xi(Q)$ (equation (8.42)) in the n_c-u plane: $\xi(Q) \to 1(\sqrt{2})$ for $u \to \infty(0)$. The dotted line denotes the value of $n_c(Q = \pi/2)$ where the dressed charge takes its maximum value for given u. Note that the line $n_c = 1$ is excluded here. There the Hubbard model falls into a different universality class – that of the isotropic Heisenberg antiferromagnet (see Section 8.2.3).

Just as in our discussion of the excitations of the fully interacting model above a classification in terms of three different processes is possible:

- states with a charge that differs from the one in the ground state by $\Delta N_\uparrow$,
- states carrying a current $2D_\uparrow k_{F\uparrow}$ due to the transfer of $D_\uparrow$ electrons from one Fermi point to the other (backscattering),
- $N^\pm$ particle-hole pairs near one of the Fermi points at $\pm k_{F\uparrow}$,

and arbitrary combinations of the above.

Due to the fact that the electrons are effectively free in this phase there is no renormalization of the dispersion due to a reordering of the Fermi sea in presence of these excitations and the finite size spectrum is easily obtained after linearizing the dispersion in the vicinity of the Fermi points:

$$\varepsilon(k) \approx \pm v_\uparrow \left(k \mp k_{F\uparrow}\right) \quad \text{for } k \approx \pm k_{F\uparrow}. \tag{8.51}$$

Now it is straightforward to see that the complete spectrum of low-lying states is of the form

$$\Delta E(\Delta N_\uparrow, D_\uparrow) = \frac{2\pi v_\uparrow}{L} \left(\frac{\Delta N_\uparrow^2}{4} + D_\uparrow^2 + N^+ + N^-\right),$$

$$\Delta P(\Delta N_\uparrow, D_\uparrow) = \frac{2\pi}{L} \left(\Delta N_\uparrow D_\uparrow + N^+ - N^-\right) + 2D_\uparrow k_{F\uparrow}. \tag{8.52}$$

The different boundary conditions for states with even and odd number of electrons lead to the constraint $D_\uparrow = \Delta N_\uparrow/2 \mod 1$, similar to (8.39).

8.2.3 The half-filled band

In Chapter 7, we have shown that in phase V, i.e. at half filling $n_c = 1$, only charge neutral excitations are gapless and consequently accessible by the techniques used in the derivation of the finite size spectra. The contributions of the spin degrees of freedom to the finite-size scaling behaviour have been found to be equivalent to those of the one-dimensional isotropic Heisenberg model [488].

The finite size scaling behaviour of the ground state energy is obtained similarly as in Section 8.1.2 (note that only the partially filled Fermi sea of spin rapidities has to be taken into account) with the result

$$E = E_0 - \frac{\pi}{6L} v_s . \tag{8.53}$$

v_s is the Fermi velocity of the magnetic excitations. The finite size scaling corrections to the energies of the low lying excitations are given by the expression

$$\Delta E(\Delta M, D_s) = \frac{2\pi v_s}{L} \left[\frac{\Delta M^2}{4Z^2} + Z^2 D^2 + N_s^+ + N_s^- \right] + o\left(\frac{1}{L}\right). \tag{8.54}$$

Again the finite size corrections are determined by a dressed charge through $Z = \xi_s(A)$. The function ξ_s is the solution to the following integral equation:

$$\xi_s(\lambda) = 1 - \int_{-A}^{A} d\mu \, a_2(\lambda - \mu) \xi_s(\mu). \tag{8.55}$$

(This is in fact just the equation for ξ_{ss} in (8.31) for $Q = \pi$). Note that the finite spectrum (8.52) in the spin polarized phase is of the form (8.54) with $Z \equiv 1$. This corresponds to the use of the 'bare' charge $\xi \equiv 1$ rather than dressed charge and reflects the absence of interaction induced renormalization of the dispersion in phase II.

For given value of u the boundaries of the integrals in (8.55) depend on the magnetic field. The Fermi-sea of spin-waves disappears for $B \geq B_0 = 2(\sqrt{u^2 + 1} - u)$. This corresponds to the transition into phase III with half filled band and completely polarized spins as discussed in Section (6.3). As the magnetic field approaches B_0 from below we have

$$A = (u^2 + 1)^{3/4} \sqrt{B_0 - B}. \tag{8.56}$$

The magnetic excitations are gapless for $B \leq B_0$, with a finite-size spectrum given as a function of the dressed charge (8.55) at $\lambda = A$

$$\xi_s(A) = 1 - \frac{(u^2 + 1)^{3/4}}{\pi u} \sqrt{B_0 - B} \tag{8.57}$$

for magnetic fields close to B_0. $\xi_s(A)$ decreases as the magnetic field becomes smaller.

The case of vanishing magnetic field, $B = 0$, has first been studied by Woynarovich and Eckle [489]. It corresponds to $A \to \infty$, and the limiting value of $\xi_s(A)$ can be computed using the Wiener Hopf method (see Appendix 8.A) giving $\lim_{A \to \infty} \xi_s(A) = 1/\sqrt{2}$.

8.2.4 Strong coupling limit

To illustrate the magnetic field dependence of the finite size spectra for a partially filled band in more detail we now consider the limit $u \to \infty$ where the Bethe Ansatz integral equations simplify significantly. The strong influence of a finite magnetic field on the spectrum was first observed by Frahm and Korepin [141].

After elimination of ρ from the integral equation (6.13) for the density of spin rapidities we obtain

$$
\sigma_1(\lambda) = \frac{1}{2\pi} \int_c dk' a_1(\lambda - \sin k')
$$

$$
- \int_s d\lambda' \left\{ a_2(\lambda - \lambda') - \int_c dk \cos k \, a_1(\lambda - \sin k) a_1(\sin k - \lambda') \right\} \sigma_1(\lambda') .
\tag{8.58}
$$

Rescaling the variables as $z = \sin k / u$ and $\eta = \lambda / u$ the kernel of this integral operator becomes

$$
\frac{1}{u} \left(k_2(\eta - \eta') - \int_{-z_0}^{z_0} dz \, k_1(\eta - z) k_1(z - \eta') \right) ,
$$

$$
k_n(x) = \frac{1}{2\pi} \frac{2n}{x^2 + n^2} .
\tag{8.59}
$$

In the strong coupling limit, i.e. large u, we can neglect the second term and $\sigma_1(\lambda) \to u \sigma_1(\eta)$ is obtained from a simple scalar Fredholm integral equation ($\Lambda_0 = A/u$)

$$
\sigma_1(\eta) = \frac{Q}{2\pi^2} k_1(\eta) - \int_{-\Lambda_0}^{\Lambda_0} d\eta' \, k_2(\eta - \eta') \sigma_1(\eta') .
\tag{8.60}
$$

Using $n_c = \int_c dk \, \rho(k) = Q/\pi + \mathcal{O}(1/u)$ we can eliminate Q from this equation.

Along the same line of arguments we can eliminate the charge parts from the integral equations for the dressed energy (6.10) and the dressed charge (8.31). To leading order in $1/u$ the resulting equations read

$$
\varepsilon_1(\eta) = 2B - B_c k_1(\eta) - \int_{-\Lambda_0}^{\Lambda_0} d\eta' \, k_2(\eta - \eta') \varepsilon_1(\eta') ,
\tag{8.61}
$$

$$
\xi_{ss}(\eta) = 1 - \int_{-\Lambda_0}^{\Lambda_0} d\eta' \, k_2(\eta - \eta') \xi_{ss}(\eta') .
\tag{8.62}
$$

B_c is the magnetic field necessary for complete spin polarization of the ground state of the less than half-filled Hubbard model (see Section 6.3). At large coupling B_c scales like $1/u$ (6.25)

$$
B_c \simeq \frac{1}{u} \left(n_c - \frac{1}{2\pi} \sin 2\pi n_c \right) + \mathcal{O}(u^{-3}), \quad \text{for } u \gg 1 .
\tag{8.63}
$$

As a consequence of the vanishing of the effective exchange coupling as t^2/u in the strong coupling limit of the t-J model (2.A.39) an infinitesimal field is sufficient to polarize the system completely. To leading order in $1/u$, however, we already obtain nontrivial results on magnetic field effects on the finite size spectra.

The remaining entries of the dressed charge matrix (8.29) are

$$Z_{cc} = 1, \quad Z_{cs} = 0,$$

$$Z_{sc} = \int_{-\Lambda_0}^{\Lambda_0} d\eta \, k_1(\eta) \xi_{ss}(\eta) = \frac{1}{n_c} \int_{-\Lambda_0}^{\Lambda_0} d\eta \, \sigma_1(\eta) = \frac{1}{2} - \frac{m}{n_c} \tag{8.64}$$

(the identity for Z_{sc} can be verified in a similar way as above by comparing the formal solutions of (8.58), (8.62) in terms of the formal von Neumann series). Hence the general form of the dressed charge matrix in the strong coupling limit is

$$Z = \begin{pmatrix} 1 & 0 \\ Z_{sc} & Z_{ss} \end{pmatrix}. \tag{8.65}$$

The B-dependence of the finite-size spectrum is in the lower two elements of Z alone.

For vanishing magnetic field $\Lambda_0 = \infty$ and the integral equations can be solved as in Section 8.2.1. The result is

$$Z = \frac{1}{2} \begin{pmatrix} 2 & 0 \\ 1 & \sqrt{2} \end{pmatrix}. \tag{8.66}$$

For small magnetic field with Λ_0 large but finite we can apply the Wiener-Hopf method and obtain the leading magnetic field dependence of the elements of Z. From the identities derived in Appendix 8.A.2 we obtain with $B_1 = B_c\sqrt{\pi^3/2e}$

$$Z_{ss} = \frac{1}{\sqrt{2}} \left(1 + \frac{1}{4\ln(B_1/B)} \right) + \mathcal{O}\left(\frac{1}{(\ln(B_1/B))^2} \right).$$

$$Z_{sc} = \frac{1}{2} - \frac{2}{\pi^2} \frac{B}{B_c} + \mathcal{O}\left(\frac{B}{B_c \ln(B_1/B)} \right). \tag{8.67}$$

Note the different functional dependence on B of Z_{ss} and Z_{sc} near $B = 0$!

In the other limit, as B approaches the saturation field (8.63) from below, i.e. near the ferromagnetic state, the boundary of integration vanishes like

$$\Lambda_0 = \sqrt{\frac{B_c - B}{B_c}}. \tag{8.68}$$

In this regime the integral equations can be solved by iteration and the charge matrix in this regime is given by

$$Z = \begin{pmatrix} 1 & 0 \\ 0 & 1 \end{pmatrix} + \frac{1}{\pi} \sqrt{\frac{B_c - B}{B_c}} \begin{pmatrix} 0 & 0 \\ 2 & -1 \end{pmatrix}. \tag{8.69}$$

In figure 8.2 we present numerical results obtained from equations (8.62) and (8.64) for the magnetic field dependence for the entire range of fields $0 \leq B < B_c$ where the electrons become spin-polarized.

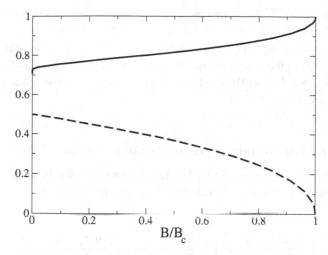

Fig. 8.2. Magnetic field dependence of the elements of the dressed charge matrix in the strong coupling limit: shown are Z_{ss} (solid curve) and Z_{sc} (dashed curve) as a function of the magnetic field in units of the field necessary for complete polarization of the system B_c. Note the different behaviour of the two functions in the limit of small B: while Z_{sc} is linear in B, Z_{ss} has a logarithmic singularity.

8.3 Finite-size spectrum of the open Hubbard chain

An important extension of the cases considered so far is that of the Hubbard chain with open rather than periodic boundary conditions. We shall focus in this section on the case of the repulsive Hubbard model below half-filling with additional local potentials coupling to the number of electrons on the boundary sites, i.e.

$$H = -\sum_{j=1}^{L-1}\sum_{\sigma}\left(c_{j\sigma}^{\dagger}c_{j+1,\sigma} + h.c.\right) + 4u\sum_{j=1}^{L}n_{j\uparrow}n_{j\downarrow} - (\mu + 2u)\hat{N} - 2BS^z$$
$$-p(n_{1\uparrow} + n_{1\downarrow}) - p'(n_{L\uparrow} + n_{L\downarrow})$$
(8.70)

It was shown first by H. Schulz [380] that the Hubbard model continues to be solvable by means of the Bethe Ansatz in the case of reflecting boundary conditions ($p = 0 = p'$). Subsequently the solution was extended to non-vanishing boundary potentials [24,49,507]. By means of the Shiba transformation (2.59) it is clear that the open-boundary Hubbard model with a magnetic field instead of the potential on the boundary site

$$-2bS_1^z - 2b'S_L^z$$
(8.71)

is also solvable [402, 403, 507]. Note that the boundary fields at the two ends of the chain can be chosen independently, hence it is possible to combine a potential p on site 1 with a boundary magnetic field b' on site L or vice versa. In general, the scattering due to the reflection at an end of the chain and that between the particles in the bulk has to satisfy so-called reflection equations to be compatible with integrability of the system [406]. For the open Hubbard chain and the boundary conditions discussed above this has been established in Ref. [507] (see also [457]). Within this approach one can rule out a solvable combination of potential and boundary magnetic field at the same end of the chain or – as far as is known to date – more complicated boundary conditions: while an integrable Hamiltonian for a Kondo spin coupled to the boundary site of the supersymmetric $t–J$ model is known [144] a similar construction for the Hubbard model still requires a deeper understanding of its algebraic structure.

8.3.1 Bethe Ansatz equations for the open Hubbard chain

For the derivation of the resulting Bethe Ansatz equations we refer the reader to the original literature, here we just state the equivalent of the Lieb-Wu equations (3.95), (3.96) for the Hamiltonian (8.70)

$$e^{i2k_j L} \frac{e^{ik_j} - p}{1 - pe^{ik_j}} \frac{e^{ik_j} - p'}{1 - p'e^{ik_j}} = \prod_{\ell=1}^{M} \frac{\sin k_j - \lambda_\ell + iu}{\sin k_j - \lambda_\ell - iu} \frac{\sin k_j + \lambda_\ell + iu}{\sin k_j + \lambda_\ell - iu} ,$$

$$j = 1, \ldots, N , \tag{8.72}$$

$$\prod_{j=1}^{N} \frac{\lambda_\ell - \sin k_j + iu}{\lambda_\ell - \sin k_j - iu} \frac{\lambda_\ell + \sin k_j + iu}{\lambda_\ell + \sin k_j - iu} = \prod_{\substack{m=1 \\ m \neq \ell}}^{M} \frac{\lambda_\ell - \lambda_m + 2iu}{\lambda_\ell - \lambda_m - 2iu} \frac{\lambda_\ell + \lambda_m + 2iu}{\lambda_\ell + \lambda_m - 2iu} ,$$

$$\ell = 1, \ldots, M .$$

Compared to the Lieb-Wu equations for the periodic chain one observes two differences: first, the geometric phase factor for the electrons is $\exp(i2kL)$ rather than $\exp(ikL)$ and the number of two-particle scattering phases on the r.h.s. of the equations is doubled: for a full path "around" the chain a particle has to be moved from its original position to site L, then to site 1 and back. Second, there are additional phases in the first set of these equations. This is the only effect of the boundary potentials p and p'.

The energy of the eigenstate of (8.70) corresponding to a solution of (8.72) is

$$E = -\sum_{j=1}^{N} \left(2\cos k_j + \mu + 2u + B\right) + 2BM . \tag{8.73}$$

There is no explicit dependence on the boundary conditions. It is present however through the dependence of the quasi momenta k_j on p, p' determined from equations (8.72).

In the thermodynamic limit $L \to \infty$ with both the particle density and magnetization kept fixed we expect that the energy density of the ground state is e_{GS} independent of the boundary conditions. The effect of the boundary potential will manifest itself in an order L^0 contribution to the energy. In addition, we expect a different finite-size scaling behaviour of the ground state energy (8.24) and the low-lying excited states (8.36).

From a physical point of view it is clear that the ground state of the model (8.70) should have up to two electrons bound to the first site for sufficiently large p (and similar to site L for sufficiently large p'). Numerical solution of the Bethe Ansatz equations (8.72) shows that this is indeed the fact: the presence of the bound states is indicated by complex quasimomenta k and rapidities λ in the configuration corresponding to the ground state for $p > 1$ [49,458].[2] These complex roots are not present in the ground state configuration for $p < 1$ [24] and are very different from the $k - \Lambda$-strings (4.13) discussed in Chapter 4.1. While p-dependent complex combinations of two k and one λ exist for any p, they are found to coincide with an η-pair in the $(SO(4)$-invariant) limit $p \to 0$. Hence, they correspond to highly excited states of the system and need not to be considered in the present context.

By careful analysis of the Bethe Ansatz equations one identifies four regions for p which have to be studied separately for the ground state [24,49]:

$p < 1$: The classification of solutions to the BAE is identical to that for the Hubbard chain with periodic boundary conditions. No additional types of complex rapidities exist.

$1 < p < p_1 = u + \sqrt{1 + u^2}$: In the ground state configuration one of the quasimomenta, say k_N, takes the value $k_N = i \ln(p)$ or, equivalently, $\sin k_N = i\,t$ where

$$ t = \frac{1}{2} \left(p - \frac{1}{p} \right) . \tag{8.74} $$

This value is realized with exponential accuracy even for systems of finite length L. Since the quasimomenta k_j parametrize the charge part of the states this solution may be interpreted as a single charge bound by the potential on site 1.

$p_1 < p < p_2 = 2u + \sqrt{1 + 4u^2}$: Upon further increasing the boundary potential one finds that the solution corresponding to the ground state contains a complex spin rapidity $\lambda_M = \sin k_N - iu = i(t - u)$ in addition to k_N. Just as the solution in region $1 < p < p_1$ this state may be interpreted as a charge bound to the edge of the chain. The *physical* excitations in the spin-sector (i.e. holes in the distribution of λ_ℓ) are still real.

$p_2 < p$: For boundary potentials larger than the Hubbard interaction $p \gtrsim 4u$ a pair of electrons forming a singlet is bound to the surface, parametrized by $\lambda_M = \sin k_N - iu = \sin k_{N-1} + iu = i(t - u)$.

Similarly, complex roots to the Bethe Ansatz equations appear as p' is varied. It can be shown however, that no additional complex solutions are possible as p or p' are increased beyond p_2 – in perfect agreement with the physical intuition.

Having characterized the configuration of Bethe Ansatz roots corresponding to the ground state of the open Hubbard chain we can proceed as in Section 8.1 to compute the finite size corrections to the energies of the ground state and the continua of low lying excitations.

Note that one may also consider eigenstates of the model where the boundary bound states discussed above are unoccupied, i.e. the corresponding complex quasi momenta or rapidities are allowed but not present in the solution considered. Above these highly excited states there arise different continua of states that can be treated in an analogous way. These

---

[2] This is similar to the situation observed in the XXZ Heisenberg chain with a boundary magnetic field [234,412] and a continuum model related to the Kondo problem [180,477].

continua have to be considered for the description of Fermi edge singularities in the presence
of boundary bound states [117].

8.3.2 Surface energy of the open Hubbard chain

In the following we restrict ourselves to the discussion of a boundary potential at site 1 of
the chain only, i.e. $p' = 0$. It is straightforward to extend the discussion to the general case,
however. For the analysis of the equations (8.72) it is convenient to double the number of
variables by identifying of $k_{-j} = -k_j$ and $\lambda_{-\ell} = -\lambda_\ell$ and setting $k_0 = 0 = \lambda_0$. After this
"symmetrization", the Bethe Ansatz equations read

$$
e^{ik_j(2L+1)} \frac{e^{ik_j} - p}{1 - pe^{ik_j}} \frac{\sin k_j + iu}{\sin k_j - iu} = \prod_{\ell=-M}^{M} \frac{\sin k_j - \lambda_\ell + iu}{\sin k_j - \lambda_\ell - iu} , \quad j = -N, \dots, N ,
$$

$$
\frac{\lambda_\ell + 2iu}{\lambda_\ell - 2iu} \prod_{j=-N}^{N} \frac{\lambda_\ell - \sin k_j + iu}{\lambda_\ell - \sin k_j - iu} = \prod_{\substack{m=-M \\ m \neq \ell}}^{M} \frac{\lambda_\ell - \lambda_m + 2iu}{\lambda_\ell - \lambda_m - 2iu} , \quad \ell = -M, \dots, M .
$$

(8.75)

In this form they depend on differences of the variables only, which permits the application
of the same methods used in the periodic boundary case for their analysis. Formally, this
doubling of variables leads to non-physical solutions to the equations which have to be
removed by hand (i.e. by considering only even solutions for quantities such as the root
densities).

To derive linear integral equations as in Section 8.1.1 above we introduce counting functions and root densities for the real roots of (8.75) after taking into account the complex
roots corresponding to the boundary bound states explicitly. This procedure implies a modification in the definition of the boundaries of integration in (8.26) to account for the complex
roots, namely:

$$
L \int_{Q_-}^{Q_+} dk\, r_c^{(L)}(k) = 2N + 1 - 2\theta_H(p - 1) - 2\theta_H(p - p_2) ,
$$

$$
L \int_{A_-}^{A_+} d\lambda\, r_s^{(L)}(\lambda) = 2M + 1 - 2\theta_H(p - p_1) .
$$

(8.76)

($\theta_H(x)$ denotes the Heaviside step function in this section). Proceeding as in Section 8.1.1
we obtain integral equations of the form (8.7) where the driving terms up to order L^{-1} are
given by

$$
r_c^{(0)}(k) = \frac{1}{\pi} + \frac{1}{L}\rho_c^{(0)}(k) + \mathcal{O}\left(\frac{1}{L}\right) , \quad r_s^{(0)}(\lambda) = \frac{1}{L}\rho_s^{(0)}(\lambda) + o\left(\frac{1}{L}\right) .
$$

(8.77)

For later use we denote the corresponding solution of the integral equations for the density
to this order as $r_i^{(\infty)}(x) + (1/L)\rho_i^{(\infty)}(x)$.

Due to the presence of complex roots the explicit form of the driving terms depends on the value of the boundary potential p in order L^{-1}:

$$\rho_c^{(0)}(k) = \frac{1}{\pi} - \cos k \, a_1(\sin k) + \frac{p \cos k - p^2}{\pi(p^2 - 2p \cos k + 1)}$$
$$+ \theta_H(p - p_1) \cos k \, (\tilde{a}_{2t}(\sin k) + \tilde{a}_{4u-2t}(\sin k)) \tag{8.78}$$

$$\rho_s^{(0)}(\lambda) = a_2(\lambda) + \begin{cases} 0 & p < 1, \\ \tilde{a}_{2u-2t}(\lambda) + \tilde{a}_{2u+2t}(\lambda) & 1 < p < p_1, \\ -\tilde{a}_{2t-2u}(\lambda) - \tilde{a}_{6u-2t}(\lambda) & p_1 < p < p_2, \\ 0 & p > p_2. \end{cases} \tag{8.79}$$

Here we have introduced the notation

$$\tilde{a}_y(x) = 2y/\pi(4x^2 + y^2). \tag{8.80}$$

Note that the index of $\tilde{a}_{4u-2t}$ in (8.78) changes sign at $p = p_2$.

The energy (3.97) of the Bethe states can be expressed in terms of the dressed energies (8.15). Being bulk-related quantities derived from the thermodynamic Bethe Ansatz, these are given by the same equations as in the case of periodic boundary conditions. To order L^0 the ground state energy is given by:

$$E_0 = Lu + Le_{GS} + f_{GS}. \tag{8.81}$$

Here e_{GS} is the bulk-energy density which does not depend on the boundary conditions. The surface contribution of order L^0 to the ground state energy is

$$f_{GS} = \frac{1}{2} \sum_{i \in \{c,s\}} \int_{J_i} dx \, e_i(x) \rho_i^{(0)}(x) + \frac{1}{2}(\mu + 2u - B + 2)$$
$$+ \theta_H(p - 1)E_1 + 2B\theta_H(p - p_1) + \theta_H(p - p_2)E_2. \tag{8.82}$$

Here $E_{1,2}$ are the energies of the bound states appearing at $p = 1$ and $p = p_2$, i.e. of the modes with complex wave numbers k_N and k_{N-1}, respectively:

$$E_1 = -p - \frac{1}{p} - \mu - 2u - B, \tag{8.83}$$
$$E_2 = -2\sqrt{1 + (t - 2u)^2} - \mu - 2u - B.$$

8.3.3 Ground-state expectation value of n_1

Due to the fact that there is a local potential coupled to the number operator of the electrons on the boundary site in (8.70) the corresponding expectation value can be computed from the identity $\langle n_1 \rangle = -\partial E_0 / \partial p$ (see Ref. [23]). With (8.82) we obtain

$$\langle n_1 \rangle = -\frac{1}{2} \sum_{i \in \{c,s\}} \int_{J_i} dx \, e_i(x) \frac{\partial \rho_i^{(0)}(x)}{\partial p} - \theta_H(p-1)\frac{\partial E_1}{\partial p} - \theta_H(p-p_2)\frac{\partial E_2}{\partial p}. \tag{8.84}$$

Without bulk magnetic field ($B = 0$) we can use the resulting simplifications in the structure of the integral equations (see Section (6.B)) to obtain

$$\langle n_1 \rangle = -\theta_H(p-1)\frac{\partial E_1}{\partial p} - \theta_H(p-p_2)\frac{\partial E_2}{\partial p} - \frac{1}{2}\int_{-Q}^{Q} dk \, \kappa(k) \, \gamma_p(k)$$

$$- \frac{1}{2}\int_{-Q}^{Q} dk \, \cos k \, \kappa(k) \begin{cases} \frac{\partial}{\partial p}\left(G_{2u}^{2u-2t}(\sin k) + G_{2u}^{2u+2t}(\sin k)\right) & 1 < p < p_2 \\ \frac{\partial}{\partial p}\left(\tilde{a}_{2t}(\sin k) - \tilde{a}_{2t-4u}(\sin k)\right) & p > p_2 \end{cases} \tag{8.85}$$

where $(y + z > 0)$

$$\gamma_p(k) = \frac{(p^2+1)\cos k - 2p}{\pi(p^2 - 2p\cos k + 1)^2},$$

$$G_y^z(x) = \int_{-\infty}^{\infty} \frac{d\omega}{2\pi} \frac{\exp(i\omega x - |\omega| z/2)}{2\cosh(\omega y/2)}. \tag{8.86}$$

In the limit of $p \to \infty$ only the first two terms survive and we obtain the expected result $\langle n_1 \rangle \to 2$ for an infinitely strong attractive boundary potential. Some numerical results on the p-dependence of the occupation of the boundary site are presented in figure 8.3.

8.3.4 Finite-size corrections to the energy of the open Hubbard chain

The calculation of the L^{-1} corrections to the energies of the ground state and low-lying excitations can be performed following the method of Woynarovich [487] as presented above. Asakawa and Suzuki [24] computed the spectrum of low-lying states of the open Hubbard chain for $p < 1$, this result is easily generalized to arbitrary strength of the boundary potential [49]. The result is a refinement of equation (8.81) for the ground state energy

$$E_0 = Lu + Le_{GS} + f_{GS} - \frac{\pi}{24L}(v_c + v_s) \tag{8.87}$$

and the following expression for the finite-size energies of the low-lying excitations

$$\Delta E(\Delta \mathbf{N}) = E - E_0$$

$$= \frac{\pi v_c}{L}\left\{\frac{1}{2\det^2(Z)}\left[(\Delta N_c - \theta_c(p))Z_{ss} - (\Delta N_s - \theta_s(p))Z_{cs}\right]^2 + N_c^+\right\}$$

$$+ \frac{\pi v_s}{L}\left\{\frac{1}{2\det^2(Z)}\left[(\Delta N_s - \theta_s(p))Z_{cc} - (\Delta N_c - \theta_c(p))Z_{sc}\right]^2 + N_s^+\right\}. \tag{8.88}$$

These expressions should be compared to the corresponding ones in equations (8.24) and (8.36) for periodic boundary conditions: The v_i are again the Fermi velocities of the low

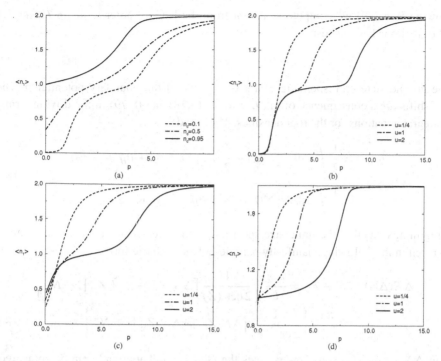

Fig. 8.3. Ground state expectation value of n_1 for the open Hubbard model in zero magnetic field as a function of the boundary potential p for (a) $u = 1$ and several electron densities; fixed density (b) $n_c = 0.1$, (c) $n_c = 0.5$ and (d) $n_c = 0.95$ and several values of u [49].

lying-charge and spin excitations. Due to the slightly different definition of the densities in the open boundary case they are given by $\pi v_i = e'_i(X_{0,i})/r_i^{(\infty)}(X_{0,i})$ here. The L^{-1}-term in (8.87) is the leading finite-size correction to the ground state energy of the Hubbard model with open boundary conditions. It does not depend on the strength of the boundary potential p but has a different universal numerical prefactor than in the periodic case (8.24).

The numbers N_i^+ in (8.88) are non-negative integers counting the number of particle hole excitations at the Fermi points (due to the symmetrization of the Bethe Ansatz roots there exists a single Fermi point for each degree of freedom). Z_{ij} are the elements of the dressed charge matrix given in terms of the integral equations (8.29), identical to the case of periodic boundary conditions, and $\Delta N_c = N - Ln_c$, $\Delta N_s = M - Lm$ are the changes in the numbers of electrons and down-spins as compared to the values obtained from the corresponding densities of the 'reference state', defined through

$$n_c = \frac{1}{2} \int_{-Q}^{Q} dk\, r_c^{(\infty)}(k), \quad m = \frac{1}{2} \int_{-A}^{A} d\lambda\, r_s^{(\infty)}(\lambda). \tag{8.89}$$

Following Ref. [49] we compute n_c and m from the leading order terms in L only. This implies that one has to chose

$$\Delta N_i \equiv \theta_i(p), \quad i \in c, s \tag{8.90}$$

in the ground state corresponding to a given choice of the boundary potential p. These phase shifts are a consequence of the L^{-1} terms (8.78) and (8.79) in the driving terms of the integral equations for the root densities. They read

$$\theta_c(p) = \frac{1}{2} \int_{-Q}^{Q} dk \, \rho_c^{(\infty)}(k) - \frac{1}{2} + \theta_H(p-1) + \theta_H(p-p_2)$$

$$\theta_s(p) = \frac{1}{2} \int_{-A}^{A} d\lambda \, \rho_s^{(\infty)}(\lambda) - \frac{1}{2} + \theta_H(p-p_1) \, . \tag{8.91}$$

Putting things together the finite-size spectrum of the open-boundary Hubbard chain can be written in the following manifestly particle-hole symmetric form

$$\Delta E(\Delta \mathbf{N}) = E - E_0 = \frac{\pi v_c}{L} \left\{ \frac{1}{2 \det^2(Z)} \left[\Delta \tilde{N}_c Z_{ss} - \Delta \tilde{N}_s Z_{cs} \right]^2 + N_c^+ \right\}$$

$$+ \frac{\pi v_s}{L} \left\{ \frac{1}{2 \det^2(Z)} \left[\Delta \tilde{N}_s Z_{cc} - \Delta \tilde{N}_c Z_{sc} \right]^2 + N_s^+ \right\} \tag{8.92}$$

where $\Delta \tilde{N}_i = \Delta N_i - \theta_i(p)$ now denotes the change in charge and spin as compared to the ground state, respectively (see also [149, 478]). There is no equivalent of the quantum numbers D_i in the finite-size spectrum of the periodic chain (8.36), the open boundaries do not allow for a non-zero current.

8.4 Relation of the dressed charge matrix to observables

As shown above, the finite size corrections in the spectrum of the one-dimensional Hubbard model are characterized in terms of the velocity of the gapless excitations and in addition the elements of the dressed charge matrix or its scalar equivalent when there is a single branch of gapless excitations only. An obvious question – in particular in view of the possible application to non integrable models – is, whether this quantity is merely a construct appearing within the Bethe Ansatz solution or can be related to other observable quantities. To address this question we first return to the discussion of the Hubbard model in a vanishing magnetic field.

8.4.1 Zero magnetic field

Following Refs. [63, 145, 190, 487] one can prove the identity

$$\frac{1}{2\rho(Q)} \frac{\partial n_c}{\partial Q} = \xi(Q). \tag{8.93}$$

This relation clearly shows that the dressed charge reveals itself in a characteristic way in the number $2\rho(Q)\partial Q$ of electrons added near the Fermi surface due to a change in the total density n_c.

A more physical interpretation of the dressed charge (which is directly applicable to models in the universality class of the Hubbard model not integrable by Bethe Ansatz) is found in its relation to certain thermodynamic coefficients [63, 191] – in this case the compressibility of the electron gas

$$\chi_c = \frac{1}{L}\frac{\partial N_c}{\partial \mu}. \tag{8.94}$$

To see this, consider the finite-size corrections (8.36) to the energies again. Using the special form of the dressed charge matrix for zero magnetic field this expression becomes

$$\Delta E(\Delta \mathbf{N}, \mathbf{D}) = \frac{2\pi}{L}\left[v_c\left(\frac{\Delta N_c^2}{4\xi(Q)^2} + \xi(Q)^2\left(D_c + \frac{D_s}{2}\right)^2 + N_c^+ + N_c^- \right) \right.$$
$$\left. + v_s\left(\frac{(\Delta N_s - \frac{1}{2}\Delta N_c)^2}{2} + \frac{D_s^2}{2} + N_c^+ + N_c^- \right) \right] + o\left(\frac{1}{L}\right). \tag{8.95}$$

Now let us change the ground state configuration under consideration by adding electrons while keeping zero magnetization, i.e. $\Delta N_c = 2\Delta N_s$ and $D_c = D_s = 0$. This state is actually the ground state of the system with $N_c + \Delta N_c$ electrons. From (8.95) we find the change in the ground state energy due to this change of the electron number to be

$$\Delta E = \frac{\pi v_c}{2L}\frac{1}{\xi(Q)^2}(\Delta N_c)^2. \tag{8.96}$$

To express this change in terms of the compressibility (8.94) we first have to separate the energy in the contribution of a micro-canonical ensemble and the contribution of the chemical potential $E = \tilde{E} - \mu N_c$. Expansion of this quantity in powers of ΔN_c and comparison to (8.96) yields the thermodynamical relation $\mu = \partial \tilde{E}/\partial N_c$ and the desired relation between the dressed charge and the compressibility:

$$\xi(Q)^2 = \pi v_c \chi_c. \tag{8.97}$$

An equivalent expression has been established for the 1D Bose gas with δ-repulsion [191].

8.4.2 Half-filling

In a completely analogous way we can express the dressed charge $\xi_s(A)$ from equation (8.55) in terms of a thermodynamical exponent, in this case the magnetic susceptibility

$$\chi(B) = \frac{\partial m}{\partial B}. \tag{8.98}$$

To do so consider the change in the ground state energy of the system due to a change in the magnetization $\Delta m = -\Delta N_s/L$. Comparison of the result from the finite size scaling analysis (8.54)

$$\Delta E = \frac{\pi v_s}{2L} \frac{1}{(\xi_s(A))^2} (\Delta N_s)^2 \tag{8.99}$$

with the one obtained from an expansion of $\Delta E = \Delta(\tilde{E} - LBm)$ with respect to Δm we find

$$\xi_s(A)^2 = \pi v_s \chi(B). \tag{8.100}$$

(Here we have used that $\partial \tilde{E}/\partial m = LB$). Again this relation is equivalent to the one found in the isotropic Heisenberg antiferromagnet [63].

8.4.3 Generic case

In the previous sections we have shown the importance of the dressed charge matrix Z (8.29) for the description of the low-energy behaviour of the Hubbard model. In addition we have obtained relations for the dressed charge that allow for a physical interpretation in several limiting cases (8.93), (8.97), (8.100). In this section we want to generalize these relations to the case with two degrees of freedom (i.e. spin and charge). First of all it is straightforward to generalize (8.93) to a matrix situation [145, 216, 487]:

$$\frac{1}{2\rho_j(X_j)} \frac{\partial n_i}{\partial X_j} = Z_{ij} . \tag{8.101}$$

As before, the dressed charge matrix Z governs the characteristic changes of the distribution functions for charges and spin waves at the Fermi surface due to changes of the density and magnetization.

Now let us try to find a generalization of equations (8.97) and (8.100) where the dressed charge was related to the compressibility and the magnetic susceptibility, respectively, to the general case. Again we want to express the finite-size corrections (8.36) with $D_c = D_s = 0$

$$\Delta E = \frac{\pi}{2L} \Delta \mathbf{N}^T (Z^{-1})^T V Z^{-1} \Delta \mathbf{N} \tag{8.102}$$

in terms of thermodynamic coefficients. To do so we proceed as in Section 8.4.1 and expand

$$E = \tilde{E} - \mu N_c - LBm \tag{8.103}$$

in small changes of the number of charges N_c and the magnetization m. To second order this gives

$$\Delta E = \left(\frac{\partial \tilde{E}}{\partial N_c} - \mu \right) \Delta N_c + \left(\frac{\partial \tilde{E}}{\partial m} - LB \right) \Delta m$$
$$+ \frac{1}{2} \frac{\partial^2 \tilde{E}}{\partial N_c^2} (\Delta N_c)^2 + \frac{1}{2} \frac{\partial^2 \tilde{E}}{\partial m^2} (\Delta m)^2 + \frac{\partial^2 \tilde{E}}{\partial N_c \partial m} \Delta N_c \Delta m . \tag{8.104}$$

Using $\Delta m = (\Delta N_c/2L - \Delta N_s/L)$ this can be compared to equation (8.102) to obtain

$$\frac{\partial \tilde{E}}{\partial N_c} = \mu, \qquad \frac{\partial \tilde{E}}{\partial m} = LB, \tag{8.105}$$

and the desired relation for the dressed charge matrix (8.29):

$$\pi (Z^{-1})^T V Z^{-1} = \begin{pmatrix} \chi_c^{-1} + \frac{1}{4}\chi^{-1} + \eta^{-1} & -(2\chi)^{-1} - \eta^{-1} \\ -(2\chi)^{-1} - \eta^{-1} & \chi^{-1} \end{pmatrix}. \tag{8.106}$$

Here the compressibility χ_c and the susceptibility χ have been given before, while

$$\eta = \frac{1}{L}\frac{\partial N_c}{\partial B} = \frac{\partial m}{\partial \mu}. \tag{8.107}$$

Note that (8.106) almost solves the problem, however, since the matrix Z enters in a symmetric combination of Z^{-1} and $(Z^{-1})^T$ only we have found just three equations to fix the four elements of Z.

A fourth equation can be obtained by considering a change in the boundary conditions. Enclosing a magnetic flux in the ring on which the electrons can move leads to an additional constant constant phase φ in the equations for the momenta k_j of the charges

$$Lk_j = 2\pi I_j + \varphi + \sum_{\beta=1}^{N_s} 2\arctan\left(\frac{\sin k_j - \lambda_\beta}{u}\right) \tag{8.108}$$

while leaving the equations for the rapidities of spin-waves unchanged. For small φ this leads to a change in the ground state energy $\Delta E \propto \varphi^2$. The momentum of this state will be

$$P = n_c \varphi. \tag{8.109}$$

This is the Aharonov-Bohm effect for this system. The current is given as $j(\varphi) = \partial E(\varphi)/\partial\varphi$.

On the other hand this change in boundary conditions corresponds to the finite-size corrections (8.36) for an excited state with $\Delta N_c = \Delta N_s = D_s = 0$ and $D_c = \varphi/2\pi$. Hence the change in energy can be given in terms of elements of the dressed charge matrix as

$$\Delta E(\varphi) = \frac{1}{2\pi L}\left(v_c Z_{cc}^2 + v_s Z_{cs}^2\right)\varphi^2. \tag{8.110}$$

This is also related to an observable quantity [39, 395, 510].

To conclude, we have derived a set of equations (8.106), (8.110) that relates the elements of the dressed charge matrix to observable quantities such as thermodynamic coefficients and velocities of spin waves and charge density waves. Hence, the complete spectrum of gapless low-lying energies of the 1D Hubbard chain can be determined by taking certain combinations of these quantities. We believe that these relations continue to hold for one-dimensional electron systems with more general interactions.

Appendices to Chapter 8

8.A Wiener Hopf calculation of the dressed charge

In the analysis of the zero temperature properties of the one-dimensional Hubbard model we are frequently faced with the problem that certain quantities of physical interest are given in terms of a function $f(x)$ which is a solution of a linear Fredholm type integral equation on some interval $[-X, X]$ on the real axis (see e.g. Appendix 5.B). The kernel of these integral operators is usually well behaved. For finite X these equations have to be solved numerically as outlined in Appendix 6.A. Only for $X \to \infty$ an analytic solution for $f(x)$ is possible by Fourier transformation which is very useful to compute quantities which depend on $f(x)$ for $|x| \ll X$. The dressed charge, however, is an example for a quantity for which we need to know the value of the function f on the boundary X of the interval. Again for finite X, this can be efficiently computed numerically. For very large X (including $X \to \infty$) an analytical calculation of $f(X)$ is possible using a perturbative scheme based on the Wiener Hopf method, introduced by Yang and Yang [495] to study the ground state properties of the XXZ model (see Chapter 17.2).

8.A.1 Weak coupling limit of the dressed charge in zero magnetic field

We now want to compute the value of $\xi(z_0)$ from (8.43) in the limit of large z_0 which determines the finite size spectrum of the Hubbard model below half-filling without external magnetic field in the weak coupling limit $u \to 0$. This equation is of the type (17.21) discussed in Chapter 17.2 with

$$K(z) = -R(z) , \qquad \bar{K}(\omega) = -\exp(-2|\omega|) \tag{8.A.1}$$

and $f_\infty(x) \equiv 2$. The Wiener-Hopf factorization (17.33) of $1 - \bar{K}(\omega)$ is

$$G^-(\omega) = G^+(-\omega) = \frac{1}{\sqrt{2\pi}} \Gamma \left(\frac{1}{2} + \frac{i\omega}{\pi} \right) \left(\frac{i\omega}{\pi e} \right)^{-i\omega/\pi} . \tag{8.A.2}$$

For z_0 sufficiently large we can solve the integral equation by means of a series expansion of $f = g_0 + g_1 + \ldots$ with g_i satisfying the Wiener-Hopf integral equations

$$g_0(z) - \int_0^\infty dz' \, \bar{K}(z - z')g_0(z') = 2,$$

$$g_1(z) - \int_0^\infty dz' \, \bar{K}(z - z')g_1(z') = \int_0^\infty dz' \, \bar{K}(2z_0 + z + z')g_0(z') \tag{8.A.3}$$

and so forth. The solution of the equation for g_0 is

$$g_0^+(\omega) = 2i \frac{G^-(0)}{\omega + i0} G^+(\omega). \tag{8.A.4}$$

Hence the result for the dressed charge $\xi(z_0)$ in the limit $z_0 \to \infty$ is

$$\lim_{z_0 \to \infty} \xi(z_0) = g_0(z = 0) = -i \lim_{\omega \to \infty} \omega g_0^+(\omega) = \sqrt{2}. \tag{8.A.5}$$

Similarly, we find for the next order in the expansion above:

$$g_1^+(\omega) = \sqrt{2}i \left[\frac{[G^-(\omega)]^2}{\omega - i0} \exp\left(-2iz_0\omega - 2|\omega|\right) \right]^+ G^+(\omega) \tag{8.A.6}$$

where

$$[f(\omega)]^+ = \frac{i}{2\pi} \int_{-\infty}^\infty d\omega' \frac{f(\omega')}{\omega - \omega' + i0}. \tag{8.A.7}$$

Combining these results we find

$$g_1(z = 0) = -i \lim_{\omega \to \infty} \omega g_1^+(\omega)$$

$$= -\frac{\sqrt{2}}{2\pi^2} \int_0^\infty \frac{dt}{t} \sin 2\pi t \, \Gamma^2\left(\frac{1}{2} + \frac{t}{e}\right)\left(\frac{t}{e}\right)^{-2t} \exp(-2\pi z_0 t)$$

$$\approx -\frac{\sqrt{2}}{2\pi z_0} + \mathcal{O}\left(\frac{1}{z_0^2}\right) \tag{8.A.8}$$

which together with (8.A.4) leads to the expression (8.49) for the dressed charge in the weak coupling limit. Application of the same method to the integral equation for the charge density (6.B.5) would allow to express the corrections to the zero coupling result in terms of the charge density rather than $z_0 = \sin Q/u$.

8.A.2 Solution of the strong coupling equations for small B

The integral equations (8.58),(8.61) and (8.62) derived in the strong coupling limit are of the form discussed in Chapter 17.2 with

$$\bar{K}(\omega) = \frac{1}{1 + \exp(2|\omega|)}. \tag{8.A.9}$$

Hence the decomposition of $1 - \bar{K}(\omega)$ is just given by the inverse of the functions $G^{\pm}$ in equation (8.A.2).

The calculation of the dressed charge $\xi_{ss}(\Lambda_0)$ is completely analogous to the weak coupling limit above, resulting in

$$\xi_{ss}(\Lambda_0) = \frac{1}{\sqrt{2}} \left(1 + \frac{1}{2\pi \Lambda_0} + \mathcal{O}\left(\frac{1}{\Lambda_0^2} \right) \right). \tag{8.A.10}$$

The B-dependence of Λ_0 can be computed from the condition $\varepsilon_1(\Lambda_0) = 0$: from (8.61) we have

$$f_{\varepsilon,\infty}(\omega) = 2\pi B \delta(\omega) - e^{-i\omega\Lambda_0} \frac{\pi B_c}{\cosh \omega},$$

giving

$$Q^+(\omega) = -\frac{B}{i} \frac{G^-(0)}{\omega + i0} + e^{-\pi\Lambda_0/2} \frac{\pi B_c G^-(-i\frac{\pi}{2})}{i(\omega + i\pi/2)} + \mathcal{O}(e^{-3\pi\Lambda_0/2}). \tag{8.A.11}$$

The second equation in (17.37) gives

$$\varepsilon_1(\Lambda_0) = -i \lim_{\omega \to \infty} \omega f_\varepsilon^+(\omega) = B\, G^-(0) - e^{-\pi\Lambda_0/2} \pi B_c G^- \left(-i\frac{\pi}{2} \right) \tag{8.A.12}$$

$$\Rightarrow \Lambda_0 = \frac{2}{\pi} \ln \left(\frac{B_1}{B} \right), \qquad B_1 = B_c \sqrt{\frac{\pi^3}{2e}}.$$

Similarly, the magnetization can be computed from (8.58) where $f_{\sigma,\infty}(\omega) = n_c \exp(-i\omega\Lambda_0)/\cosh \omega$. Proceeding as above we obtain

$$m = f_\sigma^+(\omega = 0) = 2n_c \sqrt{\frac{2\pi}{e}} e^{-\pi\Lambda_0/2} \tag{8.A.13}$$

$$= 2n_c \sqrt{\frac{2}{\pi e}} \frac{B}{B_1} + \mathcal{O}\left(\frac{B}{B_1 \ln(B_1/B)} \right) \approx \frac{2n_c}{\pi^2} \frac{B}{B_c}.$$

9

Asymptotics of correlation functions

Any experimentally measurable quantity can be expressed in terms of a suitable correlation function. As we have seen in the previous chapters, the spectrum of the one-dimensional Hubbard model can be studied in great detail, including the complete dependence on a homogeneous chemical potential and the magnetic field in a system of arbitrary size L – either by the analysis of certain integral equations for the behaviour in the thermodynamic limit and for the leading finite-size corrections or by solving the discrete Lieb-Wu equations for given L numerically. From the spectrum certain one point functions such as the overall electron density or magnetization can be computed within the framework of the Bethe Ansatz.

Expectation values of more general operators, in particular two point correlation functions

$$\langle \mathcal{O}^\dagger(\tau, x)\mathcal{O}(0, 0)\rangle \tag{9.1}$$

are not accessible by means of the methods developed above, not even if we restrict ourselves to the most interesting case of the asymptotic behaviour at large distances.

9.1 Low energy effective field theory at weak coupling

9.1.1 Continuum limit

For a well established approach to the computation of correlation functions such as (9.1) we may use the continuum limit of the Hubbard model introduced in Appendix 2.B in the limit of weak coupling. The idea is to restrict oneself to the low-energy modes of the system, which dominate the long distance asymptotics of the correlators.

For the free part of the Hubbard Hamiltonian, equation (2.12), the ground state is obtained by filling all negative energy modes, i.e. all single particle states with momenta in the interval $[-k_F, k_F]$ (we consider the case of a vanishing magnetic field, so $k_{F\uparrow} = k_{F\downarrow} = \pi n_c/2$). Hence, all low-lying excitations can be constructed by taking into account the modes with wave number close to the Fermi momenta $\pm k_F$.

As we have seen from our discussion of the exact solution above, the nature of the excitations of the one-dimensional Hubbard model is completely different from what we know from the free electron system, even for arbitrarily weak interaction. Still, following a standard perturbative approach we may assume – as long as the interaction is sufficiently weak

297

such that only states close to the Fermi points are mixed with the ground state – that only these modes will be important. Then we can decompose the Fermi fields (2.B.5) into components

$$\Psi_\sigma(x) = \lim_{a_0 \to 0} \frac{c_{n\sigma}}{\sqrt{a_0}} = \exp(ik_F x) R_\sigma(x) + \exp(-ik_F x) L_\sigma(x). \tag{9.2}$$

Here a_0 is the lattice spacing and $x = na_0$. The right- and left-moving Fermion fields R_σ and L_σ have dimension (length)$^{-1/2}$ and are slowly varying on the scale of the lattice spacing. Linearizing the single particle spectrum in the vicinity of the Fermi points and neglecting oscillating terms we obtain the following effective low-energy model (see e.g. [8]) from the continuum Hamiltonian (2.B.18)

$$H = H_0 + H_{\text{int}}$$

$$H_0 = v_F \int dx \; : \left(L_\sigma^\dagger(x)i\,\partial_x L_\sigma(x) - R_\sigma^\dagger(x)i\,\partial_x R_\sigma(x)\right) :$$

$$H_{\text{int}} = \frac{g}{2} \int dx \; \Big\{ : \left(R_\uparrow^\dagger(x)R_\uparrow(x) + L_\uparrow^\dagger(x)L_\uparrow(x)\right)\left(R_\downarrow^\dagger(x)R_\downarrow(x) + L_\downarrow^\dagger(x)L_\downarrow(x)\right) :$$

$$- : R_\uparrow^\dagger(x)R_\downarrow(x)L_\uparrow^\dagger(x)L_\uparrow(x) : - : R_\downarrow^\dagger(x)R_\uparrow(x)L_\uparrow^\dagger(x)L_\downarrow(x) : \Big\}. \tag{9.3}$$

Here H_0 describes noninteracting massless relativistic fermions with Fermi velocity

$$v_F = \left.\frac{\partial \varepsilon(k)}{\partial k}\right|_{k=k_F} = 2ta_0 \sin(k_F a_0), \tag{9.4}$$

and $: \mathcal{O} :$ denotes normal ordering of the operator $\mathcal{O}$. The coupling constant in H_{int} is given by

$$g = 2Ua_0 = 8tua_0. \tag{9.5}$$

At half filling, i.e. $k_F a_0 = \pi/2$, an additional non-oscillating interaction term describing Umklapp processes is present:

$$H_{\text{int}}^{\text{h.f.}} = \frac{g}{4} \int dx \; e^{-4ik_F x}\varepsilon_{\alpha\beta} R_\alpha^\dagger(x)R_\beta^\dagger(x)\, \varepsilon_{\gamma\delta}L_\gamma(x)L_\delta(x) + \text{h.c.} \tag{9.6}$$

This term describes scattering processes involving a finite momentum transfer equal to the reciprocal lattice vector $2\pi/a_0$, e.g. scattering of two holes located in the vicinity of $-k_F$ (at half-filling) off two particles located close to k_F. An example of an Umklapp scattering process is depicted in figure 9.1. Below half filling Umklapp processes involve high-energy degrees of freedom as is apparent from Fig.9.1 and as a result play no role in the low-energy effective theory. On the other hand, at half filling Umklapp processes invole only modes in the vicinity of the Fermi points because $4k_F a_0 = 2\pi$.

9.1.2 Bosonization and separation of spin and charge degrees of freedom

We now restrict ourselves to the discussion of the less than half-filled Hubbard model and discuss the implications of the additional interaction (9.6) later (see Chapter 10).

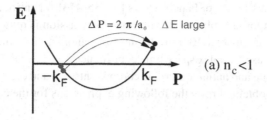

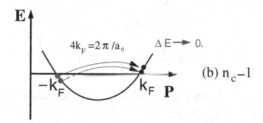

Fig. 9.1. Umklapp scattering processes below (a) and at (b) half-filling.

Examining the interaction terms in (9.3) one finds that the Fermi fields appear in certain quadratic combinations only. These bosonic operators are the chiral components of the $U(1)$ charge currents $\bar{j}$ and j[1]

$$\bar{j}(x) = \frac{1}{2} : \left(L_\uparrow^\dagger(x) L_\uparrow(x) + L_\downarrow^\dagger(x) L_\downarrow(x) \right) : ,$$
$$j(x) = \frac{1}{2} : \left(R_\uparrow^\dagger(x) R_\uparrow(x) + R_\downarrow^\dagger(x) R_\downarrow(x) \right) : , \qquad (9.7)$$

and the $SU(2)$-spin currents $\bar{\mathbf{J}}$

$$\bar{J}^z(x) = \frac{1}{2} : \left(L_\uparrow^\dagger(x) L_\uparrow(x) - L_\downarrow^\dagger(x) L_\downarrow(x) \right) : ,$$
$$\bar{J}^+(x) = \left(\bar{J}^-(x) \right)^\dagger = L_\uparrow^\dagger(x) L_\downarrow(x) . \qquad (9.8)$$

The components of the spin-currents for right movers, $\mathbf{J}$, are obtained by substituting $L \to R$.

The commutators between charge and spin currents vanish as do the commutators of currents with different chiralities. Commutators of e.g. charge currents of the same chirality, however, contain anomalous terms due to the presence of the Fermi sea[2]

$$[j(x), j(y)] = -\frac{i}{2\pi} \delta'(x - y) . \qquad (9.9)$$

The method of 'Bosonization' makes use of the fact that the same commutator relations can be realized in terms of bosonic operators alone [87, 309, 318, 321, 379]. We refer the

---

[1] ':' denotes normal ordering of point-split expressions [5].
[2] For a derivation see Appendix 10.A.

reader to the textbooks [139, 168] and reviews [7, 385, 473] for introductions and detailed treatments of this important subject. Our following discussion is quite brief; some further details can be found in Chapter 10. Introducing two canonical bosonic fields $\Phi_{c,s}(z, \bar{z}) = \varphi_{c,s}(z) + \bar{\varphi}_{c,s}(\bar{z})$ (we use complex space time coordinates $z = \tau - ix, \bar{z} = \tau + ix$) together with their holomorphic and antiholomorphic components $\varphi_{c,s}$ and $\bar{\varphi}_{c,s}$ we find that relations such as (9.9) can be obtained from the following expressions for the currents

$$j = \frac{i}{4\pi} \partial_z \Phi_c \,,$$

$$J^z = \frac{i}{4\pi} \partial_z \Phi_s \,,$$

$$J^{\pm} = \frac{1}{2\pi} \exp\{\pm i\varphi_s(z)\} \,, \tag{9.10}$$

and similar ones for $\bar{j}, \bar{\mathbf{J}}$. Furthermore, the fermionic fields can be written as bosonic exponents with $\varphi_\sigma = (\varphi_c + f_\sigma \varphi_s)/\sqrt{2}$

$$R_\sigma^\dagger = \frac{\eta_\sigma e^{if_\sigma \pi/4}}{\sqrt{2\pi}} \exp\left(\frac{i}{\sqrt{2}} \varphi_\sigma\right), \quad L_\sigma^\dagger = \frac{\eta_\sigma e^{if_\sigma \pi/4}}{\sqrt{2\pi}} \exp\left(-\frac{i}{\sqrt{2}} \bar{\varphi}_\sigma\right), \tag{9.11}$$

where $f_\uparrow = 1 = -f_\downarrow$ and η_σ are Klein factors which fulfil $\{\eta_a, \eta_b\} = 2\delta_{ab}$ and ensure anticommutation relations between the Fermi fields. An immediate consequence of this observation is that correlation functions computed for the free fermionic theory defined by H_0 in (9.3) are identical to the ones computed for the corresponding bosonic operators in the free bosonic theory

$$H_0 = \frac{v_F}{16\pi} \sum_{i=c,s} \int dx \left[(\partial_x \Theta_i)^2 + (\partial_x \Phi_i)^2\right] \,. \tag{9.12}$$

Here $\Theta_i = \varphi_i - \bar{\varphi}_i$ are known as 'dual' fields.

Furthermore, in this representation the interaction part of the Hamiltonian (9.3) separates into commuting parts depending only on charge and spin currents, respectively. Hence, in its bosonized form the continuum Hamiltonian is a sum of two contributions $H = H_c + H_s$ describing charge and spin degrees of freedom, respectively.

The charge part

$$H_c = \frac{v_F}{16\pi} \int dx \left[(\partial_x \Theta_c)^2 + (\partial_x \Phi_c)^2\right] + \frac{g}{2} \int dx : (j + \bar{j})^2 : \tag{9.13}$$

is known as the Luttinger Hamiltonian and is also obtained in the continuum limit of the XXZ model [**189**]. Bosonizing the products of U(1) currents (see Chapter 10) we obtain

$$H_c = \frac{v_F}{16\pi} \int dx \left[(\partial_x \Theta_c)^2 + \left(1 + \frac{g}{2\pi v_F}\right)(\partial_x \Phi_c)^2\right]. \tag{9.14}$$

The Hamiltonian may be simplified further by following e.g. Ref. [7]. Defining rescaled fields[3]

$$\Phi_c = \sqrt{K_c}\,\Phi'_c\,, \qquad \Theta_c = \frac{1}{\sqrt{K_c}}\Theta'_c\,, \qquad K_c = \frac{1}{\sqrt{1+g/2\pi v_F}}, \qquad (9.15)$$

we may rewrite (9.14) as

$$H_c = \frac{v_c}{16\pi}\int dx\,\left[\left(\partial_x\Theta'_c\right)^2 + \left(\partial_x\Phi'_c\right)^2\right]. \qquad (9.16)$$

Here K_c is known as the Luttinger liquid parameter and the charge velocity v_c is given by

$$v_c = \frac{v_F}{K_c} \simeq v_F + g/4\pi\,. \qquad (9.17)$$

We note that (9.17) agrees with the exact result for the charge velocity in the Hubbard model to linear order in u. The spin-part of the Hamiltonian is of the form

$$H_s = \frac{v_F}{16\pi}\int dx\,\left[\left(\partial_x\Theta_s\right)^2 + \left(\partial_x\Phi_s\right)^2\right] - \frac{g}{2}\int dx\,\left[:J^zJ^z:+:\bar{J}^z\bar{J}^z:\right]$$

$$- \frac{g}{2}\int dx\,\left[:\mathbf{J}\cdot\bar{\mathbf{J}}:+:\bar{\mathbf{J}}\cdot\mathbf{J}:\right]. \qquad (9.18)$$

It is shown in Appendix 10.A that (9.18) can be rewritten in a manifestly $SU(2)$ symmetric form

$$H_s = \int dx\,\left(\frac{2\pi v_s}{3}\,\{:\mathbf{J}\cdot\mathbf{J}:+:\bar{\mathbf{J}}\cdot\bar{\mathbf{J}}:\} - \frac{g}{2}\,\{:\mathbf{J}\cdot\bar{\mathbf{J}}:+:\bar{\mathbf{J}}\cdot\mathbf{J}:\}\right), \qquad (9.19)$$

where the velocity v_s of the spin excitations is renormalized by the interaction as

$$v_s \simeq v_F - g/4\pi\,. \qquad (9.20)$$

The terms proportional to v_s form the $SU(2)_1$ Wess-Zumino-Novikow-Witten (WZNW) model. The remaining interaction term coupling right and left currents is a marginal perturbation (i.e. has scaling dimension $d = 2$). From the renormalization group (RG) equations for the theory one finds that the relevance of the perturbation depends on the sign of the coupling constant g [12,506] (see also in Chapter 10). For repulsive interactions, i.e. $g > 0$, it is marginally irrelevant and can be neglected for the discussion of the low-lying excitations. Hence, the Hamiltonian describing the spin sector is simply

$$H_s = \frac{v_s}{16\pi}\int dx\,\left[\left(\partial_x\Theta_s\right)^2 + \left(\partial_x\Phi_s\right)^2\right]. \qquad (9.21)$$

On the other hand, the continuum limit of the attractive Hubbard model, i.e. $g < 0$ has to be described by a completely different field theory: in this case the marginal perturbation due to the current-current interaction in the spin sector is marginally relevant – in complete correspondence with the appearance of a spectral gap for spin excitations known from the Bethe Ansatz analysis of the model.

---

[3] The dual field must be rescaled in the opposite way to Φ_c in order to maintain canonical commutation relations. This is because the dual field can be expressed in terms of the momentum Π_c conjugate to Φ_c as $\Theta_c(t, x) = -8\pi\int_{-\infty}^{x} dx'\,\Pi_c(t, x')$.

Similarly, the effective field theory for the repulsive Hubbard model at half filling is different from the one discussed above: as we have found before the Hubbard model in this case has an enlarged symmetry (see Section 2.2.5) and is a Mott insulator with a spectral gap for charge excitations (denoted as phase V in Section 6.3). In the continuum limit the additional interaction term (9.6) has to be taken into account. Together with the remaining charge terms (9.13) this allows to bring the Hamiltonian for the charge degrees of freedom into an (η-) $SU(2)$-invariant form similar to (9.19) but with opposite sign of the remaining current-current interaction [8]. This results in a marginally relevant perturbation of the charge Hamiltonian which is responsible for the opening of the Mott gap. These matters are discussed in some detail in Chapter 10.

9.1.3 Bosonization results for correlation functions

For the repulsive Hubbard model below half-filling we have reformulated the model as a product of two massless bosonic theories for the charge and spin degrees of freedom with dispersions $\omega(q) = v_{c,s}|q|$, respectively. This 'spin-charge separation' holds for states near the Fermi level at weak coupling and allows to compute correlation functions of the Hubbard model. For the case of zero magnetic field this extension of the Luttinger liquid approach has been used by H. J. Schulz [382, 383] and others [70, 372] to compute the critical exponents of correlation functions for the Hubbard model.

As an example, the electronic Green's function computed in this approach reads

$$G_{\Psi\Psi^\dagger}(\tau, x) \sim \left(\frac{\exp(-ik_F x)}{\sqrt{(v_c\tau + ix)(v_s\tau + ix)}} + \text{c.c.} \right) \left[\frac{1}{v_c^2\tau^2 + x^2} \right]^{\alpha/2}. \qquad (9.22)$$

The exponent for the spin-contributions to this correlation function are independent of the interaction by virtue of the $SU(2)$ symmetry while for the charge contributions we find an interaction-dependent contribution, which is equal to $\alpha \simeq (g/8\pi v_F)^2$ at weak coupling (and thus really out of the range of the simple weak-coupling analysis performed above). Equation (9.22) is derived as follows. We start by rewriting the bosonization formulas (9.11) in terms of the rescaled fields $\Phi'_c = \varphi'_c + \bar{\varphi}'_c$, $\Theta'_c = \varphi'_c - \bar{\varphi}'_c$, see equation (9.15). We have

$$\varphi_c = \frac{1}{2}\left[\varphi'_c\left(\sqrt{K_c} + \frac{1}{\sqrt{K_c}}\right) + \bar{\varphi}'_c\left(\sqrt{K_c} - \frac{1}{\sqrt{K_c}}\right) \right],$$

$$\bar{\varphi}_c = \frac{1}{2}\left[\varphi'_c\left(\sqrt{K_c} - \frac{1}{\sqrt{K_c}}\right) + \bar{\varphi}'_c\left(\sqrt{K_c} + \frac{1}{\sqrt{K_c}}\right) \right]. \qquad (9.23)$$

The left-moving Fermi field is then bosonized as

$$L_\sigma(\tau, x) = \frac{\eta_\sigma e^{-if_\sigma\pi/4}}{\sqrt{2\pi}} e^{\frac{i}{4}\left[\varphi'_c\left(\sqrt{K_c} - \frac{1}{\sqrt{K_c}}\right) + \bar{\varphi}'_c\left(\sqrt{K_c} + \frac{1}{\sqrt{K_c}}\right)\right]} e^{\frac{if_\sigma}{2}\bar{\varphi}_s(\tau, x)}. \qquad (9.24)$$

Now $\langle L_\sigma(\tau, x) L_\sigma^\dagger(0, 0)\rangle$ and $\langle R_\sigma(\tau, x) R_\sigma^\dagger(0, 0)\rangle$ are calculated by evaluating the two-point functions of chiral vertex operators along the lines of Appendix 10.C. This leads to the result (9.22).

While the result (9.22) has been derived here for small values of the coupling constant g only, it will be shown below, that – as a consequence of the absence of spectral gaps – the result (9.22) is valid universally for the repulsive Hubbard model below half-filling. The interaction strength merely determines the numerical values of the velocities $v_{c,s}$ and the exponent α. These quantities can be calculated directly from the Bethe ansatz as we will see in what follows.

9.2 Conformal field theory and finite size scaling

In the previous section we have obtained some results for the asymptotic behaviour of correlation functions by means of an analysis of the low energy effective field theory in the limit of weak coupling. On the other hand we know from the Bethe Ansatz solution that there is no phase transition induced by the interaction in the repulsive Hubbard model below half-filling, in particular the spectrum remains gapless although the nature of the excitations in presence of the interaction is quite different from what is known for free electrons.

More general, depending on filling and external magnetic field the repulsive Hubbard model is found in several phases with one or two massless excitations at zero temperature, i.e. it has various quantum critical points at $T = 0$. In such a situation we generally expect correlation functions to decay as power laws at large distances due to *scale invariance*. Furthermore, the critical behaviour should not depend on the details of the underlying microscopic Hamiltonian but rather be a universal property shared by a large family of models driven to the same fixed point under renormalization.

Much progress in the description of critical phenomena in $1 + 1$ dimensional quantum systems has been made by the extension of simple scaling arguments through application of the concepts of conformal quantum field theory [51]. Most important in the context of an exactly solvable model such as the one-dimensional Hubbard model is the existence of one to one relations between the spectrum of low-lying excitations – which we can compute with arbitrary precision from the Bethe Ansatz – and quantities related to the universality class of the system which contain all the information needed to describe the asymptotic behaviour of correlation functions.

9.2.1 Universality classes

Due to conformal invariance, the universality class of a Lorentz-invariant theory (i.e. a model with a single 'velocity of light' v) is uniquely described by a single dimensionless number c – the central charge of the underlying Virasoro-algebra. The value of c can be extracted from the universal finite size scaling behaviour of the ground state energy of the model [6, 62, 75]

$$E_0 - L\varepsilon_0 = -\frac{\pi}{6L}vc + o\left(\frac{1}{L}\right) \tag{9.25}$$

for periodic boundary conditions and

$$E_0 - L\varepsilon_0 - f_0 = -\frac{\pi}{24L}vc + o\left(\frac{1}{L}\right) \tag{9.26}$$

for a conformal field theory subject to open boundary conditions. Here L is the size of the system, E_0 is the ground state energy of the finite system, ε_0 is the energy density of the ground state of the infinite system and f_0 is the surface energy of the system with open boundaries. v is the Fermi-velocity. The same mechanism is the origin of a universal term in the low temperature expansion of the free energy density of the model ($\beta = 1/k_B T$)

$$f = \varepsilon_0 - \frac{\pi}{6\beta^2}\frac{c}{v}. \tag{9.27}$$

Note that the expression (9.25) is exactly of the form (8.53) found for the leading finite-size corrections to the ground state energy of the repulsive Hubbard model at half filling where only the charge-neutral spinon excitations are gapless. Hence conformal invariance implies that the central charge is $c = 1$ in this phase [489]. This corresponds to a Gaussian model of a single free boson as is well established and at the heart of the Bosonization method. A similar situation arises in the attractive Hubbard model. Bogoliubov and Korepin have used this fact to compute the critical exponents from the finite-size corrections to the spectrum [65].

Our generic results (8.24) and (8.87) for the finite size scaling of the ground state energy of the repulsive Hubbard model below half-filling, however, do not fit into this simple picture. This was to be expected since we were dealing with a model having two branches of low-lying excitations with different Fermi velocities. Assuming, however, that charge and spin excitations are independent of each other we can interpret the scaling of the ground state energy in the framework of conformal quantum field theory as the result of a critical theory based on a product of two Virasoro algebras each having central charge $c = 1$ [140, 141]. The same interpretation can be drawn from the comparison of the conformal prediction (9.27) with the low-temperature expansion of the free energy of the Hubbard model (see equation (13.207) below). While this separation into two independent CFTs appears trivial for the discussion above (where we simply count critical degrees of freedom) it has dramatic consequences for correlation functions (9.1):

As we have seen in Chapter 7 the nature of the low-lying excitations in the Hubbard model changes when we modify the parameters of the model: the holons and spinons which make up the excitation spectrum at vanishing magnetic field change into fermionic quasi-particles and magnons near the saturation field. Hence the decomposition of a physical operator into its constituents from the two conformal field theories will depend continuously on these parameters. This has immediate consequences on the analytic properties of the correlation functions: instead of simple quasi particle poles we shall find branch cuts with properties depending on the electron density, magnetic field and interaction strength – as is already indicated in the perturbative result (9.22) for the Green's function at weak coupling.

Unfortunately, for a CFT with central charge $c \geq 1$ the critical exponents related to these singularities are not fixed by universality. If no additional constraints – e.g. due to symmetry – are present one has to analyze the finite size scaling of the low-lying excitations (those becoming gapless in the thermodynamic limit) of the one-dimensional quantum system in addition to that of the ground state energy to obtain information on the precise nature of these singularities.

9.2.2 Low-lying excitations and correlation functions

As a consequence of global scale invariance in a Lorentz invariant system without an internal scale e.g. due to a spectral gap, a correlation function such as (9.1) has to decay algebraically. In the simplest case the two-point correlation functions – when considered in the entire complex plane – decay as simple power laws ($z_i = v\tau_i - ix_i$)

$$\langle \mathcal{O}^{\dagger}(z_1, \bar{z}_1)\mathcal{O}(z_2, \bar{z}_2)\rangle \propto \frac{1}{(z_1 - z_2)^{2\Delta^-}(\bar{z}_1 - \bar{z}_2)^{2\Delta^+}} \equiv D(z_1, z_2)\bar{D}(\bar{z}_1, \bar{z}_2). \tag{9.28}$$

Conformal invariance extends the invariance under global scale transformations (i.e. translations and rotations) to local ones. This has two consequences: first it allows to classify all possible universality classes as discussed above, second it allows to relate geometries different from the extended complex plane to each other. Under a conformal transformation $z \to w = w(z)$ the (holomorphic part of the) correlation function (9.28) becomes

$$D(z_1, z_2) = \frac{1}{(w(z_1) - w(z_2))^{2\Delta^-}} \left(\frac{\partial w}{\partial z_1} \cdot \frac{\partial w}{\partial z_2}\right)^{\Delta^-} \tag{9.29}$$

and similarly for the antiholomorphic part $\bar{D}(\bar{z}_1, \bar{z}_2)$. The additional *local* factors appearing in the correlation functions may be interpreted as the transformation properties of the corresponding operators $\mathcal{O}$. A particularly useful transformation is

$$w(z) = \exp(2\pi z/L), \tag{9.30}$$

which maps the strip $-L < \text{Im}(z) \le 0$ to the complex plane [6, 62, 74]. Equation (9.29) enables us to compute the correlation function of finite quantum chains of length L or finite temperature correlation functions at $T = i/L$ from the simple infinite system, zero temperature expression (9.28):

$$D(z_1, z_2) = \left[\frac{\pi/L}{\sinh\left(\frac{\pi}{L}(z_1 - z_2)\right)}\right]^{2\Delta^-}. \tag{9.31}$$

For a chain of length L, i.e.

$$z = v\tau - ix, \quad -\infty < \tau < \infty, \quad -L < x \le 0, \tag{9.32}$$

we may expand equation (9.31) for large values of $\tau_{12} = \tau_1 - \tau_2$

$$D(z_1, z_2)\bar{D}(\bar{z}_1, \bar{z}_2) = \left(\frac{2\pi}{L}\right)^{2(\Delta^+ + \Delta^-)}$$

$$\times \sum_{n,m} C_{nm} \exp\left\{-\frac{2\pi v}{L}(d + n)\tau_{12}\right\} \exp\left\{-\frac{2\pi i}{L}(s + m)x_{12}\right\} \tag{9.33}$$

where we have introduced the scaling dimension $d = \Delta^+ + \Delta^-$ and conformal spin $s = \Delta^- - \Delta^+$ of the operator $\mathcal{O}$.

A similar expression for this quantity can be obtained formally for a system with periodic boundary conditions with the Hamiltonian as the generator of translations in the time direction in a transfer matrix approach. This gives ($|\Omega\rangle$ is the ground state of the Hamiltonian)

$$\langle \Omega | \mathcal{O}^\dagger(z_1, \bar{z}_1)\mathcal{O}(z_2, \bar{z}_2)|\Omega\rangle = \sum_q \langle \Omega | \mathcal{O}^\dagger(\tau_1, x_1)|q\rangle \langle q | \mathcal{O}(\tau_2, x_2)|\Omega\rangle$$

$$= \sum_q \exp\left\{-E_q \tau_{12} - i P_q x_{12}\right\} |\langle q | \mathcal{O}(0)|\Omega\rangle|^2 \quad (9.34)$$

where $|q\rangle$ are eigenstates of the Hamiltonian with energy E_q and momentum P_q. Comparing (9.33) and (9.34) we can, in principle, determine the form factors of the operator $\mathcal{O}$ and, more importantly for our subsequent analysis, obtain the following relations between the scaling dimensions and spins of the operators in the conformal field theory and the finite size corrections to the low-lying states in the spectrum of the Hamiltonian [75]

$$E^{N^+, N^-}_{\Delta^\pm} - E_0 = \frac{2\pi v}{L}(d + N^+ + N^-) + o\left(\frac{1}{L}\right),$$

$$P^{N^+, N^-}_{\Delta^\pm} - P_0 = \frac{2\pi}{L}(s + N^+ - N^-) + 2D k_F,$$

(9.35)

where N^+, N^- are non-negative integers and $2D$ is the macroscopic contribution to the momentum of the state $\mathcal{O}|\Omega\rangle$ in units of the Fermi momentum k_F.[4] Hence, each operator $\mathcal{O}$ in the conformal field theory corresponds to a tower (enumerated by $N^\pm$) of excited states in the lattice model. From the lowest energy in such a tower (corresponding to $N^+ = 0 = N^-$ in (9.35)) one can read off the scaling dimension d and conformal spin s of the corresponding so-called primary field $\phi_{\Delta^\pm}$ which has particularly simple two-point correlation functions [51, 64]

$$\langle \phi_{\Delta^\pm}(\tau, x)\phi_{\Delta^\pm}(0, 0)\rangle = \frac{\exp(2i D k_F x)}{(v\tau + ix)^{2\Delta^+}(v\tau - ix)^{2\Delta^-}}$$

$$= \frac{\exp(2i D k_F x)}{(v^2\tau^2 + x^2)^d}\left(\frac{v\tau - ix}{v\tau + ix}\right)^s. \quad (9.36)$$

Thus, provided that we have reliable results (either from an exact solution or by numerical methods) on the finite size scaling of low-lying excitations in a $1 + 1$ dimensional quantum model with a single branch of gapless excitations we can determine the asymptotic behaviour of correlation functions without the need of using perturbative methods. All the effects of interactions will lead to a renormalization of the finite-size scaling properties – or equivalently of certain thermodynamic quantities, see Chapter 8.4 – of the system. The additional step relating these quantities to the critical exponents in correlation functions relies on the principles of conformal invariance alone. What remains is the identification of

---

[4] This term is not usually discussed in the field theory literature where particles on the light cone are considered.

the operators appearing in the CFT relevant for the correlation function of a given operator in the lattice model.

9.2.3 Extension to models with several critical degrees of freedom

While the correspondence (9.35) can be directly applied to the discussion of the phases of the Hubbard model with a single critical degree of freedom, such as the Hubbard model at half-filling, the generic result (8.36), (8.38) for the spectrum of low-lying excitations of the repulsive Hubbard model below half-filling does not fit into this picture. As in the discussion of the universality classes based on the finite size scaling behaviour of the ground state energy in Section 9.2.1 a suitable generalization has to be found [140, 141].

Following our interpretation of the scaling (8.24) of the ground state energy as signature of a critical theory based on a product of two Virasoro algebras we generalize (9.35) to

$$E(\Delta \mathbf{N}, \mathbf{D}) - E_0 = \frac{2\pi}{L} \left(v_c (\Delta_c^+ + \Delta_c^-) + v_s (\Delta_s^+ + \Delta_s^-) \right) + o\left(\frac{1}{L}\right),$$

$$P(\Delta \mathbf{N}, \mathbf{D}) - P_0 = \frac{2\pi}{L} \left(\Delta_c^+ - \Delta_c^- + \Delta_s^+ - \Delta_s^- \right)$$
$$+ 2D_c k_{F,\uparrow} + 2(D_c + D_s) k_{F,\downarrow} .$$

(9.37)

Comparing equations (8.36) with (9.37) we obtain unique expressions for the conformal scaling dimensions (i.e. the sums $d_{c,s} = \Delta_{c,s}^+ - \Delta_{c,s}^-$) characteristic of the charge and spin part of the fields in this theory as functions of the elements of the dressed charge matrix (8.29). On the other hand, the comparison of the finite-size momenta (8.38) with the corresponding expression in (9.37) only provides the sum of the conformal spins $s_{c,s} = \Delta_{c,s}^+ - \Delta_{c,s}^-$. The additional requirement that all the dimensions $\Delta_{c,s}^\pm$ be non negative resolves this final problem (otherwise there would be unphysical divergences in the correlation functions). This requirement is met by writing the dimensions as complete squares [140]:

$$2\Delta_c^\pm (\Delta \mathbf{N}, \mathbf{D}) = \left(Z_{cc} D_c + Z_{sc} D_s \pm \frac{Z_{ss} \Delta N_c - Z_{cs} \Delta N_s}{2 \det Z} \right)^2 + 2N_c^\pm,$$

$$2\Delta_s^\pm (\Delta \mathbf{N}, \mathbf{D}) = \left(Z_{cs} D_c + Z_{ss} D_s \pm \frac{Z_{cc} \Delta N_s - Z_{sc} \Delta N_c}{2 \det Z} \right)^2 + 2N_s^\pm.$$

(9.38)

In general, the individual conformal spins $s_{c,s}$ defined by these expressions will depend on the system parameters and take arbitrary real values. This gives rise to the unusual analytic properties of correlation functions in correlated one-dimensional systems. Physical operators, however, cannot have arbitrary spin. Therefore, the combined conformal spin $s_c + s_s$ of the charge and spin part of a physical operator has to be integer or half-odd integer. As can be seen from (8.38) this is always the case.

Finally, we modify the conformal field theory expression for the correlation function of a primary field (9.36) to the case where these operators contain factors from two different sectors (such as charge and spin part of the model or 'holon' and 'spinon' operators):

$$\langle \phi_{\Delta^\pm}(\tau, x)\phi_{\Delta^\pm}(0, 0)\rangle$$

$$= \frac{\exp\left(2i D_c k_{F,\uparrow} x\right)\exp\left(2i(D_c + D_s)k_{F,\downarrow} x\right)}{(v_c\tau + ix)^{2\Delta_c^+}(v_c\tau - ix)^{2\Delta_c^-}(v_s\tau + ix)^{2\Delta_s^+}(v_s\tau - ix)^{2\Delta_s^-}}. \tag{9.39}$$

At small finite temperature $T > 0$ the correlation functions decay exponentially, however with a small exponent due to the vicinity to the phase transition at $T = 0$. Hence, the large distance asymptotics of correlation functions in the space like regime can still be obtained from conformal invariance. In this case expressions of type (9.39) have to be replaced by

$$\exp\left(2i D_c k_{F,\uparrow} x\right)\exp\left((2i(D_c + D_s)k_{F,\downarrow} x\right)$$

$$\times \left(\frac{\pi T}{v_c \sinh(\pi T(x - i v_c\tau)/v_c)}\right)^{2\Delta_c^+}\left(\frac{\pi T}{v_c \sinh(\pi T(x + i v_c\tau)/v_c)}\right)^{2\Delta_c^-}$$

$$\times \left(\frac{\pi T}{v_s \sinh(\pi T(x - i v_s\tau)/v_s)}\right)^{2\Delta_s^+}\left(\frac{\pi T}{v_s \sinh(\pi T(x + i v_s\tau)/v_s)}\right)^{2\Delta_s^-}. \tag{9.40}$$

This expression can be obtained from (9.39) by a conformal mapping of the complex plane without the origin (zero temperature) onto a strip of width $1/T$ in time direction.

With the expressions given above we have extended the methods developed in conformal field theory for the computation of correlation functions from finite-size spectra to models with two or more critical degrees of freedom – the case of two branches of gapless excitations considered here is easily generalized to more general situations (see e.g. [143, 145, 216]).

As a final remark we note that marginally irrelevant perturbations to a conformal field theory – such as the current current interactions which we have neglected in the perturbative analysis in Section 9.1.2 – can produce logarithmic corrections to the conformal predictions for correlation functions [10, 137]. In the spectrum they are manifest through finite size effects of order $1/(L \ln L)$ in (9.25), (9.35) [75]. In principle, these can also be calculated explicitly from the Bethe Ansatz equations, but this is beyond the scope of the present discussion.

9.3 Correlation functions of the one-dimensional Hubbard model

In the previous section we have established relations between the correlation functions of certain operators in the effective field theory describing the low-energy sector of a given microscopic model and the finite-size scaling of the low-lying states in this model. This determines the set of all possible correlation functions within the microscopic model. This is a notable achievement, but still doesn't address the problem one is usually faced with, namely to compute the asymptotic behaviour of, say, the density-density correlation function of the one-dimensional Hubbard model. The missing step would be to express a given local operator in the microscopic model in terms of the operators appearing in the conformal field theory (whose correlation functions we know to be (9.36) or (9.39)). This expansion is not known usually. Hence without additional input all one can say is that the correlation function in the microscopic model is a superposition of contributions from all operators appearing in the CFT with unknown – possible vanishing – coefficients.

Fortunately, the number of terms in such an expansion can be reduced drastically by using the selection rules for the form factors $\langle q|\mathcal{O}(x)|\Omega\rangle$ of the microscopic operator $\mathcal{O}$ appearing in (9.34).

As an illustration of this scheme let us describe how to obtain the asymptotical behaviour of the density-density correlations

$$G_{nn}(\tau, x) = \langle n(\tau, x)n(0, 0)\rangle \tag{9.41}$$

for the repulsive Hubbard model below half filling[5] in the Euclidean region: The leading term is a constant since $\langle n(\tau, x)\rangle = n_c$. Since the local density operator does not change the number of particles we have to use (9.38), (9.39) at $\Delta N_c = \Delta N_s = 0$ for the higher terms contributing to G_{nn}. The restrictions of equations (8.39) are satisfied for any integer D_c, D_s. Hence the next terms in the asymptotic expansion of G_{nn} are found to be

$$
\begin{aligned}
G_{nn}&(\tau, x) - n_c^2 \\
&\sim A_1 \cos(2k_{F,\uparrow}x + \varphi_1)/\left((x^2 + v_c^2\tau^2)^{(Z_{cc}-Z_{sc})^2}(x^2 + v_s^2\tau^2)^{(Z_{cs}-Z_{ss})^2}\right) \\
&+ A_2 \cos(2k_{F,\downarrow}x + \varphi_2)/\left((x^2 + v_c^2\tau^2)^{Z_{sc}^2}(x^2 + v_s^2\tau^2)^{Z_{ss}^2}\right) \\
&+ A_3 \cos(2(k_{F,\uparrow} + k_{F,\downarrow})x + \varphi_3)/\left((x^2 + v_c^2\tau^2)^{Z_{cc}^2}(x^2 + v_s^2\tau^2)^{Z_{cs}^2}\right) \\
&+ A_4 \frac{x^2 - v_c^2\tau^2}{(x^2 + v_c^2\tau^2)^2} + A_5 \frac{x^2 - v_s^2\tau^2}{(x^2 + v_s^2\tau^2)^2}.
\end{aligned}
\tag{9.42}
$$

Here A_k are constant coefficients, φ_k unknown phases. The general expression for the density-density correlation function reads

$$
\begin{aligned}
G_{nn}(\tau, x) = \sum A(D_c, D_s, N_c^{\pm}, N_s^{\pm}) \\
\times \frac{\exp(2i D_c k_{F,\uparrow}x)\exp(2i(D_c + D_s)k_{F,\downarrow}x)}{(v_c\tau + ix)^{2\Delta_c^+}(v_c\tau - ix)^{2\Delta_c^-}(v_s\tau + ix)^{2\Delta_s^+}(v_s\tau - ix)^{2\Delta_s^-}},
\end{aligned}
\tag{9.43}
$$

where the sum runs over all integers D_c, D_s and nonnegative integers $N_c^{\pm}$ and $N_s^{\pm}$. The scaling dimensions are

$$\Delta_c^{\pm}(\mathbf{D}, \mathbf{N}^{\pm}) = \frac{1}{2}(Z_{cc}D_c + Z_{sc}D_s)^2 + N_c^{\pm},$$

$$\Delta_s^{\pm}(\mathbf{D}, \mathbf{N}^{\pm}) = \frac{1}{2}(Z_{cs}D_c + Z_{ss}D_s)^2 + N_s^{\pm},$$

$$\tag{9.44}$$

according to (9.38).

---

[5] At half-filling there exist stronger selection rules due to the η-SU(2) symmetry which modify the resulting expression significantly, see Section 9.3.2.

Table 9.1. *Selection rules for the conformal dimensions contributing to various correlation functions of the repulsive Hubbard model below half-filling.*

	ΔN_c	ΔN_s	D_c	D_s
$G_{nn} = \langle n(\tau, x)n(0, 0)\rangle$	0	0	integer	integer
$G_{\sigma\sigma}^{z} = \langle S^z(\tau, x)S^z(0, 0)\rangle$	0	0	integer	integer
$G_{\sigma\sigma}^{\perp} = \langle S^-(\tau, x)S^+(0, 0)\rangle$	0	1	half-odd int.	integer
$G_{\psi\psi\dagger}^{\uparrow} = \langle \Psi_\uparrow(\tau, x)\Psi_\uparrow^\dagger(0, 0)\rangle$	1	0	half-odd int.	half-odd int.
$G_{\psi\psi\dagger}^{\downarrow} = \langle \Psi_\downarrow(\tau, x)\Psi_\downarrow^\dagger(0, 0)\rangle$	1	1	integer	half-odd int.

Using the fact that the Hamiltonian of the Hubbard model is invariant under the action of the parity operator R_L (2.46) the ground state is an eigenstate of R_L and, consequently,

$$G_{nn}(\tau, x) \equiv G_{nn}(\tau, -x) . \tag{9.45}$$

From this property it follows that the phases φ_k in the general expression (9.42) for the density-density correlation function actually vanish.

The general procedure outlined above is applicable for each of the phases of the Hubbard model with gapless excitations. In each case the relevant selection rules have to be determined. A listing of the selection rules of the repulsive Hubbard model below half-filling relevant to the two-point correlation functions of various operators is given in table 9.1. In general this approach gives the critical exponents as functions of the entries of the matrix Z (8.29). The numerical values of these coefficients are easily obtained by numerical solution of the integral equations (8.31). As we have seen in Chapter 8 this matrix simplifies in various special cases which we shall consider in more detail below.

9.3.1 Zero magnetic field

In Section 8.2.1 we have studied the finite size spectrum of the repulsive Hubbard model for $B = 0$ below half-filling. As a consequence of $SU(2)$ symmetry in the magnetic sector the only non-universal quantity apart from the Fermi velocities was the element $Z_{cc} = \xi(z_0)$ of the dressed charge matrix (8.42) which has been computed analytically in various limiting cases. This fact allows to simplify the expression for the conformal dimensions (9.38) significantly [140, 241]

$$\Delta_c^\pm = \frac{1}{2}\xi^2(D_c + \frac{1}{2}D_s)^2 + \frac{1}{8\xi^2}(\Delta N_c)^2 \pm \frac{1}{4}\Delta N_c(2D_c + D_s) + N_c^\pm,$$

$$\Delta_s^\pm = \frac{1}{4}(D_s)^2 + \frac{1}{4}(\Delta N_s - \frac{1}{2}\Delta N_c)^2 \pm \frac{1}{4}(2\Delta N_s - \Delta N_c)D_s + N_s^\pm. \tag{9.46}$$

Note that ξ enters the expressions for the dimensions of the charge part of the operators

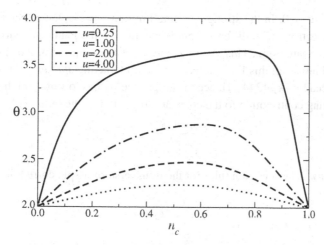

Fig. 9.2. The critical exponents of the one-dimensional Hubbard model below half-filling for vanishing magnetic field (e.g. in the density-density correlation function (9.47)) are given in terms of the number θ. Here we present θ as a function of the electron density n_c for various values of u.

only – the spin sector is independent of the system parameters due to the $SU(2)$ symmetry.[6] This also implies that the conformal spins of the holon and spinon operators, $s_{c,s} = \Delta^+_{c,s} - \Delta^-_{c,s}$, are independent of ξ. Note though, that the individual values of s_c and s_s can be different from integers or half-odd integers reflecting the fact that only the product of the corresponding holon and spinon operators are physical fields (recall our discussion following equation (9.38).

As an additional simplification for vanishing magnetic field, the Fermi momenta for spin up and spin down electrons are identical $k_{F,\uparrow} = k_{F,\downarrow} = \pi n_c/2 \equiv k_F$. This leads to various simplifications in the expressions for the correlation functions compared to the general form given for G_{nn} above. Below we express the critical exponents as functions of $\theta = 2\xi(z_0)^2$. From equation (8.50) we know that for any density $0 < n_c < 1$ the value of θ increases from 2 to 4 as the Coulomb repulsion u decreases from ∞ to 0 (see figure 9.2).

Density correlations. From (9.42) we obtain for the asymptotic behaviour of the density-density correlation function (we use the observation (9.45) on the parity of the correlator):

$$G_{nn}(\tau, x) - n_c^2 \sim A_1 \frac{\cos(2k_F x)}{|v_s \tau + ix||v_c \tau + ix|^{(\theta/4)}} + A_2 \frac{\cos(4k_F x)}{|v_c \tau + ix|^\theta}$$
$$+ A_3 \frac{x^2 - v_c^2 \tau^2}{(x^2 + v_c^2 \tau^2)^2} + A_4 \frac{x^2 - v_s^2 \tau^2}{(x^2 + v_s^2 \tau^2)^2}. \tag{9.47}$$

Because of the range of variation of θ the leading correction beyond the constant term is the term $\propto A_1$ for all finite u. Unlike these oscillations with wave number $2k_F$ which also appear in a free electron gas, $4k_F$-oscillations are a consequence of the interactions,

---

[6] More precisely, only the spin velocity depends on the system parameters.

i.e. $A_2 \to 0$ for $u \to 0$. In the strong coupling limit, at $u = \infty$, the effect of the Hubbard interaction on electrons with different spin is identical to the Pauli exclusion principle for electrons with the same spin. This allows an alternative description of the charge dynamics of the Hubbard model in this limit in terms of noninteracting *spinless* fermions where k_F has to be replaced by $2k_F$ [344]. Hence the amplitude A_1 has to vanish in this limit and the leading oscillating contribution to the correlation function is the term $\propto A_2$, namely

$$\frac{\cos(4k_F x)}{v_c^2 \tau^2 + x^2}.$$

Spin correlations. The selection rules for the transverse spin-spin correlation function (see table 9.1) imply

$$G_{\sigma\sigma}^{\perp}(\tau, x) \sim A_{2k_F} \frac{\cos(2k_F x + \varphi)}{|v_s \tau + ix||v_c \tau + ix|^{(\theta/4)}}$$

$$+ A_0 \left(\frac{1}{(v_s \tau + ix)^2} + \frac{1}{(v_s \tau - ix)^2} \right) \tag{9.48}$$

$$+ A_{4k_F} \left(\frac{\exp(4ik_F x + \varphi')}{(v_s \tau + ix)^2 |v_c \tau + ix|^{\theta}} + \text{c.c} \right) + \dots$$

As in our argument leading to (9.45) for the density-density correlation function the transformation properties of the spin operators under parity imply that

$$G_{\sigma\sigma}^{a}(\tau, x) = G_{\sigma\sigma}^{a}(\tau, -x), \quad a = \perp, z, \tag{9.49}$$

and hence $\varphi = \varphi' = 0$ in (9.48).

As for the density correlations above, the leading contribution asymptotically is that at wavenumber $2k_F$ for any non-zero u. The $k = 0$-behaviour of $G_{\sigma\sigma}^{\perp}$ is determined by the contribution due to the choice $D_c = \pm\frac{1}{2}$ and $D_s = \mp1$. This leads to the conformal dimensions $\Delta_c^{\pm} = 0$ and $(\Delta_s^+, \Delta_s^-) = (0, 1)$ and $(1, 0)$ of the $SU(2)$ spin current which is a primary field.

In absence of a magnetic field the longitudinal and the transverse spin-spin correlations should be identical due to the spin-$SU(2)$ symmetry of the Hubbard model (see Chapter 2). Simple application of the selection rules from table 9.1 for the $G_{\sigma\sigma}^{z}$ results a form identical to that of the density density correlator (9.47) – up to the numerical values of the amplitudes. That the expected symmetry is not transparent should come as no surprise since the presence of the spin-$SU(2)$ has not been used at all up to this point: in our general scheme the selection rules for the terms contributing to the correlation function of a given local operator are based only on the conservation of electron numbers with spins up and down. The spin-$SU(2)$ symmetry does lead to additional selection rules for the form factors $\langle q|S^z(0)|\Omega\rangle$ as compared to those of the local density. At zero magnetic field, $B = 0$, the ground state $|\Omega\rangle$ of the Hubbard model is an $SU(2)$-singlet. The matrix elements

$$\langle q|S_1^z|\Omega\rangle \tag{9.50}$$

determine which intermediate states $|q\rangle$ can contribute to the asymptotic behaviour of

$G^z_{\sigma\sigma}$. Using the commutation relations (2.69) between the spin operators and the fact that $\vec{S}^2_{tot}|\Omega\rangle = 0$ one easily obtains ($\vec{S} = \sum_i \vec{S}_i = (S^x, S^y, S^z)$ is the operator of the total spin, see (2.66))

$$\vec{S}^2\left(S^z_1|\Omega\rangle\right) = \left[\vec{S}^2, S^z_1\right]|\Omega\rangle = 2\,S^z_1|\Omega\rangle. \tag{9.51}$$

This shows that $S^z_1|\Omega\rangle$ is a triplet of the spin-$SU(2)$. As a consequence, only intermediate states which are spin-triplets and have $\langle S^z\rangle = 0$ can contribute to $G^z_{\sigma\sigma}$ for $B = 0$. It follows from the highest-weight property (3.99) that these states are not Bethe Ansatz states! However, due to the completeness of the Bethe Ansatz (see Chapter 4), the corresponding energies are found in the $\langle S^z\rangle = 1$-sector. Hence, the energies of the states $|q\rangle$ with non-vanishing form factors (9.50) are obtained from the finite-size spectrum (8.36) with $(\Delta N_c, \Delta N_s) = (0, -1)$ and the allowed values for D_{cs}, as given by equation (8.39). To summarize, we find that for zero magnetic field, $B = 0$, the selection rules for the conformal dimensions contributing to the longitudinal spin correlation function $G^z_{\sigma\sigma}$ are – as a consequence of the spin-$SU(2)$ symmetry of the system and the fact that the ground state of the repulsive Hubbard model is a spin singlet – identical to those for the transversal spin correlation function and consequently

$$G^z_{\sigma\sigma}(\tau, x) = \frac{1}{2}G^\perp_{\sigma\sigma}(\tau, x). \tag{9.52}$$

For $u \to \infty$ the equal time correlation functions $G_{\sigma\sigma}(x)$ decays as $x^{-3/2}$ at any filling $n_c < 1$. At half filling, where only magnetic excitations are gapless, a different exponent is found as shown in Section 9.3.2 below.

Field correlator. Finally, let us compute the Green's function $G_{\psi\psi^\dagger}$ within the conformal approach. Using the selection rules from table 9.1 we obtain

$$G_{\psi\psi^\dagger}(\tau, x) \sim A\left(\frac{\exp(-ik_F x)}{\sqrt{(v_c\tau + ix)(v_s\tau + ix)}} + c.c.\right)\left[\frac{1}{v_c^2\tau^2 + x^2}\right]^{\alpha_1/2}$$

$$+ B\left(\frac{\exp(-3ik_F x)}{\sqrt{(v_c\tau + ix)^3(v_s\tau - ix)}} + c.c.\right)\left[\frac{1}{v_c^2\tau^2 + x^2}\right]^{\alpha_3/2} \tag{9.53}$$

$$+ \dots.$$

This is an example for the effect of the conformal spins of the charge and spin part of an operator not being integer or half integer separately: both s_c and s_s are $\frac{1}{4}$ for the first term which produces the branch cut singularities in the correlation function. In the limit $u \to 0$ the velocities v_c and v_s coincide and the singularities cancel each other producing the usual quasi particle pole in the Green's function.

The anomalous exponent $\alpha_1 = 1/\theta + \theta/16 - 1/2$ which determines the singularity of the term oscillating with k_F is a monotonically growing function of u with $\alpha_1(u = 0) = 0$ and $\alpha_1(u \to \infty) = \frac{1}{8}$ (see figure 9.3). Using our result (8.49) for the small-u behaviour of $\xi(z_0)$ we find that $\alpha_1 = (u/2\pi\,\sin(Q))^2$ in agreement with the weak coupling result (9.22) from bosonization. This result shows one of the striking consequences of the interaction

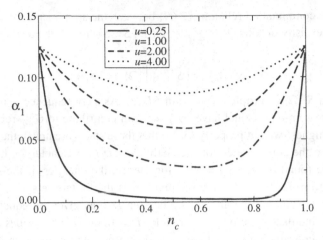

Fig. 9.3. Anomalous exponent α_1 of the k_F singularity in the Green's function (9.53) versus the electron density n_c for various values of the Hubbard interaction u.

in the one-dimensional Hubbard model – and in fact any one-dimensional system within the universality class of the so-called Tomonaga-Luttinger liquid: according to (9.A.11) the momentum distribution function of the electrons which is just the Fourier transformation of the equal time field-field correlation function (9.53) has an algebraic singularity at the Fermi points $\pm k_F$:

$$n(k) = \frac{1}{2} + |k - k_F|^{\alpha_1} \operatorname{sign}(k_F - k) \qquad \text{for } k \approx k_F . \tag{9.54}$$

This is in contrast to the usual zero temperature Fermi distribution function with a step at k_F (corresponding to $\alpha_1 = 0$): this is a consequence of the enhancement of quantum fluctuations in a one-dimensional quantum system due to correlations. Increasing the interaction strength further[7] does lead to even larger fluctuation effects (corresponding to larger values of α_1).

Similarly as with the $4k_F$ singularity in G_{nn}, the interaction gives rise to new singularities of the Green's function oscillating with odd multiples of the Fermi momentum k_F. This is an indication for the fact that the electrons are not the elementary excitations in the interacting system but rather scattering states of multiple holons and anti-holons. The exponent of the first such singularity, $\alpha_3 = 1/\theta + 9\theta/16 - 3/2$, decreases from 1 to $\frac{1}{8}$ as u goes from 0 to ∞. This result holds at any density n_c below half-filling (see figure 9.4).

9.3.2 Half-filled band

From our previous analysis we know that at half filling, $n_c = 1$, only charge-neutral excitations are gapless. Therefore only these states show up in the finite-size spectra computed

---

[7] Within the Hubbard model this is not possible beyond $u = \infty$. With an additional nearest-neighbour interaction, however, α_1 can increase beyond 1/8 [382].

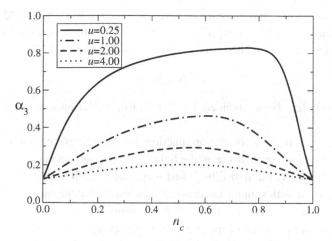

Fig. 9.4. Anomalous exponent α_3 of the $3k_F$ singularity in the Green's function (9.53) as a function of the electron density n_c for various values of the Hubbard interaction u.

in Section 8.2.3. The finite-size scaling of the ground state energy (8.53) and low-lying excitations (8.54) implies that the conformal field theory describing this critical model is a $c = 1$ Gaussian model of a free boson [489]. The operator dimensions are

$$\Delta^\pm = \frac{1}{2}\left(D_s\xi_s(A) \pm \frac{\Delta N_s}{2\xi_s(A)}\right)^2 \tag{9.55}$$

where $\xi_s(A)$ varies between $1/\sqrt{2}$ and 1 as a function of the external magnetic field. One should note, that the momentum of the intermediate state is now given by $2D_sk_{F,\downarrow}$ for states with ΔN_s even but $\pi/a_0 + 2D_sk_{F,\downarrow}$ for states with ΔN_s odd.

We start our discussion of the correlation functions at half-filling with the spin-spin correlators which we expect to decay algebraically based on our previous findings. Again we will express the critical exponents in terms of $\theta_s = 2\xi_s(A)^2$ which grows from 1 for vanishing magnetic field to 2 for $h = h_c$. Using the methods introduced above the leading contributions to the longitudinal and transverse spin-spin correlations are found to decay as

$$G_{\sigma\sigma}^z(\tau, x) \sim m^2 + A_1\frac{\cos(2k_{F,\downarrow}x)}{|v_s\tau + ix|^{\theta_s}} + A_2\frac{x^2 - v_s^2\tau^2}{(x^2 + v_s^2\tau^2)^2}, \tag{9.56a}$$

$$G_{\sigma\sigma}^\perp(\tau, x) \sim B_1\frac{\cos(\frac{\pi}{a_0}x)}{|v_s\tau + ix|^{(1/\theta_s)}}, \tag{9.56b}$$

$$+ B_2\left(\frac{\exp(i(\frac{\pi}{a_0} - 2k_{F,\downarrow})x)}{(v_s\tau - ix)^2} + \text{c.c.}\right)\left[\frac{1}{x^2 + (v_s\tau)^2}\right]^{\nu/2},$$

where $\nu = 2 - \theta_s - 1/\theta_s$ and we have used (9.49). In absence of a magnetic field the magnetization m vanishes and we have $2k_{F,\downarrow} = \pi/a_0$, $\theta_s = 1$ and $\nu = 0$. Hence the spin correlations are isotropic and the staggered part decays like $A_1\cos(\pi x/a_0)/x$ at equal times.

As mentioned above, marginally irrelevant perturbations to a conformal field theory lead to logarithmic corrections to these expressions. For example, the coefficient A_1 in the equal time spin-spin correlation functions at $h = 0$ has been found to be [137]

$$A_1 \propto \sqrt{\ln |x|}. \tag{9.57}$$

(The same result has been obtained for the isotropic Heisenberg antiferromagnet in Ref. [10].)

For the analysis of density correlation functions we have to extend our approach for the computation of critical exponents from the finite-size spectrum to make full use of the full set of symmetries present at half filling. Just using the selection rules of conservation of numbers of electrons with spin up and down, G_{nn} would show the same algebraic decay as the longitudinal spin correlator in (9.56a) with different amplitudes A_i. Using the modified selection rules in the presence of the η-$SU(2)$ in the charge sector at half-filling we will now show that all the amplitudes of such algebraically decaying contributions to G_{nn} do actually vanish [118]. Quantitative results on the dynamical correlation functions involving charge degrees of freedom can be obtained in a particular continuum limit of the Hubbard model (see Chapter 10).

At half filling the ground state $|\Omega\rangle$ of the one-dimensional Hubbard model is an $SO(4)$-singlet for magnetic field $B = 0$ [120, 293]. In the presence of a magnetic field the ground state (in the half-filled band) remains a singlet of the η-pairing $SU(2)$. From (2.80c) the electronic density operator can be expressed in terms of the local generators of the η-pairing $SU(2)$ as $\eta^z(\tau, x) = \frac{1}{2}(n(\tau, x) - 1)$. Hence the connected density correlation function is

$$G_{nn}(\tau, x) = 4\langle \eta^z(\tau, x)\eta^z(0, 0)\rangle \tag{9.58}$$

and the matrix elements

$$\langle q|\eta_1^z|\Omega\rangle \tag{9.59}$$

(η_1^z is the generator of the local η-pairing $SU(2)$ at site 1 of the lattice) determine which intermediate states $|q\rangle$ will contribute to the asymptotic behaviour of G_{nn}.

The calculation of the total η-spin quantum number of the state $\eta_1^z|\Omega\rangle$ is identical to that in (9.51): the commutation relations (2.81) between the η-pairing operators and the fact that $|\Omega\rangle$ is a singlet under the η-pairing $SU(2)$ imply that this state is a triplet of the η-pairing $SU(2)$. As a consequence, only intermediate states which are η-pairing triplets can contribute to the correlation function G_{nn} at half-filling.

In Chapter 7 we have shown that only *pure* spin excitations are gapless in the repulsive half-filled Hubbard model. These are all singlets of the η-pairing $SU(2)$. Therefore the corresponding matrix elements (9.59) vanish identically and cannot contribute in G_{nn}. We conclude that the lowest-energy intermediate states with non-zero matrix elements (9.59) are holon-antiholon scattering states with energy above the Mott-Hubbard gap. As a result the density-density correlation function exhibits exponential decay at large distances for any positive u

$$G_{nn}(\tau, x) \sim \exp(-\alpha x), \quad x \to \infty. \tag{9.60}$$

The determination of $\alpha(u)$ is an interesting open problem. On general grounds it can be expected to be proportional to the holon gap (see Section 7.2). In the scaling limit (see Chapter 10) this is indeed the case.

We emphasize that to establish the exponential decay of the density-density correlation function G_{nn} in the above way it has been essential that:[8]

(i) There is an *exact* symmetry in the charge sector of the microscopic Hamiltonian (2.22). For the Hubbard model this is the η-pairing $SU(2)$. For extended versions of the Hubbard model the existence of such a symmetry has to be established.

(ii) The ground state is a singlet of the corresponding algebra. Note that the above considerations still hold in the presence of a magnetic field as the ground state (in the half-filled band) remains a singlet of the η-pairing $SU(2)$.

(iii) All charged (non-singlet) excitations are gapped.

If one would consider correlation functions of 'point-split' densities such as $N_j^{(ps)} = \sum_\sigma c_{j\sigma}^\dagger c_{j+1,\sigma}$ the above symmetry argument does not imply the vanishing of e.g. the matrix element $\langle ST|N_1^{(ps)}|\Omega\rangle$, where $|ST\rangle$ denotes a spin-triplet excitation. Given that symmetry does not force such matrix elements to vanish we expect them to be nonzero, which then immediately implies an algebraic decay of the corresponding correlation function $G_{nn}^{(ps)}$ as in (9.56a).

Analogous conclusions can be reached for spin-spin correlations in the *attractive* Hubbard model. Application of the Shiba-transformation (2.59) changes the sign of the interaction in the Hamiltonian (2.22), whereas η-pairing and spin $SU(2)$ symmetries are interchanged. Furthermore the ground state of the attractive Hubbard model is a spin singlet. This implies that spin-spin correlation functions in the attractive Hubbard model decay exponentially at large distances

$$G_{\sigma\sigma}^z(\tau, x) \longrightarrow \exp(-\beta x), \quad x \to \infty \tag{9.61}$$

where $\beta > 0$ for any $u < 0$.

9.3.3 Magnetic field effects in the strong coupling limit

As we have seen in Section 8.2.4, the strong coupling limit of the repulsive Hubbard model allows for a detailed study of the dependence of the excitation spectrum and hence the critical exponents on the magnetic field. Apart from this, there exists an independent method for the calculation of correlation functions in this limit [344,345,353]: For $u \to \infty$, the Bethe Ansatz wave functions take a particularly simple form which allows to evaluate

---

[8] We note that these conditions are fulfilled for the half-filled Hubbard model on a bipartite lattice in *any* dimension, provided that u is larger than the critical Mott-Hubbard value [293].

correlation functions directly. The numerical and analytical (at quarter filling and $B = 0$) results obtained within this approach are in perfect agreement with the conformal field theory approach used here.

As a most remarkable result of our investigation of the finite size spectrum in the strong coupling limit we found that the dressed charge matrix exhibited two very different dependencies on the external field. This has an interesting consequence on the conformal dimensions (9.38) entering the expressions for the correlation functions [141]: For a small magnetic field we obtain from (8.67)

$$2\Delta_c^{\pm}(\Delta\mathbf{N}, \mathbf{D}) = \left(D_c + \frac{1}{2}D_s \pm \frac{1}{2}\Delta N_c \right)^2 - \frac{4B}{\pi^2 B_c} \left(D_c + \frac{1}{2}D_s \pm \frac{1}{2}\Delta N_c \right) D_s ,$$

$$2\Delta_s^{\pm}(\Delta\mathbf{N}, \mathbf{D}) = \frac{1}{2} \left(D_s \pm (\Delta N_s - \frac{1}{2}\Delta N_c) \right)^2$$

$$+ \frac{1}{4\ln(B_1/B)} \left(D_s^2 - (\Delta N_s - \frac{1}{2}\Delta N_c)^2 \right) . \tag{9.62}$$

Hence the magnetic field dependence of the critical dimensions for the charge-excitations is much weaker than that of the spin-excitations. This is not surprising since the magnetic field couples directly to the spin degree of freedom. In general, the exponents of equal time correlators (where only the sum of Δ_c and Δ_s enters) will be dominated by the latter, in time-dependent quantities, however, this effect should become observable. The (weak) B-dependence of the charge exponents shows, however, that their spin and charge are not completely independent. In fact, the conformal spins are dependent on the external magnetic field. According to our discussion above this is another signature of the mutation of holons and spinons as B varies.

The presence of a magnetic field does not change the general form of the correlators as determined within the CFT approach, i.e. they are again a sum of terms (9.39) which have to be selected according to the selection rules of the operator considered. We do have to consider the Green's functions for spin-up and spin-down electrons separately in principle. It turns out however, that for small magnetic fields there is a simple mapping between the various terms: The leading terms in $G_{\psi\psi\dagger}^{\downarrow}$ are obtained from the corresponding ones in $G_{\psi\psi\dagger}^{\uparrow}$ by interchanging the Fermi momenta $k_{F,\uparrow}$ and $k_{F,\downarrow}$ and replacing B by $-B$ in the expressions for $\Delta_c^{\pm}$.

As in the zero-field regime the leading contribution to $G_{\psi\psi}^{\uparrow}$ oscillates with wave number $k_{F,\uparrow}$. Near $B = 0$ the exponents are according to (9.62)

$$2\Delta_c^+ = \frac{1}{16} - \frac{B}{2\pi^2 B_c}, \qquad 2\Delta_c^- = \frac{9}{16} + \frac{3B}{2\pi^2 B_c},$$

$$2\Delta_s^+ = \frac{1}{2} \left(\frac{1}{4\ln(B_1/B)} \right)^2, \qquad 2\Delta_s^- = \frac{1}{2} + \frac{1}{2} \left(\frac{1}{4\ln(B_1/B)} \right)^2 . \tag{9.63}$$

The logarithmic field dependence of $\Delta_s^{\pm}$ cancels to first order. Fortunately, the next order is completely fixed by the leading correction to Z_{ss} (8.67).

There is another contribution with wavenumber $k_{F,\uparrow} + 2k_{F,\downarrow}$ (corresponding to $D_c = D_s = -1/2$) with (again the corrections of order $(\ln(B_1/B))^{-1}$ cancel)

$$2\Delta_c^+ = \frac{1}{16} - \frac{B}{2\pi^2 B_c}, \qquad 2\Delta_c^- = \frac{25}{16} - \frac{5B}{2\pi^2 B_c},$$
$$2\Delta_s^+ = \frac{1}{2} + \frac{1}{2}\left(\frac{1}{4\ln(B_1/B)}\right)^2, \qquad 2\Delta_s^- = \frac{1}{2}\left(\frac{1}{4\ln(B_1/B)}\right)^2. \tag{9.64}$$

This term becomes the $3k_F$ singularity at zero magnetic field (9.53).

The leading contributions to the asymptotics of the density correlation function G_{nn} beyond the constant term are found to have wavenumber $2k_{F,\uparrow}$ (corresponding to $D_c = -D_s = -1$). For small magnetic field the corresponding critical dimensions are

$$2\Delta_c^{\pm} = \frac{1}{4} + \frac{2B}{\pi^2 B_c}, \qquad 2\Delta_s^{\pm} = \frac{1}{2} + \frac{1}{4\ln(B_1/B)}. \tag{9.65}$$

The contribution with wavenumber $2k_{F,\downarrow}$ ($D_c = 0$ and $D_s = -1$) has the same dimensions with B in the expression for $\Delta_c^{\pm}$ replaced by $-B$. As we have discussed above the amplitudes of these contributions are known to vanish in the strong coupling limit without a magnetic field [344].

The same terms can be expected to contribute to the longitudinal spin-spin correlation functions $G_{\sigma\sigma}^z$ which – in the absence of further symmetries – has the same selection rules as G_{nn}. The logarithmic field dependence in the spinon exponents $\Delta_s^{\pm}$ is the same as the one found in the isotropic Heisenberg chain [63]. Finally, the transverse spin-spin correlation function $G_{\sigma\sigma}^{\perp}$ has its leading singularity at wavenumber $k_{F,\uparrow} + k_{F,\downarrow}$. We find for the dimensions in the strong coupling limit

$$2\Delta_c^{\pm} = \frac{1}{4}, \qquad 2\Delta_s^{\pm} = \frac{1}{2} - \frac{1}{4\ln(B_1/B)} \qquad \text{for } B \to 0. \tag{9.66}$$

In a magnetic field close to the saturation field B_c where the system becomes completely polarized the field dependence of the conformal dimensions to leading order is given by

$$2\Delta_c^{\pm}(\Delta\mathbf{N}, \mathbf{D}) = \left(D_c \pm \frac{1}{2}\Delta N_c\right)^2 + \frac{4}{\pi}\sqrt{1 - \frac{B}{B_c}}\left(D_c \pm \frac{1}{2}\Delta N_c\right)D_s,$$
$$2\Delta_s^{\pm}(\Delta\mathbf{N}, \mathbf{D}) = \left(D_s \pm \frac{1}{2}\Delta N_s\right)^2 \tag{9.67}$$
$$- \frac{2}{\pi}\sqrt{1 - \frac{B}{B_c}}\left(D_s \pm \frac{1}{2}\Delta N_s\right)\left(D_s \pm \left(\Delta N_c - \frac{1}{2}\Delta N_s\right)\right).$$

From this expression we find for the exponents of the $k_{F\uparrow}$-singularities of the Green's function $G^{\uparrow}_{\psi\psi\dagger}$

$$2\Delta^+_c = 0, \qquad\qquad 2\Delta^-_c = 1 - \frac{2}{\pi}\sqrt{1 - \frac{B}{B_c}},$$

$$2\Delta^+_s = \frac{1}{4} - \frac{3}{2\pi}\sqrt{1 - \frac{B}{B_c}}, \qquad 2\Delta^-_s = \frac{1}{4} + \frac{1}{2\pi}\sqrt{1 - \frac{B}{B_c}},$$

$$(9.68)$$

and for the corresponding $k_{F\downarrow}$-singularities of the Green's function $G^{\downarrow}_{\psi\psi\dagger}$

$$2\Delta^{\pm}_c = \frac{1}{4} \mp \frac{1}{\pi}\sqrt{1 - \frac{B}{B_c}}, \quad 2\Delta^+_s = 0, \quad 2\Delta^-_s = 1 - \frac{2}{\pi}\sqrt{1 - \frac{B}{B_c}}. \qquad (9.69)$$

(Here we have neglected contributions of order $\mathcal{O}(1 - B/B_c)$.)

At $B \geq B_c$ a phase transition similar to the one found at half filling occurs: excitations with spin develop a gap and the corresponding contributions to the correlation functions decay like exponentials asymptotically. The other excitations involve only spin-up electrons and remain massless. They can lead to algebraically decaying correlation functions. For the Hubbard model these are identical to those of non-interacting spinless fermions.

9.4 Correlation functions in momentum space

For experimental and other applications one is often interested in the Fourier transforms of two-point correlation functions. Their low-energy asymptotics can be determined by Fourier transforming the large-distance behaviour obtained above. Here we discuss the results of such a procedure for some particular examples.

9.4.1 Spectral function

The spectral function $A(\omega, k)$ is obtained from the imaginary part of the retarded single particle Green's function

$$G^{(R)}(\omega, k) = -i\theta_H(t)\langle\{c_{j+1,\sigma}(t), c^{\dagger}_{1,\sigma}\}\rangle, \qquad (9.70)$$

$$A(\omega, k) = -\frac{1}{\pi}\text{Im}\, G^{(R)}(\omega, k)$$

$$= \frac{1}{2\pi} \int_{-\infty}^{\infty} dx \int_{-\infty}^{\infty} dt\, e^{i\omega t - iqx} \left[\langle c_{j+1,\sigma}(t)\, c^{\dagger}_{1,\sigma}\rangle + \langle c^{\dagger}_{1,\sigma}\, c_{j+1,\sigma}(t)\rangle\right]. \qquad (9.71)$$

In what follows we will evaluate the spectral function in zero magnetic field by Fourier transforming the asymptotic form of the single-particle Green's function. This can be done following Refs. [309, 323, 472] and is summarized in the Appendix to Chapter 18 of the textbook [168]. We reproduce this discussion here for the sake of completeness. For momenta k in the vicinity of k_F and $3k_F$ the dominant contribution to the Fourier transform in (9.71) comes from the k_F and $3k_F$ harmonics of the asymptotic form of the Green's function (9.53). In particular, we have

$$A(\omega, k_F + q) \approx \frac{A}{2\pi} \int_{-\infty}^{\infty} dx \int_{-\infty}^{\infty} dt \; e^{i\omega t - iqx} \Big\{ [\delta + i(v_c t - x)]^{-\frac{1}{2}} [\delta + i(v_s t - x)]^{-\frac{1}{2}}$$

$$\times \left[(\delta + iv_c t)^2 + x^2\right]^{-\frac{\alpha_1}{2}} + (t, x) \to (-t, -x) \Big\}, \quad (9.72)$$

where δ is an infinitesimal regularization and $|q| \ll k_F$. It is apparent from (9.72) that the spectral function has the symmetry

$$A(\omega, k_F + q) = A(-\omega, k_F - q). \quad (9.73)$$

By virtue of (9.73) it is sufficient to determine the positive frequency behaviour of $A(\omega, k_F + q)$.

Singular behaviour. On general grounds one may expect that the singularities of $A(\omega, k_F + q)$ occur on the 'light-cones' $\omega = \pm v_{c,s} q$. The behaviour in these regions can be determined along the lines discussed in Appendix 9.A. Setting $2\Delta_c^- = \frac{1+\alpha_1}{2}$, $2\Delta_c^+ = \frac{\alpha_1}{2}$, $2\Delta_s^- = \frac{1}{2}$ and $2\Delta_s^+ = 0$ in (9.A.8) we obtain

$$A(\omega, k_F + q) \sim \begin{cases} (\omega - v_c q)^{\frac{\alpha_1 - 1}{2}} & \text{for } \omega \to v_c q, \\ (\omega + v_c q)^{\frac{\alpha_1}{2}} & \text{for } \omega \to -v_c q, \\ (\omega - v_s q)^{\alpha_1 - \frac{1}{2}} & \text{for } \omega \to v_s q, \\ (\omega + v_s q)^{\alpha_1} & \text{for } \omega \to -v_s q. \end{cases} \quad (9.74)$$

As $0 < \alpha_1 \leq \frac{1}{8}$ there are singularities for $\omega \to v_{c,s} q$, but $A(\omega, k_F + q)$ vanishes in a root like fashion for $\omega \to -v_{c,s} q$.

Explicit calculation of the integrals. After the substitution $x' = x - v_s t$ the integral over t in (9.72) may be carried out by means of the identity GR 3.384.8 of [184]

$$\int_{-\infty}^{\infty} dt \; e^{-ipt} [\beta + it]^{-\mu} [\gamma + it]^{-\nu} = \frac{2\pi}{\Gamma(\mu + \nu)} e^{\gamma p} (-p)^{\mu + \nu - 1} \theta_H(-p)$$

$$\times \Phi(\mu, \mu + \nu, [\beta - \gamma] p), \quad (9.75)$$

where $\text{Re}(\beta) > 0$, $\text{Re}(\gamma) > 0$, $\text{Re}(\mu + \nu) > 1$, $\Phi(a, b, z)$ denotes a confluent hypergeometric function and where θ_H denotes the Heaviside function. In the next step one reexpresses the confluent hypergeometric function by means of the integral representation (see 13.2.1

of [3])

$$\Phi(a, b, z) = \frac{\Gamma(b)}{\Gamma(a)\Gamma(b-a)} \int_0^1 ds\, s^{a-1}\, (1-s)^{b-a-1}\, e^{sz}\,. \tag{9.76}$$

Now one may perform the integral over x' in (9.72) by means of the identity

$$\int_{-\infty}^{\infty} dx\, e^{-ipx}\, [\beta - ix]^{-\nu} = \frac{2\pi}{\Gamma(\nu)}\, e^{-\beta p}\, p^{\nu-1}\, \theta_H(p) \tag{9.77}$$

where $\mathrm{Re}(\beta) > 0$, $\mathrm{Re}(\nu) > 0$. Sending the regularization δ to zero we arrive at

$$A(\omega, k_F + q) \approx f(\omega, q) + f(-\omega, -q)\,,$$

$$f(\omega, q) = \frac{2\pi A\, v_c^{-\frac{1}{2}} v_+^{\frac{1-\alpha_1}{2}} v_-^{-\frac{\alpha_1}{2}}}{\Gamma(\frac{1}{2})\Gamma(\frac{\alpha_1}{2})\Gamma(\frac{1+\alpha_1}{2})} \theta_H(\omega - v_s q)\, (\omega - v_s q)^{\alpha_1 - \frac{1}{2}}$$

$$\times \int_0^1 ds\, s^{\frac{\alpha_1}{2} - \frac{1}{2}} (1-s)^{\frac{\alpha_1}{2} - 1} \left[\frac{v_-}{v_c}(\omega + v_c q) - 2s(\omega - v_s q) \right]^{-\frac{1}{2}}$$

$$\times \theta_H \left(\frac{v_-}{v_c}(\omega + v_c q) - 2s(\omega - v_s q) \right), \tag{9.78}$$

where we have defined $v_{\pm} = v_c \pm v_s$. Last but not least the s-integration can be carried out by means of the identity GR 3.197.3 of [184] to give a hypergeometric function

$$\int_0^1 ds\, s^{\lambda-1}\, (1-s)^{\mu-1}\, (1 - \beta x)^{-\nu} = \frac{\Gamma(\lambda)\Gamma(\mu)}{\Gamma(\lambda+\mu)} F(\nu, \lambda, \lambda+\mu; \beta). \tag{9.79}$$

Here $\mathrm{Re}(\lambda) > 0$ and $\mathrm{Re}(\mu) > 0$. Because the Heaviside function under the s-integration must be satisfied we need to distinguish between two cases.

- Case A: $v_s q < \omega < v_c q$

$$A(\omega, k_F + q) \approx \frac{2\pi\, v_-^{-\frac{1}{2} - \frac{\alpha_1}{2}} v_+^{\frac{1}{2} - \frac{\alpha_1}{2}}}{\Gamma(\frac{1}{2})\Gamma(\alpha_1 + \frac{1}{2})} (\omega - v_s q)^{\alpha_1 - \frac{1}{2}} (\omega + v_c q)^{-\frac{1}{2}}$$

$$\times F\left(\frac{1}{2}, \frac{1}{2} + \frac{\alpha_1}{2}, \frac{1}{2} + \alpha_1; \frac{2v_c}{v_-} \frac{\omega - v_s q}{\omega + v_c q} \right). \tag{9.80}$$

- Case B: $\omega > |v_c q|$

$$A(\omega, k_F + q) \approx \frac{2\pi\, (2v_c)^{-\frac{1}{2} - \frac{\alpha_1}{2}} v_+^{\frac{1}{2} - \frac{\alpha_1}{2}}}{\Gamma(\frac{\alpha_1}{2})\Gamma(1 + \frac{\alpha_1}{2})} (\omega - v_s q)^{\frac{\alpha_1}{2} - 1} (\omega + v_c q)^{\frac{\alpha_1}{2}}$$

$$\times F\left(1 - \frac{\alpha_1}{2}, \frac{1}{2} + \frac{\alpha_1}{2}, 1 + \frac{\alpha_1}{2}; \frac{v_-}{2v_c} \frac{\omega + v_c q}{\omega - v_s q} \right). \tag{9.81}$$

The behaviour of $A(\omega, k_F + q)$ for negative frequencies $\omega < 0$ is easily obtained from (9.80) and (9.81) by means of the symmetry (9.73).

The spectral function in the vicinity of k_F for $u = 1$ at quarter-filling $n_c = 0.5$ is shown in figure 9.5. The two singularities at $\omega = v_{c,s} q$ are clearly visible. The features for $q < 0, \omega > 0$ and $q > 0, \omega < 0$ are too small to be seen. The most relevant excited states contributing to

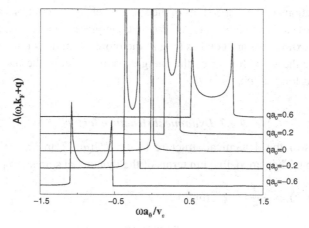

Fig. 9.5. Spectral function $A(\omega, k_F + q)$ as a function of ω for several values of qa_0. The positive (negative) frequency features for $qa_0 < 0$ ($qa_0 > 0$) are too small to be visible.

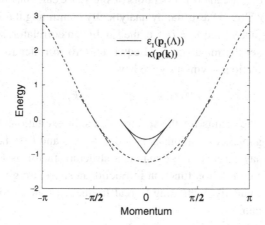

Fig. 9.6. Dressed energies $\kappa(p(k))$ and $\varepsilon_1(p_1(\Lambda))$ as functions of the dressed momenta and linear approximations to the dispersions.

$A(\omega, k_F + q)$ are the holon-spinon excitation at $\omega < 0$ and the antiholon-spinon excitation at $\omega > 0$. They are shown in figures 7.34 and 7.33 respectively.

The two peaks in $A(\omega, k_F + q)$ dispersing with velocities v_s and v_c respectively are a direct manifestation of spin-charge separation. The peaks are power-law singularities with interaction-dependent exponents (9.74) rather than δ-functions or Lorentzians. The spectral function of the Hubbard model is dramatically different from the one of noninteracting electrons. The latter is simply

$$A_{\text{free}}(\omega, k_F + q) = \delta(\omega - v_F q). \tag{9.82}$$

An important question is in which interval of momentum transfers q the results (9.80), (9.81) can be trusted. A useful criterion is obtained by considering how good the linear approximations to the dispersions of the spin and charge excitations are. In figure 9.6 we show the linear approximations to the dressed energies ε_1 and κ, see Section 7.7. We see that the approximations are good as long as the momentum does not deviate more than about $|\pi/8|$ from the value for which the energies vanish. Hence the results (9.80),(9.81) may be trusted as long as $|qa_0| < \pi/8$.

9.4.2 Dynamical structure factor

The components of the dynamical structure factor $S^{\alpha\beta}(\omega, k)$ are measured by inelastic neutron scattering. They are defined in terms of the dynamical spin susceptibility by

$$S^{\alpha\beta}(\omega, k) = -\frac{1}{\pi} \text{Im } \chi^{\alpha\beta}(\omega, k), \tag{9.83}$$

$$\chi^{\alpha\beta}(\omega, k) = \int_{-\infty}^{\infty} dx \int_{0}^{\infty} dt \, e^{i\omega t - ikx} \left[\langle S_{j+1}^{\alpha}(t) S_{1}^{\beta} \rangle - \langle S_{1,\sigma}^{\beta} S_{j+1}^{\alpha}(t) \rangle \right]. \tag{9.84}$$

Here $S_j^{\alpha} = \frac{1}{2} c_{j,\tau}^{\dagger} \sigma_{\tau,\tau'}^{\alpha} c_{j,\tau'}$ are the spin operators of site j. We calculate the retarded spin-spin correlation function $\chi^{\alpha\beta}(\omega, k)$ as usual by analytically continuing the Fourier transform of the correlator in imaginary time $\tau = it$, i.e. the Euclidean correlator, see e.g. [4].

From now on we set the magnetic field equal to zero in order to keep the discussion simple. Due to spin rotational symmetry we have

$$\chi^{\alpha\beta}(\omega, k) = \delta^{\alpha\beta} \chi^{\perp}(\omega, k). \tag{9.85}$$

As can be seen from the results for the spectrum of spin-excitations below half filling in figure 7.32 there are gapless excitations at $k \approx 0$, $k \approx 2k_F$ and $k \approx 4k_F$. In the vicinity of these momenta we may estimate the dynamical structure factor as follows. The asymptotics of the spin-spin correlation function in Euclidean space are given in (9.48). Fourier transforming and analytically continuing to real frequencies along the lines discussed in Appendix 9.A we obtain

$$S^{\alpha\alpha}(\omega, 2k_F + q) \sim \begin{cases} \left[\omega^2 - (v_s q)^2\right]^{\frac{\theta}{4} - \frac{1}{2}} & \text{for } \omega - v_s|q| \to 0^+, \\ \left[\omega^2 - (v_c q)^2\right]^{\frac{\theta}{8}} & \text{for } \omega - v_c|q| \to 0^+. \end{cases} \tag{9.86}$$

For a less than half filled band we have always $\theta > 2$ (see figure 9.2). Hence, the contribution (9.86) to the dynamical structure factor in the vicinity of $2k_F$ is always non-singular and will be modified by additional non-singular terms which are not accessible within the conformal approach. The same is found near $k = 0$ and $4k_F$. At half filling however, the structure factor is singular in the vicinity of $2k_F$. This case is discussed in more detail in Chapter 10.

9.5 Correlation functions in the open boundary Hubbard chain

In Section 8.3 we have seen how various interesting quantities such as local expectation values of certain operators could be computed from the exact solution of the Hubbard model with open boundary conditions. Furthermore, we were able to calculate the finite size corrections to the energies of low-lying excitations in this situation which – when combined with the techniques of conformal field theory – can be used to study new correlation functions. Since practically every experiment deals with systems of finite extension, either due to the geometry of the probe or due to the presence of imperfections, this opens new possibilities for the identification of the special properties of one-dimensional correlated electron systems in the real world.

Here we shall discuss two phenomena in more detail, namely Friedel oscillations [147] and Anderson's orthogonality catastrophy [17] in the open Hubbard model. Further applications of these methods to the Hubbard model, e.g. Fermi edge singularities appearing in X-ray scattering experiments or in the I–V characteristics of tunneling experiments have been discussed elsewhere [50, 117]. Similar results have been obtained for the Luttinger model using the Bosonization method [13, 114, 346, 351, 369, 501].

9.5.1 Friedel oscillations

In the presence of an impurity or boundary in a one-dimensional fermion system the translational invariance is broken leading to an inhomogeneous density distribution – so-called Friedel oscillations in the densities with wavenumber $2k_F$ [147], i.e.

$$\langle n(x) \rangle - n_c \sim \frac{\cos(2k_F x + \varphi)}{x^\gamma} \tag{9.87}$$

with an exponent γ depending on the interaction. There have been various attempts using both numerical and analytical methods to clarify the role of the interaction [114, 130, 378, 398].

It was shown by Cardy that the n-point correlation functions of a conformal field theory with a boundary are related to the $2n$-point bulk correlation functions of this system [73]. As a consequence the local density $\langle n(x) \rangle_{\text{open}}$ of the open boundary system can be extracted from the two-point density-density correlation function G_{nn} of the periodic system considered above (see also [478]).

Following Ref. [73] we have to consider the antiholomorphic part $\bar{D}(\bar{z}_1, \bar{z}_2)$ of the two-point bulk correlation function (9.28) of the primary operator $\mathcal{O}(z, \bar{z})$ in the conformal field theory. Then the one-point correlation function of this operator in the semi-infinite geometry is just

$$\langle \mathcal{O}(z, \bar{z}) \rangle \propto \bar{D}(\bar{z}, z) = (\bar{z} - z)^{-2\Delta^+} . \tag{9.88}$$

The oscillating factor has to be taken from the corresponding bulk current-current correlation function ((9.36) or (9.39)).

Using this we obtain the leading asymptotic oscillating contributions to the local density from the density-density correlation function (9.42)

$$\langle n(x) \rangle - n_c \sim A_1 \frac{\cos(2k_{F,\uparrow}x + \varphi_1)}{x^{(Z_{cc}-Z_{sc})^2+(Z_{cs}-Z_{ss})^2}} + A_2 \frac{\cos(2k_{F,\downarrow}x + \varphi_2)}{x^{Z_{sc}^2+Z_{ss}^2}}$$
$$+ A_3 \frac{\cos(2(k_{F,\uparrow} + k_{F,\downarrow})x + \varphi_3)}{x^{Z_{cc}^2+Z_{cs}^2}} . \tag{9.89}$$

The correlation function $G_{\sigma\sigma}^z(\tau, x)$ has the same functional form as the density-density correlator. Therefore, the CFT prediction for the oscillations in the local magnetization $\langle m(x) \rangle$ contains the same terms as $\langle n(x) \rangle$ but different amplitudes A_i.

The critical exponents in (9.89) depend on the elements Z_{ij} of the dressed charge. Since Z is a bulk quantity only the density of electrons, the strength of the interaction and of the magnetic field have an influence on the exponents. Similarly, the amplitudes A_i should not depend on the strength of the boundary potential. However, just as in the case of periodic boundary conditions, a prediction on the dependence of the different amplitudes A_i in the correlation function is beyond the capabilities of the CFT approach. Only in the limit of noninteracting electrons are their values known from the exact expression for this correlation function ($\bar{n}$ is the average bulk density of electrons corrected by the finite size shifts due to the boundaries: $\bar{n} = n_c - \theta/L$)

$$\langle n(x) \rangle \sim \bar{n} - \frac{1}{2\pi x} \sin(2\pi \bar{n} x) \tag{9.90}$$

from which we expect $A_1 = A_2 = \frac{1}{2\pi}$ and $A_3 = 0$ for $u \to 0$.

In Ref. [48] an extensive numerical study of this correlation function has been performed using the density matrix renormalization group (DMRG) method to compute $\langle n(x) \rangle$ and $\langle m(x) \rangle$ for chains of length up to $L = 500$ sites (see figure 9.7). This study has confirmed the predictions of conformal field theory for the critical exponents and provided information on the interaction dependence of the amplitudes A_i. The numerical results have shown that – in perfect agreement with the predictions from the Bethe Ansatz and conformal field theory – the critical exponents are determined by the dressed charge matrix, independent of the boundary conditions; in particular they do not depend on the boundary potential p. At the same time no significant p-dependence of the amplitudes A_i could be observed.

Finally, the analysis of the DMRG results has provided strong evidence for the conjecture that the finite size shift of the average bulk density of electrons due to boundary fields and, similarly, the wave numbers of the Friedel oscillations is given by the phase shifts (8.91) appearing in the analysis of the Bethe Ansatz equations of the open boundary system, i.e.

$$\bar{n}_c = n_c - \frac{1}{L}\theta_c(p), \quad \bar{m} = m - \frac{1}{L}\theta_c(p) + \frac{2}{L}\theta_s(p). \tag{9.91}$$

This gives a physical interpretation to this quantity which first appeared as a purely technical object during the finite size analysis of the Bethe Ansatz equations for the Hubbard model with open boundary conditions.

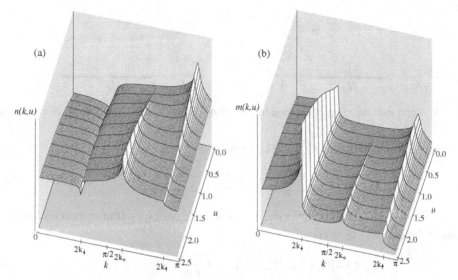

Fig. 9.7. Fourier amplitudes of (a) the density $\langle n(x) \rangle$ and (b) the magnetization $\langle m(x) \rangle$ for the Hubbard chain with reflecting ends ($p = 0 = p'$) for an electron density $n_c = 0.70$ and $n_\uparrow = 0.55$ as a function of u (from Ref. [48]). Singularities are present at $2k_{F\uparrow}$, $2k_{F\downarrow}$ and – for non vanishing interaction u – at $2k_n \equiv 2k_{F\uparrow} + 2k_{F\downarrow}$. The amplitude corresponding to $k = 0$ is not shown.

9.5.2 Orthogonality catastrophe

This problem deals with the system size dependence of the overlap of the many-particle ground states of two systems differing only in the choice of the boundary potential [17]. To study this one has to consider operators changing the boundary conditions of a system [9,13]. Let $\mathcal{O}_p(\zeta)$ be such an operator with ζ a point on the boundary of a given complex region. The action of $\mathcal{O}_p$ is to switch from one boundary condition A (parametrized by $p_A = 0$) left of ζ to a boundary condition B corresponding to a parameter $p_B \equiv p$ right of ζ.

Following Affleck [9] we consider the conformal transformation $\bar{z} \to \bar{w}(\bar{z}) = \exp(\pi\bar{z}/L)$ to obtain a relation between the correlation functions in the infinite strip $0 \leq \text{Im}(\bar{z}) \leq L$ of width L to those in the upper half-plane $\bar{w} = u + iv$, $v \geq 0$. In the latter geometry, the correlation function of the primary boundary operator $\mathcal{O}_p$ decays as a simple power law (let $0 < u_1 < u_2$)

$$\langle \mathcal{O}_p(u_1)\mathcal{O}_p^\dagger(u_2) \rangle = \frac{1}{(u_1 - u_2)^{2x_p}} \, . \tag{9.92}$$

To make contact with the integrable boundary conditions realized in the one-dimensional Hubbard model we now identify the boundary condition A with that of a reflecting end or boundary fields $p = 0$ in the Hamiltonian (8.70) and B with that of a finite boundary potential p. Applying the transformation given above to (9.92) and identifying the imaginary part of z with the spatial variable as before we obtain a correlation function in the strip geometry where due to the action of the operators $\mathcal{O}_p$ and $\mathcal{O}_p^\dagger$ the boundary conditions

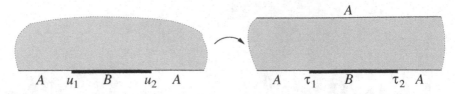

Fig. 9.8. Conformal mapping of the semi-infinite plane with boundary conditions A, B to the strip.

change as a function of time τ ($\tau_i = (L/\pi) \ln u_i$)

$$
\begin{aligned}
AA \quad p = 0 = p' \quad & \text{for } \tau < \tau_1 \text{ and } \tau > \tau_2 \\
BA \quad p \neq 0, \, p' = 0 \quad & \text{for } \tau_1 < \tau < \tau_2
\end{aligned}
\tag{9.93}
$$

(see also figure 9.8). Due to conformal invariance the corresponding correlation function reads for large $\Delta\tau = \tau_2 - \tau_1$

$$
\langle AA | \mathcal{O}_p(\tau_1) \mathcal{O}_p^\dagger(\tau_2) | AA \rangle \sim \left(\frac{\pi}{L} \right)^{2x_p} e^{-\pi x_p \Delta\tau / L} .
\tag{9.94}
$$

Here we have stated explicitly, that this correlation function is to be computed in the ground state $|AA\rangle$ of the model with reflecting ends.

To evaluate the correlation function (9.94) in a transfer matrix approach we have to insert a complete set of eigenstates $|BA; n\rangle$ of the system with boundary potentials $p \neq 0$, $p' = 0$

$$
\sum_n |\langle AA | \mathcal{O}_p | BA; n \rangle|^2 e^{-(E_n^{BA} - E_n^{AA}) \Delta\tau} \sim \left(\frac{\pi}{L} \right)^{2x_p} e^{-\pi x_p \Delta\tau / L} .
\tag{9.95}
$$

For the operator considered here the form factor $\langle AA | \mathcal{O}_p | BA; 0 \rangle$ is expected to be non-zero. Hence we can read off the orthogonality exponent x_p

$$
\langle AA | \mathcal{O}_p | BA; 0 \rangle = |\langle p | 0 \rangle| \sim L^{-x_p}
\tag{9.96}
$$

to be

$$
x_p = \frac{L}{\pi} (E_0^{BA} - E_0^{AA}) .
\tag{9.97}
$$

The key to the correct identification of the orthogonality exponent is the proper choice of the parameters ΔN_i in (8.88) (see also Ref. [49]): as has been argued in Section 8.3.4 the ground state energy E_0^{AA} is obtained by taking $\Delta N_i = \theta_i(p = 0)$ (see equation (8.90)). In the expression (9.97) for the orthogonality exponent we have to compare this energy with that of a different state, namely E_0^{BA}. It is crucial for the correct computation of the critical exponent in this approach to compute all finite-size corrections with respect to the *same* reference state. Since selection rules dictate that $|0\rangle$ and $|p\rangle$ are states with the same particle numbers N and M this implies that one has to choose $\Delta N_i = \theta_i(p = 0)$ in E_0^{BA}, too.[9]

---

[9] This argument can be verified by considering the case of the completely spin-polarized band of the Hubbard model in a sufficiently strong magnetic field corresponding to non-interacting spinless fermions [49].

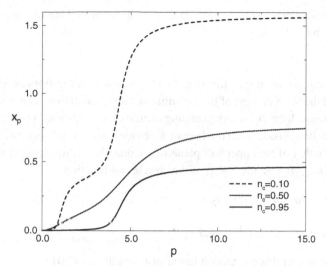

Fig. 9.9. Orthogonality exponent x_p of the open-boundary Hubbard model with $u = 1$ as a function of the boundary potential p for several electron densities (from Ref. [49]).

For vanishing bulk magnetic field $B = 0$ we have $\theta_c(p = 0) = 2\theta_s(p = 0)$ and consequently $\Delta N_c = 2\Delta N_s$. Together with the simplified expression (8.42) for the dressed charge matrix Z in this case we find

$$x_p = \frac{1}{2\xi^2}\Big(\theta_c(p = 0) - \theta_c(p)\Big) . \tag{9.98}$$

Since the parameter ξ has been identified as the contribution of the holon sector to the anomalous critical exponents of the Hubbard model this implies that there are no contributions from the spinon sector to the orthogonality exponent due to a change in the boundary chemical potential. In the limit $p \to \infty$, i.e. switching between vanishing and very strong boundary potential the orthogonality exponent becomes a function of the difference between the bulk density of electrons n_c and the occupation ($n_1 = 2$) of the boundary site in this case

$$x_p = \frac{1}{2\xi^2}(2 - n_c)^2 . \tag{9.99}$$

In figure 9.9 numerical results for the orthogonality exponent as a function of the boundary potential are shown.

Finally, we remark on the effect on a second boundary potential p' at site L of the chain: in the finite-size spectrum (8.88) this gives rise to additional shifts $\theta_i(p')$. The orthogonality exponent $x_{pp'}$

$$\langle pp'|00\rangle \sim L^{-x_{pp'}} , \tag{9.100}$$

however, cannot be obtained by simply adding the new shifts. One rather expects that the exponent is given by the sum

$$x_{pp'} = x_p + x'_p \tag{9.101}$$

of the independent contributions from the two boundaries. In the framework of boundary conformal field theory a change of the potential at *both* boundaries cannot be achieved by the action of a single local boundary changing operator $\mathcal{O}_{pp'}$. Instead, a product of two such operators has to be considered. This becomes obvious when one switches back from the strip geometry to that of the upper half plane where one obtains for the leading asymptotics (provided that $|u_1 - u_2| \ll |u_1|, |u_2|$) of the correlation function

$$\langle \mathcal{O}_{p'}(-u_2) \mathcal{O}_{p'}^{\dagger}(-u_1) \mathcal{O}_p(u_1) \mathcal{O}_p^{\dagger}(u_2) \rangle$$

$$\sim \langle \mathcal{O}_p(u_1) \mathcal{O}_p^{\dagger}(u_2) \rangle \langle \mathcal{O}_{p'}(-u_2) \mathcal{O}_{p'}^{\dagger}(-u_1) \rangle \sim \frac{1}{(u_1 - u_2)^{2x_p}} \frac{1}{(u_1 - u_2)^{2x_{p'}}}. \tag{9.102}$$

Conformal mapping of this expression to the strip results in (9.101).

Appendices to Chapter 9

9.A Singular behaviour of momentum-space correlators

The long distance asymptotics of zero temperature correlation functions are expressed as sums of terms of the form (9.39), i.e.

$$g(\tau, x) = \frac{\exp(ik_0 x)}{(v_c \tau + ix)^{2\Delta_c^+} (v_c \tau - ix)^{2\Delta_c^-} (v_s \tau + ix)^{2\Delta_s^+} (v_s \tau - ix)^{2\Delta_s^-}} . \tag{9.A.1}$$

To compute the Fourier transforms

$$\tilde{g}(\omega, k) = \frac{1}{2\pi} \int dx \int dt \, e^{i(\omega t - kx)} g(t, x) \tag{9.A.2}$$

we first have to rotate the time coordinate from Euclidean time τ to real time t. Analyticity requires

$$\tau = it + 0 \, \text{sign}(t). \tag{9.A.3}$$

Equivalently, the correct regularization of the cuts in the complex time coordinate can be obtained by replacing $\tau = it$ in (9.A.1) and giving an infinitesimal imaginary part to the velocity:

$$v_i \rightarrow v_i e^{-i0}. \tag{9.A.4}$$

All the following equations are to be understood that way. This gives

$$g(t, x) = \frac{\exp(ik_0 x)}{(v_c t + x)^{2\Delta_c^+} (v_c t - x)^{2\Delta_c^-} (v_s t + x)^{2\Delta_s^+} (v_s t - x)^{2\Delta_s^-}} . \tag{9.A.5}$$

As mentioned above this expression holds asymptotically only. Nevertheless, it allows us to calculate the behaviour of the Fourier transforms $\tilde{g}(\omega, k)$ near the singularities

$$\omega = \pm v_{c,s} k_0 . \tag{9.A.6}$$

331

Substituting $y = x + v_c t$ and $y' = qx - \omega t$ in (9.A.2) we obtain

$$
\tilde{g}(\omega, k_0 + q) = \frac{1}{2\pi} (\omega + v_c q)^{2(\Delta_c^- + \Delta_s^+ + \Delta_s^-) - 1} \int \frac{dy}{y^{2\Delta_c^+}}
$$

$$
\times \int dy' \, e^{iy'} \left\{ \left(2 v_c y' + (\omega - v_c q) y\right)^{2\Delta_c^-} \right.
$$

$$
\times \left((v_c - v_s) y' + (\omega + v_s q) y\right)^{2\Delta_s^+}
$$

$$
\left. \times \left((v_c + v_s) y' + (\omega - v_s q) y\right)^{2\Delta_s^-} \right\}^{-1} .
$$

(9.A.7)

From this expression we can read off the singular behaviour of $\tilde{g}(\omega, k_0 + q)$ near $\omega = -v_c q$ and, similarly, $\omega = +v_c q$ and $\omega = \pm v_s q$. For $q \neq 0$ we obtain

$$
\tilde{g}(\omega, k_0 + q) \sim \begin{cases} \text{const.} \, (\omega \mp v_c q)^{2(\Delta_s^+ + \Delta_s^- + \Delta_c^\pm) - 1} & \text{for } \omega \approx \pm v_c q \\ \text{const.} \, (\omega \mp v_s q)^{2(\Delta_c^+ + \Delta_c^- + \Delta_s^\pm) - 1} & \text{for } \omega \approx \pm v_s q \end{cases} .
$$

(9.A.8)

For the Fourier transform of equal time correlators, i.e. terms like

$$
\tilde{g}(k_0 + q) = \int dx \, e^{-i(k_0 + q)x} g(x, t = 0^+) = \int dx \, \frac{e^{-iqx}}{(x - i0)^{2\Delta^+} (x + i0)^{2\Delta^-}}
$$

(9.A.9)

(here $\Delta^\pm = \Delta_c^\pm + \Delta_s^\pm$), one has to consider $q > 0$ and $q < 0$ separately. Contour integration yields

$$
\frac{\tilde{g}(k_0 + q)}{\tilde{g}(k_0 - q)} = \frac{\sin 2\pi \Delta^-}{\sin 2\pi \Delta^+} = (-1)^{2s}, \quad q > 0
$$

(9.A.10)

where $s = \Delta_+ - \Delta_-$ is the (total) conformal spin of the operator under consideration. From (9.38) we know that $2s$ is always an integer. Using this we find that

$$
\tilde{g}(k_0 + q) \sim (\text{sign}(q))^{2s} |q|^\nu, \quad |q| \ll k_0
$$

$$
\nu = 2(\Delta_c^+ + \Delta_c^- + \Delta_s^+ + \Delta_s^-) - 1.
$$

(9.A.11)

The extra sign will appear in correlation functions of Fermi-fields, e.g. the field-field correlator (9.53).

10

Scaling and continuum limits at half-filling

At half-filling the repulsive Hubbard model is in a Mott insulating phase. The charge degrees of freedom are gapped, whereas the spin degrees of freedom remain gapless. At low energies the spin sector is actually scale invariant (apart from logarithmic corrections) and Conformal Field Theory (CFT) methods may be applied to determine the low-energy behaviour of correlation functions involving only the spin sector. On the other hand, the charge sector is not scale invariant and CFT does not provide any information for correlators involving the charge degrees of freedom. In this chapter we will show that there exists a particular continuum limit of the half filled Hubbard model, in which it is possible to calculate dynamical correlation functions by means of methods of integrable quantum field theory. We first construct a Lorentz invariant *scaling limit* starting from the results for the excitation spectrum and the S-matrix discussed in Chapter 7. This scaling limit is identified as the SU(2) Thirring model, which is an integrable relativistic quantum field theory. Next we discuss a *continuum limit*, which is obtained directly from the Hubbard Hamiltonian and describes the vicinity of the scaling limit.

10.1 Construction of the scaling limit

The simplest way of constructing the scaling limit is to start with the results for the dispersions of the elementary excitations and the S-matrix derived in Chapter 7 and then look for a particular limit in which Lorentz invariance is recovered. This was done for the attractive Hubbard model by E. Melzer in Ref. [324] using the results of [120, 121] for the excitation spectrum and the S-matrix of the half filled Hubbard model. A somewhat different path, based on the physical Bethe Ansatz equations (see Section 7.5), was taken by F. Woynarovich and P. Forgacs in [490, 491]. We will construct the scaling limit in the repulsive regime of the half filled Hubbard model following Melzer's work.

The basic idea behind both the scaling and continuum limits is most easily explained by considering the dispersion relations (7.8) for the elementary (anti)holon and spinon excitations. It is clear from figures 7.2 that the (anti)holon gap becomes very small for small u. Hence in the charge sector we have low-energy holon modes in the vicinity of $\mathcal{P}_h = -\frac{\pi}{2}$ and low-energy antiholon modes around $\mathcal{P}_{\bar{h}} = \frac{\pi}{2}$. In the spin sector there are gapless modes at $\mathcal{P}_s = 0, \pi$. In the scaling and continuum limits we concentrate on these low-energy modes only. It turns out that Lorentz invariance emerges as a symmetry of these low-energy degrees of freedom.

On a more formal level, the continuum limit is usually based on taking the hopping t to infinity and simultaneously the lattice spacing a_0 to zero, while keeping the Fermi velocity $v_F = 2ta_0$ fixed.[1] The scaling limit is obtained by taking in addition $u = U/4t$ to zero, while keeping the charge gap fixed

$$t \to \infty, \quad u \to 0,$$

$$M = \frac{8}{\pi} \sqrt{ut} \, \exp\left(-\frac{\pi}{2u}\right) = \text{fixed}. \tag{10.1}$$

In this limit, the (anti)holon energy and momentum become

$$\mathcal{E}_h(k) = \mathcal{E}_{\bar{h}}(k) \longrightarrow M \, \cosh\left(\frac{\pi}{2u} \sin k\right), \quad \frac{\pi}{2} < |k| \le \pi,$$

$$\mathcal{P}_h(k) = \mathcal{P}_{\bar{h}}(k) - \pi \longrightarrow -\frac{\pi}{2} + \frac{M}{2t} \sinh\left(\frac{\pi}{2u} \sin k\right), \quad \frac{\pi}{2} < |k| \le \pi. \tag{10.2}$$

Similarly, the spinon energy and momentum become

$$\mathcal{E}_s(\Lambda) \longrightarrow 4t I_1\left(\frac{\pi}{2u}\right) \exp\left(-\frac{\pi}{2u}|\Lambda|\right),$$

$$\mathcal{P}_s(\Lambda) \longrightarrow \frac{\pi}{2} - \text{sign}(\Lambda)\left[\frac{\pi}{2} - 2I_0\left(\frac{\pi}{2u}\right) \exp\left(-\frac{\pi}{2u}|\Lambda|\right)\right]. \tag{10.3}$$

The limits (10.2), (10.3) can be derived in the same way as (6.B.10), see [324]. Inspection of (10.2) suggests that in order to obtain excitations with a finite energy, it is necessary to scale k to $\pm\pi$ in a particular way

$$k \to \pm\pi, \quad u \to 0, \quad \theta = \frac{\pi}{2u} \sin(k) = \text{fixed}. \tag{10.4}$$

The holon energy and momentum (7.8) then take the following forms in the scaling limit

$$\varepsilon_h(\theta) = \lim_{u \to 0, k \to \pm\pi} \mathcal{E}_h(k) = M \cosh(\theta),$$

$$p_h(\theta) = \lim_{u \to 0, k \to \pm\pi} \frac{\mathcal{P}_h(k) + \frac{\pi}{2}}{a_0} = \frac{M}{v_F} \sinh(\theta). \tag{10.5}$$

Similarly we obtain for the antiholon

$$\varepsilon_{\bar{h}}(\theta) = \lim_{u \to 0, k \to \pm\pi} \mathcal{E}_{\bar{h}}(k) = M \cosh(\theta),$$

$$p_{\bar{h}}(\theta) = \lim_{u \to 0, k \to \pm\pi} \frac{\mathcal{P}_{\bar{h}}(k) - \frac{\pi}{2}}{a_0} = \frac{M}{v_F} \sinh(\theta). \tag{10.6}$$

In (10.5) we recognize the standard parametrization of a Lorentz invariant, massive dispersion in terms of a rapidity variable θ. By eliminating θ we obtain

$$\varepsilon_h = \sqrt{M^2 + v_F^2 p_h^2}. \tag{10.7}$$

---

[1] See however Section (11.1).

In order to obtain finite values for the energy and momentum in the spin sector we need to scale Λ as follows

$$\Lambda \to \pm 2, \quad u \to 0, \quad \beta = \pm\frac{\pi}{2u}(2 - |\Lambda|) = \text{fixed}. \tag{10.8}$$

In the scaling limit the spinon energy and momentum then become

$$\varepsilon_s(\beta) = \lim_{u \to 0, \Lambda \to \pm 2} \varepsilon_s(\Lambda) = \frac{M}{2}\exp(\pm\beta),$$

$$p_s(\beta) = \lim_{u \to 0, \Lambda \to \pm 2} \frac{\mathcal{P}_s(\Lambda) - \frac{\pi}{2}[1 \mp \text{sign}(\Lambda)]}{a_0} = \pm\frac{M}{2v_F}\exp(\pm\beta). \tag{10.9}$$

Equations (10.9) are the standard parametrization of a Lorentz invariant massless dispersion in terms of a rapidity variable β

$$\varepsilon_s = v_F|\,p_s|. \tag{10.10}$$

We note that scaling Λ to ± 2 is not a unique choice, but it is the most natural in the sense that the scale appearing in the parametrization (10.9) is a physical quantity, namely the single particle gap M.

The analysis above establishes that the scaling limit of the half filled Hubbard model is a Lorentz invariant scattering theory of massive (anti)holons and massless spinons.

10.2 The S-matrix in the scaling limit

Having obtained Lorentz invariant dispersions for (anti)holons and spinons in the scaling limit, the next step is to determine their respective S-matrices.

10.2.1 Massive charge sector

Our starting point is the SU(2) invariant S-matrix (7.134) describing scattering of holons and antiholons. Scaling the momenta of the holes to $\pm\pi$ according to (10.4), we find

$$k_{12} = \frac{\sin(k_1^h) - \sin(k_2^h)}{2u} \longrightarrow \frac{\theta_1 - \theta_2}{\pi} \equiv \frac{\theta_{12}}{\pi}. \tag{10.11}$$

This in turn yields the following SU(2) invariant form for the S-matrix in the massive charge sector in the scaling limit

$$\check{S}^{\gamma_1\gamma_2}_{\alpha_1\alpha_2}(\theta_1, \theta_2) = -\frac{\Gamma\left(\frac{1}{2} - i\frac{\theta_{12}}{2\pi}\right)\Gamma\left(1 + i\frac{\theta_{12}}{2\pi}\right)}{\Gamma\left(\frac{1}{2} + i\frac{\theta_{12}}{2\pi}\right)\Gamma\left(1 - i\frac{\theta_{12}}{2\pi}\right)}$$

$$\times \left[\delta_{\alpha_1\gamma_1}\delta_{\alpha_2\gamma_2}\frac{\theta_{12}}{\theta_{12} - i\pi} - \delta_{\alpha_1\gamma_2}\delta_{\alpha_2\gamma_1}\frac{i\pi}{\theta_{12} - i\pi}\right]. \tag{10.12}$$

Equation (10.12) is identical to the S-matrix of the massive sector of the SU(2) Thirring model [55]. We note that the SU(2) Thirring model is also known as Chiral Gross Neveu model in the literature.[2]

The fact that the S-matrix tends to $\mp i$ when the rapidity difference approaches $\pm\infty$ seems to be at odds with physical intuition, which suggests that the amplitude of purely elastic scattering ought to tend to 1 in these limits (which correponds to the Mandelstam variable s going to infinity). The unusual asymptotic behaviour of the S-matrix signifies that holons have fractional (exchange) statistics. For many calculations it is convenient to introduce Hilbert spaces of auxiliary 'in' and 'out' particles with ordinary statistics [236, 416, 428] and work with auxiliary particles instead of the physical particles with fractional statistics.

10.2.2 Massless spin sector

In the spin sector we have to scale the spectral parameters according to equation (10.8). We have to distinguish between three cases.

Scattering of right movers on right movers. Both Λ_1^h and Λ_2^h are in the vicinity of 2 and describe *right-moving* spinons. This is apparent from the fact that their momentum is always positive. The scaling limit for the difference of spectral parameters of the holes is

$$\Lambda_{12} = \frac{\Lambda_1^h - \Lambda_2^h}{2u} \longrightarrow -\frac{\beta_1 - \beta_2}{\pi} \equiv -\frac{\beta_{12}}{\pi}. \qquad (10.13)$$

The corresponding S-matrix is

$$\left(\hat{S}_{RR}\right)_{\alpha_1\alpha_2}^{\gamma_1\gamma_2} (\beta_1, \beta_2) = \frac{\Gamma\left(\frac{1}{2} - i\frac{\beta_{12}}{2\pi}\right)\Gamma\left(1 + i\frac{\beta_{12}}{2\pi}\right)}{\Gamma\left(\frac{1}{2} + i\frac{\beta_{12}}{2\pi}\right)\Gamma\left(1 - i\frac{\beta_{12}}{2\pi}\right)}$$
$$\times \left[\delta_{\alpha_1\gamma_1}\delta_{\alpha_2\gamma_2}\frac{\beta_{12}}{\beta_{12} - i\pi} - \delta_{\alpha_1\gamma_2}\delta_{\alpha_2\gamma_1}\frac{i\pi}{\beta_{12} - i\pi}\right]. \qquad (10.14)$$

Scattering of left movers on left movers. Here both Λ_1^h and Λ_2^h are in the vicinity of -2 and describe *left-moving* spinons; their momenta (10.9) are negative. The S-matrix as a function of the rapidity difference is identical to the one in the RR sector

$$\left(\hat{S}_{LL}\right)_{\alpha_1\alpha_2}^{\gamma_1\gamma_2} (\beta_1, \beta_2) = \left(\hat{S}_{RR}\right)_{\alpha_1\alpha_2}^{\gamma_1\gamma_2} (\beta_1, \beta_2). \qquad (10.15)$$

Scattering of right movers on left movers. The scattering between the left and right sectors is diagonal and rapidity independent

$$\left(\hat{S}_{RL}\right)_{\alpha_1\alpha_2}^{\gamma_1\gamma_2} (\beta_1, \beta_2) = -i\,\delta_{\alpha_1\gamma_1}\delta_{\alpha_2\gamma_2}. \qquad (10.16)$$

---

[2] In the literature the massive sector in the SU(2) Thirring model is often referred to as the 'spin' sector. This should not be confused with the spin sector in the underlying Hubbard model.

The S-matrices (10.14)–(10.16) agree with the ones derived for the $SU_1(2)$ Wess-Zumino-Novikov-Witten model in Ref. [504].[3]

10.2.3 Scattering between spin and charge

The scattering between spin and charge is diagonal and rapidity independent

$$\tilde{S}^{\gamma_1\gamma_2}_{\alpha_1\alpha_2}(\theta, \beta) = -i\,\delta_{\alpha_1\gamma_1}\delta_{\alpha_2\gamma_2}. \tag{10.17}$$

Hence spin and charge degrees of freedom are essentially decoupled in the scaling limit.

10.3 Continuum limit

In the previous section we have seen that there exists a particular scaling limit, in which the half filled Hubbard model becomes equivalent to an integrable quantum field theory, the SU(2) Thirring model. An obvious question is whether this relationship can be established directly on the level of the Hamiltonian, by taking an appropriate continuum limit. As we will see this indeed turns out to be the case. As the interaction strength U/t goes to zero in the scaling limit, we may use the noninteracting case $U = 0$ in order to establish the region in momentum space where low-energy degrees of freedom reside. We start by decomposing the Hamiltonian into free and interacting parts

$$\mathcal{H} = \mathcal{H}_0 + U\sum_{j=1}^{L} n_{j,\uparrow} n_{j,\downarrow}. \tag{10.18}$$

The free part is given by (5.4)

$$\mathcal{H}_0 = -t\sum_{j,\sigma}\left[c^\dagger_{j,\sigma}c_{j+1,\sigma} + c^\dagger_{j+1,\sigma}c_{j,\sigma}\right]$$

$$= \frac{1}{L}\sum_{l,\sigma} -2t\cos(k_l)\,\tilde{n}_\sigma(k_l), \tag{10.19}$$

where $\tilde{n}_\sigma(k_l) = \tilde{c}^\dagger_{k_l,\sigma}\tilde{c}_{k_l,\sigma}$ is the number operator for spin σ electrons in momentum space. We note that we have dropped the constant $-\frac{U}{2}\sum_{j,\sigma} n_{j,\sigma} = -\frac{UL}{2}$ in the definition of the Hamiltonian (10.18). The ground state of $\mathcal{H}_0$ is obtained by filling all negative energy modes, i.e. all single-electron states with momenta in the interval $[-k_F, k_F]$, where the Fermi momentum is $k_F = \frac{\pi}{2a_0}$. All low-energy modes are found in the vicinity of the two Fermi points $\pm k_F$ and these are the degrees of freedom of interest from the point of view of taking a continuum limit. The above considerations suggest the following decomposition of the lattice Fermion annihilation operators

$$c_{l,\sigma} \longrightarrow \sqrt{a_0}\left[\exp(ik_F x)\,R_\sigma(x) + \exp(-ik_F x)\,L_\sigma(x)\right]. \tag{10.20}$$

---

[3] The phase-shift for left-right scattering was chosen as -1 'in the infrared' in [504].

Here a_0 is the lattice spacing and $x = la_0$. The right and left moving fermionic quantum fields R_σ and L_σ have dimension (length)$^{-1/2}$ and are slowly varying on the scale of the lattice spacing. Inserting the prescription (10.20) into the Hamiltonian (10.18) one obtains after some straightforward calculations

$$\mathcal{H} = \sum_\sigma v_F \int dx \; : \left[L_\sigma^\dagger i\partial_x L_\sigma - R_\sigma^\dagger i\partial_x R_\sigma \right] : + g \int dx \; : \left[\mathbf{I} \cdot \bar{\mathbf{I}} - \mathbf{J} \cdot \bar{\mathbf{J}} \right] :$$

$$+ \frac{g}{6} \int dx \; : \left[\mathbf{I} \cdot \mathbf{I} + \bar{\mathbf{I}} \cdot \bar{\mathbf{I}} - \mathbf{J} \cdot \mathbf{J} - \bar{\mathbf{J}} \cdot \bar{\mathbf{J}} \right] : , \qquad (10.21)$$

where $v_F = 2ta_0$ is the Fermi velocity, $g = 2Ua_0$ and $: \mathcal{O} :$ denotes normal ordering of the operator $\mathcal{O}$. Here $\mathbf{J}$ and $\mathbf{I}$ are the chiral components of SU(2) spin and pseudospin currents

$$\bar{I}^z = \frac{1}{2} : \left(L_\uparrow^\dagger L_\uparrow + L_\downarrow^\dagger L_\downarrow \right) : , \quad \bar{I}^+ = (\bar{I}^-)^\dagger = L_\uparrow^\dagger L_\downarrow^\dagger ,$$

$$I^z = \frac{1}{2} : \left(R_\uparrow^\dagger R_\uparrow + R_\downarrow^\dagger R_\downarrow \right) : , \quad I^+ = (I^-)^\dagger = R_\uparrow^\dagger R_\downarrow^\dagger ,$$

$$\bar{J}^z = \frac{1}{2} : \left(L_\uparrow^\dagger L_\uparrow - L_\downarrow^\dagger L_\downarrow \right) : , \quad \bar{J}^+ = (\bar{J}^-)^\dagger = L_\uparrow^\dagger L_\downarrow ,$$

$$J^z = \frac{1}{2} : \left(R_\uparrow^\dagger R_\uparrow - R_\downarrow^\dagger R_\downarrow \right) : , \quad J^+ = (J^-)^\dagger = R_\uparrow^\dagger R_\downarrow . \qquad (10.22)$$

Here ':' denotes normal ordering of point-split expressions [5]. The 'kinetic' terms in the Hamiltonian (10.21) can be expressed as normal ordered bilinears of currents as well [5, 7, 11, 93, 168]

$$\frac{2\pi}{3} \int dx \; : [\mathbf{I} \cdot \mathbf{I} + \mathbf{J} \cdot \mathbf{J}] : = - \int dx \left[\sum_\sigma : R_\sigma^\dagger i\partial_x R_\sigma : \right],$$

$$\frac{2\pi}{3} \int dx \; : [\bar{\mathbf{I}} \cdot \bar{\mathbf{I}} + \bar{\mathbf{J}} \cdot \bar{\mathbf{J}}] : = \int dx \left[\sum_\sigma : L_\sigma^\dagger i\partial_x L_\sigma : \right]. \qquad (10.23)$$

A derivation of (10.23) is outlined in Appendix 10.A, another can be found in the Appendix of [26]. Using (10.23) the Hamiltonian (10.21) can now be split into two parts, corresponding to the spin and charge sectors respectively

$$\mathcal{H} = \mathcal{H}_c + \mathcal{H}_s ,$$

$$\mathcal{H}_c = \frac{2\pi v_c}{3} \int dx \; : [\mathbf{I} \cdot \mathbf{I} + \bar{\mathbf{I}} \cdot \bar{\mathbf{I}}] : + g \int dx \; : \mathbf{I} \cdot \bar{\mathbf{I}} : ,$$

$$\mathcal{H}_s = \frac{2\pi v_s}{3} \int dx \; : [\mathbf{J} \cdot \mathbf{J} + \bar{\mathbf{J}} \cdot \bar{\mathbf{J}}] : - g \int dx \; : \mathbf{J} \cdot \bar{\mathbf{J}} : . \qquad (10.24)$$

Here $v_s = v_F - Ua_0/2\pi$ and $v_c = v_F + Ua_0/2\pi$. These values agree with the small-U approximations of (7.21) and (7.26). We note that the Hamiltonian (10.24) displays the required SO(4) symmetry of the half filled Hubbard model. The continuum theory corresponding to the Hamiltonian (10.24) is not Lorentz invariant as the spin and charge velocities are different. Lorentz invariance is only recovered in the scaling limit, where

$U a_0 \to 0$ and hence

$$v_{s,c} \longrightarrow v_F . \tag{10.25}$$

Renormalization group equations. The spin and charge Hamiltonians (10.24) are both of the form

$$\mathcal{H} = \frac{2\pi v}{3} \int dx : [\mathbf{L} \cdot \mathbf{L} + \bar{\mathbf{L}} \cdot \bar{\mathbf{L}}] : -\lambda v \int dx : \mathbf{L} \cdot \bar{\mathbf{L}} :, \tag{10.26}$$

where L^α, $\bar{L}^\alpha$ fulfil the $SU_1(2)$ Kac-Moody algebra

$$[L^\alpha(x), L^\beta(x')] = i\varepsilon_{\alpha\beta\gamma} L^\gamma(x) \delta(x - x') - \frac{i\delta^{\alpha\beta}}{4\pi} \delta'(x - x'), \tag{10.27}$$

$$[\bar{L}^\alpha(x), \bar{L}^\beta(x')] = i\varepsilon_{\alpha\beta\gamma} \bar{L}^\gamma(x) \delta(x - x') - \frac{i\delta^{\alpha\beta}}{4\pi} \delta'(x - x'), \tag{10.28}$$

$$[L^\alpha(x), \bar{L}^\beta(x')] = 0 . \tag{10.29}$$

The relations (10.27)–(10.29) can be established along the lines of Appendix 10.A, see [5]. The Renormalization Group (RG) equations for the theory (10.26) can be derived directly from the current algebra (10.27)–(10.29) as is shown in [12]. Under an appropriate choice of renormalization scheme the RG equations (to all orders in the coupling constant) can be cast in the form [506] (see also [115, 164, 302])

$$r \frac{\partial \lambda}{\partial r} = -\frac{2\lambda^2}{4\pi - \lambda}, \tag{10.30}$$

where r is the RG length scale. The RG equations (10.30) imply that λ diminishes under renormalization. This means that if we start with $\lambda > 0$ as is the case for the spin sector of (10.24), then the current-current interaction flows to zero. In other words, the interaction of spin currents in $\mathcal{H}_s$ is marginally irrelevant and hence we will ignore it in what follows. Taking it into account would generate extra logarithms in certain formulas below. On the other hand, if initially $\lambda < 0$ as in the charge sector of (10.24), then the interaction grows under renormalization: it is marginally relevant.

Let us now show that in the scaling limit the model (10.24) is equivalent to the SU(2) Thirring model. We define a metric

$$g_{\mu\nu} = \begin{pmatrix} -1 & 0 \\ 0 & 1 \end{pmatrix}, \tag{10.31}$$

two-dimensional Gamma matrices $\{\gamma^\mu, \gamma^\nu\} = 2g^{\mu\nu}$,

$$\gamma^0 = i\sigma^y = \begin{pmatrix} 0 & 1 \\ -1 & 0 \end{pmatrix}, \quad \gamma^1 = \sigma^x = \begin{pmatrix} 0 & 1 \\ 1 & 0 \end{pmatrix}, \tag{10.32}$$

and two spinor fields

$$\Psi_1(t, x) = \begin{pmatrix} R_\uparrow(t, x) \\ L_\uparrow(t, x) \end{pmatrix}, \quad \Psi_2(t, x) = \begin{pmatrix} R_\downarrow^\dagger(t, x) \\ L_\downarrow^\dagger(t, x) \end{pmatrix}. \tag{10.33}$$

In terms of these spinor fields the Hamiltonian (10.24) without the marginally irrelevant interaction of spin currents can be expressed in the scaling limit $U a_0 \to 0^+$ as

$$\mathcal{H} = \int dx \; i v_F \sum_{a=1}^{2} \bar{\Psi}_a(t, x) \, \gamma_1 \partial_x \Psi_a(t, x) - \frac{g}{4} \int dx \sum_{\alpha=1}^{3} J_\mu^\alpha(t, x) \, J^{\alpha\mu}(t, x), \quad (10.34)$$

where

$$J_\mu^\alpha(t, x) = \frac{1}{2} \bar{\Psi}_a(t, x) \, \gamma_\mu \, \sigma_{ab}^\alpha \, \Psi_b(t, x) . \quad (10.35)$$

In (10.34) we recognize the standard expression for the Hamiltonian of the SU(2) Thirring model [93, 186, 327]. We note that the SU(2) Thirring model as well as its U(1) generalization are Bethe Ansatz solvable [110, 221, 455].

Continuum limit of operators In order to study correlation functions of physical operators, we need to know what their respective continuum limits are. This is easily done using the prescription (10.20). Let $\mathcal{O}_j$ be an operator defined on site j in the lattice model. We denote its continuum limit by $\mathcal{O}(x)$, where $x = j a_0$.

1. Current Operator: The current operator in the lattice model is given by (1.A.4)

$$J_{j,j+1} = -i t \sum_\sigma \left[c_{j,\sigma}^\dagger c_{j+1,\sigma} - c_{j+1,\sigma}^\dagger c_{j,\sigma} \right] . \quad (10.36)$$

In the continuum limit it becomes

$$J(x) = v_F \sum_\sigma : \left[R_\sigma^\dagger(x) R_\sigma(x) - L_\sigma^\dagger(x) L_\sigma(x) \right] :, \quad (10.37)$$

where we have dropped terms that contain higher powers of the lattice spacing a_0. This is justified as these terms vanish when we take $a_0 \to 0$. We recall that the electric current is related to $J(x)$ by (1.A.5)

$$J_{\text{el}}(x) = -e a_0 J(x) , \quad (10.38)$$

where $-e$ is the electron charge.

2. Electron density: The electron number operator on site j is

$$n_j = \sum_\sigma c_{j,\sigma}^\dagger c_{j,\sigma} . \quad (10.39)$$

In the continuum limit it turns into (we again set $x = j a_0$)

$$n(x) = a_0 \sum_\sigma : \left[R_\sigma^\dagger(x) R_\sigma(x) + L_\sigma^\dagger(x) L_\sigma(x) \right] :$$

$$+ a_0 (-1)^j \sum_\sigma \left[R_\sigma^\dagger(x) L_\sigma(x) + L_\sigma^\dagger(x) R_\sigma(x) \right]$$

$$\equiv n_0(x) + (-1)^j n_{2k_F}(x). \quad (10.40)$$

We see that in the continuum limit only the Fourier components with $k \approx 0$ and

$k \approx 2k_F = \frac{\pi}{a_0}$ remain. For general band fillings, Fourier components with $k \approx 4k_F, k \approx 6k_F$ *etc.* would be present as well [168]. We note that such terms cannot be derived by means of the simple procedure we have employed here as their coefficients are proportional to the interaction U. What one needs to do to capture such contributions is to integrate out the high-energy terms in the path-integral representation of the density-density correlation function $\langle n(x)\, n(0)\rangle$ perturbatively. In this way terms proportional to e.g. $\cos(4k_F x)$ get generated.

3. Spin operators: The lattice spin operators are defined by ($\alpha = x, y, z$)

$$S_j^\alpha = \frac{1}{2} c_{j,\tau}^\dagger\, \sigma_{\tau\tau'}^\alpha\, c_{j,\tau'}\,, \tag{10.41}$$

and in the continuum limit have the following decomposition

$$
\begin{aligned}
S^\alpha(x) &- J^\alpha(x) + (-1)^j\, n^\alpha(x)\,, \\
J^\alpha(x) &= \frac{a_0}{2}\left[R_\tau^\dagger(x)\,\sigma_{\tau\tau'}^\alpha\, R_{\tau'}(x) + L_\tau^\dagger(x)\,\sigma_{\tau\tau'}^\alpha\, L_{\tau'}(x)\right], \\
n^\alpha(x) &= \frac{a_0}{2}\left[R_\tau^\dagger(x)\,\sigma_{\tau\tau'}^\alpha\, L_{\tau'}(x) + L_\tau^\dagger(x)\,\sigma_{\tau\tau'}^\alpha\, R_{\tau'}(x)\right].
\end{aligned} \tag{10.42}
$$

Like in the expansion of the density operator only Fourier components with $k \approx 0$ ('smooth magnetization') and $k \approx 2k_F$ ('staggered magnetization') remain.

10.3.1 Bosonization

The SU(2) Thirring model (as well as the theory (10.24)) is equivalent to a theory of canonical Bose fields with nonlinear interactions. The bosonic theory may be constructed by standard 'Bosonization' methods [87, 309, 318, 321, 379]. We refer the reader to the textbooks [139, 168] and reviews [7, 385, 473] for introductions and detailed discussions of this important subject. In what follows it is often convenient to work in Euclidean space with imaginary time $\tau = it$. The SU(2) Thirring model is bosonized in terms of two canonical Bose fields $\Phi_s(\tau, x)$ and $\Phi_c(\tau, x)$, corresponding to collective spin and charge degrees of freedom respectively. These Bose fields have chiral decompositions

$$\Phi_a(\tau, x) = \varphi_a(z) + \bar\varphi_a(\bar z)\,, \qquad a = s, c\,, \tag{10.43}$$

where the chiral boson fields φ_a and $\bar\varphi_a$ fulfil the following commutation relations[4]

$$[\varphi_a(\tau, x), \bar\varphi_b(\tau, y)] = 2\pi i \delta_{ab}, \qquad a, b = c, s. \tag{10.44}$$

The chiral bosons $\varphi(z)$ and $\bar\varphi(\bar z)$ depend on τ and x only through the combinations

$$z = v_F\tau - ix\,, \qquad \bar z = v_F\tau + ix\,. \tag{10.45}$$

It is convenient to define so-called dual fields $\Theta_{c,s}(\tau, x)$ by

$$\Theta_a(\tau, x) = \varphi_a(z) - \bar\varphi_a(\bar z)\,, \qquad a = s, c. \tag{10.46}$$

---

[4] We consider quantization on an infinite line with vanishing boundary conditions at $x = \pm\infty$. In a periodic quantization scheme the chiral Bose fields commute.

The canonical Bose fields and their dual fields are related by

$$v_F \partial_x \Theta(\tau, x) = -i \partial_\tau \Phi(\tau, x) , \quad \partial_\tau \Theta(\tau, x) = i v_F \partial_x \Phi(\tau, x) . \tag{10.47}$$

The creation operators of left and right moving fermions with spin σ are expressed in terms of the chiral Bose fields as

$$L_\sigma^\dagger(\tau, x) = \frac{\eta_\sigma}{\sqrt{2\pi}} e^{if_\sigma \pi/4} \exp\left(-\frac{i}{2} \bar{\varphi}_c(\tau, x)\right) \exp\left(-\frac{if_\sigma}{2} \bar{\varphi}_s(\tau, x)\right),$$

$$R_\sigma^\dagger(\tau, x) = \frac{\eta_\sigma}{\sqrt{2\pi}} e^{if_\sigma \pi/4} \exp\left(\frac{i}{2} \varphi_c(\tau, x)\right) \exp\left(\frac{if_\sigma}{2} \varphi_s(\tau, x)\right). \tag{10.48}$$

Here $f_\uparrow = 1 = -f_\downarrow$ and η_a are Klein factors[5] fulfilling the anticommutation relations

$$\{\eta_a, \eta_b\} = 2\delta_{ab} . \tag{10.49}$$

The role of the Klein factors is to ensure the appropriate anticommutation relations of the Fermi fields. They may be represented by Pauli matrices

$$\eta_\uparrow = \sigma^x , \qquad \eta_\downarrow = \sigma^y . \tag{10.50}$$

In general, the Hilbert space of states should be thought of as containing a 'Klein' piece, in which the Klein factors act. The factors of $e^{if_\sigma \pi/4}$ in (10.48) are introduced in order to obtain the usual bosonized expressions for the staggered components of the spin operators.

We choose a normalisation of operators such that for $|\mathbf{x} - \mathbf{y}| \longrightarrow 0$

$$\exp\left(i\alpha \Phi_c(\mathbf{x})\right) \exp\left(i\beta \Phi_c(\mathbf{y})\right) \longrightarrow |\mathbf{x} - \mathbf{y}|^{4\alpha\beta} \exp\left(i\alpha \Phi_c(\mathbf{x}) + i\beta \Phi_c(\mathbf{y})\right). \tag{10.51}$$

Here we have used a vector notation for the coordinates in Euclidean space $\mathbf{x} = (v_F \tau, x)$. We note that the normalisation (10.51) is standard in Conformal Field Theory [102] and is often used in the literature [304, 307, 506]. It implies that the operators $\exp\left(i\alpha \Phi_c(\mathbf{x})\right)$ are dimensionful objects

$$\dim\left[\exp\left(i\alpha \Phi_c(\mathbf{x})\right)\right] = \text{length}^{-2\alpha^2}. \tag{10.52}$$

Applying the bosonization identities we obtain the following bosonic form of the low energy effective Hamiltonian in the continuum limit

$$H = \mathcal{H}_c + \mathcal{H}_s ,$$

$$\mathcal{H}_c = \frac{v_c}{16\pi} \int dx \left[(\partial_x \Phi_c)^2 + (\partial_x \Theta_c)^2\right]$$

$$- \frac{g}{(2\pi)^2} \int dx \left[\cos(\Phi_c) + \frac{1}{16}\left\{(\partial_x \Theta_c)^2 - (\partial_x \Phi_c)^2\right\}\right],$$

$$\mathcal{H}_s = \frac{v_s}{16\pi} \int dx \left[(\partial_x \Phi_s)^2 + (\partial_x \Theta_s)^2\right]$$

$$+ \frac{g}{(2\pi)^2} \int dx \left[\cos(\Phi_s) + \frac{1}{16}\left\{(\partial_x \Theta_s)^2 - (\partial_x \Phi_s)^2\right\}\right]. \tag{10.53}$$

---

[5] By virtue of (10.44) we only need to intoduce two Klein factors. In a periodic quantization scheme four Klein factors are required.

It is important to recall that the theory corresponding to the Hamiltonian (10.53) is not Lorentz invariant and describes the vicinity of the scaling limit. The scaling limit itself corresponds to setting $v_c = v_s = v_F$ and dropping the marginally irrelevant interaction in the spin sector.

Bosonization of operators. Products of the Fermi fields may be bosonized by using the expressions (10.48) and the normalisation condition (10.51), which governs the short-distance operator product expansions.

1. *Current operator*

$$J(\tau, x) = \frac{i}{2\pi} \partial_\tau \, \Phi_c(\tau, x) \, . \tag{10.54}$$

2. *Electron density*

$$n(\tau, x) = -\frac{a_0}{2\pi} \partial_x \, \Phi_c(\tau, x)$$

$$+ (-1)^{\frac{x}{a_0}} \frac{2a_0}{\pi} \, \sin\left[\frac{\Phi_c(\tau, x)}{2}\right] \cos\left[\frac{\Phi_s(\tau, x)}{2}\right] \tag{10.55}$$

By virtue of the normalization (10.52) the units in (10.55) work out correctly: Φ_c is dimensionless, whereas $\sin(\Phi_c/2)$ and $\cos(\Phi_s/2)$ have dimensions of length$^{-1/2}$.

3. *Spin operators*

$$S^z(\tau, x) = -\frac{a_0}{4\pi} \partial_x \, \Phi_s(\tau, x)$$

$$+ (-1)^{\frac{x}{a_0}} \frac{a_0}{\pi} \, \cos\left[\frac{\Phi_c(\tau, x)}{2}\right] \sin\left[\frac{\Phi_s(\tau, x)}{2}\right], \tag{10.56}$$

$$S^+(\tau, x) = S^x(\tau, x) + i S^y(\tau, x)$$

$$= \frac{i a_0 \eta_\uparrow \eta_\downarrow}{2\pi} \left[\exp\left(i\varphi_s(\tau, x)\right) + \exp\left(-i\bar{\varphi}_s(\tau, x)\right)\right]$$

$$+ \frac{i a_0 \eta_\uparrow \eta_\downarrow}{\pi} (-1)^{\frac{x}{a_0}} \, \cos\left[\frac{\Phi_c(\tau, x)}{2}\right] \exp\left(i\frac{\Theta_s(\tau, x)}{2}\right). \tag{10.57}$$

The general structure of the bosonization identities for a local operator involving lattice Fermi creation and annihilation operators on several adjacent sites $n, \ldots, n+l$ is

$$\mathcal{O}_{n,n+1,\ldots,n+l}(\tau) \longrightarrow \sum_j a_j \mathcal{C}_j(\tau, x) + (-1)^{x/a_0} b_j \mathcal{D}_j(\tau, x), \tag{10.58}$$

where $\mathcal{C}_j$ and $\mathcal{D}_j$ are (quasi)local operators in the bosonic theory (10.53). In (10.54), (10.55) and (10.56) we have written the 'naive' U-independent results for the amplitudes a_j and b_j obtained by bosonizing in the free theory. In the continuum limit, the amplitudes a_j and b_j generally depend on the interaction strength U as we have mentioned before. The reason is that the continuum theory should be thought of as the result of tracing out high-energy degrees of freedom in the Hubbard model. In a path-integral formulation one would decompose the Fermi operators into high- and low-energy pieces and then integrate out the high-energy pieces. Such a procedure produces (10.53) as the leading term for the Hamiltonian at small U/t. However, there are additional contributions of higher order in U/t. The same holds true on the level of correlation functions.

10.4 Correlation functions in the scaling limit

Let us now turn to the calculation of the large-distance asymptotics of correlation functions in the scaling limit. An important simplification arises due to spin charge separation.

10.4.1 Spin-charge factorization of correlation functions

The bosonized expression for the two pieces (10.53) of the Hamiltonian commute with one another. $\mathcal{H}_s$ and $\mathcal{H}_c$ correspond to collective spin and charge degrees of freedom respectively. This means that we can choose an eigenbasis of the Hamitonian in which this spin-charge separation is manifest and eigenstates are represented as products

$$|\text{eigenstate}\rangle = |\text{spin}\rangle \otimes |\text{charge}\rangle. \tag{10.59}$$

In particular, the ground state can be represented as

$$|0\rangle = |0\rangle_s \otimes |0\rangle_c, \tag{10.60}$$

where $|0\rangle_{s,c}$ are the ground states of the spin and charge Hamiltonians (10.24), (10.53). Similarly, physical observables like spin (10.56), current (10.54) and electron density (10.55) are expressed as (sums over) products of commuting operators acting in the spin and charge sectors respectively. Let us consider a physical operator $\mathcal{O}$ and represent this factorization by

$$\mathcal{O}(\tau, x) = \mathcal{O}_s(\tau, x) \otimes \mathcal{O}_c(\tau, x). \tag{10.61}$$

It now follows, that correlation functions can be represented as products over correlation functions in the spin and charge sectors respectively

$$\langle 0|\mathcal{O}(\tau, x)\, \bar{\mathcal{O}}(0, 0)|0\rangle = {}_s\langle 0|\mathcal{O}_s(\tau, x)\, \bar{\mathcal{O}}_s(0, 0)|0\rangle_s$$
$$\times {}_c\langle 0|\mathcal{O}_c(\tau, x)\, \bar{\mathcal{O}}_c(0, 0)|0\rangle_c. \tag{10.62}$$

The representation (10.62) is very useful, because the Hamiltonians (10.53) describing the spin and charge sectors respectively are both exactly solvable. Furthermore, the spin sector is a simple Gaussian model and correlation functions are easily evaluated (see Appendix 10.C). This leaves us with the task of determining correlation functions of local operators in the gapped charge sector. The latter is integrable and as a result the elementary excitations have the property of factorizable scattering. In massive quantum field theories with factorizable scattering it is possible to determine two-point correlation functions of local operators by a method known as 'Form Factor Bootstrap Approach' (FFBA). The method was conceived in the late seventies by M. Karowski and P. Weisz [53, 238]. It was developed further in a series of seminal papers by F. Smirnov and coworkers [248–250, 374, 414, 415, 417–419]. Detailed accounts of the method are given in F. Smirnov's book [420] and in papers

[32–34, 37, 72, 148, 262, 289, 303, 500]. Here we summarize the essential steps and refer the reader to the aforementioned references for further explanations. The basic idea underlying the FFBA is to express two-point functions in a spectral (Lehmann) representation in terms of the elementary excitations with factorizable scattering. The matrix elements (or form factors) of a given local operator between the ground state and an excited state can then be inferred from the knowledge of the exact S-matrix.

10.4.2 Spectral representation of two-point functions in the charge sector

A basis of scattering states of the elementary holon and antiholon excitations may be constructed by means of the Faddeev-Zamolodchikov (FZ) algebra. We define rapidity dependent creation and annihilation operators subject to the following algebra, which may be thought of as 'generalized commutation relations'

$$Z^{\varepsilon_1}(\theta_1)Z^{\varepsilon_2}(\theta_2) = \tilde{S}^{\varepsilon_1,\varepsilon_2}_{\varepsilon'_1,\varepsilon'_2}(\theta_1 - \theta_2)Z^{\varepsilon'_2}(\theta_2)Z^{\varepsilon'_1}(\theta_1) , \tag{10.63}$$

$$Z^{\dagger}_{\varepsilon_1}(\theta_1)Z^{\dagger}_{\varepsilon_2}(\theta_2) = Z^{\dagger}_{\varepsilon'_2}(\theta_2)Z^{\dagger}_{\varepsilon'_1}(\theta_1)\tilde{S}^{\varepsilon'_1,\varepsilon'_2}_{\varepsilon_1,\varepsilon_2}(\theta_1 - \theta_2) , \tag{10.64}$$

$$Z^{\varepsilon_1}(\theta_1)Z^{\dagger}_{\varepsilon_2}(\theta_2) = Z^{\dagger}_{\varepsilon'_2}(\theta_2)\tilde{S}^{\varepsilon'_2,\varepsilon_1}_{\varepsilon_2,\varepsilon'_1}(\theta_2 - \theta_1)Z^{\varepsilon'_1}(\theta_1)$$
$$+ 2\pi \, \delta^{\varepsilon_1}_{\varepsilon_2}\delta(\theta_1 - \theta_2) . \tag{10.65}$$

Here $\tilde{S}^{\varepsilon_1,\varepsilon_2}_{\varepsilon'_1,\varepsilon'_2}(\theta)$ is the factorizable two-particle S-matrix (10.12) of (anti)holons and $\varepsilon_j = h, \bar{h}$. Our notations are such that e.g. the operator $Z^{\dagger}_h(\theta)$ creates a holon with rapidity θ.[6] Using the ZF operators a Fock space of states can be constructed as follows. The vacuum is defined by

$$Z_{\varepsilon_i}(\theta)|0\rangle_c = 0 . \tag{10.66}$$

Multiparticle states are then obtained by acting with strings of creation operators $Z^{\dagger}_\varepsilon(\theta)$ on the vacuum

$$|\theta_n \dots \theta_1\rangle_{\varepsilon_n \dots \varepsilon_1} = Z^{\dagger}_{\varepsilon_n}(\theta_n) \dots Z^{\dagger}_{\varepsilon_1}(\theta_1)|0\rangle_c. \tag{10.67}$$

In terms of this basis the resolution of the identity is given by

$$1 = |0\rangle_c \, _c\langle 0| + \sum_{n=1}^{\infty} \sum_{\varepsilon_i=h,\bar{h}} \int_{-\infty}^{\infty} \frac{d\theta_1 \dots d\theta_n}{(2\pi)^n n!} |\theta_n \dots \theta_1\rangle_{\varepsilon_n \dots \varepsilon_1} \, ^{\varepsilon_1 \dots \varepsilon_n}\langle \theta_1 \dots \theta_n| . \tag{10.68}$$

Inserting (10.68) between the operators in a two-point correlation function, we obtain the following spectral representation

$$\langle \mathcal{O}_c(t,x)\mathcal{O}^{\dagger}_c(0,0)\rangle_c = \sum_{n=0}^{\infty} \sum_{\varepsilon_i=h,\bar{h}} \int \frac{d\theta_1 \dots d\theta_n}{(2\pi)^n n!} \exp\left(i \sum_{j=1}^{n} P(\theta_j)x - E(\theta_j)t\right)$$
$$\times |_c\langle 0|\mathcal{O}_c(0,0)|\theta_n \dots \theta_1\rangle_{\varepsilon_n \dots \varepsilon_1}|^2. \tag{10.69}$$

---

[6] The existence of the FZ operators is a postulate in the FFBA approach. They are not strictly necessary but greatly simplify the notations in what follows. For zero electron density an explicit representation of the FZ operators is derived in Chapter 15.

Here the $n = 0$ term just gives the absolute value squared of the vacuum expectation value of $\mathcal{O}_c$ and

$$P(\theta) = \frac{M}{v_F} \sinh\theta, \qquad E(\theta) = M \cosh\theta \,. \tag{10.70}$$

Finally, the form factors of the operator $\mathcal{O}_c$ between the vacuum and excited states are defined by

$$f^{\mathcal{O}_c}(\theta_1 \ldots \theta_n)_{\varepsilon_1 \ldots \varepsilon_n} \equiv {}_c\langle 0|\mathcal{O}_c(0,0)|\theta_n \ldots \theta_1\rangle_{\varepsilon_n \ldots \varepsilon_1} \,. \tag{10.71}$$

It is often useful to work with imaginary time $\tau = it$. The τ-evolution of operators is given by

$$\mathcal{O}(\tau, x) = e^{H\tau} \, \mathcal{O}(0) \, e^{-H\tau} \,,$$
$$\bar{\mathcal{O}}(\tau, x) = e^{H\tau} \, \mathcal{O}^\dagger(0) \, e^{-H\tau} \,, \tag{10.72}$$

where H is the Hamiltonian. As usual $\bar{\mathcal{O}}$ is not the hermitian conjugate of $\mathcal{O}$. The spectral representation for $\tau > 0$ reads

$$\langle \mathcal{O}_c(\tau, x)\bar{\mathcal{O}}_c(0,0)\rangle_c = \sum_{n=0}^{\infty} \sum_{\varepsilon_i} \int \frac{d\theta_1 \ldots d\theta_n}{(2\pi)^n n!} \exp\left(\sum_{j=1}^{n} i P(\theta_j)x - E(\theta_j)\tau\right)$$
$$\times |_c\langle 0|\mathcal{O}_c(0,0)|\theta_n \ldots \theta_1\rangle_{\varepsilon_n \ldots \varepsilon_1}|^2. \tag{10.73}$$

10.4.3 Form factors

In the FFBA the form factors are determined from a set of axiomatized assumptions. For the case of the SU(2) Thirring model they have been formulated by A.N. Kirillov and F. Smirnov in Refs. [248,249] and by S. Lukyanov in [303]. The basic assumption is that the form factor $f^{\mathcal{O}}(\theta_1, \ldots, \theta_n)_{\varepsilon_1, \ldots, \varepsilon_n}$ is a meromorphic function of θ_n in the strip $0 \leq \mathrm{Im}(\theta_n) \leq 2\pi$, whose only singularities are simple poles. In addition the form factors are subject to the following conditions [420]:

1. The form factors have the 'Symmetry Property'

$$f^{\mathcal{O}}(\ldots, \theta_j, \theta_{j+1}, \ldots)_{\ldots, \varepsilon_j, \varepsilon_{j+1}, \ldots} = f^{\mathcal{O}}(\ldots, \theta_{j+1}, \theta_j, \ldots)_{\ldots, \varepsilon'_{j+1}, \varepsilon'_j, \ldots}$$
$$\times S^{\varepsilon'_{j+1}\varepsilon'_j}_{\varepsilon_{j+1}\varepsilon_j}(\theta_{j+1} - \theta_j). \tag{10.74}$$

The symmetry property is a direct consequence of the definition of the form factor and is easily established by use of the FZ algebra.

2. The form factors fulfil 'Smirnov's Axiom', which for charge neutral operators reads

$$f^{\mathcal{O}}(\theta_1, \ldots, \theta_n + 2\pi i)_{\varepsilon_1, \ldots, \varepsilon_n} = i^n \exp(2\pi i\omega[\mathcal{O}, \Psi]) \, f^{\mathcal{O}}(\theta_n, \theta_1, \ldots, \theta_{n-1})_{\varepsilon_n, \varepsilon_1, \ldots, \varepsilon_{n-1}}. \tag{10.75}$$

Here $\omega[\mathcal{O}, \Psi]$ is the 'Locality Index' [303, 420, 500] of the operator $\mathcal{O}$ with respect to the 'elementary' field Ψ, which is any field with nonzero matrix elements between the vacuum and one-particle states. The locality index between two operators $\mathcal{A}$ and $\mathcal{B}$ is calculated by considering the operator product $\mathcal{A}(\mathbf{x})\mathcal{B}(\mathbf{y})$, where $\mathbf{x} = (v\tau, x)$. Denoting by $\mathcal{A}_C$ the analytical continuation in $\mathbf{x}$ along a counterclockwise contour C around the point $\mathbf{y}$, the quantity $\omega[\mathcal{A}, \mathcal{B}]$ is defined as

$$\mathbf{A}_C \left[\mathcal{A}(\mathbf{x})\mathcal{B}(\mathbf{y}) \right] = \exp(2\pi i \omega[\mathcal{A}, \mathcal{B}]) \mathcal{A}(\mathbf{x})\mathcal{B}(\mathbf{y}) . \tag{10.76}$$

3. The form factors have simple poles at the points $\theta_n = \theta_j + i\pi$. The residues at these poles are subject to the 'Annihilation Pole Condition'. For charge-neutral operators $\mathcal{O}$ it reads

$$i \text{Res} \Big|_{\theta_n = \theta_{n-1} + i\pi} \ f^{\mathcal{O}}(\theta_1, \dots, \theta_n)_{\varepsilon_1, \dots, \varepsilon_n} = f^{\mathcal{O}}(\theta_1, \dots, \theta_{n-2})_{\varepsilon'_1, \dots, \varepsilon'_{n-2}}$$
$$\times C_{\varepsilon_n, \varepsilon'_1} \Big[\delta^{\varepsilon'_1}_{\varepsilon_1} \cdots \delta^{\varepsilon'_{n-1}}_{\varepsilon_{n-1}} + i^n e^{2\pi i \omega[\mathcal{O}, \Psi]} S^{\varepsilon'_{n-1}\varepsilon'_1}_{\gamma_1 \varepsilon_1}(\theta_{n-1} - \theta_1)$$
$$\times S^{\gamma_1 \varepsilon'_2}_{\gamma_2 \varepsilon_2}(\theta_{n-1} - \theta_2) \cdots S^{\gamma_{n-3}\varepsilon'_{n-2}}_{\varepsilon_{n-1}\varepsilon_{n-2}}(\theta_{n-1} - \theta_{n-2}) \Big]. \tag{10.77}$$

Here $C = i\sigma^y$ is the charge conjugation matrix. The annihilation pole condition relates form factors with different numbers of particles.

4. The form factors behave under Lorentz transformations as

$$f^{\mathcal{O}}(\theta_1 + u, \dots, \theta_n + u)_{\varepsilon_1, \dots, \varepsilon_n} = \exp(su) \ f^{\mathcal{O}}(\theta_1, \dots, \theta_n)_{\varepsilon_1, \dots, \varepsilon_n} , \tag{10.78}$$

where s is the Lorentz spin of the operator $\mathcal{O}$.

We note that the form factor axioms for the SU(2) Thirring model are of a slightly different form compared to e.g. the sine-Gordon model. As explained in [420], the reason for this is that the elementary excitations in the SU(2) Thirring model have Lorentz spin-$\frac{1}{4}$ and hence possess unusual (fractional) statistics.

10.4.4 Optical conductivity

The real part of the optical conductivity is related to the retarded current-current correlation function by [315]

$$\text{Re} \ \sigma(\omega) = -\frac{\text{Im} \chi^J(\omega)}{\omega} , \quad \omega > 0 ,$$

$$\chi^J(\omega) = -\frac{ie^2}{a_0^2} \int_0^\infty dt \ \exp(i\omega t) \int_{-\infty}^\infty dx \ \langle 0|[J(t, x), J(0, 0)]|0\rangle . \tag{10.79}$$

It is obvious from the bosonized expression (10.54) that the current operator couples only to the charge degrees of freedom. This implies that spinons do not contribute to the optical conductivity in the scaling limit. When expressing (10.79) in a spectral representation, we only have to consider scattering states of holons and antiholons. Furthermore, the current operator is neutral, which implies that only intermediate states with equal numbers of holons and antiholons will contribute to the current-current correlation function. Taking these considerations into account, we arrive at the following spectral representation for the optical conductivity

$$\text{Re}\,\sigma(\omega) = \frac{2\pi^2 e^2}{a_0^2 \omega} \sum_{n=1}^{\infty} \sum_{\varepsilon_i = h, \bar{h}} \int \frac{d\theta_1 \dots d\theta_n}{(2\pi)^n n!} \left| f^J(\theta_1 \dots \theta_n)_{\varepsilon_1 \dots \varepsilon_n} \right|^2$$

$$\times \delta(\sum_k \frac{M}{v_F} \sinh \theta_k) \, \delta(\omega - \sum_k M \cosh \theta_k)$$

$$= \sigma_{h\bar{h}}(\omega) + \sigma_{hh\bar{h}\bar{h}}(\omega) + \sigma_{hhh\bar{h}\bar{h}\bar{h}}(\omega) + \cdots \qquad (10.80)$$

Here $\sigma_{h\bar{h}}(\omega)$ is the contribution of intermediate states with one holon and one antiholon, $\sigma_{hh\bar{h}\bar{h}}(\omega)$ is the contribution of two-holon–two-antiholon states and so on. The most important piece is $\sigma_{h\bar{h}}(\omega)$, where the appropriate intermediate state is the $\eta^z = 0$ state of the charge-triplet excitation constructed in Section 7.2, see figure 7.9.

As a direct consequence of the delta-functions in (10.80) an intermediate state with n holons and n antiholons will contribute to the real part of $\sigma(\omega)$ only above a frequency of $2nM$. Hence $\sigma_{h\bar{h}}(\omega)$ gives the full optical conductivity in the frequency interval $[0, 4M]$. The four-particle contribution $\sigma_{hh\bar{h}\bar{h}}(\omega)$ has been analyzed in Ref. [90], where it was shown that it is negligible at low frequencies and becomes comparable to the two-particle contribution $\sigma_{h\bar{h}}(\omega)$ only around $\omega \approx 100M$. This suggests that the form factor expansion converges rather quickly at low energies and this has indeed been observed in many cases, see e.g. Refs. [37, 72, 97, 289], and can be understood in terms of phase-space arguments [72, 338].

In order to determine $\sigma_{h\bar{h}}(\omega)$ we need the form factor of the current operator between the vacuum and a holon-antiholon intermediate state. The latter can be found for example in [32, 238, 304, 420], but it is instructive to work it out directly from the Axioms above. This is done in Appendix 10.B. The result is

$$\left| f^J(\theta_1, \theta_2)_{h\bar{h}} \right|^2 = \left| f^J(\theta_1, \theta_2)_{\bar{h}h} \right|^2 = \left[M \cosh\left(\frac{\theta_1 + \theta_2}{2}\right) \right]^2 |g(\theta_1 - \theta_2)|^2 , \qquad (10.81)$$

$$g(\theta) = \sinh\left(\frac{\theta}{2}\right) \exp\left[-\int_0^{\infty} \frac{dx}{x} \frac{\sin^2[(\theta + i\pi)x/\pi] \exp(-x)}{\cosh(x) \sinh(2x)} \right] . \qquad (10.82)$$

Inserting (10.81) and (10.82) into (10.80) we arrive at the following result for the holon-antiholon contribution to the real part of the optical conductivity

$$\sigma_{h\bar{h}}(\omega) = \frac{e^2 t}{a_0} \frac{\sqrt{\omega^2 - 4M^2}}{\omega^2} \theta_H(\omega - 2M)$$

$$\times \exp\left[-\int_0^{\infty} \frac{dx}{x} \frac{\exp(-x)[1 - \cos(4x\theta_0/\pi) \cosh 2x]}{\cosh(x) \sinh(2x)} \right] , \qquad (10.83)$$

where t is the hopping matrix element, $\theta_H(x)$ is the Heaviside function and

$$\theta_0 = \text{arcosh}\left(\frac{\omega}{2M}\right) . \qquad (10.84)$$

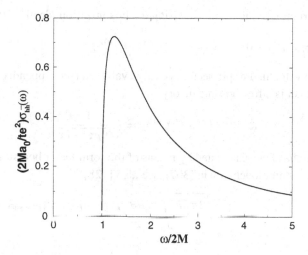

Fig. 10.1. Holon-antiholon contribution to the optical conductivity.

The result (10.83) was first obtained in [224]. In figure 10.1 we plot $\sigma_{h\bar{h}}(\omega)$ as a function of frequency. We see that $\sigma_{h\bar{h}}(\omega)$ vanishes below the optical gap $2M$, then increases as $\sqrt{\omega - 2M}$, peaks around $\omega \approx 2.5M$ and then decays at large frequencies. As the full real part of the optical conductivity $\mathrm{Re}\,\sigma(\omega)$ is equal to $\sigma_{h\bar{h}}(\omega)$ in the frequency interval $[0, 4M]$ (10.83) gives the exact threshold behaviour of $\sigma(\omega)$.

The square-root increase of $\mathrm{Re}\,\sigma(\omega)$ above the optical gap should be contrasted to the results for a Peierls insulator[7] (noninteracting tight-binding electrons with alternating hopping), where there is a square root *singularity* [159]

$$\mathrm{Re}\,\sigma_{\mathrm{PI}}(\omega) \propto \frac{1}{\sqrt{\omega - 2M}} . \qquad (10.85)$$

The behaviour of $\mathrm{Re}\,\sigma(\omega)$ in the 'perturbative' regime $\omega \gg M$ can be analyzed by RG improved perturbation theory [90, 165].

10.4.5 Single particle Green's function

In order to determine the single particle Green's function it is convenient to work with Euclidean (imaginary) time $\tau = it$. In the field theory limit the imaginary time Green's function is expanded as

$$\langle 0|c_{n+1,\sigma}(\tau)\, c_{1,\sigma}^{\dagger}(0)|0\rangle \approx a_0\, \langle 0|R_{\sigma}(\tau, x)\, R_{\sigma}^{\dagger}(0, 0)|0\rangle\, \exp(ik_F x)$$
$$+ a_0\, \langle 0|L_{\sigma}(\tau, x)\, L_{\sigma}^{\dagger}(0, 0)|0\rangle \exp(-ik_F x) . \qquad (10.86)$$

Cross-terms such as $\langle L_{\sigma}(\tau, x)\, R_{\sigma}^{\dagger}(0, 0)\rangle$ vanish because of the chiral symmetry of the Gaussian model describing the spin sector.[8] Let us now turn to the single particle Green's

---

[7] Rigorous results for the Peierls instability were obtained in [295, 297].
[8] In other words, the spin pieces of such correlation functions vanish because $\langle e^{i\alpha\varphi(z)}\, e^{i\beta\bar{\varphi}(\bar{z})}\rangle = 0$ in the Gaussian model.

function of the right moving fermions. Using the bosonization identity (10.48) we obtain

$$\langle 0|R_\sigma(\tau, x)R_\sigma^\dagger(0,0)|0\rangle = \frac{1}{2\pi} \langle e^{-\frac{i}{2}\varphi_c(\tau,x)} e^{\frac{i}{2}\varphi_c(0)}\rangle_c \ \langle e^{-\frac{i}{2}f_\sigma\varphi_s(\tau,x)} e^{\frac{i}{2}f_\sigma\varphi_s(0)}\rangle_s \ . \quad (10.87)$$

The expectation value in the spin sector is easily evaluated (see Appendix 10.C) as the spin sector Hamiltonian is a free bosonic theory

$$\langle e^{-\frac{i}{2}f_\sigma\varphi_s(\tau,x)} e^{\frac{i}{2}f_\sigma\varphi_s(0)}\rangle_s = \frac{1}{\sqrt{v_F\tau - ix}} \ . \quad (10.88)$$

The charge piece has been calculated by means of the form factor bootstrap approach by S. Lukyanov and A. Zamolodchikov in [307], see also [128]

$$\langle e^{-\frac{i}{2}\varphi_c(\tau,x)} e^{\frac{i}{2}\varphi_c(0)}\rangle_c = Z_0\sqrt{\frac{2\pi M}{v_F}} \int_{-\infty}^{\infty} \frac{d\theta}{2\pi} e^{\theta/2} e^{-M\tau\cosh\theta + i\frac{Mx}{v_F}\sinh\theta} + \cdots , \quad (10.89)$$

where

$$Z_0 = \frac{\Gamma(1/4)}{2^{\frac{5}{4}}\sqrt{\pi}} \exp\left[\int_0^\infty \frac{dt}{t} \frac{\sinh^2(t/2)\exp(-t)}{\sinh(2t)\cosh(t)}\right] \approx 0.921862. \quad (10.90)$$

We note that the knowledge of the normalization Z_0 is very useful for comparisons with numerical computations of e.g. the momentum distributions function or the tunneling density of states. The leading term on the r.h.s. of equation (10.89) is the contribution of a one antiholon intermediate state. The θ-dependence of the form factor of $\mathcal{O}_c = \exp(i\varphi_c/2)$ between the vacuum and 1-antiholon states is fixed by the Lorentz spin $s = 1/4$ of $\mathcal{O}_c$. There are subleading contributions due to intermediate states with two antiholons and one holon, three antiholons and two holons and so on. It has been shown in [105] that their contribution is negligible at large distances. Carrying out the θ-integration we obtain

$$\langle 0|R_\sigma(\tau, x)R_\sigma^\dagger(0,0)|0\rangle \simeq \frac{Z_0}{2\pi} \frac{\exp[-M\sqrt{\tau^2 + x^2 v_F^{-2}}]}{v_F\tau - ix} \ . \quad (10.91)$$

The leading corrections to (10.91) due to intermediate states with two antiholons and one holon are of order $\mathcal{O}(\exp(-3mr))$, where $r^2 = \tau^2 + x^2/v_F^2$. Similarly we have

$$\langle 0|L_\sigma(\tau, x)L_\sigma^\dagger(0,0)|0\rangle \simeq \frac{Z_0}{2\pi} \frac{\exp[-M\sqrt{\tau^2 + x^2 v_F^{-2}}]}{v_F\tau + ix} \ . \quad (10.92)$$

The result (10.91) was first written down by P. B. Wiegmann in [480] and independently conjectured by J. Voit in [474]. We note that the expression (10.91) looks quite natural, as it reconciles the known behaviour at short distances $v_F r \to 0$, $(v_F\tau - ix)^{-1}$, with the presence of a spectral gap M leading to the exponential decay at large distances $v_F r \to \infty$. However, the fact that such a simple ansatz gives the correct answer is really a coincidence. In the case of the U(1) Thirring model [307] it would lead to an incorrect result.

It is now straightforward to Fourier transform (10.91) and (10.92) and then analytically continue to real frequencies. The single particle Green's function of right-moving fermions in Euclidean space is (to ease notations we suppress the spin index σ in the formulas below)

$$
\begin{aligned}
G_E^R(\tau, x) &= -\langle 0|\ T_\tau\ R(\tau, x)\ R^\dagger(0, 0)|0\rangle \\
&= -\theta_H(\tau)\langle 0|R(\tau, x)\ R^\dagger(0, 0)|0\rangle + \theta_H(-\tau)\langle 0|R^\dagger(0, 0)\ R(\tau, x)|0\rangle \\
&\simeq -\frac{Z_0}{2\pi} \frac{\exp[-M\sqrt{\tau^2 + x^2 v_F^{-2}}]}{v_F\tau - ix}.
\end{aligned} \tag{10.93}
$$

The Fourier transform of (10.93) may be calculated by going to polar coordinates

$$
\tau = r\cos(\varphi + \varphi_0), \quad x = v_F r\sin(\varphi + \varphi_0), \quad \tan(\varphi_0) = \bar{\omega}/v_F q. \tag{10.94}
$$

Carrying out the the integrals (we use 6.611 (1.) of [184]) we obtain

$$
G_E^R(\bar{\omega}, k_F + q) \simeq \frac{Z_0}{v_F q - i\bar{\omega}} \left[\frac{M}{\sqrt{M^2 + v_F^2 q^2 + \bar{\omega}^2}} - 1 \right]. \tag{10.95}
$$

Analytically continuing (10.95) as well as the analogous formula for the Green's function of left-moving fermions to real frequencies $\bar{\omega} \to \varepsilon - i\omega$, we obtain the following result for the retarded Green's function in the vicinity of the Fermi points $\pm k_F$

$$
G^{(R)}(\omega, \pm k_F + q) \simeq -\frac{Z_0}{\omega \mp v_F q} \left[\frac{M}{\sqrt{M^2 + v_F^2 q^2 - \omega^2}} - 1 \right]. \tag{10.96}
$$

The result (10.96) is valid at low energies where $|q| \ll k_F$ We note that the Green's function (10.96) has a branch cut but no poles. This is a direct reflection of the fact that there are no coherent single particle excitations with the quantum numbers of an electron. We also note that in the approximation (10.96)

$$
G^{(R)}(0, \pm k_F) = 0, \tag{10.97}
$$

i.e. the Green's function vanishes at the Fermi 'surface' $\pm k_F$. We expect that (10.97) continues to hold when the contributions of multiparticle intermediate states are taken into account and is an exact property of the Green's function. This has an important consequence [129]. Luttinger's Theorem relates the density of electrons N_e/V to the single particle Green's function $G(\omega, k)$ in the following way [4, 112]

$$
\frac{N_e}{V} = 2 \int_{G(0,k)>0} \frac{d^D k}{(2\pi)^D}, \tag{10.98}
$$

where the integration is over the interior of the region defined by either singularities or zeroes of the single particle Green's function. The former is the case for a Fermi liquid whereas the Green's function (10.96) fulfils (10.98) by virtue of having zeroes at the position of non-interacting Fermi surface, i.e. $\pm k_F$. Luttinger's Theorem (10.98) is interesting, because it implies that the integral on the right hand side of (10.98) is independent of electron-electron interactions. In particular this means that the volume of the Fermi surface of a Fermi liquid

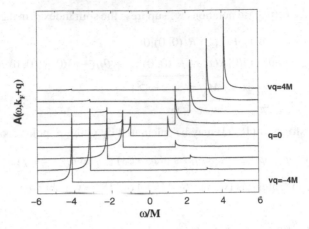

Fig. 10.2. Spectral function in the vicinity of $k_F = \frac{\pi}{2a_0}$. The curves are constant-q scans with $\frac{v_F q}{M} = -4, -3, \ldots, 4$ and have been offset along the y-axis by a constant with respect to one another. We have smoothed the square root singularity above the threshold as described in the text.

is unaffected by electron-electron interactions. In the half filled Hubbard model Luttinger's theorem appears to hold despite the absence of a Fermi surface.

10.4.6 Spectral function

The spectral function is defined as

$$A(\omega, k_F \pm q) = -\frac{1}{\pi} \text{Im} G^{(R)}(\omega, k_F \pm q) . \tag{10.99}$$

We note that as a consequence of the particle-hole symmetry at half filling we have

$$A(-\omega, k_F - q) = A(\omega, k_F + q) . \tag{10.100}$$

The spectral function is of direct experimental interest as it can be measured by angle-resolved photoemission spectroscopy (ARPES), see Refs. [71, 91] and references therein. Taking the imaginary part of (10.96) we arrive at the following simple result for the spectral function in the vicinity of $\pm k_F$ ($|q| \ll k_F$)

$$A(\omega, \pm k_F + q) \simeq \frac{Z_0 M}{\pi |\omega \mp v_F q|} \frac{\theta_H(|\omega| - \sqrt{M^2 + v_F^2 q^2})}{\sqrt{\omega^2 - M^2 - v_F^2 q^2}} . \tag{10.101}$$

In figure 10.2 we plot the spectral function in the vicinity of $k_F = \frac{\pi}{2a_0}$ in a series of constant-q scans. For presentational purposes we have smoothed the square root singularities of $A(\omega, \pm k_F + q)$ above the thresholds at $\omega = \pm\sqrt{M^2 + v_F^2 q^2}$ by giving ω a small imaginary part and then taking the real part of the resulting function. The spectral function is completely incoherent, there are no poles corresponding to the coherent propagation of electrons or holes. Instead there is a scattering continuum that is singular at the spectral gap.

Let us relate these results to our discussion of the excitation spectrum at half filling in Section 7.2. For positive frequencies the relevant excitation is an antiholon-spinon excitation. The latter is obtained by acting with $\eta^\dagger$ on the spin-charge scattering state. The resulting boundaries of the scattering continuum are obtained by shifting the momentum figure 7.11 by π. In the scaling limit there is an exponentially small gap at momenta $\pm\frac{\pi}{2}$ and in the vicinity of these points the spectral function is given by (10.101). For negative frequencies we need to consider a holon-spinon excited state i.e. the spin-charge scattering state of Section 7.2.

10.4.7 Density response function

The two-point function of electron densities is given by

$$G_{nn}(\tau, x) = \langle 0|n(\tau, x)\, n(0)|0\rangle$$
$$= G_{nn}^0(\tau, x) + (-1)^{x/a_0}\, G_{nn}^{2k_F}(\tau, x)\,, \tag{10.102}$$

where

$$G_{nn}^0(\tau, x) = \langle 0|n_0(\tau, x)\, n_0(0, 0)|0\rangle\,,$$
$$G_{nn}^{2k_F}(\tau, x) = \langle 0|n_{2k_F}(\tau, x)\, n_{2k_F}(0, 0)|0\rangle\,. \tag{10.103}$$

Here the smooth n_0 and staggered n_{2k_F} components of the electron density have been defined in (10.40). We note that there are no 'mixed terms' as is most easily seen in the bosonic representation (10.55). As the spin sector is gapless we have

$$\langle 0|\cos\left[\frac{\Phi_s(\tau, x)}{2}\right]|0\rangle_s = 0\,, \tag{10.104}$$

and as a result

$$\langle 0|n_0(\tau, x)\, n_{2k_F}(0, 0)|0\rangle = 0\,. \tag{10.105}$$

We first consider density-density correlation functions in momentum space and determine the large (Euclidean) distance asymptotics of $G^0(\tau, x)$ and $G^{2k_F}(\tau, x)$ afterwards, see equations (10.137) and (10.139). Of particular physical interest is minus the imaginary part of the Fourier transform of the retarded dynamical density-density correlation function

$$L(\omega, k) = -\mathrm{Im}\left\{\chi''(\omega, k)\right\}\,,$$

$$\chi''(\omega, k) = -i\int_{-\infty}^{\infty} dx \int_0^\infty dt\, e^{i\omega t - ikx}\left[G_{nn}(t, x) - G_{nn}(-t, -x)\right]. \tag{10.106}$$

The function $L(\omega, k)$ is experimentally measurable by Electron Energy Loss Spectroscopy

(EELS) [282]. We evaluate $L(\omega, k)$ in the vicinity of the low-energy modes at $k = 0, 2k_F = \pi/a_0$ by Fourier transforming the large-distance asymptotics of the density-density correlation function (10.102). The problem clearly splits into two parts: the dominant contributions in the vicinity of $k \approx 0$ and $k \approx 2k_F$ come from the Fourier transforms of G_{nn}^0 and $G_{nn}^{2k_F}$ respectively. We will calculate $\chi^n(\omega, q)$ by analytically continuing the Euclidean correlation function in frequency

$$\chi^n(\omega, k) = \chi_E^n(\bar{\omega}, k)\Big|_{\bar{\omega} \to \varepsilon - i\omega},$$

$$\chi_E^n(\bar{\omega}, k) = - \int_{-\infty}^{\infty} dx \int_{-\infty}^{\infty} d\tau\, e^{i\bar{\omega}\tau - ikx} \langle T_\tau \{n(\tau, x)\, n(0, 0)\}\rangle . \qquad (10.107)$$

Here T_τ denotes the usual τ-ordering

$$T_\tau \{n(\tau, x)\, n(0, 0)\} = \theta_H(\tau)n(\tau, x)\, n(0, 0) + \theta_H(-\tau)n(0, 0)\, n(\tau, x), \qquad (10.108)$$

where θ_H is the Heaviside function.

Small k behaviour. In the vicinity of $k = 0$ the dynamical density response is dominated by the contribution from G_{nn}^0. It is obvious from the bosonized expression of $n(\tau, x)$ (10.55) that this contribution does not involve the spin sector. We may therefore proceed in complete analogy to our calculation of the optical conductivity. Employing a spectral representation in terms of scattering states of holons and antiholons we obtain

$$\chi^n(\omega, q \approx 0) = \sum_{n=1}^{\infty} \sum_{\varepsilon_i} \int \frac{d\theta_1 \ldots d\theta_n}{(2\pi)^{n-1} n!} |f^{n_0}(\theta_1 \ldots \theta_n)_{\varepsilon_1 \ldots \varepsilon_n}|^2$$

$$\times \left[\frac{\delta(q - \sum_j M_{\varepsilon_j} \sinh\theta_j / v_F)}{\omega - \sum_j M_{\varepsilon_j} \cosh\theta_j + i\eta} - \frac{\delta(q + \sum_j M_{\varepsilon_j} \sinh\theta_j / v_F)}{\omega + \sum_j M_{\varepsilon_j} \cosh\theta_j + i\eta} \right],$$
$$(10.109)$$

where η is a positive infinitesimal. As n_0 is charge neutral, the leading contribution at small energies is due to intermediate states with one holon and one antiholon. More precisely, the relevant intermediate state is the $\eta^z = 0$ state of the charge-triplet excitation constructed in Section 7.2, see figure 7.9. The corresponding form factors can be determined along the same lines as the two-particle form factor of the current operator (see Appendix 10.B), the only difference coming from the transformation properties under Lorentz transformations. The result is [32, 238, 304, 420]

$$|f^{n_0}(\theta_1, \theta_2)_{h\bar{h}}|^2 = |f^{n_0}(\theta_1, \theta_2)_{\bar{h}h}|^2$$

$$= \left[\frac{a_0 M}{v_F} \sinh\left(\frac{\theta_1 + \theta_2}{2} \right) \right]^2 |g(\theta_1 - \theta_2)|^2 , \qquad (10.110)$$

where the function $g(\theta)$ is given by (10.82). Using (10.110) to evaluate the two-particle contribution to (10.109) and then taking the imaginary part, we obtain for $\omega > 0$

$$L(\omega, q \approx 0) = \frac{v_F}{8t^2} \frac{v_F^2 q^2 \sqrt{\omega^2 - v_F^2 q^2 - 4M^2}}{\left(\omega^2 - v_F^2 q^2\right)^{3/2}} \theta_H(\omega^2 - v_F^2 q^2 - 4M^2)$$

$$\times \exp\left[-\int_0^\infty \frac{dx}{x} \frac{\exp(-x)\left[1 - \cos(4x\theta_0/\pi)\cosh 2x\right]}{\cosh(x)\sinh(2x)}\right]$$

$$+ \text{contributions from 4,6,8 ... particles,} \tag{10.111}$$

where

$$\theta_0 = \text{arcosh}\left(\frac{\sqrt{\omega^2 - v_F^2 q^2}}{2M}\right). \tag{10.112}$$

The contributions of intermediate states with four or more particles vanish due to energy-momentum conservation as long as

$$\omega^2 - v_F^2 q^2 \leq 16M^2. \tag{10.113}$$

In this range of frequencies and momenta the holon-antiholon contribution gives the exact result. Above the two-particle threshold at $\omega = \sqrt{v_F^2 q^2 + 4M^2}$, $L(\omega, q)$ (as a function of ω for fixed q) increases from zero in a universal square root fashion. This is due to the momentum dependence of the form factors. We note that because of the proportionality to q^2, $L(\omega, q)$ is always small for small momenta.

From the Heisenberg equations of motions for the lattice density operator

$$\frac{\partial n_j}{\partial t} = -\left[J_{j,j+1} - J_{j-1,j}\right], \tag{10.114}$$

one can derive a relation between $L(\omega, q)$ and the real part of the optical conductivity

$$\text{Re } \sigma(\omega) = \lim_{q \to 0} \frac{e^2 \omega}{a_0^4 q^2} L(\omega, q). \tag{10.115}$$

A direct comparison of (10.111) with (10.83) confirms this relation.

Behaviour around $k = 2k_F$. In the vicinity of $k = 2k_F$ the dynamical density response is dominated by the contribution from $G_{nn}^{2k_F}(\tau, x)$ and involves both the spin and the charge sector. This is most easily seen by considering the bosonized form of n_{2k_F}

$$n_{2k_F} = \frac{2a_0}{\pi} \sin\left[\frac{\Phi_c}{2}\right] \cos\left[\frac{\Phi_s}{2}\right]. \tag{10.116}$$

Due to spin-charge separation the correlation functions in (10.103) factorize into spin and charge pieces. In the spin sector (10.53) we are dealing with a simple Gaussian model and considerations along the lines of Appendix 10.C give

$$_s\langle 0| \cos\left(\frac{\Phi_s(\tau, x)}{2}\right) \cos\left(\frac{\Phi_s(0, 0)}{2}\right) |0\rangle_s = \frac{1}{2}\left[x^2 + v_F^2\tau^2\right]^{-\frac{1}{2}}, \tag{10.117}$$

where $|0\rangle_s$ denotes the vacuum in the spin sector. The correlator in the charge sector can again be determined by using the spectral representation (10.73). In order to carry out the x and τ integrations in the Fourier transforms, it is convenient to set $q = 0$. As $\sin(\Phi_c/2)$ is a Lorentz scalar, the q-dependence can be easily restored in the end by taking

$$\bar{\omega} \to \sqrt{\bar{\omega}^2 + v_F^2 q^2} . \tag{10.118}$$

The Euclidean correlator is then expressed as

$$\chi_E^n(\bar{\omega}, 2k_F) = -\frac{1}{2}\left[\frac{2a_0}{\pi}\right]^2 \int_{-\infty}^{\infty} dx \int_{-\infty}^{\infty} d\tau \, e^{i\bar{\omega}\tau}[x^2 + v_F^2\tau^2]^{-\frac{1}{2}}$$

$$\times \sum_{n=2}^{\infty} \int \frac{d\theta_1 \ldots d\theta_n}{(2\pi)^n n!} |f^{\sin(\Phi_c/2)}(\theta_1, \ldots, \theta_n)_{\varepsilon_1,\ldots,\varepsilon_n}|^2 e^{-\sum_{j=1}^{n} E_j|\tau| - i P_j x}, \tag{10.119}$$

where $E_j = M\cosh(\theta_j)$ and $P_j = M\sinh(\theta_j)/v_F$. The leading contribution in the spectral sum comes from the terms with $n = 2$, i.e. intermediate states with one holon and one antiholon. The relevant excitation is the $\eta^z = 0$ state of the charge triplet constructed in Section 7.2. The corresponding form factors can be determined in analogy with Appendix 10.B. An important difference compared to the current operator is that the locality index of $\sin(\Phi_c/2)$ is 1/2. The form factors are [32, 304, 420]

$$|f^{\sin(\Phi_c/2)}(\theta_1, \theta_2)_{+-}|^2 = |f^{\sin(\Phi_c/2)}(\theta_1, \theta_2)_{-+}|^2$$

$$= \frac{ZM}{v_F}|g(\theta_1 - \theta_2)|^2 , \tag{10.120}$$

where $g(\theta)$ is given by (10.82) and Z is an unknown constant. Concentrating on the holon-antiholon contribution, we may change variables to

$$\theta_\pm = \frac{\theta_1 \pm \theta_2}{2} , \tag{10.121}$$

and then carry out the integral over θ_+. This results in

$$\chi_E^n(\bar{\omega}, 2k_F) \approx -\frac{1}{2\pi}\left[\frac{2a_0}{\pi}\right]^2 \int_{-\infty}^{\infty} dx \int_{-\infty}^{\infty} d\tau \, e^{i\bar{\omega}\tau}[x^2 + v_F^2\tau^2]^{-\frac{1}{2}}$$

$$\times \sum_{\varepsilon_1,\varepsilon_2} \int \frac{d\theta_-}{2\pi}|f^{\sin(\Phi_c/2)}(2\theta_-)_{\varepsilon_1\varepsilon_2}|^2 K_0(2M\sqrt{\tau^2 + (x/v_F)^2}\cosh\theta_-) . \tag{10.122}$$

Introducing polar coordinates

$$\tau = r\sin\varphi , \quad \frac{x}{v_F} = r\cos\varphi , \quad r^2 = \tau^2 + x^2/v_F^2 , \tag{10.123}$$

we may integrate over φ first, giving a Bessel function $J_0(\bar{\omega}r)$, and then carry out the r integral by means of the identity 6.576 (3.) of [184]

$$\int_0^\infty dx\, x^{-\lambda}\, K_\mu(ax)\, J_\nu(bx) = \frac{b^\nu \Gamma\left(\frac{\nu-\lambda+\mu+1}{2}\right)\Gamma\left(\frac{\nu-\lambda-\mu+1}{2}\right)}{2^{\lambda+1}a^{\nu-\lambda+1}\Gamma(1+\nu)}$$
$$\times F\left(\frac{\nu-\lambda+\mu+1}{2},\frac{\nu-\lambda-\mu+1}{2};\nu+1,-\frac{b^2}{a^2}\right). \tag{10.124}$$

Here $F(a,b;c,z)$ is a hypergeometric function. This gives

$$\chi_E^n(\bar{\omega},2k_F) \approx -\left[\frac{a_0}{\pi}\right]^2 \sum_{\varepsilon_1,\varepsilon_2}\int d\theta\, \frac{|f^{\sin(\Phi_c/2)}(2\theta)_{\varepsilon_1\varepsilon_2}|^2}{c(\theta)}F\left(\tfrac{1}{2},\tfrac{1}{2};1,-\tfrac{\bar{\omega}^2}{c(\theta)^2}\right), \tag{10.125}$$

where we have introduced the shorthand notation

$$c(\theta) = 2M\cosh(\theta). \tag{10.126}$$

We may now restore the q-dependence by using (10.118), analytically continue to real frequencies and then take the imaginary part. This leads to the following result for the function $L(\omega, 2k_F+q)$ for small $|q| \ll k_F$ and $\omega > 0$

$$L(\omega, 2k_F+q) \approx \left[\frac{a_0}{\pi}\right]^2 \sum_{\varepsilon_1,\varepsilon_2}\int d\theta\, \frac{|f^{\sin(\Phi_c/2)}(2\theta)_{\varepsilon_1\varepsilon_2}|^2}{c(\theta)}$$
$$\times \mathrm{Im}\left\{F\left(\tfrac{1}{2},\tfrac{1}{2};1,\tfrac{\omega^2-v_F^2q^2}{c(\theta)^2}\right)\right\}. \tag{10.127}$$

The imaginary part of the hypergeometric function is zero unless

$$c(\theta) < \sqrt{\omega^2 - v_F^2 q^2}, \tag{10.128}$$

which implies that $\omega^2 > v_F^2q^2 + 4M^2$. In order to extract the behaviour just above the threshold $(\omega^2 - v_F^2q^2 - 4M^2 \ll 4M^2)$ we may use the transformation formulas for hypergeometric functions (see Ref. [3] equation 15.3.10) to obtain

$$\mathrm{Im}\left\{F\left(\tfrac{1}{2},\tfrac{1}{2};1,\tfrac{\omega^2-v_F^2q^2}{c(\theta)^2}\right)\right\} = F\left(\tfrac{1}{2},\tfrac{1}{2};1,1-\tfrac{\omega^2-v_F^2q^2}{c^2(\theta)}\right)$$
$$\times \theta_H(\omega^2 - v_F^2q^2 - c^2(\theta)). \tag{10.129}$$

The remaining θ-integral in (10.127) is therefore over a very small interval and can be taken by Taylor-expanding the integrand. The leading contribution to the behaviour just above the threshold is

$$L(\omega, 2k_F+q) \propto \left(\frac{\sqrt{\omega^2-v_F^2q^2}-2M}{M}\right)^{\frac{3}{2}}, \quad \frac{\sqrt{\omega^2-v_F^2q^2}}{2M} \to 1^+. \tag{10.130}$$

The important result is that $L(\omega, 2k_F+q)$ vanishes as the threshold is approached from above. There are no threshold singularities! Instead $L(\omega, 2k_F+q)$ vanishes in a power-law fashion. The same holds true in presence of a magnetic field as has been shown in [89].

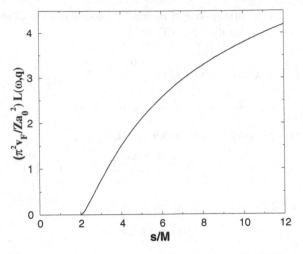

Fig. 10.3. Density response function $L(\omega, q)$ as a function of $s = \sqrt{\omega^2 - v_F^2 q^2}$ in the vicinity of $2k_F$.

The behaviour of (10.127) for large frequencies $\omega \gg \sqrt{v_F^2 q^2 + 4M^2}$ (but necessarily $\omega \ll t$ for field theory to apply) is

$$L(\omega, 2k_F + q) \longrightarrow \text{const.} \times \ln(\omega/m). \qquad (10.131)$$

As n_{2k_F} is a Lorentz scalar, the function $L(\omega, 2k_F + q)$ only depends on the Mandelstam variable

$$s = \sqrt{\omega^2 - v_F^2 q^2}. \qquad (10.132)$$

It is convenient to plot $L(\omega, 2k_F + q)$ as a function of s, which is done in figure 10.3. At first sight it looks quite strange that $L(\omega, q)$ does not go to zero at large ω and fixed q. However, we recall that the field theory calculations cease to be applicable at energies comparable to the bandwidth $4t$. In order to understand why $L(\omega, q)$ increases with frequency it is useful to determine the density-density correlations in the tight-binding model of noninteracting electrons ($U = 0$). A simple calculation gives

$$L_{u=0}(\omega, k) = \frac{4a_0}{\pi} \text{Im} \left\{ \frac{\text{artanh}\left[\frac{4t \sin^2(ka_0/2)}{(4t \sin^2(ka_0/2))^2 + (\varepsilon - i\omega)^2} \right]}{\sqrt{(4t \sin^2(ka_0/2))^2 + (\varepsilon - i\omega)^2}} \right\}. \qquad (10.133)$$

A density plot of $L_{u=0}(\omega, k)$ is shown in figure 10.4. Most of the spectral weight is located at high energies and $L_{u=0}(\omega, k)$ increases with frequency. At finite interaction strength u the low-energy behaviour is changed by the dynamical generation of the Mott gap. However, a small value of u will not alter the overall distribution of spectral weight. Hence one would expect that $L(\omega, q)$ (eventually) will increase with frequency, in accordance with the field theory result.

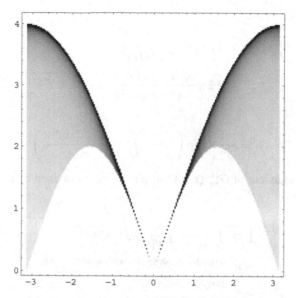

Fig. 10.4. Density plot of the density response function $L_{u=0}(\omega, k)$ for noninteracting electrons as a function of ka_0 (x-axis) and ω/t (y-axis).

The density response function for the half filled Hubbard model at weak coupling was analyzed by the FFBA in [89].

Large distance asymptotics. Let us now derive explicit expressions for the large (Euclidean) distance asymptotics of $G_{nn}(\tau, x)$. Employing a spectral representation (10.73) and retaining only the leading term, which is due to intermediate states with one holon and one antiholon, we obtain

$$G_{nn}^0(\tau, x) \approx \int \frac{d\theta_1 d\theta_2}{(2\pi)^2} e^{-M\tau(\cosh\theta_1 + \cosh\theta_2) + i(Mx/v_F)(\sinh\theta_1 + \sinh\theta_2)}$$
$$\times |f^{n_0}(\theta_1, \theta_2)_{h\bar{h}}|^2 \,, \tag{10.134}$$

where $|f^{n_0}(\theta_1, \theta_2)_{h\bar{h}}|^2 = |f^{n_0}(\theta_1, \theta_2)_{\bar{h}h}|^2$ is given by (10.110). We now change integration variables to $\theta_\pm = (\theta_1 \pm \theta_2)/2$. The θ_+ dependence enters via the factor

$$\sinh^2(\theta_+) \, e^{2i(Mx/v_F)\sinh(\theta_+)\cosh(\theta_-)} = -\frac{\partial^2}{\partial x^2} \frac{e^{2i(Mx/v_F)\sinh(\theta_+)\cosh(\theta_-)}}{4M^2 v_F^{-2} \cosh^2(\theta_-)} \,. \tag{10.135}$$

The integration over θ_+ can now be carried out

$$G_{nn}^0(\tau, x) \approx -a_0^2 \frac{\partial^2}{\partial x^2} \int \frac{d\theta_-}{4\pi^2} \left| \frac{g(2\theta_-)}{\cosh\theta_-} \right|^2 K_0(2Mr\cosh\theta_-) \,, \tag{10.136}$$

where $r = \sqrt{\tau^2 + (x/v_F)^2}$. The remaining integral is dominated by the saddle points at $\theta \approx \pm(Mr)^{-1/2} \ll 1$. In the saddle-point approximation we obtain the following result in the limit $Mr \to \infty$

$$G^0_{nn}(\tau, x) \approx -a_0^2 \, \mathcal{D}_0 \, \partial_x^2 \frac{e^{-2Mr}}{(2Mr)^2}$$

$$\approx -\mathcal{D}_0 \frac{a_0^2[x^2 - v_F^2 r/2M]}{(x^2 + v_F^2 \tau^2)^2} e^{-2M\sqrt{\tau^2 + (x/v_F)^2}} , \tag{10.137}$$

where

$$\mathcal{D}_0 = \frac{1}{\sqrt{2\pi}} \exp\left(-1 + \int_0^\infty \frac{dx}{x} \frac{e^{-x}}{\,} \frac{\sinh x}{\cosh^2 x}\right) . \tag{10.138}$$

The asymptotic behaviour of $G^{2k_F}_{nn}(\tau, x)$ for $Mr \to \infty$ is obtained by analogous considerations. We obtain

$$G^{2k_F}_{nn}(\tau, x) \approx \left[\frac{2a_0}{\pi}\right]^2 \frac{1}{2\sqrt{x^2 + v_F^2 \tau^2}} \int \frac{d\theta_1 d\theta_2}{(2\pi)^2} |f^{\sin(\Phi_c/2)}(\theta_1, \theta_2)_{h\bar{h}}|^2$$

$$\times e^{-M\tau(\cosh\theta_1 + \cosh\theta_2) + i(Mx/v_F)(\sinh\theta_1 + \sinh\theta_2)}$$

$$\approx \mathcal{D} \frac{e^{-2M\sqrt{\tau^2 + (x/v_F)^2}}}{(x^2 + v_F^2 \tau^2)^{\frac{3}{2}}}, \tag{10.139}$$

where

$$\mathcal{D} = \frac{a_0^2 v_F}{M} \frac{\sqrt{2}Z}{\pi^3} \exp\left(-1 + \int_0^\infty \frac{dx}{x} \frac{e^{-x}}{\,} \frac{\sinh x}{\cosh^2 x}\right) . \tag{10.140}$$

10.4.8 Spin correlation functions

Due to spin-rotational invariance we have

$$\langle S^x(\tau, x) S^x(0, 0) \rangle = \langle S^y(\tau, x) S^y(0, 0) \rangle = \langle S^z(\tau, x) S^z(0, 0) \rangle , \tag{10.141}$$

so that it is sufficient to consider $\langle S^z(\tau, x) S^z(0, 0) \rangle$. Using the bosonized expression (10.56) for the spin-operator, we obtain

$$\langle S^z(\tau, x) S^z(0, 0) \rangle = \left[\frac{a_0}{4\pi}\right]^2 \langle \partial_x \Phi_s(\tau, x) \, \partial_x \Phi_s(0, 0) \rangle_s$$

$$+ (-1)^{x/a_0} \left[\frac{a_0}{\pi}\right]^2 \left\langle \sin\left(\frac{\Phi_s(\tau, x)}{2}\right) \sin\left(\frac{\Phi_s(0, 0)}{2}\right)\right\rangle_s$$

$$\times \left\langle \cos\left(\frac{\Phi_c(\tau, x)}{2}\right) \cos\left(\frac{\Phi_c(0, 0)}{2}\right)\right\rangle_c . \tag{10.142}$$

The expectation values in the spin sector are easily evaluated

$$\langle S^z(\tau, x) S^z(0, 0) \rangle = \frac{a_0^2}{8\pi^2} \left[\frac{1}{(v_F\tau + ix)^2} + \frac{1}{(v_F\tau - ix)^2}\right]$$

$$+ (-1)^{x/a_0} \frac{a_0^2}{2\pi^2} \frac{1}{(v_F^2 \tau^2 + x^2)^{\frac{1}{2}}} \chi^{\cos(\Phi_s/2)}(\tau, x) , \tag{10.143}$$

where we have defined

$$\chi^{\cos(\Phi_s/2)}(\tau, x) = \left\langle \cos\left(\frac{\Phi_c(\tau, x)}{2}\right) \cos\left(\frac{\Phi_c(0, 0)}{2}\right)\right\rangle_c. \qquad (10.144)$$

The two-point function in the charge sector (10.144) can be evaluated in the spectral representation (10.73). The first two terms in the spectral sum are

$$\chi^{\cos(\Phi_s/2)}(\tau, x) \approx C^2 + \frac{1}{2} \sum_{\varepsilon_1, \varepsilon_2} \int \frac{d\theta_1 d\theta_2}{(2\pi)^2} |f^{\cos(\Phi_c/2)}(\theta_1 - \theta_2)_{\varepsilon_1 \varepsilon_2}|^2$$

$$\times e^{-M|\tau|(\cosh\theta_1 + \cosh\theta_2)) - i(Mx/v_F)(\sinh\theta_1 + \sinh\theta_2)} + \cdots \qquad (10.145)$$

where

$$C = \left\langle \cos\left(\frac{\Phi_c(0, 0)}{2}\right)\right\rangle. \qquad (10.146)$$

The contribution to (10.145) due to holon-antiholon intermediate states is exponentially suppressed at large τ. Hence, the large-τ asymptotics of the spin-spin correlation functions at half-filling are given by the 'conformal' result

$$\langle S^z(\tau, x) S^z(0, 0)\rangle \approx \frac{a_0^2}{8\pi^2}\left[\frac{1}{(v_F\tau + ix)^2} + \frac{1}{(v_F\tau - ix)^2}\right]$$

$$+ (-1)^{x/a_0} \frac{C^2 a_0^2}{2\pi^2} \frac{1}{(v_F^2\tau^2 + x^2)^{\frac{1}{2}}}. \qquad (10.147)$$

The leading corrections to (10.147) are proportional to e^{-2Mr} and can be evaluated by the same method we used to obtain the large-distance asymptotics of the density-density correlation functions. In the dynamical structure factor these corrections contribute only for energies higher than $2M$, i.e. twice the single particle gap.

10.5 Correlation functions in the continuum limit

As we have mentioned before, the continuum limit describes the vicinity of the scaling limit of the half filled Hubbard model. This is a very important regime. If one is interested in physical properties for weak repulsion, say $U \approx t$, one cannot rely on the results obtained in the scaling limit but one may hope that this regime is captured by the continuum limit. The reason is that the window of applicability of the results in the continuum limit is ultimately determined by the requirement that the ratio of the charge gap Δ (7.18) to the electronic bandwidth $4t$ is small, i.e.

$$\Delta \ll 4t. \qquad (10.148)$$

For $U = t$ we have $\Delta \approx 0.002\,512\,35t$ and this criterion is satisfied. As we have mentioned before, the Hamiltonian in the continuum limit (10.24) is the sum of two integrable quantum field theories and correlation functions can be obtained in a way analogous to the scaling limit. The main differences between scaling and continuum limits are

- spin and charge velocities are different in the continuum limit but are the same in the scaling limit;
- the charge gap is given by (7.18) in the continuum limit and by (10.1) in the scaling limit;
- the spin Hamiltonian (10.24) in the continuum limit contains a marginally irrelevant current-current interaction;
- the amplitudes of the bosonized expressions of local operators depend on the interaction U in the continuum limit and are not generally known.

The most important difference is that $v_c \neq v_s$ in the continuum limit. In what follows we will ignore the marginally irrelevant current-current interaction in the spin sector. Taking it into account leads to logarithmic corrections, which are generally small. We also will neglect the U-dependence of the amplitudes of the bosonized expressions for local operators. Furthermore, we will use the exact result (7.18) for the charge gap rather than (10.1), which holds only in the scaling limit.

10.5.1 Optical conductivity

In the scaling limit the current operator does not involve the spin degrees of freedom and this continues to hold true in the continuum limit. Hence we may proceed as in the scaling limit and employ a spectral representation in terms of scattering states of holons and antiholons. The main difference between the scaling limit and the continuum limit are the replacements of the Fermi velocity by the charge velocity and of the charge gap M by Δ as mentioned above. Retaining the bosonized expression obtained in the scaling limit

$$J(\tau, x) = \frac{i}{2\pi} \partial_\tau \Phi_c , \tag{10.149}$$

and then going through the same steps as in the calculation for the scaling limit, we obtain the following result for the holon-antiholon contribution to the real part of the optical conductivity

$$\sigma_{h\bar{h}}(\omega) = \frac{e^2 t}{a_0} \frac{v_c}{v_F} \frac{\sqrt{\omega^2 - 4\Delta^2}}{\omega^2} \theta_H(\omega - 2\Delta)$$
$$\times \exp\left[-\int_0^\infty \frac{dx}{x} \frac{\exp(-x)\left[1 - \cos(4x\theta_0/\pi)\cosh 2x\right]}{\cosh(x)\sinh(2x)} \right], \tag{10.150}$$

where $\theta_0 = \text{arcosh}\left(\frac{\omega}{2\Delta}\right)$. The result (10.150) may be compared with the results of numerical computations of the real part of the optical conductivity for the half filled Hubbard model. The numerical method used in Ref. [224] is the Dynamical Density Matrix Renormalization Group (DDMRG), see [223, 271] and references therein. The method allows for the calculation of $\text{Re}\,\sigma(\omega)$ on large lattices of 128 sites, provided a broadening of the energy levels of intermediate states is introduced. The numerical results for $U = 3t$ presented in [224] are in excellent agreement with (10.150) despite the fact that at $U = 3t$ the charge gap is already sizeable $\Delta(U = 3t) = 0.315\,687t$. The velocity ratio entering (10.150) is obtained

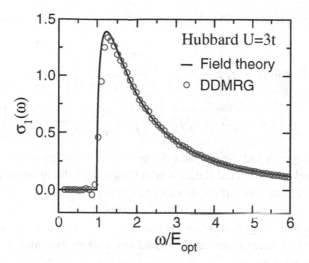

Fig. 10.5. Comparison of the holon-antiholon contribution in units of e^2/a_0 to the optical conductivity to DDMRG computations.

from (7.26) as

$$\left.\frac{v_c}{v_F}\right|_{U=3t} \approx 1.22371 . \tag{10.151}$$

In figure 10.5 we show a comparison between equation (10.150) and DDMRG results for the Hubbard model with $U = 3t$, for which the aforementioned broadening has been removed by means of a numerical deconvolution [222]. The agreement is quite satisfactory. We note that the corrections to (10.150) due to intermediate states with more than one holon and one antiholon are extremely small in the relevant range of frequencies.

10.5.2 Single particle Green's function

The single particle Green's functions of left- and right-moving fermions in Euclidean space in the continuum limit are obtained by introducing spin and charge velocities in the expressions (10.88) and (10.89) for the spin and charge pieces and replacing the expression M for the charge gap by Δ. This gives

$$\langle R_\sigma(\tau, x)R_\sigma^\dagger(0, 0)\rangle \simeq \frac{Z_0}{2\pi}\frac{\exp[-\Delta\sqrt{\tau^2 + x^2 v_c^{-2}}]}{\sqrt{v_s\tau - ix}\sqrt{v_c\tau - ix}} , \tag{10.152}$$

$$\langle L_\sigma(\tau, x)L_\sigma^\dagger(0, 0)\rangle \simeq \frac{Z_0}{2\pi}\frac{\exp[-\Delta\sqrt{\tau^2 + x^2 v_c^{-2}}]}{\sqrt{v_s\tau + ix}\sqrt{v_c\tau + ix}} , \tag{10.153}$$

where the constant Z_0 is given by (10.90). Like in the scaling limit, the leading corrections to (10.152) and (10.153) involve intermediate states containing two antiholons and one holon and are thus of order $\mathcal{O}(\exp(-3\Delta r))$. After Fourier transformation and analytic continuation to real frequencies we obtain the following result for the retarded Green's function (for spin σ) [128]

$$G^{(R)}(\omega, k_F + q) \simeq -Z_0 \sqrt{\frac{2v_c}{v_c + v_s}} \frac{\omega + v_c q}{\sqrt{\Delta^2 + v_c^2 q^2 - \omega^2}}$$

$$\times \left[\left(\Delta + \sqrt{\Delta^2 + v_c^2 q^2 - \omega^2} \right)^2 - \frac{v_c - v_s}{v_c + v_s} (\omega + v_c q)^2 \right]^{-\frac{1}{2}}. \quad (10.154)$$

Here the charge gap Δ and the spin and charge velocities are given by (7.18), (7.21) and (7.26) respectively. The spectral function is obtained from the imaginary part of $G^{(R)}$ by (10.99) and we refrain from writing down its rather lengthy explicit expression. In order to interpret the spectral function, it is useful to determine its threshold. The intermediate state with lowest possible energy that couples to the fermion annihilation operator in the vicinity of k_F is a scattering state of one antiholon and one spinon. The energy and dispersion of such states in general are

$$P = a_0^{-1} \left[\mathcal{P}_s(\Lambda) + \mathcal{P}_{\bar{h}}(k) \right],$$
$$E = t \left[\mathcal{E}_s(\Lambda) + \mathcal{E}_{\bar{h}}(k) \right], \quad (10.155)$$

where $\mathcal{P}_{s,\bar{h}}$ and $\mathcal{E}_{s,\bar{h}}$ are given by (7.8) and where we have reinstituted the lattice spacing a_0 and the hopping integral t. In the continuum limit we must recover Lorentz invariance in the spin and charge sectors separately,[9] so that

$$P = k_F + p_c + p_s,$$
$$E = v_s p_s + \sqrt{\Delta^2 + v_c^2 p_c^2}, \quad (10.156)$$

where the spinon momentum is positive, because $\mathcal{P}_s(\Lambda) \in [0, \pi]$ and the low-energy modes occur for $\mathcal{P}_{\bar{h}} \approx \frac{\pi}{2}$. In other words we are dealing with a right-moving spinon. The threshold (for positive frequencies $\omega > 0$) is obtained by minimizing the energy at fixed total momentum with respect to $p_{s,c}$

$$E_{\text{thres}}(k_F + q) = \begin{cases} \sqrt{\Delta^2 + v_c^2 q^2} & \text{if } q \le Q \\ v_s q + \Delta \sqrt{1 - \alpha^2} & \text{if } q \ge Q \end{cases}, \quad (10.157)$$

where

$$\alpha = \frac{v_s}{v_c}, \quad Q = \frac{v_s \Delta}{v_c \sqrt{v_c^2 - v_s^2}}. \quad (10.158)$$

Inspection of (10.157) shows that the threshold is not symmetric around k_F. For momenta smaller than $k_F + Q$ the threshold is equal to the antiholon dispersion. Furthermore, at the threshold the entire momentum is carried by the antiholon whereas the spinon momentum is zero. On the other hand, for momenta larger than $k_F + Q$, the 'excess' momentum $P - k_F - Q$ is carried by the spin degrees of freedom.

---

[9] We recall that as $v_s \ne v_c$ the full theory is not Lorentz invariant.

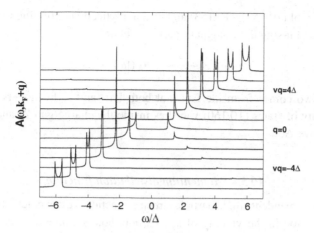

Fig. 10.6. Spectral function for $U = t$ in the vicinity of $k_F = \frac{\pi}{2a_0}$. The curves are constant-q scans with $\frac{v_c q}{\Delta} = -6, -3, \dots, 6$ and have been offset along the y-axis by a constant with respect to one another. We have smoothed the singularities by making ω in the expression (10.154) for the Green's function slightly complex and then taking the imaginary part.

The spectral function corresponding to $U = t$ is shown in a series of constant-q scans in figure 10.6. The most important difference between the spectral functions in the continuum and scaling limits is readily seen by comparing figures 10.6 and 10.2. In the continuum limit and at $\omega > 0$, the single peak above the threshold splits into two peaks if q is larger than a critical value Q. This splitting is a direct manifestation of spin-charge separation. The first peak occurs at the threshold and follows the spinon dispersion, whereas the higher energy peak disperses like an antiholon, i.e. occurs at $\omega = \sqrt{\Delta^2 + v_c^2 q^2}$. A peak splitting of this type is known to exist from numerical computations of the spectral function for the half filled Hubbard model [38, 136, 365, 390]. Several experiments have reported observations of related features in the ARPES spectra of the quasi one-dimensional charge transfer insulators $SrCuO_2$ and Sr_2CuO_3 [150, 245].

10.5.3 Tunneling density of states

The tunneling density of states is defined as

$$\rho(\omega) = -\frac{1}{\pi}\text{Im} \int_{-\pi}^{\pi} \frac{dk}{2\pi} G^{(R)}(\omega, k)$$
$$= -\frac{1}{\pi}\text{Im} \int_{-\infty}^{\infty} dt \, e^{i\omega t} \, G^{(R)}(t, 0) \,, \qquad (10.159)$$

where $G^{(R)}(t, x)$ is the retarded single particle Green's function. In the continuum limit we can determine $\rho(\omega)$ for small frequencies from the expressions (10.152), (10.153) for the Green's function at $x = 0$. Calculating the Fourier transformation with respect to the imaginary time τ of (10.152), (10.153) and then analytically continuing to real frequencies $\omega \longrightarrow \varepsilon - i\omega$ and taking the imaginary part we obtain

$$\rho(\omega) \simeq \frac{Z_0}{\pi \sqrt{v_s v_c}} \theta_H(|\omega| - \Delta) .$$ (10.160)

The factor of two comes from the fact that both left- and right-movers contribute. The tunneling density of states (10.160) vanishes inside the charge gap Δ and is constant for frequencies above Δ.

10.5.4 Momentum distribution function

From the time-independent single particle Green's function we can calculate the momentum distribution function in the vicinity of k_F. Particle-hole symmetry implies that $n_\sigma(k_F + q) = 1 - n_\sigma(k_F - q)$, which fixes $n(k_F) = \frac{1}{2}$. Assuming that momentum dependence in the vicinity of k_F is mainly due to the leading term (10.152) in the expansion of the Green's function we arrive at the following approximate result [128]

$$n_\sigma(k_F + q) = \frac{1}{2} - \frac{Z_0}{\pi} \arctan\left(\frac{v_c q}{\Delta}\right) .$$ (10.161)

The momentum distribution function in the half-filled Hubbard model has been determined numerically using the Density Matrix Renormalisation Group in [342]. For small values of u the numerical results are in agreement with (10.161).

10.5.5 Density-density response function

Last but not least, let us consider the density-density response function $L(\omega, k)$ defined in (10.106). The generalization of the result (10.111) for small $q \approx 0$ to the case $v_c \neq v_s$ works in the same way as for the optical conductivity and we will not dwell on it. The determination of the response function in the continuum limit for $q \approx 2k_F$ is somewhat more complicated. The threshold of $L(\omega > 0, k \approx 2k_F)$ can be determined by considering the lowest intermediate state that couples to the density operator at $k = 2k_F$, which is a scattering state of one holon, one antiholon and two spinons. The energy and dispersion of such states in general are

$$P = a_0^{-1}\left[\mathcal{P}_s(\Lambda_1) + \mathcal{P}_s(\Lambda_2) + \mathcal{P}_{\bar{h}}(k_1) + \mathcal{P}_h(k_2)\right] ,$$
$$E = t\left[\mathcal{E}_s(\Lambda_1) + \mathcal{E}_s(\Lambda_2) + \mathcal{E}_{\bar{h}}(k_1) + \mathcal{E}_h(k_2)\right] ,$$ (10.162)

where $\mathcal{P}_{s,\bar{h}}$ and $\mathcal{E}_{s,\bar{h}}$ are given by (7.8). In the continuum limit we only consider low-energy states, which occur at momenta close to $2k_F = \frac{\pi}{a_0}$. We choose a parametrization such that

$$\mathcal{P}_s(\Lambda_1) = p_1 a_0 \geq 0 , \qquad \mathcal{P}_s(\Lambda_2) = \pi - p_2 a_0 \leq \pi ,$$
$$\mathcal{P}_h(k_1) = -\frac{\pi}{2} + q_1 a_0, \qquad \mathcal{P}_{\bar{h}}(k_2) = \frac{\pi}{2} + q_2 a_0.$$ (10.163)

This leads to the following expression for the total momentum and energy in the continuum limit

$$P = [2k_F + p_1 - p_2 + q_1 + q_2], \quad p_{1,2} \geq 0,$$

$$E = v_s[p_1 + p_2] + \sum_{j=1}^{2} \sqrt{\Delta^2 + v_c^2 q_j^2}. \tag{10.164}$$

Here we have used that the low-energy modes occur for $\mathcal{P}_h \approx -\frac{\pi}{2}, \mathcal{P}_{\bar{h}} \approx \frac{\pi}{2}$ and $\mathcal{P}_s = 0, \pi$. The threshold in the vicinity of $2k_F$ is obtained by minimizing the energy at fixed total momentum with respect to p_1, p_2, q_1, q_2

$$E_{\text{thres}}(2k_F + q) = \begin{cases} \sqrt{4\Delta^2 + v_c^2 q^2} & \text{if } |q| \leq 2Q \\ v_s|q| + 2\Delta\sqrt{1 - \alpha^2} & \text{if } |q| \geq 2Q \end{cases}, \tag{10.165}$$

where Q and α are given by (10.158). The threshold is symmetric around $2k_F$. Working with a spectral representation in real time (10.69) and carrying out the integrals over x and t in the Fourier transforms by means of the identity 3.382 of [184]

$$\int_{-\infty}^{\infty} dx \frac{\exp(-ipx)}{(\beta + ix)^\nu} = \theta_H(-p) \frac{2\pi}{\Gamma(\nu)} (-p)^{\nu-1} \exp(\beta p), \tag{10.166}$$

one may derive the following expression for the function $L(\omega, q)$ [89]

$$L(\omega, 2k_F + q) \approx 2 \left[\frac{a_0}{\pi}\right]^2 \int_{-\infty}^{\infty} \frac{d\theta_+ d\theta_-}{\pi} |f^{\sin(\Phi_c/2)}(2\theta_-)_{h,\bar{h}}|^2$$

$$\times (\Omega\Omega')^{-\frac{1}{2}} \Theta(\Omega) \Theta(\Omega'), \tag{10.167}$$

where

$$\Omega = \omega - v_s q - 2\Delta \cosh(\theta_-) [\cosh(\theta_+) - \alpha \sinh(\theta_+)],$$

$$\Omega' = \omega + v_s q - 2\Delta \cosh(\theta_-) [\cosh(\theta_+) + \alpha \sinh(\theta_+)]. \tag{10.168}$$

One may check that (10.167) leads to a threshold described by (10.165) and reduces to (10.127) in the limit $\alpha \to 1$, in which the θ_+-integral can be carried out easily. The remaining integrals in (10.167) need to be evaluated numerically. The function $L(\omega, 2k_F + q)$ vanishes at the threshold and is free of singularities. As a matter of fact there are no significant differences between the continuum limit and the scaling limit. The fact that no pronounced features associated with v_s and v_c appear in $L(\omega, 2k_F + q)$ may be understood intuitively by recalling that the relevant excitations involve at least four particles (two spinons and one holon/antiholon each). Fixing the total energy and momentum of the state still leaves us with two free momenta over which we need to average.

10.6 Finite temperatures

A natural question is whether the calculations of dynamical correlation functions in the scaling and continuum limits can be extended to finite temperatures. Proposals on how to calculate one and two point functions for integrable Quantum Field Theories at finite temperatures exist [284, 285, 306, 375], but the problem is quite difficult in general. In [261] a method was proposed that allows for the calculation of two point functions in massive integrable quantum field theories at temperatures (much) smaller than the spectral gap. This approach was used in [129] to determine the single particle spectral function of the half filled Hubbard model in the continuum limit at low temperatures. The main finding of [129] is that even a small temperature leads to a significant smearing out of the (anti)holon peak in the spectral function. This is easily understood: as the spin sector remains gapless in the half filled Hubbard model a temperature that is small compared to the charge gap may still significantly alter the dynamical response of the spin degrees of freedom. On the other hand, the tunneling density of states for frequencies below the charge gap is exponentially small at low temperatures.

Appendices to Chapter 10

10.A Current algebra

In this appendix we summarize some properties of the SU(2) spin and pseudospin currents (10.22). Our discussion closely follows Affleck's work [5]. Let us consider spinless fermions with Hamiltonian density (to ease notations we set the velocity equal to 1 in what follows)

$$\mathcal{H} = i : \left[L^\dagger \, \partial_x \, L - R^\dagger \, \partial_x \, R \right] : . \tag{10.A.1}$$

The corresponding Lagrangian density is

$$\mathcal{L} = i : L^\dagger \left[\partial_t - \partial_x \right] L : + i : R^\dagger \left[\partial_t + \partial_x \right] R : . \tag{10.A.2}$$

Introducing light-cone coordinates

$$x_\pm = t \pm x \,, \qquad \partial_\pm = \frac{1}{2} (\partial_t \pm \partial_x) \tag{10.A.3}$$

we may rewrite $\mathcal{L}$ as

$$\mathcal{L} = 2i : \left[L_\sigma^\dagger \, \partial_- \, L_\sigma + R_\sigma^\dagger \, \partial_+ \, R_\sigma \right] : . \tag{10.A.4}$$

The Euler-Lagrange equations are

$$\partial_- L = 0 = \partial_+ R \,. \tag{10.A.5}$$

From now on we concentrate on the right-moving fermions. The corresponding formulas for left-movers are obtained in the same way. The fermion anticommutation relations read

$$\{R^\dagger(x_-), R(y_-)\} = \delta(x_- - y_-) \,. \tag{10.A.6}$$

We note that (10.A.6) holds not only for equal times, but for general x_-, y_- because R depends on t and x only through x_- by virtue of the Euler-Lagrange equations. The fermion propagator is given by

$$\langle R(x_-) \, R^\dagger(y_-) \rangle = \frac{1}{2\pi i (x_- - y_-)} \,. \tag{10.A.7}$$

Let us now introduce a current

$$J(x_-) = \lim_{\varepsilon \to 0} : R^\dagger(x_- - \varepsilon) \, R(x_- + \varepsilon) : \, , \qquad (10.A.8)$$

where ':' denotes normal ordering. The commutator of two currents can now be evaluated by using (10.A.6). It is sufficient to point-split one of the currents, so that

$$[J(x_-), J(y_-)] = \lim_{\varepsilon \to 0} [J(x_-), : R^\dagger(y_- - \varepsilon)R(y_- + \varepsilon) :]$$

$$= \lim_{\varepsilon \to 0} [\delta(x_- - y_- + \varepsilon) - \delta(x_- - y_- - \varepsilon)] \, R^\dagger(y_- - \varepsilon)R(y_- + \varepsilon)$$

$$= \lim_{\varepsilon \to 0} [\delta(x_- - y_- + \varepsilon) - \delta(x_- - y_- - \varepsilon)] \frac{i}{4\pi\varepsilon}$$

$$= \frac{i}{2\pi} \, \delta'(x_- - y_-) \, . \qquad (10.A.9)$$

Here we have used that

$$R^\dagger(y_- - \varepsilon)R(y_- + \varepsilon) = : R^\dagger(y_- - \varepsilon)R(y_- + \varepsilon) : + \langle R^\dagger(y_- - \varepsilon)R(y_- + \varepsilon) \rangle$$

$$= : R^\dagger(y_- - \varepsilon)R(y_- + \varepsilon) : + \frac{i}{4\pi\varepsilon} \, , \qquad (10.A.10)$$

and that the normal ordered piece does not have a singularity for $\varepsilon \to 0$. The derivative of a delta-function generated by the commutator of currents is known as a 'Schwinger term' [387].

Let us now consider the operator product of two currents

$$J(x_- - \varepsilon) \, J(x_- + \varepsilon) = \lim_{\delta \to 0} : R^\dagger(x_- - \varepsilon - \delta)R(x_- - \varepsilon + \delta) :$$

$$\times : R^\dagger(x_- + \varepsilon - \delta)R(x_- + \varepsilon + \delta) : . \qquad (10.A.11)$$

Removing the normal orderings by (10.A.10) and then normal ordering the resulting term quartic in Fermi operators leads to

$$J(x_- - \varepsilon) \, J(x_- + \varepsilon) = : R^\dagger(x_- - \varepsilon)R(x_- - \varepsilon)R^\dagger(x_- + \varepsilon)R(x_- + \varepsilon) :$$

$$+ \frac{i}{4\pi\varepsilon} \, : R^\dagger(x_- - \varepsilon)R(x_- + \varepsilon) :$$

$$- \frac{i}{4\pi\varepsilon} \, : R^\dagger(x_- + \varepsilon)R(x_- - \varepsilon) : - \frac{1}{(4\pi\varepsilon)^2} \, . \qquad (10.A.12)$$

Hence the normal-ordered operator product of two currents is

$$: J(x_-) \, J(x_-) : = \frac{i}{2\pi} \left[: R^\dagger(x_-) \, \partial_- R(x_-) : - : \partial_- R^\dagger(x_-) \, R(x_-) : \right]. \qquad (10.A.13)$$

Equation (10.A.13) implies that

$$\int dx \, : R^\dagger(x) \, i\partial_x \, R(x) : = -\pi \int dx \, : J(x)J(x) : . \qquad (10.A.14)$$

Let us now turn to spinful fermions. We want to determine the normal-ordered operator products of the spin and pseudospin currents (10.22). A calculation completely analogous to the one we just did in the spinless case gives

$$: J^z \, J^z : + : I^z \, I^z : \; = \frac{i}{4\pi} \sum_\sigma \left[: R_\sigma^\dagger \, \partial_- R_\sigma : - : \partial_- R_\sigma^\dagger \, R_\sigma : \right]. \tag{10.A.15}$$

Similarly we find

$$\frac{1}{2} : \left[J^+ J^- + I^+ I^- + \text{h.c.} \right] : \; = \frac{i}{2\pi} \sum_\sigma \left[: R_\sigma^\dagger \, \partial_- R_\sigma : - : \partial_- R_\sigma^\dagger \, R_\sigma : \right]. \tag{10.A.16}$$

Putting everything together, we obtain after integration by parts

$$\int dx \sum_\sigma : R_\sigma^\dagger \, i \partial_x \, R_\sigma : \; = -\frac{2\pi}{3} \int dx \; [: \mathbf{J} \cdot \mathbf{J} : + : \mathbf{I} \cdot \mathbf{I} :]. \tag{10.A.17}$$

10.B Two-particle form factors

In this appendix we discuss how to determine two-particle form factors using the Axioms stated in Section 10.4.3.

10.B.1 Current operator

Combining the symmetry property with Smirnov's axiom, one obtains a Riemann-Hilbert Problem in the variable θ_2

$$f^J(\theta_1, \theta_2 + 2\pi i)_{\varepsilon_1, \varepsilon_2} = -f^J(\theta_1, \theta_2)_{\varepsilon_1', \varepsilon_2'} \, S_{\varepsilon_1 \varepsilon_2}^{\varepsilon_1' \varepsilon_2'}(\theta_1 - \theta_2) \,. \tag{10.B.1}$$

Here we have used that the current operator has mutual locality index zero, which is most easily established in the bosonic representation. The current operator is odd under charge conjugation, which implies that it couples only to the triplet state. This is because the charge conjugation matrix $C = i\sigma^y$ is antisymmetric.[1] The Riemann-Hilbert Problem (10.B.1) can be 'scalarized' by working in a basis in which the S-matrix is diagonal. Symmetrizing in the indices ε_1 and ε_2 we obtain

$$f^J(\theta_1, \theta_2 + 2\pi i)_{(\varepsilon_1, \varepsilon_2)} = -f^J(\theta_1, \theta_2)_{\varepsilon_1', \varepsilon_2'} \, S_{(\varepsilon_1 \varepsilon_2)}^{\varepsilon_1' \varepsilon_2'}(\theta_1 - \theta_2)$$

$$= f^J(\theta_1, \theta_2)_{(\varepsilon_1, \varepsilon_2)} \frac{\Gamma\left(\frac{1}{2} - i\,\frac{\theta_{12}}{2\pi}\right) \Gamma\left(1 + i\,\frac{\theta_{12}}{2\pi}\right)}{\Gamma\left(\frac{1}{2} + i\,\frac{\theta_{12}}{2\pi}\right) \Gamma\left(1 - i\,\frac{\theta_{12}}{2\pi}\right)}, \tag{10.B.2}$$

where $f_{(a,b)} = f_{a,b} + f_{b,a}$. The symmetry property by itself reads

$$f^J(\theta_1, \theta_2)_{(\varepsilon_1, \varepsilon_2)} = -f^J(\theta_2, \theta_1)_{(\varepsilon_1, \varepsilon_2)} \frac{\Gamma\left(\frac{1}{2} - i\,\frac{\theta_{21}}{2\pi}\right) \Gamma\left(1 + i\,\frac{\theta_{21}}{2\pi}\right)}{\Gamma\left(\frac{1}{2} + i\,\frac{\theta_{21}}{2\pi}\right) \Gamma\left(1 - i\,\frac{\theta_{21}}{2\pi}\right)}. \tag{10.B.3}$$

---

[1] As explained in [420], the situation is reversed in the Sine-Gordon model, where the charge conjugation matrix is symmetric $C = \sigma^x$.

The properties of the current operator under Lorentz transformations imply that the two-particle form factor of the current operator must be of the form

$$f^J(\theta_1, \theta_2)_{(\varepsilon_1, \varepsilon_2)} = \cosh\left(\frac{\theta_1 + \theta_2}{2}\right) f(\theta_1 - \theta_2) . \tag{10.B.4}$$

Combining (10.B.4) with (10.B.2) we see that the function $f(\theta)$ fulfils the scalar Riemann-Hilbert Problem

$$f(\theta - 2\pi i) = -f(\theta) \frac{\Gamma\left(\frac{1}{2} - i\frac{\theta}{2\pi}\right)\Gamma\left(1 + i\frac{\theta}{2\pi}\right)}{\Gamma\left(\frac{1}{2} + i\frac{\theta}{2\pi}\right)\Gamma\left(1 - i\frac{\theta}{2\pi}\right)} \equiv f(\theta) \, \tilde{K}(\theta). \tag{10.B.5}$$

A special solution of (10.B.5) is easily constructed

$$f_0(\theta) = \exp\left(-\int_{-\infty}^{\infty} dx \, \frac{1 - \exp(i\theta x)}{\exp(2\pi x) - 1} K(x)\right) ,$$

$$K(x) = \int_{-\infty}^{\infty} \frac{d\theta}{2\pi} \exp(-i\theta x) \ln(\tilde{K}(\theta)). \tag{10.B.6}$$

Carrying out the Fourier transformation we obtain two special solutions

$$f_0^{\pm}(\theta) = C \exp\left(\pm\frac{\theta}{2} - \int_0^{\infty} \frac{dx}{x} \frac{\sin^2([\theta + i\pi]x/\pi)\exp(-x)}{\sinh(2x)\cosh(x)}\right) , \tag{10.B.7}$$

where C is a normalization constant. The symmetry property together with the definition (10.B.4) yields the condition

$$f(0) = 0 , \tag{10.B.8}$$

which means that we need to take $2f(\theta) = f_0^+(\theta) - f_0^-(\theta)$, or

$$f(\theta) = C \sinh(\theta/2) \exp\left(-\int_0^{\infty} \frac{dx}{x} \frac{\sin^2([\theta + i\pi]x/\pi)\exp(-x)}{\sinh(2x)\cosh(x)}\right). \tag{10.B.9}$$

The constant C is determined in [32, 238, 304, 420]

$$|C| = 2M . \tag{10.B.10}$$

Putting everything together we arrive at the following expression for the soliton – anti-soliton form factor of the current operator in the SU(2) Thirring model

$$f^J(\theta_1, \theta_2)_{+-} = f^J(\theta_1, \theta_2)_{-+} = \frac{1}{2} \cosh\left(\frac{\theta_1 + \theta_2}{2}\right) f(\theta_1 - \theta_2) . \tag{10.B.11}$$

10.C Correlation functions in the Gaussian model

In this appendix we summarize some facts about the Gaussian model in two dimensions. For a more detailed treatment we refer the reader to the books [102, 168, 461] or the review [389].

Consider a Gaussian model with Hamiltonian

$$\mathcal{H} = \frac{v}{16\pi} \int dx \left[(\partial_x \Phi)^2 + (\partial_x \Theta)^2 \right] , \tag{10.C.1}$$

where the dual field $\Theta(t, x)$ is related to the canonical Bose field Φ by

$$\partial_x \Theta(t, x) = -\frac{1}{v} \partial_t \Phi(t, x) , \quad \partial_t \Theta(t, x) = -v \partial_x \Phi(t, x) . \tag{10.C.2}$$

The canonical momentum is

$$\Pi(t, x) = \frac{\partial H}{\partial \partial_t \Phi(t, x)} = \frac{1}{8\pi v} \partial_t \Phi(t, x) ,$$
$$[\Pi(t, x), \Phi(t, y)] = -i\delta(x - y) . \tag{10.C.3}$$

The Lagrangian is

$$\mathcal{L} = \frac{1}{16\pi} \int dx \left[\frac{1}{v} (\partial_t \Phi)^2 - v(\partial_x \Phi)^2 \right] . \tag{10.C.4}$$

Finally, the action of the Gaussian model in a two-dimensional Euclidean space is

$$S = \frac{1}{16\pi} \int dx \, d\tau \left[\frac{1}{v} (\partial_\tau \Phi)^2 + v(\partial_x \Phi)^2 \right] , \tag{10.C.5}$$

The generating functional of multi-point correlation functions is defined in the usual way as a path integral

$$Z[J] = \int \mathcal{D}\Phi \, e^{-S - \int d\tau \, dx \, \Phi(\tau, x) \, J(\tau, x)} , \tag{10.C.6}$$

where $J(\tau, x)$ is a source. Fourier-transforming gives

$$Z[J] = \int \mathcal{D}\Phi \, e^{-\int \frac{d\omega \, dq}{(2\pi)^2} \left[\Phi(\omega, q) \frac{\omega^2 + v^2 q^2}{16\pi v} \Phi(-\omega, -q) + \Phi(\omega, q) \, J(-\omega, -q) \right]} . \tag{10.C.7}$$

Making the change of variables

$$\Phi'(\omega, q) = \Phi(\omega, q) + \frac{8\pi v}{\omega^2 + v^2 q^2} J(\omega, q) , \tag{10.C.8}$$

in the path integral, we arrive at

$$Z[J] = Z[0] \exp \left[\int \frac{d\omega \, dq}{(2\pi)^2} J(\omega, q) \frac{4\pi v}{\omega^2 + v^2 q^2} J(-\omega, -q) \right] . \tag{10.C.9}$$

Fourier-transforming back we obtain

$$\frac{Z[J]}{Z[0]} = \exp \left[\frac{1}{2} \int d\tau \, dx \, d\tau' \, dx' \, J(\tau, x) \, G(\tau - \tau', x - x') \, J(\tau', x') \right] , \tag{10.C.10}$$

where the propagator $G(\tau, x)$ is given by

$$G(\tau, x) = \int \frac{d\omega \, dq}{(2\pi)^2} e^{-i\omega\tau + iqx} \frac{8\pi v}{\omega^2 + v^2 q^2} . \tag{10.C.11}$$

The propagator fulfils the differential equation

$$\left(\frac{1}{v}\partial_\tau^2 + v\partial_x^2\right) G(\tau, x) = -8\pi\,\delta(\tau)\,\delta(x)\,. \tag{10.C.12}$$

Regularizing the ultraviolet and infrared divergences in the integrals (10.C.11) by long and short distance cutoffs, we obtain

$$G(\tau, x) = 2\ln\left[\frac{R^2}{x^2 + v^2\tau^2 + a_0^2}\right]. \tag{10.C.13}$$

We are particularly interested in correlation functions of vertex operators

$$\langle T_\tau\, e^{i\alpha\Phi(\tau_1,x_1)}\, e^{i\alpha'\Phi(0,0)}\rangle = \frac{1}{Z[0]}\int \mathcal{D}\Phi\, e^{i\alpha\Phi(\tau_1,x_1)}\, e^{i\alpha'\Phi(0,0)}\, e^{-S}\,. \tag{10.C.14}$$

These can be obtained from the generating functional $Z[J]$ by making the following choice for the source

$$J(\tau, x) = -i\alpha\,\delta(\tau - \tau_1)\,\delta(x - x_1) - i\alpha'\,\delta(\tau)\,\delta(x)\,. \tag{10.C.15}$$

Inserting (10.C.15) into (10.C.10) we obtain

$$\langle T_\tau\, e^{i\alpha\Phi(\tau_1,x_1)}\, e^{i\alpha'\Phi(0,0)}\rangle = \left[\frac{a_0^2}{R^2}\right]^{(\alpha+\alpha')^2} \left[\frac{x_1^2 + v^2\tau_1^2}{a_0^2}\right]^{2\alpha\alpha'} \tag{10.C.16}$$

As $a_0 \ll R$, the r.h.s. of (10.C.16) actually vanishes unless

$$\alpha' = -\alpha\,. \tag{10.C.17}$$

The condition (10.C.17) is known as 'electroneutrality' condition. In Conformal Field Theory it is customary to use a normalization that is different from (10.C.13), namely

$$G(\tau, x) = -2\ln\left[x^2 + v^2\tau^2\right]. \tag{10.C.18}$$

Correlation functions of vertex operators in the CFT normalization are

$$\langle T_\tau\, e^{i\alpha\Phi(\tau,x)}\, e^{-i\alpha\Phi(0,0)}\rangle = \left[x^2 + v^2\tau^2\right]^{-2\alpha^2}\,. \tag{10.C.19}$$

In the CFT normalization (10.C.19), vertex operators have a physical dimension of

$$\text{length}^{-2\alpha^2}\,. \tag{10.C.20}$$

Correlation functions of vertex operators of the dual field Θ can be calculated analogously. The Minkowski space Lagrangian for the dual field is

$$\mathcal{L} = \frac{1}{16\pi}\int dx\left[\frac{1}{v}(\partial_t\Theta)^2 - v(\partial_x\Theta)^2\right]\,. \tag{10.C.21}$$

Following through exactly the same steps as before, we obtain the following result for vertex operators of the dual field

$$\langle T_\tau \, e^{i\alpha\Theta(\tau,x)} \, e^{-i\alpha\Theta(0,0)} \rangle = \left[x^2 + v^2\tau^2 \right]^{-2\alpha^2} . \tag{10.C.22}$$

Last but not least we need to know correlation functions of so-called chiral vertex operators. Let us define complex coordinates z and $\bar{z}$ by

$$z = v\tau - ix , \qquad \bar{z} = v\tau + ix ,$$
$$\partial_\tau = v(\partial_z + \partial_{\bar{z}}) , \qquad \partial_x = -i(\partial_z - \partial_{\bar{z}}) . \tag{10.C.23}$$

Next we define chiral Bose fields by

$$\Phi(\tau, x) = \varphi(z) + \bar{\varphi}(\bar{z}) , \qquad \Theta(\tau, x) = \varphi(z) - \bar{\varphi}(\bar{z}) . \tag{10.C.24}$$

The chiral fields have propagators

$$\langle T_\tau \, \varphi(z) \, \varphi(0) \rangle = -2 \ln [z] ,$$
$$\langle T_\tau \, \bar{\varphi}(\bar{z}) \, \bar{\varphi}(0) \rangle = -2 \ln [\bar{z}] . \tag{10.C.25}$$

Correlation functions of chiral vertex operators are then given by

$$\langle T_\tau \, e^{i\alpha\varphi(z)} \, e^{-i\alpha\varphi(0)} \rangle = z^{-2\alpha^2} ,$$
$$\langle T_\tau \, e^{i\alpha\bar{\varphi}(\bar{z})} \, e^{-i\alpha\bar{\varphi}(0)} \rangle = \bar{z}^{-2\alpha^2} . \tag{10.C.26}$$

The results (10.C.26) follow from the identity

$$\langle T_\tau \, e^{i\alpha\varphi(z)} \, e^{-i\alpha\varphi(0)} \rangle = e^{\alpha^2 \langle T_\tau \, \varphi(z) \, \varphi(0) \rangle} , \tag{10.C.27}$$

which is a general property of free bosonic field theories.

11

Universal correlations at low density

In Chapter 9 we developed the picture of the asymptotics of the correlation functions of the Hubbard model for the phases with gapless excitations, i.e., for the phases with ground states belonging to regions II and IV of the ground state phase diagram discussed in Chapter 6.3 (see figures 6.4, 6.5). Our results relied on the predictions of conformal field theory which are expected to hold for a whole universality class of models related to the Hubbard model and which only need the finite size data calculated in Chapter 8 as input parameters. Correlation functions at half-filling (phase V) were considered in the previous chapter on the basis of a special continuum limit and the predictions of certain integrable quantum field theories.

Here we shall consider correlation functions in the phase whose ground state is the empty band (phase I in the ground state phase diagram, figures 6.4, 6.5). At zero temperature the boundary of this phase in the μ-B plane (see figure 6.5) is determined by equation (6.20): $\mu < \mu_0(B)$. For small finite temperature a small number of particles is populating the system. They form a dilute, thermodynamically ideal gas with pressure proportional to the temperature (see below). We shall say the system is in the gas phase [176] and shall give a more precise meaning to this statement later.

At zero temperature it costs the energy $\mu_0(B) - \mu > 0$ (see equation (6.20)) to add a single electron to the phase. Thus, the lowest lying possible excitation has a gap, and conformal field theory cannot be applied. We shall see below that instead the non-relativistic Fermi gas with infinite point-like repulsion becomes the universal model for the long-wavelength, low-temperature physics of one-dimensional electrons. This model, which we call the impenetrable electron gas, allows for a rigorous mathematical treatment by means of the Bethe ansatz. In particular, a so-called determinant representation [270] for the two-point Green function could be derived [217,218]. This determinant representation is related to an integrable classical evolution equation and to a Riemann-Hilbert problem, that enable the exact calculation of the asymptotics of the two-point Green functions [173,174].

In Section 11.1 we shall explain our ideas about the universality of the long-wavelength, low-temperature asymptotics of electronic correlations in the gas phase [171,176]. We shall start with the paradigmatic Hubbard model, and then argue that certain modifications of the interaction do not change the long-wavelength, low-temperature physics of the model. In the appropriate scaling limit all modified Hamiltonians lead to the same effective model: the impenetrable electron gas model. A summary of the calculation [173,174] of the asymptotics of the two-point Green functions of this model is presented in Section 11.2.

11.1 The Hubbard model in the gas phase

11.1.1 The gas phase

For simplicity let us neglect the magnetic field in this section. The density D of non-interacting, one-dimensional spin-$\frac{1}{2}$ Fermions on a lattice is given by the integral over the Fermi weight,

$$D = \frac{2}{\pi} \int_0^\pi dp \, \frac{1}{e^{(\varepsilon(p)-\mu)/T} + 1}. \tag{11.1}$$

Here $\varepsilon(p)$ is the dispersion of the Fermions, T denotes the temperature and μ the chemical potential. Let us assume $\varepsilon(p)$ to be monotonically increasing and bounded from below. If the chemical potential is smaller than a critical value, $\mu < \mu_c = \min_{p>0} \varepsilon(p)$, then D vanishes in the zero temperature limit $T \to 0+$. For $\mu > \mu_c$, on the other hand, the density D approaches a finite positive value as $T \to 0+$. This means that the system undergoes a phase transition at $T = 0$ as a function of the chemical potential. The critical point is at $\mu = \mu_c$. Assuming that $\varepsilon(p) = \mu_c + p^2 + \mathcal{O}(p^4)$, we obtain

$$D = \frac{2}{\pi} \sqrt{\mu - \mu_c} \tag{11.2}$$

for $\mu_c < \mu < \mu_c + \delta, \delta \ll 1$. Clearly, the phase transition is a transition from a band insulator for $\mu < \mu_c$ to a conductor for $\mu > \mu_c$.

Let us come back to the Hubbard model as an example for an interacting electron system. In Chapter 6 we found for the transition from phase I to phase IV (see equation (6.41) and figure 6.7)

$$D = n_c = \frac{1}{\pi} \sqrt{\mu - \mu_0(0)} \tag{11.3}$$

for μ close to but larger than $\mu_0(0)$. Thus, the qualitative picture is the same for free and for interacting electrons. There is a phase transition from a zero density phase to a phase with a finite density of particles which in the interacting case may be interpreted as a transition from a correlated insulator to a correlated metal.[1]

Let us complete our picture by considering small positive temperatures. For fixed chemical potential we expect a small number of thermally excited particles in phase I forming a dilute gas whose properties can be studied by means of the thermodynamic Bethe ansatz equations (see Chapter 5). For this purpose it is convenient to express the Gibbs free energy $f = -P$ (P pressure), equation (5.57), in terms of the dressed energies $\kappa(k)$, $\varepsilon_n(\Lambda)$, $\varepsilon'_n(\Lambda)$, equations (5.68), of elementary excitations at finite temperature. $\kappa(k)$ is the dressed energy

---

[1] Comparing (11.2) and (11.3), however, we see that the results for the electron density close to the critical point $\mu = \mu_0(0)$ differ by a factor of two. This difference may be interpreted as a ground state signature of the spin-charge separation discussed in Chapter 7: The elementary charge excitations of the Hubbard model at finite positive u are spinless due to the interaction, hence the density is smaller by a factor of two.

of particle (or hole) excitations, $\varepsilon_n(\Lambda)$ describes spin excitations and $\varepsilon'_n(\Lambda)$ so-called k-Λ strings [95]. All k-Λ strings are gapped (see (5.80), (5.82)). They do not contribute to the low-temperature thermodynamic properties of the Hubbard model [436] and drop out of the equation for the pressure for sufficiently small T, which then simplifies to

$$P = \frac{T}{2\pi} \int_{-\pi}^{\pi} dk \, \ln\left(1 + e^{-\frac{\kappa(k)}{T}}\right). \tag{11.4}$$

Similarly, the integral equations (5.54), (5.55) for the dressed energies at low temperature become

$$\kappa(k) = -\mu - 2u - 2\cos k - T \sum_{n=1}^{\infty} \left([n]\ln\left(1 + e^{-\frac{\varepsilon_n}{T}}\right)\right)(\sin k), \tag{11.5}$$

$$\ln\left(1 + e^{\frac{\varepsilon_n(\Lambda)}{T}}\right) = -\int_{-\pi}^{\pi} dk \, \cos k \, a_n(\Lambda - \sin k) \ln\left(1 + e^{-\frac{\kappa(k)}{T}}\right)$$

$$+ \sum_{m=1}^{\infty} \left(A_{nm} \ln\left(1 + e^{-\frac{\varepsilon_n}{T}}\right)\right)(\Lambda), \tag{11.6}$$

where $n = 1, 2, 3, \ldots$ in equation (11.6), and

$$a_n(\Lambda) = \frac{nu/\pi}{(nu)^2 + \Lambda^2}. \tag{11.7}$$

$[n]$ and A_{nm} are integral operators defined by

$$([0]f)(\Lambda) = f(\Lambda), \tag{11.8}$$

$$([n]f)(\Lambda) = \int_{-\infty}^{\infty} d\Lambda' \, a_n(\Lambda - \Lambda') f(\Lambda'), \quad n = 1, 2, \ldots \tag{11.9}$$

$$A_{nm} = \sum_{j=1}^{\min\{n,m\}} \left([|n - m| + 2(j - 1)] + [|n - m| + 2j]\right). \tag{11.10}$$

The gas phase is characterized by the absence of a Fermi surface for $\kappa(k)$. Thus, $\kappa(k)$ is positive in the zero temperature limit, and the first term on the right-hand side of (11.6) becomes exponentially small in T. Dropping this term, the equations (11.6) decouple from (11.5). Since the equations become independent of Λ, it is not hard to solve them. The solution, $\exp\{\varepsilon_n(\Lambda)/T\} = n(n + 2)$, is the same as in the infinite coupling limit $u \to \infty$ (see e.g. [432]). Inserting this solution into (11.5) we obtain

$$\kappa(k) = -\mu - 2u - 2\cos k - T \ln 2. \tag{11.11}$$

Our initial assumption, that $\lim_{T \to 0} \kappa(k) > 0$ self-consistently holds for all k if $\mu + 2u + 2 < 0$, which is precisely the same as the condition $\mu < \mu_0(0)$ for being in the gas phase stated above. With (11.11) the low temperature expression for the pressure becomes

$$P = \frac{T}{2\pi} \int_{-\pi}^{\pi} dk \, \ln\left(1 + 2e^{\frac{\mu + 2u + 2\cos k}{T}}\right) \approx \sqrt{\frac{T}{\pi}} \, e^{\frac{\mu + 2u + 2}{T}}, \tag{11.12}$$

and we see that the density $D = \partial P/\partial \mu$ and the pressure P are related by the ideal gas law,

$$P = TD. \tag{11.13}$$

There are two more important lessons to learn from our simple calculation. First, the low temperature limit in the gas phase works the same way as the strong coupling limit at finite temperatures. Second, the low temperature Gibbs free energy $f = -P$ in the gas phase shows no signature of the discreteness of the lattice. It is the same as for the impenetrable electron gas (see below), which is a continuum model. This agrees well with our intuitive understanding of the gas phase at low temperature: (i) the mean free path ($= 1/D$) of the electrons is large compared to the lattice spacing (which we set equal to unity so far); (ii) their kinetic energy is of the order T. Hence, the effective repulsion is large for $T \ll u$; (iii) the ideal gas law holds at low temperature.

11.1.2 Scaling

The above arguments show that only electrons with small momenta, corresponding to long wavelengths, contribute to the low-temperature properties of the Hubbard model in the gas phase. Thus, the Hubbard model in the gas phase at low temperature is effectively described by its continuum limit (see appendix 2.B). Recall that in order to perform the continuum limit we have to introduce the lattice spacing a_0 and coordinates $x = a_0 n$ connected with the nth lattice site. The total length of the system is $\ell = a_0 L$. The continuum limit is the limit $a_0 \to 0$ for fixed ℓ. In this limit we obtain canonical field operators $\Psi_\sigma(x)$ for electrons of spin σ as

$$\Psi_\sigma(x) = \lim_{a_0 \to 0} c_{n,\sigma}/\sqrt{a_0}. \tag{11.14}$$

Let us perform the rescaling

$$T_H = a_0^2 T, \quad \mu_H + 2u + 2 = a_0^2 \mu, \quad k_H = a_0 k, \quad t_H = t/a_0^2, \quad B_H = a_0^2 B, \tag{11.15}$$

where k denotes the momentum, t the time and B the magnetic field, which we shall incorporate below. The index 'H' refers to the Hubbard model. Then, in the limit $a_0 \to 0$, we find

$$H_H/T_H = H/T. \tag{11.16}$$

Here H is the Hamiltonian for continuum electrons with delta interaction,

$$H = \int_{-\ell/2}^{\ell/2} dx \left\{ (\partial_x \Psi_\alpha^+(x)) \partial_x \Psi_\alpha(x) + \frac{4u}{a_0} \Psi_\uparrow^+(x) \Psi_\downarrow^+(x) \Psi_\downarrow(x) \Psi_\uparrow(x) \right.$$
$$\left. - \mu \Psi_\alpha^+(x) \Psi_\alpha(x) \right\}. \tag{11.17}$$

Note that the coupling $c_1 = u/a_0$ of the continuum model goes to infinity! This is a peculiarity of the one-dimensional system. The effective interaction in the low-density phase becomes large. Similar scaling arguments lead to an effective coupling $c_2 = u$ in two dimensions and to $c_3 = a_0 u$ in three dimensions, i.e., unlike one-dimensional electrons three-dimensional electrons in the gas phase are free.

11.1.3 Universality

What happens to more general Hamiltonians in the continuum limit? Let us consider Hamiltonians of the form $H_G = H_H + V$, where H_H is the Hubbard Hamiltonian and V contains additional short-range interactions. We shall assume that V is a sum of local terms V_j which preserve the particle number. Then V_j contains as many creation as annihilation operators, and the number of field operators in V_j is even. We shall further assume that V_j is Hermitian and space parity invariant.

According to equation (11.14) every field $c_{j,\sigma}$ on the lattice contributes a factor of $a_0^{1/2}$ in the continuum limit. One factor of a_0 is absorbed by the volume element $dx = a_0$, when turning from summation to integration. Thus, if V_j contains 8 or more field operators, then $V \sim a_0^3$ and V/T_H vanishes. If V_j contains 6 field operators, then at least two of the creation operators and two of the annihilation operators must belong to different lattice sites, since otherwise $V_j = 0$. A typical term is, for instance, $V_j = c_{j,\uparrow}^+ c_{j,\downarrow}^+ c_{j+1,\uparrow}^+ c_{j+1,\uparrow} c_{j,\downarrow} c_{j,\uparrow}$. In the continuum limit we have $c_{j+1,\uparrow} = a_0^{1/2} \Psi_\uparrow(x) + a_0^{3/2} \partial_x \Psi_\uparrow(x) + O(a_0^{5/2})$. Hence, the leading term vanishes due to the Pauli principle. The next to leading term acquires an additional power of a_0. We conclude that $V \sim a_0^3$ and thus $V/T_H \to 0$.

If V_j contains 4 fields, then

$$V \sim a_0^2 \Psi_\uparrow^+(x) \Psi_\downarrow^+(x) \Psi_\downarrow(x) \Psi_\uparrow(x) + O(a_0^4). \tag{11.18}$$

Here the first term on the right-hand side is the density-density interaction of the electron gas. In order to arrive at the impenetrable electron gas model the coefficient in front of this term has to be positive. Note that there are no terms of the order of a_0^3 on the right hand side of (11.18) and thus no other terms than the first one in the continuum limit. Terms of the order of a_0^3 would contain precisely one spatial derivative. They are ruled out, since they would break space parity.

Considering the case, when V_j contains 2 fields, we find, except for the kinetic energy and the chemical potential term, terms which correspond to a coupling to an external magnetic field B_H. For these terms to be finite in the continuum limit we have to rescale the magnetic field as $B_H = a_0^2 B$ (see equation (11.15)).

Our considerations show that the impenetrable electron gas model with magnetic field,

$$H_B = H + B \int_{-\ell/2}^{\ell/2} dx \, \Psi_\alpha^+(x) \sigma_{\alpha\beta}^z \Psi_\beta(x), \tag{11.19}$$

is indeed the universal model (for small T) for the gas phase of one-dimensional lattice electrons with repulsive short-range interaction.

11.1.4 Asymptotics of correlation functions in the gas phase

The impenetrable electron gas is the infinite coupling limit of the electron gas with repulsive delta interaction ($a_0 \to 0$ in (11.19)), which was the first model solved by nested Bethe ansatz [154,493]. The pressure of the system as a function of T, μ and B is known explicitly [432],

$$P = \frac{T}{2\pi} \int_{-\infty}^{\infty} dk \, \ln\left(1 + e^{\frac{\mu+B-k^2}{T}} + e^{\frac{\mu-B-k^2}{T}}\right), \tag{11.20}$$

and may serve as thermodynamic potential. The expression (11.20) is formally the same as for a gas of free spinless Fermions with effective (temperature dependent) chemical potential $\mu_{eff} = \mu + T \ln(2 \cosh B/T)$. Hence the Fermi surface vanishes for $\lim_{T \to 0} \mu_{eff} = \mu + |B| < 0$. The finite temperature correlation functions of the impenetrable electron gas depend crucially on the sign of μ_{eff}. This allows us to give precise meaning to the gas phase at finite temperature by the condition $\mu_{eff} < 0$, which is also sufficient for deriving the ideal gas law (11.13) from the low temperature limit of (11.20). Note that for zero magnetic field and small temperature equation (11.20) coincides with the right hand side of (11.12).

The time- and temperature-dependent (two-point) Green functions are defined as

$$G_{\uparrow\uparrow}^+(x, t) = \frac{\text{tr}\{e^{-H_B/T} \Psi_\uparrow(x, t)\Psi_\uparrow^+(0, 0)\}}{\text{tr}\{e^{-H_B/T}\}}, \tag{11.21a}$$

$$G_{\uparrow\uparrow}^-(x, t) = \frac{\text{tr}\{e^{-H_B/T} \Psi_\uparrow^+(x, t)\Psi_\uparrow(0, 0)\}}{\text{tr}\{e^{-H_B/T}\}}. \tag{11.21b}$$

For the impenetrable electron gas these correlation functions were represented as determinants of Fredholm integral operators in [217,218]. The determinant representation provides a powerful tool to study their properties analytically.

In [173,174] the determinant representation was used to derive a nonlinear partial differential equation for two classical auxiliary fields, which determine the correlation functions. This partial differential equation is closely related to the Heisenberg equation of the quantum Hamiltonian (11.17). It is called the separated nonlinear Schrödinger equation. Together with a corresponding Riemann-Hilbert problem it determines the large-time, long-distance asymptotics of the correlators (11.21a), (11.21b) (for details see the following sections). In [173,174] the asymptotics $x, t \to \infty$ was calculated for fixed ratio $k_0 = x/2t$. The crucial parameter for the asymptotics is the average number of particles xD in the interval $[0, x]$. If x is large but $xD \ll 1$ (i.e., T small), an electron propagates freely from 0 to x, and the correlation functions (11.21a), (11.21b) are those of free Fermions,

$$G_f^+(x, t) = \frac{e^{-\frac{i\pi}{4}}}{2\sqrt{\pi}} t^{-\frac{1}{2}} e^{it(\mu-B)} e^{\frac{ix^2}{4t}}, \tag{11.22a}$$

$$G_f^-(x, t) = \frac{e^{\frac{i\pi}{4}}}{2\sqrt{\pi}} e^{\frac{(\mu-B-k_0^2)}{T}} t^{-\frac{1}{2}} e^{-it(\mu-B)} e^{-\frac{ix^2}{4t}}. \tag{11.22b}$$

The true asymptotic region is characterized by a large number xD of particles in the interval $[0, x]$, specifically, $xD \gg z_c^{-1}$, where $z_c = (T^{3/4}e^{-k_0^2/2T})/(2\pi^{1/4}k_0^{3/2})$. If the latter condition is satisfied, the correlation functions decay due to multiple scattering. The cases $B > 0$ and $B \leq 0$ have to be treated separately. For $B > 0$ a critical line, $x = 4t\sqrt{B}$, separates the x-t plane into a time and a space like regime. The asymptotics (for small T) in these respective regimes are:

Time like regime ($x < 4t\sqrt{B}$):

$$G_{\uparrow\uparrow}^{\pm}(x, t) = G_f^{\pm}(x, t) \frac{t^{\mp i\nu(z_c)}e^{-xD_\downarrow}}{\sqrt{4\pi z_c x D_\downarrow}},$$ (11.23)

where

$$\nu(z_c) = -\frac{2D_\downarrow k_0^{3/2}e^{-k_0^2/2T}}{\pi^{1/4}T^{5/4}}, \quad D_\downarrow = \frac{1}{2}\sqrt{\frac{T}{\pi}}e^{(\mu+B)/T}.$$ (11.24)

$D_\downarrow = \partial P/\partial(\mu + B)$ is the low temperature expression for the density of down-spin electrons.

Space like regime ($x > 4t\sqrt{B}$):

$$G_{\uparrow\uparrow}^{\pm}(x, t) = G_f^{\pm}(x, t) t^{\mp i\nu(\gamma^{-1})}e^{-xD_\downarrow},$$ (11.25)

where

$$\nu(\gamma^{-1}) = -\frac{e^{(3B+\mu-k_0^2)/T}}{2\pi}.$$ (11.26)

For $B \leq 0$ there is no distinction between time- and space-like regimes. The asymptotics is given by (11.25).

It seems fair to mention here that the calculation of the asymptotics (11.23), (11.25) is rather hard and lengthy. The required technics are presented in detail in parts III and IV of the book [270]. We can only sketch the calculations in the remaining sections of this chapter. We wish to emphasize that the terms on the right-hand side of equations (11.23) and (11.25) are the three leading terms of an asymptotic expansions in t and that the method employed in [173] allows for a systematic calculation of the next, subleading orders.

The leading exponential factor in (11.23) and (11.25), has a clear physical interpretation: because of the specific form of the infinite repulsion in (11.17), up-spin electrons are only scattered by down-spin electrons. This is reflected in the fact that the correlation length is $1/D_\downarrow$. The expression $1/D_\downarrow$ may be interpreted as the mean free path of the up-spin electrons. Thus, the correlation length for up-spin electrons is equal to their mean free path. The exponential decay of the two-point Green functions means that, due to the strong interaction, an up-spin electron is confined by the cloud of surrounding down-spin electrons. Thus, we are facing an interesting situation: although at small distances the electrons look like free Fermions, they are confined on a 'macroscopic scale' set by the mean free path $1/D_\downarrow$.

11.1.5 Outline of the derivation

The derivation of the above results on the asymptotics of the two-point Green functions at low temperature is based on the fact that the impenetrable electron gas model is exactly solvable by Bethe ansatz. The Bethe ansatz eigenfunctions [154, 493] and the thermodynamics of the model [432] are long since known. But only more recently a determinant representation for the two-point Green functions was derived by Izergin and Pronko [217, 218]. Their derivation includes the following steps:

(i) A change of basis for the spin part of the Bethe ansatz wave function from inhomogeneous XXX to XX spin chain eigenfunctions, which is possible at infinite repulsion.
(ii) The calculation of form factors in the finite volume.
(iii) A summation of the form factors.
(iv) The thermodynamic limit.

The details of the calculation can be found in the article [218].

The asymptotic analysis of the correlation functions was performed in [173, 174]. Then starting point was the determinant representation of Izergin and Pronko (Section 11.2.1), which is valid for all x and t. The non-trivial ingredients of the determinant representation are certain auxiliary functions b_{++} and B_{--} and the Fredholm determinant $\det(\hat{I} + \hat{V})$ of an integral operator $\hat{V}$. A direct yet lengthy calculation shows that b_{++} and B_{--} satisfy the separated non-linear Schrödinger equation (Section 11.2.2), which is a well-known integrable partial differential equation. The logarithm of the Fredholm determinant plays the role of its tau-function (Section 11.2.3). Moreover, a Riemann-Hilbert problem that fixes b_{++} and B_{--} as solutions of the separated non-linear Schrödinger equation can be derived from the determinant representation (Section 11.2.4). The Riemann-Hilbert problem is the appropriate starting point for the asymptotic analysis of the correlation functions $G^{\pm}_{\uparrow\uparrow}$ (Section 11.2.5).

Luckily, the differential equation and the Riemann-Hilbert problem turn out to be of the same form as in case of the impenetrable (spinless) Bose gas [211, 270]. Therefore a theorem obtained in the asymptotic analysis of the impenetrable Bose gas [212] could be applied to the impenetrable electrons as well. In contrast to the bosonic case, there is, however, an additional external integration in the determinant representation of the impenetrable electron gas. This integration can be carried out in the low-temperature limit, by the method of steepest descent (Section 11.2.6).

11.2 Correlation functions of the impenetrable electron gas

11.2.1 Determinant representation

Let us now recall the determinant representation for the correlation functions $G^{\pm}_{\uparrow\uparrow}(x, t)$, which was derived in [217, 218]. We shall basically follow the account of [174]. Yet, it turns out to be useful to rescale the variables and the correlation functions.

The rescaling

$$x_r = -\sqrt{T}x/2, \quad t_r = Tt/2, \tag{11.27}$$

$$g^{\pm} = G_{\uparrow\uparrow}^{\pm}/\sqrt{T}, \tag{11.28}$$

$$\beta = \mu_{eff}/T, \quad h = B/T \tag{11.29}$$

removes the explicit temperature dependence from all expressions. Furthermore, it will allow us to make close contact with results which were obtained for the impenetrable Bose gas [211,212,270]. The index 'r' in (11.27) stands for 'rescaled'. For the sake of simplicity we shall suppress this index in the following sections. We shall come back to physical space and time variables only in the last section, where we consider the low temperature limit.

The rescaled correlation functions g^+ and g^- in the rescaled variables can be expressed as [173,174],

$$g^+(x,t) = \frac{-e^{2it(\beta - h - \ln(2\mathrm{ch}(h)))}}{2\pi} \int_{-\pi}^{\pi} d\eta \, \frac{F(\gamma,\eta)}{1 - \cos(\eta)} b_{++} \det\left(\hat{I} + \hat{V}\right), \tag{11.30}$$

$$g^-(x,t) = \frac{e^{-2it(\beta - h - \ln(2\mathrm{ch}(h)))}}{4\pi\gamma} \int_{-\pi}^{\pi} d\eta \, F(\gamma,\eta) B_{--} \det\left(\hat{I} + \hat{V}\right). \tag{11.31}$$

Here γ and $F(\gamma, \eta)$ are elementary functions,

$$\gamma = 1 + e^{2h}, \tag{11.32}$$

$$F(\gamma,\eta) = 1 + \frac{e^{i\eta}}{\gamma - e^{i\eta}} + \frac{e^{-i\eta}}{\gamma - e^{-i\eta}}. \tag{11.33}$$

$\det(\hat{I} + \hat{V})$ is the Fredholm determinant of the integral operator $\hat{I} + \hat{V}$, where $\hat{I}$ is the identity operator, and $\hat{V}$ is defined by its kernel $V(\lambda, \mu)$. λ and μ are complex variables, and the path of integration is the real axis. In order to define $V(\lambda, \mu)$ we have to introduce certain auxiliary functions. Let us define

$$\tau(\lambda) = i(\lambda^2 t + \lambda x), \tag{11.34}$$

$$\vartheta(\lambda) = \frac{1}{1 + e^{\lambda^2 - \beta}}, \tag{11.35}$$

$$E(\lambda) = \mathrm{p.v.} \int_{-\infty}^{\infty} d\mu \, \frac{e^{-2\tau(\mu)}}{\pi(\mu - \lambda)}, \tag{11.36}$$

$$e_-(\lambda) = \sqrt{\frac{\vartheta(\lambda)}{\pi}} e^{\tau(\lambda)}, \tag{11.37}$$

$$e_+(\lambda) = \frac{1}{2}\sqrt{\frac{\vartheta(\lambda)}{\pi}} e^{-\tau(\lambda)} \left\{(1 - \cos(\eta))e^{2\tau(\lambda)}E(\lambda) + \sin(\eta)\right\}. \tag{11.38}$$

Note that $\vartheta(\lambda)$ is the Fermi weight. $V(\lambda, \mu)$ can be expressed in terms of e_+ and e_-,

$$V(\lambda, \mu) = \frac{e_+(\lambda)e_-(\mu) - e_+(\mu)e_-(\lambda)}{\lambda - \mu}. \tag{11.39}$$

Denote the resolvent of $\hat{V}$ by $\hat{R}$,

$$\left(\hat{I} + \hat{V}\right)\left(\hat{I} - \hat{R}\right) = \left(\hat{I} - \hat{R}\right)\left(\hat{I} + \hat{V}\right) = \hat{I}. \tag{11.40}$$

Then $\hat{R}$ is an integral operator with symmetric kernel [211],

$$R(\lambda, \mu) = \frac{f_+(\lambda)f_-(\mu) - f_+(\mu)f_-(\lambda)}{\lambda - \mu}, \tag{11.41}$$

which is of the same form as $V(\lambda, \mu)$. The functions $f_\pm$ are obtained as the solutions of the integral equations

$$f_\pm(\lambda) + \int_{-\infty}^{\infty} d\mu \, V(\lambda, \mu) f_\pm(\mu) = e_\pm(\lambda). \tag{11.42}$$

We may now define the 'potentials'

$$B_{ab} = \int_{-\infty}^{\infty} d\lambda \, e_a(\lambda) f_b(\lambda), \quad C_{ab} = \int_{-\infty}^{\infty} d\lambda \, \lambda e_a(\lambda) f_b(\lambda) \tag{11.43}$$

for $a, b = \pm$. B_{--} enters the definition of $g^-(x, t)$, equation (11.31). b_{++} in (11.30) is defined as

$$b_{++} = B_{++} - G(x, t), \tag{11.44}$$

where

$$G(x, t) = \frac{(1 - \cos(\eta))e^{-i\pi/4}}{2\sqrt{2\pi t}} e^{ix^2/2t}. \tag{11.45}$$

The remaining potentials B_{ab} and C_{ab} will be needed later.

It is instructive to compare the determinant representation (11.30) for the correlation function $g^+(x, t)$ with the corresponding expression for impenetrable bosons (see e.g. page 345 of [270]). The main formal differences are the occurrence of the η-integral in (11.30) and the occurrence of η in the definition of e_+. As can be seen from the derivation of (11.30) in [218], the η-integration is related to the spin degrees of freedom. For $\eta = \pm\pi$ the expression $-\frac{1}{2}e^{2i\beta t}b_{++} \det(\hat{I} + \hat{V})$ agrees with the field-field correlator for impenetrable Bosons (recall, however, the different physical meaning of β).

11.2.2 Differential equations

As in case of impenetrable Bosons [211, 270] it is possible to derive a set of integrable nonlinear partial differential equations for the potentials b_{++} and B_{--} and to express the logarithmic derivatives of the Fredholm determinant $\det(\hat{I} + \hat{V})$ in terms of the potentials B_{ab} and C_{ab}.

The functions $f_\pm$ satisfy linear differential equations with respect to the variables x, t, and β,

$$\hat{L}\begin{pmatrix} f_+ \\ f_- \end{pmatrix} = \hat{M}\begin{pmatrix} f_+ \\ f_- \end{pmatrix} = \hat{N}\begin{pmatrix} f_+ \\ f_- \end{pmatrix} = 0, \tag{11.46}$$

The Lax operators $\hat{L}$, $\hat{M}$ and $\hat{N}$ are given as

$$\hat{L} = \partial_x + i\lambda\sigma^z - 2iQ, \tag{11.47}$$

$$\hat{M} = \partial_t + i\lambda^2\sigma^z - 2i\lambda Q + \partial_x U, \tag{11.48}$$

$$\hat{N} = 2\lambda\partial_\beta + \partial_\lambda + 2it\lambda\sigma^z + ix\sigma^z - 4itQ - 2\partial_\beta U, \tag{11.49}$$

where the matrices Q and U are defined according to

$$Q = \begin{pmatrix} 0 & b_{++} \\ B_{--} & 0 \end{pmatrix}, \quad U = \begin{pmatrix} -B_{+-} & b_{++} \\ -B_{--} & B_{+-} \end{pmatrix}. \tag{11.50}$$

Mutual compatibility of the linear differential equations (11.46) leads to a set of nonlinear partial differential equations for the potentials b_{++} and B_{--} [173]. In particular, the space and time evolution is driven by the separated non-linear Schrödinger equation,

$$i\partial_t b_{++} = -\tfrac{1}{2}\partial_x^2 b_{++} - 4b_{++}^2 B_{--}, \tag{11.51}$$

$$i\partial_t B_{--} = \tfrac{1}{2}\partial_x^2 B_{--} + 4B_{--}^2 b_{++}. \tag{11.52}$$

11.2.3 Connection between Fredholm determinant and potentials

To describe the correlation functions (11.30) and (11.31) one has to relate the Fredholm determinant $\det(\hat{I} + \hat{V})$ and the potentials B_{ab} and C_{ab}. Let us use the abbreviation $\sigma(x, t, \beta) = \ln\det(\hat{I} + \hat{V})$. The logarithmic derivatives of the Fredholm determinant with respect to x, t and β are

$$\partial_x\sigma = -2iB_{+-}, \tag{11.53}$$

$$\partial_t\sigma = -2i(C_{+-} + C_{-+} + G(x, t)B_{--}), \tag{11.54}$$

$$\partial_\beta\sigma = -2it\partial_\beta(C_{+-} + C_{-+} + G(x, t)B_{--}) - 2ix\partial_\beta B_{+-} - 2(\partial_\beta B_{+-})^2$$
$$\quad - 2it(B_{--}\partial_\beta b_{++} - b_{++}\partial_\beta B_{--}) + 2(\partial_\beta b_{++})(\partial_\beta B_{--}). \tag{11.55}$$

For the calculation of the asymptotics of the Fredholm determinant we further need the second derivatives of σ with respect to space and time,

$$\partial_x^2\sigma = 4B_{--}b_{++}, \tag{11.56}$$

$$\partial_x\partial_t\sigma = 2i(B_{--}\partial_x b_{++} - b_{++}\partial_x B_{--}), \tag{11.57}$$

$$\partial_t^2\sigma = 2i(B_{--}\partial_t b_{++} - b_{++}\partial_t B_{--}) + 8B_{--}^2 b_{++}^2 + 2(\partial_x B_{--})(\partial_x b_{++}). \tag{11.58}$$

Note that

$$\lim_{\beta\to-\infty}\sigma = 0. \tag{11.59}$$

This follows from $\lim_{\beta\to-\infty}\vartheta(\lambda) = 0$ and is important for fixing the integration constant in the calculation of the asymptotics of the determinant.

11.2.4 The Riemann-Hilbert problem

From now on we will restrict ourselves to the case of negative effective chemical potential, $\beta < 0$. Recall that this is the condition for the system to be in the gas phase. For negative β the logarithmic derivatives $\partial_x \sigma$ and $\partial_t \sigma$ of the Fredholm determinant and the potentials b_{++} and B_{--} are determined by the following matrix Riemann-Hilbert problem, which was derived from the determinant representation (see Section 11.2.1) in [174].

(i) $\phi : \mathbb{C} \to \text{End}(\mathbb{C}^2)$ is analytic in $\mathbb{C} \setminus \mathbb{R}$.
(ii) $\lim_{\lambda \to \infty} \phi(\lambda) = I_2$.
(iii) ϕ has a discontinuity across the real axis described by the condition

$$\phi_-(\lambda) = \phi_+(\lambda) \begin{pmatrix} 1 & p(\lambda)e^{-2\tau(\lambda)} \\ q(\lambda)e^{2\tau(\lambda)} & 1 + p(\lambda)q(\lambda) \end{pmatrix} \tag{11.60}$$

for all $\lambda \in \mathbb{R}$.

Here I_2 denotes the 2×2 unit matrix. The functions $p(\lambda)$ and $q(\lambda)$ are defined as

$$p(\lambda) = i(\cos(\eta) - 1)(1 - \vartheta(\lambda))\alpha_+(\lambda)\alpha_-(\lambda), \tag{11.61}$$

$$q(\lambda) = -\frac{2i\vartheta(\lambda)}{\alpha_+(\lambda)\alpha_-(\lambda)}, \tag{11.62}$$

where

$$\alpha(\lambda) = \exp\left\{ -\frac{1}{2\pi i} \int_{-\infty}^{\infty} \frac{d\mu}{\mu - \lambda} \ln\left(1 + \vartheta(\mu)(e^{-i\eta} - 1)\right) \right\}. \tag{11.63}$$

The functions $\partial_x \sigma$, $\partial_t \sigma$, b_{++} and B_{--} can be expressed through the coefficients in the asymptotic expansions of $\phi(\lambda)$ and $\ln(\alpha(\lambda))$ for large spectral parameter λ. Let

$$\phi(\lambda) = I_2 + \frac{\phi^{(1)}}{\lambda} + \frac{\phi^{(2)}}{\lambda^2} + \mathcal{O}\left(\frac{1}{\lambda^3}\right) \tag{11.64}$$

and

$$\ln(\alpha(\lambda)) = \frac{\alpha_1}{\lambda} + \frac{\alpha_2}{\lambda^2} + \mathcal{O}\left(\frac{1}{\lambda^3}\right). \tag{11.65}$$

Then

$$\partial_x \sigma = 2i\alpha_1 + i \, \text{tr}\{\phi^{(1)}\sigma^z\}, \quad \partial_t \sigma = 4i\alpha_2 + 2i \, \text{tr}\{\phi^{(2)}\sigma^z\}, \tag{11.66}$$

$$b_{++} = \phi_{12}^{(1)}, \quad B_{--} = -\phi_{21}^{(1)}. \tag{11.67}$$

The Riemann-Hilbert problem is the appropriate starting point for the asymptotic analysis of the potentials b_{++} and B_{--} which determine the asymptotics of the two-point Green functions $G_{\uparrow\uparrow}^{\pm}$. For impenetrable Bosons a similar analysis was carried out in [212]. Fortunately, the result of [212] depends only on some general properties of the functions $p(\lambda)$ and $q(\lambda)$ entering the conjugation matrix in (11.60), and also applies in the present case. Alternatively, the non-linear steepest descent method of Deift and Zhou [96] could be applied.

11.2.5 Derivation of the asymptotics

The direct asymptotic analysis of the Riemann-Hilbert problem yields the leading order asymptotics $(x, t \to \infty$ for fixed ratio $\lambda_0 = -2x/t)$ of the functions $\partial_x \sigma$, $\partial_t \sigma$, b_{++} and B_{--} [173, 212]. It turns out, in particular, that b_{++} and B_{--} are a decaying solution of the separated nonlinear Schrödinger equation (11.51), (11.52). Now the form of the complete asymptotic decomposition of the decaying solutions of the separated nonlinear Schrödinger equation is known [2, 388].

$$b_{++} = t^{-\frac{1}{2}} \left[u_0 + \sum_{n=1}^{\infty} \sum_{k=0}^{2n} \frac{\ln^k 4t}{t^n} u_{nk} \right] \exp\left\{ \frac{ix^2}{2t} - iv \ln 4t \right\}, \tag{11.68a}$$

$$B_{--} = t^{-\frac{1}{2}} \left[v_0 + \sum_{n=1}^{\infty} \sum_{k=0}^{2n} \frac{\ln^k 4t}{t^n} v_{nk} \right] \exp\left\{ -\frac{ix^2}{2t} + iv \ln 4t \right\}, \tag{11.68b}$$

where u_0, v_0, u_{nk}, v_{nk} and v are functions of $\lambda_0 = -x/2t$ and of β and η. Inserting the asymptotic expansions for B_{--} and b_{++} into the differential equations (11.51), (11.52) we obtain expressions for u_{nk}, v_{nk} and v in terms of u_0 and v_0, i.e., the two unknown functions u_0 and v_0 determine the whole asymptotic expansion (11.68a), (11.68b). But u_0 and v_0 are obtained from the asymptotic analysis of the Riemann-Hilbert problem (for the explicit expressions see [173]). Hence we know, in principle, the complete asymptotic decomposition of the potentials b_{++} and B_{--}.

In order to obtain the asymptotics of the two-point Green functions we still need the asymptotics of the Fredholm determinant. The Fredholm determinant is related to b_{++} and B_{--} through equations (11.56)–(11.58) and (11.53)–(11.55). We may integrate (11.56)–(11.58) to obtain the asymptotic expansions of $\partial_x \sigma$ and $\partial_t \sigma$. The integration constant is a function of β. It is fixed by the leading asymptotics, which, using (11.66), can be obtained from the direct asymptotic analysis of the Riemann-Hilbert problem. Then, integrating (11.53)–(11.55) yields σ up to a numerical constant, which follows from the asymptotic condition (11.59). The calculation is the same as for the impenetrable Bose gas and can be found on pages 455–457 of [270].

Finally, we obtain the following expressions for the leading asymptotics of the correlation function,

$$g^+(x, t) = e^{ix^2/2t + 2it\beta} e^{-2it(h + \ln(2\mathrm{ch}(h)))} \int_{-\pi}^{\pi} d\eta \, \frac{F(\gamma, \eta)}{1 - \cos(\eta)} \cdot$$
$$\times \, C^+(\lambda_0, \beta, \eta) (4t)^{\frac{1}{2}(\nu - i)^2} \exp\left\{ \frac{1}{\pi} \int_{-\infty}^{\infty} d\lambda \, |x + 2\lambda t| \ln(\varphi(\lambda, \beta)) \right\}, \tag{11.69}$$

$$g^-(x, t) = e^{-ix^2/2t - 2it\beta} e^{2it(h + \ln(2\mathrm{ch}(h)))} \int_{-\pi}^{\pi} d\eta \, \frac{F(\gamma, \eta)}{2\gamma} \cdot$$
$$\times \, C^-(\lambda_0, \beta, \eta) (4t)^{\frac{1}{2}(\nu + i)^2} \exp\left\{ \frac{1}{\pi} \int_{-\infty}^{\infty} d\lambda \, |x + 2\lambda t| \ln(\varphi(\lambda, \beta)) \right\}, \tag{11.70}$$

where

$$\varphi(\lambda, \beta) = 1 + \vartheta(\lambda)\left(e^{-i\eta\,\mathrm{sign}(\lambda-\lambda_0)} - 1\right),\tag{11.71}$$

$$\nu = -\frac{1}{2\pi}\ln\left(1 - 2(1 - \cos(\eta))\vartheta(\lambda_0)(1 - \vartheta(\lambda_0))\right),\tag{11.72}$$

$$C^+(\lambda_0, \beta, \eta) = -|\sin(\eta/2)|\frac{\sqrt{\nu}}{2\pi}\exp\left\{\frac{1}{2}(\lambda_0^2 - \beta) + i\Psi_0 + \frac{\nu^2}{2}\right.$$

$$-\int_{-\infty}^{\beta} d\beta\,(i\nu/2 + \nu\partial_\beta\Psi_0)\tag{11.73}$$

$$\left.+\frac{1}{2\pi^2}\int_{-\infty}^{\beta} d\beta\left(\partial_\beta\int_{-\infty}^{\infty} d\lambda\,\mathrm{sign}(\lambda - \lambda_0)\ln(\varphi(\lambda, \beta))\right)^2\right\},$$

$$C^-(\lambda_0, \beta, \eta) = C^+(\lambda_0, \beta, \eta)\exp(-(\lambda_0^2 - \beta) - 2i\Psi_0)/\sin^2(\eta/2).\tag{11.74}$$

$\lambda_0 = -x/2t$ is the stationary point of the phase $\tau(\lambda)$ (i.e., $\tau'(\lambda_0) = 0$), and the functions Ψ_0 and Ψ_1 are defined as

$$\Psi_0 = -\frac{3\pi}{4} + \arg\Gamma(i\nu) + \Psi_1,\tag{11.75}$$

$$\Psi_1 = -\frac{1}{\pi}\int_{-\infty}^{\infty} d\lambda\,\mathrm{sign}(\lambda - \lambda_0)\ln|\lambda - \lambda_0|\partial_\lambda\ln(\varphi(\lambda, \beta)).\tag{11.76}$$

Equations (11.69) and (11.70) are valid for large t and fixed finite ratio $\lambda_0 = -x/2t$. Correlations in the pure space direction $t = 0$ were discussed in [56]. We would like to emphasize that (11.69) and (11.70) still hold for arbitrary temperatures. The low temperature limit will be discussed in the next section. Note that there is no pole of the integrand at $\eta = 0$, since $\sqrt{\nu} \sim |\eta|$ for small η and thus $C^+(\lambda_0, \beta, \eta) \sim \eta^2$.

11.2.6 Asymptotics in the low-temperature limit

For the following steepest descent calculation we transform the η-integrals in (11.69), (11.70) into complex contour integrals over the the unit circle, setting $z = e^{i\eta}$. Since we would like to consider low temperatures, we have to restore the explicit temperature dependence by scaling back to the physical space and time variables x and t and to the physical correlation functions $G^\pm_{\uparrow\uparrow}$. Recall that in the previous sections we have suppressed an index 'r' referring to 'rescaled'. Let us restore this index in order to define $k_0 = \lambda_0\sqrt{T} = x/2t$, $\vartheta(k) = \vartheta_r(k/\sqrt{T})$, $\varphi(k, \beta) = \varphi_r(k/\sqrt{T}, \beta)$, $C^\pm(k_0, \beta, z) = C^\pm_r(\lambda_0, \beta, \eta)$, $F(\gamma, z) = F_r(\gamma, \eta)$. Then

$$G^+_{\uparrow\uparrow}(x, t) = 2i\sqrt{T}\,e^{ix^2/4t + it(\mu - B)}\oint dz\,\frac{F(\gamma, z)}{(z-1)^2}\,C^+(k_0, \beta, z)(2Tt)^{\frac{1}{2}(\nu(z)-i)^2}e^{tS(z)},\tag{11.77}$$

$$G^-_{\uparrow\uparrow}(x, t) = -i\sqrt{T}\,e^{-ix^2/4t - it(\mu - B)}\oint dz\,\frac{F(\gamma, z)}{2\gamma z}\,C^-(k_0, \beta, z)(2Tt)^{\frac{1}{2}(\nu(z)+i)^2}e^{tS(z)},$$

$$\tag{11.78}$$

where

$$S(z) = \frac{1}{\pi} \int_{-\infty}^{\infty} dk \, |k - k_0| \ln(\varphi(k, \beta)). \tag{11.79}$$

We would like to calculate the contour integrals (11.77), (11.78) by the method of steepest descent. For this purpose we have to consider the analytic properties of the integrands. Let us assume that $k_0 \geq 0$, and let us cut the complex plane along the real axis from $-\infty$ to $-e^{-\beta}$ and from $-e^{\beta - k_0^2/T}$ to 0. The integrands in (11.77) and (11.78) can be analytically continued as functions of z into the cut plane with the only exception of the two simple poles of $F(\gamma, z)$ at $z = \gamma^{\pm 1}$. We may therefore deform the contour of integration as long as we never cross the cuts and take into account the pole contributions, if we cross $z = \gamma$ or $z = \gamma^{-1}$.

The saddle point equation $\partial S/\partial z = 0$ can be represented in the form

$$\int_0^{\infty} \frac{dk \, k}{1 + z^{-1} e^{-\beta} e^{(k-k_0)^2/T}} = \int_0^{\infty} \frac{dk \, k}{1 + z e^{-\beta} e^{(k+k_0)^2/T}}. \tag{11.80}$$

This equation was discussed in the appendix of [174]. In [174] it was shown that (11.80) has exactly one real positive solution which is located in the interval [0, 1]. It was argued that this solution gives the leading saddle point contribution to (11.77) and (11.78). At small temperatures (11.80) can be solved explicitly. There are two solutions $z_{\pm} = \pm z_c$, where

$$z_c = \frac{T^{3/4}}{2\pi^{1/4} k_0^{3/2}} e^{-k_0^2/2T}. \tag{11.81}$$

In the derivation of (11.81) we assumed that $k_0 \neq 0$. The case $k_0 = 0$ has to be treated separately (see below).

The phase $t S(z)$ has the low-temperature approximation

$$t S(z) = -2k_0 Dt \left\{ \left(1 - \frac{1}{z}\right) z_c^2 + 1 - z \right\}. \tag{11.82}$$

Here $D = \partial P/\partial \mu$ is the density of the electron gas. The low-temperature expansion (11.82) is valid in an annulus $e^{\beta - k_0^2/T} \ll |z| \ll e^{-\beta}$, which lies in our cut plane. The unit circle and the circle $|z| = z_c$ are inside this annulus. We may thus first apply (11.82) and then deform the contour of integration from the unit circle to the small circle $|z| = z_c$. Let us parameterize the small circle as $z = z_c e^{i\alpha}$, $\alpha \in [-\pi, \pi]$. Then $S(z(\alpha)) = -2k_0 D((z_c - 1)^2 + 2z_c(1 - \cos(\alpha)))$, which implies that the small circle is a steepest descent contour and that on this contour $S(z_-) \leq S(z) \leq S(z_+)$. The maximum of $S(z)$ on the steepest descent contour at $z = z_+$ is unique and therefore provides the leading saddle point contribution to (11.77), (11.78) as $t \to \infty$. The saddle point approximation becomes good when $t S(z(\alpha)) = -2k_0 Dt((z_c - 1)^2 + z_c\alpha^2 + \mathcal{O}(\alpha^4))$ becomes sharply peaked around $\alpha = 0$. Hence, the relevant parameter for the calculation of the asymptotics of $G_{\uparrow\uparrow}^{\pm}$ is $2k_0 Dt = xD$ rather than t. xD has to be large compared to z_c^{-1}. The parameter xD has a simple interpretation. It is the average number of particles in the interval $[0, x]$. Let us consider two different limiting cases.

(i) $xD \to 0$, the number of electrons in the interval $[0, x]$ vanishes. In this regime the interaction of the electrons is negligible. An electron propagates freely from 0 to x. $G_{\uparrow\uparrow}^{\pm}$ cannot be calculated by the method of steepest descent. We have to use the integral representation (11.69), (11.70) instead. Since $t S(z)$ and $\nu(z)$ tend to zero on the contour of integration, the integrals in (11.69) and (11.70) are easily calculated. We find $G_{\uparrow\uparrow}^{\pm} = G_f^{\pm}$ (see (11.22a), (11.22b)), which is the well known result for free Fermions.

(ii) $xD \gg z_c^{-1}$, the average number of electrons in the interval $[0, x]$ is large. This is the *true asymptotic region*, $x \to \infty$. In this region the interaction becomes important. At the same time the method of steepest descent can be used to calculate $G_{\uparrow\uparrow}^{\pm}$. This case will be studied below.

In the process of deformation of the contour from the unit circle to the small circle of radius z_c we may cross the pole of the function $F(\gamma, z)$ at $z = \gamma^{-1}$. Then we obtain a contribution of the pole to the asymptotics of $G_{\uparrow\uparrow}^{\pm}$. It turns out that the pole contributes to $G_{\uparrow\uparrow}^{\pm}$, when the magnetic field is below a critical positive value, $B_c = k_0^2/4$. Below this value the pole contribution always dominates the contribution of the saddle point. Hence, we have to distinguish two different asymptotic regions, $B > B_c$ and $B < B_c$. On the other hand, if we consider the asymptotics for fixed magnetic field, we have to treat the cases $B > 0$ and $B \le 0$ separately. For $B > 0$ we have to distinguish between a time like regime ($k_0^2 < 4B$) and a space-like regime ($k_0^2 > 4B$). In these respective regimes we obtain the asymptotics (11.23), (11.25).

In the limit $B \to -\infty$, $\mu \to -\infty$, $\mu - B$ fixed there are no $\downarrow$-spin electrons left in the system, $D_\downarrow \to 0$. This is the free Fermion limit. In the free Fermion limit $B < B_c$, and the asymptotics of $G_{\uparrow\uparrow}^+(x, t)$ and $G_{\uparrow\uparrow}^-(x, t)$ are given by the equations (11.25), which turn into the expressions (11.22a), (11.22b) for free Fermions.

The pure time direction $k_0 = 0$ requires a separate calculation. For $k_0 = 0$ the saddle point equation (11.80) has the solutions $z = \pm 1$ for all temperatures. The unit circle is a steepest-descent contour with unique maximum of $S(z)$ at $z = 1$, which gives the leading asymptotic contribution to the integrals in (11.77) and (11.78). We find algebraically decaying correlations,

$$G_{\uparrow\uparrow}^+(0, t) = C_0^+ t^{-1} e^{it(\mu - B)}, \quad G_{\uparrow\uparrow}^-(0, t) = C_0^- t^{-1} e^{-it(\mu - B)}, \tag{11.83}$$

where

$$C_0^+ = \frac{e^{-i\frac{\pi}{4}}}{2\sqrt{2\pi T}}(1 + 2e^{-2B/T})$$
$$\times \left[(e^{(\mu+B)/T} + e^{(\mu-B)/T})(1 + e^{(\mu+B)/T} + e^{(\mu-B)/T}) \right]^{-\frac{1}{2}}, \tag{11.84a}$$

$$C_0^- = \frac{e^{i\frac{\pi}{4}}}{2\sqrt{2\pi T}} \frac{1 + 2e^{-2B/T}}{1 + e^{2B/T}} \left[\frac{e^{(\mu+B)/T} + e^{(\mu-B)/T}}{1 + e^{(\mu+B)/T} + e^{(\mu-B)/T}} \right]^{\frac{1}{2}}. \tag{11.84b}$$

These formulae are valid at any temperature.

11.3 Conclusions

In this chapter we have considered dynamical correlations of the Hubbard model in the gas phase and have obtained explicit expressions for the asymtotics in space and time of the Green function at small temperature. The results do not rely on field theoretical methods, but rather came out of a direct Bethe ansatz calculation.

12

The algebraic approach to the Hubbard model

12.1 Introduction to the quantum inverse scattering method

The quantum inverse scattering method is the modern algebraic theory of exactly solvable quantum systems. It arose [404, 410, 411] as an attempt to carry over the concepts of the inverse scattering method for classical non-linear evolution equations [2, 134] into quantum mechanics. As a result, our understanding of both the theory of integrable partial differential equations and the theory of exactly solvable quantum systems changed, and the algebraic roots of the exact solvability became apparent. These roots originate from the Yang-Baxter equation and its classical counterpart.

Before turning to our actual subject, which is the application of the quantum inverse scattering method to the Hubbard model, we give a brief general introduction. We shall limit our exposition basically to the material which is needed later for the understanding of the algebraic structure of the Hubbard model. The reader who is interested in the general scope of the method and in the history of its development is referred to the excellent books and review articles [131, 270, 276, 277, 407].

12.1.1 Integrability

As a motivation for the definition of the Yang-Baxter algebra in the following subsection we shall first recall the concept of integrability in classical mechanics. Then, by considering the elementary example of the harmonic oscillator, we shall see that this concept does not directly apply to quantum mechanical systems and needs to be extended.

Consider N classical point particles described by their canonical momenta $\mathbf{p} = (p_1, \ldots, p_N)$ and position variables $\mathbf{q} = (q_1, \ldots, q_N)$. Suppose their dynamics is generated by a Hamiltonian $H(\mathbf{p}, \mathbf{q})$ through the canonical equations of motion

$$\dot{\mathbf{p}} = -\frac{\partial H}{\partial \mathbf{q}}, \quad \dot{\mathbf{q}} = \frac{\partial H}{\partial \mathbf{p}}. \tag{12.1}$$

Suppose further that there are N independent, single-valued and analytic functions $I_j(\mathbf{p}, \mathbf{q})$, $j = 1, \ldots, N$, in involution,

$$\{I_j, I_k\} = 0 \tag{12.2}$$

(here $\{\cdot,\cdot\}$ denotes the Poisson bracket), and let H be among the I_j. Then Liouville's theorem [22] states that the equations of motion (12.1) are 'solvable by quadratures'. More precisely, starting from (12.2) a canonical transformation can be constructed, such that the equations of motion are trivial in the new (action-angle) variables. No further input apart from the involutive integrals of motion is needed.

The simplest case where the Liouville theorem applies, is the case of conservative systems with only one degree of freedom. For these systems the Hamiltonian is a constant of motion. The equation $H(p,q) = E$ can be solved for $p = p(q, E)$. Thus, $dt = (\partial H(p(q, E), q)/\partial p)^{-1} dq$, which gives $t = t(q)$ upon integration.

It is probably fair to say that no concept of integrability in quantum mechanics exists which is as general as Liouville's theorem. By analogy with the classical case discussed above let us assume we are given a Hamiltonian H contained in a set of mutually commuting quantum integrals of motion $I_j, j = 1, \ldots, N$. Then there is no general theorem which would explain how to obtain the spectrum and the eigenfunctions of H from the commutation relations

$$[I_j, I_k] = 0 \tag{12.3}$$

alone. The construction of action-angle variables does not easily translate into quantum mechanics. Additional information is required.

Thinking of those textbook examples which can be algebraically solved by elementary means, like the harmonic oscillator, the Coulomb problem, the Morse oscillator or the two-particle Sutherland system, we can get an idea of what kind of additional information is necessary. We should look for an embedding of the commutative algebra (12.3) of quantum integrals of motion into some larger algebra, with the space of states of our system playing the role of the representation space of this algebra. For the examples mentioned above the embedding algebra is a quadratic algebra, either a Lie algebra or a Lie super algebra.

Let us recall, for instance, the solution of the harmonic oscillator [198]. It relies on the Heisenberg Lie algebra

$$[a, a^\dagger] = 1. \tag{12.4}$$

Suppose we are given a highest weight representation of the algebra (12.4), i.e., a representation containing a highest weight state $|0\rangle$, such that $a|0\rangle = 0$. Suppose further that a and $a^\dagger$ are mutually adjoint. Let $H = a^\dagger a + \frac{1}{2}$. Equation (12.4) implies that $[a, \cdot]$ acts as a derivative on functions of $a^\dagger$, i.e., $[a, (a^\dagger)^n] = n(a^\dagger)^{n-1}$. It follows that

$$\begin{aligned}
H(a^\dagger)^n|0\rangle &= ((a^\dagger)^n H + [H, (a^\dagger)^n])|0\rangle \\
&= (\tfrac{1}{2}(a^\dagger)^n + a^\dagger[a, (a^\dagger)^n])|0\rangle \\
&= (n + \tfrac{1}{2})(a^\dagger)^n|0\rangle.
\end{aligned} \tag{12.5}$$

Hence, the states $|n\rangle = (a^\dagger)^n|0\rangle$ are eigenstates of the Hamiltonian $H = a^\dagger a + \frac{1}{2}$.

The connection to physics comes through the infinite dimensional highest weight representation

$$a = \tfrac{1}{\sqrt{2}}(x + \partial_x), \quad a^\dagger = \tfrac{1}{\sqrt{2}}(x - \partial_x) \tag{12.6}$$

acting on the space of square integrable functions on the real line. The highest weight state is the unique normalized solution

$$\psi(x) = \frac{e^{-\frac{x^2}{2}}}{\sqrt{\pi}} \qquad (12.7)$$

of the differential equation

$$a\psi(x) = 0 \qquad (12.8)$$

and is the well-known ground state of the harmonic oscillator.

Note that the above scheme depends only on (i) the algebra (12.4) and (ii) the existence of the highest-weight state $|0\rangle$ (see (12.7), (12.8)).

12.1.2 The Yang-Baxter algebra

The quantum inverse scattering method deals with systems which are based on an associative quadratic algebra $\mathcal{T}_R$ defined in terms of its generators $T_\beta^\alpha(\lambda)$, $\alpha, \beta = 1, \ldots, d$; $\lambda \in \mathbb{C}$, by the relation

$$R(\lambda, \mu) T_1(\lambda) T_2(\mu) = T_2(\mu) T_1(\lambda) R(\lambda, \mu). \qquad (12.9)$$

Here the following notation has been employed

$$T(\lambda) = \begin{pmatrix} T_1^1(\lambda) & \cdots & T_d^1(\lambda) \\ \vdots & & \vdots \\ T_1^d(\lambda) & \cdots & T_d^d(\lambda) \end{pmatrix}, \qquad (12.10)$$

$$T_1(\lambda) = T(\lambda) \otimes I_d, \qquad (12.11)$$
$$T_2(\lambda) = I_d \otimes T(\lambda), \qquad (12.12)$$

where I_d is the $d \times d$ unit matrix. $R(\lambda, \mu) \in \mathrm{End}(\mathbb{C}^d \otimes \mathbb{C}^d)$ is a numerical $d^2 \times d^2$ matrix, called the R-matrix. The R-matrix fixes the structure of the quadratic algebra $\mathcal{T}_R$ in a similar manner as the tensor of structure constants does in case of a Lie algebra. We assume that R is invertible for almost all $\lambda, \mu \in \mathbb{C}$.

The algebra $\mathcal{T}_R$ thus defined has a rich commutative subalgebra. Multiplying equation (12.9) by $R^{-1}(\lambda, \mu)$ from the right and taking the trace we obtain

$$\mathrm{tr}\big(R(\lambda,\mu)T_1(\lambda)T_2(\mu)R^{-1}(\lambda,\mu)\big)$$

$$= R^{\alpha\beta}_{\gamma\delta}(\lambda,\mu)T_1{}^{\gamma\delta}_{\alpha'\beta'}(\lambda)T_2{}^{\alpha'\beta'}_{\gamma'\delta'}(\mu)R^{-1}{}^{\gamma'\delta'}_{\alpha\beta}(\lambda,\mu)$$

$$= R^{-1}{}^{\gamma'\delta'}_{\alpha\beta}(\lambda,\mu)R^{\alpha\beta}_{\gamma\delta}(\lambda,\mu)T^\delta_{\alpha'}(\lambda)\delta^\gamma_{\beta'}\delta^{\alpha'}_{\gamma'}T^{\beta'}_{\delta'}(\mu)$$

$$= \delta^{\gamma'}_\gamma \delta^\delta_{\delta'} T^\gamma_{\gamma'}(\lambda)T^\delta_\delta(\mu) = T^\gamma_\gamma(\lambda)T^\delta_\delta(\mu)$$

$$= T_2{}^{\alpha\beta}_{\gamma\delta}(\mu)T_1{}^{\gamma\delta}_{\alpha\beta}(\lambda) = \delta^\alpha_\gamma T^\beta_\delta(\mu)T^\gamma_\alpha(\lambda)\delta^\delta_\beta$$

$$= T^\delta_\delta(\mu)T^\gamma_\gamma(\lambda). \qquad (12.13)$$

Here and in the following implicit summation with respect to doubly occurring indices is, understood. With the definition

$$t(\lambda) = T_\gamma^\gamma(\lambda) = \mathrm{tr}(T(\lambda)) \tag{12.14}$$

we have the important result

$$[t(\lambda), t(\mu)] = 0. \tag{12.15}$$

It means that $t(\lambda)$ is a generating function of a commutative subalgebra of $\mathcal{T}_R$, e.g., if $t(\lambda) = I_0 + \lambda I_1 + \lambda^2 I_2 + \ldots$, then (12.15) implies that $[I_j, I_k] = 0$.

Let us assume we are given a representation of $\mathcal{T}_R$ on the space of states of some physical system. Then $t(\lambda)$ generates a set of mutually commuting operators which by construction are embedded into the quadratic algebra $\mathcal{T}_R$. Thus, on the one hand we have a chance to meet the requirements of Liouville's theorem in the classical limit (if it exists), while on the other hand, the quadratic relations of the algebra $\mathcal{T}_R$ may provide means to simultaneously diagonalize the quantum integrals of motion, generated by $t(\lambda)$, in a similar manner as in our example above of the harmonic oscillator.

Here is the standard terminology for the notions introduced so far: The associative quadratic algebra $\mathcal{T}_R$ is the Yang-Baxter algebra. $T(\lambda)$ is the monodromy matrix and $t(\lambda)$ its associated transfer matrix. Their complex argument λ is the spectral parameter. The space $\mathbb{C}^d$ is called the auxiliary space, while the name for the representation space of the Yang-Baxter algebra usually is quantum space.

Remark. We would like to emphasize that

$$\left(I_d \otimes T(\mu)\right)\left(T(\lambda) \otimes I_d\right) \neq T(\lambda) \otimes T(\mu), \tag{12.16}$$

since the matrix elements $T_\beta^\alpha(\lambda)$ and $T_\delta^\gamma(\mu)$ of the monodromy matrix, in general, do not commute.

An alternative way to write the defining relations (12.9) of the Yang-Baxter algebra is by introduction of a matrix $\check{R}(\lambda, \mu)$ with matrix elements

$$\check{R}_{\gamma\delta}^{\alpha\beta}(\lambda, \mu) = R_{\gamma\delta}^{\beta\alpha}(\lambda, \mu). \tag{12.17}$$

It is easy to see that (12.9) is equivalent to

$$\check{R}(\lambda, \mu)\left(T(\lambda) \otimes T(\mu)\right) = \left(T(\mu) \otimes T(\lambda)\right)\check{R}(\lambda, \mu). \tag{12.18}$$

This formulation is sometimes more convenient for practical calculations.

12.1.3 The Yang-Baxter equation

The Yang-Baxter equation is a sufficient condition for the consistency of the Yang-Baxter algebra $\mathcal{T}_R$. At the same time it guarantees the existence of non-trivial representations of $\mathcal{T}_R$.

We embed the algebra $\mathcal{T}_R$ into the tensor product $\mathbb{C}^d \otimes \mathbb{C}^d \otimes \mathbb{C}^d$ of auxiliary spaces, $T_1(\lambda) = T(\lambda) \otimes I_d \otimes I_d$, $T_2(\lambda) = I_d \otimes T(\lambda) \otimes I_d$, $T_3(\lambda) = I_d \otimes I_d \otimes T(\lambda)$, and denote the three possible canonical embeddings of the R-matrix $R(\lambda, \mu)$ into the space of endomorphisms on $\mathbb{C}^d \otimes \mathbb{C}^d \otimes \mathbb{C}^d$ by $R_{12}(\lambda, \mu)$, $R_{13}(\lambda, \mu)$ and $R_{23}(\lambda, \mu)$. Then $R_{12}(\lambda, \mu) = R(\lambda, \mu) \otimes I_d$ and $R_{23}(\lambda, \mu) = I_d \otimes R(\lambda, \mu)$. The third possibility cannot be written in such a simple form. From the requirement that $(R_{13}(\lambda, \mu)\mathbf{x} \otimes \mathbf{y} \otimes \mathbf{z})^{\alpha\beta\gamma} = R_{\delta\varphi}^{\alpha\gamma}(\lambda, \mu)x^\delta y^\beta z^\varphi$ we conclude that $R_{13}{}_{\delta\varepsilon\varphi}^{\alpha\beta\gamma} = R_{\delta\varphi}^{\alpha\gamma}(\lambda, \mu)\delta_\varepsilon^\beta$.

Now comes an argument that is very similar to the argument in appendix 3.B.3, where we encountered the Yang-Baxter equation for the first time. From the definition (12.9) of the Yang-Baxter algebra it can be seen that there are two different ways to reverse the order of monodromy matrices in the triple product $T_1(\lambda)T_2(\mu)T_3(\nu)$ by application of R-matrices,

$$R_{12}(\lambda, \mu)R_{13}(\lambda, \nu)R_{23}(\mu, \nu)T_1(\lambda)T_2(\mu)T_3(\nu)$$
$$= T_3(\nu)T_2(\mu)T_1(\lambda)R_{12}(\lambda, \mu)R_{13}(\lambda, \nu)R_{23}(\mu, \nu), \tag{12.19}$$

$$R_{23}(\mu, \nu)R_{13}(\lambda, \nu)R_{12}(\lambda, \mu)T_1(\lambda)T_2(\mu)T_3(\nu)$$
$$= T_3(\nu)T_2(\mu)T_1(\lambda)R_{23}(\mu, \nu)R_{13}(\lambda, \nu)R_{12}(\lambda, \mu). \tag{12.20}$$

Obviously, these two equations are always compatible, if the R-matrix satisfies the equation

$$R_{12}(\lambda, \mu)R_{13}(\lambda, \nu)R_{23}(\mu, \nu) = R_{23}(\mu, \nu)R_{13}(\lambda, \nu)R_{12}(\lambda, \mu). \tag{12.21}$$

This is the famous Yang-Baxter equation. It is not only a sufficient condition for the consistency of $\mathcal{T}_R$ as an associative algebra, but also provides its so-called fundamental representation whose construction is the subject of the following subsections.

Most of the known solutions of the Yang-Baxter equation that are connected to applications in physics depend on the spectral parameters only through their difference. In these cases there exists a matrix $R(\lambda)$ of a single argument, such that $R(\lambda, \mu) = R(\lambda - \mu)$ solves the Yang-Baxter equation (12.21). One says $R(\lambda)$ is a solution of difference form of the Yang-Baxter equation. Shastry's R-matrix which connects the Hubbard model to a solution of the Yang-Baxter equation is not of difference form. Therefore we introduced the Yang-Baxter equation in its more general form (12.21).

12.1.4 The standard basis

In order to obtain convenient expressions for our final formulae we first have to introduce some more notation. We shall need the standard basis on the space of endomorphisms, $\text{End}(\mathbb{C}^d)$, on $\mathbb{C}^d$. In the following we denote by $e_\gamma \in \mathbb{C}^d$, $\gamma = 1, \ldots, d$, a column vector with only non-vanishing entry 1 in row γ. The set $\{e_\gamma \in \mathbb{C}^d | \gamma = 1, \ldots, d\}$ is a basis of $\mathbb{C}^d$. Let $e_\alpha^\beta \in \text{End}(\mathbb{C}^d)$, such that $e_\alpha^\beta e_\gamma = \delta_\gamma^\beta e_\alpha$. Then $\{e_\alpha^\beta \in \text{End}(\mathbb{C}^d) | \alpha, \beta = 1, \ldots, d\}$ is a basis of $\text{End}(\mathbb{C}^d)$. e_α^β is a $d \times d$ matrix with only non-vanishing entry 1 in row α and column β.

Hence, the e_α^β multiply according to the rule

$$e_\alpha^\beta e_\gamma^\delta = \delta_\gamma^\beta e_\alpha^\delta \tag{12.22}$$

(compare appendix 3.D.1).

Expressing the same facts in different words, we may also say that the basis $\{e_\alpha^\beta\}$ is a basis of the space of operators on the auxiliary space $\mathbb{C}^d$. We shall now consider representations of the Yang-Baxter algebra, where the quantum space is an L-fold tensor product of auxiliary spaces $\mathbb{C}^d$. Thus, for $d = 2$ and $L = 1$, the quantum space is equal to $\mathbb{C}^2$ and can be interpreted as the space of states of a spin $\frac{1}{2}$. For $d = 2$ and arbitrary L it is the space of states of a spin-$\frac{1}{2}$ quantum spin chain of length L. For $d > 2$ the spin is replaced by a generalized so-called su(d) spin. In order to construct operators on the spin chain space of states we consider the canonical embedding of the basis $\{e_\alpha^\beta\}$ into End$(\mathbb{C}^d)^{\otimes L}$,

$$e_{j\,\alpha}^{\ \beta} = I_d^{\otimes(j-1)} \otimes e_\alpha^\beta \otimes I_d^{\otimes(L-j)} \,. \tag{12.23}$$

The index $j = 1, \ldots, L$ will be called the site index. From (12.22) and (12.23) we infer the local multiplication rule

$$e_{j\,\alpha}^{\ \beta} e_{j\,\gamma}^{\ \delta} = \delta_\gamma^\beta e_{j\,\alpha}^{\ \delta} \tag{12.24}$$

and the commutation relations

$$[e_{j\,\alpha}^{\ \beta}, e_{k\,\gamma}^{\ \delta}] = 0 \tag{12.25}$$

for $j \neq k$.

With the aid of the basic operators $e_{j\,\alpha}^{\ \beta}$ we can conveniently express arbitrary more complicated operators of interest. We can, for instance, expand the canonical embeddings $R_{jk}(\lambda, \mu)$ of the R-matrix in the Yang-Baxter equation (12.21) in terms of the $e_{j\,\alpha}^{\ \beta}$. Let $L = 3$. Then

$$R_{jk}(\lambda, \mu) = R_{\beta\delta}^{\alpha\gamma} e_{j\,\alpha}^{\ \beta} e_{k\,\gamma}^{\ \delta} \tag{12.26}$$

for $jk = 12, 13, 23$.

We also obtain a very useful expression for the transposition operators P_{jk} which play an important role below in the construction of integrable lattice models with local interactions,

$$P_{jk} = e_{j\,\alpha}^{\ \beta} e_{k\,\beta}^{\ \alpha} \,. \tag{12.27}$$

The following properties of the transposition operators are easily verified. They follow from (12.24) and (12.25),

$$P_{kj} = P_{jk} \,, \tag{12.28a}$$

$$P_{jj} = d \cdot \mathrm{id} \,, \tag{12.28b}$$

$$P_{jk}^2 = \mathrm{id} \,, \quad j \neq k \,, \tag{12.28c}$$

$$P_{jk} e_{k\,\alpha}^{\ \beta} = e_{j\,\alpha}^{\ \beta} P_{jk} \,, \tag{12.28d}$$

$$P_{jk} e_{l\,\alpha}^{\ \beta} = e_{l\,\alpha}^{\ \beta} P_{jk} \,, \quad l \neq j, k \,. \tag{12.28e}$$

We see from these formulae that the P_{jk} induce the action of the symmetric group on the site indices of the matrices $e_{j\alpha}^{\beta}$. Because of (12.28d) and (12.28e), the P_{jk} generate a faithful representation of the symmetric group $\mathfrak{S}^L$,

$$P_{jk}P_{kl} = P_{jl}P_{jk} = P_{kl}P_{jl}. \tag{12.29}$$

Combining the definitions (12.17) and (12.27) with (12.26) and using (12.24) and (12.25) we obtain $\check{R}_{jk}(\lambda, \mu) = P_{jk}R_{jk}(\lambda, \mu)$. This allows us to rewrite the Yang-Baxter equation in an alternative form, for multiplication of (12.21) by $P_{23}P_{12}P_{23} = P_{12}P_{23}P_{12}$ yields

$$\check{R}_{23}(\lambda, \mu)\check{R}_{12}(\lambda, \nu)\check{R}_{23}(\mu, \nu) = \check{R}_{12}(\mu, \nu)\check{R}_{23}(\lambda, \nu)\check{R}_{12}(\lambda, \mu). \tag{12.30}$$

Written in this form the Yang-Baxter equation is of striking similarity with the braid relation (3.C.2a). In fact, every representation of the braid relation (3.C.2a) in $\text{End}(\mathbb{C}^d \otimes \mathbb{C}^d)$ gives a spectral parameter independent solution of the Yang-Baxter equation (12.30). Again, if $\check{R}(\lambda, \mu) = \check{R}(\lambda - \mu)$, then $\check{R}(\lambda)$ is said to be a solution of difference form of (12.30).

12.1.5 Fundamental models

We shall now explain how solutions the Yang-Baxter equation (12.21) give rise to representations of the Yang-Baxter algebra (12.18). First, using (12.24) and (12.25), we rewrite the Yang-Baxter equation in components,

$$R_{\alpha'\beta'}^{\alpha\beta}(\lambda, \mu)R_{\alpha''\gamma'}^{\alpha'\gamma}(\lambda, \nu)R_{\beta''\gamma''}^{\beta'\gamma'}(\mu, \nu) = R_{\beta'\gamma'}^{\beta\gamma}(\mu, \nu)R_{\alpha'\gamma''}^{\alpha\gamma'}(\lambda, \nu)R_{\alpha''\beta''}^{\alpha'\beta'}(\lambda, \mu). \tag{12.31}$$

Next, we introduce the L-matrix at site j by defining its matrix elements

$$L_{j\beta}^{\alpha}(\lambda, \mu) = R_{\beta\delta}^{\alpha\gamma}(\lambda, \mu)e_{j\gamma}^{\delta}. \tag{12.32}$$

These matrix elements are operators in $\left(\text{End}(\mathbb{C}^d)\right)^{\otimes L}$. Multiplication of the Yang-Baxter equation (12.31) by $e_{j\gamma}^{\gamma''}$ implies that

$$\check{R}(\lambda, \mu)\left(L_j(\lambda, \nu) \otimes L_j(\mu, \nu)\right) = \left(L_j(\mu, \nu) \otimes L_j(\lambda, \nu)\right)\check{R}(\lambda, \mu). \tag{12.33}$$

Thus, $L_j(\lambda, \nu)$ is a representation of the Yang-Baxter algebra (12.18). This representation is called the fundamental representation.

From (12.25) and (12.32) we know that $[L_{j+1\beta}^{\alpha}(\lambda, \nu_{j+1}), L_{j\delta}^{\gamma}(\lambda, \nu_j)] = 0$. It follows that

$$\left(L_{j+1}(\lambda, \nu_{j+1}) \otimes L_{j+1}(\mu, \nu_{j+1})\right)\left(L_j(\lambda, \nu_j) \otimes L_j(\mu, \nu_j)\right)$$
$$= L_{j+1}(\lambda, \nu_{j+1})L_j(\lambda, \nu_j) \otimes L_{j+1}(\mu, \nu_{j+1})L_j(\mu, \nu_j). \tag{12.34}$$

Thus, (12.33) and (12.34) imply that the product $L_{j+1}(\lambda, \nu_{j+1})L_j(\lambda, \nu_j)$ of two L-matrices as well is a representation of the Yang-Baxter algebra (12.18). This representation may be interpreted as a tensor product representation of two fundamental representations. The property of the Yang-Baxter algebra, that a tensor product of two representations is a representation, is called the co-multiplication property. By iterated co-multiplication we can easily verify that the L-fold ordered product of L-matrices

$$T(\lambda) = L_L(\lambda, \nu_L) \dots L_1(\lambda, \nu_1) \tag{12.35}$$

is a representation of the Yang-Baxter algebra (12.18). The trace of this monodromy matrix, $t(\lambda) = \mathrm{tr}(T(\lambda))$, belongs by construction to a commuting family of transfer matrices, $[t(\lambda), t(\mu)] = 0$.

A solution $R(\lambda, \mu)$ of the Yang-Baxter equation (12.31) is called regular, if there are values λ_0, ν_0 of the spectral parameters, such that $R^{\alpha\beta}_{\gamma\delta}(\lambda_0, \nu_0) = \delta^\alpha_\delta \delta^\beta_\gamma$. For a regular R-matrix it follows from the definition (12.32) that

$$L_{j\,\beta}^{\ \alpha}(\lambda_0, \nu_0) = e_{j\,\beta}^{\ \alpha}. \tag{12.36}$$

Setting $\nu_j = \nu_0$ for $j = 1, \dots, L$ in (12.35) and using the definition of the transposition operator (12.27) we obtain

$$
\begin{aligned}
t(\lambda_0) &= L_L{}^{\beta_L}_{\beta_{L-1}}(\lambda_0, \nu_0) \dots L_2{}^{\beta_2}_{\beta_1}(\lambda_0, \nu_0) L_1{}^{\beta_1}_{\beta_L}(\lambda_0, \nu_0) \\
&= e_1{}^{\beta_1}_{\beta_L} e_2{}^{\beta_2}_{\beta_1} e_3{}^{\beta_3}_{\beta_2} \dots e_L{}^{\beta_L}_{\beta_{L-1}} \\
&= P_{12} P_{23} \dots P_{L-1 L} = \hat{U},
\end{aligned} \tag{12.37}
$$

where the operator $\hat{U}$ defined by the last equation (12.37) is the right-shift operator, which is the generator of cyclic shifts of the site index.

Along with the transfer matrix $t(\lambda)$ itself every appropriately chosen differentiable function of $t(\lambda)$ may be used as a generating function of a set of mutually commuting operators. A particularly useful choice of a generating function is $\tau(\lambda) = \ln(\hat{U}^{-1} t(\lambda))$. Its expansion around $\lambda = \lambda_0$ is

$$\tau(\lambda) = (\lambda - \lambda_0) \hat{U}^{-1} t'(\lambda_0) + \mathcal{O}\big((\lambda - \lambda_0)^2\big). \tag{12.38}$$

It can be shown [308] that the coefficients in the series expansion are local in the sense, that the nth coefficient is a sum over local densities acting non-trivially at $n + 1$ neighbouring sites at most. The two-site term

$$H = \hat{U}^{-1} t'(\lambda_0) = \sum_{j=1}^{L} H_{j-1,j}, \tag{12.39}$$

where $H_{0,1} = H_{L,1}$ by definition and the prime denotes differentiation with respect to the argument, may be interpreted as local Hamiltonian. For the 'densities' $H_{j-1,j}$ we obtain the explicit expression

$$H_{j-1,j} = \partial_\lambda \check{R}_{j-1,j}(\lambda, \nu_0)\Big|_{\lambda=\lambda_0}. \tag{12.40}$$

It is not difficult to verify this equation. First note that

$$
\begin{aligned}
t'(\lambda_0) &= \partial_\lambda L_1{}^{\beta_1}_{\beta_L}(\lambda, \nu_0)\Big|_{\lambda=\lambda_0} e_2{}^{\beta_2}_{\beta_1} \dots e_L{}^{\beta_L}_{\beta_{L-1}} + \dots \\
&\quad \dots + e_1{}^{\beta_1}_{\beta_L} \dots e_{L-1}{}^{\beta_{L-1}}_{\beta_{L-2}} \partial_\lambda L_L{}^{\beta_L}_{\beta_{L-1}}(\lambda, \nu_0)\Big|_{\lambda=\lambda_0}.
\end{aligned} \tag{12.41}
$$

Then, because of

$$e_{j-1}{}^{\beta_{j-1}}_{\beta_{j-2}} e_j{}^{\beta_j}_{\beta_{j-1}} H_{j-1,j}$$

$$= e_{j-1}{}^{\beta_{j-1}}_{\beta_{j-2}} e_j{}^{\beta_j}_{\beta_{j-1}} \left. \partial_\lambda \check{R}^{\alpha\beta}_{\gamma\delta}(\lambda, v_0)\right|_{\lambda=\lambda_0} e_{j-1}{}^{\gamma}_{\alpha} e_j{}^{\delta}_{\beta}$$

$$= e_{j-1}{}^{\gamma}_{\beta_{j-2}} \delta^{\beta_{j-1}}_\alpha \delta^{\beta_j}_\beta \left. \partial_\lambda R^{\beta\alpha}_{\gamma\delta}(\lambda, v_0)\right|_{\lambda=\lambda_0} e_j{}^{\delta}_{\beta_{j-1}}$$

$$= e_{j-1}{}^{\beta_{j-1}}_{\beta_{j-2}} \left. \partial_\lambda R^{\beta_j\gamma}_{\beta_{j-1}\delta}(\lambda, v_0)\right|_{\lambda=\lambda_0} e_j{}^{\delta}_{\gamma}$$

$$= e_{j-1}{}^{\beta_{j-1}}_{\beta_{j-2}} \left. \partial_\lambda L_j{}^{\beta_j}_{\beta_{j-1}}(\lambda, v_0)\right|_{\lambda=\lambda_0}, \tag{12.42}$$

equation (12.40) follows from (12.37), (12.39) and (12.41).

The next coefficients in the series expansion of $\tau(\lambda)$ can be calculated in a similar manner. The reader may verify that the second order term is

$$\tfrac{1}{2}\tau''(\lambda_0) = \tfrac{1}{2} \sum_{j=1}^{L} \left\{ \left. \partial_\lambda^2 \check{R}_{j-1,j}(\lambda, v_0)\right|_{\lambda=\lambda_0} - H^2_{j-1,j} - [H_{j-1,j}, H_{j,j+1}] \right\}, \tag{12.43}$$

where periodic boundary conditions on the indices are again implied. Note that no closed, explicit formula for the nth order term is known. The calculation of the higher order derivatives of $\tau(\lambda)$ is cumbersome. A more efficient way of calculating higher local commuting operators is by means of a recursion relation generated by the so-called boost operator[1] [453]. The higher order commuting operators usually do not have a simple intuitive interpretation in physical terms, and their explicit form does not matter much for applications.[2] They are mostly interesting for their bare existence which says something about the mathematical structure of the model.

One says that equation (12.35) defines the fundamental model associated with the R-matrix $R(\lambda, v)$. If all the v_j, $j = 1, \ldots, L$, are equal, the model is called homogeneous, otherwise inhomogeneous. Only the homogeneous model leads to the local Hamiltonian H, equation (12.39).

We have encountered a particular example of a fundamental inhomogeneous model in appendix 3.B, where the Bethe ansatz wave function of the Hubbard model was constructed. The 'spin problem' obtained after inserting the ansatz for the wavefunction into the Schrödinger equation led us to consider the transfer matrix of the inhomogeneous XXX model (see appendix 3.B.5). In the following subsection we will recall and partly generalize our former results.

Before closing this subsection let us add a comment. Fundamental models are only a small subclass of all models solvable by the quantum inverse scattering method. This subclass, however, contains many models which are interesting from the point of view

---

[1] For a boost operator related to the Hubbard model see [301].

[2] There is, however, an interesting conclusion that can be drawn from (12.43): if the R-matrix is unitary (see (12.46) below) and of difference form, then the first two terms under the sum cancel each other and $\sum_{j=1}^{L}[H_{j-1,j}, H_{j,j+1}]$ commutes with the Hamiltonian. This gives us a simple criterion to test whether a local Hamiltonian $H_{j-1,j}$ can be connected to a fundamental model with unitary R-matrix of difference form.

of applications in physics. If we allow for a slight generalization to so-called fundamental *graded* models [179] which will be discussed in Section 12.3 below, then the models which interest us the most (i.e., the isotropic Heisenberg chain, the t-0 model, the supersymmetric t-J model and the Hubbard model) are all contained in this class. This is the reason why we restrict our discussion to fundamental models here. For additional introductory information about non-fundamental models the reader is referred to the literature (e.g. [270, 277]).

12.1.6 An example – the XXX models

We now reconsider and generalize the inhomogeneous XXX model we have encountered in appendix 3.B.5. Having in mind the construction of another class of fundamental models in the next section we offer a slightly more abstract point of view. Classes of solutions of the Yang-Baxter equation can often be constructed by first considering an appropriate algebra, which provides a solution of the Yang-Baxter equation without spectral parameter, and, in a second step, attempting to introduce a spectral parameter into the equations. This procedure has the advantage that every representation of the algebra at hand gives a new solution of the Yang-Baxter equation.

Let us, for instance, consider the associative algebra with unity 'id' defined in terms of its generators A_i, $i = 1, \dots, L$, by the relations

$$A_i A_{i+1} A_i = A_{i+1} A_i A_{i+1}, \tag{12.44a}$$

$$A_i A_j = A_j A_i \qquad \text{for } |i - j| > 1, \tag{12.44b}$$

$$A_i^2 = \text{id}, \tag{12.44c}$$

where $A_{L+1} = A_1$ by definition. Comparing with (3.C.2) we see that this algebra is isomorphic to the group algebra of the symmetric group $\mathfrak{S}^L$. Note that the braid relation (12.44a) is very similar to a spectral parameter independent form of the Yang-Baxter equation (12.30).

A short calculation, similar to the one in appendix (3.C.1) shows that

$$(\text{id} + \lambda A_2)(\text{id} + (\lambda + \mu)A_1)(\text{id} + \mu A_2) = (\text{id} + \mu A_1)(\text{id} + (\lambda + \mu)A_2)(\text{id} + \lambda A_1).$$
$$\tag{12.45}$$

Clearly, the transposition operators P_{ii+1}, equation (12.27), provide a representation $A_i = P_{ii+1}$ of (12.44). Let $P = e_\alpha^\beta \otimes e_\beta^\alpha$. Then, by (12.45), the matrix $\check{R}(\lambda) = \alpha(\lambda)(\text{id} + \lambda P)$, with an arbitrary function $\alpha(\lambda)$, is a solution of difference form of the Yang-Baxter equation (12.30). Note that since $d = 2, 3, 4, \dots$ is arbitrary, we have obtained solutions of the Yang-Baxter equation of arbitrarily high dimension. The arbitrary function $\alpha(\lambda)$ could be introduced because of the homogeneity of the Yang-Baxter equation. It is still at our disposal. We may choose it in such a way that the matrix $\check{R}(\lambda)$ has a convenient normalization. Choosing, for instance, $\alpha(\lambda) = 1/(\lambda + 1)$ we find

$$\check{R}(\lambda)\check{R}(-\lambda) = \text{id} \tag{12.46}$$

which follows from $P^2 = \text{id}$. An R-matrix having this property is called unitary.

In order to make contact with our notation in appendix 3.B.5 we rescale the spectral parameter, $\lambda \to \lambda/ic$, where c is some coupling constant. It follows that the matrix

$$\check{R}(\lambda) = \frac{ic \cdot \mathrm{id} + \lambda P}{ic + \lambda} \tag{12.47}$$

is a solution of difference form of the Yang-Baxter equation (12.30). Upon introducing functions

$$b(\lambda) = \frac{ic}{\lambda + ic}, \quad c(\lambda) = \frac{\lambda}{\lambda + ic}, \tag{12.48}$$

the matrix $\check{R}(\lambda)$ takes the form

$$\check{R}(\lambda) = b(\lambda)\mathrm{id} + c(\lambda)P. \tag{12.49}$$

For the particular case $d = 2$, for instance, this is the 4×4 matrix

$$\check{R}(\lambda) = \begin{pmatrix} 1 & & & \\ & b(\lambda) & c(\lambda) & \\ & c(\lambda) & b(\lambda) & \\ & & & 1 \end{pmatrix} \tag{12.50}$$

and agrees with equation (3.B.58), if we put $c = 2u$.

Having obtained a set of solutions of the Yang-Baxter equation we may now apply the formalism of the previous section and construct the corresponding fundamental models. We should remark that the construction of the fundamental models in the previous section is not unique and may be altered by a number of rather trivial transformations, like shifts or rescaling of the spectral parameter, or multiplication of, for instance, the Hamiltonian by a number. We shall freely use this possibility.

Let us write down the R-matrix $R(\lambda, \mu) = P\check{R}(\lambda, \mu)$ with $\check{R}(\lambda, \mu)$ obtained from (12.49) in components. We have

$$\begin{aligned} R(\lambda, \mu) &= c(\lambda - \mu)\mathrm{id} + b(\lambda - \mu)P \\ &= c(\lambda - \mu)e_\beta^\alpha \otimes e_\alpha^\beta + b(\lambda - \mu)e_\alpha^\beta \otimes e_\beta^\alpha \\ &= \left(c(\lambda - \mu)\delta_\beta^\alpha \delta_\delta^\gamma + b(\lambda - \mu)\delta_\delta^\alpha \delta_\beta^\gamma\right)e_\alpha^\beta \otimes e_\gamma^\delta, \end{aligned} \tag{12.51}$$

and the components of the R-matrix can be read off from the right hand side of this equation. It follows (see (12.32)) that

$$L_{j\beta}^\alpha(\lambda, \mu) = c(\lambda - \mu)\delta_\beta^\alpha + b(\lambda - \mu)e_{j\beta}^\alpha. \tag{12.52}$$

Since $b(0) = 1$ and $c(0) = 0$, the R-matrix (12.51) is regular. The monodromy matrix of the corresponding fundamental model is given by equation (12.35) with $L_{j\beta}^\alpha(\lambda, \nu_j)$ taken from (12.52). As the Hamiltonian density in the homogeneous case we choose (compare

(12.40))

$$H_{j-1,j} = i\partial_\lambda \check{R}_{j-1,j}(\lambda, 0)\big|_{\lambda=0}$$
$$= i\big(b'(0) + c'(0)P_{j-1j}\big)$$
$$= (P_{j-1j} - 1)/c, \tag{12.53}$$

where a factor of 'i' has been supplied in order to make the expression hermitian.

The operator $H_{j-1,j}$ is acting on a tensor product of d-dimensional local quantum spaces $\mathbb{C}^d$. For $d = 2$ we may use the well-known formula $P_{j-1j} = \frac{1}{2}(1 + \sigma^\alpha_{j-1}\sigma^\alpha_j)$ and obtain the Hamiltonian

$$\hat{H} = \frac{1}{2c} \sum_{j=1}^{L} \big(\sigma^\alpha_{j-1}\sigma^u_j - 1\big) \tag{12.54}$$

of the isotropic (or **XXX**) Heisenberg chain. If the 'exchange coupling' c is positive we have the antiferromagnetic chain, if c is negative the ferromagnetic chain. For general d the Hamiltonian

$$\hat{H} = \frac{1}{c} \sum_{j=1}^{L} (P_{j-1j} - 1) \tag{12.55}$$

defines the so-called su(d)-**XXX** chain.

The transfer matrix of the su(d)-**XXX** chain can be diagonalized by using the Yang-Baxter algebra (12.18). The procedure is called the nested algebraic Bethe ansatz and is a generalization of the method presented in appendix 3.B.5. The nested algebraic Bethe ansatz for the su(d)-**XXX** chain was constructed by Kulish and Reshetikhin [273]. We shall not discuss it at this point, since it would lead us too far away from our actual subject. Let us only reconsider the arguments for the case $d = 2$ in a different, more general form. This is sufficient for our purpose to give an example of how the algebraic Bethe ansatz works and will moreover be useful for the algebraic Bethe ansatz for the Hubbard model in Section 12.6 and for the quantum transfer matrix approach to the thermodynamics of the **XXX** chain in Chapter 13.

12.1.7 Algebraic Bethe ansatz for the gl(2) generalized model

The models connected to the Yang-Baxter algebra are interesting for physicists mainly, because powerful methods are available for the solution of their spectral problem. One of these methods is the algebraic Bethe ansatz. It relies on the direct use of the quadratic commutation relations (12.18) defining the Yang-Baxter algebra. In order for the algebraic Bethe ansatz to work it must be possible to identify the elements of the monodromy matrix as 'particle' creation and annihilation operators. In particular, a pseudo vacuum state must exist which is annihilated by all the annihilation operators. In all known cases, where an algebraic Bethe ansatz has been successful so far, the elements of the monodromy matrix could be arranged in such a way, that the monodromy matrix acts as an upper triangular matrix on the pseudo vacuum, and the pseudo vacuum is an eigenstate of its diagonal elements.

The paradigmatic example of an algebraic Bethe ansatz is the algebraic Bethe ansatz for models with the R-matrix (12.50) of the spin-$\frac{1}{2}$ XXX chain [132]. The monodromy matrices of these models are 2×2. Thus, they have only a single element below the diagonal which must annihilate the pseudo vacuum for the algebraic Bethe ansatz to work. It turns out that the particular form of the vacuum eigenvalues of the diagonal elements is irrelevant for the construction of the algebraic Bethe ansatz. These vacuum eigenvalues, usually denoted $a(\lambda)$ and $d(\lambda)$, enter the algebraic Bethe ansatz solution as functional parameters. Thus, the algebraic Bethe ansatz can be simultaneously constructed for a whole class of models with the same R-matrix. One may even think of a model *defined* by the functional parameters and the triangular action of the monodromy matrix on the pseudo vacuum. The question whether or not such kind of 'generalized model' exists for arbitrary parameters $a(\lambda)$ and $d(\lambda)$ was first addressed in [267] and after some refinement was answered in the affirmative in [449,450]. The model has been termed the gl(2) generalized model, because of the gl(2) invariance of the R-matrix (12.50).

The algebraic Bethe ansatz solution of the spin-$\frac{1}{2}$ XXX chain presented in appendix 3.B.5 is a special case of the algebraic Bethe ansatz of the gl(2) generalized model. In principle, the arguments given in appendix 3.B.5 also apply to the gl(2) generalized model. Here, however, we shall take the opportunity to present an alternative derivation of the algebraic Bethe ansatz solution. We shall start by making the notion of the gl(2) generalized model more precise.

Let us consider the Yang-Baxter algebra (12.18) with R-matrix (12.50). The corresponding monodromy matrix is a 2×2 matrix, say,

$$T(\lambda) = \begin{pmatrix} A(\lambda) & B(\lambda) \\ C(\lambda) & D(\lambda) \end{pmatrix}. \tag{12.56}$$

The gl(2) generalized model is the set of all (linear) representations of the Yang-Baxter algebra (12.18) with R-matrix (12.50), for which a pseudo vacuum $|0\rangle$ exists, such that $T(\lambda)$ acts triangularly on $|0\rangle$,

$$A(\lambda)|0\rangle = a(\lambda)|0\rangle, \qquad D(\lambda)|0\rangle = d(\lambda)|0\rangle, \tag{12.57}$$

$$C(\lambda)|0\rangle = 0. \tag{12.58}$$

The complex valued functions $a(\lambda)$ and $d(\lambda)$ are called the parameters of the generalized model. These parameters characterize the representation in much the same way as the highest-weight vector characterizes a highest-weight representation of a Lie algebra.

Let us denote the representation space of a given representation of the generalized model by $\mathcal{H}$. It is clear from the quadratic commutation relations contained in (12.18) and from

(12.57), (12.58) that we may assume $\mathcal{H}$ to be spanned by all vectors of the form

$$|\mu_1, \ldots, \mu_M\rangle = B(\mu_1) \ldots B(\mu_M)|0\rangle . \tag{12.59}$$

This assumption is at least sensible if $\mathcal{H}$ is finite dimensional.

The family of transfer matrices we want to diagonalize is given by

$$t(\lambda) = \text{tr}(T(\lambda)) = A(\lambda) + D(\lambda) . \tag{12.60}$$

This is a commuting family of transfer matrices, $[t(\lambda), t(\mu)] = 0$, by construction (see Section 12.1.2). Therefore $t(\lambda)$ and $t(\mu)$ have a common system of eigenfunctions, which means that the eigenvectors of $t(\lambda)$ are independent of the spectral parameter λ. The task of the algebraic Bethe ansatz for the generalized model is to diagonalize $t(\lambda)$, i.e., to solve the eigenvalue problem

$$t(\lambda)|\Psi\rangle = \Lambda(\lambda)|\Psi\rangle . \tag{12.61}$$

This task can be accomplished by solely resorting to the Yang-Baxter algebra (12.18) and the properties (12.57), (12.58) of the pseudo vacuum state.

Out of the 16 quadratic relations contained in (12.18) we select the following three,

$$A(\lambda)B(\mu) = \frac{B(\mu)A(\lambda)}{c(\mu - \lambda)} - \frac{b(\mu - \lambda)}{c(\mu - \lambda)}B(\lambda)A(\mu) , \tag{12.62a}$$

$$D(\lambda)B(\mu) = \frac{B(\mu)D(\lambda)}{c(\lambda - \mu)} - \frac{b(\lambda - \mu)}{c(\lambda - \mu)}B(\lambda)D(\mu) , \tag{12.62b}$$

$$B(\lambda)B(\mu) = B(\mu)B(\lambda) . \tag{12.62c}$$

We are interested in the commutation relation of a product $B(\mu_1) \ldots B(\mu_M)$ with $A(\lambda)$ and $D(\lambda)$. These commutation relations can be obtained by iterated use of (12.62). We claim that

$$A(\lambda)\prod_{k=1}^{M} B(\mu_k) = \left[\prod_{k=1}^{M} B(\mu_k)\right]A(\lambda)\prod_{k=1}^{M} \frac{1}{c(\mu_k - \lambda)}$$
$$- \sum_{k=1}^{M}\left[B(\lambda)\prod_{\substack{l=1 \\ l\neq k}}^{M} B(\mu_l)\right]A(\mu_k)\frac{b(\mu_k - \lambda)}{c(\mu_k - \lambda)}\prod_{\substack{l=1 \\ l\neq k}}^{M} \frac{1}{c(\mu_l - \mu_k)} . \tag{12.63a}$$

$$D(\lambda)\prod_{k=1}^{M} B(\mu_k) = \left[\prod_{k=1}^{M} B(\mu_k)\right]D(\lambda)\prod_{k=1}^{M} \frac{1}{c(\lambda - \mu_k)}$$
$$- \sum_{k=1}^{M}\left[B(\lambda)\prod_{\substack{l=1 \\ l\neq k}}^{M} B(\mu_l)\right]D(\mu_k)\frac{b(\lambda - \mu_k)}{c(\lambda - \mu_k)}\prod_{\substack{l=1 \\ l\neq k}}^{M} \frac{1}{c(\mu_k - \mu_l)} . \tag{12.63b}$$

Proof. Equations (12.63) can be proven by induction over M. Let us concentrate on (12.63a). This equation is satisfied for $M = 1$, since it reduces to (12.62a). Let us assume that (12.63a) is true for some $M \in \mathbb{N}$. Multiplication of (12.63a) by $B(\mu_{M+1})$ from the right and use of (12.62) leads to

$$
A(\lambda) \prod_{k=1}^{M+1} B(\mu_k) = \left[\prod_{k=1}^{M+1} B(\mu_k) \right] A(\lambda) \prod_{k=1}^{M+1} \frac{1}{c(\mu_k - \lambda)}
$$

$$
- \sum_{k=1}^{M} \left[B(\lambda) \prod_{\substack{l=1 \\ l \neq k}}^{M+1} B(\mu_l) \right] A(\mu_k) \frac{b(\mu_k - \lambda)}{c(\mu_k - \lambda)} \prod_{\substack{l=1 \\ l \neq k}}^{M+1} \frac{1}{c(\mu_l - \mu_k)}
$$

$$
- \left[B(\lambda) \prod_{l=1}^{M} B(\mu_l) \right] A(\mu_{M+1}) \left\{ \frac{b(\mu_{M+1} - \lambda)}{c(\mu_{M+1} - \lambda)} \prod_{k=1}^{M} \frac{1}{c(\mu_k - \lambda)} \right.
$$

$$
\left. - \sum_{k=1}^{M} \frac{b(\mu_{M+1} - \mu_k)}{c(\mu_{M+1} - \mu_k)} \frac{b(\mu_k - \lambda)}{c(\mu_k - \lambda)} \prod_{\substack{l=1 \\ l \neq k}}^{M} \frac{1}{c(\mu_l - \mu_k)} \right\}. \tag{12.64}
$$

The right hand side of this equation reduces to the right-hand side of (12.63a) with M replaced by $M + 1$ thanks to the identity

$$
\frac{b(\mu_{M+1} - \lambda)}{c(\mu_{M+1} - \lambda)} \prod_{l=1}^{M} \frac{1}{c(\mu_l - \mu_{M+1})} = \frac{b(\mu_{M+1} - \lambda)}{c(\mu_{M+1} - \lambda)} \prod_{k=1}^{M} \frac{1}{c(\mu_k - \lambda)}
$$

$$
- \sum_{k=1}^{M} \frac{b(\mu_{M+1} - \mu_k)}{c(\mu_{M+1} - \mu_k)} \frac{b(\mu_k - \lambda)}{c(\mu_k - \lambda)} \prod_{\substack{l=1 \\ l \neq k}}^{M} \frac{1}{c(\mu_l - \mu_k)}, \tag{12.65}
$$

which, for mutually distinct μ_k, can be easily proven by means of Liouville's theorem.[3] It follows that (12.63a) is true for all $M \in \mathbb{N}$. The proof of (12.63b) is almost literally the same and is left as an exercise to the reader. $\qquad \square$

We may now add equations (12.63a) and (12.63b). This gives us a commutation relation between $t(\lambda)$ and the multiple product $B(\mu_1) \ldots B(\mu_M)$. Using the fact that $b(\lambda)/c(\lambda)$ is an odd function of λ we obtain

$$
t(\lambda) \prod_{k=1}^{M} B(\mu_k) = \left[\prod_{k=1}^{M} B(\mu_k) \right] \left\{ A(\lambda) \prod_{k=1}^{M} \frac{1}{c(\mu_k - \lambda)} + D(\lambda) \prod_{k=1}^{M} \frac{1}{c(\lambda - \mu_k)} \right\}
$$

$$
+ \sum_{k=1}^{M} \left[B(\lambda) \prod_{\substack{l=1 \\ l \neq k}}^{M} B(\mu_l) \right] \frac{b(\lambda - \mu_k)}{c(\lambda - \mu_k)} \prod_{\substack{l=1 \\ l \neq k}}^{M} \frac{1}{c(\mu_k - \mu_l)}
$$

$$
\times \left\{ A(\mu_k) \prod_{\substack{l=1 \\ l \neq k}}^{M} \frac{c(\mu_k - \mu_l)}{c(\mu_l - \mu_k)} - D(\mu_k) \right\}. \tag{12.66}
$$

By hypothesis, $|0\rangle$ is a joint eigenvector of $A(\lambda)$ and $D(\lambda)$ with eigenvalues $a(\lambda)$ and $d(\lambda)$ (see (12.57)). When we act with equation (12.66) on the pseudo vacuum we can therefore replace the operators $A(\lambda)$ and $D(\lambda)$ by their pseudo vacuum eigenvalues. Then the first

---

[3] A function which is bounded and analytic everywhere in the complex plane must be a constant.

curly bracket on the right-hand side of (12.66) turns into

$$\Lambda(\lambda) = a(\lambda) \prod_{k=1}^{M} \frac{1}{c(\mu_k - \lambda)} + d(\lambda) \prod_{k=1}^{M} \frac{1}{c(\lambda - \mu_k)} . \tag{12.67}$$

The second curly bracket vanishes, provided that

$$\frac{d(\mu_k)}{a(\mu_k)} = \prod_{\substack{l=1 \\ l \neq k}}^{M} \frac{c(\mu_k - \mu_l)}{c(\mu_l - \mu_k)} \tag{12.68}$$

for $k = 1, \ldots, M$. Thus, $|\Psi\rangle = B(\mu_1) \ldots B(\mu_M)|0\rangle$ is an eigenvector of the transfer matrix $t(\lambda)$ with eigenvalue $\Lambda(\lambda)$ if the equations (12.68) are satisfied. These equations are, of course, nothing but the Bethe ansatz equations.

As an application of our abstract Bethe ansatz solution let us again consider the spin-$\frac{1}{2}$ XXX chain. The monodromy matrix (12.35) is then an L-fold product of elementary L-matrices (12.52). It defines a representation of the Yang-Baxter algebra with R-matrix (12.50). An appropriate pseudo vacuum is the ferromagnetic state $|0\rangle = \binom{1}{0}^{\otimes L}$. The only thing we need to do if we want to apply our general result (12.67), (12.68) is to calculate the parameters $a(\lambda)$ and $d(\lambda)$. We note that

$$L_j(\lambda, \nu_j)|0\rangle = \begin{pmatrix} 1 & b(\lambda - \nu_j)e_j^1 \\ 0 & c(\lambda - \nu_j) \end{pmatrix} |0\rangle . \tag{12.69}$$

It follows that

$$a(\lambda) = 1 , \quad d(\lambda) = \prod_{j=1}^{L} c(\lambda - \nu_j) . \tag{12.70}$$

In order to compare with our previous result obtained in appendix 3.B.5 we have to adjust the notation. Replacing $\Lambda(\lambda)$ with $\tau(\lambda)$, λ with $\lambda - ic/2$, μ_k with $\lambda_k - ic/2$, ν_j with $\sin(k_j)$, c with $2u$, and L with N, we reproduce equations (3.B.83) and (3.B.84) from (12.67), (12.68), and (12.70).

In the homogeneous case, $\nu_j = 0$, $j = 1, \ldots, L$, equations (12.67), (12.68) and (12.70) provide the eigenvalues E of the spin-$\frac{1}{2}$ XXX Hamiltonian (12.54). Comparing (12.39), (12.40), and (12.53) we find (for $L \geq 2$)

$$E = i \frac{\Lambda'(0)}{\Lambda(0)} = i \sum_{k=1}^{M} \frac{c'(\mu_k)}{c(\mu_k)} = - \sum_{k=1}^{M} \frac{c}{\mu_k(\mu_k + ic)} \tag{12.71}$$

where the μ_k, $k = 1, \ldots, M$, are subject to the Bethe ansatz equations (12.68) with $a(\lambda)$ and $d(\lambda)$ according to (12.70).

Let us set $\mu_k = \lambda_k - ic/2$ in equations (12.68) and (12.71) and $c = 2$ in (12.54), (12.68) and (12.71). Then we arrive at the following classical result due to Bethe [60].

Lemma 9. *The antiferromagnetic, isotropic spin-$\frac{1}{2}$ Heisenberg Hamiltonian*

$$\hat{H} = \sum_{j=1}^{L} \left(S_{j-1}^\alpha S_j^\alpha - \tfrac{1}{4} \right) \tag{12.72}$$

has eigenvalues

$$E = -\sum_{k=1}^{M} \frac{2}{\lambda_k^2 + 1},\tag{12.73}$$

where the Bethe ansatz roots λ_k have to be calculated from the Bethe ansatz equations

$$\left(\frac{\lambda_k - i}{\lambda_k + i}\right)^L = \prod_{\substack{l=1 \\ l \neq k}}^{M} \frac{\lambda_k - \lambda_l - 2i}{\lambda_k - \lambda_l + 2i}, \quad k = 1, \dots, M.\tag{12.74}$$

A number of comments are in order here. The isotropic spin-$\frac{1}{2}$ Heisenberg chain is one of the best understood models that can be solved by algebraic Bethe ansatz. Lemma 9 is only the starting point for the study of its physical properties. Its thermodynamics, its excitation spectrum in the thermodynamic limit, its S-matrix and its finite size corrections which determine the long-distance behaviour of correlations, can be obtained in much the same way as discussed with the example of the Hubbard model in the first part of this book. In Chapter 13 where we treat the quantum transfer matrix approach to the thermodynamics of the Hubbard model we shall come back to the isotropic Heisenberg chain in order to introduce the method with a sufficiently simple example. For further reading we refer to the article [132] and to the books [157, 270, 439]. In [132] it is shown that, due to the su(2) symmetry, we should have imposed the restriction $2M \leq L$ in lemma 9.

The algebraic Bethe ansatz also provides a relatively simple expression for the eigenvectors and is therefore a powerful tool for the calculation of local properties of the spin-$\frac{1}{2}$ XXX chain. The norm of the eigenvectors was obtained in [268]. Expressions for expectation values of local operators in terms of the Bethe ansatz roots have been derived in [413] (see also [251]).

It is interesting to compare the two derivations of the algebraic Bethe ansatz in appendix 3.B.5 and in this section. The approach of appendix 3.B.5 is constructive but non-rigorous, since we have implicitly assumed the linear independence of vectors of the form $B(\mu_1) \dots B(\mu_M)|0\rangle$. The approach of this section is rigorous but non-constructive. Induction over M is mathematically elegant, but only helpful once the result is known. Thus, it is useful to be aware of both approaches.

The algebraic Bethe ansatz is not the only approach to solve the spectral problem of models connected with the Yang-Baxter algebra. It is only applicable if a pseudo vacuum exists. For models like the Toda chain [405], which has the same R-matrix as the spin-$\frac{1}{2}$ XXX chain, but has no pseudo vacuum, the algebraic Bethe ansatz fails. For these types of models another powerful technique, the 'method of separation of variables' was devised by E. K. Sklyanin [408].

12.1.8 Graphical representation of the Yang-Baxter equation

So far we have developed the formalism of the quantum inverse scattering method in a purely algebraic language. In this subsection we will slightly detour from this route and give a complementary view on the method. R-matrices can be represented by graphs. Relations between products of R-matrices then become relations between graphs or graphical rules for calculations. The use of these graphical rules sometimes simplifies complicated algebraic proofs. The origin of the graphical representation is the statistical mechanics of vertex models [45]. For many purposes it is mainly a matter of personal taste or habits whether one prefers to work with the algebraic or with the graphical representation. We shall use the graphical representation in our next chapter on the quantum transfer matrix approach to thermodynamics. But here is the appropriate place to introduce it.

Let us represent $R(\lambda, \mu)$ by two crossing arrows, one of which is thought of to 'carry' the spectral parameter λ, the other one the spectral parameter μ. Such kind of symbol may alternatively be understood as a (directed) vertex. Let us attach indices α, β, γ, δ to the four ends of the vertex. The indices at the tips of the arrows will be called outgoing indices and the indices at the tails incoming indices. The attachment of the indices to the vertex is unique if we agree to put them in clockwise order, starting with α as 'leftmost' outgoing index. This way we also establish a unique correspondence between the outgoing indices α and β and the spectral parameters λ and μ. Thus, we may uniquely identify an R-matrix element $R_{\gamma\delta}^{\alpha\beta}(\lambda, \mu)$ with a labeled vertex,

$$R_{\gamma\delta}^{\alpha\beta}(\lambda,\mu) \quad = \quad \alpha \longleftarrow \underset{\lambda}{\overset{\beta}{\underset{\mu}{\big|}}} \gamma \quad . \tag{12.75}$$

The key point now that makes the graphical representation suitable for calculations is to symbolize contraction of indices by connection of lines. Then we have, for instance,

$$R_{\alpha'\alpha''}^{\alpha\beta}(\lambda,\mu)R_{\beta''\gamma''}^{\alpha'\gamma}(\lambda,\nu) \quad = \quad \alpha \longleftarrow \underset{\alpha''}{\overset{\beta}{\underset{\mu}{\big|}}} \underset{\gamma''}{\overset{\gamma}{\underset{\nu}{\big|}}} \underset{\lambda}{\,} \beta'' \quad . \tag{12.76}$$

In this equation it is implied that the arrow from β'' to α carries the same spectral parameter λ throughout. It is consistent with our sum convention to define

$$\alpha \longleftarrow \underset{\lambda}{\,} \beta \quad = \quad \delta_\beta^\alpha \quad . \tag{12.77}$$

With the aid of our rules we may express various identities involving the R-matrix in graphical form. The Yang-Baxter equation is shown in figure 12.1. In its graphical form it is probably most easily memorized. Attaching labels to the graphs and using (12.75) we can easily reconstruct equation (12.31) from figure 12.1. The unitarity condition is shown

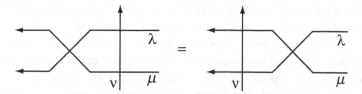

Fig. 12.1. Graphical representation of the Yang-Baxter equation (12.31).

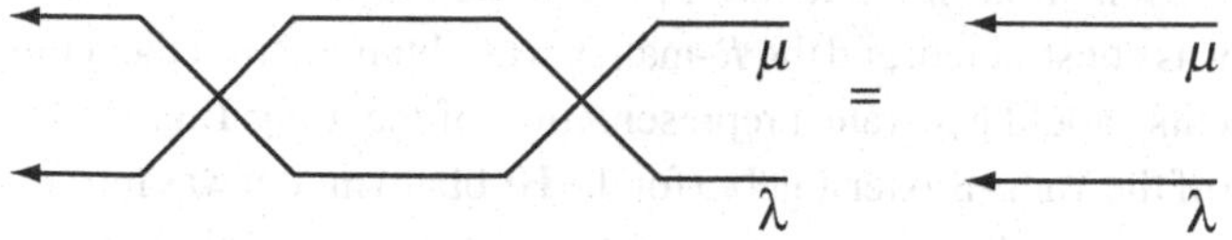

Fig. 12.2. Graphical representation of the unitarity condition (12.46).

in figure 12.2. Finally, a regular R-matrix with 'shift point' (λ_0, μ_0) satisfies

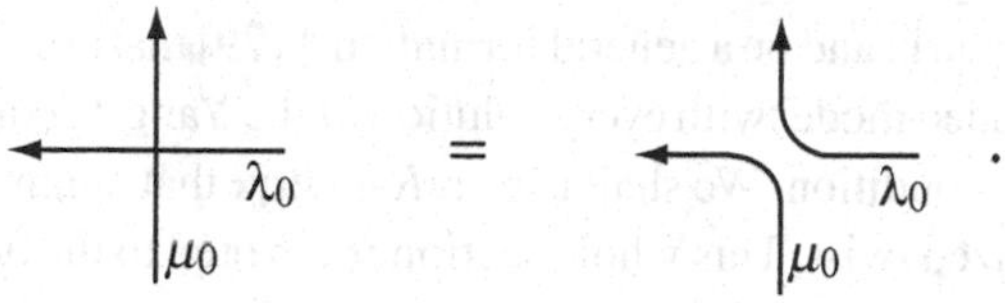

$$(12.78)$$

Examples of the usage and the usefulness of the graphical representation will be given in Chapter 13.

12.2 Shastry's R-matrix

Soon after the quantum inverse scattering method was created it became apparent that most of the models solvable by coordinate Bethe ansatz can be connected to the Yang-Baxter algebra. Yet it turned out to be difficult to find an R-matrix associated with the Hubbard model. The reason is that due to an argument of Reshetikhin [277] the Hubbard model cannot be constructed as a fundamental model with an R-matrix of difference form. Another source of obstruction is provided by the fact that the Hubbard model is formulated in terms of fermionic rather than spin degrees of freedom. This problem can be circumvented by considering the spin model related to the Hubbard model by a Jordan-Wigner transformation [227]. An R-matrix for this spin model was obtained by B. S. Shastry in 1986 [391, 392].

As we shall see, Shastry's R-matrix is rather peculiar as compared to the R-matrices of most of the other prominent exactly solvable models. In particular, it is not of difference form. Let us emphasize that the algebraic structure of the Hubbard model is less simple and less well understood than, for instance, the algebraic structure of the isotropic Heisenberg chain. From the point of view of its algebraic structure the Hubbard model is the subject of recent and still ongoing work. Therefore, the results presented in this part of the book are less complete than the results of the first parts. We expect further interesting developments in the future in which the reader is invited to participate.

Shastry's original work [391, 392] relies on rather brute force calculations involving extensive computer algebra. Later on, in an elegant paper [393], he introduced a novel algebraic means, the so-called 'decorated star triangle relation' that enabled to derive the R-matrix simultaneously with a peculiar representation of the Yang-Baxter algebra. This construction, however, did not include a proof showing that the R-matrix thus constructed would satisfy the Yang-Baxter equation. An algebraic proof based on Korepanov's representation [264] of the tetrahedral Zamolodchikov algebra was given by Shiroishi and Wadati [401].

Another derivation of Shastry's R-matrix was obtained in [349, 475], where a (quantum) Lax pair [270] was constructed, and the R-matrix was obtained from the requirement that the L-matrix elements should generate a representation of the Yang-Baxter algebra. In [349] a fermionic form of the Yang-Baxter algebra for the Hubbard model was introduced for the first time. The corresponding R-matrix turned out to be slightly modified as compared to Shastry's R-matrix. As we shall see below (see lemma 12) the two R-matrices are related through a 'generalized twist transformation' which leaves the Yang-Baxter equation invariant.

Our account of the algebraic structure of the Hubbard model does not rely on [349]. It is based on Shastry's R-matrix and on a general formalism [179] that associates a fermionic, so-called fundamental graded model with every solution of the Yang-Baxter equation satisfying a certain compatibility condition. We shall use an R-matrix that again connects to Shastry's R-matrix by a generalized twist. This whole section is devoted to the derivation of Shastry's R-matrix and to the discussion of similarity transformations which leave the Yang-Baxter equation invariant. Fundamental graded model are introduced in the following Section 12.3.

In the derivation of Shastry's R-matrix we partly follow the article [312], where a purely algebraic version of Shastry's arguments [393] has been presented. The derivation of [312] gives rise to a whole family of fundamental models with su(n)-spin degrees of freedom, the simplest of which ($n = 2$) is related to the Hubbard model through a Jordan-Wigner transformation. The proof that these R-matrices actually solve the Yang-Baxter equation is shown in appendix 12.A. It relies on the work of Korepanov [263] and is due to Shiroishi and Wadati [401]. A generalized twist transformation transforms Shastry's R-matrix into an R-matrix that generates the Hubbard model as a fundamental graded model. It is this R-matrix that we shall use as the starting point for the exploration of the algebraic properties of the Hubbard model in Section 12.4.

12.2.1 The XX models

We start our considerations by introducing the su(d)-XX models which will serve as building blocks for the construction of Shastry's R-matrix.

The most general anisotropic spin-$\frac{1}{2}$ model with nearest-neighbour interactions is defined by the Hamiltonian

$$H_{XYZ} = \sum_{j=1}^{L} \left(J_x \sigma_{j-1}^x \sigma_j^x + J_y \sigma_{j-1}^y \sigma_j^y + J_z \sigma_{j-1}^z \sigma_j^z \right), \tag{12.79}$$

where periodic boundary conditions, $\sigma_0^\alpha = \sigma_L^\alpha$, $\alpha = x, y, z$, are implied and J_x, J_y and J_z are three generally distinct 'exchange couplings'. H_{XYZ} is called the XYZ Hamiltonian. The XYZ Hamiltonian is connected to Baxter's famous eight-vertex model solution of the Yang-Baxter equation [43–45]. For $J_x = J_y$ the Hamiltonian describes the partially anisotropic so-called XXZ model. The completely isotropic case $J_x = J_y = J_z$ is called the XXX model and was considered along with its su(d) generalizations in the previous section. Here we shall discuss another interesting family of solutions of the Yang-Baxter equation connected to a special choice of exchange couplings in (12.79). For $J_x = J_y = J/2$ and $J_z = 0$ the XYZ Hamiltonian becomes

$$H_{XX} = \frac{J}{2} \sum_{j=1}^{L} \left(\sigma_{j-1}^x \sigma_j^x + \sigma_{j-1}^y \sigma_j^y \right) = J \sum_{j=1}^{L} \left(\sigma_{j-1}^+ \sigma_j^- + \sigma_{j-1}^- \sigma_j^+ \right). \tag{12.80}$$

We call this the XX Hamiltonian. Other common names are XY Hamiltonian or XX0 Hamiltonian, respectively. As in case of the XXX model there is a whole family of su(d) generalizations ($d = 3, 4, \ldots$) of the XX model. This family of models will be constructed below. The XX model relates to the tight binding model of free spinless fermions through a Jordan-Wigner transformation (see Section 12.3). As mentioned above it will serve as a building block for the construction of Shastry's R-matrix in the next subsection.

Let us consider the associative algebra with unity defined in terms of its generators A_i, $i = 1, \ldots, L$, by the relations

$$A_i A_{i+1} A_i = 0 = A_{i+1} A_i A_{i+1}, \tag{12.81a}$$

$$A_i A_j = A_j A_i \qquad \text{for } |i - j| > 1, \tag{12.81b}$$

$$A_i^3 = A_i, \tag{12.81c}$$

$$\{A_i^2, A_{i\pm1}\} = A_{i\pm1}, \tag{12.81d}$$

where we impose periodic boundary conditions on the generators. The curly brackets in the last line denote the anticommutator. The algebra defined by (12.81) was introduced by Maassarani [312] and was termed 'free fermion algebra'. Note that (12.81a) means that the A_i satisfy the braid relation. Using (12.81c) and (12.81d) we see that the same is also true for A_i^2, $A_i^2 A_{i+1}^2 A_i^2 = A_{i+1}^2 A_i^2 A_{i+1}^2$.

Using the free fermion algebra (12.81) and the addition theorems for trigonometric functions it is easily verified (see equations (12.A.11) and (12.A.12) in appendix 12.A) that

$$\begin{aligned}
&\left(\cos(\lambda) + A_2^2(1 - \cos(\lambda)) + A_2 \sin(\lambda) \right) \\
&\times \left(\cos(\lambda + \mu) + A_1^2(1 - \cos(\lambda + \mu)) + A_1 \sin(\lambda + \mu) \right) \\
&\times \left(\cos(\mu) + A_2^2(1 - \cos(\mu)) + A_2 \sin(\mu) \right) \\
&= \left(\cos(\mu) + A_1^2(1 - \cos(\mu)) + A_1 \sin(\mu) \right) \\
&\quad \times \left(\cos(\lambda + \mu) + A_2^2(1 - \cos(\lambda + \mu)) + A_2 \sin(\lambda + \mu) \right) \\
&\quad \times \left(\cos(\lambda) + A_1^2(1 - \cos(\lambda)) + A_1 \sin(\lambda) \right).
\end{aligned} \tag{12.82}$$

The latter equation constitutes an abstract solution of the Yang-Baxter equation. Let $Q \in \text{End}(\mathbb{C}^d \otimes \mathbb{C}^d)$, such that $A_i = Q_{ii+1} = Q_{\beta\delta}^{\alpha\gamma} e_{i\alpha}^{\beta} e_{i+1\gamma}^{\delta}$ is a representation of the free fermion algebra. Let $P^{(1)} = Q^2$ and $P^{(2)} = \text{id} - P^{(1)}$. Then (12.82) implies that

$$\check{R}(\lambda) = P^{(1)} + P^{(2)} \cos(\lambda) + Q \sin(\lambda) \tag{12.83}$$

is a solution of difference form of the Yang-Baxter equation (12.30). This solution is regular, since

$$\check{R}(0) = P^{(1)} + P^{(2)} = \text{id}. \tag{12.84}$$

From the defining relations of the free fermion algebra we conclude that $P^{(1)}$ and $P^{(2)}$ form a complete set of projection operators,

$$P^{(i)} P^{(j)} = \delta_{ij} P^{(j)}, \quad i, j = 1, 2. \tag{12.85}$$

The operator Q, being the square root of $P^{(1)}$, satisfies

$$P^{(1)} Q = Q P^{(1)} = Q, \tag{12.86a}$$

$$P^{(2)} Q = Q P^{(2)} = 0. \tag{12.86b}$$

Using (12.85) and (12.86) it follows that

$$\check{R}(\lambda)\check{R}(-\lambda) = \cos^2(\lambda) \cdot \text{id}. \tag{12.87}$$

Thus $\check{R}(\lambda)/\cos(\lambda)$ is unitary. According to equation (12.40) the Hamiltonian density of the fundamental homogeneous model generated by $\check{R}(\lambda)$ is

$$H_{j-1,j} = \partial_\lambda \check{R}_{j-1j}(\lambda)\Big|_{\lambda=0} = Q_{j-1j}. \tag{12.88}$$

The work that remains to be done is to construct representations of the free fermion algebra (12.81). We have claimed that we would construct the R-matrix of the XX model and its su(d) generalizations. Then, by comparing (12.80) and (12.88) the Hamiltonian density $\frac{1}{2}(\sigma_{j-1}^x \sigma_j^x + \sigma_{j-1}^y \sigma_j^y) = e_{j-1}^2{}_1 e_{j}^1{}_2 + e_{j-1}^1{}_2 e_{j}^2{}_1$ should generate the free fermion algebra (12.81). This is indeed the case. We have, for instance,

$$(e_1{}_1^2 e_2{}_2^1 + e_1{}_2^1 e_2{}_1^2)(e_2{}_1^2 e_3{}_2^1 + e_2{}_2^1 e_3{}_1^2)(e_1{}_1^2 e_2{}_2^1 + e_1{}_2^1 e_2{}_1^2)$$
$$= (e_1{}_1^2 e_2{}_2^2 e_3{}_2^1 + e_1{}_2^1 e_2{}_1^1 e_3{}_1^2)(e_1{}_1^2 e_2{}_2^1 + e_1{}_2^1 e_2{}_1^2) = 0, \tag{12.89}$$

and the first relation (12.81a) is verified. The remaining relations (12.81) are verified by similar calculations. Hence,

$$Q = e_1^2 \otimes e_2^1 + e_2^1 \otimes e_1^2 \tag{12.90}$$

generates a representation of the free fermion algebra (12.81).

This result straightforwardly generalizes. One verifies by direct calculation that $A_i = Q_{ii+1}$ with

$$Q = \sum_{\alpha=2}^{d} \left(x e_1^\alpha \otimes e_\alpha^1 + x^{-1} e_\alpha^1 \otimes e_1^\alpha \right) \tag{12.91}$$

and $d = 2, 3, 4, \ldots$ generates a representation of the algebra (12.81). Here $x \in \mathbb{C}$ is a free parameter. Thus, because of (12.83), we have obtained the following solution of difference form of the Yang-Baxter equation (12.30),

$$
\check{R}(\lambda) = \sum_{\alpha=2}^{d} \left(e_1^1 \otimes e_\alpha^\alpha + e_\alpha^\alpha \otimes e_1^1 \right) + \left[e_1^1 \otimes e_1^1 + \sum_{\alpha,\beta=2}^{d} e_\alpha^\alpha \otimes e_\beta^\beta \right] \cos(\lambda)
$$
$$
+ \sum_{\alpha=2}^{d} \left(x e_1^\alpha \otimes e_\alpha^1 + x^{-1} e_\alpha^1 \otimes e_1^\alpha \right) \sin(\lambda) . \tag{12.92}
$$

This solution was originally obtained in [313]. Our derivation follows [312]. The fundamental models generated by $R(\lambda) = P \check{R}(\lambda)$ are called the su(d)-XX models. Note that these models (for $d > 2$) are not just degenerate special cases of the higher rank XXZ models (see e.g. [47]).

Later, when we derive an explicit 16×16-matrix expression of Shastry's R-matrix for the Hubbard model, we shall need the explicit form of $\check{R}(\lambda)$ in the case of a two-dimensional auxiliary space, $d = 2$,

$$
\check{R}(\lambda) = \begin{pmatrix} \cos(\lambda) & & & \\ & 1 & x \sin(\lambda) & \\ & x^{-1} \sin(\lambda) & 1 & \\ & & & \cos(\lambda) \end{pmatrix} . \tag{12.93}
$$

In this case, the same R-matrix (with $x = 1$) can also be obtained as an appropriate limit of Baxter's R-matrix of the eight-vertex model [45].

In the following subsection we shall use the R-matrices (12.92) as building blocks for the construction of a family of R-matrices related to the Hubbard model. Note, however, that the XX models are also interesting on their own right. The su(3)-XX R-matrix, for instance, generates the t-0 model [179] (see appendix 2.A.6).

12.2.2 Conjugation matrix and decorated Yang-Baxter equation

Shastry's R-matrix is built by gluing together two copies of the XX models by means of a 'conjugation matrix' $C \in \text{End}(\mathbb{C}^d)$. Following [312] we characterize it by its properties:

$$
C^2 = I_d, \qquad \{C_i, Q_{12}\} = 0, \quad i = 1, 2, \tag{12.94a}
$$
$$
C_1 Q_{12} = Q_{12} C_2, \qquad Q_{12}^2 = \tfrac{1}{2}(I_{d^2} - C_1 C_2), \tag{12.94b}
$$

where $Q \in \text{End}(\mathbb{C}^d \otimes \mathbb{C}^d)$ is a representation of the free fermion algebra (12.81). Coming back to our peculiar representation (12.91) we easily verify that

$$
C = e_1^1 - \sum_{\alpha=2}^{d} e_\alpha^\alpha , \tag{12.95}
$$

is a corresponding conjugation matrix.

What is the use of this conjugation matrix? (12.94a) implies $[C_i, P_{12}^{(j)}] = 0$ for $i, j = 1, 2$. Hence (see (12.83)),

$$C_i \check{R}_{12}(\lambda) = \check{R}_{12}(-\lambda)C_i, \quad i = 1, 2. \tag{12.96}$$

The conjugation matrix thus changes the sign of the argument of $\check{R}(\lambda)$. It can be used to show that the XX-model R-matrices satisfy another functional equation besides the Yang-Baxter equation. For this purpose let us first rewrite the Yang-Baxter equation (12.21) in difference form,

$$R_{12}(\lambda - \mu)R_{13}(\lambda)R_{23}(\mu) = R_{23}(\mu)R_{13}(\lambda)R_{12}(\lambda - \mu). \tag{12.97}$$

Here we shifted λ and μ by ν. Then we multiply (12.97) by $C_1 C_2$ from the left and by C_3 from the right and reverse the sign of μ. We obtain the so-called decorated Yang-Baxter equation,

$$R_{12}(\lambda + \mu)C_1 R_{13}(\lambda)R_{23}(\mu) = R_{23}(\mu)R_{13}(\lambda)C_1 R_{12}(\lambda + \mu). \tag{12.98}$$

An equivalent form of this equation was introduced by Shastry in [393] and termed 'decorated star triangle relation'.

12.2.3 Constructing the R-matrix

Now we are prepared to introduce Shastry's R-matrix in the generalized form constructed by Maassarani [311, 312]. It is composed of two copies of the XX-model R-matrix (12.92). Its construction is based on equations (12.97), (12.98) and on the properties (12.94) of the conjugation matrix.

Let us first consider a system consisting of two non-interacting XX-models. We define

$$\check{r}_\uparrow(\lambda) = \check{R}_{\beta\delta}^{\alpha\gamma}(\lambda) e_\alpha^\beta \otimes I_d \otimes e_\gamma^\delta \otimes I_d, \tag{12.99a}$$

$$\check{r}_\downarrow(\lambda) = \check{R}_{\beta\delta}^{\alpha\gamma}(\lambda) I_d \otimes e_\alpha^\beta \otimes I_d \otimes e_\gamma^\delta, \tag{12.99b}$$

where $\check{R}(\lambda)$ is the XX-model R-matrix (12.92), and the associated transposition matrices

$$P_\uparrow = e_\alpha^\beta \otimes I_d \otimes e_\beta^\alpha \otimes I_d, \quad P_\downarrow = I_d \otimes e_\alpha^\beta \otimes I_d \otimes e_\beta^\alpha. \tag{12.100}$$

Then, by construction, both matrices, $r_\uparrow(\lambda) = P_\uparrow \check{r}_\uparrow(\lambda)$ and $r_\downarrow(\lambda) = P_\downarrow \check{r}_\downarrow(\lambda)$, satisfy the Yang-Baxter equation in its difference form (12.97). Since they commute the same is trivially true for their product

$$r(\lambda) = r_\uparrow(\lambda)r_\downarrow(\lambda), \tag{12.101}$$

it satisfies (12.97). $r(\lambda)$ is acting on the tensor product $(\mathbb{C}^d \otimes \mathbb{C}^d) \otimes (\mathbb{C}^d \otimes \mathbb{C}^d)$ of two auxiliary spaces $\mathbb{C}^d \otimes \mathbb{C}^d$. The vector space $\mathbb{C}^d \otimes \mathbb{C}^d$ is isomorphic to $\mathbb{C}^{d^2}$. Hence, we can interpret $r(\lambda)$ as a matrix in $\text{End}(\mathbb{C}^{d^2} \otimes \mathbb{C}^{d^2})$ if we employ the usual conventions $e_1^1 \otimes e_1^1 \rightarrow e_1^1, e_1^1 \otimes e_2^1 \rightarrow e_2^1$ etc. for the tensor product of two matrices. Then $r(\lambda)$ is a $d^4 \times d^4$-matrix.

The transposition operator P which interchanges the two tensor factors in $\mathbb{C}^{d^2} \otimes \mathbb{C}^{d^2}$ is given by the product $P = P_\uparrow P_\downarrow$.

Let us briefly consider the fundamental model associated with the R-matrix $r(\lambda)$. We have to follow the steps in Section (12.1.5). The L-matrix with components

$$\ell_{j\,\beta}^{\,\alpha}(\lambda) = r_{\beta\delta}^{\alpha\gamma}(\lambda) e_j{}_\gamma^\delta, \qquad (12.102)$$

where all Greek indices run from 1 to d^2 and summation over γ and δ is implied, generates a fundamental representation of the Yang-Baxter algebra,

$$\check{r}(\lambda - \mu)\big(\ell_j(\lambda) \otimes \ell_j(\mu)\big) = \big(\ell_j(\mu) \otimes \ell_j(\lambda)\big)\check{r}(\lambda - \mu). \qquad (12.103)$$

Due to the regularity of the XX-model R-matrix we have $\check{r}'(0) = \check{r}'_\uparrow(0) + \check{r}'_\downarrow(0)$. Therefore (see equation (12.40)) the Hamiltonian of the corresponding homogeneous model is equal to

$$H = \sum_{j=1}^{L} \big(Q_{j-1,j}^\uparrow + Q_{j-1,j}^\downarrow\big), \qquad (12.104)$$

with $Q^\sigma = \check{r}'_\sigma(0)$, $\sigma = \uparrow, \downarrow$. It is clear by construction that $Q^\uparrow$ and $Q^\downarrow$ commute. These operators correspond to two independent embeddings of Q into $\mathbb{C}^d \otimes \mathbb{C}^d \otimes \mathbb{C}^d \otimes \mathbb{C}^d$. The Hamiltonian (12.104) may thus be understood as a direct sum of two XX Hamiltonians.

In order to couple the two XX models we shall need two commuting copies, $C^\uparrow = C \otimes I_d$ and $C^\downarrow = I_d \otimes C$, of the conjugation matrix (12.95). The matrix $r(\lambda)$ then satisfies (12.98) with C replaced by $C^\uparrow C^\downarrow$. It follows that

$$\check{r}(\lambda + \mu)\big(C^\uparrow C^\downarrow \otimes I_{d^2}\big)\big(\ell_j(\lambda) \otimes \ell_j(\mu)\big)$$
$$= \big(\ell_j(\mu) \otimes \ell_j(\lambda)\big)\big(I_{d^2} \otimes C^\uparrow C^\downarrow\big)\check{r}(\lambda + \mu). \qquad (12.105)$$

It was Shastry's original observation [393] that (12.103) and (12.105) can be used to construct a representation of the Yang-Baxter algebra which is related to the Hubbard model together with its R-matrix. Later, Maassarani [312] noticed that Shastry's construction can be generalized to the family of XX models. Shastry proceeded as follows.

(i) He constructed a higher conserved operator I_2 of the Hubbard model, i.e., he constructed an operator which commutes with the Hubbard Hamiltonian.

(ii) By means of a Jordan-Wigner transformation he obtained 'spin chain operators', $H^{(s)}$, $I_2^{(s)}$, related to the Hubbard Hamiltonian and the conserved operator I_2.

(iii) He guessed an L-matrix, such that the series of conserved operators generated by the associated transfer matrix $t(\lambda)$ started as

$$\ln(U^{-1} t(\lambda)) = \lambda H^{(s)} + \lambda^2 I_2^{(s)} + \ldots$$

(iv) He constructed an R-matrix, such that his L-matrix appeared as the fundamental representation of the Yang-Baxter algebra defined by that R-matrix.

We shall not repeat all steps of Shastry's derivation here. What we are really interested in is the R-matrix. We shall need it in the construction of a fermionic version of the Yang-Baxter algebra that is directly connected to the Hubbard model in the next sections and in the construction of the thermodynamics of the Hubbard model within the so-called quantum transfer matrix approach in Chapter 13. Shastry did not present an analytic proof that his R-matrix satisfies the Yang-Baxter equation. Such proof was given later by Shiroishi and Wadati [401]. Still, point (iv) above of Shastry's argument is rather simple and beautiful. We therefore reproduce it below. The technically more involved proof of Shiroishi and Wadati is presented in appendix 12.A.

Let us define a matrix $G(h) \in \text{End}(\mathbb{C}^d \otimes \mathbb{C}^d)$ by

$$G(h) = \exp\left\{h\, C^{\uparrow}C^{\downarrow}/2\right\} = \text{ch}(h/2) + C^{\uparrow}C^{\downarrow}\, \text{sh}(h/2). \qquad (12.106)$$

$G(h)$ is acting on the auxiliary space of $r(\lambda)$. Next we define matrices $L_j(\lambda)$ and $L_j(\mu)$ as

$$L_j(\lambda) = G(h)\ell_j(\lambda)G(h), \quad L_j(\mu) = G(l)\ell_j(\mu)G(l). \qquad (12.107)$$

It follows from (12.103) and (12.105) that

$$\left(G(l) \otimes G(h)\right)\check{r}(\lambda - \mu)\left(G(-h) \otimes G(-l)\right)\left(L_j(\lambda) \otimes L_j(\mu)\right)$$
$$= \left(L_j(\mu) \otimes L_j(\lambda)\right)\left(G(-l) \otimes G(-h)\right)\check{r}(\lambda - \mu)\left(G(h) \otimes G(l)\right),$$

$$\left(G(l) \otimes G(h)\right)\check{r}(\lambda + \mu)\left(C^{\uparrow}C^{\downarrow} \otimes I_{d^2}\right)\left(G(-h) \otimes G(-l)\right)\left(L_j(\lambda) \otimes L_j(\mu)\right)$$
$$= \left(L_j(\mu) \otimes L_j(\lambda)\right)\left(G(-l) \otimes G(-h)\right)\left(I_{d^2} \otimes C^{\uparrow}C^{\downarrow}\right)\check{r}(\lambda + \mu)\left(G(h) \otimes G(l)\right).$$

Let us consider an arbitrary linear combination of these equations with coefficients α, β. Then

$$\left\{\left(G(l) \otimes G(h)\right)\left[\beta\check{r}(\lambda - \mu) + \alpha\check{r}(\lambda + \mu)\left(C^{\uparrow}C^{\downarrow} \otimes I_{d^2}\right)\right]\left(G(-h) \otimes G(-l)\right)\right\}$$
$$\times \left(L_j(\lambda) \otimes L_j(\mu)\right) = \left(L_j(\mu) \otimes L_j(\lambda)\right)$$
$$\times \left\{\left(G(-l) \otimes G(-h)\right)\left[\beta\check{r}(\lambda - \mu) + \alpha\left(I_{d^2} \otimes C^{\uparrow}C^{\downarrow}\right)\check{r}(\lambda + \mu)\right]\left(G(h) \otimes G(l)\right)\right\}.$$

This equation takes the form of the defining relations of a Yang-Baxter algebra if we can find coefficients α, β, such that the terms in curly bracket on the left- and right-hand side of this equation agree, i.e., we have to determine α and β such that

$$\left(G(2l) \otimes G(2h)\right)\left[\beta\check{r}(\lambda - \mu) + \alpha\check{r}(\lambda + \mu)\left(C^{\uparrow}C^{\downarrow} \otimes I_{d^2}\right)\right]$$
$$= \left[\beta\check{r}(\lambda - \mu) + \alpha\left(I_{d^2} \otimes C^{\uparrow}C^{\downarrow}\right)\check{r}(\lambda + \mu)\right]\left(G(2h) \otimes G(2l)\right).$$

Using (12.106) and (12.96) this can be seen to be equivalent to

$$\beta\left[\text{sh}(h)\text{ch}(l)\left(C^{\uparrow}C^{\downarrow} \otimes I_{d^2}\right) + \text{ch}(h)\text{sh}(l)\left(I_{d^2} \otimes C^{\uparrow}C^{\downarrow}\right), r(\lambda - \mu)\right]$$
$$- \alpha\left[\text{ch}(h)\text{ch}(l)\left(C^{\uparrow}C^{\downarrow} \otimes I_{d^2}\right) + \text{sh}(h)\text{sh}(l)\left(I_{d^2} \otimes C^{\uparrow}C^{\downarrow}\right), r(\lambda + \mu)\right] = 0.$$

By a straightforward calculation based on (12.94) this equation further reduces to

$$\{\beta \cos(\lambda - \mu)\text{sh}(h - l) - \alpha \cos(\lambda + \mu)\text{ch}(h - l)\}\left(I_{d^2} \otimes C^{\uparrow}C^{\downarrow} - C^{\uparrow}C^{\downarrow} \otimes I_{d^2}\right)$$
$$+ \{\beta \sin(\lambda - \mu)\text{sh}(h + l) - \alpha \sin(\lambda + \mu)\text{ch}(h + l)\}$$
$$\times \left[Q^{\uparrow}(C^{\uparrow} \otimes C^{\downarrow} - C^{\uparrow}C^{\downarrow} \otimes I_{d^2}) + Q^{\downarrow}(C^{\downarrow} \otimes C^{\uparrow} - C^{\uparrow}C^{\downarrow} \otimes I_{d^2})\right] = 0.$$

The latter equation is satisfied if and only if the terms in curly brackets both vanish. Thus

$$\frac{\alpha}{\beta} = \frac{\cos(\lambda - \mu)\text{sh}(h - l)}{\cos(\lambda + \mu)\text{ch}(h - l)} = \frac{\sin(\lambda - \mu)\text{sh}(h + l)}{\sin(\lambda + \mu)\text{ch}(h + l)}. \tag{12.108}$$

These are two algebraic equations. Using the addition theorems for trigonometric and hyperbolic functions the second equation (12.108) is shown to be equivalent to

$$\frac{\text{sh}(2h)}{\sin(2\lambda)} = \frac{\text{sh}(2l)}{\sin(2\mu)} = u. \tag{12.109}$$

Here the parameter u is a free parameter which will turn out to be the coupling constant of the Hubbard model. Equation (12.109) defines a function that expresses h in terms of λ and u and l in terms of μ and u. In order to define this function uniquely we choose the principal branch of the inverse sh-function. Then $h(\lambda = 0) = l(\mu = 0) = 0$. The first equation (12.108) fixes the ratio α/β as a function of λ, μ and u. Since only the ratio is fixed, β remains a free function. This, of course, is due to the homogeneity of the Yang-Baxter algebra.

Let us summarize. We have found that the L-matrix (12.107) is a representation of the Yang-Baxter algebra with R-matrix

$$\check{R}(\lambda, \mu) = \beta\left(G(l) \otimes G(h)\right)$$
$$\times \left[\check{r}(\lambda - \mu) + (\alpha/\beta)\check{r}(\lambda + \mu)(C^{\uparrow}C^{\downarrow} \otimes I_{d^2})\right]\left(G(-h) \otimes G(-l)\right) \tag{12.110}$$

provided that h and l are fixed as functions of λ, μ and u by (12.109), and α/β is given by (12.108). The matrix (12.110) is Shastry's R-matrix in its generalized form obtained by Maassarani. Our derivation was based on the properties (12.94) of the conjugation matrix and on the decorated Yang-Baxter equation (12.98).

The R-matrix (12.110) is rather special in that it is not of difference form. The second spectral parameter μ is a true independent parameter. For $\mu = 0$ the R-matrix $R(\lambda, \mu) = P\check{R}(\lambda, \mu)$ takes its simplest form

$$R(\lambda, 0) = \frac{\beta}{\text{ch}(h)}(G(h) \otimes I_{d^2})r(\lambda)(G(h) \otimes I_{d^2}). \tag{12.111}$$

According to our general formula (12.32) the associated L-matrix is

$$L_j(\lambda, 0) = \frac{\beta}{\text{ch}(h)}G(h)\ell_j(\lambda)G(h) = \frac{\beta}{\text{ch}(h)}L_j(\lambda) \tag{12.112}$$

with $L_j(\lambda)$ from (12.107). This is an interesting result. Fixing either β to ch(h) in (12.112) or changing the definition (12.107) we can interpret $L_j(\lambda)$ as a fundamental representation of the Yang-Baxter algebra. On the other hand, (12.112) also implies that $R(\lambda, \mu)$ satisfies the Yang-Baxter equation (12.21) for $\nu = 0$.

In fact, the R-matrix (12.110) satisfies the Yang-Baxter equation (12.21) for all λ, μ and ν. For the case $d = 2$, which corresponds to the Hubbard model, this can, for instance, be shown with the aid of some computer algebra [391]. For general d a deeper understanding of the algebraic properties of the XX models is necessary. The decorated Yang-Baxter equation (12.98) is part of a whole algebra of similar relations. It is this so-called tetrahedral Zamolodchikov algebra that will be used in appendix 12.A to prove that $R(\lambda, \mu)$ satisfies the Yang-Baxter equation (12.21) also for $\nu \neq 0$.

The ratio α/β, equation (12.108) vanishes for $\mu = \lambda$. It follows that

$$\check{R}(\lambda, \lambda) = \beta(\lambda, \lambda) \cdot \mathrm{id}. \tag{12.113}$$

Thus $R(\lambda, \mu)$ is regular for an appropriate choice of the function $\beta(\lambda, \mu)$ which is not fixed by (12.108). We further infer from the unitarity of the XX model R-matrix and from the fact that $[\check{r}(\lambda - \mu), \check{r}(\lambda + \mu)] = 0$ that

$$\check{R}(\lambda, \mu)\check{R}(\mu, \lambda) = \beta(\lambda, \mu)\beta(\mu, \lambda)$$
$$\times \cos^2(\lambda - \mu)\left(\cos^2(\lambda - \mu) - \cos^2(\lambda + \mu)\mathrm{th}^2(h - l)\right). \tag{12.114}$$

Thus $\check{R}(\lambda, \mu)$ becomes unitary for either of the two choices

$$\beta(\lambda, \mu) = \frac{\mathrm{ch}(h - l)}{\cos(\lambda - \mu)}\left[\cos(\lambda - \mu)\mathrm{ch}(h - l) \pm \cos(\lambda + \mu)\mathrm{sh}(h - l)\right]^{-1}. \tag{12.115}$$

Unless otherwise stated we assume in the following that $\beta(\lambda, \mu)$ is given by (12.115) with the plus sign. With this choice of $\beta(\lambda, \mu)$ we have $\beta(\lambda, \lambda) = 1$, and $R(\lambda, \mu)$ is also regular.

For the interpretation of the fundamental model generated by $\check{R}(\lambda, \mu)$ we still have to derive the Hamiltonian in the homogeneous case. We shall restrict ourselves to $\mu = 0$. Then

$$\partial_\lambda \check{R}(\lambda, 0)\Big|_{\lambda=0} = Q^\uparrow + Q^\downarrow + \frac{u}{2}(C^\uparrow C^\downarrow \otimes I_{d^2} + I_{d^2} \otimes C^\uparrow C^\downarrow) - u. \tag{12.116}$$

Here we used (12.111), (12.115) and (12.109). For its interpretation it is most appropriate to conceive the local quantum space as $\mathbb{C}^d \otimes \mathbb{C}^d$ rather than $\mathbb{C}^{d^2}$. With the definition

$$e_{j,\uparrow\alpha}^{\beta} = I_{d^2}^{\otimes(j-1)} \otimes (e_\alpha^\beta \otimes I_d) \otimes I_{d^2}^{\otimes(L-j)}, \tag{12.117a}$$

$$e_{j,\downarrow\alpha}^{\beta} = I_{d^2}^{\otimes(j-1)} \otimes (I_d \otimes e_\alpha^\beta) \otimes I_{d^2}^{\otimes(L-j)}, \tag{12.117b}$$

$\alpha, \beta = 1, \ldots, d$, the Hamiltonian reads

$$H = \sum_{j=1}^{L}\sum_{a=\uparrow,\downarrow}\sum_{\alpha=2}^{d}\left(e_{j-1,a}{}^{\alpha}_{1}e_{j,a}{}^{1}_{\alpha} + e_{j-1,a}{}^{1}_{\alpha}e_{j,a}{}^{\alpha}_{1}\right) + u\sum_{j=1}^{L}(C_j^\uparrow C_j^\downarrow - 1). \tag{12.118}$$

For $d = 2$ we may express the various matrices under the sums in terms of Pauli matrices.

We have $e_1^2 = \sigma^+$, $e_2^1 = \sigma^-$ and $C = \sigma^z$. Thus, H turns into

$$H = \sum_{j=1}^{L} \sum_{a=\uparrow,\downarrow} \left(\sigma_{j-1,a}^+ \sigma_{j,a}^- + \sigma_{j-1,a}^- \sigma_{j,a}^+ \right) + u \sum_{j=1}^{L} (\sigma_{j,\uparrow}^z \sigma_{j,\downarrow}^z - 1) \qquad (12.119)$$

which is Shastry's original spin Hamiltonian. The Hamiltonian is, up to a twist of the boundary conditions, equivalent to the Hubbard Hamiltonian by a Jordan-Wigner transformation [393]. The case of general μ for $d = 2$ and the inhomogeneous vertex model were discussed in [401].

12.2.4 Explicit form of the R-matrix (d = 2)

The R-matrix (12.110) with $d = 2$ is called Shastry's R-matrix. It is related to the Hubbard model. Later, when we construct the algebraic Bethe ansatz, and also in Chapter 15 where we algebraically construct the eigenstates on the infinite interval, we shall need its explicit form. Recall that the R-matrix (12.110) is a $d^4 \times d^4$-matrix. Thus, for $d = 2$ its dimension is 16×16. In order to obtain its explicit form we insert the R-matrix (12.93) of the su(2)-XX model, the conjugation matrix (12.95), and the expressions (12.108) for α/β and (12.115) for β into equation (12.110),

$$\check{R}(\lambda, \mu) = \frac{1}{\rho_4} \cdot$$

$$
\begin{pmatrix}
\rho_1 & & & & & & & & & & & & & & & \\
 & 1 & & & \rho_9 & & & & & & & & & & & \\
 & & 1 & & & & & \rho_9 & & & & & & & & \\
 & & & \rho_3 & & & \rho_6 & & & \rho_6 & & & -\rho_8 & & & \\
 & \rho_{10} & & & 1 & & & & & & & & & & & \\
 & & & & & \rho_4 & & & & & & & & & & \\
 & & & \rho_6 & & & \rho_5 & & & -\rho_7 & & & \rho_6 & & & \\
 & & & & & & & 1 & & & & & \rho_{10} & & & \\
 & & \rho_{10} & & & & & & 1 & & & & & & & \\
 & & & \rho_6 & & & -\rho_7 & & & \rho_5 & & & \rho_6 & & & \\
 & & & & & & & & & & \rho_4 & & & & & \\
 & & & & & & & & & & & 1 & & & \rho_{10} & \\
 & & & -\rho_8 & & & \rho_6 & & & \rho_6 & & & \rho_3 & & & \\
 & & & & & & & & & & & & & 1 & & \\
 & & & & & & & \rho_9 & & & & & & & 1 & \\
 & & & & & & & & & & & & \rho_9 & & & 1 \\
 & & & & & & & & & & & & & & & \rho_1
\end{pmatrix}
\qquad (12.120)
$$

We omitted the matrix elements that are equal to zero. The lines inside the matrix are guides to the eye.

The R-matrix (12.120) is constructed from two independent copies of the XX-model R-matrix. In principle, each of these copies carries one more free parameter (the parameter x

in (12.92)). Yet, the two 'twist parameters', say x and y, that come with the representations Q of the free fermion algebra may be set equal to unity, since the dependence of $\check{R}(\lambda, \mu)$ on x and y can be restored by a so-called twist transformation, to be discussed in the next subsection.

Following the terminology of statistical mechanics we shall call the functions $\rho_j = \rho_j(\lambda, \mu)$ the Boltzmann weights. Our notation is taken from [349]. The explicit expressions for the Boltzmann weights are

$$\rho_1(\lambda, \mu) = \cos(\lambda)\cos(\mu)e^{h-l} + \sin(\lambda)\sin(\mu)e^{l-h}, \tag{12.121a}$$

$$\rho_4(\lambda, \mu) = \cos(\lambda)\cos(\mu)e^{l-h} + \sin(\lambda)\sin(\mu)e^{h-l}, \tag{12.121b}$$

$$\rho_3(\lambda, \mu) = \frac{\cos(\lambda)\cos(\mu)e^{h-l} - \sin(\lambda)\sin(\mu)e^{l-h}}{\cos^2(\lambda) - \sin^2(\mu)}, \tag{12.121c}$$

$$\rho_5(\lambda, \mu) = \frac{\cos(\lambda)\cos(\mu)e^{l-h} - \sin(\lambda)\sin(\mu)e^{h-l}}{\cos^2(\lambda) - \sin^2(\mu)}, \tag{12.121d}$$

$$\rho_6(\lambda, \mu) = \frac{\mathrm{sh}(2(h-l))}{2u(\cos^2(\lambda) - \sin^2(\mu))}, \tag{12.121e}$$

$$\rho_7(\lambda, \mu) = \rho_4(\lambda, \mu) - \rho_5(\lambda, \mu), \tag{12.121f}$$

$$\rho_8(\lambda, \mu) = \rho_1(\lambda, \mu) - \rho_3(\lambda, \mu), \tag{12.121g}$$

$$\rho_9(\lambda, \mu) = \sin(\lambda)\cos(\mu)e^{l-h} - \cos(\lambda)\sin(\mu)e^{h-l}, \tag{12.121h}$$

$$\rho_{10}(\lambda, \mu) = \sin(\lambda)\cos(\mu)e^{h-l} - \cos(\lambda)\sin(\mu)e^{l-h}. \tag{12.121i}$$

The parameters λ, μ, h and l are subject to the constraints (12.109). We note the following relations [349] that are often useful in calculations,

$$\rho_1\rho_4 + \rho_9\rho_{10} = 1, \tag{12.122a}$$

$$\rho_1\rho_5 + \rho_3\rho_4 = 2, \tag{12.122b}$$

$$\rho_3\rho_5 - \rho_6^2 = 1. \tag{12.122c}$$

Yet another set of useful identities describes the behaviour of the Boltzmann weights under exchange of the arguments λ and μ. Let $\bar{\rho}_j(\lambda, \mu) = \rho_j(\mu, \lambda)$. Then

$$\bar{\rho}_1 = \rho_4, \quad \bar{\rho}_3 = \rho_5, \quad \bar{\rho}_7 = \rho_8, \tag{12.123a}$$

$$\bar{\rho}_6 = -\rho_6, \quad \bar{\rho}_9 = -\rho_9, \quad \bar{\rho}_{10} = -\rho_{10}. \tag{12.123b}$$

12.2.5 Invariances of the Yang-Baxter equation

The Yang-Baxter equation is invariant under various kinds of transformations. We exhibit three simple lemmata that will be needed for our further considerations. In what follows we shall always assume that $\check{R}(\lambda, \mu) \in \mathrm{End}(\mathbb{C}^n \otimes \mathbb{C}^n)$ is a solution of the Yang-Baxter equation (12.30).

Lemma 10. *Gauge transformations. Let $V(\lambda) \in \text{End}(\mathbb{C}^n)$ be an invertible $n \times n$-matrix. Then $\left(V(\mu) \otimes V(\lambda)\right) \check{R}(\lambda, \mu)\left(V^{-1}(\lambda) \otimes V^{-1}(\mu)\right)$ is a solution of the Yang-Baxter equation (12.30).*

Two solutions of the Yang-Baxter equation which are connected by a gauge transformation are sometimes called gauge equivalent. For example, Shastry's R-matrix (12.110) is gauge equivalent to

$$\check{R}(\lambda, \mu) = \beta \, \check{r}(\lambda - \mu) + \alpha \, \check{r}(\lambda + \mu)(C^{\uparrow} C^{\downarrow} \otimes I_{d^2}). \tag{12.124}$$

This fact is used in appendix 12.A, where we prove that Shastry's R-matrix satisfies the Yang-Baxter equation.

Lemma 11. *Twist transformations.[4] Suppose that $V \in \text{End}(\mathbb{C}^n)$ is invertible and satisfies the condition*

$$[\check{R}(\lambda, \mu), V \otimes V] = 0. \tag{12.125}$$

Then the matrices $\left(V \otimes I_n\right) \check{R}(\lambda, \mu)\left(V^{-1} \otimes I_n\right)$ and $\left(I_n \otimes V\right) \check{R}(\lambda, \mu)\left(I_n \otimes V^{-1}\right)$ satisfy the Yang-Baxter equation (12.30).

Twist transformations can be used to couple the electrons in the Hubbard model to an external electro-magnetic field (see Section (1.3)).

Lemma 11 is a special case of the following more general

Lemma 12. *Generalized twists [169, 280]. Let $V \in \text{End}(\mathbb{C}^n \otimes \mathbb{C}^n)$ be an invertible solution of the Yang-Baxter equation*

$$V_{12} V_{13} V_{23} = V_{23} V_{13} V_{12}, \tag{12.126}$$

such that

$$[\check{R}_{23}(\lambda, \mu), V_{13} V_{12}] = [\check{R}_{12}(\lambda, \mu), V_{13} V_{23}] = 0. \tag{12.127}$$

Then $V \check{R}(\lambda, \mu) V^{-1}$ is a solution of the Yang-Baxter equation (12.30).

Let us consider the important special case of V being a diagonal matrix. Then (12.126) is trivially satisfied, and the only remaining restrictions on V come from (12.127). It turns out to be useful to write V as

$$V = \begin{pmatrix} A & & & \\ & B & & \\ & & C & \\ & & & D \end{pmatrix}, \tag{12.128}$$

where

$$A = \text{diag}(a_1, a_2, a_3, a_4), \ldots, D = \text{diag}(d_1, d_2, d_3, d_4). \tag{12.129}$$

---

[4] This lemma was kindly communicated to us by S. Murakami.

We further introduce the matrices

$$\tilde{A} = \mathrm{diag}(a_1, b_1, c_1, d_1), \ldots, \tilde{D} = \mathrm{diag}(a_4, b_4, c_4, d_4). \tag{12.130}$$

A simple calculation shows that (12.127) is equivalent to

$$[\check{R}(\lambda, \mu), X \otimes X] = [\check{R}(\lambda, \mu), \tilde{X} \otimes \tilde{X}] = 0 \tag{12.131}$$

for $X = A, B, C, D$. On the other hand, Shastry's R-matrix (12.120) has the following property [178]: let $\alpha, \beta, \gamma, \delta \in \mathbb{C}$. Then

$$[\check{R}(\lambda, \mu), \mathrm{diag}(\alpha, \beta, \gamma, \delta) \otimes \mathrm{diag}(\alpha, \beta, \gamma, \delta)] = 0 \quad \Leftrightarrow \quad \alpha\delta = \beta\gamma. \tag{12.132}$$

This means that (12.127) is satisfied if and only if

$$\begin{aligned}
x_1 x_4 &= x_2 x_3 \quad \text{for } x = a, b, c, d, \\
a_j d_j &= b_j c_j \quad \text{for } j = 1, 2, 3, 4.
\end{aligned} \tag{12.133}$$

In general, our considerations allow us to introduce additional parameters into the R-matrix which may appear as additional coupling constants in the Hamiltonian [280]. Here we do not work out the consequences of general diagonal twists, but rather give two simple but important examples. First of all, the matrix

$$V = \mathrm{diag}(1, -1, -1, 1 | 1, -1, -1, 1 | 1, 1, 1, 1 | 1, 1, 1, 1) \tag{12.134}$$

obviously satisfies (12.133). Thus, we have shown that

$$V \check{R}(\lambda, \mu) V^{-1} = \frac{1}{\rho_4} \cdot$$

$$\left(\begin{array}{cccc|cccc|cccc|cccc}
\rho_1 & & & & & & & & & & & & & & & \\
 & 1 & & & -\rho_9 & & & & & & & & & & & \\
 & & 1 & & & & & & -\rho_9 & & & & & & & \\
 & & & \rho_3 & & & -\rho_6 & & & \rho_6 & & & -\rho_8 & & & \\
\hline
 & -\rho_{10} & & & 1 & & & & & & & & & & & \\
 & & & & & \rho_4 & & & & & & & & & & \\
 & & & -\rho_6 & & & \rho_5 & & & \rho_7 & & & -\rho_6 & & & \\
 & & & & & & & 1 & & & & & & \rho_{10} & & \\
\hline
 & & -\rho_{10} & & & & & & 1 & & & & & & & \\
 & & & \rho_6 & & & \rho_7 & & & \rho_5 & & & \rho_6 & & & \\
 & & & & & & & & & & \rho_4 & & & & & \\
 & & & & & & & & & & & 1 & & & \rho_{10} & \\
\hline
 & & & -\rho_8 & & & -\rho_6 & & & \rho_6 & & & \rho_3 & & & \\
 & & & & & & & \rho_9 & & & & & & 1 & & \\
 & & & & & & & & & & & \rho_9 & & & 1 & \\
 & & & & & & & & & & & & & & & \rho_1
\end{array}\right) \tag{12.135}$$

is a solution of the Yang-Baxter equation. We shall see below that the R-matrix (12.135) generates the Hubbard model within the formalism of the *graded* quantum inverse scattering method, which is our next subject.

Another example of a diagonal twist is generated by the matrix

$$V_{OWA} = \text{diag}(1, 1, -i, -i| - i, -i, 1, 1| - 1, -1, i, i|i, i, -1, -1). \qquad (12.136)$$

The corresponding R-matrix $V_{OWA} \check{R} V_{OWA}^{-1}$ was constructed by different means by Olmedilla, Wadati and Akutsu [349].

12.3 Graded quantum inverse scattering method

In Section 12.1 we encountered so-called fundamental models. They were constructed by attaching the operators $e_{j\alpha}^{\beta}$, equation (12.23), to a solution $R(\lambda, \mu)$ of the Yang-Baxter equation. For a lack of any better name, we henceforth call them 'local projection operators'. The local projection operators were used as local quantum space operators. They are characterized by two properties: the 'projection property' (12.24) and their commutativity (12.25).

We shall now introduce Fermi operators into the formalism. We shall proceed slightly indirectly. Clearly, models containing fermions cannot be fundamental models of the kind presented in Section 12.1, since all local quantum space operators constructed from local projection operators $e_{j\alpha}^{\beta}$, equation 12.23, commute for different site indices. In order to introduce fermions we therefore seek for a generalization of the local projection operators, such that they still satisfy (12.24), but commute or anticommute depending on the values of their 'matrix indices' α, β. Such kind of construction requires the introduction of graded vector spaces and graded algebras. Accordingly, the modified operators $e_{j\alpha}^{\beta}$ will be called 'graded local projection operators'.

With the aid of graded local projection operators we can construct fundamental *graded* models in much the same way as in Section 12.1, namely, by appropriately attaching graded local projection operators to a solution of the Yang-Baxter equation. This works for all R-matrices that satisfy a rather weak compatibility condition due to Kulish and Sklyanin [276].

Fermi operators come into the play by the observation that the graded local projection operators are matrix representations of fermionic projection operators. The transformation from fermionic projection operators to graded local projection operators can be interpreted as a generalization of the Jordan-Wigner transformation [227] to fermions with an arbitrary number of internal degrees of freedom ($su(n)$ fermions). In fact, for spinless fermions the Jordan-Wigner transformation is recovered. The material developed below is taken from the articles [177, 179].

12.3.1 Graded vector spaces

In this subsection we shall recall the basic concepts of graded vector spaces and graded associative algebras. In the context of the quantum inverse scattering method these concepts were first used by Kulish and Sklyanin [272, 276]. We shall further recall the notions of 'graded local projection operators' and graded transposition operators. Graded local projection operators were introduced in the article [179]. They enable the definition of fundamental graded representations of the Yang-Baxter algebra, which will be given in the following subsection.

Graded vector spaces are vector spaces equipped with a notion of odd and even, that allows us to treat fermions within the formalism of the quantum inverse scattering method. Let us start with a finite dimensional local space of states V, on which we impose an additional structure, the parity, from the outset. Let $V = V_0 \oplus V_1$, $\dim V_0 = m$, $\dim V_1 = n$. We shall call $\mathbf{v}_0 \in V_0$ even and $\mathbf{v}_1 \in V_1$ odd. The subspaces V_0 and V_1 are called the homogeneous components of V. The parity p is a function $V_i \to \mathbb{Z}_2$ defined on the homogeneous components of V,

$$p(\mathbf{v}_i) = i, \quad i = 0, 1, \quad \mathbf{v}_i \in V_i. \tag{12.137}$$

The vector space V endowed with this structure is called a graded vector space or super space. Let us fix a basis $\{e_1, \ldots, e_{m+n}\}$ of definite parity and let us define $p(\alpha) := p(e_\alpha)$.

In order to introduce Fermi operators into the formalism of the quantum inverse scattering method we have to construct an algebra of commuting and anticommuting *operators*. For this purpose the concept of parity must be extended to operators in End(V) and to tensor products of these operators. Let $e_\alpha^\beta \in \text{End}(V)$, $e_\alpha^\beta e_\gamma = \delta_\gamma^\beta e_\alpha$. The set $\{e_\alpha^\beta \in \text{End}(V) | \alpha, \beta = 1, \ldots m + n\}$ is a basis of End(V). Hence, the definition

$$p(e_\alpha^\beta) = p(\alpha) + p(\beta) \tag{12.138}$$

induces a grading on End(V) regarded as a vector space.

It is easy to see that an element $A = A_\beta^\alpha e_\alpha^\beta \in \text{End}(V)$ is homogeneous with parity $p(A)$, if and only if

$$(-1)^{p(\alpha)+p(\beta)} A_\beta^\alpha = (-1)^{p(A)} A_\beta^\alpha. \tag{12.139}$$

The latter equation implies for two homogeneous elements $A, B \in \text{End}(V)$ that their product AB is homogeneous with parity

$$p(AB) = p(A) + p(B). \tag{12.140}$$

In other words, multiplication of matrices in End(V) preserves homogeneity, and therefore, End(V) endowed with the grading (12.138) is a graded associative algebra [276].

Let us consider the L-fold tensorial power $\mathcal{H} = (\text{End}(V))^{\otimes L}$ of End(V). The definition (12.138) has a natural extension to $\mathcal{H}$, namely,

$$p(e_{\alpha_1}^{\beta_1} \otimes \cdots \otimes e_{\alpha_L}^{\beta_L}) = p(\alpha_1) + p(\beta_1) + \cdots + p(\alpha_L) + p(\beta_L). \tag{12.141}$$

From this formula it can be seen in a similar way as before, that homogeneous elements

$A = A^{\alpha_1...\alpha_L}_{\beta_1...\beta_L} e^{\beta_1}_{\alpha_1} \otimes \cdots \otimes e^{\beta_L}_{\alpha_L}$ of $\mathcal{H}$ with parity $p(A)$ are characterized by the equation

$$(-1)^{\sum_{j=1}^{L}(p(\alpha_j)+p(\beta_j))} A^{\alpha_1...\alpha_L}_{\beta_1...\beta_L} = (-1)^{p(A)} A^{\alpha_1...\alpha_L}_{\beta_1...\beta_L}, \qquad (12.142)$$

which generalizes (12.139). Again the product AB is homogeneous with parity $p(AB) = p(A) + p(B)$, if A and B are homogeneous. Thus the definition (12.141) induces the structure of a graded associative algebra on $\mathcal{H}$.

Let us define the super-bracket

$$[X, Y]_\pm = XY - (-1)^{p(X)p(Y)} YX \qquad (12.143)$$

for X, Y taken from the homogeneous components of $\text{End}(V)$, and let us extend it linearly to $\text{End}(V)$ in both of its arguments. Then, $\text{End}(V)$ endowed with the super-bracket becomes the Lie-super algebra $\text{gl}(m|n)$. Note that the above definition of a super-bracket makes sense in any graded algebra and is particularly valid in $\mathcal{H}$.

The following definition of 'graded local projection operators' [179] will be crucial for our definition of fundamental graded representations of the Yang-Baxter algebra in the next subsection. Define the matrices

$$e^{\beta}_{j\alpha} = (-1)^{(p(\alpha)+p(\beta))\sum_{k=j+1}^{L} p(\gamma_k)} I^{\otimes(j-1)}_{m+n} \otimes e^{\beta}_{\alpha} \otimes e^{\gamma_{j+1}}_{\gamma_{j+1}} \otimes \cdots \otimes e^{\gamma_L}_{\gamma_L}, \qquad (12.144)$$

where I_{m+n} is the $(m+n) \times (m+n)$ unit matrix, and summation over double *tensor indices* (i.e., over $\gamma_{j+1}, \ldots, \gamma_L$) is understood. We shall keep this sum convention throughout the remainder of this chapter. The index j on the left hand side of (12.144) will later refer to the jth site of a physical lattice model and is called the site index. A simple consequence of the definition (12.144) for $j \neq k$ are the commutation relations

$$e^{\beta}_{j\alpha} e^{\delta}_{k\gamma} = (-1)^{(p(\alpha)+p(\beta))(p(\gamma)+p(\delta))} e^{\delta}_{k\gamma} e^{\beta}_{j\alpha}. \qquad (12.145)$$

It further follows from equation (12.144) that $e^{\beta}_{j\alpha}$ is homogeneous with parity

$$p(e^{\beta}_{j\alpha}) = p(\alpha) + p(\beta). \qquad (12.146)$$

Hence, in agreement with intuition, equation (12.145) says that odd matrices with different site indices mutually anticommute, whereas even matrices commute with each other as well as with the odd matrices. For products of matrices $e^{\beta}_{j\alpha}$ which act on the same site (12.144) implies the projection property

$$e^{\beta}_{j\alpha} e^{\delta}_{j\gamma} = \delta^{\beta}_{\gamma} e^{\delta}_{j\alpha}. \qquad (12.147)$$

Equations (12.145) and (12.147) justify our terminology. The $e^{\beta}_{j\alpha}$ are graded analogues of local projection operators. We call them graded local projection operators or projection operators, for short. Using the super-bracket (12.143), equations (12.145) and (12.147) can be combined into

$$[e^{\beta}_{j\alpha}, e^{\delta}_{k\gamma}]_\pm = \delta_{jk} \left(\delta^{\beta}_{\gamma} e^{\delta}_{j\alpha} - (-1)^{(p(\alpha)+p(\beta))(p(\gamma)+p(\delta))} \delta^{\delta}_{\alpha} e^{\beta}_{j\gamma} \right). \qquad (12.148)$$

The right-hand side of the latter equation with $j = k$ gives the structure constants of the Lie super algebra $\text{gl}(m|n)$ with respect to the basis $\{e^{\beta}_{j\alpha}\}$.

Since any $m + n$-dimensional vector space over the complex numbers is isomorphic to $\mathbb{C}^{m+n}$, we may simply set $V = \mathbb{C}^{m+n}$. We may further assume that our homogeneous basis $\{e_\alpha \in \mathbb{C}^{m+n} | \alpha = 1, \ldots, m + n\}$ is canonical, i.e., we may represent the vector e_α by a column vector having the only non-zero entry $+1$ in row α. Our basic matrices e_α^β are then $(m + n) \times (m + n)$-matrices with a single non-zero entry $+1$ in row α and column β, and we recover (12.23) from (12.144) for $m = d$ and $n = 0$.

Remark. The meaning of (12.144) becomes more evident by considering a simple example. Let $m = n = 1$ and $p(1) = 0$, $p(2) = 1$. Then, using (12.148), we obtain

$$[e_{j1}^2, e_{k1}^2]_\pm = \{e_{j1}^2, e_{k1}^2\} = 0, \tag{12.149}$$

$$[e_{j2}^1, e_{k2}^1]_\pm = \{e_{j2}^1, e_{k2}^1\} = 0, \tag{12.150}$$

$$[e_{j1}^2, e_{k2}^1]_\pm = \{e_{j2}^1, e_{k1}^2\} = \delta_{jk}(e_{j1}^1 + e_{j2}^2) = \delta_{jk} \tag{12.151}$$

for $j, k = 1, \ldots, L$. The curly brackets in (12.149), (12.150) denote the anticommutator. The matrices e_{j1}^2 and e_{k2}^1 satisfy the canonical anticommutation relations for spinless Fermi operators. We can therefore identify $e_{j1}^2 \to c_j$ and $e_{k2}^1 \to c_k^\dagger$. Introducing Pauli matrices $\sigma^+ = e_1^2, \sigma^- = e_2^1$ and $\sigma^z = e_1^1 - e_2^2$ we obtain, by carrying out the summation, the following explicit matrix representation from our basic definition (12.144):

$$c_j = I_2^{\otimes(j-1)} \otimes \sigma^+ \otimes (\sigma^z)^{\otimes(L-j)}, \tag{12.152}$$

$$c_k^\dagger = I_2^{\otimes(k-1)} \otimes \sigma^- \otimes (\sigma^z)^{\otimes(L-k)}. \tag{12.153}$$

This is the well-known Jordan-Wigner transformation [227] expressing Fermi operators for spinless fermions in terms of Pauli matrices. We may thus interpret equation (12.144) as a generalization of the Jordan-Wigner transformation. In general, equation (12.144) provides matrix representations not of Fermi operators but, more generally, of fermionic projection operators. Representations of Fermi operators can be obtained be taking appropriate linear combinations of matrices $e_{j\alpha}^\beta$. This issue will be explained below.

The transposition operator plays an important role in the construction of local integrable lattice models. It enters the expression for the shift operator on homogeneous lattices. In the graded case the definition of the transposition operator requires the following modification of signs,

$$P_{jk} = (-1)^{p(\beta)} e_{j\alpha}^\beta e_{k\beta}^\alpha. \tag{12.154}$$

As indicated by its name, this operator induces the action of the symmetric group $\mathfrak{S}^L$ on the site indices of the matrices $e_{j\alpha}^\beta$. The properties of P_{jk} (for $j \neq k$) are the same as in the non-graded case (see (12.28)). They are easily derived from (12.145) and (12.147).

In the next subsection the graded associative algebra $\mathcal{H}$ will be considered as the space of states of a lattice model associated with a solution of the Yang-Baxter equation. We will define a monodromy matrix whose entries are elements of $\mathcal{H}$. The following definitions will prove to be useful.

Consider $(m + n) \times (m + n)$-matrices $A, B, C, \ldots$ with entries in $\mathcal{H}$, such that $p(A^\alpha_\beta) = p(B^\alpha_\beta) = p(C^\alpha_\beta) = \cdots = p(\alpha) + p(\beta)$ for $\alpha, \beta = 1, \ldots, m + n$. These matrices form an associative algebra, say $\mathcal{A}$, since $p(A^\alpha_\beta B^\beta_\gamma) = p(\alpha) + p(\gamma)$. For $A, B \in \mathcal{A}$ define a super tensor product (or graded tensor product)

$$(A \otimes_s B)^{\alpha\gamma}_{\beta\delta} = (-1)^{(p(\alpha)+p(\beta))p(\gamma)} A^\alpha_\beta B^\gamma_\delta. \tag{12.155}$$

This definition has an interesting consequence. Let $A, B, C, D \in \mathcal{A}$, such that

$$[B^\alpha_\beta, C^\gamma_\delta]_\pm = 0. \tag{12.156}$$

Then

$$(A \otimes_s B)(C \otimes_s D) = AC \otimes_s BD. \tag{12.157}$$

For later use we further define the super trace of a matrix $A \in \mathcal{A}$ by

$$\mathrm{str}(A) = (-1)^{p(\alpha)} A^\alpha_\alpha. \tag{12.158}$$

12.3.2 Fundamental graded models

In this subsection we shall introduce the notion of *fundamental graded representations* of the Yang-Baxter algebra [179]. For a given grading we shall associate a fundamental model with every solution of the Yang-Baxter equation (12.31) that satisfies the compatibility condition of Kulish and Sklyanin [276],

$$R^{\alpha\beta}_{\gamma\delta}(\lambda, \mu) = (-1)^{p(\alpha)+p(\beta)+p(\gamma)+p(\delta)} R^{\alpha\beta}_{\gamma\delta}(\lambda, \mu). \tag{12.159}$$

This compatibility condition simply means that certain matrix elements vanish. For R-matrices satisfying (12.159) we define a graded L-matrix at site j,

$$\mathcal{L}_j{}^\alpha_\beta(\lambda, \mu) = (-1)^{p(\alpha)p(\gamma)} R^{\alpha\gamma}_{\beta\delta}(\lambda, \mu) e_{j\,\gamma}^{\ \delta}. \tag{12.160}$$

Its properties are summarized in the following.

Lemma 13. *Properties of the graded L-matrix.*
 (i) *Homogeneity. The matrix elements of the graded L-matrix are homogeneous with parity*

$$p\big(\mathcal{L}_j{}^\alpha_\beta(\lambda, \mu)\big) = p(\alpha) + p(\beta). \tag{12.161}$$

(ii) *Commutativity. The entries of the graded L-matrix super commute for different site indices,*

$$[\mathcal{L}_j{}^\alpha_\beta(\lambda, \mu), \mathcal{L}_k{}^\gamma_\delta(\nu, \rho)]_\pm = 0, \tag{12.162}$$

for $j \neq k$.
(iii) *Bilinear relation. The entries of the graded L-matrix at the same lattice site satisfy the bilinear relation*

$$\check{R}(\lambda, \mu)\big(\mathcal{L}_j(\lambda, \nu) \otimes_s \mathcal{L}_j(\mu, \nu)\big) = \big(\mathcal{L}_j(\mu, \nu) \otimes_s \mathcal{L}_j(\lambda, \nu)\big)\check{R}(\lambda, \mu), \tag{12.163}$$

where, as in the non-graded case (12.17), the matrix $\check{R}(\lambda, \mu)$ is defined by $\check{R}^{\alpha\beta}_{\gamma\delta}(\lambda, \mu) = R^{\beta\alpha}_{\gamma\delta}(\lambda, \mu)$.

The lemma follows from the Yang-Baxter equation (12.31) and from equation (12.159). Equation (12.163) may be interpreted as defining a graded Yang-Baxter algebra with R-matrix $\check{R}(\lambda, \mu)$. We call $\mathcal{L}_j(\lambda, \mu)$ its fundamental graded representation.

Starting from (12.163) we can construct integrable lattice models as in the non-graded case (see Section 12.1.5). Let us briefly recall the construction with emphasis on the modifications that appear due to the grading. Define a monodromy matrix $T(\lambda)$ as an L-fold ordered product of fundamental L-matrices,

$$T(\lambda) = \mathcal{L}_L(\lambda, \nu_L) \ldots \mathcal{L}_1(\lambda, \nu_1) \tag{12.164}$$

Due to equation (12.140) the matrix elements of $T(\lambda)$ are homogeneous with parity $p(T^{\alpha}_{\beta}(\lambda)) = p(\alpha) + p(\beta)$. Repeated application of (12.163) and (12.157) shows that this monodromy matrix is a representation of the graded Yang-Baxter algebra,

$$\check{R}(\lambda, \mu)\big(T(\lambda) \otimes_s T(\mu)\big) = \big(T(\mu) \otimes_s T(\lambda)\big)\check{R}(\lambda, \mu). \tag{12.165}$$

It follows from (12.159) and (12.165) that

$$\big[\mathrm{str}(T(\lambda)), \mathrm{str}(T(\mu))\big] = 0, \tag{12.166}$$

which is in complete analogy with the non-graded case. We see that the transfer matrix is now given by $t(\lambda) = \mathrm{str}(T(\lambda))$.

The construction of a local lattice Hamiltonian as well is very similar to the non-graded case. Suppose that $R(\lambda, \mu)$ is a regular solution of the Yang-Baxter equation, $R^{\alpha\beta}_{\gamma\delta}(\lambda_0, \nu_0) = \delta^{\alpha}_{\delta}\delta^{\beta}_{\gamma}$ for some $\lambda_0, \nu_0 \in \mathbb{C}$. Then (12.160) implies that

$$\mathcal{L}_{j\,\beta}^{\ \alpha}(\lambda_0, \nu_0) = (-1)^{p(\alpha)p(\beta)}e_{j\,\beta}^{\ \alpha}, \tag{12.167}$$

and we can see (compare (12.154)) as in the non-graded case that we obtain the right-shift operator for $\nu_1 = \cdots = \nu_L = \nu_0$ and $\lambda = \lambda_0$,

$$t(\lambda_0) = P_{12}P_{23} \ldots P_{L-1L} = \hat{U}. \tag{12.168}$$

It follows that $\tau(\lambda) = \ln(\hat{U}^{-1}t(\lambda))$ generates a sequence of local operators [308],

$$\tau(\lambda) = (\lambda - \lambda_0)\hat{U}^{-1}t'(\lambda_0) + \mathcal{O}\big((\lambda - \lambda_0)^2\big), \tag{12.169}$$

which, as a consequence of (12.166), mutually commute. As expected, the local terms $H_{j-1,j}$ in the Hamiltonian

$$\hat{H} = \hat{U}^{-1}t'(\lambda_0) = \sum_{j=1}^{L} H_{j-1,j} \tag{12.170}$$

(where $H_{0,1} = H_{L,1}$) now come with a certain number of minus signs,

$$H_{j-1,j} = (-1)^{p(\gamma)(p(\alpha)+p(\gamma))} \partial_\lambda \check{R}^{\alpha\beta}_{\gamma\delta}(\lambda, \nu_0)\Big|_{\lambda=\lambda_0} e_{j-1\,\alpha}^{\ \ \gamma}e_{j\,\beta}^{\ \delta}. \tag{12.171}$$

We would like to emphasize the following points. (i) The R-matrix $\check{R}(\lambda, \mu)$ in equation (12.163) does *not* undergo a modification due to the grading. (ii) The only necessary compatibility condition which has to be satisfied in order to introduce a fundamental graded representation of the Yang-Baxter algebra associated with a solution of the Yang-Baxter equation is equation (12.159), which was introduced in [276].

The role of the matrix $\check{R}(\lambda, \mu)$ in the graded Yang-Baxter algebra (12.163) is to switch the order of the two auxiliary spaces. The definition of an operator that similarly switches the order of quantum spaces in a product of two L-matrices requires appropriate use of the grading. Such an operator was introduced for several important models in [99,466,467] and was called fermionic R-operator. A general definition of the fermionic R-operator associated with a solution $R(\lambda, \mu)$ of the Yang-Baxter equation (12.31) was obtained in [177]. For a given grading and a solution $R(\lambda, \mu)$ of the Yang-Baxter equation (12.31) that is compatible with this grading (see (12.159)) we define (following [177]) the fermionic R-operator

$$\mathcal{R}^f_{jk}(\lambda, \mu) = (-1)^{p(\gamma)+p(\alpha)(p(\beta)+p(\gamma))} R^{\alpha\beta}_{\gamma\delta}(\lambda, \mu) e_{j}{}^{\gamma}_{\alpha} e_{k}{}^{\delta}_{\beta}. \tag{12.172}$$

Let us summarize its properties in the following

Lemma 14. *Properties of the fermionic R-operator.*
  (i) Evenness. The fermionic R-operator is even,

$$p(\mathcal{R}^f_{jk}(\lambda, \mu)) = 0. \tag{12.173}$$

  (ii) Bilinear relation. The fermionic R-operator satisfies

$$\mathcal{R}^f_{jk}(v_j, v_k)\mathcal{L}_k(\lambda, v_k)\mathcal{L}_j(\lambda, v_j) = \mathcal{L}_j(\lambda, v_j)\mathcal{L}_k(\lambda, v_k)\mathcal{R}^f_{jk}(v_j, v_k). \tag{12.174}$$

  (iii) Yang-Baxter equation. The fermionic R-operator satisfies the following form of the Yang-Baxter equation,

$$\mathcal{R}^f_{12}(\lambda, \mu)\mathcal{R}^f_{13}(\lambda, \nu)\mathcal{R}^f_{23}(\mu, \nu) = \mathcal{R}^f_{23}(\mu, \nu)\mathcal{R}^f_{13}(\lambda, \nu)\mathcal{R}^f_{12}(\lambda, \mu). \tag{12.175}$$

  (iv) Regularity. If $R(\lambda, \mu)$ is regular, say $R^{\alpha\beta}_{\gamma\delta}(\lambda_0, \nu_0) = \delta^{\alpha}_{\delta}\delta^{\beta}_{\gamma}$, then

$$\mathcal{R}^f_{jk}(\lambda_0, \nu_0) = P_{jk}, \tag{12.176}$$

  where P_{jk} is the graded permutation operator (12.154).
  (v) Unitarity. If $R(\lambda, \mu)$ is unitary, i.e., if

$$R^{\alpha\beta}_{\gamma\delta}(\lambda, \mu)R^{\delta\gamma}_{\alpha'\beta'}(\mu, \lambda) = \delta^{\alpha}_{\beta'}\delta^{\beta}_{\alpha'}, \tag{12.177}$$

  then $\mathcal{R}^f_{jk}(\lambda, \mu)$ is unitary in the sense that

$$\mathcal{R}^f_{jk}(\lambda, \mu)\mathcal{R}^f_{kj}(\mu, \lambda) = \mathrm{id}. \tag{12.178}$$

The fermionic R-operator has at least three interesting applications. First of all, we shall need it below to prove an inversion theorem [177] that allows us to express the graded local projection operators $e_{j}{}^{\beta}_{\alpha}$ in terms of the elements of the monodromy matrix (12.164). Second, the form (12.175) of the Yang-Baxter equation is sometimes more convenient for the construction of generic fermionic models. We shall illustrate this point below with an example. Finally, it is possible to define a 'monodromy operator' [466, 467] and to use (12.175) as a starting point for an algebraic Bethe ansatz. For this purpose one has to introduce two fermionic auxiliary sites, say a and b, and has to define

$$T_a^f(\lambda) = \mathcal{R}_{a,L}^f(\lambda, \nu_L) \ldots \mathcal{R}_{a,1}^f(\lambda, \nu_1). \tag{12.179}$$

Then

$$\mathcal{R}_{ab}^f(\lambda, \mu) T_a^f(\lambda) T_b^f(\mu) = T_b^f(\mu) T_a^f(\lambda) \mathcal{R}_{ab}^f(\lambda, \mu). \tag{12.180}$$

The monodromy operator is connected to the monodromy matrix (12.164) by the formula

$$T_a^f(\lambda) = (-1)^{p(\beta)+p(\alpha)p(\beta)} e_{a\alpha}^{\beta} T_\beta^\alpha(\lambda). \tag{12.181}$$

12.3.3 Global symmetries from local symmetries

Symmetries of fundamental models solvable by the quantum inverse scattering method are most naturally understood in terms of the symmetries of the corresponding R-matrix. Some of the readers may be familiar with the 'non-graded' case: Suppose an R-matrix $R(\lambda, \mu) \in \mathrm{End}(\mathbb{C}^d \otimes \mathbb{C}^d)$ satisfies

$$[R(\lambda, \mu), x \otimes I_d + I_d \otimes x] = 0 \tag{12.182}$$

for some $x = x_\beta^\alpha e_\alpha^\beta \in gl(d)$ and for all $\lambda, \mu \in \mathbb{C}$. Then the transfer matrix of the corresponding inhomogeneous model commutes with

$$X = \sum_{j=1}^{L} x_\beta^\alpha e_{j\alpha}^\beta = \sum_{j=1}^{L} x_j. \tag{12.183}$$

We are familiar with such kind of symmetry from our proof of the spin-su(2) symmetry of the Hubbard model in Section (3.D).

A natural starting point for a generalization to the graded case is the invariance equation

$$[\mathcal{R}_{12}^f(\lambda, \mu), x_1 + x_2] = 0 \tag{12.184}$$

for the fermionic R-operator, which turns into (12.182) in the non-graded case. Here we assume that $\mathcal{R}_{12}^f(\lambda, \mu)$ is constructed from a given solution of the Yang-Baxter equation (12.31) compatible with some grading $p : \{1, \ldots, m+n\} \to \mathbb{Z}_2$. We shall further assume that $x = x_\beta^\alpha e_\alpha^\beta$ is homogeneous with parity $p(x)$ in $gl(m|n)$. Inserting the definition (12.172) of the fermionic R-operator into (12.184) and comparing the coefficients in front of $e_{1\alpha}^\beta e_{2\gamma}^\delta$ we obtain

$$\tilde{R}_{\beta'\delta}^{\alpha\gamma}(\lambda, \mu) x_\beta^{\beta'} - x_{\alpha'}^{\alpha} \tilde{R}_{\beta\delta}^{\alpha'\gamma}(\lambda, \mu)$$
$$= (-1)^{p(x)p(\alpha)} x_{\gamma'}^{\gamma} \tilde{R}_{\beta\delta}^{\alpha\gamma'}(\lambda, \mu) - (-1)^{p(x)p(\beta)} \tilde{R}_{\beta\delta'}^{\alpha\gamma}(\lambda, \mu) x_\delta^{\delta'}, \tag{12.185}$$

where $\tilde{R}_{\beta\delta}^{\alpha\gamma}(\lambda,\mu)=(-1)^{p(\alpha)p(\gamma)}R_{\beta\delta}^{\alpha\gamma}(\lambda,\mu)$. Note that we used equations (12.139) and (12.159) to rearrange the minus signs in (12.185). Equation (12.185) is the basic invariance equation for the R-matrix that replaces (12.182) in the graded case and obviously turns into (12.182) when the grading is trivial. The corresponding equation for the L-matrix is

$$\mathcal{L}_{j\beta'}^{\alpha}(\lambda,\mu)x_{\beta}^{\beta'}-x_{\alpha'}^{\alpha}\mathcal{L}_{j\beta}^{\alpha'}(\lambda,\mu)$$
$$=(-1)^{p(x)p(\alpha)}x_{j}\mathcal{L}_{j\beta}^{\alpha}(\lambda,\mu)-(-1)^{p(x)p(\beta)}\mathcal{L}_{j\beta}^{\alpha}(\lambda,\mu)x_{j} \qquad (12.186)$$

and is obtained from (12.185) by multiplication by $e_{j\gamma}^{\delta}$. A simple induction argument leads from (12.186) to the invariance equation

$$T_{\beta'}^{\alpha}(\lambda)x_{\beta}^{\beta'}-x_{\alpha'}^{\alpha}T_{\beta}^{\alpha'}(\lambda)=(-1)^{p(x)p(\alpha)}XT_{\beta}^{\alpha}(\lambda)-(-1)^{p(x)p(\beta)}T_{\beta}^{\alpha}(\lambda)X \qquad (12.187)$$

for the monodromy matrix. Again $X=x_{1}+\cdots+x_{L}$. Using the definition (12.143) of the super-bracket equation (12.187) can be equivalently written as

$$\left[T_{\beta}^{\alpha}(\lambda),X\right]_{\pm}=(-1)^{p(x)p(\beta)}\left(x_{\alpha'}^{\alpha}T_{\beta}^{\alpha'}(\lambda)-T_{\beta'}^{\alpha}(\lambda)x_{\beta}^{\beta'}\right). \qquad (12.188)$$

The latter equation is useful for studying the highest weight properties of states within the algebraic Bethe ansatz. Multiplying (12.188) by $(-1)^{p(\alpha)}$, setting $\beta=\alpha$, and summing over α we conclude that

$$\left[\mathrm{str}(T(\lambda)),X\right]=0. \qquad (12.189)$$

In many cases the symmetry of the R-matrix is evident by construction, e.g., when the R-matrix is an intertwiner of representations of quantum groups. Yet there are examples, as Shastry's R-matrix of the Hubbard model, where the symmetries are less obvious. Moreover, as can be seen from the above derivation, the symmetries of the transfer matrix are determined by the symmetries of $\tilde{R}$ rather than R and therefore depend on the choice of the grading.

It may be argued that, in the presence of a grading, the matrix $\tilde{R}$ is more fundamental than R, since $\tilde{R}$ determines the L-matrix (12.160), the symmetries of the model and (if it exists) the semi-classical limit [276]. Substituting $\tilde{R}$ into the Yang-Baxter equation (12.31), we obtain the so-called graded Yang-Baxter equation, which equivalently might have been taken as the starting point of our section on the graded Yang-Baxter algebra. Since it is the non-graded matrix $\check{R}$, however, which fixes the structure of the Yang-Baxter algebra, equation (12.163), we did not adopt this point of view.

12.3.4 Fermi operators

Let us now explain how the various graded objects introduced in the previous subsections can be expressed in terms of Fermi operators. The key observation is that, as far as the matrices $e_{j\alpha}^{\beta}$ are concerned, all calculations of the previous subsections solely rely on the commutation relations (12.145) and on the projection property (12.147). Fermionic projection operators satisfy the same equations. We may thus say that the matrices $e_{j\alpha}^{\beta}$

are matrix representations of fermionic projection operators. As we have seen, the matrices $e_{j\alpha}^{\beta}$ are suitable for formulating a graded version of the quantum inverse scattering method. For the physical interpretation of a Hamiltonian constructed from a given solution of the Yang-Baxter equation, however, it is useful to introduce Fermi operators into the formalism.

To begin with, let us consider spinless fermions on a ring of L lattice sites,

$$\{c_j, c_k\} = \{c_j^\dagger, c_k^\dagger\} = 0, \quad \{c_j, c_k^\dagger\} = \delta_{jk}, \quad j, k = 1, \dots, L. \tag{12.190}$$

It is easy to verify that the entries $(X_j)_\beta^\alpha$ of the matrix

$$X_j = \begin{pmatrix} 1 - n_j & c_j \\ c_j^\dagger & n_j \end{pmatrix} \tag{12.191}$$

are fermionic projection operators: define $X_{j\alpha}^{\beta} = (X_j)_\beta^\alpha$; then

$$X_{j\alpha}^{\beta} X_{j\gamma}^{\delta} = \delta_\gamma^\beta X_{j\alpha}^{\delta}. \tag{12.192}$$

The operators $X_{j\alpha}^{\beta}$ carry parity, induced by the anti-commutation rule (12.190) for the Fermi operators. For $j \neq k$ $X_{j\alpha}^{\beta}$ and $X_{k\gamma}^{\delta}$ anticommute, if both are built up of an odd number of Fermi operators, and otherwise commute. This fact can be expressed as follows. Let $p(1) = 0$, $p(2) = 1$ and $p(X_{j\alpha}^{\beta}) = p(\alpha) + p(\beta)$. Then $X_{j\alpha}^{\beta}$ is odd (contains an odd number of Fermi operators), if $p(X_{j\alpha}^{\beta}) = 1$, and even, if $p(X_{j\alpha}^{\beta}) = 0$. The commutation rules for the projectors $X_{j\alpha}^{\beta}$ are thus

$$X_{j\alpha}^{\beta} X_{k\gamma}^{\delta} = (-1)^{(p(\alpha)+p(\beta))(p(\gamma)+p(\delta))} X_{k\gamma}^{\delta} X_{j\alpha}^{\beta}. \tag{12.193}$$

Now (12.192) and (12.193) are of the same form as (12.147) and (12.145), respectively. Since the calculations in the previous section relied solely on (12.145) and (12.147), we may simply replace $e_{j\alpha}^{\beta} \to X_{j\alpha}^{\beta}$ in equations (12.160) and (12.171).

Fermionic representations compatible with an arbitrary grading can be constructed by considering several species of fermions and graded products of projection operators. We shall explain this for the case of two species first. This is the most interesting case in applications, since we may interpret the two species as up- and down-spin electrons. We have to attach a spin index to the Fermi operators, $c_j \to c_{j\sigma}, \sigma = \uparrow, \downarrow, \{c_{j\sigma}, c_{k\tau}^\dagger\} = \delta_{jk}\delta_{\sigma\tau}$. Accordingly, there are two species of projection operators, $X_{j\alpha}^{\beta} \to X_{j\alpha}^{\sigma\beta}$.

Let us define projection operators for electrons by the tensor products

$$X_{j\alpha\gamma}^{\beta\delta} = (-1)^{(p(\alpha)+p(\beta))p(\gamma)} X_{j\alpha}^{\downarrow\beta} X_{j\gamma}^{\uparrow\delta} = \left(X_j^\downarrow \otimes_s X_j^\uparrow \right)_{\beta\delta}^{\alpha\gamma}. \tag{12.194}$$

Then

$$X_{j\alpha\gamma}^{\beta\delta} X_{j\alpha'\gamma'}^{\beta'\delta'} = \delta_{\alpha'}^{\beta} \delta_{\gamma'}^{\delta} X_{j\alpha\gamma}^{\beta'\delta'}. \tag{12.195}$$

$X_{j\,\alpha\gamma}^{\ \beta\delta}$ inherits the parity from $X_j^{\downarrow\beta}{}_{\!\alpha}$ and $X_j^{\uparrow\delta}{}_{\!\gamma}$. The number of Fermi operators contained in $X_{j\,\alpha\gamma}^{\ \beta\delta}$ is the sum of the number of Fermi operators in $X_j^{\downarrow\beta}{}_{\!\alpha}$ and $X_j^{\uparrow\delta}{}_{\!\gamma}$. Hence $p(X_{j\,\alpha\gamma}^{\ \beta\delta}) = p(X_j^{\downarrow\beta}{}_{\!\alpha}) + p(X_j^{\uparrow\delta}{}_{\!\gamma}) = p(\alpha) + \cdots + p(\delta)$, and the analogue of (12.193) holds for $X_{j\,\alpha\gamma}^{\ \beta\delta}$ as well. Again we present all projection operators in form of a matrix $(X_j)_{\beta\delta}^{\alpha\gamma} = X_{j\,\alpha\gamma}^{\ \beta\delta}$,

$$X_j = X_j^{\downarrow} \otimes_s X_j^{\uparrow}$$

$$= \begin{pmatrix} (1-n_{j\downarrow})(1-n_{j\uparrow}) & (1-n_{j\downarrow})c_{j\uparrow} & c_{j\downarrow}(1-n_{j\uparrow}) & c_{j\downarrow}c_{j\uparrow} \\ (1-n_{j\downarrow})c_{j\uparrow}^\dagger & (1-n_{j\downarrow})n_{j\uparrow} & -c_{j\downarrow}c_{j\uparrow}^\dagger & -c_{j\downarrow}n_{j\uparrow} \\ c_{j\downarrow}^\dagger(1-n_{j\uparrow}) & c_{j\downarrow}^\dagger c_{j\uparrow} & n_{j\downarrow}(1-n_{j\uparrow}) & n_{j\downarrow}c_{j\uparrow} \\ -c_{j\downarrow}^\dagger c_{j\uparrow}^\dagger & -c_{j\downarrow}^\dagger n_{j\uparrow} & n_{j\downarrow}c_{j\uparrow}^\dagger & n_{j\downarrow}n_{j\uparrow} \end{pmatrix}. \qquad (12.196)$$

Here we used the standard ordering of matrix elements of tensor products, corresponding to a renumbering $(11) \to 1$, $(12) \to 2$, $(21) \to 3$, $(22) \to 4$. Within this convention $X_{j\,\alpha\gamma}^{\ \beta\delta}$ is replaced by $X_{j\,\alpha}^{\ \beta}$, $\alpha, \beta = 1, \ldots, 4$, which then satisfies (12.192) and (12.193) with grading $p(1) = p(4) = 0$, $p(2) = p(3) = 1$.

Note that Fermi operators can be recovered as linear combinations of projection operators. By inspection of equation (12.196) we obtain

$$c_{j\uparrow}^\dagger = X_{j2}^1 + X_{j4}^3, \qquad\qquad c_{j\uparrow} = X_{j1}^2 + X_{j3}^4, \qquad (12.197a)$$

$$c_{j\downarrow}^\dagger = X_{j3}^1 - X_{j4}^2, \qquad\qquad c_{j\downarrow} = X_{j1}^3 - X_{j2}^4. \qquad (12.197b)$$

The identification of the fermionic projection operators $X_{j\,\alpha}^{\ \beta}$ with the matrices $e_{j\,\alpha}^{\ \beta}$ in turns then gives us matrix representations of the Fermi operators:

$$c_{j,\uparrow}^\dagger = e_{j2}^1 + e_{j4}^3 = I_4^{\otimes(j-1)} \otimes (e_2^1 + e_4^3) \otimes e_{\gamma_{j+1}}^{\gamma_{j+1}}(-1)^{p(\gamma_{j+1})} \otimes \cdots \otimes e_{\gamma_L}^{\gamma_L}(-1)^{p(\gamma_L)}$$

$$= I_4^{\otimes(j-1)} \otimes (I_2 \otimes \sigma^-) \otimes (\sigma^z \otimes \sigma^z)^{\otimes(L-j)}$$

$$= I_2^{\otimes(2j-1)} \otimes \sigma^- \otimes (\sigma^z)^{\otimes 2(L-j)}, \qquad (12.198)$$

and similarly

$$c_{j,\downarrow}^\dagger = e_{j3}^1 - e_{j4}^2 = I_2^{\otimes 2(j-1)} \otimes \sigma^- \otimes (\sigma^z)^{\otimes(2L-2j+1)}, \qquad (12.199)$$

$$c_{j,\uparrow} = e_{j1}^2 + e_{j3}^4 = I_2^{\otimes(2j-1)} \otimes \sigma^+ \otimes (\sigma^z)^{\otimes 2(L-j)}, \qquad (12.200)$$

$$c_{j,\downarrow} = e_{j1}^3 - e_{j2}^4 = I_2^{\otimes 2(j-1)} \otimes \sigma^+ \otimes (\sigma^z)^{\otimes(2L-2j+1)}. \qquad (12.201)$$

This is, of course, a familiar generalization of the Jordan-Wigner transformation to Fermions with spin degrees of freedom.

So far we have considered the case of spinless fermions with two-dimensional local space of states and grading $m = n = 1$, and the case of electrons with four-dimensional space of states and grading $m = n = 2$. There are four different possibilities to realize (12.145) and (12.147) in the case of a three-dimensional local space of states, $m + n = 3$. They can be obtained by deleting the α's row and column of the matrix X_j in equation

(12.196), $\alpha = 1, 2, 3, 4$. (12.192) and (12.193) remain valid, since the operators $X_{j\alpha}^{\beta}$ are projectors.

It should be clear by now how to generalize the above considerations to an arbitrary number of species of fermions. In the case of N species we may define

$$X_{j\alpha_1...\alpha_N}^{\beta_1...\beta_N} = \left(X_j^N \otimes_s \cdots \otimes_s X_j^1\right)_{\beta_1...\beta_N}^{\alpha_1...\alpha_N} . \tag{12.202}$$

Then

$$X_{j\alpha_1...\alpha_N}^{\beta_1...\beta_N} X_{j\gamma_1...\gamma_N}^{\delta_1...\delta_N} = \delta_{\gamma_1}^{\beta_1} \cdots \delta_{\gamma_N}^{\beta_N} X_{j\alpha_1...\alpha_N}^{\delta_1...\delta_N} , \tag{12.203}$$

$$X_{j\alpha_1...\alpha_N}^{\beta_1...\beta_N} X_{k\gamma_1...\gamma_N}^{\delta_1...\delta_N} = (-1)^{\sum_{j,k=1}^{N}(p(\alpha_j)+p(\beta_j))(p(\gamma_k)+p(\delta_k))} X_{k\gamma_1...\gamma_N}^{\delta_1...\delta_N} X_{j\alpha_1...\alpha_N}^{\beta_1...\beta_N} \tag{12.204}$$

which can be shown by induction over the number of species. Here the grading is $m = n = 2^{N-1}$. The most general case is obtained deleting rows and columns from X_j, equation (12.202), in analogy with the example considered above.

Let us note that the operators $X_{j\alpha}^{\beta}$ for two species of fermions appear under the name Hubbard projection operators in the literature.

Remark. An alternative way [92, 349, 367] of introducing Fermi operators into the quantum inverse scattering method is by applying a (generalized) Jordan-Wigner transformation as formulated for creation and annihilation operators [92] to the non-graded L-matrix and then pulling out the non-local factors. This approach was of primary importance, for instance, for a fermionic formulation of the Yang-Baxter algebra of the Hubbard model [349] and led to the discovery of a SO(4)-invariant form of the monodromy matrix of the Hubbard model [178, 349, 399]. Still, we prefer the method presented above for conceptual clarity. The approach of [92, 349, 367] so far has not led to general formulae such as (12.160) or (12.171) and does not enable us to control the boundary conditions, which lead in turns to unwanted twists and to the appearance of numerous factors of 'i' in the equations.

12.3.5 Examples

A few simple examples will make us more familiar with the formalism developed so far. The simplest non-trivial example we can offer is the su(2)-XX model with R-matrix (12.93) and grading $p(1) = 0$, $p(2) = 1$. This choice of the grading is compatible with the R-matrix (see (12.159)). The corresponding local Hamiltonian is obtained from equation (12.171),

$$H_{j-1,j} = x^{-1} e_{j-1}{}_2^1 e_j{}_1^2 - x\, e_{j-1}{}_1^2 e_j{}_2^1 . \tag{12.205}$$

We may use the projection operators $X_{j\alpha}^{\beta}$ defined in the line below (12.191) to express $H_{j-1,j}$ in terms of Fermi operators. Specializing the free parameter x in (12.93) to $x = -1$, we obtain

$$H = -\sum_{j=1}^{L} \left(c_j^\dagger c_{j+1} + c_{j+1}^\dagger c_j \right), \qquad (12.206)$$

which is the Hamiltonian of the spinless tight-binding model. The corresponding L-matrix follows from the definition (12.160). It is of the form $\mathcal{L}_j(\lambda, \mu) = \mathcal{L}_j(\lambda - \mu)$ with

$$\mathcal{L}_j(\lambda) = \begin{pmatrix} \cos(\lambda)X_{j1}^1 - \sin(\lambda)X_{j2}^2 & X_{j2}^1 \\ X_{j1}^2 & -\sin(\lambda)X_{j1}^1 - \cos(\lambda)X_{j2}^2 \end{pmatrix} \qquad (12.207)$$

and satisfies the bilinear relation (12.163) with R-matrix (12.93) and $x = -1$. The algebraic Bethe ansatz for this model is left as an exercise to the reader.

The next example in this line is generated by the R-matrix of the su(3)-XX model, i.e., by the R-matrix (12.92) with $d = 3$. This R-matrix is compatible with the grading $p(1) = 0$, $p(2) = p(3) = 1$. Our general formula (12.171) yields the expression

$$H_{j-1,j} = -x\left(e_{j-1}{}_1^2 e_{j2}^1 + e_{j-1}{}_1^3 e_{j3}^1\right) + x^{-1}\left(e_{j-1}{}_2^1 e_{j1}^2 + e_{j-1}{}_3^1 e_{j1}^3\right) \qquad (12.208)$$

for the Hamiltonian density in terms of the graded local projection operators $e_{j\alpha}^\beta$. We are free to fermionize it in different ways. We may, for instance, use the set of fermionic projection operators obtained from the matrix X_j in (12.196) by deleting the fourth row and column. The elements $(X_j)_\beta^\alpha$, $\alpha, \beta = 1, 2, 3$, of the reduced matrix

$$X_j = \begin{pmatrix} (1 - n_{j\downarrow})(1 - n_{j\uparrow}) & (1 - n_{j\downarrow})c_{j\uparrow} & c_{j\downarrow}(1 - n_{j\uparrow}) \\ (1 - n_{j\downarrow})c_{j\uparrow}^\dagger & (1 - n_{j\downarrow})n_{j\uparrow} & -c_{j\downarrow}c_{j\uparrow}^\dagger \\ c_{j\downarrow}^\dagger(1 - n_{j\uparrow}) & c_{j\downarrow}^\dagger c_{j\uparrow} & n_{j\downarrow}(1 - n_{j\uparrow}) \end{pmatrix} \qquad (12.209)$$

thus obtained from a complete set of projection operators on the space of states locally spanned by the basis vectors $|0\rangle, c_{j\uparrow}^\dagger |0\rangle, c_{j\downarrow}^\dagger |0\rangle$. Double occupancy of lattice sites is forbidden on this space. Let $X_{j\alpha}^\beta = (X_j)_\beta^\alpha$, $\alpha, \beta = 1, 2, 3$. The operator

$$X_{j\alpha}^\alpha = 1 - n_{j\uparrow}n_{j\downarrow} \qquad (12.210)$$

projects the local space of lattice electrons onto the space from which double occupancy is excluded. The corresponding global projection operator is (recall appendix 2.A)

$$P_0 = \prod_{j=1}^{L}(1 - n_{j\uparrow}n_{j\downarrow}). \qquad (12.211)$$

Replacing the graded local projection operators $e_{j\alpha}^\beta$ by $X_{j\alpha}^\beta$ and setting $x = -1$ in (12.208) we obtain the local Hamiltonian

$$\begin{aligned} H_{j-1,j} = -\{ & (c_{j,\uparrow}^\dagger c_{j-1,\uparrow} + c_{j-1,\uparrow}^\dagger c_{j,\uparrow})(1 - n_{j-1,\downarrow})(1 - n_{j,\downarrow}) \\ & + (c_{j,\downarrow}^\dagger c_{j-1,\downarrow} + c_{j-1,\downarrow}^\dagger c_{j,\downarrow})(1 - n_{j-1,\uparrow})(1 - n_{j,\uparrow}) \}. \end{aligned} \qquad (12.212)$$

Because it acts on the space with no double occupancy, we may replace it by $H_{j-1,j}P_0$. This trick leads us to the compact form

$$H = -P_0 \sum_{j=1}^{L} \left(c_{j,a}^\dagger c_{j+1,a} + c_{j+1,a}^\dagger c_{j,a}\right) P_0 \qquad (12.213)$$

of the Hamiltonian. We identify it as the t-0 Hamiltonian encountered earlier in appendix 2.A as an effective strong coupling approximation to the Hubbard Hamiltonian below half-filling. The L-matrix (12.160) that generates the t-0 model as a fundamental graded model is of the form $\mathcal{L}_j(\lambda, \mu) = \mathcal{L}_j(\lambda - \mu)$ with

$$\mathcal{L}_j(\lambda) = \begin{pmatrix} \cos(\lambda)X_{j1}^1 - \sin(\lambda)(X_{j2}^2 + X_{j3}^3) & X_{j2}^1 & X_{j3}^1 \\ X_{j1}^2 & -\sin(\lambda)X_{j1}^1 - \cos(\lambda)X_{j2}^2 & -\cos(\lambda)X_{j3}^2 \\ X_{j1}^3 & -\cos(\lambda)X_{j2}^3 & -\sin(\lambda)X_{j1}^1 - \cos(\lambda)X_{j3}^3 \end{pmatrix}$$

$$(12.214)$$

This L-matrix can be used as a starting point for an algebraic Bethe ansatz solution of the t-0 model (see [179]).

Our last example in this section is a family of R-matrices closely related to the XXX models considered in Section 12.1.6. We observed that the graded transposition operator (12.154) shares the properties (12.28) (except for (12.28b)) with its non-graded counterpart. In particular, setting $A_i = P_{i,i+1}$, we see that the graded transposition operators form a representation of the group algebra of the symmetric group (12.44). Let us define

$$\mathcal{R}_{jk}^f(\lambda, \mu) = c(\lambda - \mu) + b(\lambda - \mu)P_{jk} \qquad (12.215)$$

with $b(\lambda)$ and $c(\lambda)$ taken from (12.48). Then, by (12.45), $\mathcal{R}_{jk}^f(\lambda, \mu)$ satisfies the Yang-Baxter equation in the form (12.175). Since $c(0) = 0$ and $b(0) = 1$, we conclude that $\mathcal{R}_{jk}^f(\lambda, \lambda) = P_{jk}$ and thus is regular in the sense of (12.176). It is moreover unitary in the sense of (12.178). Hence $\mathcal{R}_{jk}^f(\lambda, \mu)$ may be interpreted as the fermionic R-operator of a model to be identified. Rewriting it as

$$\begin{aligned} \mathcal{R}_{jk}^f(\lambda, \mu) &= c(\lambda - \mu) + b(\lambda - \mu)(-1)^{p(\beta)}e_{j\alpha}{}^\beta e_{k\beta}{}^\alpha \\ &= \left[c(\lambda - \mu)\delta_\gamma^\alpha \delta_\delta^\beta + b(\lambda - \mu)(-1)^{p(\gamma)}\delta_\delta^\alpha \delta_\gamma^\beta\right]e_{j\alpha}{}^\gamma e_{k\beta}{}^\delta \\ &= (-1)^{p(\gamma)+p(\alpha)(p(\beta)+p(\gamma))} \\ &\quad \times \left[c(\lambda - \mu)(-1)^{p(\alpha)p(\beta)}\delta_\gamma^\alpha \delta_\delta^\beta + b(\lambda - \mu)\delta_\delta^\alpha \delta_\gamma^\beta\right]e_{j\alpha}{}^\gamma e_{k\beta}{}^\delta \end{aligned} \qquad (12.216)$$

we can compare it with the definition (12.172) of the fermionic R-operator. It follows that the term in square brackets,

$$R_{\gamma\delta}^{\alpha\beta}(\lambda, \mu) = c(\lambda - \mu)(-1)^{p(\alpha)p(\beta)}\delta_\gamma^\alpha \delta_\delta^\beta + b(\lambda - \mu)\delta_\delta^\alpha \delta_\gamma^\beta, \qquad (12.217)$$

satisfies the Yang-Baxter equation (12.31).

Our example shows that the fermionic R-operator is a useful tool for the construction of solutions of the Yang-Baxter equation. The peculiar feature of the family (12.217) of solutions of the Yang-Baxter equation is its dependence on a grading. In fact, by construction, (12.217) satisfies the Yang-Baxter equation for arbitrary grading $p : \{1, \ldots m + n\} \to \mathbb{Z}_2$. It is therefore natural to call such kind of solutions graded R-matrices. The family (12.217) of rational graded R-matrices is in Kulish and Sklyanin's early list [276] of known solutions of the Yang-Baxter equation.

We may now associate a fundamental graded model with the solution (12.217) of the Yang-Baxter equation. As we have learned the first step then is to choose a grading which is compatible with the R-matrix. Note that this grading need not be identical with the grading which enters the definition of the R-matrix. Let us choose an arbitrary grading $q : \{1, \ldots, m + n\} \to \mathbb{Z}_2$. Then, because of the Kronecker deltas in (12.217),

$$R_{\gamma\delta}^{\alpha\beta}(\lambda, \mu) = (-1)^{q(\alpha)+q(\beta)+q(\gamma)+q(\delta)} R_{\gamma\delta}^{\alpha\beta}(\lambda, \mu), \tag{12.218}$$

i.e., the compatibility condition (12.159) is satisfied for arbitrary p and q.

Let us elaborate on the case $p = q$. Since $c(0) = 0$ and $b(0) = 1$, the R-matrix (12.217) is regular. Furthermore,

$$-ic\partial_\lambda \check{R}_{\gamma\delta}^{\alpha\beta}(\lambda, 0)\Big|_{\lambda=0} = \delta_\gamma^\alpha \delta_\delta^\beta - (-1)^{p(\alpha)p(\beta)}\delta_\delta^\alpha \delta_\gamma^\beta. \tag{12.219}$$

Thus, using $p = q$ in (12.171) we obtain the Hamiltonian

$$H = -\sum_{j=1}^{L}(P_{j,j+1} - 1) \tag{12.220}$$

of the homogeneous, fundamental graded model associated with the R-matrix (12.217). Here $P_{j,j+1}$ is the graded transposition operator (12.154).

The family (12.220) of Hamiltonians based on graded permutations includes a number of models that are interesting for applications in physics. In the 'non-graded' case $n = 0$ we recover the XXX models considered in Section 12.1.6. The case $m = n = 1$ provides another realization of the tight-binding model of spinless Fermions. For $m = n = 2$ we obtain the so-called EKS model [126, 127]. Here we shall have a closer look only at the case $m = 1, n = 2$ which leads to the supersymmetric t-J model.

In order to see this we employ the same fermionization scheme as above where we considered the su(3)-XX model, i.e., we replace the graded local projection operators $e_{j\alpha}^{\beta}$ with the fermionic projection operators $X_{j\alpha}^{\beta}$ taken from (12.209). The summation in (12.154) is again over three values, $\alpha, \beta = 1, 2, 3$, and the grading is $p(1) = 0$, $p(2) = p(3) = 1$. An elegant way of taking into account the simplifications arising from the restriction to the Hilbert space of electrons with no double occupancy is by considering $P_{jk}P_0$ with P_0 from (12.211) instead of P_{jk}. Since $n_{j,\uparrow}n_{j,\downarrow}P_0 = 0$, we obtain

$$(P_{jk} - 1)P_0 = P_0(c_{ja}^\dagger c_{ka} + c_{ka}^\dagger c_{ja})P_0 - 2(S_j^\alpha S_k^\alpha - \tfrac{1}{4}n_j n_k)P_0 - (n_j + n_k)P_0. \tag{12.221}$$

Here we have denoted the density of electrons by $n_j = n_{j,\uparrow} + n_{j,\downarrow}$ and the spin densities by $S_j^\alpha = \frac{1}{2}\sigma_{ab}^\alpha c_{ja}^\dagger c_{jb}$. Comparing (12.221) with (2.A.39) we see that we have indeed generated the local Hamiltonian of the supersymmetric t-J model. Let us finally write down the corresponding L-matrix, which follows from equation (12.160):

$$\mathcal{L}_j(\lambda, \mu) = c(\lambda - \mu) + b(\lambda - \mu) \begin{pmatrix} X_{j1}^1 & X_{j2}^1 & X_{j3}^1 \\ X_{j1}^2 & -X_{j2}^2 & -X_{j3}^2 \\ X_{j1}^3 & -X_{j2}^3 & -X_{j3}^3 \end{pmatrix}. \tag{12.222}$$

This L-matrix can be used to perform an algebraic Bethe ansatz for the supersymmetric t-J model which was originally obtained in [119, 138, 272] (for the algebraic Bethe ansatz of the corresponding generalized model see [170, 181]).

At this point it becomes also evident, why the model is called supersymmetric t-J model. Being a sum over transposition operators (see (12.220)) the Hamiltonian obviously commutes with the generators

$$E_\beta^\alpha = \sum_{j=1}^L X_{j\beta}^\alpha, \tag{12.223}$$

$\alpha, \beta = 1, 2, 3$, of the Lie superalgebra gl(1|2). This symmetry is a consequence of the gl(1|2) invariance of the transfer matrix, which follows from the fact that the fermionic R-operator (12.215) commutes with $X_{j\beta}^\alpha + X_{k\beta}^\alpha$ (see Section 12.3.3).

12.4 The Hubbard model as a fundamental graded model

We are now going to show that the Hubbard model can be interpreted as a fundamental graded model. This is an important fact. It means that the general theory developed in the previous section can be applied. Furthermore, in the next section it will allow us to re-express the local Fermi operators in terms of the elements of the monodromy matrix. We would like to emphasize that the choice of the appropriate R-matrix turns out to be crucial for our considerations. This appropriate R-matrix is not Shastry's original R-matrix (12.120) but our modified version (12.135).

Due to various subtleties the history of the fermionic formulation of the Yang-Baxter algebra connected with the Hubbard model is slightly involved. The pioneering work was done by Wadati, Olmedilla and Akutsu [347, 349, 475]. Their work basically contains everything what is necessary to discuss the symmetries and to perform an algebraic Bethe ansatz [320, 371]. The main difference to our account here is of conceptional nature. Similar to Shastry in his work on the spin model, Wadati, Olmedilla and Akutsu did not derive the L-matrix from the R-matrix but obtained L-matrix and R-matrix simultaneously. This way it was impossible to control the possible twists compatible with the Yang-Baxter equation (see lemma 12 and below). Consequentially, the supertrace of the monodromy matrix evaluated at spectral parameter equal to zero did not give the shift operator but the shift operator multiplied by a global gauge transformation [178]. Another disadvantage of the original approach of Wadati *et al.* is that it does allows one to treat neither the inhomogeneous model nor the homogeneous model with the second spectral parameter μ different from zero [401]. The latter case is of interest for models with R-matrix not of difference form, since it may happen that the second spectral parameter generates an additional non-trivial interaction in the Hamiltonian. The disadvantages of the approach of Wadati, Olmedilla and Akutsu were overcome in the article [466], where the Hubbard model was treated within the fermionic R-operator approach. The monodromy operator in [466] is a product of fermionic R-operators which satisfy the Yang-Baxter equation (12.175) and are regular in the sense of (12.176). Thus, the monodromy operator of [466] is free of twists and covers the homogeneous as well as the inhomogeneous case. Its 'supertrace' reduces to the shift operator in the homogeneous case when all spectral parameters (λ and the μ's) assume the same value. It was therefore possible in [466] to derive a local fermionic Hamiltonian which contains μ as an additional coupling constant. This Hamiltonian is related to the corresponding generalization of Shastry's spin Hamiltonian [401] by a Jordan-Wigner transformation.

Our account below is in many respects equivalent to the account of [466]. We stay more on the conservative side, however, in that we express operations on the auxiliary space by usual matrices rather than Fermi operators. Both approaches are connected by the general arguments of Section 12.3, since our L-matrix as well as the fermionic R-operator of [466] relate to the R-matrix (12.135) by the general formulae (12.160) and (12.172) (see appendix 12.E).

In Section 12.4.1 we apply the general formula (12.171) to the R-matrix, equation (12.135), and obtain the Hubbard Hamiltonian if we set the second spectral parameter μ of the R-matrix equal to zero. Keeping this second spectral parameter, on the other hand, we obtain a generalized Hamiltonian [466] containing the Hubbard Hamiltonian as a limiting case. Section 12.4.2 is devoted to a discussion of the symmetries of the Hubbard model on the level of the transfer matrix [178, 399, 464, 466]. It turns out that the η-pairing symmetry is peculiar, since for our form (12.135) of the R-matrix it is not a symmetry in the sense of Section 12.3.3. The η-pairing symmetry can be obtained from the rotational symmetry by means of a Shiba transformation [178]. In our discussion of the η-pairing symmetry in Section 12.4.2, however, we start from a local argument given in [466].

12.4.1 Hamiltonian and L-matrix

In the following we denote the R-matrix on the right hand side of (12.135) by $\check{R}(\lambda, \mu)$. Let us choose the grading $p(1) = p(4) = 0$, $p(2) = p(3) = 1$. Our R-matrix is compatible with this choice.

We want to show that equation (12.171) generates the Hamiltonian density of the Hubbard model after appropriate fermionization. First of all, differentiating all the Boltzmann weights

(12.121) with respect to λ, we obtain

$$
\partial_\lambda \check{R}(\lambda,0)\Big|_{\lambda=0} = u\,I_{16} +
$$

$$
\left(
\begin{array}{cccc|cccc|cccc|cccc}
u & & & & & & & & & & & & & & & \\
 & 0 & & & & -1 & & & & & & & & & & \\
 & & 0 & & & & & -1 & & & & & & & & \\
 & & & u & & & -1 & & & 1 & & & & & & \\
\hline
 & -1 & & & 0 & & & & & & & & & & & \\
 & & & & & -u & & & & & & & & & & \\
 & & -1 & & & & -u & & & & & -1 & & & & \\
 & & & & & & & 0 & & & & & & 1 & & \\
\hline
 & & & -1 & & & & & 0 & & & & & & & \\
 & & & 1 & & & & & & -u & & & & 1 & & \\
 & & & & & & & & & & -u & & & & & \\
 & & & & & & & & & & & 0 & & & 1 & \\
\hline
 & & & & & & -1 & & & & & 1 & u & & & \\
 & & & & & & & 1 & & & & & & 0 & & \\
 & & & & & & & & & -1 & & 1 & & & 0 & \\
 & & & & & & & & & & & & & & & u
\end{array}
\right)
\qquad (12.224)
$$

Here u is the parameter entering the relation (12.109) between λ and h, and the omitted matrix elements are all zero. Using (12.171) we conclude that

$$
H_{j-1,j} = \left(e_{j-1}{}_1^2 + e_{j-1}{}_3^4\right)\left(e_{j}{}_2^1 + e_{j}{}_4^3\right) + \left(e_{j}{}_1^2 + e_{j}{}_3^4\right)\left(e_{j-1}{}_2^1 + e_{j-1}{}_4^3\right)
$$

$$
+ \left(e_{j-1}{}_1^3 - e_{j-1}{}_2^4\right)\left(e_{j}{}_3^1 - e_{j}{}_4^2\right) + \left(e_{j}{}_1^3 - e_{j}{}_2^4\right)\left(e_{j-1}{}_3^1 - e_{j-1}{}_4^2\right) \qquad (12.225)
$$

$$
+ u\left[\left(e_{j-1}{}_1^1 + e_{j-1}{}_4^4\right)\left(e_{j}{}_1^1 + e_{j}{}_4^4\right) - \left(e_{j}{}_2^2 + e_{j}{}_3^3\right)\left(e_{j-1}{}_2^2 + e_{j-1}{}_3^3\right) + 1\right].
$$

This expression can be fermionized by replacing the graded local projection operators $e_{j}{}_\alpha^\beta$ with the Hubbard projection operators listed in (12.196). Recall that according to our convention we have to use the transpose of (12.196). Then

$$
H_{j-1,j} = -c_{j-1,a}^\dagger c_{j,a} - c_{j,a}^\dagger c_{j-1,a}
$$

$$
+ \frac{u}{2}\left[(1 - 2n_{j-1,\uparrow})(1 - 2n_{j-1,\downarrow}) + (1 - 2n_{j,\uparrow})(1 - 2n_{j,\downarrow})\right] + u, \quad (12.226)
$$

which is the Hamiltonian density of the Hubbard model with coupling constant u.

Note that we put the second spectral parameter μ equal to zero in our above derivation of the Hamiltonian density from (12.135). The corresponding homogeneous monodromy matrix is given by (12.164) with $v_j = 0$, and L-matrix (see equation (12.160))

$$\mathcal{L}_j(\lambda, 0) = \frac{e^h}{\cos^2(\lambda)}$$

$$\times \begin{pmatrix} e^h f_{j,\downarrow} f_{j,\uparrow} & -f_{j,\downarrow} c_{j,\uparrow}^\dagger & -f_{j,\uparrow} c_{j,\downarrow}^\dagger & -e^h c_{j,\downarrow}^\dagger c_{j,\uparrow}^\dagger \\ -f_{j,\downarrow} c_{j,\uparrow} & e^{-h} f_{j,\downarrow} g_{j,\uparrow} & -e^{-h} c_{j,\downarrow}^\dagger c_{j,\uparrow} & -g_{j,\uparrow} c_{j,\downarrow}^\dagger \\ -f_{j,\uparrow} c_{j,\downarrow} & -e^{-h} c_{j,\uparrow}^\dagger c_{j,\downarrow} & e^{-h} g_{j,\downarrow} f_{j,\uparrow} & g_{j,\downarrow} c_{j,\uparrow}^\dagger \\ -e^h c_{j,\uparrow} c_{j,\downarrow} & -g_{j,\uparrow} c_{j,\downarrow} & g_{j,\downarrow} c_{j,\uparrow} & e^h g_{j,\downarrow} g_{j,\uparrow} \end{pmatrix} . \quad (12.227)$$

where we introduced the shorthand notation

$$f_{j,a} = n_{j,a} \sin(\lambda) - (1 - n_{j,a}) \cos(\lambda), \quad (12.228a)$$
$$g_{j,a} = n_{j,a} \cos(\lambda) + (1 - n_{j,a}) \sin(\lambda) \quad (12.228b)$$

for $a = \downarrow, \uparrow$.

The graded Yang-Baxter algebra (12.165) with R-matrix (12.135) and monodromy matrix (12.164) built up from an L-fold ordered product of L-matrices (12.227) is the starting point for the algebraic Bethe ansatz solution of the Hubbard model which we shall discuss below in Section 12.6.

In general, we may keep distinct all the v_j entering the definition of the monodromy matrix (12.164). The resulting inhomogeneous model will be useful in the next chapter, where we consider the quantum transfer matrix approach to the thermodynamics of the Hubbard model.

The fact that the R-matrix (12.135) is not of difference form has an interesting conse-quence [401]: the second spectral parameter μ is an additional independent parameter of the model. Setting all the v_j, $j = 1, \ldots, L$, in (12.164) equal to μ we obtain a homogeneous model which depends on μ and reduces to the Hubbard model in the special case $\mu = 0$. The generalized model generates a family of local Hamiltonians, because the regularity of the R-matrix (12.135) extends to all $\lambda = \mu$. Using once more equation (12.171), this time with $\lambda_0 = v_0 = \mu$, and the same fermionization scheme as above, we obtain the Hamiltonian

$$H = -\frac{1}{\text{ch}(2l)} \sum_{j=1}^{L} \sum_{a=\uparrow,\downarrow} \left(t_{j,-a}^{(-)} c_{j,a}^\dagger c_{j-1,a} + t_{j,-a}^{(+)} c_{j-1,a}^\dagger c_{j,a} \right)$$

$$+ \frac{u}{\text{ch}(2l)} \sum_{j=1}^{L} \left[(1 - 2n_{j,\uparrow})(1 - 2n_{j,\downarrow}) + \cos^2(2\mu) \right]$$

$$\quad (12.229)$$

$$+ \frac{u \sin^2(2\mu)}{\text{ch}(2l)} \sum_{j=1}^{L} \left[(c_{j,\downarrow}^\dagger c_{j-1,\downarrow} - c_{j-1,\downarrow}^\dagger c_{j,\downarrow})(c_{j,\uparrow}^\dagger c_{j-1,\uparrow} - c_{j-1,\uparrow}^\dagger c_{j,\uparrow}) \right.$$

$$\left. - (1 - n_{j-1,\downarrow} - n_{j,\downarrow})(1 - n_{j-1,\uparrow} - n_{j,\uparrow}) \right].$$

Here we used the convention $- \uparrow = \downarrow$ and $- \downarrow = \uparrow$. The transition amplitudes $t_{j,a}^{(\pm)}$ depend on whether or not sites $j - 1$ and j are occupied:

$$t_{j,a}^{(\pm)} = a_\pm + (n_{j-1,a} + n_{j,a})(1 - a_\pm) + n_{j-1,a} n_{j,a}(a_+ + a_- - 2) \qquad (12.230)$$

for $a = \uparrow, \downarrow$ with

$$a_\pm = \text{ch}(2l) \pm \text{sh}(2l) \cos(2\mu). \qquad (12.231)$$

Such kinds of amplitudes, depending on the particle densities, are sometimes called correlated hopping amplitudes.

The Hamiltonian (12.229) is hermitian for purely imaginary μ, $\text{Re } \mu = 0$. The additional interactions in the third sum on the right hand side of (12.229) couple the local currents of up- and down-spin electrons and the average densities of up- and down-spin electrons at neighbouring lattice sites. Note that the Hamiltonian (12.229) is spin reversal invariant and, up to a trivial, overall shift, transforms into $H(-u)$ under a Shiba transformation (2.59).

The model defined by (12.229) was first obtained in [401, 466]. It is exactly solvable by construction. So far it has not attracted much attention, but may be worth studying since unlike many other so-called 'generalized Hubbard Hamiltonians', it is a true generalization of the Hubbard Hamiltonian containing it as a limiting case. The model with non-zero μ played a role in the construction of a long-range interacting model related to Shastry's R-matrix [334] and in the construction of the boost operator for the Hubbard model [301].

We omit the lengthy expression for the L-matrix associated with the Hamiltonian (12.229). The reader may readily generate it by him- or herself using (12.135) and (12.160). An explicit expression for the fermionic R-operator can be obtained from the general definition (12.172). For the sake of completeness we present it in appendix 12.E. The fermionic R-operator is an alternative tool to perform many of the calculations of the following Sections (see [464, 466]). In our account, however, we shall never make use of its explicit form.

12.4.2 Symmetries

We now consider the symmetries of the Hubbard model from the point of view of the underlying Yang-Baxter algebra. In Section 12.3.3 we have learned that it is not the R-matrix $\check{R}(\lambda, \mu)$ that determines the symmetries of a fundamental graded model, but the 'graded R-matrix' $\tilde{R}(\lambda, \mu)$ with matrix elements

$$\tilde{R}_{\beta\delta}^{\alpha\gamma}(\lambda, \mu) = (-1)^{p(\alpha)p(\gamma)} R_{\beta\delta}^{\alpha\gamma}(\lambda, \mu). \qquad (12.232)$$

For the discussion of the two su(2) symmetries we need its explicit form. Thus, we first have to multiply the right hand side of (12.135) by the permutation matrix $P = e_\alpha^\beta \otimes e_\gamma^\delta \otimes e_\beta^\alpha \otimes e_\delta^\gamma$, $\alpha, \ldots, \delta = 1, 2$, and then have to reverse the signs in the 6th, 7th, 10th and 11th row. The resulting matrix is

$$
\tilde{R}(\lambda, \mu) = \frac{1}{\rho_4} \cdot
$$

$$
\begin{pmatrix}
\rho_1 & & & & & & & & & & & & & & & \\
 & -\rho_{10} & & & 1 & & & & & & & & & & & \\
 & & -\rho_{10} & & & & & & 1 & & & & & & & \\
 & & -\rho_8 & & & -\rho_6 & & & \rho_6 & & & & \rho_3 & & & \\
 & 1 & & & -\rho_9 & & & & & & & & & & & \\
 & & & & & -\rho_4 & & & & & & & & & & \\
 & & -\rho_6 & & & -\rho_7 & & & -\rho_5 & & & & -\rho_6 & & & \\
 & & & & & & & \rho_9 & & & & 1 & & & & \\
 & & & & & 1 & & & -\rho_9 & & & & & & & \\
 & & \rho_6 & & & -\rho_5 & & & -\rho_7 & & & & \rho_6 & & & \\
 & & & & & & & & & & -\rho_4 & & & & & \\
 & & & & & & & \rho_9 & & & & 1 & & & & \\
 & & \rho_3 & & & -\rho_6 & & & \rho_6 & & & & -\rho_8 & & & \\
 & & & & & & & & & & & 1 & & \rho_{10} & & \\
 & & & & & & & & & & & & & 1 & \rho_{10} & \\
 & & & & & & & & & & & & & & & \rho_1
\end{pmatrix}
\qquad (12.233)
$$

Note that $\tilde{R}(\lambda, \mu)$ is symmetric and, by construction, satisfies the graded Yang-Baxter equation [276],

$$
(-1)^{p(\beta')(p(\gamma)+p(\gamma'))} \tilde{R}^{\alpha\beta}_{\alpha'\beta'}(\lambda, \mu) \tilde{R}^{\alpha'\gamma}_{\alpha''\gamma'}(\lambda, \nu) \tilde{R}^{\beta'\gamma'}_{\beta''\gamma''}(\mu, \nu)
$$
$$
= (-1)^{p(\beta')(p(\gamma')+p(\gamma''))} \tilde{R}^{\beta\gamma}_{\beta'\gamma'}(\mu, \nu) \tilde{R}^{\alpha\gamma'}_{\alpha'\gamma''}(\lambda, \nu) \tilde{R}^{\alpha'\beta'}_{\alpha''\beta''}(\lambda, \mu). \qquad (12.234)
$$

The two su(2) symmetries of the Hubbard model are related to the two su(2) subalgebras of gl(2|2). We introduce the notation

$$
\Sigma_s^+ = e_2^3 = \sigma^+ \otimes \sigma^-, \qquad \Sigma_s^- = e_3^2 = \sigma^- \otimes \sigma^+, \qquad (12.235)
$$
$$
\Sigma_s^z = e_2^2 - e_3^3 = \tfrac{1}{2}(\sigma^z \otimes I_2 - I_2 \otimes \sigma^z) \qquad (12.236)
$$

and

$$
\Sigma_\eta^+ = e_1^4 = \sigma^+ \otimes \sigma^+, \qquad \Sigma_\eta^- = e_4^1 = \sigma^- \otimes \sigma^-, \qquad (12.237)
$$
$$
\Sigma_\eta^z = e_1^1 - e_4^4 = \tfrac{1}{2}(\sigma^z \otimes I_2 + I_2 \otimes \sigma^z) \qquad (12.238)
$$

for their generators. The labels 's' and 'η' refer to spin and η-spin, respectively. Setting $\Sigma_j^x = \Sigma_j^+ + \Sigma_j^-$ and $\Sigma_j^y = -i(\Sigma_j^+ - \Sigma_j^-)$ for $j = s, \eta$ we find the su(2) commutation relations

$$
[\Sigma_j^\alpha, \Sigma_j^\beta] = 2i\varepsilon^{\alpha\beta\gamma} \Sigma_j^\gamma \qquad (12.239)
$$

for $j = s, \eta$.

It turns out that spin and η-spin symmetry cannot be treated on the same footing. The invariance of the Hubbard model under rotations in spin space is a symmetry in the sense of Section 12.3.3, i.e., it is a consequence of the Lie-algebra invariance of the matrix $\tilde{R}(\lambda, \mu)$. The invariance (for an even number of lattice sites) of the Hamiltonian and the higher conserved quantities under rotations of the η-spin, on the other hand, follows by a somewhat different reasoning.

Let us first consider the spin. We claim that

$$[\tilde{R}(\lambda, \mu), \Sigma_s^\alpha \otimes I_4 + I_4 \otimes \Sigma_s^\alpha] = 0, \tag{12.240}$$

for $\alpha = x, y, z$. This claim is easily verified. One may, for instance, first check it for Σ_s^+ by explicit multiplication of 16×16-matrices. The corresponding equation for Σ_s^- then follows by taking the transposed of the invariance equation for Σ_s^+ and taking into account that $\tilde{R}(\lambda, \mu)$ is symmetric. Finally, the equation for Σ_s^z follows using $[\Sigma_s^+, \Sigma_s^-] = \Sigma_s^z$ and the Jacobi identity.

We thus have established the invariance of $\tilde{R}(\lambda, \mu)$ in the sense of equation (12.185), with respect to the su(2) representation generated by the Σ_s^α and may now pursue the general reasoning of Section 12.3.3. The global quantum space operators corresponding to the matrices Σ_s^α are two times the spin operators (2.66). This follows from the graded analogue of (12.183) upon fermionization. Then (12.188) tells us that

$$[\mathcal{T}(\lambda), \tfrac{1}{2}\Sigma_s^\alpha + S^\alpha] = 0, \tag{12.241}$$

$\alpha = x, y, z$. Note that Σ_s^α acts in the auxiliary space, but S^α acts in the quantum space. Equation (12.241) encodes the commutation relations of the elements of the monodromy matrix with the spin operators S^α. It is valid in the homogeneous as well as in the inhomogeneous case. According to our general formula (12.189) it follows that

$$[\text{str}(\mathcal{T}(\lambda)), S^\alpha] = 0, \tag{12.242}$$

for $\alpha = x, y, z$. Consequentially, in the homogeneous case the Hamiltonian and all the mutually commuting operators generated by $\tau(\lambda) = \ln(\hat{U}^{-1}\text{str}(\mathcal{T}(\lambda)))$ preserve the spin.

The matrices $\Sigma_\eta^\pm$ do not enter invariance equations like (12.240), but the somewhat different relations

$$\{\tilde{R}(\lambda, \mu), \Sigma_\eta^\pm \otimes I_4 - I_4 \otimes \Sigma_\eta^\pm\} = 0. \tag{12.243}$$

These relations, too, are easily verified. One of them has to be proven by explicit calculation, the other one then follows by taking the transpose and using the symmetry of $\tilde{R}(\lambda, \mu)$.

A first implication of (12.243) is the invariance equation

$$[\tilde{R}(\lambda, \mu), \Sigma_\eta^z \otimes I_4 + I_4 \otimes \Sigma_\eta^z] = 0. \tag{12.244}$$

The simple proof relies on the identity $[A, [B, C]] = \{\{A, B\}, C\} - \{\{A, C\}, B\}$, which holds for arbitrary matrices, and on the fact that $[\Sigma_\eta^+, \Sigma_\eta^-] = \Sigma_\eta^z$. The quantum space operator corresponding to Σ_η^z is $\sum_{j=1}^{L}(X_{j1}^1 - X_{j4}^4) = L - \hat{N} = -2\eta^z$. Using equation (12.188) we obtain the commutation relations between the elements of the monodromy matrix and the particle number operator,

$$[T(\lambda), \Sigma_\eta^z - \hat{N}] = 0. \tag{12.245}$$

Taking the super trace of the latter equation we arrive at

$$[\text{str}(T(\lambda)), \hat{N}] = 0 \tag{12.246}$$

and thus have shown the invariance of all the higher conserved quantities of the Hubbard model under global gauge transformations.

Next, we work out the consequences of equation (12.243). Let us write x for either Σ_η^+ or Σ_η^-. Then, spelled out in components, (12.243) reads

$$\tilde{R}^{\alpha\gamma}_{\beta'\delta}(\lambda, \mu) x^{\beta'}_\beta + x^\alpha_{\alpha'} \tilde{R}^{\alpha'\gamma}_{\beta\delta}(\lambda, \mu) = \tilde{R}^{\alpha\gamma}_{\beta\delta'}(\lambda, \mu) x^{\delta'}_\delta + x^\gamma_{\gamma'} \tilde{R}^{\alpha\gamma'}_{\beta\delta}(\lambda, \mu). \tag{12.247}$$

It follows that

$$\mathcal{L}_{j\,\beta'}^\alpha(\lambda, \mu) x^{\beta'}_\beta + x^\alpha_{\alpha'} \mathcal{L}_{j\,\beta}^{\alpha'}(\lambda, \mu) = \mathcal{L}_{j\,\beta}^\alpha(\lambda, \mu) x_j + x_j \mathcal{L}_{j\,\beta}^\alpha(\lambda, \mu) \tag{12.248}$$

and, for an even number of lattice sites,

$$T_{\beta'}^\alpha(\lambda) x^{\beta'}_\beta - x^\alpha_{\alpha'} T_\beta^{\alpha'}(\lambda) = -T_\beta^\alpha(\lambda) \sum_{j=1}^L (-1)^j x_j - \sum_{j=1}^L (-1)^j x_j\, T_\beta^\alpha(\lambda). \tag{12.249}$$

The proof is rather elementary. One first uses (12.248) to show (12.249) for two sites and then proceeds by induction.

For Σ_η^+ and Σ_η^- we obtain the local quantum space operators $x_j = c_{j,\downarrow} c_{j,\uparrow}$ and $x_j = c^\dagger_{j,\uparrow} c^\dagger_{j,\downarrow}$, respectively, and (12.249) turns into

$$[T(\lambda), \Sigma_\eta^\pm] = \{T(\lambda), \eta^\mp\}. \tag{12.250}$$

This equation comprises all commutation relations between the elements of the monodromy matrix and $\eta^\pm$. Taking the super trace we arrive at

$$\{\text{str}(T(\lambda)), \eta^\pm\} = 0. \tag{12.251}$$

Thus, unlike the other su(2) generators, S^α, η^z, the operators $\eta^\pm$ anticommute with the transfer matrix. Specializing to $\lambda = 0$ in the homogeneous case we recover our former result (see equation (2.88)) $\{\hat{U}, \eta^\pm\} = 0$. It follows that the generating function of local conserved quantities, $\tau(\lambda)$, commutes with $\eta^\pm$,

$$[\tau(\lambda), \eta^\pm] = 0. \tag{12.252}$$

Thus, the whole set of mutually commuting operators generated by $\tau(\lambda)$ is invariant under rotations of the η-spin.

Let us now turn to the discussion of the discrete symmetries of the monodromy matrix (compare Section 2.2). Application of the spin flip operator (2.57) to the L-matrix (12.227) yields

$$J^{(s)} \mathcal{L}_j(\lambda, 0) J^{(s)} = M \mathcal{L}_j(\lambda, 0) M, \tag{12.253}$$

where

$$M = \begin{pmatrix} 1 & & & \\ & & 1 & \\ & 1 & & \\ & & & -1 \end{pmatrix}. \tag{12.254}$$

Since $(J^{(s)})^2 = \mathrm{id}$ and $M^2 = I_4$, we conclude that the monodromy matrix of the homogeneous model satisfies

$$J^{(s)} T(\lambda) J^{(s)} = M T(\lambda) M. \tag{12.255}$$

The latter equation describes the behaviour of the monodromy matrix elements under spin flips. We obtain, in particular,

$$[t(\lambda), J^{(s)}] = 0. \tag{12.256}$$

Thus, not only the Hamiltonian but all the higher conserved operators generated by $\tau(\lambda)$ are spin reversal invariant.

It is slightly more tricky to figure out the behaviour of the L-matrix (12.227) under the Shiba transformation generated by $J_a^{(sh)}$, equation (2.58). We shall consider an even number of lattice sites L from the outset. Let us start with $a = \downarrow$. Then the functions $f_{j,\downarrow}, g_{j,\downarrow}$ defined in (12.228) transform as

$$J_\downarrow^{(sh)} f_{j,\downarrow} (J_\downarrow^{(sh)})^\dagger = (1 - 2n_{j,\downarrow}) g_{j,\downarrow}, \tag{12.257}$$

$$J_\downarrow^{(sh)} g_{j,\downarrow} (J_\downarrow^{(sh)})^\dagger = -(1 - 2n_{j,\downarrow}) f_{j,\downarrow}. \tag{12.258}$$

The L-matrix (12.227) turns into

$$J_\downarrow^{(sh)} \mathcal{L}_j(\lambda, 0|u) (J_\downarrow^{(sh)})^\dagger$$
$$= e^{2h} (1 - 2n_{j,\downarrow}) ((\sigma^z)^j \otimes I_2) (\sigma^y \otimes \sigma^z) \mathcal{L}_j(\lambda, 0| - u) (\sigma^y \otimes \sigma^z) ((\sigma^z)^{j-1} \otimes I_2). \tag{12.259}$$

Here we indicated explicitly the dependence of the L-matrix on the coupling u. Taking into account that $1 - 2n_{j,\downarrow} = e^{i\pi n_{j,\downarrow}}$ and that L is even, we obtain the transformation rule for the monodromy matrix in the form

$$J_\downarrow^{(sh)} T(\lambda|u) (J_\downarrow^{(sh)})^\dagger = e^{2hL + i\pi \hat{N}_\downarrow} (\sigma^y \otimes \sigma^z) T(\lambda| - u) (\sigma^y \otimes \sigma^z). \tag{12.260}$$

From here it is easy to determine how the generating function $\tau(\lambda)$ behaves under the Shiba transformation. Observing that $\mathrm{str}\{(\sigma^y \otimes \sigma^z) T(\lambda|u) (\sigma^y \otimes \sigma^z)\} = -\mathrm{str}\{T(\lambda|u)\}$ we find

$$J_\downarrow^{(sh)} t(\lambda|u) (J_\downarrow^{(sh)})^\dagger = -e^{2hL + i\pi \hat{N}_\downarrow} t(\lambda| - u), \tag{12.261}$$

$$J_\downarrow^{(sh)} \hat{U}^{-1} (J_\downarrow^{(sh)})^\dagger = -\hat{U}^{-1} e^{-i\pi \hat{N}_\downarrow}. \tag{12.262}$$

It follows that

$$J_\downarrow^{(sh)} \tau(\lambda|u) (J_\downarrow^{(sh)})^\dagger = \tau(\lambda| - u) + 2h(\lambda|u)L. \tag{12.263}$$

Thus $\tau(\lambda|u)$ is invariant under the Shiba transformation up to a change of the sign of the coupling and a trivial overall shift.

Remark. The unpleasant term $2h(\lambda|u)L$ can be easily avoided by changing the normalization of the L-matrix (12.227). If we replace the prefactor $e^h/\cos^2(\lambda)$ in equation (12.227) by 1, then the term $2h(\lambda|u)L$ disappears from equations (12.260) and (12.263). Another possibility to make (12.263) more symmetric is by redefinition of the generating function $\tau(\lambda|u)$. Setting $\tilde{\tau}(\lambda|u) = \tau(\lambda|u) - h(\lambda|u)L$ we obtain

$$J_{\downarrow}^{(sh)}\tilde{\tau}(\lambda|u)\big(J_{\downarrow}^{(sh)}\big)^{\dagger} = \tilde{\tau}(\lambda| - u). \tag{12.264}$$

It should be clear from our discussion that the Shiba transformation relates the Hubbard model with positive u with the Hubbard model with negative u. Instead of working with the Shiba transformations for down-spin electrons we could have equivalently worked with the Shiba transformation for up-spin electrons. The corresponding operators $J_{\downarrow}^{(sh)}$ and $J_{\uparrow}^{(sh)}$ are related by a spin reversal transformation. Hence, we can apply the transformation (12.255) to equation (12.260) to obtain the transformation rule for the monodromy matrix under a Shiba transformation which affects the up-spins. We multiply equation (12.260) by $J^{(s)}M$ from the left and from the right and use (12.255) and $M(\sigma^y \otimes \sigma^z)M = I_2 \otimes \sigma^y$. Then

$$J_{\uparrow}^{(sh)}T(\lambda|u)\big(J_{\uparrow}^{(sh)}\big)^{\dagger} = e^{2hL+i\pi\hat{N}_{\uparrow}}(I_2 \otimes \sigma^y)T(\lambda| - u)(I_2 \otimes \sigma^y). \tag{12.265}$$

Because of (12.256), equation (12.263) is invariant under spin reversal.

The simultaneous action of Shiba transformations for up- and down-spins on the monodromy matrix is obtained by combining (12.260) and (12.265),

$$J_{\downarrow}^{(sh)}J_{\uparrow}^{(sh)}T(\lambda|u)\big(J_{\downarrow}^{(sh)}J_{\uparrow}^{(sh)}\big)^{\dagger} = e^{i\pi\hat{N}}(\sigma^y \otimes \sigma^x)T(\lambda|u)(\sigma^y \otimes \sigma^x). \tag{12.266}$$

As a consequence we obtain the following invariance equation for the generating function $\tau(\lambda)$,

$$J_{\downarrow}^{(sh)}J_{\uparrow}^{(sh)}\tau(\lambda|u)\big(J_{\downarrow}^{(sh)}J_{\uparrow}^{(sh)}\big)^{\dagger} = \tau(\lambda|u). \tag{12.267}$$

As an exercise for the reader we propose to use the Shiba transformation in order to re-derive the commutation relations (12.245), (12.250) between the elements of the monodromy matrix and $\eta^{\pm}$, η^z out of the commutation relations (12.241) between $T(\lambda)$ and $S^{\pm}$, S^z. This way the η-pairing symmetry of the transfer matrix was originally established in [178]. Our derivation here of the η-pairing symmetry (see (12.243) and below) closely follows in spirit the article [466]. For related work with emphasis on different aspects of the problem the reader is referred to [399, 464].

The alert reader may have noticed that we have discussed the discrete symmetries only for the special case of the homogeneous model with the second spectral parameter of the L-matrix equal to zero. Still, we wish to note that our account can easily be extended to the completely inhomogeneous case.

12.5 Solution of the quantum inverse problem

The quantum inverse scattering method relies on a transformation from a set of local 'field operators' $\{e_{j\alpha}^{\beta}\}$ to a set of non-local operators, the elements of the monodromy matrix (see figure 12.3). It is natural to ask: What is the inverse transformation? This question has a remarkably simple answer [177, 251, 317]. Here we shall present the result of [177], where the inverse transformation was constructed for fundamental graded models.

We shall assume we are given a solution of the Yang-Baxter equation (12.31) which is regular and unitary. Let p be a grading that is compatible with the R-matrix in the sense of equation (12.159), and let $\mathcal{T}(\lambda)$ be the corresponding inhomogeneous monodromy matrix (12.164). Then the following inversion formula holds,

$$e_{n\alpha}^{\beta} = (-1)^{p(\alpha)p(\beta)} \prod_{j=1}^{n-1} \text{str}(\mathcal{T}(v_j)) \cdot \mathcal{T}_{\alpha}^{\beta}(v_n) \cdot \prod_{j=n+1}^{L} \text{str}(\mathcal{T}(v_j)). \tag{12.268}$$

Note that because of (12.166), no ordering is required for the products on the right-hand side of (12.268).

The proof of the inversion formula (12.268) for graded inhomogeneous models is given in appendix 12.B. In the homogeneous case the inversion formula takes a particularly simple form. Without any loss of generality we may assume that the 'shift point' where the L-matrix elements are proportional to $e_{j\alpha}^{\beta}$ (see (12.167)) is at $\lambda_0 = v_0 = 0$. Then $\text{str}(\mathcal{T}(0)) = \hat{U}$, where $\hat{U}$ is the shift operator (12.168), and equation (12.268) turns into

$$e_{n\alpha}^{\beta} = (-1)^{p(\alpha)p(\beta)} \hat{U}^{n-1} \mathcal{T}_{\alpha}^{\beta}(0) \hat{U}^{L-n}. \tag{12.269}$$

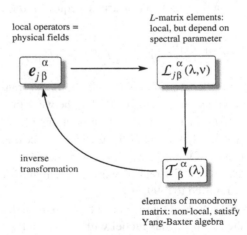

Fig. 12.3. The inverse transformation.

Note that $\hat{U}^L = \text{id}$. Hence, the latter equation is equivalent to

$$e_1{}^\beta_\alpha = (-1)^{p(\alpha)p(\beta)} T^\beta_\alpha(0)\, \hat{U}^{-1}, \tag{12.270}$$

by a similarity transformation with $\hat{U}^{1-n}$.

The proof of (12.270) is considerably more simple than the proof of the inversion formula (12.268) for the inhomogeneous model. We just evaluate the elements of the monodromy matrix at the shift point, say $\lambda_0 = \nu_0 = 0$, where (12.167) is valid. Then

$$
\begin{aligned}
T^\alpha_\beta(0) &= (-1)^{p(\alpha)p(\beta_{L-1})+p(\beta_{L-1})p(\beta_{L-2})+\cdots+p(\beta_1)p(\beta)} e_{L}{}^\alpha_{\beta_{L-1}} \cdots e_2{}^{\beta_2}_{\beta_1} e_1{}^{\beta_1}_\beta \\
&= (-1)^{p(\alpha)p(\beta)+p(\beta_1)+\cdots+p(\beta_{L-1})} e_1{}^{\beta_1}_\beta e_2{}^{\beta_2}_{\beta_1} \cdots e_L{}^\alpha_{\beta_{L-1}} \\
&= (-1)^{p(\alpha)p(\beta)} e_1{}^\alpha_\beta P_{12} \dots P_{L\,1L} \\
&= (-1)^{p(\alpha)p(\beta)} e_1{}^\alpha_\beta \hat{U},
\end{aligned}
\tag{12.271}
$$

and the proof of (12.270) is complete.

The analogous calculation for the inhomogeneous case is presented in appendix 12.B in the proof of lemma 19. The complications that occur in the inhomogeneous case are due to the more complicated structure of the shift operator for inhomogeneous models (see lemma 17 below).

We illustrate equation (12.269) with the example of the Hubbard model. It is rather natural to divide the 4×4-monodromy matrix of the Hubbard model into four 2×2 blocks [335, 336],

$$
T(\lambda) = \begin{pmatrix}
D^1_1(\lambda) & C^1_1(\lambda) & C^1_2(\lambda) & D^1_2(\lambda) \\
B^1_1(\lambda) & A^1_1(\lambda) & A^1_2(\lambda) & B^1_2(\lambda) \\
B^2_1(\lambda) & A^2_1(\lambda) & A^2_2(\lambda) & B^2_2(\lambda) \\
D^2_1(\lambda) & C^2_1(\lambda) & C^2_2(\lambda) & D^2_2(\lambda)
\end{pmatrix}. \tag{12.272}
$$

This block structure reflects the discrete symmetries and the SO(4) symmetry, which is connected to the blocks $A(\lambda)$ and $D(\lambda)$ of the monodromy matrix.

Using the same fermionization scheme as in Section 12.4 we obtain

$$c^\dagger_{n,\uparrow} = \hat{U}^{n-1}\big(C^1_1(0) + B^2_2(0)\big)\hat{U}^{L-n}, \tag{12.273a}$$

$$c_{n,\uparrow} = \hat{U}^{n-1}\big(B^1_1(0) + C^2_2(0)\big)\hat{U}^{L-n}, \tag{12.273b}$$

$$c^\dagger_{n,\downarrow} = \hat{U}^{n-1}\big(C^1_2(0) - B^2_1(0)\big)\hat{U}^{L-n}, \tag{12.273c}$$

$$c_{n,\downarrow} = \hat{U}^{n-1}\big(B^2_1(0) - C^2_1(0)\big)\hat{U}^{L-n}. \tag{12.273d}$$

Here the operators on the right hand side are elements of the monodromy matrix (12.272) evaluated at $\lambda = 0$. Similar formulae are easily written down for the local spin operators. In the inhomogeneous case the terms $\hat{U}^{n-1}$ and $\hat{U}^{L-n}$ are replaced by products of the super trace of $T(\lambda)$ evaluated at the inhomogeneities.

The equations (12.273) are expected to be useful for the calculation of matrix elements of local operators, which is one of the most interesting open problems for the one-dimensional Hubbard model.

12.6 On the algebraic Bethe ansatz for the Hubbard model

Performing an algebraic Bethe ansatz entails the use of the commutation relations between the elements of the monodromy matrix, collected in the Yang-Baxter algebra, for the diagonalization of the transfer matrix. A necessary requirement for this type of procedure is that the monodromy matrix acts triangularly on an appropriate pseudo vacuum state, such that the pseudo vacuum is an eigenstate of the operators on its diagonal. As we shall see below, this necessary requirement is fulfilled for the Hubbard model. Let us stress, however, that no general recipe for an algebraic Bethe ansatz is known. Depending on the structure of the R-matrix an algebraic Bethe ansatz, if possible at all, may be a demanding task.

In the case of the Hubbard model the construction of an algebraic Bethe ansatz remained a notorious problem for quite a while. Let us comment on the reasons: (i) The Yang-Baxter algebra for a 16×16 R-matrix comprises the large number of 256 commutation relations and is not really a convenient tool for calculations as long as no sub-structure of the commutation relations is identified. (ii) Moreover, the parameterization of the Boltzmann weights of our R-matrix (12.135) in terms of λ and μ is inconvenient as well, because of the constraint (12.109), and does not fall into the usual classification scheme of rational, trigonometric or elliptic R-matrices. The R-matrix is not of difference form. We know, however, from the coordinate Bethe ansatz solution in Chapter 3 that the 'spin problem' of the Hubbard model is intimately related to the rational R-matrix of the XXX spin chain (or the 6-vertex model with rational Boltzmann weights). Thus, there should be a 'hidden 6-vertex structure' inside the Yang-Baxter algebra generated by the R-matrix (12.135). (iii) The fact that the Lieb-wu equations may be formally generated by the Yang-Baxter algebra connected with gl(1|2) invariant 3×3 R-matrix (12.217) (see [169, 170]) seems to leave the possibility that there is an algebraic Bethe ansatz not based on Shastry's R-matrix and yet to be discovered. (iv) As we shall see below, the monodromy matrix contains too many creation operators which makes the identification of the operators relevant for an algebraic Bethe ansatz a delicate matter.

The difficulties connected with the involved structure of the Yang-Baxter algebra for the Hubbard model were eventually overcome by Martins and Ramos who constructed an algebraic Bethe ansatz in [320, 371] (see also [319]). The key points in their analysis were the discovery of a hidden 6-vertex structure inside the Yang-Baxter algebra and a clever recursive construction of the eigenvectors, which was inspired by Tarasov's work [451] on the algebraic Bethe ansatz for the Izergin-Korepin model. Martins and Ramos were led to the hidden 6-vertex structure through the analysis of a certain rational vertex model generated by the intertwiner of two typical four-dimensional representations of Y(gl(1|2)) [370]. This intertwiner is a 16×16 matrix with 36 non-vanishing Boltzmann weights at the same positions as in our R-matrix (12.135).

In this section we first classify the entries of the monodromy matrix as creation or annihilation operators, or operators which do not change the particle number. Then we work out the block structure of the Yang-Baxter algebra generated by the R-matrix (12.135). After calculating the action of the monodromy matrix on the Fock vacuum $|0\rangle$, starting from Subsection 12.6.4, we review Martins' and Ramos' construction of the one- and two-particle eigenstates of the transfer matrix. We obtain the one- and two-particle eigenvalues of the transfer matrix which will be enough to guess the generalization to the N-particle case.

12.6.1 Classification of the monodromy matrix elements with respect to change of particle number

The two invariance equations (12.241) for $\alpha = z$ and (12.245) contain the information about how the action of the elements of the monodromy matrix affects the numbers of up- and down-spin electrons. For the classification of the elements of the monodromy matrix we need more explicit expressions. Recall that $\hat{N} = \hat{N}_\uparrow + \hat{N}_\downarrow$ and $2S^z = \hat{N}_\uparrow - \hat{N}_\downarrow$, where $\hat{N}_\uparrow$ and $\hat{N}_\downarrow$ are the particle number operators for electrons with spin up and spin down. Then (12.241), (12.245) imply

$$[\mathcal{T}(\lambda), \tfrac{1}{2}(\Sigma_\eta^z - \Sigma_s^z) - \hat{N}_\uparrow] = [\mathcal{T}(\lambda), \tfrac{1}{2}(I_2 \otimes \sigma^z) - \hat{N}_\uparrow] = 0, \qquad (12.274a)$$

$$[\mathcal{T}(\lambda), \tfrac{1}{2}(\Sigma_\eta^z + \Sigma_s^z) - \hat{N}_\downarrow] = [\mathcal{T}(\lambda), \tfrac{1}{2}(\sigma^z \otimes I_2) - \hat{N}_\downarrow] = 0. \qquad (12.274b)$$

Performing the matrix multiplications in the auxiliary space explicitly we obtain

$$[\hat{N}_\uparrow, \mathcal{T}(\lambda)] = \begin{pmatrix} 0 & C_1^1(\lambda) & 0 & D_2^1(\lambda) \\ -B_1^1(\lambda) & 0 & -A_2^1(\lambda) & 0 \\ 0 & A_1^2(\lambda) & 0 & B_2^2(\lambda) \\ -D_1^2(\lambda) & 0 & -C_2^2(\lambda) & 0 \end{pmatrix}, \qquad (12.275a)$$

$$[\hat{N}_\downarrow, \mathcal{T}(\lambda)] = \begin{pmatrix} 0 & 0 & C_2^1(\lambda) & D_2^1(\lambda) \\ 0 & 0 & A_2^1(\lambda) & B_2^1(\lambda) \\ -B_1^2(\lambda) & -A_1^2(\lambda) & 0 & 0 \\ -D_1^2(\lambda) & -C_1^2(\lambda) & 0 & 0 \end{pmatrix}. \qquad (12.275b)$$

Here the notation for the elements of the monodromy matrix is taken from equation (12.272). In (12.275) the 32 commutation relations between the elements of the monodromy matrix and the particle number operators $\hat{N}_\uparrow$, $\hat{N}_\downarrow$ are written in compact form. Let us pick out the commutators involving $C_1^1(\lambda)$, $B_2^2(\lambda)$, $C_2^1(\lambda)$, and $B_2^1(\lambda)$,

$$\hat{N}_\uparrow C_1^1(\lambda) = C_1^1(\lambda)(\hat{N}_\uparrow + 1), \quad [\hat{N}_\downarrow, C_1^1(\lambda)] = 0, \tag{12.276a}$$

$$\hat{N}_\uparrow B_2^2(\lambda) = B_2^2(\lambda)(\hat{N}_\uparrow + 1), \quad [\hat{N}_\downarrow, B_2^2(\lambda)] = 0, \tag{12.276b}$$

$$\hat{N}_\downarrow C_2^1(\lambda) = C_2^1(\lambda)(\hat{N}_\downarrow + 1), \quad [\hat{N}_\uparrow, C_2^1(\lambda)] = 0, \tag{12.276c}$$

$$\hat{N}_\downarrow B_2^1(\lambda) = B_2^1(\lambda)(\hat{N}_\downarrow + 1), \quad [\hat{N}_\uparrow, B_2^1(\lambda)] = 0. \tag{12.276d}$$

Equations (12.276a), (12.276b) mean that $C_1^1(\lambda)$ and $B_2^2(\lambda)$, for instance, add an up-spin electron to a state and do not change the number of down-spin electrons. Thus, $C_1^1(\lambda)$ and $B_2^2(\lambda)$ are one-particle creation operators of up-spin electrons. Similarly, $C_2^1(\lambda)$ and $B_2^1(\lambda)$ can be interpreted as one-particle creation operators of down-spin electrons.

We leave to the reader the simple task of writing explicitly all the relations contained in (12.276). It turns out that $B_1^1(\lambda)$ and $C_2^2(\lambda)$ are one-particle annihilation operators of up-spin electrons, while $C_1^2(\lambda)$ and $B_1^1(\lambda)$ annihilate a down-spin electron. The operators $A_2^1(\lambda)$ and $A_1^2(\lambda)$ preserve the total number of particles, but flip the spin. $D_2^1(\lambda)$ and $D_1^2(\lambda)$ create or annihilate a pair of electrons of opposite spin. The diagonal elements of the monodromy matrix leave spin and particle number unchanged.

It is a striking fact that we have identified five creation operators $C_1^1(\lambda)$, $C_2^1(\lambda)$, $B_2^1(\lambda)$, $B_2^2(\lambda)$ and $D_2^1(\lambda)$ inside the monodromy matrix. By way of contrast there are only two local creation operators, $c_{j,\uparrow}^\dagger$, $c_{j,\downarrow}^\dagger$, of electrons. This means that the choice of creation operators for the algebraic Bethe ansatz is not obvious. Presumably there is more than one way of constructing an algebraic Bethe ansatz (compare the construction of eigenstates over the empty lattice in Chapter 15).

The operator entries of the submatrices $A(\lambda), \ldots, D(\lambda)$ of the monodromy matrix each are related by adjungation in quantum space,

$$A(\lambda)^\dagger = \mathrm{ctg}^{2L}\left(\tfrac{\pi}{2} - \bar{\lambda}\right) e^{i\pi \hat{N}} \sigma^y A\left(\tfrac{\pi}{2} - \bar{\lambda}\right) \sigma^y, \tag{12.277a}$$

$$B(\lambda)^\dagger = (-i)\,\mathrm{ctg}^{2L}\left(\tfrac{\pi}{2} - \bar{\lambda}\right) e^{i\pi \hat{N}} \sigma^y B\left(\tfrac{\pi}{2} - \bar{\lambda}\right) \sigma^x, \tag{12.277b}$$

$$C(\lambda)^\dagger = (-i)\,\mathrm{ctg}^{2L}\left(\tfrac{\pi}{2} - \bar{\lambda}\right) e^{i\pi \hat{N}} \sigma^x C\left(\tfrac{\pi}{2} - \bar{\lambda}\right) \sigma^y, \tag{12.277c}$$

$$D(\lambda)^\dagger = \mathrm{ctg}^{2L}\left(\tfrac{\pi}{2} - \bar{\lambda}\right) e^{i\pi \hat{N}} \sigma^x D\left(\tfrac{\pi}{2} - \bar{\lambda}\right) \sigma^x. \tag{12.277d}$$

Here no adjungation of matrices is implied with the daggers on the left of hand side of these equations. The bar means complex conjugation. Note that the prefactors of $\mathrm{ctg}^{2L}\left(\tfrac{\pi}{2} - \bar{\lambda}\right)$ could have been avoided if we had chosen a different normalization of the L-matrix (12.227). Equations (12.277) can easily be proven by induction over L.

12.6.2 Yang-Baxter algebra in block form

The structure of the Yang-Baxter algebra (12.165) generated by the R-matrix (12.135) becomes clearer after performing a similarity transformation which reveals the block structure

(12.272) of the monodromy matrix. A similarity transformation X transforms (12.165) into

$$X\check{R}(\lambda, \mu)X^{-1}X\big(T(\lambda) \otimes_s T(\mu)\big)X^{-1} = X\big(T(\mu) \otimes_s T(\lambda)\big)X^{-1}X\check{R}(\lambda, \mu)X^{-1}. \quad (12.278)$$

Choosing a matrix X with only non-zero entries

$$\begin{aligned}
X_6^1 &= X_7^2 = X_{10}^3 = X_{11}^4 = X_5^5 = X_8^6 = X_9^7 = X_{12}^8 = 1, \\
X_2^9 &= X_3^{10} = X_{14}^{11} = X_{15}^{12} = X_1^{13} = X_4^{14} = X_{13}^{15} = X_{16}^{16} = -1
\end{aligned} \quad (12.279)$$

we can write (12.278) more explicitly as

$$
\begin{pmatrix}
\rho_4\check{r}_1 & 0 & 0 & \rho_6 K \\
0 & I_4 & \rho_{10}J & 0 \\
0 & \rho_9 J^t & I_4 & 0 \\
\rho_6 K^t & 0 & 0 & \rho_1\check{r}_2
\end{pmatrix}
\begin{pmatrix}
A \otimes \bar{A} & A \otimes \bar{B} & B \otimes \bar{A} & B \otimes \bar{B} \\
A \otimes \bar{C} & A \otimes \bar{D} & -B \otimes \bar{C} & -B \otimes \bar{D} \\
C \otimes \bar{A} & C \otimes \bar{B} & D \otimes \bar{A} & D \otimes \bar{B} \\
-C \otimes \bar{C} & -C \otimes \bar{D} & D \otimes \bar{C} & D \otimes \bar{D}
\end{pmatrix}
$$

$$
=
\begin{pmatrix}
\bar{A} \otimes A & \bar{A} \otimes B & \bar{B} \otimes A & \bar{B} \otimes B \\
\bar{A} \otimes C & \bar{A} \otimes D & -\bar{B} \otimes C & -\bar{B} \otimes D \\
\bar{C} \otimes A & \bar{C} \otimes B & \bar{D} \otimes A & \bar{D} \otimes B \\
-\bar{C} \otimes C & -\bar{C} \otimes D & \bar{D} \otimes C & \bar{D} \otimes D
\end{pmatrix}
\begin{pmatrix}
\rho_4\check{r}_1 & 0 & 0 & \rho_6 K \\
0 & I_4 & \rho_{10}J & 0 \\
0 & \rho_9 J^t & I_4 & 0 \\
\rho_6 K^t & 0 & 0 & \rho_1\check{r}_2
\end{pmatrix}.
$$

$$(12.280)$$

The entries of the various matrices in equation (12.280) are 4×4 matrices themselves. For the formula to fit on the line we suppressed the arguments. The 2×2 matrices $A, \dots, D$ depend on λ, and a bar means here that λ is replaced with μ. The 4×4 matrices $\check{r}_1(\lambda, \mu)$ and $\check{r}_2(\lambda, \mu)$ depend non-trivially on the Boltzmann weights,

$$
\check{r}_1 =
\begin{pmatrix}
1 & & & \\
& \frac{\rho_5}{\rho_4} & 1 - \frac{\rho_5}{\rho_4} & \\
& 1 - \frac{\rho_5}{\rho_4} & \frac{\rho_5}{\rho_4} & \\
& & & 1
\end{pmatrix}, \quad
\check{r}_2 =
\begin{pmatrix}
1 & & & \\
& \frac{\rho_3}{\rho_1} & \frac{\rho_3}{\rho_1} - 1 & \\
& \frac{\rho_3}{\rho_1} - 1 & \frac{\rho_3}{\rho_1} & \\
& & & 1
\end{pmatrix}. \quad (12.281)
$$

The matrices J and K are constant matrices, most conveniently expressed in terms of the permutation matrix P,

$$J = P(\sigma^z \otimes I_2), \quad (12.282)$$

$$K = (I_4 - P)(\sigma^z \otimes I_2). \quad (12.283)$$

We would like to note that the structure of the Yang-Baxter algebra, when written in block form (12.280), resembles the Yang-Baxter algebra of an (asymmetric) 8-vertex model with grading $p(1) = 0$, $p(2) = 1$. A complete list of the relations contained in equation (12.280) is given in appendix 12.C.

12.6.3 Action of the monodromy matrix on the vacuum

As for any algebraic Bethe ansatz calculation we have to determine the action of the monodromy matrix on the vacuum. Using the definitions (12.228) we obtain

$$f_{j,a}|0\rangle = -\cos(\lambda)|0\rangle, \quad g_{j,a}|0\rangle = \sin(\lambda)|0\rangle, \tag{12.284}$$

for $a = \uparrow, \downarrow$. Thus,

$$\mathcal{L}_j(\lambda, 0)|0\rangle = \begin{pmatrix} e^{2h} & * & * & * \\ 0 & -\operatorname{tg}(\lambda) & 0 & * \\ 0 & 0 & -\operatorname{tg}(\lambda) & * \\ 0 & 0 & 0 & \operatorname{tg}^2(\lambda)e^{2h} \end{pmatrix}|0\rangle, \tag{12.285}$$

where the asterisks denote some operators whose specific form does not matter here. Denoting the eigenvalues on the diagonal by

$$\omega_1(\lambda) = e^{2h}, \tag{12.286a}$$

$$\omega_2(\lambda) = -\operatorname{tg}(\lambda), \tag{12.286b}$$

$$\omega_3(\lambda) = \operatorname{tg}^2(\lambda)e^{2h}, \tag{12.286c}$$

we obtain the following result for the monodromy matrix,

$$\mathcal{T}(\lambda)|0\rangle = \begin{pmatrix} \omega_1^L(\lambda) & * & * & * \\ 0 & \omega_2^L(\lambda) & 0 & * \\ 0 & 0 & \omega_2^L(\lambda) & * \\ 0 & 0 & 0 & \omega_3^L(\lambda) \end{pmatrix}|0\rangle. \tag{12.287}$$

It follows that the Fock vacuum is an eigenstate of the transfer matrix, $t(\lambda)|0\rangle = \Lambda(\lambda)|0\rangle$, with eigenvalue

$$\Lambda(\lambda) = \omega_1^L(\lambda) - 2\omega_2^L(\lambda) + \omega_3^L(\lambda). \tag{12.288}$$

12.6.4 Hidden six-vertex structure and one-particle states

In this subsection we shall construct one-particle eigenstates of the transfer matrix $t(\lambda) = \operatorname{str}(\mathcal{T}(\lambda)) = \operatorname{tr}(D(\lambda)) - \operatorname{tr}(A(\lambda))$. We shall use a notation that may appear artificial at first reading, but will turn out to be useful in the two- and more-particle cases. First of all we introduce the vector notation

$$\mathbf{C}^n(\lambda) = \left(C_1^n(\lambda), C_2^n(\lambda)\right), \quad \mathbf{B}_n(\lambda) = \begin{pmatrix} B_n^1(\lambda) \\ B_n^2(\lambda) \end{pmatrix} \tag{12.289}$$

for $n = 1, 2$. We have seen above in Section (12.6.1) that $\mathbf{C}^1(\lambda)$ and $\mathbf{B}_2(\lambda)$ are vectors of creation operators, and $\mathbf{C}^2(\lambda)$ and $\mathbf{B}_1(\lambda)$ are vectors of annihilation operators, respectively. Let $\phi^{(1)}(\lambda) = \mathbf{C}^1(\lambda)$. We shall seek for eigenvectors of the transfer matrix of the form

$$|\phi^{(1)}\rangle = \phi^{(1)}(\lambda_1)|\hat{F}\rangle, \tag{12.290}$$

where $|\hat{F}\rangle = \begin{pmatrix} f_1 \\ f_2 \end{pmatrix} \otimes |0\rangle = \begin{pmatrix} f_1|0\rangle \\ f_2|0\rangle \end{pmatrix}$; $f_1, f_2 \in \mathbb{C}$. In order to calculate the action of the transfer matrix on $|\phi^{(1)}\rangle$ we have to commute $A_1^1(\lambda), A_2^2(\lambda), D_1^1(\lambda)$, and $D_2^2(\lambda)$ through $\mathbf{C}^1(\lambda_1)$. The appropriate commutation relations can be extracted from the list (12.C.1) in appendix 12.C, which is a more explicit form of (12.280). Let us pick out equation (12.C.1n),

$$\rho_6 K^t(A \otimes \bar{B}) - \rho_1 \check{r}_2(C \otimes \bar{D}) = -\bar{C} \otimes D + \rho_9(\bar{D} \otimes C)J^t. \tag{12.291}$$

We multiply this equation by $(1, 0) \otimes (1, 0)$ from the left and by $I_2 \otimes \binom{1}{0}$ from the right and use $[(1, 0) \otimes (1, 0)]K^t = 0$, $[(1, 0) \otimes (1, 0)]\check{r}_2 = (1, 0) \otimes (1, 0)$, and $J^t(I_2 \otimes \binom{1}{0}) = P(I_2 \otimes \sigma^z)(I_2 \otimes \binom{1}{0}) = \binom{1}{0} \otimes I_2$. We further use the relations (12.123) to interchange the arguments λ and μ. It follows that

$$D_1^1(\lambda)\mathbf{C}^1(\mu) = \frac{\rho_4}{\rho_9}\mathbf{C}^1(\mu)D_1^1(\lambda) - \frac{1}{\rho_9}\mathbf{C}^1(\lambda)D_1^1(\mu). \tag{12.292}$$

For the commutation relation of $D_2^2(\lambda)$ with $\mathbf{C}^1(\mu)$ we start from (12.C.1o),

$$\rho_6 K^t(B \otimes \bar{A}) + \rho_1 \check{r}_2(D \otimes \bar{C}) = -\rho_{10}(\bar{C} \otimes D)J + (\bar{D} \otimes C). \tag{12.293}$$

We observe that $\rho_1 \check{r}_2 = \rho_3 I_4 + \rho_8(\sigma^z \otimes \sigma^z)P$, and thus $[(1, 0) \otimes (0, 1)]\rho_1 \check{r}_2 = \rho_3(1, 0) \otimes (0, 1) - \rho_8(0, 1) \otimes (1, 0)$. Similarly, $J(\binom{0}{1} \otimes I_2) = P(\sigma^z \otimes I_2)(\binom{0}{1} \otimes I_2) = -I_2 \otimes \binom{0}{1}$. We further define the vector

$$\boldsymbol{\xi} = [(1, 0) \otimes (0, 1)]K^t = (0, 1, -1, 0), \tag{12.294}$$

which will frequently appear in what follows. Multiplication of equation (12.293) by $(1, 0) \otimes (0, 1)$ from the left and by $\binom{0}{1} \otimes I_2$ from the right then gives us the desired formula,

$$D_2^2(\lambda)\mathbf{C}^1(\mu) = -\frac{\rho_{10}}{\rho_8}\mathbf{C}^1(\mu)D_2^2(\lambda) + \frac{\rho_6}{\rho_8}\boldsymbol{\xi}\big(\mathbf{B}_2(\lambda) \otimes A(\mu)\big)$$
$$+\frac{\rho_3}{\rho_8}D_2^1(\lambda)\mathbf{C}^2(\mu) - \frac{1}{\rho_8}D_2^1(\mu)\mathbf{C}^2(\lambda). \tag{12.295}$$

We now turn to the problem of finding the appropriate commutation relations of $A(\lambda)$ with $\mathbf{C}^1(\mu)$. This is the point where the hidden 6-vertex structure comes into play. Let us consider equations (12.C.1i) and (12.C.1l),

$$\rho_9 J^t(A \otimes \bar{C}) + C \otimes \bar{A} = \rho_4(\bar{C} \otimes A)\check{r}_1 + \rho_6(\bar{D} \otimes B)K^t, \tag{12.296}$$

$$-\rho_9 J^t(B \otimes \bar{D}) + D \otimes \bar{B} = \rho_6(\bar{C} \otimes A)K + \rho_1(\bar{D} \otimes B)\check{r}_2. \tag{12.297}$$

Equation (12.296) contains the commutation relations between $A(\lambda)$ and $\mathbf{C}^1(\mu)$, but still is not of the appropriate form. Instead, Ramos and Martins considered a certain linear combination of (12.296) and (12.297). They were led to this idea by analogy with their calculations for the gl(1|2) vertex model mentioned above (see [370]), where all Boltzmann weights are parameterized by rational functions and thus are easier to handle.

Let us introduce the shorthand notation $L = I_4 - P$. Then $\check{r}_1$ can be written as

$$\check{r}_1 = I_4 - \frac{\rho_7}{\rho_4}L. \tag{12.298}$$

Can this matrix be equivalent to the rational R-matrix (12.50)? If yes, then a reparameterization $v(\lambda)$ should exist, such that $v(\lambda) - v(\mu) = \rho_7/\rho_5$. This cannot be the case, since

ρ_7/ρ_5 is not odd. Let us consider, however, the combination of Boltzmann weights

$$a(\lambda, \mu) = \frac{\rho_7}{\rho_4} - \frac{\rho_6^2}{\rho_4 \rho_8} = -\frac{\rho_9 \rho_{10}}{\rho_4 \rho_8}. \tag{12.299}$$

Inserting the function

$$v(\lambda) = -i \, \mathrm{ctg}\,(2\lambda)\mathrm{ch}(2h) \tag{12.300}$$

it can be rewritten as

$$a(\lambda, \mu) = \frac{v(\lambda) - v(\mu)}{v(\lambda) - v(\mu) + 2iu}. \tag{12.301}$$

The matrix

$$\check{r} = \check{r}_1 + \frac{\rho_6^2}{\rho_4 \rho_8} L = I_4 - a(\lambda, \mu)L \tag{12.302}$$

is therefore equivalent to the R-matrix (12.50) of the spin-$\frac{1}{2}$ XXX chain and, in fact, encodes the hidden 6-vertex structure discovered by Ramos and Martins. The matrix $\check{r}$ may appear as an ad hoc notion at this point. We wish to stress, however, that it naturally arises in the work [335, 336] on the Hubbard model on the infinite interval (see Chapter 15).

We shall now use equation (12.302) in order to introduce the matrix $\check{r}$ into the commutation relation between $A(\lambda)$ and $\mathbf{C}^1(\mu)$. We note the relations

$$K K^t = 2L, \qquad \rho_1 \check{r}_2 K^t = (2\rho_3 - \rho_1)K^t. \tag{12.303}$$

We multiply (12.297) by $\rho_6 K^t/2\rho_8$ from the right and, using (12.303), add it to (12.296). The resulting equation is equivalent to

$$A \otimes \bar{C} = \frac{\rho_4}{\rho_9} J(\bar{C} \otimes A)\check{r} - \frac{1}{\rho_9} J(C \otimes \bar{A})$$
$$+ \frac{\rho_6}{2\rho_8}(B \otimes \bar{D})K^t - \frac{\rho_6}{2\rho_8 \rho_9} J(D \otimes \bar{B})K^t + \frac{\rho_1 \rho_6}{2\rho_8 \rho_9} J(\bar{D} \otimes B)K^t. \tag{12.304}$$

This equation can be further simplified. Let us multiply (12.297) by J from the left and by $L = K K^t/2$ from the right. Taking into account that $K^2 = 0$ and $\check{r}_2 K = K$, we obtain

$$-\rho_9(B \otimes \bar{D})L + J(D \otimes \bar{B})L - \rho_1 J(\bar{D} \otimes B)L = 0. \tag{12.305}$$

We multiply this equation by $\rho_6/2\rho_8 \rho_9$ and add it to (12.304). Then

$$A \otimes \bar{C} = \frac{\rho_4}{\rho_9} J(\bar{C} \otimes A)\check{r} - \frac{1}{\rho_9} J(C \otimes \bar{A}) + \frac{\rho_6}{\rho_8}(B \otimes \bar{D})\frac{K^t - L}{2}$$
$$- \frac{\rho_6}{\rho_8 \rho_9} J(D \otimes \bar{B})\frac{K^t - L}{2} + \frac{\rho_1 \rho_6}{\rho_8 \rho_9} J(\bar{D} \otimes B)\frac{K^t - L}{2}. \tag{12.306}$$

Finally, to project out the commutation relations with the vector $\mathbf{C}^1(\mu)$ of creation operators we multiply (12.306) by $I_2 \otimes (1, 0)$ from the left. We further take advantage of the relation

$(K^t - L)/2 = \left[\binom{0}{1} \otimes \binom{1}{0}\right] \otimes \boldsymbol{\xi}$ and end up with

$$A(\lambda) \otimes \mathbf{C}^1(\mu) = \frac{\rho_4}{\rho_9}\left(\mathbf{C}^1(\mu) \otimes A(\lambda)\right)\check{r} - \frac{1}{\rho_9}\left(\mathbf{C}^1(\lambda) \otimes A(\mu)\right) + \frac{\rho_6}{\rho_8}\left(\boldsymbol{\xi} \otimes \mathbf{B}_2(\lambda)\right)D_1^1(\mu)$$

$$- \frac{\rho_6}{\rho_8\rho_9}D_2^1(\lambda)\left(\boldsymbol{\xi} \otimes \mathbf{B}_1(\mu)\right) + \frac{\rho_1\rho_6}{\rho_8\rho_9}D_2^1(\mu)\left(\boldsymbol{\xi} \otimes \mathbf{B}_1(\lambda)\right). \qquad (12.307)$$

Equations (12.292), (12.295) and (12.307) will turn out to be sufficient to diagonalize the transfer matrix in the one-particle sector.

It is equation (12.307) where the auxiliary spin problem we are familiar with from the coordinate Bethe ansatz appears in the form of the inconspicuous matrix $\check{r}$ on the right-hand side. Having in mind the generalization to two and more particles we introduce a convenient notation. The definition

$$A_j(\lambda) = I_2^{\otimes(j-1)} \otimes A(\lambda) \otimes I_2^{\otimes(N-j)} \qquad (12.308)$$

embeds the submatrix $A(\lambda)$ of the monodromy matrix into $\left(\text{End}(\mathbb{C})\right)^{\otimes N} \otimes \mathcal{H}$, where $\mathcal{H}$ is the electronic space of states of an L-site chain. N will be the number of particles. We shall write $A_0^{(0)}(\lambda) = A(\lambda) \otimes I_2^{\otimes N}$ and $A_j^{(0)}(\lambda) = I_2 \otimes A_j(\lambda)$. The index 'zero' here refers to an auxiliary space, which may be imagined as an additional site. We set $r = P\check{r}$, where P is the 4×4 permutation matrix with matrix elements $P_{\beta\delta}^{\alpha\gamma} = \delta_\delta^\alpha \delta_\beta^\gamma$ and introduce the L-matrix of the auxiliary spin problem

$$L_j(\lambda) = r_{\beta\delta}^{\alpha\gamma}(\lambda, \lambda_j)\, e_\alpha^\beta \otimes e_{j\gamma}^\delta \qquad (12.309)$$

and the corresponding inhomogeneous monodromy matrix

$$T^{(1)}(\lambda) = L_N(\lambda)\ldots L_1(\lambda) = \begin{pmatrix} A^{(1)}(\lambda) & B^{(1)}(\lambda) \\ C^{(1)}(\lambda) & D^{(1)}(\lambda) \end{pmatrix}. \qquad (12.310)$$

The L-matrix acts triangularly on the auxiliary pseudo vacuum $\binom{1}{0}^{\otimes N}$. As can be seen from (12.302) and (12.309) we have

$$L_j(\lambda)\binom{1}{0}^{\otimes N} = \begin{pmatrix} 1 & * \\ 0 & a(\lambda, \lambda_j) \end{pmatrix}\binom{1}{0}^{\otimes N}, \qquad (12.311)$$

where the asterisk stands for $(1 - a(\lambda, \lambda_j))e_{j2}^1$. Hence, $\binom{1}{0}^{\otimes N}$ is an appropriate pseudo vacuum for the monodromy matrix $T^{(1)}(\lambda)$,

$$T^{(1)}(\lambda)\binom{1}{0}^{\otimes N} = \begin{pmatrix} 1 & B^{(1)}(\lambda) \\ 0 & \prod_{j=1}^N a(\lambda, \lambda_j) \end{pmatrix}\binom{1}{0}^{\otimes N}, \qquad (12.312)$$

and we can apply the general formulae for the solution of the gl(2) generalized model obtained in Section 12.1.7. From (12.312), (12.67), (12.68) it follows that the auxiliary transfer matrix $\mathrm{tr}_0\big(T^{(1)}(\lambda)\big) = A^{(1)}(\lambda) + D^{(1)}(\lambda)$ has eigenvalues

$$\Lambda^{(1)}(\lambda) = \prod_{k=1}^{M} \frac{1}{a(\mu_k, \lambda)} + \prod_{j=1}^{N} a(\lambda, \lambda_j) \prod_{k=1}^{M} \frac{1}{a(\lambda, \mu_k)}, \qquad (12.313)$$

where the Bethe ansatz roots μ_k are solutions of the Bethe ansatz equations

$$\prod_{j=1}^{N} \frac{1}{a(\mu_k, \lambda_j)} = \prod_{\substack{l=1 \\ l \neq k}}^{M} \frac{a(\mu_l, \mu_k)}{a(\mu_k, \mu_l)}, \qquad k = 1, \dots, M. \qquad (12.314)$$

The corresponding eigenvectors are $|\mu_1, \dots, \mu_M\rangle = B^{(1)}(\mu_1) \dots B^{(1)}(\mu_M)\binom{1}{0}^{\otimes N}$.

How can this be utilized for the diagonalization of the transfer matrix of the Hubbard model in the one-particle sector? We first rewrite equation (12.307) in the form

$$A(\lambda) \otimes \mathbf{C}^1(\lambda_1) = \frac{\rho_4(\lambda, \lambda_1)}{\rho_9(\lambda, \lambda_1)} \big(I_2 \otimes \mathbf{C}^1(\lambda_1)\big) A_0^{(0)}(\lambda) T^{(1)}(\lambda)$$

$$- \frac{1}{\rho_9(\lambda, \lambda_1)} \big(I_2 \otimes \mathbf{C}^1(\lambda)\big) A_0^{(0)}(\lambda_1) T^{(1)}(\lambda_1)$$

$$+ \frac{\rho_6(\lambda, \lambda_1)}{\rho_8(\lambda, \lambda_1)} \big(\boldsymbol{\xi} \otimes \mathbf{B}_2(\lambda)\big) D_1^1(\lambda_1) + \dots \qquad (12.315)$$

Here μ was replaced by λ_1. The dots denote terms which annihilate the pseudo vacuum $|0\rangle$. $T^{(1)}(\lambda) = L_1(\lambda, \lambda_1)$ is the monodromy matrix for the auxiliary spin problem for a chain of length $N = 1$. Taking the trace in space zero of (12.315) we obtain

$$\mathrm{tr}_0\big(A_0^{(0)}(\lambda)\big) \mathbf{C}^1(\lambda_1) = \frac{\rho_4(\lambda, \lambda_1)}{\rho_9(\lambda, \lambda_1)} \mathbf{C}^1(\lambda_1) \, \mathrm{tr}_0\big(A_0^{(0)}(\lambda) T^{(1)}(\lambda)\big)$$

$$- \frac{1}{\rho_9(\lambda, \lambda_1)} \mathbf{C}^1(\lambda) \, \mathrm{tr}_0\big(A_0^{(0)}(\lambda_1) T^{(1)}(\lambda_1)\big)$$

$$+ \frac{\rho_6(\lambda, \lambda_1)}{\rho_8(\lambda, \lambda_1)} \boldsymbol{\xi} \big(\mathbf{B}_2(\lambda) \otimes I_2\big) D_1^1(\lambda_1) + \dots \qquad (12.316)$$

A similar commutation relation for $\mathrm{tr}\big(D(\lambda)\big)$ and $\mathbf{C}^1(\lambda_1)$ is obtained by adding (12.292) and (12.295),

$$\mathrm{tr}_0\big(D_0^{(0)}(\lambda)\big) \mathbf{C}^1(\lambda_1) = \mathbf{C}^1(\lambda_1) \Big[D_1^1(\lambda) \frac{\rho_4(\lambda, \lambda_1)}{\rho_9(\lambda, \lambda_1)} - D_2^2(\lambda) \frac{\rho_{10}(\lambda, \lambda_1)}{\rho_8(\lambda, \lambda_1)} \Big]$$

$$- \frac{1}{\rho_9(\lambda, \lambda_1)} \mathbf{C}^1(\lambda) D_1^1(\lambda_1)$$

$$+ \frac{\rho_6(\lambda, \lambda_1)}{\rho_8(\lambda, \lambda_1)} \boldsymbol{\xi} \big(\mathbf{B}_2(\lambda) \otimes I_2\big) \mathrm{tr}_0\big(A_0^{(0)}(\lambda_1) T^{(1)}(\lambda_1)\big) + \dots \qquad (12.317)$$

Again the dots denote terms which annihilate the pseudo vacuum. We now subtract (12.316) from (12.317), insert the definition $\phi^{(1)}(\lambda) = \mathbf{C}^1(\lambda)$, and use (12.299). Then

$$t(\lambda)\phi^{(1)}(\lambda_1) = -\phi^{(1)}(\lambda_1)\left[\left\{D_1^1(\lambda) - \text{tr}_0\big(A_0^{(0)}(\lambda)T^{(1)}(\lambda)\big)\right\}\frac{1}{a(\lambda,\lambda_1)} + D_2^2(\lambda)\right]\frac{\rho_{10}(\lambda,\lambda_1)}{\rho_8(\lambda,\lambda_1)}$$
$$+\left[\frac{1}{\rho_9(\lambda,\lambda_1)}\phi^{(1)}(\lambda) + \frac{\rho_6(\lambda,\lambda_1)}{\rho_8(\lambda,\lambda_1)}\xi\big(\mathbf{B}_2(\lambda)\otimes I_2\big)\right]\left\{\text{tr}_0\big(A_0^{(0)}(\lambda)T^{(1)}(\lambda)\big) - D_1^1(\lambda)\right\} + \dots$$

$$\tag{12.318}$$

Recall our ansatz (12.290) for the one-particle eigenvector. At this point we shall see that it actually works. For any eigenvector $\binom{f_1}{f_2}$ of the auxiliary transfer matrix we find, using (12.287),

$$\text{tr}_0\big(A_0^{(0)}(\lambda)T^{(1)}(\lambda)\big)|\hat{F}\rangle = \omega_2^L(\lambda)\text{tr}_0\big(T^{(1)}(\lambda)\big)\binom{f_1}{f_2}\otimes|0\rangle$$
$$= \omega_2^L(\lambda)\Lambda^{(1)}(\lambda)|\hat{F}\rangle,\tag{12.319}$$

$$D_1^1(\lambda)|\hat{F}\rangle = \omega_1^L(\lambda)|\hat{F}\rangle,\quad D_2^2(\lambda)|\hat{F}\rangle = \omega_3^L(\lambda)|\hat{F}\rangle.\tag{12.320}$$

Thus, if $\binom{f_1}{f_2}$ is an eigenvector of the auxiliary transfer matrix $A^{(1)}(\lambda) + D^{(1)}(\lambda)$, then $|\phi^{(1)}\rangle = \phi^{(1)}(\lambda_1)|\hat{F}\rangle$ is a one-particle eigenvector of the transfer matrix $t(\lambda)$ of the Hubbard model with eigenvalue

$$\Lambda(\lambda) = -\left[\left\{\omega_1^L(\lambda) - \omega_2^L(\lambda)\Lambda^{(1)}(\lambda)\right\}\frac{1}{a(\lambda,\lambda_1)} + \omega_3^L(\lambda)\right]\frac{\rho_{10}(\lambda,\lambda_1)}{\rho_8(\lambda,\lambda_1)},\tag{12.321}$$

provided that the Bethe ansatz equation

$$\left(\frac{\omega_1(\lambda_1)}{\omega_2(\lambda_1)}\right)^L = \Lambda^{(1)}(\lambda_1)\tag{12.322}$$

is satisfied.

The last step is to insert the explicit form of the eigenvalue $\Lambda^{(1)}(\lambda)$ of the auxiliary transfer matrix into (12.321) and (12.322) and to write explicitly the eigenvector $|\phi^{(1)}\rangle$. Since $N = 1$, the problem is rather trivial, and we could have avoided the machinery of the algebraic Bethe ansatz for the solution of the auxiliary problem. We included it already here in order to facilitate the comparison with the two-particle case. From the definitions (12.309), (12.310) we have $A^{(1)}(\lambda) + D^{(1)}(\lambda) = (1 + a(\lambda,\lambda_1))I_2$. Thus, $A^{(1)}(\lambda) + D^{(1)}(\lambda)$ is proportional to the unit matrix (acting in the 'quantum space of the auxiliary problem'). It follows that

$$\Lambda^{(1)} = 1 + a(\lambda,\lambda_1),\tag{12.323}$$

and that every vector in $\mathbb{C}^2$ is a corresponding eigenvector. These findings are in agreement with our algebraic Bethe ansatz result. Because of the su(2) symmetry of the auxiliary monodromy matrix we have to impose the restriction $N \geq 2M$ (see appendix 3.D.3). Thus, $M = 0$ and, using the convention that a product has to be replaced with 1, once its upper limit is smaller than its lower limit, (12.313) implies the correct result $\Lambda^{(1)}(\lambda) = 1 + a(\lambda,\lambda_1)$.

The only Bethe ansatz vector in the sector $M = 0$ is the pseudo vacuum $\binom{1}{0}$. A second eigenvector $\binom{0}{1}$ follows by application of σ^-.

Let us insert (12.323) into (12.321), (12.322), and let us summarize. A set of one-particle algebraic Bethe ansatz eigenvectors of the transfer matrix $t(\lambda)$ of the Hubbard model is given by

$$|\phi^{(1)}\rangle = \mathbf{C}^1(\lambda_1)\binom{1}{0} \otimes |0\rangle = C_1^1(\lambda_1)|0\rangle . \tag{12.324}$$

The corresponding eigenvalue is

$$\Lambda(\lambda) = -\left[\omega_1^L(\lambda)\frac{1}{a(\lambda, \lambda_1)} - \omega_7^L(\lambda)\frac{1}{a(\lambda, \lambda_1)} - \omega_2^L(\lambda) + \omega_3^L(\lambda)\right]\frac{\rho_{10}(\lambda, \lambda_1)}{\rho_8(\lambda, \lambda_1)} , \tag{12.325}$$

where the Bethe ansatz root λ_1 has to be calculated from the Bethe ansatz equation

$$\left(\frac{\omega_1(\lambda_1)}{\omega_2(\lambda_1)}\right)^L = 1 . \tag{12.326}$$

For a full understanding of the one-particle states (12.324) we have to study their behaviour under the action of the spin and η-spin operators. Let us start with the spin. The commutation relations of the monodromy matrix elements with the spin operators are compactly combined in equation (12.241). Those equations contain, in particular, the relations

$$[S^\pm, \mathbf{C}^1(\lambda)] = \mathbf{C}^1(\lambda)\sigma^\pm , \tag{12.327a}$$

$$[S^z, \mathbf{C}^1(\lambda)] = \mathbf{C}^1(\lambda)\tfrac{1}{2}\sigma^z . \tag{12.327b}$$

The spin operators annihilate the pseudo vacuum $|0\rangle$. Hence, we infer from (12.327) that

$$S^+|\phi^{(1)}\rangle = 0 , \tag{12.328a}$$

$$S^z|\phi^{(1)}\rangle = \tfrac{1}{2}|\phi^{(1)}\rangle , \tag{12.328b}$$

$$S^-|\phi^{(1)}\rangle = C_2^1(\lambda_1)|0\rangle . \tag{12.328c}$$

Thus, the vector $|\phi^{(1)}\rangle = C_1^1(\lambda_1)|0\rangle$ is a spin-$\frac{1}{2}$ highest weight state, the vectors $C_1^1(\lambda_1)|0\rangle$ and $C_2^1(\lambda_1)|0\rangle$ form a spin-$\frac{1}{2}$ doublet.

Similarly, we conclude from (12.245) and (12.250) that

$$\{\eta^-, \mathbf{C}^1(\lambda)\} = -\mathbf{C}^2(\lambda) , \tag{12.329a}$$

$$[\eta^z, \mathbf{C}^1(\lambda)] = \tfrac{1}{2}\mathbf{C}^1(\lambda) , \tag{12.329b}$$

$$\{\eta^+, \mathbf{C}^1(\lambda)\} = 0 . \tag{12.329c}$$

It follows that

$$\eta^-|\phi^{(1)}\rangle = 0 , \tag{12.330a}$$

$$\eta^z|\phi^{(1)}\rangle = \tfrac{1}{2}(1 - L)|\phi^{(1)}\rangle . \tag{12.330b}$$

Thus, $|\phi^{(1)}\rangle$ is the lowest weight state in a multiplet of η-spin L, the remaining states being generated by consecutive action of η^+ on $|\phi^{(1)}\rangle$.

By comparison of the one-particle Bethe ansatz equation (12.326) with the corresponding one-particle Lieb-Wu equation $e^{ik_1 L} = 1$ (see (3.95)) we are led to the reparameterization [499]

$$e^{ik(\lambda)} = \frac{\omega_1(\lambda)}{\omega_2(\lambda)} = -\operatorname{ctg}(\lambda)e^{2h} \tag{12.331}$$

of equation (12.326), which relates the spectral parameter λ to the pseudo momentum k of the coordinate Bethe ansatz. Let us further define

$$e^{ip(\lambda)} = \frac{\omega_2(\lambda)}{\omega_3(\lambda)} = -\operatorname{ctg}(\lambda)e^{-2h} . \tag{12.332}$$

The functions $k(\lambda)$, $p(\lambda)$ and $v(\lambda)$, equation (12.300), are then simply related through

$$v(\lambda) = -\sin(k(\lambda)) + iu = -\sin(p(\lambda)) - iu . \tag{12.333}$$

Hence, the constraint (12.109) connecting λ and h for given u turns into the relation

$$\sin(k) - \sin(p) = 2iu \tag{12.334}$$

between k and p.

12.6.5 The two-particle states

The generalization of the one-particle calculation of the previous section to two particles is not straightforward. We expect that the rather general ansatz

$$\begin{aligned}|\phi^{(2)}\rangle &= \phi^{(2)}(\lambda_1, \lambda_2)|\hat{F}\rangle , \\ |\hat{F}\rangle &= \mathbf{f} \otimes |0\rangle , \quad \mathbf{f} \in \mathbb{C}^2 \otimes \mathbb{C}^2\end{aligned} \tag{12.335}$$

should work. But what is the appropriate choice of the vector $\phi^{(2)}(\lambda_1, \lambda_2)$ of creation operators? It turns out that a choice of the form $\mathbf{C}^1(\lambda_1) \otimes \mathbf{C}^1(\lambda_2)$, which works in the construction of the nested algebraic Bethe ansatz for, say, the supersymmetric t-J model [119], fails in the case at hand.

In the construction of the one-particle eigenstates we saw that we did not need to consider the one-particle creation operators $B_2^1(\lambda)$, $B_2^2(\lambda)$. It was sufficient to work only with $C_1^1(\lambda)$, $C_2^1(\lambda)$. For the two-particle states, however, there is yet another candidate for a creation operator, namely $D_2^1(\lambda)$. As we shall see, two-particle eigenstates of the Hubbard model transfer matrix can be constructed by choosing $\phi^{(2)}(\lambda_1, \lambda_2)$ as an appropriate combination of $\mathbf{C}^1(\lambda_1) \otimes \mathbf{C}^1(\lambda_2)$ and $D_2^1(\lambda_1)$. An idea for the construction of such an appropriate combination comes from the observation that a relation of the form

$$\phi^{(2)}(\lambda_1, \lambda_2) \sim \phi^{(2)}(\lambda_2, \lambda_1)\check{r}(\lambda_1, \lambda_2) \tag{12.336}$$

is crucial for all known nested algebraic Bethe ansatz calculations. Let us therefore seek to introduce the matrix $\check{r}$ into the commutation relation between two matrices $C(\lambda)$ in a similar manner as in the derivation of equation (12.306) from (12.296) and (12.297).

We take equations (12.C.1m) and (12.C.1p) from appendix 12.C,

$$\rho_6 K^t(A \otimes \bar{A}) - \rho_1 \check{r}_2(C \otimes \bar{C}) = -\rho_4(\bar{C} \otimes C)\check{r}_1 + \rho_6(\bar{D} \otimes D)K^t, \tag{12.337}$$

$$\rho_6 K^t(B \otimes \bar{B}) + \rho_1 \check{r}_2(D \otimes \bar{D}) = -\rho_6(\bar{C} \otimes C)K + \rho_1(\bar{D} \otimes D)\check{r}_2. \tag{12.338}$$

Proceeding similarly as in the derivation of (12.306) we multiply (12.338) by $\rho_6 K^t / 2\rho_8$ from the right and add the result to (12.337). We obtain an equation which is equivalent to

$$\check{r}_2(C \otimes \bar{C}) - \frac{\rho_6}{2\rho_8} \check{r}_2(D \otimes \bar{D})K^t - \frac{\rho_6}{\rho_1} K^t(A \otimes \bar{A}) - \frac{\rho_6^2}{2\rho_1\rho_8} K^t(B \otimes \bar{B})K^t$$

$$= \frac{\rho_4}{\rho_1}(\bar{C} \otimes C)\check{r} - \frac{\rho_6}{2\rho_8}(\bar{D} \otimes D)K^t. \tag{12.339}$$

Next, we multiply (12.338) by $\rho_6 L/2\rho_1\rho_8$ from the right and add the result to (12.339). Since $K^2 = 0$ and $\check{r}_2 K = K$, we obtain

$$\check{r}_2(C \otimes \bar{C}) - \frac{\rho_6}{\rho_8} \check{r}_2(D \otimes \bar{D})\frac{K^t - L}{2} - \frac{\rho_6}{\rho_1} K^t(A \otimes \bar{A}) - \frac{\rho_6^2}{\rho_1\rho_8} K^t(B \otimes \bar{B})\frac{K^t - L}{2}$$

$$= \frac{\rho_4}{\rho_1}(\bar{C} \otimes C)\check{r} - \frac{\rho_6}{\rho_8}(\bar{D} \otimes D)\frac{K^t - L}{2}. \tag{12.340}$$

Finally, to project out the commutation relation between $\mathbf{C}^1(\lambda)$ and $\mathbf{C}^1(\mu)$ we multiply by $(1, 0) \otimes (1, 0)$ from the left. We also use once more that $(K^t - L)/2 = \left[\binom{0}{1} \otimes \binom{1}{0}\right] \otimes \boldsymbol{\xi}$. Then

$$\mathbf{C}^1(\lambda) \otimes \mathbf{C}^1(\mu) - \frac{\rho_6(\lambda, \mu)}{\rho_8(\lambda, \mu)} D_2^1(\lambda)D_1^1(\mu)\boldsymbol{\xi} = \frac{\rho_4(\lambda, \mu)}{\rho_1(\lambda, \mu)}\left(\mathbf{C}^1(\mu) \otimes \mathbf{C}^1(\lambda)\right)\check{r}(\lambda, \mu)$$

$$- \frac{\rho_6(\lambda, \mu)}{\rho_8(\lambda, \mu)} D_2^1(\mu)D_1^1(\lambda)\boldsymbol{\xi}. \tag{12.341}$$

Now $\boldsymbol{\xi}$ is an eigenvector of $\check{r}(\lambda, \mu)$,

$$\boldsymbol{\xi} = -\frac{\rho_4\rho_8}{\rho_1\rho_7}\boldsymbol{\xi}\,\check{r}. \tag{12.342}$$

Inserting (12.342) into the right hand side of (12.341) and using (12.123) we end up with

$$\mathbf{C}^1(\lambda) \otimes \mathbf{C}^1(\mu) - \frac{\rho_6(\lambda, \mu)}{\rho_8(\lambda, \mu)} D_2^1(\lambda)D_1^1(\mu)\boldsymbol{\xi}$$

$$= \left[\mathbf{C}^1(\mu) \otimes \mathbf{C}^1(\lambda) - \frac{\rho_6(\mu, \lambda)}{\rho_8(\mu, \lambda)} D_2^1(\mu)D_1^1(\lambda)\boldsymbol{\xi}\right]\frac{\rho_4(\lambda, \mu)}{\rho_1(\lambda, \mu)}\check{r}(\lambda, \mu). \tag{12.343}$$

This equation is of the form (12.336). Thus, the combination of creation operators

$$\phi^{(2)}(\lambda_1, \lambda_2) = \mathbf{C}^1(\lambda_1) \otimes \mathbf{C}^1(\lambda_2) - \frac{\rho_6(\lambda_1, \lambda_2)}{\rho_8(\lambda_1, \lambda_2)} D_2^1(\lambda_1)D_1^1(\lambda_2)\boldsymbol{\xi}, \tag{12.344}$$

originally introduced by Ramos and Martins [371], provides a promising ansatz for the two-particle eigenvectors (12.335).

To show that this ansatz actually works one has to commute the submatrix $A(\lambda)$ of the monodromy matrix and the monodromy matrix elements $D_1^1(\lambda)$ and $D_2^2(\lambda)$ through $\phi^{(2)}(\lambda_1, \lambda_2)$. Unfortunately, this is a cumbersome task. We shall omit the details of the calculation here. One has to use (12.292), (12.295), and (12.307) and a number of additional commutation relations listed in (12.C.1). The coefficients in the resulting formulae are rational functions of the Boltzmann weights ρ_j and also contain the R-matrix $\check{r}$ of the auxiliary spin problem. These coefficients have to be appropriately rearranged using certain identities, involving three different arguments λ, μ, ν, between the Boltzmann weights and certain elementary properties of the R-matrix $\check{r}$. There is one guiding principle in the calculations: as a consequence of (12.343) the coefficients of the different terms are related by interchange of λ_1 and λ_2 and multiplication by $\frac{\rho_4}{\rho_1}\check{r}$. This fact allows one to figure out the required identities between the Boltzmann weights. In appendix 12.D we provide a complete list of the additional commutation relations and the identities we have used. With the aid of appendix 12.D the reader should be able to verify the following formulae,

$$D_1^1(\lambda)\,\phi^{(2)}(\lambda_1, \lambda_2) = \phi^{(2)}(\lambda_1, \lambda_2)\,D_1^1(\lambda)\prod_{j=1}^{2}\frac{\rho_4(\lambda, \lambda_j)}{\rho_9(\lambda, \lambda_j)}$$

$$-\sum_{j=1}^{2}\phi^{(2)}(\lambda, \hat{\lambda}_j)\,S_{j-1}D_1^1(\lambda_j)\frac{1}{\rho_9(\lambda, \lambda_j)}\prod_{\substack{k=1\\k\neq j}}^{2}\frac{\rho_4(\lambda_j, \lambda_k)}{\rho_9(\lambda_j, \lambda_k)}$$

$$-D_2^1(\lambda)D_1^1(\lambda_1)D_1^1(\lambda_2)\,\boldsymbol{\xi}\,\frac{\rho_6(\lambda_1, \lambda_2)}{\rho_8(\lambda_1, \lambda_2)}\left[\prod_{j=1}^{2}\frac{\rho_{10}(\lambda, \lambda_j)}{\rho_8(\lambda, \lambda_j)}+1\right], \tag{12.345}$$

$$D_2^2(\lambda)\,\phi^{(2)}(\lambda_1, \lambda_2) = \phi^{(2)}(\lambda_1, \lambda_2)\,D_2^2(\lambda)\prod_{j=1}^{2}\frac{\rho_{10}(\lambda, \lambda_j)}{\rho_8(\lambda, \lambda_j)}$$

$$+\sum_{j=1}^{2}\left[\left[\boldsymbol{\xi}\left(\mathbf{B}_2(\lambda)\otimes I_2\right)\right]\otimes\phi^{(1)}(\hat{\lambda}_j)\right]S_{j-1}\,\mathrm{tr}_0\left(A_0^{(0)}(\lambda_j)T^{(1)}(\lambda_j)\right)$$

$$\times\frac{\rho_6(\lambda, \lambda_j)}{\rho_8(\lambda, \lambda_j)}\prod_{\substack{k=1\\k\neq j}}^{2}\frac{\rho_4(\lambda_j, \lambda_k)}{\rho_9(\lambda_j, \lambda_k)}$$

$$+D_2^1(\lambda)\,\boldsymbol{\xi}\left[A(\lambda_1)\otimes A(\lambda_2)\right]\frac{\rho_6(\lambda_1, \lambda_2)}{\rho_8(\lambda_1, \lambda_2)}\left[\prod_{j=1}^{2}\frac{\rho_{10}(\lambda, \lambda_j)}{\rho_8(\lambda, \lambda_j)}+1\right]+\dots, \tag{12.346}$$

$$A(\lambda)\otimes\phi^{(2)}(\lambda_1, \lambda_2) = \left[I_2\otimes\phi^{(2)}(\lambda_1, \lambda_2)\right]A_0^{(0)}(\lambda)T^{(1)}(\lambda)\prod_{j=1}^{2}\frac{\rho_4(\lambda, \lambda_j)}{\rho_9(\lambda, \lambda_j)}$$

$$-\sum_{j=1}^{2}\left[\left[I_2\otimes\phi^{(2)}(\lambda, \hat{\lambda}_j)\right]S_{j-1}A_0^{(0)}(\lambda_j)T^{(1)}(\lambda_j)\frac{1}{\rho_9(\lambda, \lambda_j)}\right.$$

$$-\left[\boldsymbol{\xi}\otimes\mathbf{B}_2(\lambda)\otimes\phi^{(1)}(\hat{\lambda}_j)\right]S_{j-1}D_1^1(\lambda_j)\frac{\rho_6(\lambda_1, \lambda_2)}{\rho_8(\lambda_1, \lambda_2)}\right]\prod_{\substack{k=1\\k\neq j}}^{2}\frac{\rho_4(\lambda_j, \lambda_k)}{\rho_9(\lambda_j, \lambda_k)}+\dots \tag{12.347}$$

As before the dots in (12.346) and (12.347) denote terms which annihilate the pseudo vacuum. We introduced two new notations. First,

$$\hat{\lambda}_j = \begin{cases} \lambda_2 & \text{if } j = 1 \\ \lambda_1 & \text{if } j = 2 \,, \end{cases} \tag{12.348}$$

and second

$$S_{j-1} = \begin{cases} \text{id} & \text{if } j = 1 \\ \frac{\rho_4(\lambda_1, \lambda_2)}{\rho_1(\lambda_1, \lambda_2)} \, \check{r}^{(0)}_{1,2}(\lambda_1, \lambda_2) & \text{if } j = 2 \,, \end{cases} \tag{12.349}$$

where $\check{r}^{(0)}_{1,2}(\lambda_1, \lambda_2) = I_2 \otimes \check{r}(\lambda_1, \lambda_2)$ acts on a triple tensor product of auxiliary spaces $\mathbb{C}^2$, the first one being interpreted as the auxiliary space of the auxiliary spin problem, the tensor product of the other two as the corresponding quantum space.

The commutation relations of $t(\lambda) = \text{str}(\mathcal{T}(\lambda))$ with $\phi^{(2)}(\lambda_1, \lambda_2)$ are now easily obtained from (12.345)–(12.347). We take the trace of (12.347) in space zero and subtract the result from the sum of equations (12.345) and (12.346). Then

$$t(\lambda) \, \phi^{(2)}(\lambda_1, \lambda_2) = \phi^{(2)}(\lambda_1, \lambda_2)$$

$$\times \left[\left\{ D_1^1(\lambda) - \text{tr}_0\big(A_0^{(0)}(\lambda) T^{(1)}(\lambda)\big) \right\} \prod_{j=1}^2 \frac{1}{a(\lambda, \lambda_j)} + D_2^2(\lambda) \right] \prod_{j=1}^2 \frac{\rho_{10}(\lambda, \lambda_j)}{\rho_8(\lambda, \lambda_j)}$$

$$+ \sum_{j=1}^2 \left[\frac{1}{\rho_9(\lambda, \lambda_j)} \, \phi^{(2)}(\lambda, \hat{\lambda}_j) + \frac{\rho_6(\lambda, \lambda_j)}{\rho_8(\lambda, \lambda_j)} \left[\boldsymbol{\xi} \, (\mathbf{B}_2(\lambda) \otimes I_2) \right] \otimes \phi^{(1)}(\hat{\lambda}_j) \right] S_{j-1}$$

$$\times \left\{ \text{tr}_0\big(A_0^{(0)}(\lambda_j) T^{(1)}(\lambda_j)\big) - D_1^1(\lambda_j) \right\} \prod_{\substack{k=1 \\ k \neq j}}^2 \frac{\rho_4(\lambda_j, \lambda_k)}{\rho_9(\lambda_j, \lambda_k)}$$

$$+ \frac{\rho_6(\lambda_1, \lambda_2)}{\rho_8(\lambda_1, \lambda_2)} \left[\prod_{j=1}^2 \frac{\rho_{10}(\lambda, \lambda_j)}{\rho_8(\lambda, \lambda_j)} + 1 \right] D_2^1(\lambda)$$

$$\times \boldsymbol{\xi} \left\{ \left[A(\lambda_1) \otimes A(\lambda_2) \right] - D_1^1(\lambda_1) D_1^1(\lambda_2) \right\} + \dots \tag{12.350}$$

Let us compare (12.350) with the corresponding one-particle result (12.318). We see that the first terms on the right-hand sides of both equations are similar to each other. The third term on the right-hand side of (12.350), however, has a novel structure, not observed in the one-particle calculation. It is clear from the comparison of (12.318) and (12.350) that the second term on the right-hand side of (12.350) can be canceled by the same trick as before. We assume that the 'spin part' $\mathbf{f} \in \mathbb{C}^2 \otimes \mathbb{C}^2$ of our two-particle ansatz (12.335) is an eigenvector with eigenvalue $\Lambda^{(1)}(\lambda)$ of the auxiliary transfer matrix $\text{tr}_0\big(T^{(1)}(\lambda)\big)$. Then, as in equations (12.319), (12.320),

$$\mathrm{tr}_0\big(A_0^{(0)}(\lambda)T^{(1)}(\lambda)\big)|\hat{F}\rangle = \omega_2^L(\lambda)\Lambda^{(1)}(\lambda)|\hat{F}\rangle\,, \tag{12.351}$$

$$D_1^1(\lambda)|\hat{F}\rangle = \omega_1^L(\lambda)|\hat{F}\rangle\,, \quad D_2^2(\lambda)|\hat{F}\rangle = \omega_3^L(\lambda)|\hat{F}\rangle\,, \tag{12.352}$$

and the second term on the right-hand side of (12.350) vanishes, if the Bethe ansatz equations

$$\left(\frac{\omega_1(\lambda_j)}{\omega_2(\lambda_j)}\right)^L = \Lambda^{(1)}(\lambda_j)\,, \quad j = 1, 2\,, \tag{12.353}$$

are satisfied. Amazingly, the Bethe ansatz equations are also sufficient for the third term on the right-hand side of (12.350) to vanish. This becomes evident after transforming the term $\xi\big[A(\lambda_1) \otimes A(\lambda_2)\big]$ as follows,

$$
\begin{aligned}
\xi\big[A(\lambda_1) \otimes A(\lambda_2)\big] &= \xi\, A_1(\lambda_1)A_2(\lambda_2) \\
&= \xi\, A_1(\lambda_1)\check{r}_{1,2}(\lambda_1, \lambda_2)P_{12}P_{12}\check{r}_{1,2}(\lambda_2, \lambda_1)A_2(\lambda_2) \\
&= \xi\, A_1(\lambda_1)r_{1,2}(\lambda_1, \lambda_2)r_{1,2}(\lambda_2, \lambda_1)A_2(\lambda_2) \tag{12.354} \\
&= \xi\, \mathrm{tr}_0\big(A_0^{(0)}(\lambda_1)T^{(1)}(\lambda_1)\big)\mathrm{tr}_0\big(A_0^{(0)}(\lambda_2)T^{(2)}(\lambda_2)\big)\,.
\end{aligned}
$$

Thus, the vector $|\phi^{(2)}\rangle$, equation (12.335), is an eigenvector of the transfer matrix $t(\lambda)$ with eigenvalue

$$\Lambda(\lambda) = \left[\left\{\omega_1^L(\lambda) - \omega_2^L(\lambda)\Lambda^{(1)}(\lambda)\right\}\prod_{j=1}^{2}\frac{1}{a(\lambda, \lambda_j)} + \omega_3^L(\lambda)\right]\prod_{j=1}^{2}\frac{\rho_{10}(\lambda, \lambda_j)}{\rho_8(\lambda, \lambda_j)}\,, \tag{12.355}$$

if $\mathbf{f}$ is an eigenvector with eigenvalue $\Lambda^{(1)}(\lambda)$ of the auxiliary transfer matrix $\mathrm{tr}_0\big(T^{(1)}(\lambda)\big)$, and if the Bethe ansatz equations (12.353) are satisfied.

For the eigenvector $\mathbf{f}$ and the corresponding eigenvalue $\Lambda^{(1)}(\lambda)$ we can resort to our general algebraic Bethe ansatz solution (12.313), (12.314) of the spin problem. Inserting (12.313) for $N = 2$ into (12.353) and (12.355) we can summarize our result: the vector $|\phi^{(2)}\rangle$ is a two-particle eigenvector of the transfer matrix $t(\lambda)$ with eigenvalue

$$
\begin{aligned}
\Lambda(\lambda) = \Bigg[\omega_1^L(\lambda)\prod_{j=1}^{2}\frac{1}{a(\lambda, \lambda_j)} &- \omega_2^L(\lambda)\prod_{j=1}^{2}\frac{1}{a(\lambda, \lambda_j)}\prod_{k=1}^{M}\frac{1}{a(\mu_k, \lambda)} \\
&- \omega_2^L(\lambda)\prod_{k=1}^{M}\frac{1}{a(\lambda, \mu_k)} + \omega_3^L(\lambda)\Bigg]\prod_{j=1}^{2}\frac{\rho_{10}(\lambda, \lambda_j)}{\rho_8(\lambda, \lambda_j)}\,, \tag{12.356}
\end{aligned}
$$

if the Bethe ansatz equations

$$\left(\frac{\omega_1(\lambda_j)}{\omega_2(\lambda_j)}\right)^L = \prod_{k=1}^{M}\frac{1}{a(\mu_k, \lambda_j)}\,, \quad j = 1, 2\,, \tag{12.357}$$

$$\prod_{j=1}^{2}\frac{1}{a(\mu_k, \lambda_j)} = \prod_{\substack{l=1 \\ l \neq k}}^{M}\frac{a(\mu_l, \mu_k)}{a(\mu_k, \mu_l)}\,, \quad k = 1, \ldots, M\,, \tag{12.358}$$

are satisfied and if $\mathbf{f} = B^{(1)}(\mu_1) \ldots B^{(1)}(\mu_M) {\binom{1}{0}}^{\otimes 2}$. Because of the su(2) symmetry of the auxiliary spin problem, the allowed values of M are restricted to $2M \leq N$, i.e., $M = 0, 1$.

To complete our picture of the two-particle states just constructed we have to inspect their behaviour under the action of spin- and η-spin operators. For this purpose we must commute the spin- and η-spin operators through $\phi^{(2)}(\lambda_1, \lambda_2)$. The commutation relations with the vector of creation operators $\mathbf{C}^1(\lambda)$ are collected in equations (12.327), (12.329). In addition, we need the commutation relations with $D_1^1(\lambda)$ and $D_2^1(\lambda)$, which can be extracted from equations (12.241), (12.245), and (12.250),

$$[S^\alpha, (D_1^1(\lambda), D_2^1(\lambda))] = 0, \quad \alpha = x, y, z, \tag{12.359a}$$

$$\{\eta^-, (D_1^1(\lambda), D_2^1(\lambda))\} = (-D_1^2(\lambda), D_1^1(\lambda) - D_2^2(\lambda)), \tag{12.359b}$$

$$[\eta^z, (D_1^1(\lambda), D_2^1(\lambda))] = (0, D_2^1(\lambda)), \tag{12.359c}$$

$$\{\eta^+, (D_1^1(\lambda), D_2^1(\lambda))\} = (D_2^1(\lambda), 0). \tag{12.359d}$$

Let us consider the spin first. We observe that

$$\boldsymbol{\xi}(\sigma_1^\alpha + \sigma_2^\alpha) = 0, \quad \alpha = x, y, z. \tag{12.360}$$

Hence, using (12.327), (12.359a), we conclude that

$$[S^\alpha, \phi^{(2)}(\lambda_1, \lambda_2)] = \phi^{(2)}(\lambda_1, \lambda_2) \tfrac{1}{2}(\sigma_1^\alpha + \sigma_2^\alpha) \tag{12.361}$$

for $\alpha = x, y, z$. The operators $\tfrac{1}{2}(\sigma_1^\alpha + \sigma_2^\alpha)$ are the operators of the components of the total spin for the two-site auxiliary spin problem. Let us define $\mathbf{f} = B^{(1)}(\mu_1) \ldots B^{(1)}(\mu_M) {\binom{1}{0}}^{\otimes 2}$, $M = 0, 1$. Then we know from equations (3.D.19) and (3.D.20) of appendix 3.D that

$$(\sigma_1^+ + \sigma_2^+)\mathbf{f} = 0, \tag{12.362}$$

if the Bethe ansatz equations (12.358) are satisfied, and that

$$S^z \mathbf{f} = \tfrac{1}{2}(N - 2M)\mathbf{f}, \tag{12.363}$$

irrespective of the Bethe ansatz equations. It follows from (12.361) and from (12.362), (12.363) that

$$S^+|\phi^{(2)}\rangle = 0, \tag{12.364a}$$

$$S^z|\phi^{(2)}\rangle = \tfrac{1}{2}(N - 2M)|\phi^{(2)}\rangle. \tag{12.364b}$$

The algebraic Bethe ansatz states $|\phi^{(2)}\rangle$ are spin $\tfrac{1}{2}(N - 2M)$ highest weight states.

The equations that encode the lowest weight properties of the two-particle Bethe ansatz states with respect to η-spin are similar to (12.364),

$$\eta^-|\phi^{(2)}\rangle = 0, \tag{12.365a}$$

$$\eta^z|\phi^{(2)}\rangle = \tfrac{1}{2}(2 - L)|\phi^{(2)}\rangle. \tag{12.365b}$$

The two-particle algebraic Bethe ansatz state $|\phi^{(2)}\rangle$ is the lowest weight state of a multiplet of η-spin $\frac{1}{2}(L-2)$. Equation (12.365b) is valid for arbitrary spectral parameters λ_1, λ_2. For (12.365a) to be true the Bethe ansatz equations (12.357) must be satisfied.

In order to show (12.365b) one uses (12.329b) and (12.359c). These equations imply

$$[\eta^z, \phi^{(2)}(\lambda_1, \lambda_2)] = \phi^{(2)}(\lambda_1, \lambda_2). \tag{12.366}$$

Recalling that $\eta^z|0\rangle = -\frac{L}{2}|0\rangle$, we see that (12.365b) follows from (12.366). For the proof of (12.365a), except for (12.329a), (12.359b), one has to use equation (12.D.2c) from appendix 12.D. When combined, these equations imply that

$$\eta^-|\phi^{(2)}\rangle = \frac{\rho_6(\lambda_1, \lambda_2)}{\rho_8(\lambda_1, \lambda_2)} \xi\{[A(\lambda_1) \otimes A(\lambda_2)] - D_1^1(\lambda_1)D_1^1(\lambda_2)\}|\hat{F}\rangle. \tag{12.367}$$

Thus (12.365a) follows, if λ_1 and λ_2 satisfy the Bethe ansatz equations (12.357).

At this point it is interesting to note that we could have obtained the Bethe ansatz equations (12.357), (12.358) by taking the ansatz (12.335), (12.344) and merely stipulating that it obey the two su(2) highest weight equations (12.364a) and (12.365a).

12.6.6 The N-particle transfer matrix eigenvalue

Comparing the expressions (12.288), (12.325), and (12.356) for the transfer matrix eigenvalues in the zero-, one-, and two-particle sectors it is natural to guess the following N-particle generalization:

$$\Lambda(\lambda) = \left[\omega_1^L(\lambda) \prod_{j=1}^{N} \frac{1}{a(\lambda, \lambda_j)} - \omega_2^L(\lambda) \prod_{j=1}^{N} \frac{1}{a(\lambda, \lambda_j)} \prod_{k=1}^{M} \frac{1}{a(\mu_k, \lambda)} \right.$$
$$\left. - \omega_2^L(\lambda) \prod_{k=1}^{M} \frac{1}{a(\lambda, \mu_k)} + \omega_3^L(\lambda)\right](-1)^N \prod_{j=1}^{N} \frac{\rho_{10}(\lambda, \lambda_j)}{\rho_8(\lambda, \lambda_j)}, \tag{12.368}$$

where the λ_j and μ_k have to satisfy the Bethe ansatz equations

$$\left(\frac{\omega_1(\lambda_j)}{\omega_2(\lambda_j)}\right)^L = \prod_{k=1}^{M} \frac{1}{a(\mu_k, \lambda_j)}, \quad j = 1, \ldots, N, \tag{12.369}$$

$$\prod_{j=1}^{N} \frac{1}{a(\mu_k, \lambda_j)} = \prod_{\substack{l=1 \\ l \neq k}}^{M} \frac{a(\mu_l, \mu_k)}{a(\mu_k, \mu_l)}, \quad k = 1, \ldots, M. \tag{12.370}$$

These formulae pass all significant tests. First of all, it is easy to see that equations (12.369) and (12.370) turn into the Lieb-Wu equations (3.95), (3.96) after an appropriate definition of charge momenta k_j and spin rapidities Λ_k. The transformation that connects the spectral parameter λ with a charge momentum $k(\lambda)$ was identified in (12.331). We set $k_j = k(\lambda_j)$, $j = 1, \ldots, N$. Then (12.331) implies that $(\omega_1(\lambda_j)/\omega_2(\lambda_j))^L = e^{ik_jL}$. In order to identify the products in equations (12.369), (12.370) with the corresponding products in the Lieb-Wu equations we define the spin rapidities $\Lambda_k = -v(\mu_k)$, $k = 1, \ldots, M$, and use (12.333) and (12.301).

It is also easy to reproduce our former results (3.97) for the eigenvalues of the shift operator and for the energy eigenvalues from (12.368). We obtain

$$\omega = \Lambda(0) = e^{i(k_1 + \cdots + k_N)}, \tag{12.371}$$

$$E = \frac{\Lambda'(0)}{\Lambda(0)} = -2 \sum_{j=1}^{N} \cos(k_j) + 2u(L - N). \tag{12.372}$$

Here the expression for the energy eigenvalues is in accordance with the local Hamiltonian (12.226), constructed from the transfer matrix, which differs from (2.31) by a constant shift uL in the energy.

Finally, as a last argument in favour of (12.368)–(12.370), we wish to note that the Bethe ansatz equations (12.369), (12.370) are the conditions for the residua at the simple poles of the rational functions $1/a(\lambda, \mu)$ in (12.368) to vanish.

12.6.7 On the construction of N-particle states

Ramos and Martins [371] proposed a generalization of the two-particle eigenstates (12.335), (12.344) of the form

$$|\phi^{(N)}\rangle = \phi^{(N)}(\lambda_1, \ldots, \lambda_N)|\hat{F}\rangle,$$

$$|\hat{F}\rangle = \mathbf{f} \otimes |0\rangle, \quad \mathbf{f} \in \left(\mathbb{C}^2\right)^{\otimes N}, \tag{12.373}$$

where $\phi^{(N)}(\lambda_1, \ldots, \lambda_N)$ obeys a second order recursion relation, i.e., $\phi^{(N)}$ depends on $\phi^{(N-1)}$ and on $\phi^{(N-2)}$. Since this is a complicated matter, however, we do not go into the details and refer the reader to the original work [320, 371] at this point.

12.7 Conclusions

In this chapter we have discussed the algebraic approach to the Hubbard model. Our account was based on Shastry's R-matrix [312, 391, 393, 401] and on the quantum inverse scattering method for fundamental graded models [177, 179, 276]. In the section on the algebraic Bethe ansatz we closely followed the work of Martins and Ramos [320, 371]. We worked, however, within a fermionic formulation of the problem and used the slightly modified 'fermionic version' (12.135) of Shastry's R-matrix. This brings out more clearly the structure of the transfer matrix eigenvalue which was presented here for the first time in the form (12.368) reflecting the grading $(+, -, -, +)$ and decaying into a product of a 'rational part' (term in square brackets in (12.368)) with a 'non-rational' part. Written in this form the eigenvalue of the transfer matrix looks similar to the one obtained in [362] for a gl(2|1) based vertex-model.

In a sense the results of this chapter explain why the Hubbard model is solvable by Bethe ansatz. Of course, this is not the only achievement of the algebraic approach. In the next chapter we shall see that it not only reproduces the results of the coordinate Bethe ansatz, but enables us to go further. The results of this chapter will turn out to be crucial for the calculation of the largest eigenvalue of the quantum transfer matrix, the analysis of which leads to an alternative approach of the thermodynamics of the Hubbard model.

We are convinced that the algebraic approach to the Hubbard model bears some potential for future applications. Further developments are expected from a better understanding of the algebraic Bethe ansatz eigenstates, which may finally lead to a proof of the norm formula (3.120) and to explicit formulae for form factors in the finite volume. We also expect that Shastry's R-matrix will eventually be understood as a special case of a more general solution of the Yang-Baxter equation that generates the two-parametric model proposed by Alcaraz and Bariev in [15].

The appendices to this chapter contain technical details left out in the main text and a collection of formulae needed in the algebraic Bethe ansatz calculation in Section 12.6.

Appendices to Chapter 12

12.A A proof that Shastry's R-matrix satisfies the Yang-Baxter equation

We show below that Shastry's R-matrix satisfies the Yang-Baxter equation. Our account closely follows Shiroishi and Wadati [401] who carried out explicitly an idea of Korepanov [263] to use the so-called tetrahedral Zamolodchikov algebra [263–265] for coupling two su(2)-XX models into a 'two-layer' interacting model. Shiroishi and Wadati discovered [401] that this model is nothing but Shastry's spin model (related to the Hubbard model through a Jordan-Wigner transformation) and [400] that the tetrahedral Zamolodchikov algebra may be interpreted as a generalization of Shastry's decorated star triangle relation.

Here we will be slightly more general than the original literature. We will couple two su(d)-XX models for arbitrary $d \geq 2$, i.e., we will show that the general R-matrix (12.110) satisfies the Yang-Baxter equation. In fact, the proof of Shiroishi and Wadati goes through without any modification, because the 'S-matrix' (see below) occurring in the tetrahedral Zamolodchikov algebra constructed from the su(d)-XX model R-matrices (12.92) is independent of d. This was first noticed (though not emphasized) in [360].

12.A.1 The tetrahedral Zamolodchikov algebra

Let us first of all introduce the tetrahedral Zamolodchikov algebra. We define

$$R^0_{jk} = R_{jk}(\theta_j - \theta_k), \qquad R^1_{jk} = R_{jk}(\theta_j + \theta_k)C_j \qquad (12.A.1)$$

for $jk = 12, 13, 23$. Here R is the R-matrix of the su(d)-XX model defined below (12.92) and C is the conjugation matrix (12.95).

Lemma 15. *The matrices R^0_{jk} and R^1_{jk} introduced in equation (12.A.1) satisfy the defining relations*

$$R^a_{12} R^b_{13} R^c_{23} = S^{abc}_{def} R^f_{23} R^e_{13} R^d_{12}, \qquad a, b, c = 0, 1, \qquad (12.A.2)$$

of the tetrahedral Zamolodchikov algebra if S has the only non-vanishing elements

$$S^{000}_{000} = S^{110}_{110} = S^{101}_{101} = S^{011}_{011} = 1, \qquad (12.A.3)$$

472

and

$$S_{100}^{111} = U(\theta_1, \theta_2, \theta_3), \qquad S_{010}^{111} = V(\theta_1, \theta_2, \theta_3), \qquad S_{001}^{111} = W(\theta_1, \theta_2, \theta_3),$$

$$S_{111}^{100} = U(\theta_1, \theta_2, -\theta_3), \qquad S_{010}^{100} = W(\theta_1, \theta_2, -\theta_3), \qquad S_{001}^{100} = V(\theta_1, \theta_2, -\theta_3),$$

$$S_{111}^{010} = V(\theta_1, -\theta_2, \theta_3), \qquad S_{100}^{010} = W(\theta_1, -\theta_2, \theta_3), \qquad S_{001}^{010} = U(\theta_1, -\theta_2, \theta_3),$$

$$S_{111}^{001} = W(-\theta_1, \theta_2, \theta_3), \qquad S_{100}^{001} = V(-\theta_1, \theta_2, \theta_3), \qquad S_{010}^{001} = U(-\theta_1, \theta_2, \theta_3), \qquad (12.A.4)$$

where the functions U, V, W are defined as

$$U(\theta_1, \theta_2, \theta_3) = -\frac{\cos(\theta_1 + \theta_3)\sin(\theta_2 + \theta_3)}{\sin(\theta_1 - \theta_3)\cos(\theta_2 - \theta_3)},$$

$$V(\theta_1, \theta_2, \theta_3) = -\frac{\sin(\theta_1 + \theta_2)\sin(\theta_2 + \theta_3)}{\cos(\theta_1 - \theta_2)\cos(\theta_2 - \theta_3)}, \qquad (12.A.5)$$

$$W(\theta_1, \theta_2, \theta_3) = \frac{\sin(\theta_1 + \theta_2)\cos(\theta_1 + \theta_3)}{\cos(\theta_1 - \theta_2)\sin(\theta_1 - \theta_3)}.$$

Proof. Equation (12.A.2) is a set of eight identities, since the indices a, b, c take values 0, 1 each. The eight identities come in two groups of four, the first group corresponding to $(a, b, c) = (0, 0, 0), (1, 1, 0), (1, 0, 1), (0, 1, 1)$ the second one to $(a, b, c) = (1, 1, 1), (1, 0, 0), (0, 1, 0), (0, 0, 1)$. We first show that due to the properties (12.94), (12.96) of the conjugation matrix C the identities in the respective groups are mutually equivalent. Then it is sufficient to explicitly verify one identity in every group. For the first group the identity $R_{12}^0 R_{13}^0 R_{23}^0 = R_{23}^0 R_{13}^0 R_{23}^0$ is just the Yang-Baxter equation which we know is satisfied by the R-matrix of the su(d)-XX model.

Replacing λ by θ_1, μ by θ_2 and ν by $-\theta_3$ in the Yang-Baxter equation (12.21) and multiplying by $C_1 C_2$ from the right we obtain

$$R_{12}(\theta_1 - \theta_2)R_{13}(\theta_1 + \theta_3)C_1 R_{23}(\theta_2 + \theta_3)C_2$$
$$= R_{23}(\theta_2 + \theta_3)R_{13}(\theta_1 + \theta_3)R_{12}(\theta_1 - \theta_2)C_1 C_2 \qquad (12.A.6)$$
$$= R_{23}(\theta_2 + \theta_3)C_2 R_{13}(\theta_1 + \theta_3)C_1 R_{12}(\theta_1 - \theta_2),$$

which means that

$$R_{12}^0 R_{13}^1 R_{23}^1 = R_{23}^1 R_{13}^1 R_{12}^0. \qquad (12.A.7)$$

Thus, (12.A.2) is satisfied for $(a, b, c) = (0, 1, 1)$. The second equation (12.A.6) follows from (12.96). The proof for $(a, b, c) = (1, 0, 1)$ and $(1, 1, 0)$ is similar.

Next, let us assume that (12.A.2) is satisfied for $(a, b, c) = (1, 1, 1)$,

$$R_{12}(\theta_1 + \theta_2)C_1 R_{13}(\theta_1 + \theta_3)C_1 R_{23}(\theta_2 + \theta_3)C_2$$
$$= U(\theta_1, \theta_2, \theta_3)\, R_{23}(\theta_2 - \theta_3)R_{13}(\theta_1 - \theta_3)R_{12}(\theta_1 + \theta_2)C_1$$
$$\quad + V(\theta_1, \theta_2, \theta_3)\, R_{23}(\theta_2 - \theta_3)R_{13}(\theta_1 + \theta_3)C_1 R_{12}(\theta_1 - \theta_2) \qquad (12.A.8)$$
$$\quad + W(\theta_1, \theta_2, \theta_3)\, R_{23}(\theta_2 + \theta_3)C_2 R_{13}(\theta_1 - \theta_3)R_{12}(\theta_1 - \theta_2).$$

Replacing θ_3 by $-\theta_3$, multiplying by $C_1 C_2$ from the right and using (12.94) and (12.96) the latter equation turns into

$$R_{12}(\theta_1 + \theta_2)C_1 R_{13}(\theta_1 - \theta_3)R_{23}(\theta_2 - \theta_3)$$
$$= U(\theta_1, \theta_2, -\theta_3)\, R_{23}(\theta_2 + \theta_3)C_2 R_{13}(\theta_1 + \theta_3)C_1 R_{12}(\theta_1 + \theta_2)C_1$$
$$+ V(\theta_1, \theta_2, -\theta_3)\, R_{23}(\theta_2 + \theta_3)C_2 R_{13}(\theta_1 - \theta_3)R_{12}(\theta_1 - \theta_2)$$
$$+ W(\theta_1, \theta_2, -\theta_3)\, R_{23}(\theta_2 - \theta_3)R_{13}(\theta_1 + \theta_3)C_1 R_{12}(\theta_1 - \theta_2)\,. \tag{12.A.9}$$

Thus, (12.A.2) is satisfied for $(a, b, c) = (1, 0, 0)$ if and only if it is satisfied for $(a, b, c) = (1, 1, 1)$. One shows in a similar way that the cases $(a, b, c) = (0, 1, 0)$, $(0, 0, 1)$ as well are equivalent to $(a, b, c) = (1, 1, 1)$.

The remaining step in the proof of our lemma is the verification of equation (12.A.8). First of all, using (12.94), (12.96) we deduce that (12.A.8) is equivalent to

$$\check{R}_{23}(\theta_1 + \theta_2)\check{R}_{12}(-\theta_1 - \theta_3)\check{R}_{23}(-\theta_2 - \theta_3) = U\,\check{R}_{12}(\theta_2 - \theta_3)\check{R}_{23}(\theta_1 - \theta_3)\check{R}_{12}(\theta_1 + \theta_2)$$
$$+ V\,\check{R}_{12}(\theta_2 - \theta_3)\check{R}_{23}(\theta_1 + \theta_3)\check{R}_{12}(-\theta_1 + \theta_2)(I_{d^2} - 2Q_{12}^2)$$
$$+ W\,\check{R}_{12}(\theta_2 + \theta_3)\check{R}_{23}(\theta_1 - \theta_3)\check{R}_{12}(-\theta_1 + \theta_2)\,. \tag{12.A.10}$$

Here we dropped the arguments of the functions U, V, W and used the second equation (12.94b) in order to replace $C_1 C_2$ by $I_{d^2} - 2Q_{12}^2$. Inserting the definition (12.83) of the matrix $\check{R}$ into (12.A.10), the equation turns into a polynomial identity in Q_{12} and Q_{23} with trigonometric coefficients depending on $\theta_1, \theta_2, \theta_3$. Now Q is a representation of the free fermion algebra (12.81). This suggest that upon substituting $A_1 = Q_{12}$, $A_2 = Q_{23}$ equation (12.A.10) may be valid as an identity in the free fermion algebra.

To proceed with the proof we exploit (12.81) to obtain the useful formula

$$\left(c_1 + s_1 A_2 + (1 - c_1)A_2^2\right)\left(c_2 + s_2 A_1 + (1 - c_2)A_1^2\right)\left(c_3 + s_3 A_2 + (1 - c_3)A_2^2\right)$$
$$= c_1 c_2 c_3 + A_1 s_2 c_3 + A_2(s_1 c_2 + s_3) + A_1 A_2\, c_1 s_2 s_3 + A_2 A_1\, s_1 s_2 c_3$$
$$+ A_1^2 c_1(1 - c_2)c_3 + A_2^2\left(c_2(1 - c_1 c_3) + s_1 s_3\right)$$
$$+ A_1 A_2^2 s_2(c_1 - c_3) + A_2 A_1^2(1 - c_2)(s_1 - s_3)$$
$$+ A_1^2 A_2^2(1 - c_2)(1 - c_1 c_3 - s_1 s_3)\,. \tag{12.A.11}$$

Here the coefficients $s_1, c_1, s_2, \ldots$ are still arbitrary. Interchanging the indices of A_1 and A_2 we obtain

$$\left(c_1 + s_1 A_1 + (1 - c_1)A_1^2\right)\left(c_2 + s_2 A_2 + (1 - c_2)A_2^2\right)\left(c_3 + s_3 A_1 + (1 - c_3)A_1^2\right)$$
$$= c_1 c_2 c_3 + A_1(s_1 c_2 + s_3) + A_2 s_2 c_3 + A_1 A_2\, s_1 s_2 c_3 + A_2 A_1\, c_1 s_2 s_3$$
$$+ A_1^2\left(c_2(1 - c_1 c_3) + s_1 s_3\right) + A_2^2 c_1(1 - c_2)c_3$$
$$+ A_1 A_2^2(1 - c_2)(s_1 - s_3) + A_2 A_1^2 s_2(c_1 - c_3)$$
$$+ A_2^2 A_1^2(1 - c_2)(1 - c_1 c_3 - s_1 s_3)\,. \tag{12.A.12}$$

Multiplying by $(1 - 2A_1^2)$ from the right and using once more the free fermion algebra

(12.81) gives us

$$
\begin{aligned}
\big(c_1 + s_1 A_1 &+ (1 - c_1)A_1^2\big)\big(c_2 + s_2 A_2 + (1 - c_2)A_2^2\big)\big(c_3 + s_3 A_1 + (1 - c_3)A_1^2\big) \cdot (1 - 2A_1^2) \\
&= c_1 c_2 c_3 - A_1(s_1 c_2 + s_3) + A_2 s_2 c_3 + A_1 A_2 s_1 s_2 c_3 - A_2 A_1 c_1 s_2 s_3 \\
&\quad - A_1^2\big(c_2(1 + c_1 c_3) + s_1 s_3\big) + A_2^2 c_1(1 - c_2)c_3 \\
&\quad - A_1 A_2^2(1 - c_2)(s_1 - s_3) - A_2 A_1^2 s_2(c_1 + c_3) \\
&\quad + A_1^2 A_2^2(1 - c_2)(s_1 s_3 - c_1 c_3 - 1).
\end{aligned}
$$

(12.A.13)

Let us replace Q_{12} by A_1 and Q_{23} by A_2 in equation (12.A.10). Then the term on the left-hand side of (12.A.10) is of the form (12.A.11), the first and the third terms on the right-hand side are of the form (12.A.12), and the second term on the right-hand side is of the form (12.A.13). Upon appropriately specifying the coefficients $s_1, c_1, s_2, \ldots$ and comparing the terms in front of the ten independent monomials $1, A_1, A_2, \ldots, A_1^2 A_2^2$ we end up with ten linear equations for U, V, W. The terms multiplying $1, A_2, A_1 A_2$, for instance, yield

$$
\begin{aligned}
\cos(\theta_1 + \theta_2)\cos(\theta_1 + \theta_3)\cos(\theta_2 + \theta_3) &= U \, \cos(\theta_2 - \theta_3)\cos(\theta_1 - \theta_3)\cos(\theta_1 + \theta_2) \\
&+ V \, \cos(\theta_2 - \theta_3)\cos(\theta_1 + \theta_3)\cos(\theta_1 - \theta_2) \\
&+ W \, \cos(\theta_2 + \theta_3)\cos(\theta_1 - \theta_3)\cos(\theta_1 - \theta_2),
\end{aligned}
$$

$$
\begin{aligned}
\sin(\theta_1 + \theta_2)\cos(\theta_1 + \theta_3) - \sin(\theta_2 + \theta_3) &= U \, \sin(\theta_1 - \theta_3)\cos(\theta_1 + \theta_2) \\
&+ V \, \sin(\theta_1 + \theta_3)\cos(\theta_1 - \theta_2) \\
&+ W \, \sin(\theta_1 - \theta_3)\cos(\theta_1 - \theta_2),
\end{aligned}
$$

$$
\begin{aligned}
\cos(\theta_1 + \theta_2)\sin(\theta_1 + \theta_3)\sin(\theta_2 + \theta_3) &= U \, \sin(\theta_2 - \theta_3)\sin(\theta_1 - \theta_3)\cos(\theta_1 + \theta_2) \\
&+ V \, \sin(\theta_2 - \theta_3)\sin(\theta_1 + \theta_3)\cos(\theta_1 - \theta_2) \\
&+ W \, \sin(\theta_2 + \theta_3)\sin(\theta_1 - \theta_3)\cos(\theta_1 - \theta_2).
\end{aligned}
$$

(12.A.14)

These equations have the unique solution (12.A.5). The reader may easily generate the remaining 7 linear equations and may verify that they are solved by U, V, W, equation (12.A.5), which completes the proof of the lemma. $\qquad\square$

Korepanov [264, 265] studied the case $d = 2$ in a more general context. Instead of the trigonometric R-matrix (12.93) he considered a more general free-fermion R-matrix parameterized by Jacobi-elliptic functions. For this more general R-matrix (in the case $d = 2$) he could show that the linear space spanned by the products $R_{23}^a R_{13}^b R_{12}^c$ for $a, b, c, = 0, 1$ and fixed values of the spectral parameters is eight-dimensional. However, in the trigonometric limit considered here these products are no longer linearly independent [263].

Lemma 16. *The products $R_{23}^a R_{13}^b R_{12}^c$ (see equation (12.A.1)) satisfy the linear relations*

$$
R_{23}^1 R_{13}^1 R_{12}^1 = X \, R_{23}^0 R_{13}^0 R_{12}^1 + Y \, R_{23}^0 R_{13}^1 R_{12}^0 + Z \, R_{23}^1 R_{13}^0 R_{12}^0,
$$

(12.A.15a)

$$
R_{23}^0 R_{13}^0 R_{12}^0 = X^{-1} R_{23}^1 R_{13}^1 R_{12}^0 + Y^{-1} R_{23}^1 R_{13}^0 R_{12}^1 + Z^{-1} R_{23}^0 R_{13}^1 R_{12}^1
$$

(12.A.15b)

with X, Y, Z defined by

$$X(\theta_1, \theta_2, \theta_3) = -\frac{\sin(\theta_1 + \theta_3)\cos(\theta_2 + \theta_3)}{\sin(\theta_1 - \theta_3)\cos(\theta_2 - \theta_3)},$$

$$Y(\theta_1, \theta_2, \theta_3) = \frac{\cos(\theta_1 + \theta_2)\cos(\theta_2 + \theta_3)}{\cos(\theta_1 - \theta_2)\cos(\theta_2 - \theta_3)}, \qquad (12.A.16)$$

$$Z(\theta_1, \theta_2, \theta_3) = \frac{\cos(\theta_1 + \theta_2)\sin(\theta_1 + \theta_3)}{\cos(\theta_1 - \theta_2)\sin(\theta_1 - \theta_3)}.$$

Proof. The proof relies on the fact that the equations (12.A.15) are equivalent to

$$\check{R}_{12}(\theta_2 + \theta_3)\check{R}_{23}(\theta_1 + \theta_3)\check{R}_{12}(\theta_1 + \theta_2)$$
$$= X\,\check{R}_{12}(\theta_2 - \theta_3)\check{R}_{23}(\theta_1 - \theta_3)\check{R}_{12}(\theta_1 + \theta_2)(I_{d^2} - 2Q_{12}^2)$$
$$+ Y\,\check{R}_{12}(\theta_2 - \theta_3)\check{R}_{23}(\theta_1 + \theta_3)\check{R}_{12}(-\theta_1 + \theta_2) \qquad (12.A.17)$$
$$+ Z\,\check{R}_{12}(\theta_2 + \theta_3)\check{R}_{23}(\theta_1 - \theta_3)\check{R}_{12}(-\theta_1 + \theta_2)(I_{d^2} - 2Q_{12}^2)$$

and

$$\check{R}_{12}(\theta_2 - \theta_3)\check{R}_{23}(\theta_1 - \theta_3)\check{R}_{12}(\theta_1 - \theta_2)$$
$$= X^{-1}\,\check{R}_{12}(\theta_2 + \theta_3)\check{R}_{23}(\theta_1 + \theta_3)\check{R}_{12}(\theta_1 - \theta_2)(I_{d^2} - 2Q_{12}^2)$$
$$+ Y^{-1}\,\check{R}_{12}(\theta_2 + \theta_3)\check{R}_{23}(\theta_1 - \theta_3)\check{R}_{12}(-\theta_1 - \theta_2) \qquad (12.A.18)$$
$$+ Z^{-1}\,\check{R}_{12}(\theta_2 - \theta_3)\check{R}_{23}(\theta_1 + \theta_3)\check{R}_{12}(-\theta_1 - \theta_2)(I_{d^2} - 2Q_{12}^2),$$

respectively. Replacing Q_{12} by A_1 and Q_{23} by A_2 and comparing the resulting equations with (12.A.12) and (12.A.13) we obtain two sets of ten linear equations which have the unique solution (12.A.16) each. $\qquad\qquad\Box$

12.A.2 The proof

The proof of Shiroishi and Wadati is based on the lemmata 15 and 16. Quite generally, these lemmata can be used to study the question whether a function $\alpha(\lambda, \mu)$ exists such that

$$\check{R}(\lambda, \mu) = \check{r}(\lambda - \mu) + \alpha(\lambda, \mu)\check{r}(\lambda + \mu)(C^\uparrow C^\downarrow \otimes I_{d^2}) \qquad (12.A.19)$$

solves the Yang-Baxter equation (12.30). Here the notation refers to Section (12.2.3). The matrix $\check{r}(\lambda)$ is a product of two R-matrices of the su(d)-XX model embedded into End($\mathbb{C}^d \otimes \mathbb{C}^d \otimes \mathbb{C}^d \otimes \mathbb{C}^d$) $\cong$ End($\mathbb{C}^{d^2} \otimes \mathbb{C}^{d^2}$) (see (12.99)). Furthermore, $C^\uparrow = C \otimes I_d$ and $C^\downarrow = I_d \otimes C$, where C is the conjugation matrix (12.95).

We shall show below that $\check{R}$, equation (12.A.19), is a solution of the Yang-Baxter equation (12.30) if

$$\alpha(\lambda, \mu) = \frac{\cos(\lambda - \mu)\text{sh}(h(\lambda) - h(\mu))}{\cos(\lambda + \mu)\text{ch}(h(\lambda) - h(\mu))} \qquad (12.A.20)$$

where $h(\lambda)$ is implicitly defined by the constraint equation

$$\text{sh}(2h) = u \sin(2\lambda). \tag{12.A.21}$$

The R-matrix (12.A.19) with (12.A.20) and (12.A.21) is related to the su(d) version (12.110) of Shastry's R-matrix through a gauge transformation (see below lemma 10) and through multiplication by $\beta(\lambda, \mu)$, equation (12.115). Thus, the su(d) version (12.110) of Shastry's R-matrix satisfies the Yang-Baxter equation.

We begin with adjusting our notation to the previous subsection where the tetrahedral Zamolodchikov algebra was introduced. Instead of λ, μ, ν we shall write $\theta_1, \theta_2, \theta_3$. The matrices

$$r^0_{\sigma jk} = r_{\sigma jk}(\theta_j - \theta_k), \quad r^1_{\sigma ik} = r_{\sigma ik}(\theta_j + \theta_k)C^\sigma_j \tag{12.A.22}$$

for $\sigma = \uparrow, \downarrow$ and $jk = 12, 13, 23$ generate two commuting representations of the tetrahedral Zamolodchikov algebra (12.A.2) with S-matrix (12.A.3), (12.A.4) and, in addition, satisfy (12.A.15). Let us consider the ansatz

$$R_{jk} = r^0_{\uparrow jk} r^0_{\downarrow jk} + \alpha_{jk} r^1_{\uparrow jk} r^1_{\downarrow jk} \tag{12.A.23}$$

and let us insert it into the Yang-Baxter equation $R_{12}R_{13}R_{23} = R_{23}R_{13}R_{12}$. Using (12.A.3) we obtain

$$
\begin{aligned}
\alpha_{12}&\left(r^1_{\uparrow12}r^0_{\uparrow13}r^0_{\uparrow23}\,r^1_{\downarrow12}r^0_{\downarrow13}r^0_{\downarrow23} - r^0_{\uparrow23}r^0_{\uparrow13}r^1_{\uparrow12}\,r^0_{\downarrow23}r^0_{\downarrow13}r^1_{\downarrow12}\right) \\
+ \alpha_{13}&\left(r^0_{\uparrow12}r^1_{\uparrow13}r^0_{\uparrow23}\,r^0_{\downarrow12}r^1_{\downarrow13}r^0_{\downarrow23} - r^0_{\uparrow23}r^1_{\uparrow13}r^0_{\uparrow12}\,r^0_{\downarrow23}r^1_{\downarrow13}r^0_{\downarrow12}\right) \\
+ \alpha_{23}&\left(r^0_{\uparrow12}r^0_{\uparrow13}r^1_{\uparrow23}\,r^0_{\downarrow12}r^0_{\downarrow13}r^1_{\downarrow23} - r^1_{\uparrow23}r^0_{\uparrow13}r^0_{\uparrow12}\,r^1_{\downarrow23}r^0_{\downarrow13}r^0_{\downarrow12}\right) \\
+ \alpha_{12}\alpha_{13}\alpha_{23}&\left(r^1_{\uparrow12}r^1_{\uparrow13}r^1_{\uparrow23}\,r^1_{\downarrow12}r^1_{\downarrow13}r^1_{\downarrow23} - r^1_{\uparrow23}r^1_{\uparrow13}r^1_{\uparrow12}\,r^1_{\downarrow23}r^1_{\downarrow13}r^1_{\downarrow12}\right) = 0.
\end{aligned}
\tag{12.A.24}
$$

We prefer to continue with the more handy notation

$$
\begin{aligned}
a_\sigma &= r^1_{\sigma12}r^0_{\sigma13}r^0_{\sigma23}, & \bar{a}_\sigma &= r^0_{\sigma23}r^0_{\sigma13}r^1_{\sigma12}, \\
b_\sigma &= r^0_{\sigma12}r^1_{\sigma13}r^0_{\sigma23}, & \bar{b}_\sigma &= r^0_{\sigma23}r^1_{\sigma13}r^0_{\sigma12}, \\
c_\sigma &= r^0_{\sigma12}r^0_{\sigma13}r^1_{\sigma23}, & \bar{c}_\sigma &= r^1_{\sigma23}r^0_{\sigma13}r^0_{\sigma12}, \\
d_\sigma &= r^1_{\sigma12}r^1_{\sigma13}r^1_{\sigma23}, & \bar{d}_\sigma &= r^1_{\sigma23}r^1_{\sigma13}r^1_{\sigma12},
\end{aligned}
\tag{12.A.25}
$$

$\sigma = \uparrow, \downarrow$. Then (12.A.24) turns into

$$
\alpha_{12}(a_\uparrow a_\downarrow - \bar{a}_\uparrow \bar{a}_\downarrow) + \alpha_{13}(b_\uparrow b_\downarrow - \bar{b}_\uparrow \bar{b}_\downarrow) + \alpha_{23}(c_\uparrow c_\downarrow - \bar{c}_\uparrow \bar{c}_\downarrow)
$$
$$
+ \alpha_{12}\alpha_{13}\alpha_{23}(d_\uparrow d_\downarrow - \bar{d}_\uparrow \bar{d}_\downarrow) = 0. \tag{12.A.26}
$$

From the tetrahedral Zamolodchikov algebra (12.A.2) we extract the relations

$$
\begin{aligned}
a_\sigma &= S^{100}_{010}\,\bar{b}_\sigma + S^{100}_{001}\,\bar{c}_\sigma + S^{100}_{111}\,\bar{d}_\sigma, \\
b_\sigma &= S^{010}_{001}\,\bar{c}_\sigma + S^{010}_{111}\,\bar{d}_\sigma + S^{010}_{100}\,\bar{a}_\sigma, \\
c_\sigma &= S^{001}_{111}\,\bar{d}_\sigma + S^{001}_{100}\,\bar{a}_\sigma + S^{001}_{010}\,\bar{b}_\sigma, \\
d_\sigma &= S^{111}_{100}\,\bar{a}_\sigma + S^{111}_{010}\,\bar{b}_\sigma + S^{111}_{001}\,\bar{c}_\sigma,
\end{aligned}
\tag{12.A.27}
$$

$\sigma = \uparrow, \downarrow$. Furthermore, equation (12.A.15a) gives us

$$\bar{d}_\sigma = X\bar{a}_\sigma + Y\bar{b}_\sigma + Z\bar{c}_\sigma , \qquad (12.A.28)$$

$\sigma = \uparrow, \downarrow$. Inserting (12.A.28) into (12.A.27) we obtain

$$
\begin{aligned}
a_\sigma &= X S_{111}^{100} \bar{a}_\sigma + \left(S_{010}^{100} + Y S_{111}^{100}\right)\bar{b}_\sigma + \left(S_{001}^{100} + Z S_{111}^{100}\right)\bar{c}_\sigma , \\
b_\sigma &= \left(S_{100}^{010} + X S_{111}^{010}\right)\bar{a}_\sigma + Y S_{111}^{010} \bar{b}_\sigma + \left(S_{001}^{010} + Z S_{111}^{010}\right)\bar{c}_\sigma , \\
c_\sigma &= \left(S_{100}^{001} + X S_{111}^{001}\right)\bar{a}_\sigma + \left(S_{010}^{001} + Y S_{111}^{001}\right)\bar{b}_\sigma + Z S_{111}^{001} \bar{c}_\sigma , \\
d_\sigma &= S_{100}^{111} \bar{a}_\sigma + S_{010}^{111} \bar{b}_\sigma + S_{001}^{111} \bar{c}_\sigma ,
\end{aligned} \qquad (12.A.29)
$$

$\sigma = \uparrow, \downarrow$. Finally, we insert (12.A.28) and (12.A.29) into (12.A.26). We obtain a sum over six independent terms, $\bar{a}_\uparrow \bar{a}_\downarrow$, $\bar{b}_\uparrow \bar{b}_\downarrow$, $\bar{c}_\uparrow \bar{c}_\downarrow$, $\bar{a}_\uparrow \bar{b}_\downarrow + \bar{a}_\downarrow \bar{b}_\uparrow$, $\bar{b}_\uparrow \bar{c}_\downarrow + \bar{b}_\downarrow \bar{c}_\uparrow$, $\bar{c}_\uparrow \bar{a}_\downarrow + \bar{c}_\downarrow \bar{a}_\uparrow$. The R-matrix (12.A.23) satisfies the Yang-Baxter equation if the coefficients in front of these terms vanish, i.e., if

$$
\begin{aligned}
&\alpha_{12}\left(\left(X S_{111}^{100}\right)^2 - 1\right) + \alpha_{13}\left(S_{100}^{010} + X S_{111}^{010}\right)^2 + \alpha_{23}\left(S_{100}^{001} + X S_{111}^{001}\right)^2 \\
&\quad + \alpha_{12}\alpha_{13}\alpha_{23}\left(\left(S_{100}^{111}\right)^2 - X^2\right) = 0 ,
\end{aligned} \qquad (12.A.30)
$$

$$
\begin{aligned}
&\alpha_{12}\left(S_{010}^{100} + Y S_{111}^{100}\right)^2 + \alpha_{13}\left(\left(Y S_{111}^{010}\right)^2 - 1\right) + \alpha_{23}\left(S_{010}^{001} + Y S_{111}^{001}\right)^2 \\
&\quad + \alpha_{12}\alpha_{13}\alpha_{23}\left(\left(S_{010}^{111}\right)^2 - Y^2\right) = 0 ,
\end{aligned} \qquad (12.A.31)
$$

$$
\begin{aligned}
&\alpha_{12}\left(S_{001}^{100} + Z S_{111}^{100}\right)^2 + \alpha_{13}\left(S_{001}^{010} + Z S_{111}^{010}\right)^2 + \alpha_{23}\left(\left(Z S_{111}^{001}\right)^2 - 1\right) \\
&\quad + \alpha_{12}\alpha_{13}\alpha_{23}\left(\left(S_{001}^{111}\right)^2 - Z^2\right) = 0 ,
\end{aligned} \qquad (12.A.32)
$$

$$
\begin{aligned}
&\alpha_{12} X S_{111}^{100}\left(S_{010}^{100} + Y S_{111}^{100}\right) + \alpha_{13} Y S_{111}^{010}\left(S_{100}^{010} + X S_{111}^{010}\right) \\
&\quad + \alpha_{23}\left(S_{100}^{001} + X S_{111}^{001}\right)\left(S_{010}^{001} + Y S_{111}^{001}\right) + \alpha_{12}\alpha_{13}\alpha_{23}\left(S_{100}^{111} S_{010}^{111} - XY\right) = 0 , \quad (12.A.33)
\end{aligned}
$$

$$
\begin{aligned}
&\alpha_{12}\left(S_{010}^{100} + Y S_{111}^{100}\right)\left(S_{001}^{100} + Z S_{111}^{100}\right) + \alpha_{13} Y S_{111}^{010}\left(S_{001}^{010} + Z S_{111}^{010}\right) \\
&\quad + \alpha_{23} Z S_{111}^{001}\left(S_{010}^{001} + Y S_{111}^{001}\right) + \alpha_{12}\alpha_{13}\alpha_{23}\left(S_{010}^{111} S_{001}^{111} - YZ\right) = 0 ,
\end{aligned} \qquad (12.A.34)
$$

$$
\begin{aligned}
&\alpha_{12} X S_{111}^{100}\left(S_{001}^{100} + Z S_{111}^{100}\right) + \alpha_{13}\left(S_{100}^{010} + X S_{111}^{010}\right)\left(S_{001}^{010} + Z S_{111}^{010}\right) \\
&\quad + \alpha_{23} Z S_{111}^{001}\left(S_{100}^{001} + X S_{111}^{001}\right) + \alpha_{12}\alpha_{13}\alpha_{23}\left(S_{100}^{111} S_{001}^{111} - XZ\right) = 0 .
\end{aligned} \qquad (12.A.35)
$$

Here we have to insert the explicit expressions (12.A.4), (12.A.5) of the 'S-matrix' elements and (12.A.16) of the coefficients X, Y and Z. The resulting equations simplify by application of the addition theorems for trigonometric functions. All in all there remain six equations of the form

$$\alpha_{12} A + \alpha_{13} B + \alpha_{23} C + \alpha_{12}\alpha_{13}\alpha_{23} D = 0 . \qquad (12.A.36)$$

By taking appropriate linear combinations of these equations they can be reduced to just two linearly independent equations, e.g.,

$$\alpha_{12} \sin(2(\theta_1 - \theta_3)) - \alpha_{13} \sin(2(\theta_1 - \theta_2)) + \alpha_{12}\alpha_{13}\alpha_{23} \sin(2(\theta_2 + \theta_3)) = 0 , \qquad (12.A.37)$$

$$\alpha_{12} \sin(2(\theta_1 + \theta_2)) - \alpha_{13} \sin(2(\theta_1 + \theta_3)) + \alpha_{23} \sin(2(\theta_2 + \theta_3)) = 0 , \qquad (12.A.38)$$

or, equivalently,

$$\alpha_{23}\sin(2(\theta_2+\theta_3))=-\alpha_{12}\sin(2(\theta_1+\theta_2))+\alpha_{13}\sin(2(\theta_1+\theta_3))$$
$$=\frac{1}{\alpha_{12}}\sin(2(\theta_1-\theta_2))-\frac{1}{\alpha_{13}}\sin(2(\theta_1-\theta_3))\,. \qquad (12.A.39)$$

It is easy to see that the latter equation is solved by $\alpha_{jk}=\alpha(\theta_j,\theta_k)$, $jk=12,13,23$ with $\alpha(\lambda,\mu)$ according to (12.A.20) and (12.A.21), which completes the proof.

Remark. It should be clear that also two su(d)-XX models with different values of d, say d_1 and d_2, can be coupled to a generalization of Shastry's R-matrix.

12.B A proof of the inversion formula

In this appendix we supply a proof originally given in [177] of the general inversion formula (12.268) for fundamental graded and inhomogeneous models. The proof relies on the properties of an appropriate generalization of the shift operator whose construction is our first concern here.

The inhomogeneous monodromy matrix as defined in (12.164) is an *ordered* product of L-matrices. In the following we shall indicate the order of the factors by supplying subscripts to the monodromy matrix

$$\mathcal{T}_{1\dots L}(\lambda;v_1,\dots,v_L)=\mathcal{T}(\lambda;v_1,\dots,v_L)=\mathcal{L}_L(\lambda,v_L)\dots\mathcal{L}_1(\lambda,v_1)\,. \qquad (12.B.1)$$

As can be seen from equations (12.173) and (12.174) the fermionic R-operator $\mathcal{R}^f_{jj+1}(v_j,v_{j+1})$ interchanges the two neighbouring factors $\mathcal{L}_{j+1}(\lambda,v_{j+1})$ and $\mathcal{L}_j(\lambda,v_j)$ in the monodromy matrix. Since the symmetric group $\mathfrak{S}^L$ is generated by the transpositions of nearest neighbours, the L-matrices on the right hand side of (12.B.1) can be arbitrarily reordered by application of an appropriate product of fermionic R-operators. This means that for every $\tau\in\mathfrak{S}^L$ there exists an operator $\mathcal{R}^\tau_{1\dots L}(v_1,\dots,v_L)$, which is a product of fermionic R-operators and induces the action of the permutation $\tau\in\mathfrak{S}^L$ on the inhomogeneous monodromy matrix,

$$\mathcal{R}^\tau_{1\dots L}(v_1,\dots,v_L)\mathcal{T}_{1\dots L}(\lambda;v_1,\dots,v_L)$$
$$=\mathcal{T}_{\tau(1)\dots\tau(L)}(\lambda;v_{\tau(1)},\dots,v_{\tau(L)})\mathcal{R}^\tau_{1\dots L}(v_1,\dots,v_L)\,. \qquad (12.B.2)$$

The non-graded analogue of this operator was introduced in [316]. The generalized shift operator we are looking for is a special case of $\mathcal{R}^\tau_{1\dots L}(v_1,\dots,v_L)$ for τ being a cyclic permutation.

Let us first of all construct the operator $\mathcal{R}^\tau_{1\dots L}(v_1,\dots,v_L)$ explicitly for arbitrary $\tau\in\mathfrak{S}^L$. We shall use the shorthand notations $\mathcal{R}^\tau_{1\dots L}=\mathcal{R}^\tau_{1\dots L}(v_1,\dots,v_L)$, $\mathcal{T}_{1\dots L}(\lambda)=\mathcal{T}_{1\dots L}(\lambda;v_1,\dots,v_L)$, and $\mathcal{R}^f_{jk}=\mathcal{R}^f_{jk}(v_j,v_k)$ whenever the order of the inhomogeneities $v_1,\dots,v_L$ is the same as the order of the corresponding lattice sites. For $j=1,\dots,L-1$

define $\pi_j \in \mathfrak{S}^L$ by

$$
\pi_j(k) = \begin{cases} j+1 & \text{if } k = j, \\ j & \text{if } k = j+1, \\ k & \text{else.} \end{cases} \tag{12.B.3}
$$

The $\pi_j \in \mathfrak{S}^L$ are transpositions of nearest neighbours. It follows from (12.173), (12.174) that

$$
\mathcal{R}^f_{jj+1} \, \mathcal{T}_{1...L}(\lambda) = \mathcal{T}_{\pi_j(1)...\pi_j(L)}(\lambda) \, \mathcal{R}^f_{jj+1} . \tag{12.B.4}
$$

This means that $\mathcal{R}^{\pi_j}_{1...L} = \mathcal{R}^f_{jj+1}$. Choose $\tau \in \mathfrak{S}^L$ arbitrarily. Then

$$
\mathcal{R}^f_{\tau(j),\tau(j+1)} \, \mathcal{T}_{\tau(1)...\tau(L)}(\lambda) = \mathcal{T}_{\tau\pi_j(1)...\tau\pi_j(L)}(\lambda) \, \mathcal{R}^f_{\tau(j),\tau(j+1)} . \tag{12.B.5}
$$

Since the transpositions of nearest neighbours π_j, $j = 1, \ldots, L-1$, generate the symmetric group $\mathfrak{S}^L$, there is a finite sequence $(j_p)_{p=1}^n$, such that $\tau = \pi_{j_1} \ldots \pi_{j_n}$. Let $\tau_p = \pi_{j_1} \ldots \pi_{j_p}$, $p = 1, \ldots, n$ and $\tau_0 = \mathrm{id}$. Then $\tau = \tau_n$, and, using (12.B.5), we conclude that

$$
\mathcal{R}^f_{\tau_{p-1}(j_p),\tau_{p-1}(j_p+1)} \, \mathcal{T}_{\tau_{p-1}(1)...\tau_{p-1}(L)}(\lambda) = \mathcal{T}_{\tau_p(1)...\tau_p(L)}(\lambda) \, \mathcal{R}^f_{\tau_{p-1}(j_p),\tau_{p-1}(j_p+1)} , \tag{12.B.6}
$$

for $p = 1, \ldots, n$. By iteration of the latter equation we obtain

$$
\mathcal{R}^f_{\tau_{n-1}(j_n),\tau_{n-1}(j_n+1)} \cdots \mathcal{R}^f_{\tau_1(j_2),\tau_1(j_2+1)} \mathcal{R}^f_{j_1,j_1+1} \, \mathcal{T}_{1...L}(\lambda)
$$
$$
= \mathcal{T}_{\tau(1)...\tau(L)}(\lambda) \, \mathcal{R}^f_{\tau_{n-1}(j_n),\tau_{n-1}(j_n+1)} \cdots \mathcal{R}^f_{\tau_1(j_2),\tau_1(j_2+1)} \mathcal{R}^f_{j_1,j_1+1} . \tag{12.B.7}
$$

Thus, we have constructed an explicit expression for the operator $\mathcal{R}^\tau_{1...L}$,

$$
\mathcal{R}^\tau_{1...L} = \mathcal{R}^f_{\tau_{n-1}(j_n),\tau_{n-1}(j_n+1)} \cdots \mathcal{R}^f_{\tau_1(j_2),\tau_1(j_2+1)} \mathcal{R}^f_{j_1,j_1+1} . \tag{12.B.8}
$$

Let us now specify to the case, when τ is equal to the cyclic permutation $\gamma = \pi_1 \ldots \pi_{L-1}$. Then $j_p = p$, $p = 1, \ldots, L-1$, in our above construction, and $\gamma_p = \pi_1 \ldots \pi_p$. Thus, $\gamma_{p-1}(j_p) = \gamma_{p-1}(p) = 1$, and $\gamma_{p-1}(j_p + 1) = \gamma_{p-1}(p+1) = p+1$. Using (12.B.8) we obtain

$$
\mathcal{R}^\gamma_{1...L} = \mathcal{R}^f_{1L} \mathcal{R}^f_{1L-1} \cdots \mathcal{R}^f_{12} . \tag{12.B.9}
$$

The operator $\mathcal{R}^\gamma_{1...L}$ induces a shift by one site on the inhomogeneous monodromy matrix. Now (12.B.5) implies that

$$
\mathcal{R}^\gamma_{\gamma(1)...\gamma(L)} \mathcal{T}_{\gamma(1)...\gamma(L)}(\lambda) = \mathcal{T}_{\gamma^2(1)...\gamma^2(L)}(\lambda) \mathcal{R}^\gamma_{\gamma(1)...\gamma(L)} . \tag{12.B.10}
$$

It follows by multiplication by $\mathcal{R}^\gamma_{1...L}$ from the right, that

$$
\mathcal{R}^{\gamma^2}_{1...L} = \mathcal{R}^\gamma_{\gamma(1)...\gamma(L)} \mathcal{R}^\gamma_{1...L} . \tag{12.B.11}
$$

Iterating the above steps we arrive at the following lemma.

Lemma 17. *The operator*

$$
\mathcal{R}^{\gamma^n}_{1...L} = \mathcal{R}^\gamma_{\gamma^{n-1}(1)...\gamma^{n-1}(L)} \mathcal{R}^\gamma_{\gamma^{n-2}(1)...\gamma^{n-2}(L)} \cdots \mathcal{R}^\gamma_{1...L} , \tag{12.B.12}
$$

where

$$\mathcal{R}^{\gamma}_{\gamma^{p-1}(1)\dots\gamma^{p-1}(L)} = \mathcal{R}^{f}_{pp-1}\dots\mathcal{R}^{f}_{p1}\mathcal{R}^{f}_{pL}\dots\mathcal{R}^{f}_{pp+1}, \tag{12.B.13}$$

generates a shift by n sites on the inhomogeneous lattice, i.e.,

$$\mathcal{R}^{\gamma^n}_{1\dots L}\,\mathcal{T}_{1\dots L}(\lambda) = \mathcal{T}_{n+1\dots L1\dots n}(\lambda)\,\mathcal{R}^{\gamma^n}_{1\dots L}. \tag{12.B.14}$$

Since $\gamma^L = \mathrm{id}$, we conclude from (12.B.14) that

$$\mathcal{R}^{\gamma^L}_{1\dots L}\,\mathcal{T}_{1\dots L}(\lambda) = \mathcal{T}_{1\dots L}(\lambda)\,\mathcal{R}^{\gamma^L}_{1\dots L}. \tag{12.B.15}$$

If $\mathcal{R}^{f}_{jk}$ is unitary, we have the following stronger result.

Lemma 18. *Let* $\mathcal{R}^{f}_{jk}$ *be unitary (cf equation (12.178)). Then*

$$\mathcal{R}^{\gamma^L}_{1\dots L} = \mathrm{id}. \tag{12.B.16}$$

Proof. Let us first prove the case $L = 2$. Then $\mathcal{R}^{\gamma}_{12} = \mathcal{R}^{f}_{12}$ and $\mathcal{R}^{\gamma^2}_{12} = \mathcal{R}^{\gamma}_{\gamma 1\gamma 2}\mathcal{R}^{\gamma}_{12} = \mathcal{R}^{\gamma}_{21}\mathcal{R}^{\gamma}_{12} = \mathcal{R}^{f}_{21}\mathcal{R}^{f}_{12} = \mathrm{id}$. The last equation holds, since by hypothesis, $\mathcal{R}^{f}_{12}$ is unitary.
For the case $L > 2$ we start from the Yang-Baxter equation (12.175),

$$\mathcal{R}^{f}_{L,L-n}\mathcal{R}^{f}_{L,j}\mathcal{R}^{f}_{L-n,j} = \mathcal{R}^{f}_{L-n,j}\mathcal{R}^{f}_{L,j}\mathcal{R}^{f}_{L,L-n}. \tag{12.B.17}$$

By iterated use of (12.B.17) we obtain

$$\mathcal{R}^{f}_{L,L-n}(\mathcal{R}^{f}_{L,L-n-1}\dots\mathcal{R}^{f}_{L,1})(\mathcal{R}^{f}_{L-n,L-n-1}\dots\mathcal{R}^{f}_{L-n,1})$$
$$= (\mathcal{R}^{f}_{L-n,L-n-1}\dots\mathcal{R}^{f}_{L-n,1})(\mathcal{R}^{f}_{L,L-n-1}\dots\mathcal{R}^{f}_{L,1})\mathcal{R}^{f}_{L,L-n}, \tag{12.B.18}$$

for $n = 1, \dots, L - 2$.
Let us introduce the truncated cyclic permutations $\gamma_p = \pi_1\dots\pi_{p-1}$, for $p = 2, \dots, L$, as above. γ_p induces a cyclic shift on the p-tuple $(1, \dots, p)$ and leaves the $(L - p)$-tuple $(p+1, \dots, L)$ invariant. Using (12.B.18), it follows that

$$\mathcal{R}^{f}_{L,L-n}\dots\mathcal{R}^{f}_{L,1}\,\mathcal{R}^{\gamma}_{\gamma^{L-n-1}(1)\dots\gamma^{L-n-1}(L)}$$
$$= \mathcal{R}^{f}_{L,L-n}(\mathcal{R}^{f}_{L,L-n-1}\dots\mathcal{R}^{f}_{L,1})(\mathcal{R}^{f}_{L-n,L-n-1}\dots\mathcal{R}^{f}_{L-n,1})$$
$$\times (\mathcal{R}^{f}_{L-n,L}\dots\mathcal{R}^{f}_{L-n,L-n+1})$$
$$= (\mathcal{R}^{f}_{L-n,L-n-1}\dots\mathcal{R}^{f}_{L-n,1})(\mathcal{R}^{f}_{L,L-n-1}\dots\mathcal{R}^{f}_{L,1})$$
$$\times \underbrace{\mathcal{R}^{f}_{L,L-n}\mathcal{R}^{f}_{L-n,L}}_{=\ \mathrm{id}}(\mathcal{R}^{f}_{L-n,L-1}\dots\mathcal{R}^{f}_{L-n,L-n+1}) \tag{12.B.19}$$
$$= (\mathcal{R}^{f}_{L-n,L-n-1}\dots\mathcal{R}^{f}_{L-n,1})(\mathcal{R}^{f}_{L-n,L-1}\dots\mathcal{R}^{f}_{L-n,L-n+1})$$
$$\times (\mathcal{R}^{f}_{L,L-n-1}\dots\mathcal{R}^{f}_{L,1})$$
$$= \mathcal{R}^{\gamma_{L-1}}_{\gamma^{L-n-1}_{L-1}(1),\dots,\gamma^{L-n-1}_{L-1}(L-1),L}\,\mathcal{R}^{f}_{L,L-n-1}\dots\mathcal{R}^{f}_{L,1}.$$

Hence,

$$\mathcal{R}^{\gamma^L}_{1...L} = \mathcal{R}^{\gamma}_{\gamma^{L-1}(1)...\gamma^{L-1}(L)}\, \mathcal{R}^{\gamma}_{\gamma^{L-2}(1)...\gamma^{L-2}(L)} \cdots \mathcal{R}^{\gamma}_{1...L}$$

$$= \mathcal{R}^{f}_{L,L-1} \cdots \mathcal{R}^{f}_{L,1}\, \mathcal{R}^{\gamma}_{\gamma^{L-2}(1)...\gamma^{L-2}(L)}\, \mathcal{R}^{\gamma}_{\gamma^{L-3}(1)...\gamma^{L-3}(L)} \cdots \mathcal{R}^{\gamma}_{1...L}$$

$$= \mathcal{R}^{\gamma_{L-1}}_{\gamma^{L-2}_{L-1}(1),...,\gamma^{L-2}_{L-1}(L-1),\,L}\, \mathcal{R}^{f}_{L,L-2} \cdots \mathcal{R}^{f}_{L,1}\, \mathcal{R}^{\gamma}_{\gamma^{L-3}(1)...\gamma^{L-3}(L)} \cdots \mathcal{R}^{\gamma}_{1...L}$$

$$= \mathcal{R}^{\gamma^{L-1}_{L-1}}_{1...L} = \mathcal{R}^{\gamma^{L-2}_{L-2}}_{1...L} = \cdots = \mathcal{R}^{\gamma_2^2}_{1...L}\,. \tag{12.B.20}$$

Since $\gamma_2 = \pi_1$ and $\mathcal{R}^{\pi_1}_{1...L} = \mathcal{R}^{f}_{12}$, the latter equation reduces the proof of lemma 18 for $L > 2$ to the case $L = 2$, which was proven above. $\square$

Our next lemma establishes a connection between the inhomogeneous monodromy matrix (12.B.1) and the shift operator (12.B.12).

Lemma 19. *Let* $X = X^{\alpha}_{\beta} e^{\beta}_{\alpha} \in \mathrm{End}(\mathbb{C}^{m+n})$ *and let the* R-*matrix* $R(\lambda, \mu)$ *be regular, say,* $R^{\alpha\beta}_{\gamma\delta}(\mu, \mu) = \delta^{\alpha}_{\delta}\delta^{\beta}_{\gamma}$. *Then*

$$\mathrm{str}(X\mathcal{T}_{n...L1...n-1}(\nu_n)) = (-1)^{p(\alpha)+p(\alpha)p(\beta)} X^{\alpha}_{\beta} e^{\beta}_{\alpha}\, \mathcal{R}^{\gamma}_{\gamma^{n-1}1...\gamma^{n-1}L}\,. \tag{12.B.21}$$

Proof.

$$\mathrm{str}(X\mathcal{T}_{n...L1...n-1}(\nu_n))$$

$$= (-1)^{p(\alpha)} X^{\alpha}_{\beta}\, \mathcal{L}_{n-1}{}^{\beta}_{\beta_{n-1}}(\nu_n, \nu_{n-1}) \ldots \mathcal{L}_1{}^{\beta_2}_{\beta_1}(\nu_n, \nu_1)\mathcal{L}_L{}^{\beta_1}_{\beta_L}(\nu_n, \nu_L)$$
$$\times \ldots \mathcal{L}_{n+1}{}^{\beta_{n+2}}_{\beta_{n+1}}(\nu_n, \nu_{n+1})(-1)^{p(\alpha)p(\beta_{n+1})}\, e^{\beta_{n+1}}_{n\alpha}$$

$$= (-1)^{\left\{p(\alpha)+p(\alpha)p(\beta)+\sum_{\substack{j=1 \\ j\neq n}}^{L}(p(\beta_j)+p(\alpha_j)p(\beta_j))\right\}}$$
$$\times X^{\alpha}_{\beta}\, \delta^{\beta}_{\alpha_{n-1}} \delta^{\beta_{n-1}}_{\alpha_{n-2}} \ldots \delta^{\beta_2}_{\alpha_1} \delta^{\beta_1}_{\alpha_L} \delta^{\beta_L}_{\alpha_{L-1}} \ldots \delta^{\beta_{n+2}}_{\alpha_{n+1}} e^{\beta_{n+1}}_{n\alpha}$$
$$\times \mathcal{L}_{n+1}{}^{\alpha_{n+1}}_{\beta_{n+1}}(\nu_n, \nu_{n+1}) \ldots \mathcal{L}_L{}^{\alpha_L}_{\beta_L}(\nu_n, \nu_L)\mathcal{L}_1{}^{\alpha_1}_{\beta_1}(\nu_n, \nu_1)$$
$$\times \ldots \mathcal{L}_{n-1}{}^{\alpha_{n-1}}_{\beta_{n-1}}(\nu_n, \nu_{n-1})$$

$$= (-1)^{\left\{p(\alpha)+p(\alpha)p(\beta)+\sum_{\substack{j=1 \\ j\neq n,n+1}}^{L}(p(\beta_j)+p(\alpha_j)p(\beta_j))\right\}}$$
$$\times X^{\alpha}_{\beta}\, e^{\beta}_{n\alpha} e^{\beta_{n-1}}_{n\alpha_{n-1}} e^{\beta_{n-2}}_{n\alpha_{n-2}} \ldots e^{\beta_1}_{n\alpha_1} e^{\beta_L}_{n\alpha_L} \ldots e^{\beta_{n+2}}_{n\alpha_{n+2}}$$
$$\times (-1)^{p(\beta_{n+1})+p(\alpha_{n+1})p(\beta_{n+1})} e^{\beta_{n+1}}_{n\alpha_{n+1}} \mathcal{L}_{n+1}{}^{\alpha_{n+1}}_{\beta_{n+1}}(\nu_n, \nu_{n+1})$$
$$\times \mathcal{L}_{n+2}{}^{\alpha_{n+2}}_{\beta_{n+2}}(\nu_n, \nu_{n+2}) \ldots \mathcal{L}_L{}^{\alpha_L}_{\beta_L}(\nu_n, \nu_L)\mathcal{L}_1{}^{\alpha_1}_{\beta_1}(\nu_n, \nu_1)$$
$$\times \ldots \mathcal{L}_{n-1}{}^{\alpha_{n-1}}_{\beta_{n-1}}(\nu_n, \nu_{n-1})$$

$$= (-1)^{\left\{p(\alpha)+p(\alpha)p(\beta)+\sum_{\substack{j=1 \\ j\neq n,n+1}}^{L}(p(\beta_j)+p(\alpha_j)p(\beta_j))\right\}}$$
$$\times X^{\alpha}_{\beta}\, e^{\beta}_{n\alpha} e^{\beta_{n-1}}_{n\alpha_{n-1}} e^{\beta_{n-2}}_{n\alpha_{n-2}} \ldots e^{\beta_1}_{n\alpha_1} e^{\beta_L}_{n\alpha_L} \ldots e^{\beta_{n+2}}_{n\alpha_{n+2}}$$
$$\times \mathcal{L}_{n+2}{}^{\alpha_{n+2}}_{\beta_{n+2}}(\nu_n, \nu_{n+2}) \ldots \mathcal{L}_L{}^{\alpha_L}_{\beta_L}(\nu_n, \nu_L)\mathcal{L}_1{}^{\alpha_1}_{\beta_1}(\nu_n, \nu_1)$$
$$\times \ldots \mathcal{L}_{n-1}{}^{\alpha_{n-1}}_{\beta_{n-1}}(\nu_n, \nu_{n-1})\, \mathcal{R}^{f}_{n,n+1}$$

$$= (-1)^{p(\alpha)+p(\alpha)p(\beta)} X_\beta^\alpha e_{n\alpha}^{\beta} \mathcal{R}_{n,n-1}^f \cdots \mathcal{R}_{n,1}^f \mathcal{R}_{n,L}^f \cdots \mathcal{R}_{n,n+1}^f$$

$$= (-1)^{p(\alpha)+p(\alpha)p(\beta)} X_\beta^\alpha e_{n\alpha}^{\beta} \mathcal{R}_{\gamma^{n-1}1...\gamma^{n-1}L}^\gamma . \tag{12.B.22}$$

Here we used the regularity in the first equation. In the second equation we reversed the order of factors and introduced a product of Kronecker deltas. In the third equation we used the identity

$$\delta_{\alpha_{n-1}}^\beta \delta_{\alpha_{n-2}}^{\beta_{n-1}} \cdots \delta_{\alpha_1}^{\beta_2} \delta_{\alpha_L}^{\beta_1} \delta_{\alpha_{L-1}}^{\beta_L} \cdots \delta_{\alpha_{n+1}}^{\beta_{n+2}} e_{n\alpha}^{\beta_{n+1}} = e_{n\alpha}^{\beta} e_{n\alpha_{n-1}}^{\beta_{n-1}} e_{n\alpha_{n-2}}^{\beta_{n-2}} \cdots e_{n\alpha_1}^{\beta_1} e_{n\alpha_L}^{\beta_L} \cdots e_{n\alpha_{n+1}}^{\beta_{n+1}}, \tag{12.B.23}$$

which follows from (12.147). In the fourth equation we used that

$$\mathcal{R}_{jk}^f = (-1)^{p(\beta)+p(\alpha)p(\beta)} e_{j\alpha}^\beta \mathcal{L}_{k\,\beta}^\alpha(v_j, v_k) \tag{12.B.24}$$

and the fact that $\mathcal{R}_{jk}^f$ is even. In the fifth equation we iterated the two previous steps of our calculation. Finally in the sixth equation the formula (12.B.13) entered. $\qquad\square$

Setting $X_\beta^\alpha = \delta_\beta^\alpha$ in (12.B.21) and using the cyclic invariance of the super trace we obtain the following corollary to lemma 19.

Corollary.

$$\mathcal{R}_{\gamma^{n-1}1...\gamma^{n-1}L}^\gamma = \mathrm{str}(\mathcal{T}_{1...L}(v_n)). \tag{12.B.25}$$

Equation (12.B.25) is the inhomogeneous analogue of equation (2.41).

Lemma 20. *We have the following expression for the shift operator in terms of the elements of the monodromy matrix,*

$$\mathcal{R}_{1...L}^{\gamma^n} = \prod_{j=1}^n \mathrm{str}(\mathcal{T}_{1...L}(v_j)). \tag{12.B.26}$$

If $R(\lambda, \mu)$ is unitary (cf equation (3.B.26)), then $\mathcal{R}_{1...L}^{\gamma^n}$ is invertible with inverse

$$\left(\mathcal{R}_{1...L}^{\gamma^n}\right)^{-1} = \prod_{j=n+1}^L \mathrm{str}(\mathcal{T}_{1...L}(v_j)). \tag{12.B.27}$$

Proof. The lemma follows from lemma 17, lemma 18 and the corollary to lemma 19. $\quad\square$

We are now prepared to prove our main result, equation (12.268).

Proof of equation (12.268). Using lemma 17, lemma 19, the corollary to lemma 19 and lemma 20 we obtain

$$\mathrm{str}(X\mathcal{T}_{n...L1...n-1}(v_n)) = \mathcal{R}_{1...L}^{\gamma^{n-1}} \mathrm{str}(X\mathcal{T}_{1...L}(v_n)) \left(\mathcal{R}_{1...L}^{\gamma^{n-1}}\right)^{-1}$$

$$= \prod_{j=1}^{n-1} \mathrm{str}(\mathcal{T}_{1...L}(v_j)) \cdot \mathrm{str}(X\mathcal{T}_{1...L}(v_n)) \cdot \prod_{j=n}^L \mathrm{str}(\mathcal{T}_{1...L}(v_j)) \tag{12.B.28}$$

$$= (-1)^{p(\alpha)+p(\alpha)p(\beta)} X_\beta^\alpha e_{n\alpha}^\beta \, \mathrm{str}(\mathcal{T}_{1...L}(v_n)).$$

It follows that

$$(-1)^{p(\alpha')+p(\alpha')p(\beta')} X^{\alpha'}_{\beta'} e_n{}^{\beta'}_{\alpha'}$$

$$= \prod_{j=1}^{n-1} \mathrm{str}(\mathcal{T}_{1\ldots L}(\nu_j)) \cdot \mathrm{str}(X\mathcal{T}_{1\ldots L}(\nu_n)) \cdot \prod_{j=n+1}^{L} \mathrm{str}(\mathcal{T}_{1\ldots L}(\nu_j)). \qquad (12.\mathrm{B}.29)$$

Finally, upon specifying $X^{\alpha'}_{\beta'} = (-1)^{p(\alpha')+p(\alpha')p(\beta')} \delta^{\alpha'}_{\alpha} \delta^{\beta}_{\beta'}$, we arrive at equation (12.268). □

12.C A list of commutation relations

Here is a list of the commutation relations contained in equation (12.280):

$$\rho_4 \check{r}_1(A \otimes \bar{A}) - \rho_6 K(C \otimes \bar{C}) = \rho_4(\bar{A} \otimes A)\check{r}_1 + \rho_6(\bar{B} \otimes B)K^t \qquad (12.\mathrm{C}.1\mathrm{a})$$

$$\rho_4 \check{r}_1(A \otimes \bar{B}) - \rho_6 K(C \otimes \bar{D}) = \bar{A} \otimes B + \rho_9(\bar{B} \otimes A)J^t \qquad (12.\mathrm{C}.1\mathrm{b})$$

$$\rho_4 \check{r}_1(B \otimes \bar{A}) + \rho_6 K(D \otimes \bar{C}) = \rho_{10}(\bar{A} \otimes B)J + \bar{B} \otimes A \qquad (12.\mathrm{C}.1\mathrm{c})$$

$$\rho_4 \check{r}_1(B \otimes \bar{B}) + \rho_6 K(D \otimes \bar{D}) = \rho_6(\bar{A} \otimes A)K + \rho_1(\bar{B} \otimes B)\check{r}_2 \qquad (12.\mathrm{C}.1\mathrm{d})$$

$$A \otimes \bar{C} + \rho_{10} J(C \otimes \bar{A}) = \rho_4(\bar{A} \otimes C)\check{r}_1 - \rho_6(\bar{B} \otimes D)K^t \qquad (12.\mathrm{C}.1\mathrm{e})$$

$$A \otimes \bar{D} + \rho_{10} J(C \otimes \bar{B}) = \bar{A} \otimes D - \rho_9(\bar{B} \otimes C)J^t \qquad (12.\mathrm{C}.1\mathrm{f})$$

$$-B \otimes \bar{C} + \rho_{10} J(D \otimes \bar{A}) = \rho_{10}(\bar{A} \otimes D)J - \bar{B} \otimes C \qquad (12.\mathrm{C}.1\mathrm{g})$$

$$-B \otimes \bar{D} + \rho_{10} J(D \otimes \bar{B}) = \rho_6(\bar{A} \otimes C)K - \rho_1(\bar{B} \otimes D)\check{r}_2 \qquad (12.\mathrm{C}.1\mathrm{h})$$

$$\rho_9 J^t(A \otimes \bar{C}) + C \otimes \bar{A} = \rho_4(\bar{C} \otimes A)\check{r}_1 + \rho_6(\bar{D} \otimes B)K^t \qquad (12.\mathrm{C}.1\mathrm{i})$$

$$\rho_9 J^t(A \otimes \bar{D}) + C \otimes \bar{B} = \bar{C} \otimes B + \rho_9(\bar{D} \otimes A)J^t \qquad (12.\mathrm{C}.1\mathrm{j})$$

$$-\rho_9 J^t(B \otimes \bar{C}) + D \otimes \bar{A} = \rho_{10}(\bar{C} \otimes B)J + \bar{D} \otimes A \qquad (12.\mathrm{C}.1\mathrm{k})$$

$$-\rho_9 J^t(B \otimes \bar{D}) + D \otimes \bar{B} = \rho_6(\bar{C} \otimes A)K + \rho_1(\bar{D} \otimes B)\check{r}_2 \qquad (12.\mathrm{C}.1\mathrm{l})$$

$$\rho_6 K^t(A \otimes \bar{A}) - \rho_1 \check{r}_2(C \otimes \bar{C}) = -\rho_4(\bar{C} \otimes C)\check{r}_1 + \rho_6(\bar{D} \otimes D)K^t \qquad (12.\mathrm{C}.1\mathrm{m})$$

$$\rho_6 K^t(A \otimes \bar{B}) - \rho_1 \check{r}_2(C \otimes \bar{D}) = -\bar{C} \otimes D + \rho_9(\bar{D} \otimes C)J^t \qquad (12.\mathrm{C}.1\mathrm{n})$$

$$\rho_6 K^t(B \otimes \bar{A}) + \rho_1 \check{r}_2(D \otimes \bar{C}) = -\rho_{10}(\bar{C} \otimes D)J + (\bar{D} \otimes C) \qquad (12.\mathrm{C}.1\mathrm{o})$$

$$\rho_6 K^t(B \otimes \bar{B}) + \rho_1 \check{r}_2(D \otimes \bar{D}) = -\rho_6(\bar{C} \otimes C)K + \rho_1(\bar{D} \otimes D)\check{r}_2. \qquad (12.\mathrm{C}.1\mathrm{p})$$

The notation is explained below equation (12.280).

12.D Some identities needed in the construction of the two-particle algebraic Bethe ansatz states

This appendix contains a number of formulae which are needed in the construction of the two-particle algebraic Bethe ansatz states and which can be extracted from (12.C.1) using the same projection technique as in Section 12.6.4. Since the ansatz for the two-particle eigenvector contains a term $D^1_2(\lambda_1)D^1_1(\lambda_2)$, we need first of all the following commutation relations in addition to (12.292), (12.295), and (12.307),

$$D_1^1(\lambda)D_1^1(\mu) = D_1^1(\mu)D_1^1(\lambda)$$

$$D_1^1(\lambda)D_2^1(\mu) = -\frac{\rho_4}{\rho_7}D_2^1(\mu)D_1^1(\lambda) + \frac{\rho_5}{\rho_7}D_2^1(\lambda)D_1^1(\mu) + \frac{\rho_6}{\rho_7}\left(\mathbf{C}^1(\lambda) \otimes \mathbf{C}^1(\mu)\right)\boldsymbol{\xi}^t$$

$$D_2^2(\lambda)D_1^1(\mu) = D_1^1(\mu)D_2^2(\lambda) + \frac{\rho_3}{\rho_8}\left[D_2^1(\lambda)D_1^2(\mu) - D_2^1(\mu)D_1^2(\lambda)\right]$$

$$+ \frac{\rho_6}{\rho_8}\left[\boldsymbol{\xi}\left(\mathbf{B}_2(\lambda) \otimes \mathbf{B}_1(\mu)\right) + \left(\mathbf{C}^1(\mu) \otimes \mathbf{C}^2(\lambda)\right)\boldsymbol{\xi}^t\right]$$

$$D_2^2(\lambda)D_2^1(\mu) = -\frac{\rho_1}{\rho_8}D_2^1(\mu)D_2^2(\lambda) + \frac{\rho_3}{\rho_8}D_2^1(\lambda)D_2^2(\mu) + \frac{\rho_6}{\rho_8}\boldsymbol{\xi}\left(\mathbf{B}_2(\lambda) \otimes \mathbf{B}_2(\mu)\right)$$

$$A(\lambda)D_1^1(\mu) = D_1^1(\mu)A(\lambda) + \frac{1}{\rho_9}\left[\mathbf{C}^1(\mu) \otimes \mathbf{B}_1(\lambda) - \mathbf{C}^1(\lambda) \otimes \mathbf{B}_1(\mu)\right]$$

$$A(\lambda)D_2^1(\mu) = -D_2^1(\mu)A(\lambda) + \frac{1}{\rho_{10}}\left[\mathbf{B}_2(\lambda) \otimes \mathbf{C}^1(\mu) - \mathbf{B}_2(\mu) \otimes \mathbf{C}^1(\lambda)\right]. \tag{12.D.1}$$

We further need

$$\mathbf{C}^1(\lambda) \otimes \mathbf{B}_2(\mu) = \frac{\rho_9}{\rho_{10}}\mathbf{B}_2(\mu) \otimes \mathbf{C}^1(\lambda) - \frac{1}{\rho_{10}}\left[D_2^1(\lambda)A(\mu) - D_2^1(\mu)A(\lambda)\right] \tag{12.D.2a}$$

$$\mathbf{B}_1(\lambda) \otimes \mathbf{C}^1(\mu) = -\frac{\rho_{10}}{\rho_9}\mathbf{C}^1(\mu) \otimes \mathbf{B}_1(\lambda) + \frac{1}{\rho_9}\left[D_1^1(\lambda)A(\mu) - D_1^1(\mu)A(\lambda)\right] \tag{12.D.2b}$$

$$\mathbf{C}^2(\lambda) \otimes \mathbf{C}^1(\mu) = -\frac{\rho_4}{\rho_8}\left(\mathbf{C}^1(\mu) \otimes \mathbf{C}^2(\lambda)\right)\check{r}_2 + \frac{\rho_3}{\rho_8}\left(\mathbf{C}^1(\lambda) \otimes \mathbf{C}^2(\mu)\right)$$

$$- \frac{\rho_6}{\rho_8}\boldsymbol{\xi}\left(A(\lambda) \otimes A(\mu)\right) + \frac{\rho_6}{\rho_8}\left[D_1^1(\mu)D_2^2(\lambda) + D_2^1(\mu)D_1^2(\lambda)\right]\boldsymbol{\xi}.$$

$$\tag{12.D.2c}$$

Equations (12.292), (12.295), (12.307), (12.341), (12.D.1), and (12.D.2) are sufficient to arrange the operators on the right hand sides of (12.345)–(12.347) in the appropriate order, which is the first step to be done in the algebraic Bethe ansatz for two particles. In a second step the coefficients multiplying the operators have to be simplified. For this task we use the identity

$$(\boldsymbol{\xi} \otimes M)P_{12} = \boldsymbol{\xi} \otimes M + M \otimes \boldsymbol{\xi}, \tag{12.D.3}$$

which holds for any 2×2 matrix M, and certain relations between the Boltzmann weights involving three different arguments. Most of the simplifications can be achieved by means of the following four relations in conjunction with equation (12.123),

$$\rho_4(\lambda, \mu)\rho_9(\lambda, \nu) - \rho_4(\lambda, \nu)\rho_9(\lambda, \mu) = \rho_9(\mu, \nu), \tag{12.D.4a}$$

$$\rho_8(\lambda, \mu)\rho_{10}(\lambda, \nu) - \rho_8(\lambda, \nu)\rho_{10}(\lambda, \mu) = \rho_6(\lambda, \mu)\rho_6(\lambda, \nu)\rho_9(\mu, \nu), \tag{12.D.4b}$$

$$\rho_6(\lambda, \mu)\left(\rho_9(\lambda, \nu)\rho_9(\mu, \nu) + \rho_8(\lambda, \nu)\rho_1(\mu, \nu)\right) = \rho_8(\lambda, \mu)\rho_6(\lambda, \nu), \tag{12.D.4c}$$

$$\rho_6(\lambda, \mu)\left(\rho_{10}(\lambda, \nu)\rho_{10}(\mu, \nu) + \rho_7(\lambda, \nu)\rho_4(\mu, \nu)\right) = \rho_7(\lambda, \mu)\rho_6(\lambda, \nu). \tag{12.D.4d}$$

Another more complicated relation, which was used in the final step of the derivation of equation (12.345), is

$$\frac{1}{\rho_9(\lambda, \lambda_1)} \frac{\rho_6(\lambda, \lambda_2)}{\rho_8(\lambda, \lambda_2)} \frac{\rho_4(\lambda_1, \lambda_2)}{\rho_9(\lambda_1, \lambda_2)} + \frac{1}{\rho_9(\lambda, \lambda_2)} \frac{\rho_6(\lambda, \lambda_1)}{\rho_8(\lambda, \lambda_1)} \frac{\rho_1(\lambda_1, \lambda_2)}{\rho_9(\lambda_1, \lambda_2)} \frac{\rho_7(\lambda_1, \lambda_2)}{\rho_8(\lambda_1, \lambda_2)}$$

$$= \frac{\rho_6(\lambda_1, \lambda_2)}{\rho_8(\lambda_1, \lambda_2)} \left[\prod_{j=1}^{2} \frac{\rho_4(\lambda, \lambda_j)}{\rho_9(\lambda, \lambda_j)} + \prod_{j=1}^{2} \frac{\rho_{10}(\lambda, \lambda_j)}{\rho_8(\lambda, \lambda_j)} + 2 \right]. \tag{12.D.5}$$

This equation was verified by means of Mathematica. We expect it to break up into simpler relations like (12.D.4).

We would like to note a transformation of the Boltzmann weights, obtained in [335], which is sometimes useful in calculations. Keeping λ fixed and shifting μ by $\pi/2$ the Boltzmann weights transform as

$$\rho_1 \to -\frac{\rho_7}{\rho_6}, \quad \rho_4 \to -\frac{\rho_8}{\rho_6}, \quad \rho_9 \to -\frac{\rho_{10}}{\rho_6}, \quad \rho_{10} \to -\frac{\rho_9}{\rho_6},$$

$$\rho_3 \to \frac{\rho_5}{\rho_6}, \quad \rho_5 \to \frac{\rho_3}{\rho_6}, \quad \rho_6 \to -\frac{1}{\rho_6}. \tag{12.D.6}$$

This transformation, combined with (12.123), connects equation (12.D.4a) with (12.D.4b) and equation (12.D.4c) with (12.D.4d).

12.E An explicit expression for the fermionic R-operator of the Hubbard model

In [464, 466] an explicit expression for the fermionic R-operator was derived without recourse to Shastry's R-matrix (see, however, the appendix of [466]). Here we shall show that this fermionic R-operator is equivalent to our R-matrix (12.135). As before we shall denote the right-hand side of (12.135) by $\check{R}(\lambda, \mu)$ and the permutation matrix $e_\gamma^\alpha \otimes e_\delta^\beta \otimes e_\alpha^\gamma \otimes e_\beta^\delta$ by P. We introduce the matrix

$$W(h, l) = \exp\left\{ -\frac{1}{2} \left[h(\sigma^z \otimes \sigma^z) \otimes I_4 + l I_4 \otimes (\sigma^z \otimes \sigma^z) \right] \right\}. \tag{12.E.1}$$

According to lemma 10 the transformed R-matrix

$$R(\lambda, \mu) = \frac{\cos(\lambda - \mu)}{\mathrm{ch}(h - l)} \rho_4(\lambda, \mu) W(h, l) P \check{R}(\lambda, \mu) W^{-1}(h, l) \tag{12.E.2}$$

is a solution of the Yang-Baxter equation (12.31). A corresponding fermionic R-operator can be obtained by using the general formula (12.172) with grading $p(1) = p(4) = 0$, $p(2) = p(3) = 1$ and the fermionization scheme (12.196). After a straightforward but slightly cumbersome calculation we arrive at

$$\mathcal{R}_{12}^f(\lambda, \mu) = \mathcal{R}_{12}^\uparrow(\lambda - \mu) \mathcal{R}_{12}^\downarrow(\lambda - \mu)$$

$$+ \frac{\cos(\lambda - \mu)}{\cos(\lambda + \mu)} \mathrm{th}(h - l) \mathcal{R}_{12}^\uparrow(\lambda + \mu) \mathcal{R}_{12}^\downarrow(\lambda + \mu)(1 - 2n_{1,\uparrow})(1 - 2n_{1,\downarrow}), \tag{12.E.3}$$

where

$$\mathcal{R}^a_{12}(\lambda) = \cos(\lambda)\big((1 - n_{1,a})(1 - n_{2,a}) - n_{1,a}n_{2,a}\big)$$
$$- \sin(\lambda)\big((1 - n_{1,a})n_{2,a} + n_{1,a}(1 - n_{2,a})\big) + c^\dagger_{1,a}c_{2,a} + c^\dagger_{2,a}c_{1,a} \quad (12.E.4)$$

for $a = \uparrow, \downarrow$. By construction $\mathcal{R}^f_{12}(\lambda, \mu)$ satisfies the Yang-Baxter equation in the form (12.175). Equations (12.E.3) and (12.E.4) agree with the corresponding result of [464].[1]

Let us note that the fermionic R-operator (12.E.3) was obtained before our R-matrix (12.135). In fact, we first derived (12.135) by applying equation (12.172) to (12.E.3) which we took from [464]. The result encouraged us to search for the twist lemma, lemma 12, which directly connects (12.135) with Shastry's R-matrix.

---

[1] Note that the expressions for the fermionic R-operator in [464] and [466] slightly differ.

13

The path integral approach to thermodynamics

A very curious situation arises in the context of the calculation of the partition function from the spectrum of an integrable Hamiltonian. Despite the validity of the Bethe ansatz equations for all energy eigenvalues of the model the direct evaluation of the partition function is rather difficult. In contrast to ideal quantum gases the eigenstates are not explicitly known: the Bethe ansatz equations provide just implicit descriptions that pose problems of their own kind. Yet, knowing the behaviour of quantum chains at finite temperatures is important for many reasons. As a matter of fact, the groundstate is strictly inaccessible due to the very fundamentals of thermodynamics. Therefore the study of finite temperatures is relevant for theoretical as well as experimental reasons. At high temperatures, quantum systems show only trivial static properties without correlations. Lowering the temperature, the systems enter a large regime with non-universal correlations and finally approach the quantum critical point at exactly zero temperature showing universal, non-trivial properties with divergent correlation lengths governed by conformal field theory [51].

In Chapter 5 of this book we have discussed the traditional Thermodynamical Bethe Ansatz (TBA) as developed for the Heisenberg model and the Hubbard model [155, 433–435] on the basis of a method [496] invented for the Bose gas. Here, the partition function was evaluated in the thermodynamic limit by identifying the dominant energy states. The macro-state for a given temperature T is described by a set of root densities (Section 5.2) satisfying integral equations obtained from the Bethe ansatz equations. In terms of the density functions expressions for the energy and the entropy are derived. The minimization of the free energy functional yields what are nowadays known as the TBA equations.

There are 'loose ends' in the above sketched procedure. Most importantly, the description of the spectrum of the Heisenberg model was built on the so-called 'string hypothesis' according to which admissible Bethe ansatz patterns of roots are built from regular building blocks. This hypothesis was criticized a number of times and led to activities providing alternative access to the finite temperature properties [42, 253, 254, 261, 425–427, 438, 456, 459]. The central idea of these works was a lattice path-integral formulation of the partition function of the Hamiltonian and the definition of a suitable 'quantum transfer matrix' (QTM), cf. also Section 13.1. At this point we would just like to mention that the two different approaches yield completely different equations; however, both are correct! This is understood in detail for the Heisenberg model [253, 281] as the TBA results (originally obtained in a combinatorial manner) can be obtained from the QTM procedure in an algebraic way by use of the fusion hierarchy described in appendix 13.A. For the Hubbard model the analogous statement is still in the state of conjecture.

The purpose of this chapter is to introduce the concepts and techniques of the analysis of eigenvalues of the QTM. In order to familiarize the reader with this alternative approach we use the Heisenberg chain as a warm-up exercise in Section 13.2. The Hubbard model poses some additional problems as it is based on a non-difference type solution to the Yang-Baxter equation. We therefore collect some essential properties of the Hamiltonian in Section 13.3, present the Bethe ansatz results for the Hubbard QTM in Section 13.4 and map these to an auxiliary problem that enjoys the property of being of difference type, see Section 13.5. The Bethe ansatz equations of the latter system are transformed into a finite set of coupled non-linear integral equations (NLIE) in Section 13.6 and the largest eigenvalue of the QTM is expressed in terms of integrals of the solution functions in Section 13.7.

The main strength of the finite set of NLIEs is its usefulness in the entire temperature range from high to extremely low temperatures. As a demonstration of this we present in Section 13.8 a numerical analysis of the specific heat, as well as magnetic and charge susceptibilities of the Hubbard chain. In Section 13.9 we analytically solve or simplify the NLIEs in various limiting cases. Notably in the low-temperature limit we find the structure of the dressed energy formalism known from finite-size analysis of the Hamiltonian at exactly $T = 0$.

13.1 The quantum transfer matrix and integrability

In this section we approach the problem of quantum systems at finite temperatures in terms of classical systems on lattices in one dimension higher. In many applications, the quantum systems are considered as the original objects and the classical systems as derived objects. In standard treatments of quantum systems on chains this leads to classical models on chequerboard lattices. Our viewpoint is slightly different. We consider the classical systems as primary and the quantum system as secondary as it is derived in the Hamiltonian limit from a suitable transfer matrix. This will lead us to classical systems on lattices that are partially staggered with alternating rows, but identical columns.

We first review some notations and basic properties of R-matrices (as collections of local Boltzmann weights) and the associated L-matrices, see Chapter 12. The elements of the L-matrix at site j are operators acting in the local Hilbert space $h_j \simeq \mathbb{C}^d$ with dimension d. The L-matrix' element in row α and column β is given in terms of the R-matrix

$$L_j{}^\alpha_\beta(\lambda, \mu) = R^{\alpha\gamma}_{\beta\delta}(\lambda, \mu) e_j{}^\delta_\gamma \qquad [\Leftrightarrow L^\alpha_\beta(\lambda, \mu) \cdot e_\delta = R^{\alpha\gamma}_{\beta\delta}(\lambda, \mu) e_\gamma], \qquad (13.1)$$

where e^δ_γ is a $d \times d$ matrix with only non-vanishing entry 1 in row γ and column δ. Furthermore, $e_\gamma \in \mathbb{C}^d$, $\gamma = 1, \ldots, d$, are the basis vectors and summation over repeated indices is performed. A direct consequence of the Yang-Baxter equation satisfied by the

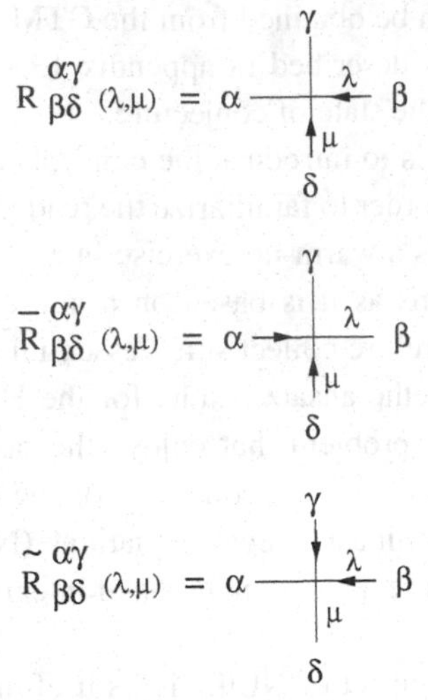

$$R\,^{\alpha\gamma}_{\beta\delta}\,(\lambda,\mu) \;=\; \alpha \underset{\delta}{\overset{\gamma}{\rule{0pt}{1em}}} \beta$$

$$\overline{R}\,^{\alpha\gamma}_{\beta\delta}\,(\lambda,\mu) \;=\; \alpha \underset{\delta}{\overset{\gamma}{\rule{0pt}{1em}}} \beta$$

$$\widetilde{R}\,^{\alpha\gamma}_{\beta\delta}\,(\lambda,\mu) \;=\; \alpha \underset{\delta}{\overset{\gamma}{\rule{0pt}{1em}}} \beta$$

Fig. 13.1. Graphical depiction of the fundamental R-matrix R and the associated $\overline{R}$ and $\widetilde{R}$.

R- and L-matrices is the commutativity of the row-to-row transfer matrix,[1]

$$T(\lambda) := \mathrm{Tr}_{\mathrm{aux}}\,[R(\lambda,\mu_L) \otimes \ldots \otimes R(\lambda,\mu_1)] = \mathrm{Tr}_{\mathrm{aux}}\,[L_L(\lambda,\mu_L)\ldots L_1(\lambda,\mu_1)]. \tag{13.2}$$

For all systems we are going to consider, we find for small spectral parameters λ and μ the Hamiltonian limit

$$R_{1,2}(\lambda,\mu) = P[1 - (\lambda - \mu)H_{1,2}] + O(\lambda^2,\lambda\mu,\mu^2), \tag{13.3}$$

for difference type as well as non-difference type R-matrices (P denotes the permutation operator, $P(x \otimes y) = y \otimes x$). This defines the local interaction $H_{1,2}$ in the space $h_1 \otimes h_2$ (here a trivial coefficient may have to be introduced as in (12.53) in order to achieve hermiticity of the operator.)

A direct consequence of (13.3) is the expansion of $T(\lambda)$ for small λ (and zero μ_i parameters)

$$T(\lambda) = e^{i\Pi - \lambda H_L + O(\lambda^2)}, \tag{13.4}$$

where Π denotes the momentum operator and H_L is the Hamiltonian on the lattice of length L.

We consider $R^{\alpha\gamma}_{\beta\delta}(\lambda,\mu)$ as the local Boltzmann weight associated with a vertex configuration $\alpha, \beta, \gamma, \delta$ on the left, right, upper, and lower bond (see figure 13.1) where the spectral parameters λ and μ 'live' on the horizontal and vertical bonds, respectively. For later use we

---

[1] Throughout this chapter we try to follow the notation of the original publication [233] when not in conflict with the general use in this book. Any deviation should be obvious and in any case rather irrelevant.

introduce $\overline{R}(\lambda, \mu)$ and $\widetilde{R}(\lambda, \mu)$ (λ and μ associated with the horizontal and vertical bond) by clockwise and anticlockwise 90° rotations of R, or in matrix notation

$$\overline{R}^{\alpha\gamma}_{\beta\delta}(\lambda, \mu) = R^{\gamma\beta}_{\delta\alpha}(\mu, \lambda), \qquad \widetilde{R}^{\alpha\gamma}_{\beta\delta}(\lambda, \mu) = R^{\delta\alpha}_{\gamma\beta}(\mu, \lambda). \tag{13.5}$$

We further introduce an auxiliary transfer matrix $\overline{T}(\lambda)$ made of Boltzmann weights $\overline{R}(-\lambda, 0)$. From the Hamiltonian limit (13.3) we see that

$$\overline{T}(\lambda) = e^{-i\Pi - \lambda H_L + O(\lambda^2)}, \tag{13.6}$$

so that the partition function is given by

$$Z_L = \operatorname{Tr} e^{-\beta H_L} = \lim_{N\to\infty} \operatorname{Tr}\left[T(\tau)\overline{T}(\tau)\right]^{N/2}\big|_{\tau=\beta/N}. \tag{13.7}$$

We regard the resulting system as a fictitious two-dimensional model on a $L \times N$ square lattice, cf. figure 13.2, where N is the extension in the fictitious (imaginary time) direction, sometimes referred to as the Trotter number. The lattice consists of alternating rows, each of which is a product of only R weights or of only $\overline{R}$ weights, respectively. All columns are identical and are made up of alternating R and $\overline{R}$ weights. This formulation realizes a lattice path integral of the quantum system in the sense that the trace of an exponential of the Hamiltonian is replaced by a summation over all configurations ('paths') of a classical model.

It is therefore very natural to introduce a different transfer matrix concept based on the transfer direction along the horizontal axis (chain) and to investigate the column-to-column

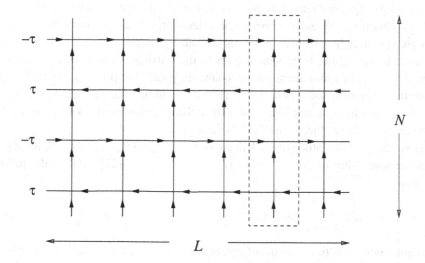

Fig. 13.2. Depiction of the two-dimensional classical model onto which the quantum chain at finite temperature is mapped. The square lattice has width L equal to the chain length, and height identical to the Trotter number N. The alternating rows of the lattice correspond to the transfer matrices $T(\tau)$ and $\overline{T}(\tau)$, $\tau = \beta/N$. The column-to-column transfer matrix (quantum transfer matrix) is of particular importance to the treatment of the thermodynamic limit.

transfer matrix being the contribution of all Boltzmann weights of a particular column to the total partition function. In the remainder of this chapter we will refer to the column-to-column transfer matrix (in particular in the limit of Trotter number $N \to \infty$) as the 'quantum transfer matrix' of the quantum chain, and denote it by T^{QTM}, because it is the closest analogue to the transfer matrix of a classical spin chain.

Now, by looking at the system in a $90°$ rotated frame which turns $\overline{R}$ and R weights into R and $\widetilde{R}$ weights, it is natural to define a general QTM with arbitrary spectral parameter λ on the (vertical) line

$$T^{\mathrm{QTM}}(\lambda, \tau) := \mathrm{Tr}_{\mathrm{aux}}\left[\bigotimes^{N/2} R(\lambda, -\tau) \otimes \widetilde{R}(\lambda, \tau)\right] \tag{13.8}$$

$$= \mathrm{Tr}_{\mathrm{aux}}\left[L_N^{\mathrm{QTM}}(\lambda, -\tau) L_{N-1}^{\mathrm{QTM}}(\lambda, \tau) \dots L_2^{\mathrm{QTM}}(\lambda, -\tau) L_1^{\mathrm{QTM}}(\lambda, \tau)\right]$$

which in the case $\lambda = 0$ is identical to the physically relevant $T^{\mathrm{QTM}} = T^{\mathrm{QTM}}(0, \tau)$.

The L-matrices are defined differently for even and odd indices

$$L_j^{\mathrm{QTM}\,\alpha}{}_\beta(\lambda, \mu) = \begin{cases} R^{\alpha\gamma}_{\beta\delta}(\lambda, \mu) e_{j\gamma}^{\delta}, & \text{for } j \text{ even,} \\ \widetilde{R}^{\alpha\gamma}_{\beta\delta}(\lambda, \mu) e_{j\gamma}^{\delta} = R^{\delta\alpha}_{\gamma\beta}(\mu, \lambda) e_{j\gamma}^{\delta}, & \text{for } j \text{ odd.} \end{cases} \tag{13.9}$$

Using the QTM we may express the partition function as

$$Z_{L,N} := \mathrm{Tr}\left[T(\tau)\overline{T}(\tau)\right]^{N/2} = \mathrm{Tr}\left(T^{\mathrm{QTM}}\right)^L. \tag{13.10}$$

Here a note on our general terminology is in order. Throughout this book we use the concept of the monodromy matrix and its associated transfer matrix. The elements of the monodromy matrix are operators in a Hilbert space that we call 'quantum space'. The transfer matrix is obtained by taking the trace with respect to the 'auxiliary space' of the monodromy matrix, yielding an operator acting on the 'quantum space'. The procedure of taking a trace in quantum/auxiliary space is denoted by $\mathrm{Tr}/\mathrm{Tr}_{\mathrm{aux}}$. Note that the auxiliary/quantum space of the row-to-row transfer matrix corresponds to the quantum/auxiliary space of the QTM when embedding these objects into the 2d lattice!

The free energy f per lattice site is defined by $f = -k_B T \lim_{L\to\infty} \lim_{N\to\infty} \log Z_{L,N}/L$. The interchangeability of the two limits $(L, N \to \infty)$ [425, 427] leads to the following expression

$$f = -k_B T \lim_{N\to\infty} \lim_{L\to\infty} \frac{1}{L} \log \mathrm{Tr}\left[T^{\mathrm{QTM}}(0, \tau)\right]^L, \qquad \tau = \frac{\beta}{N}. \tag{13.11}$$

Of particular interest is the spectrum of eigenvalues $\Lambda(\lambda, \tau)$ of $T^{\mathrm{QTM}}(\lambda, \tau)$. There is a gap between the largest and the second largest eigenvalue of the QTM for finite β [425, 427]. Therefore, the free energy per site is given just in terms of the largest eigenvalue $\Lambda_{\max}$

$$f = -k_B T \lim_{N\to\infty} \log \Lambda_{\max}\left(0, \tau = \frac{\beta}{N}\right). \tag{13.12}$$

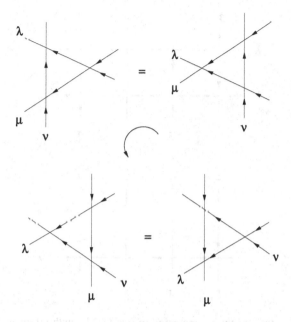

Fig. 13.3. Graphical depiction of the fundamental Yang-Baxter equation for R and the associated one for R and $\widetilde{R}$ obtained through rotation. Hence the intertwiner for R vertices is identical to the intertwiner for $\widetilde{R}$ vertices.

Now the evaluation of the free energy is reduced to that of the single eigenvalue $\Lambda_{\max}$. Of course, a sophisticated treatment is necessary in taking the Trotter limit $N \to \infty$ as $\tau = \beta/N$ now explicitly depends on it. The following sections are devoted to this analysis.

A general comment is in order. It seems redundant to define $T^{\text{QTM}}(\lambda, \tau)$ for arbitrary λ as we only need the value $\lambda = 0$. The general case, however, manifests the integrability structure and the existence of infinitely many conserved quantities. This is best seen in the commutativity of the matrices with different λ's

$$[T^{\text{QTM}}(\lambda, \tau), T^{\text{QTM}}(\lambda', \tau)] = 0, \tag{13.13}$$

with fixed τ. One can prove this by showing that two QTMs are intertwined by the same R operator as for the row-to-row case, graphically demonstrated in figure 13.3. The final step in the proof of commutativity is the standard railroad argument demonstrated for the QTM in figure 13.4. The existence of the parameter labeling the family of commuting matrices makes the subsequent analysis much more transparent. We will investigate the eigenvalues $\Lambda(\lambda, \tau)$ in dependence on λ where distributions of zeroes occur along lines parallel to the imaginary axis. For the sake of convenience we perform a rotation in the complex plane by substituting λ by iv, i.e. we will investigate $\Lambda(iv, \tau)$.

Next we want to comment on the study of the thermodynamics of the quantum chain in the presence of an external field that couples to a conserved quantity, e.g. a magnetic field h acting on the spin $S = \sum_{j=1}^{L} S_j$, where S_j denotes a certain component of the jth spin

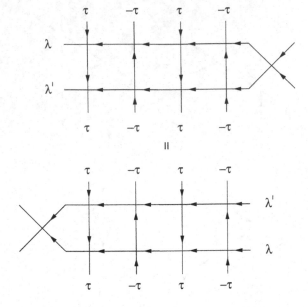

Fig. 13.4. 'Railroad proof' for the commutation of two QTM's with arbitrary spectral parameters λ and λ'. Due to the reasoning graphically depicted in figure 13.3 the intertwiner for R vertices is identical to the intertwiner for $\widetilde{R}$ vertices.

for instance S_j^z. This of course changes (13.7) only trivially

$$Z_L = \text{Tr}\, e^{-\beta(H_L - hS)}$$
$$Z_{L,N} = \text{Tr}\left([T(\tau)\overline{T}(\tau)]^{N/2} e^{\beta h S}\right). \tag{13.14}$$

The equivalent two-dimensional $L \times N$ lattice is modified in a simple way by a horizontal seam. Each vertical bond of this seam carries an individual Boltzmann weight $e^{\pm \beta h/2}$ if $S_j = \pm 1/2$, describing the action of the operator

$$e^{\beta h S} = \prod_j e^{\beta h S_j}. \tag{13.15}$$

Consequently, the QTM is modified by a field dependent boundary operator D

$$T^{\text{QTM}}(\lambda, \tau) = \text{Tr}_{\text{aux}}\left[D \cdot L_N^{\text{QTM}}(\lambda, -\tau) L_{N-1}^{\text{QTM}}(\lambda, \tau) \dots L_2^{\text{QTM}}(\lambda, -\tau) L_1^{\text{QTM}}(\lambda, \tau)\right], \tag{13.16}$$

where $D = \exp(\beta h/2 \cdot S_{\text{aux}})$ acts only in the auxiliary space. In the case of a spin-$\frac{1}{2}$ model with $S_{\text{aux}} = S_{\text{aux}}^z$ we have $D = \text{diag}(\exp(\beta h/2), \exp(-\beta h/2))$.

It will turn out that these modifications can still be treated exactly as the additional operators acting on the bonds belong to symmetries of the model. Therefore, the properties of many-particle systems can be studied within a grand canonical ensemble for general magnetic fields and chemical potentials.

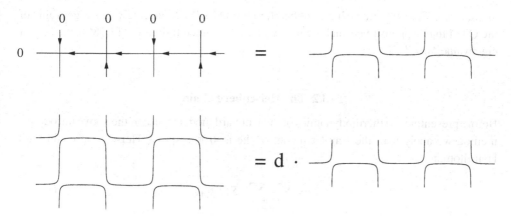

Fig. 13.5. Upper row: graphical illustration of the QTM for infinite temperature ($\tau = 0$). Note that the only non-zero matrix elements are equal to 1 and are realized for pairwise identical local indices. Lower row: the product of two QTM's for infinite temperature reproduces the QTM with a scale factor equal to the dimension d of the local Hilbert space.

Infinite Temperature

Let us now discuss the infinite temperature case with $\tau = \beta/N = 0$. Here the QTM simplifies enormously as the R-matrix (13.3) with spectral parameters $\lambda = \mu = 0$ reduces to P. With $T := T^{\mathrm{QTM}}(0, \tau = 0)$ we find

$$T \cdot T = d \cdot T,$$
$$\operatorname{Tr} T = d, \tag{13.17}$$

with d being the dimension of the local Hilbert space. The derivation of the first property is illustrated in figure 13.5, the second property is derived in a completely analogous manner. From the first line we see that the eigenvalues are 0 or d, the second line implies that exactly one eigenvalue takes the value d and all remaining eigenvalues are equal to 0. From

$$\Lambda_{\max} = d, \tag{13.18}$$

and a view to (13.12) we find the expected high-temperature asymptotics $f = -k_B T \log d$. This limit is actually realized by the free energy of the Hubbard model obtained in the TBA approach where (5.69) corresponds to (13.18) with $d = 4$. (The slightly more general case of infinite temperature with finite ratios μ/T and B/T is also found by the above argument: (13.17) still holds with d replaced by $1 + \exp((\mu + B)/T) + \exp((\mu - B)/T) + \exp(2\mu/T)$.)

However, the main application of (13.18) is the identification of the largest eigenvalue of the QTM: very often we will restrict our analysis to a special solution to the eigenvalue equation that 'happens' to yield the largest eigenvalue. A detailed study of all other eigenvalues and the formal proof that those are smaller is technically rather difficult. Fortunately, the largest eigenvalue of the QTM is known to be unique and separated by a gap from the rest of the spectrum [425, 427]. Furthermore this eigenvalue is an analytic function of the temperature $T(\neq 0)$ with the high-temperature limit (13.18). Conversely, any eigenvalue of the QTM that happens to be analytic in T with high-temperature limit (13.18) is the largest eigenvalue *for all T*!

13.2 The Heisenberg chain

Before presenting the thermodynamics of the Hubbard model by use of the above introduced method we apply it to the simpler model of the isotropic spin-$\frac{1}{2}$ Heisenberg chain with Hamiltonian

$$H = \sum_{j=1}^{L} \sum_{\alpha=x,y,z} S_j^\alpha S_{j+1}^\alpha, \tag{13.19}$$

where S_j^α is the α-component of the spin operator acting in the jth local Hilbert space.

The operator (13.19) is obtained in the Hamiltonian limit of the six-vertex model which is a difference type solution $R(\lambda, \mu) = R(\lambda - \mu)$ to the YBE. The Hamiltonian limit (13.3) yields the desired (hermitian) local interaction if we choose c in (12.48) imaginary, actually $c = 2i$, instead of real as done in Chapter 12.1.

The diagonalization of the QTM is achieved by the algebraic Bethe ansatz very much like to the homogeneous case of the row-to-row transfer matrix. In the latter case the L-matrix is given in (12.52). In our 'staggered' case (13.9) we find

$$L_j^{\text{QTM}\alpha}{}_\beta(\lambda, \mu) = \begin{cases} c(\mu - \lambda)\delta_\beta^\alpha + b(\mu - \lambda)e_{j\alpha}^\beta, & \text{for } j \text{ odd}, \\ c(\lambda - \mu)\delta_\beta^\alpha + b(\lambda - \mu)e_{j\beta}^\alpha, & \text{for } j \text{ even}. \end{cases} \tag{13.20}$$

We note that the two expressions for $L_{j\,\text{even}}^{\text{QTM}}(\lambda, \mu)$ [$=: L(\lambda - \mu)$], and $L_{j\,\text{odd}}^{\text{QTM}}(\lambda, \mu)$ differ only slightly. A direct calculation shows that up to some scalar factor the second expression $L_{j\,\text{odd}}(\lambda, \mu)$ is equal to $r^{-1} \cdot L(\lambda - \mu - ic) \cdot r$ where the operator r acts in the local Hilbert space with $r \cdot e_1 = -e_2, r \cdot e_2 = +e_1$. (This is based on a relation of the R- and $\tilde{R}$-matrices known in S-matrix theory as 'crossing symmetry'.) Therefore, the QTM is equivalent to a staggered row-to-row transfer matrix with alternating spectral parameters $\lambda - \mu$ and $\lambda - \mu - ic$. Due to this property the diagonalization of the QTM may be reduced to that of a seemingly simpler system of L-operators of the same type with only shifts of the spectral parameters. However, a direct diagonalization of the QTM is not more involved. Hence we will not use the special 'crossing symmetry' satisfied by the Heisenberg model.

An appropriate pseudo-vacuum is the staggered state $|0\rangle = |12...12\rangle = \binom{1}{0} \otimes \binom{0}{1} \otimes \ldots \otimes \binom{1}{0} \otimes \binom{0}{1}$. We note that

$$L_j(\lambda, \mu)|0\rangle = \begin{cases} \begin{pmatrix} c(\mu - \lambda) & b(\mu - \lambda)e_{j1}^2 \\ 0 & 1 \end{pmatrix} |0\rangle, & \text{for } j \text{ odd}, \\[3mm] \begin{pmatrix} 1 & b(\lambda - \mu)e_{j2}^1 \\ 0 & c(\lambda - \mu) \end{pmatrix} |0\rangle, & \text{for } j \text{ even}. \end{cases} \tag{13.21}$$

From this we see that the monodromy matrix corresponding to $T^{\text{QTM}}(\lambda, \tau)$ is upper-triangular when being applied to $|0\rangle$. Hence the entire reasoning of Section 12.1.7 holds with result (12.67) for the eigenvalue provided the vacuum functions (12.70) are replaced by

$$a(\lambda) \longrightarrow [c(\tau - \lambda)]^{N/2} \quad , \quad d(\lambda) \longrightarrow [c(\tau + \lambda)]^{N/2} \quad . \tag{13.22}$$

The eigenvalue $\Lambda(\lambda, \tau)$ of $T^{\text{QTM}}(\lambda, \tau)$ is now given by (12.67). After some substitutions $\lambda \to iv$, $\mu_j \to iv_j$, $c = 2i$ and factorization of common terms of the vacuum functions we find the following expression

$$\Lambda(iv, \tau) = \frac{\Lambda(v)}{[(v - i(2 - \tau))(v + i(2 - \tau))]^{N/2}}, \tag{13.23}$$

with numerator $\Lambda(v)$ given by

$$\Lambda(v) := \lambda_1(v) + \lambda_2(v) \tag{13.24}$$

where $\lambda_{1,2}(v)$ are shorthand notations for

$$\lambda_1(v) := e^{+\beta h/2}\phi(v - i)\frac{q(v + 2i)}{q(v)},$$

$$\lambda_2(v) := e^{-\beta h/2}\phi(v + i)\frac{q(v - 2i)}{q(v)}. \tag{13.25}$$

The function $\phi(v)$ is known explicitly

$$\phi(v) := [(v - i(1 - \tau))(v + i(1 - \tau))]^{N/2}, \quad \tau := \beta/N, \tag{13.26}$$

and $q(v)$ is defined by

$$q(v) := \prod_j^m (v - v_j). \tag{13.27}$$

Here $m \, (= 0, \ldots, N/2)$ is an integer and is identical to $m = N/2$ in the case of the largest eigenvalue. Above we have collected the results for the general case of the Heisenberg chain in the presence of a magnetic field h. The field term results into a twisted boundary condition of the QTM with (imaginary) twist angle βh.

The 'unknown' zeroes of $q(v)$ are the Bethe ansatz rapidities and are determined by the Bethe ansatz equations

$$\mathfrak{a}(v_j) = -1, \tag{13.28}$$

where the function $a(v)$ is defined by

$$\mathfrak{a}(v) = \frac{\lambda_1(v)}{\lambda_2(v)} = e^{\beta h}\frac{\phi(v - i)q(v + 2i)}{\phi(v + i)q(v - 2i)}. \tag{13.29}$$

From an algebraic point of view, we are dealing with a set of coupled non-linear equations similar to those which already occurred in the study of the eigenvalues of the Hamiltonian.

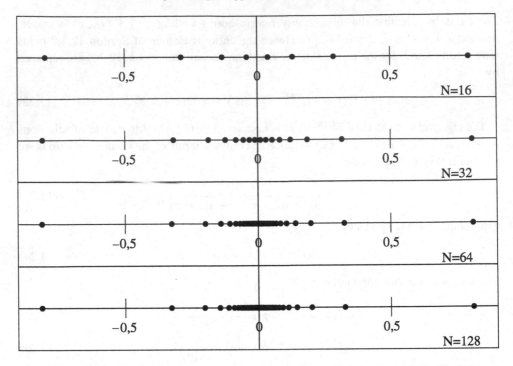

Fig. 13.6. Depiction of Bethe ansatz rapidities v_j for $\beta = 3.2$ and finite $N = 16, 32, 64, 128$ ($\tau = \beta/N$). Note that the distribution remains discrete in the limit of $N \to \infty$. The positions of the outermost rapidities hardly change for increasing N, the additional rapidities are distributed towards the origin which turns into an accumulation point for $N \to \infty$.

As far as analytical properties are concerned, there is a profound difference as in (13.29) the ratio of ϕ-functions possesses zeroes and poles that converge to the real axis in the limit $N \to \infty$. As a consequence, the distribution of Bethe ansatz rapidities is *discrete* and shows an *accumulation point* at the origin, cf. figure 13.6. This prevents the definition of meaningful root densities. Hence, in contrast to the Hamiltonian case, the treatment of the problem by means of linear integral equations is not possible. For $N = \infty$ the number of rapidities inside/outside of any (arbitrarily small) open neighbourhood of the origin is infinite/finite. In the limit of temperature $T \to \infty$ ($\beta \to 0$) all rapidities v_j converge to 0. For $T \to 0$ ($\beta \to \infty$) the rapidities move to infinity. This is illustrated in figure 13.7.

13.2.1 Derivation of non-linear integral equations I

The next step in the treatment of the thermodynamics of the Heisenberg chain is the derivation of a set of integral equations for the function $\mathfrak{a}(v)$. These equations will turn out to be non-linear. A major ingredient in our reasoning will be certain analytical properties of the functions.

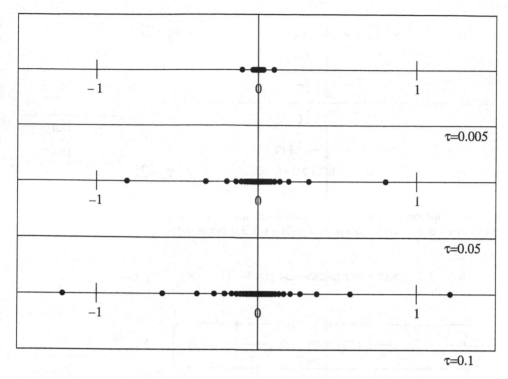

Fig. 13.7. Depiction of Bethe ansatz rapidities v_j for $N = 64$ and different temperatures $\tau = \beta/N = 0.005, 0.05, 0.1$. Note that for lower temperature, i.e. higher value of τ, the rapidities increase.

In figure 13.8 the distribution of zeroes and poles of $\mathfrak{a}(v)$ is shown resulting from the explicit factorization in the definition (13.29). Next we define the associated auxiliary function $\mathfrak{A}(v)$ by

$$\mathfrak{A}(v) = 1 + \mathfrak{a}(v). \tag{13.30}$$

Of course, the set of poles of $\mathfrak{A}(v)$ is identical to the set of poles of $\mathfrak{a}(v)$. However, the set of zeroes is different. From (13.28) we find that the Bethe ansatz rapidities are zeroes of $\mathfrak{A}(v)$ and are depicted by open circles in figure 13.9. There are additional zeroes off the real axis with imaginary parts close to ± 2. These zeroes are depicted in figure 13.9 by open squares. They are, however, not of further interest to our immediate reasoning. In the remainder of this section we are going to formulate a linear integral expression for the function $\log \mathfrak{a}(v)$ in terms of $\log \mathfrak{A}(v)$. The idea underlying our calculation is the observation that

- all functions we are dealing with are rational functions and hence are determined by the positions of their zeroes and poles and the asymptotic value for the argument approaching ∞,
- all zeroes and poles of $\mathfrak{a}(v)$ are explicitly known or directly related to the zeroes of $\mathfrak{A}(v)$ at the real axis (i.e. v_j).

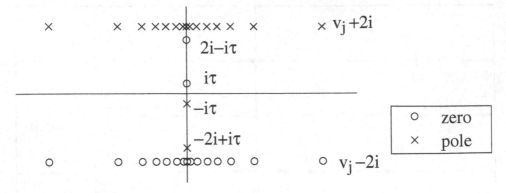

Fig. 13.8. Distribution of zeroes and poles of the auxiliary function $\mathfrak{a}(v)$. All zeroes and poles $v_j \mp 2i$ are of first order, the zeroes and poles at $\pm(2i - i\tau)$, $\pm i\tau$ are of order $N/2$.

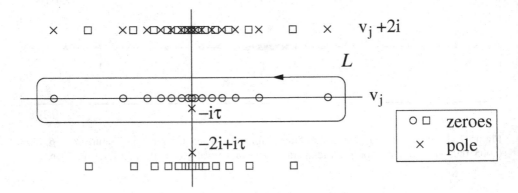

Fig. 13.9. Distribution of zeroes and poles of the auxiliary function $\mathfrak{A}(v) = 1 + \mathfrak{a}(v)$. Note that the positions of zeroes ($\circ$) and poles ($\times$) are directly related to those occurring in the function $\mathfrak{a}(v)$, or they are 'far' away from the real axis ($\square$). The closed contour $\mathcal{L}$ by definition surrounds the real axis as well as the pole at $-i\tau$.

We can exploit these ideas by considering the integral

$$\frac{1}{2\pi i} \int_{\mathcal{L}} \frac{1}{v - w} \log \mathfrak{A}(w) dw \tag{13.31}$$

defined for a closed contour $\mathcal{L}$ surrounding the real axis and the point $-i\tau$ in anticlockwise manner. Note that the number of zeroes of $\mathfrak{A}(v)$ surrounded by this contour, i.e. the number of rapidities v_j, is equal to $N/2$ and hence equal to the order of the pole at $-i\tau$. Therefore the integrand $\log \mathfrak{A}(w)$ does not show any non-zero winding number on the contour and as a result the integral is well defined. We may evaluate the integral (13.31) rather easily by performing an integration by parts, observing that the 'surface term' vanishes and obtain

$$(13.31) = \frac{1}{2\pi i} \int_{\mathcal{L}} \log(v - w)[\log \mathfrak{A}(w)]' dw. \tag{13.32}$$

Inside the contour $\mathcal{L}$ lie the simple zeroes v_j of $\mathfrak{A}$ and the pole $-i\tau$ of order $N/2$. Hence by use of Cauchy's theorem, we obtain

$$\frac{1}{2\pi i} \int_{\mathcal{L}} \frac{1}{v-w} \log \mathfrak{A}(w)dw = \sum_{j=1}^{N/2} \log(v-v_j) - \frac{N}{2}\log(v+i\tau) = \log \frac{q(v)}{(v+i\tau)^{N/2}}.$$

$$(13.33)$$

Alternative derivation

Identity (13.33) was derived by direct computations, because the function $[\log \mathfrak{A}]'$ is meromorphic, or more precisely analytic apart from simple poles. Hence by use of Cauchy's theorem the explicit expression for the integral could be derived. In the case of the Hubbard model we would like to apply a similar strategy. However, the functions involved in the latter analysis possess isolated as well as non-isolated singularities. We therefore present a proof for the identity (13.33) without resorting to the use of a particular type of singularity.

Let $g(v)$ be an analytic function with certain singularities in the complex plane and $\mathcal{L}$ be a contour surrounding the (sub)set of singularities S. We define the function $f(v)$ for complex v outside the contour $\mathcal{L}$ by

$$f(v) := \frac{1}{2\pi i} \int_{\mathcal{L}} \frac{1}{v-w} g(w)dw. \qquad (13.34)$$

Obviously, the function $f(v)$ is analytic everywhere outside $\mathcal{L}$, it has asymptotics 0 at infinity and it can be analytically continued across the border of the contour $\mathcal{L}$. This gives

$$f(v) = \frac{1}{2\pi i} \int_{\mathcal{L}} \frac{1}{v-w} g(w)dw + g(v), \qquad (13.35)$$

for v inside $\mathcal{L}$. Apparently the integral expression on the r.h.s. is analytic inside the contour $\mathcal{L}$. Hence the total r.h.s., i.e. the analytically continued function $f(v)$, has the same singularities as $g(v)$ at all points in the set S. To summarize, $f(v)$ is defined on the entire complex plane, it has asymptotics 0 and shares the same singularities with $g(v)$ at the set S.

If we use $g(v) = \log \mathfrak{A}(v)$, S is the set of singularities $\{v_j | j = 1, ..., m\} \cup \{-i\tau\}$, and $\mathcal{L}$ is the path surrounding S (see Fig.13.9) then $f(v)$ as defined above is a function with the same analyticity properties as

$$\log \frac{q(v)}{(v+i\tau)^{N/2}}. \qquad (13.36)$$

From this follows immediately that

$$f(v) = \log \frac{q(v)}{(v+i\tau)^{N/2}}, \qquad (13.37)$$

which is proved by noting the three properties for the difference function of l.h.s. and r.h.s.:

- analyticity on the complex plane with a possible exception on the set S,
- continuity on the set S, because of cancellation of all singularities ($\Rightarrow$ analyticity everywhere),
- zero asymptotics ($\Rightarrow$ boundedness $\Rightarrow$ constant due to Liouville's theorem $\Rightarrow$ constant $= 0$).

Because of (13.33) we have a linear integral representation of $\log q(v)$ in terms of $\log \mathfrak{A}(v)$. Due to the definition (13.29) the function $\log \mathfrak{a}(v)$ is a linear combination of $\log q$ and explicitly known functions. Hence we find

$$\log \mathfrak{a}(v) = \beta h + \log \left(\frac{(v - i\tau)(v + 2i + i\tau)}{(v + i\tau)(v + 2i - i\tau)} \right)^{N/2}$$
$$+ \underbrace{\frac{1}{2\pi i} \int_{\mathcal{L}} \left[\frac{1}{v - w + 2i} - \frac{1}{v - w - 2i} \right] \log \mathfrak{A}(w) dw.}_{-\frac{2}{\pi} \int_{\mathcal{L}} \frac{1}{(v-w)^2 + 4}} \qquad (13.38)$$

This expression is remarkable as it is a non-linear integral equation (NLIE) of convolution type for $\mathfrak{a}(v)$. It is valid for any value of the Trotter number N which only enters in the driving term, i.e. the first term on the r.h.s. of (13.38). This term shows a well defined limiting behaviour for $N \to \infty$

$$\frac{N}{2} \log \left(\frac{(v - i\tau)(v + 2i + i\tau)}{(v + i\tau)(v + 2i - i\tau)} \right) \to -\frac{i\beta}{v} + \frac{i\beta}{v + 2i} = \frac{2\beta}{v(v + 2i)}, \qquad (13.39)$$

leading to the NLIE for $\mathfrak{a}(v)$ in the limit $N \to \infty$

$$\log \mathfrak{a}(v) = \beta h + \frac{2\beta}{v(v + 2i)} - \frac{2}{\pi} \int_{\mathcal{L}} \frac{1}{(v - w)^2 + 4} \log \mathfrak{A}(w) dw. \qquad (13.40)$$

From this NLIE we can calculate the function $\mathfrak{a}(v)$ on the axes $\text{Im}(v) = \pm 1$ by means of numerical iterations.

13.2.2 Integral expressions for the eigenvalue I

In (13.38) and (13.40) we have found integral equations determining the function $\mathfrak{a}$ for finite and infinite Trotter number N, respectively. These equations are useful if and only if we manage to extract the eigenvalue function (13.24) in terms of $\mathfrak{a}$ or $\mathfrak{A}$. In the following paragraphs we want to show how to do this.

From (13.24) we see that $\Lambda(v)$ is a rational function, and courtesy of the BA equations without poles. Hence, $\Lambda(v)$ is a polynomial of degree N. Any polynomial is determined by its zeroes and the asymptotic behaviour, i.e. the coefficient of the leading monomial. The zeroes of $\Lambda(v) = \lambda_1(v) + \lambda_2(v)$ are 'additional' solutions to the equation $\mathfrak{a}(v) = \lambda_1(v)/\lambda_2(v) = -1$, i.e. solutions to the BA equations or zeroes of $\mathfrak{A}(v) = 1 + \mathfrak{a}(v)$ that do not coincide with BA roots! These zeroes are so-called hole-type solutions to the BA equations which we label by w_l, $l = 1, \ldots, N$. These zeroes are located in the complex plane close to the axes with imaginary parts ± 2, see zeroes in Fig. 13.9 depicted by $\square$. In terms of w_l the function $\Lambda(v)$ reads

$$\Lambda(v) = \left(e^{+\beta h/2} + e^{-\beta h/2} \right) \prod_{l=1}^{N} (v - w_l). \qquad (13.41)$$

By application of Cauchy's theorem we find for v sufficiently close to the real axis (such that $v - 2i$ is outside the contour $\mathcal{L}$)

$$\frac{1}{2\pi i} \int_{\mathcal{L}} \frac{1}{v - w - 2i} [\log \mathfrak{A}(w)]' dw = \sum_j \frac{1}{v - v_j - 2i} - \frac{N/2}{v + i\tau - 2i} \qquad (13.42)$$

as the only singularities of the integrand surrounded by the contour $\mathcal{L}$ are the simple zeroes v_j and the pole $-i\tau$ of order $N/2$ of the function $\mathfrak{A}$. Also, we obtain

$$\frac{1}{2\pi i} \int_{\mathcal{L}} \frac{1}{v - w} [\log \mathfrak{A}(w)]' dw = \sum_j \frac{1}{v - v_j - 2i} - \sum_l \frac{1}{v - w_l} + \frac{N/2}{v + 2i - i\tau}, \qquad (13.43)$$

where here the integral is evaluated by use of the singularities outside the contour. To this end we deform the contour such that the upper (lower) part of $\mathcal{L}$ is closed into the upper (lower) half-plane and the relevant singularities are the simple poles $v_j + 2i$, the zeroes w_l, and the pole $i\tau - 2i$ of order $N/2$ of the function $\mathfrak{A}$.

Next, we take the difference of (13.42) and (13.43), perform an integration by parts with respect to w, and finally integrate with respect to v

$$\frac{1}{2\pi i} \int_{\mathcal{L}} \left[\frac{1}{v - w} - \frac{1}{v - w - 2i} \right] \log \mathfrak{A}(w) dw$$

$$= \log \frac{[(v - i(2 - \tau))(v + i(2 - \tau))]^{N/2}}{\prod_l (v - w_l)} + \text{constant}. \qquad (13.44)$$

The constant is determined from the asymptotic behaviour at $v \to \infty$ with the result: constant $= -\log \mathfrak{A}(\infty) = -\log(1 + \exp(\beta h))$. Combining (13.41) and (13.44) we find

$$\log \Lambda(v) = -\beta h/2 + \frac{N}{2} \log[(v - i(2 - \tau))(v + i(2 - \tau))]$$

$$- \frac{1}{2\pi i} \int_{\mathcal{L}} \left[\frac{1}{v - w} - \frac{1}{v - w - 2i} \right] \log \mathfrak{A}(w) dw \qquad (13.45)$$

and from (13.23)

$$\log \Lambda(iv, \tau) = -\beta h/2 + \frac{1}{\pi} \int_{\mathcal{L}} \frac{\log \mathfrak{A}(w)}{(v - w)(v - w - 2i)} dw. \qquad (13.46)$$

Formulas (13.46) and (13.40) are the basis of an efficient analytical and numerical treatment of the thermodynamics of the Heisenberg chain. There are, however, variants of these integral equations that are somewhat more convenient for this purpose, especially for magnetic fields close to 0.

An equivalent formulation is obtained by means of the 'particle-hole' transformation $\bar{\mathfrak{a}}$ of the function $\mathfrak{a}$

$$\bar{\mathfrak{a}}(v) := \frac{1}{\mathfrak{a}(v)}, \qquad \bar{\mathfrak{A}}(v) := 1 + \bar{\mathfrak{a}}(v). \qquad (13.47)$$

In terms of these functions (13.40) and (13.46) read

$$\log \bar{a}(v) = -\beta h + \frac{2\beta}{v(v - 2i)} + \frac{2}{\pi} \int_{\mathcal{L}} \frac{\log \overline{\mathfrak{A}}(w)}{(v - w)^2 + 4} dw, \tag{13.48}$$

and

$$\log \Lambda(iv, \tau) = \beta h/2 - \frac{1}{\pi} \int_{\mathcal{L}} \frac{\log \overline{\mathfrak{A}}(w)}{(v - w)(v - w + 2i)} dw. \tag{13.49}$$

For many purposes it turns out to be advantageous to perform a partial 'particle-hole' transformation of the function $a(v)$ on the axis $\text{Im}(v) = +1$ only. Replacing $\log \mathfrak{A} = \log \overline{\mathfrak{A}} + \log a$ on the upper part of $\mathcal{L}$ in (13.40) leads to an equation involving convolution type integrals with $\log \mathfrak{A}$, $\log \overline{\mathfrak{A}}$ and $\log a$. This equation can be resolved explicitly for $\log a$ by straightforward calculations. We will however skip these as in subsection 13.2.3 an alternative method of derivation of this set of NLIEs will be given.

13.2.3 Derivation of non-linear integral equations II

Here we want to describe another method for deriving the integral equations based on solutions to functional equations. If we define functions $q^{\pm}(v)$ by

$$q^+(v) = \prod_{v_k^+} (v - v_k^+), \quad q^-(v) = \prod_{v_k^-} (v - v_k^-), \tag{13.50}$$

where $v_k^{\pm}$ are the zeroes of $\mathfrak{A}(v)$ with imaginary parts close to ± 2, respectively, then we can write down the explicit factorization of $\mathfrak{A}$ and $\overline{\mathfrak{A}}$

$$\mathfrak{A}(v) = \left(1 + e^{\beta h}\right) \frac{q^+(v)q^-(v)q(v)}{\phi(v + i)q(v - 2i)},$$

$$\overline{\mathfrak{A}}(v) = \left(1 + e^{-\beta h}\right) \frac{q^+(v)q^-(v)q(v)}{\phi(v - i)q(v + 2i)}. \tag{13.51}$$

From this set of equations we will derive explicit expressions for $q(v)$ in terms of the functions $\mathfrak{A}(v)$ and $\overline{\mathfrak{A}}(v)$. At first glance such a task looks impossible since the functions $q_{\pm}(v)$ are unknown and the above set of equations seems underdetermined. However, this gap is filled by use of suitable additional analyticity properties of the involved functions, namely $q^+(v)$, $q(v)$, and $q^-(v)$ having zeroes with imaginary parts close to $+2$, 0, and -2, respectively.

We transform the multiplicative functional equations (13.51) into additive form by taking the logarithm. Subsequently we take the derivative, so the additive form of the functional equations is kept, but each of the individual terms is a function with zero asymptotics and hence admits the Fourier transform for integration contours with real part ranging from $-\infty$ to $+\infty$. As the Fourier transform of a function with shifted argument is identical to the 'old' Fourier transform times a known coefficient we obtain algebraic equations for the Fourier coefficients of the functions involved.

For definiteness, let us denote the Fourier pair for a function $f(v)$ by

$$\mathrm{FT}_k[f] = \frac{1}{2\pi} \int_{-\infty}^{\infty} f(v) e^{-ikv} dv,$$

$$f(v) = \int_{-\infty}^{\infty} \mathrm{FT}_k[f] e^{ikv} dk, \tag{13.52}$$

i.e. starting with a function $f(v)$ we obtain the Fourier transform $\mathrm{FT}_k[f]$ for which the inverse transform reproduces $f(v)$. We note that in our applications $f(v)$ is often analytic, so the integration contour may differ from the real axis, but still gives the same integral thanks to Cauchy's theorem as long as the real part of the contour ranges from $-\infty$ to $+\infty$. If the function $f(v)$ possesses different 'analyticity strips' that are separated by singularities, then there exist Fourier transforms that are well defined in each strip, but differ for different strips.

For the convolution $f * g$ of functions f and g

$$f * g(v) := \int_{-\infty}^{\infty} f(v - w) g(w) dw \tag{13.53}$$

we note the 'theorem'

$$\mathrm{FT}_k[f * g] = 2\pi \, \mathrm{FT}_k[f] \, \mathrm{FT}_k[g]. \tag{13.54}$$

In most cases we will apply the Fourier transform to logarithmic derivatives of functions like $q_\pm, q, \mathfrak{A} \ldots$. For any function $h(v)$ we therefore introduce the shorthand notation

$$h_k := \mathrm{FT}_k\left[\frac{d}{dv} \log h(v)\right], \tag{13.55}$$

i.e. h_k is the Fourier transform of the logarithmic derivative of $h(v)$.

As an explicit example we treat the case of a simple linear function $p(v)$

$$p(v) := v - v_0, \quad \frac{d}{dv} \log p(v) = \frac{1}{v - v_0}. \tag{13.56}$$

For this function the Fourier integral can be evaluated explicitly by closing the integration contour in the lower (upper) half-plane for positive (negative) values of the argument k. By use of Cauchy's theorem we then find

$$
\begin{aligned}
k > 0 \quad p_k &= \begin{cases} 0, & \mathrm{Im}(v_0) > 0, \\ -\mathrm{i} e^{-ikv_0}, & \mathrm{Im}(v_0) < 0, \end{cases} \\
k < 0 \quad p_k &= \begin{cases} +\mathrm{i} e^{-ikv_0}, & \mathrm{Im}(v_0) > 0, \\ 0, & \mathrm{Im}(v_0) < 0. \end{cases}
\end{aligned}
\tag{13.57}
$$

In the general case where the Fourier transform is computed for an integration path $\mathcal{L}$ different from the real axis the above result still holds if the condition '$\mathrm{Im}(v_0) > (<) 0$' is replaced by 'v_0 above (below) $\mathcal{L}$'.

Before tackling (13.51) we note the lemma

$$\text{FT}_k\left[f(v - v_0)\right] = e^{-ikv_0}\text{FT}_k\left[f(v)\right] \tag{13.58}$$

relating the two Fourier integrals provided that the integration contour $\mathcal{L}$ and $\mathcal{L} - v_0$ belong to the same analyticity strip.

We apply the Fourier transform to the logarithmic derivative of (13.51) with a contour $\mathcal{L}(\overline{\mathcal{L}})$ along a straight line slightly below (above) the real axis for the function $\mathfrak{A}$ $(\overline{\mathfrak{A}})$. Let us study the case $k > 0$ first. We note for the first equation (with path $\mathcal{L}$) that the transforms corresponding to $q^+(v)$, $q(v)$, and $\phi(v - i)$ vanish, because the zeroes of these functions are located above $\mathcal{L}$. Likewise for the second equation (with path $\overline{\mathcal{L}}$) the transforms corresponding to $q^+(v)$ and $q(v - 2i)$ vanish. We keep the remaining terms where the transforms corresponding to $q(v + 2i)$ and $\phi(v + i)$, $q(v - 2i)$ are reduced to those of $q(v)$ and $\phi(v)$ by virtue of (13.58).

$$k > 0 \quad \mathfrak{A}_k = q_k^- - e^{-2k}q_k$$
$$\overline{\mathfrak{A}}_k = -e^{-k}\phi_k + q_k^- + q_k$$
$$\mathfrak{A}_k - \overline{\mathfrak{A}}_k = e^{-k}\phi_k - (1 + e^{-2k})q_k. \tag{13.59}$$

Applying an analogous reasoning to the case $k < 0$ we finally arrive at

$$k < 0 \quad \mathfrak{A}_k = -e^k\phi_k + q_k^+ + q_k,$$
$$\overline{\mathfrak{A}}_k = q_k^+ - e^{2k}q_k$$
$$\mathfrak{A}_k - \overline{\mathfrak{A}}_k = -e^k\phi_k + (1 + e^{2k})q_k. \tag{13.60}$$

These equations are explicit expressions for $q(v)$. Note however, that the notation used above is slightly ambiguous. We have used the same symbol q_k for two different functions: the Fourier transforms of the logarithmic derivative of $q(v)$ in the upper and the lower half-plane, respectively. For practical calculations this does not matter as for $q(v)$ in the upper (lower) half-plane, the Fourier coefficient vanishes for $k < 0$ $(k > 0)$. A non-vanishing Fourier coefficient is obtained for $k > 0$ $(k < 0)$ and is then identical to q_k. In other words, q_k for $k > 0$ $(k < 0)$ is the only non-vanishing Fourier coefficient for $q(v)$ in the upper (lower) half-plane.

From (13.29) we find an explicit expression for the Fourier coefficient of $\mathfrak{a}(v)$ and $\overline{\mathfrak{a}}(v)$ in terms of $q(v)$:

$$k > 0 \quad \mathfrak{a}_k = e^{-k}\phi_k - e^{-2k}q_k$$
$$= \frac{\phi_k}{e^k + e^{-k}} + \frac{e^{-k}}{e^k + e^{-k}}(\mathfrak{A}_k - \overline{\mathfrak{A}}_k)$$

$$k < 0 \quad \mathfrak{a}_k = -e^k\phi_k + e^{2k}q_k$$
$$= -\frac{\phi_k}{e^k + e^{-k}} + \frac{e^k}{e^k + e^{-k}}(\mathfrak{A}_k - \overline{\mathfrak{A}}_k) \tag{13.61}$$

where we have finally inserted the explicit expression in terms of $\mathfrak{A}$ and $\overline{\mathfrak{A}}$. The function ϕ_k is known explicitly. A calculation using (13.56,13.57) yields

$$\phi_k = -\text{sign}(k)\text{i}\frac{N}{2}\left(e^{k(1-\tau)} + e^{-k(1-\tau)}\right). \tag{13.62}$$

We insert this into (13.61) and obtain for both signs of k

$$\mathfrak{a}_k = -\text{i}\frac{N}{2}\frac{e^{k(1-\tau)} + e^{-k(1-\tau)}}{e^k + e^{-k}} + \frac{e^{-|k|}}{e^k + e^{-k}}(\mathfrak{A}_k - \overline{\mathfrak{A}}_k). \tag{13.63}$$

Next we apply the inverse Fourier transform to the latter equation (13.63). The first term on the r.h.s. is evaluated by means of the identity

$$\int_{-\infty}^{\infty}\frac{e^{\varepsilon k}}{e^k + e^{-k}}e^{\text{i}kv}dk = \frac{\frac{\pi}{2}}{\cosh\frac{\pi}{2}(v - \text{i}\varepsilon)}. \tag{13.64}$$

The second term on the r.h.s. of (13.63) turns into a convolution by virtue of (13.54) and we arrive at

$$[\log\mathfrak{a}(v)]' = -\text{i}\frac{N}{2}[e(v - \text{i}(1 - \tau)) + e(v + \text{i}(1 - \tau))] + \kappa * [(\log\mathfrak{A})' - (\log\overline{\mathfrak{A}})'], \tag{13.65}$$

where $e(v)$ and the kernel $\kappa(v)$ take the form

$$e(v) := \frac{\frac{\pi}{2}}{\cosh\frac{\pi}{2}v}, \qquad \kappa(v) := \frac{1}{2\pi}\int_{-\infty}^{\infty}\frac{e^{-|k|}}{e^k + e^{-k}}e^{\text{i}kv}dk. \tag{13.66}$$

The equation (13.65) is almost in its desired form. After integrating (13.65) with respect to v we obtain a very similar equation where the primes are dropped and the first term on the r.h.s. is replaced by its integral plus an integration constant which has to be fixed. This constant is determined in the limit $v \to \infty$ for which the integral equation turns into an algebraic equation for the asymptotics of $\log\mathfrak{a}$ and $\log\mathfrak{A}$, $\log\overline{\mathfrak{A}}$. These asymptotics are known from (13.29) and are $\mathfrak{a} = \exp(\beta h)$, $\mathfrak{A} = 1 + \exp(\beta h)$, and $\overline{\mathfrak{A}} = 1 + \exp(-\beta h)$ yielding the integration constant $\beta h/2$. Finally, taking the limit $N \to \infty$ and noting that $\tau = \beta/N$ we find

$$\log\mathfrak{a}(v) = +\frac{\beta h}{2} - \beta e(v + \text{i}) + \kappa * [\log\mathfrak{A} - \log\overline{\mathfrak{A}}]. \tag{13.67}$$

This integral equation was derived for arguments v with imaginary part between $+2$ and -2, the integration contours in the convolutions with $\log\mathfrak{A}$ and $\log\overline{\mathfrak{A}}$ being $\mathcal{L}$ and $\overline{\mathcal{L}}$, i.e. straight lines below and above the real axis, respectively. An integral expression for the function $\log\overline{\mathfrak{a}}$ is easily obtained as $\overline{\mathfrak{a}}$ and $\mathfrak{a}$ are related in a simple way by virtue of (13.47).

$$\log\overline{\mathfrak{a}}(v) = -\frac{\beta h}{2} - \beta e(v - \text{i}) + \kappa * [\log\overline{\mathfrak{A}} - \log\mathfrak{A}]. \tag{13.68}$$

The two integral equations can be solved by numerical integration (along the contours $\mathcal{L}$ and $\overline{\mathcal{L}}$) and iteration. In what follows we consider the functions $\mathfrak{a}$, $\overline{\mathfrak{a}}$ etc. as explicitly known.

13.2.4 Integral expressions for the eigenvalue II

From (13.24) we find two expressions for the eigenvalue Λ of the QTM

$$\Lambda(v) = \phi(v - i)\frac{q(v + 2i)}{q(v)}\mathfrak{A}(v)$$
$$= \phi(v + i)\frac{q(v - 2i)}{q(v)}\overline{\mathfrak{A}}(v). \tag{13.69}$$

Either of these expressions is suitable for an explicit calculation of Λ by use of the Fourier transforms of the logarithmic derivatives of the respective functions involved. The computations are done most elegantly if both expressions are combined yielding the following functional equation

$$\Lambda(v + i)\Lambda(v - i) = \phi(v - 2i)\phi(v + 2i)\mathfrak{A}(v - i)\overline{\mathfrak{A}}(v + i), \tag{13.70}$$

where the functions on the r.h.s. are known functions. The functional equation admits a unique solution for Λ

$$\log \Lambda(v) = -\beta e_0(v) + \frac{1}{2\pi}\int_{-\infty}^{\infty} e(v - w)\log[\mathfrak{A}(w - i)\overline{\mathfrak{A}}(w + i)]dw, \tag{13.71}$$

where the first term on the r.h.s. $e_0(v)$ is a convolution of e with $\log \phi$ similar to the second term. We do not give the explicit expression of $e_0(v)$ as we are interested in the free energy (13.12). Hence $e_0(v = 0)$ is the groundstate energy of the Heisenberg chain with magnetic field $h = 0$. Any T and h dependence enters only in the second term in (13.71).

The NLIEs (13.67,13.68) and the expression (13.71) for the eigenvalue $\Lambda(v)$ are the main result of this chapter and completely determine the thermoynamics of the spin-$\frac{1}{2}$ Heisenberg chain.

Considering the historical developement of the method described above, we note that NLIEs very similar to (13.40) were derived for the row-to-row transfer matrix in [256,257]. These equations were then generalized to the related cases of QTMs (staggered transfer matrices) of the Heisenberg and RSOS chains [253,254] and the sine-Gordon model [100].

Very recently [440], in addition to the TBA approach (Chapter 5) and the QTM approach (this Section) a third formulation of the thermodynamics of the Heisenberg chain has been developed (see Appendix 13.B). At the heart of this formulation is a single NLIE with a structure very different from that of the two sets of NLIEs discussed above. Nevertheless, this new equation has been derived from the 'old' NLIEs [440, 443] and is certainly equivalent to them. (For an algebraic derivation of the TBA equations and the 'third formulation' of the thermodynamics see Appendices 13.A and 13.B.) In the first applications of the new NLIE, numerical calculations of the free energy have been performed with excellent agreement with the 'Yang-Yang' TBA and QTM results. Also, analytical high temperature expansions up to 100th order (!) have been carried out on the basis of the new formulation [441]. The 'third formulation' of thermodynamics of integrable quantum chains has been extended meanwhile to higher rank models like $sl(r + 1)$ Uimin-Sutherland models.

13.3 Shastry's model as a classical analogue of the 1d Hubbard model

Here we recall the essential properties of the Hamiltonian of the Hubbard model and its exactly solvable classical counterpart in 2d. The Hubbard model describes a lattice fermion system with electron-hopping term and on-site Coulomb repulsion with Hamiltonian

$$H_{\text{Hubbard},L} = \sum_{i=1}^{L} H_{i,i+1} + H_{\text{external}}, \tag{13.72}$$

$$H_{i,i+1} = - \sum_{a=\uparrow,\downarrow} (c_{i+1,a}^{\dagger} c_{i,a} + c_{i,a}^{\dagger} c_{i+1,a}) + U(n_{i,\uparrow} - \tfrac{1}{2})(n_{i,\downarrow} - \tfrac{1}{2}).$$

The external field term $H_{\text{external}} = - \sum_i [\mu(n_{i,\uparrow} + n_{i,\downarrow}) + B(n_{i,\uparrow} - n_{i,\downarrow})]$ will be omitted for the time being. According to [393], it is easier to find a classical analogue after performing the Jordan-Wigner transformation for electrons in 1d. The resulting spin Hamiltonian is

$$H_{i,i+1} = (\sigma_{i\uparrow}^{+}\sigma_{i+1\uparrow}^{-} + \sigma_{i+1\uparrow}^{+}\sigma_{i\uparrow}^{-}) + (\sigma_{i\downarrow}^{+}\sigma_{i+1\downarrow}^{-} + \sigma_{i+1\downarrow}^{+}\sigma_{i\downarrow}^{-}) + \frac{U}{4}\sigma_{i\uparrow}^{z}\sigma_{i\downarrow}^{z}, \tag{13.73}$$

where L denotes the length of the chain. Note that we are now imposing periodic boundary conditions for the spin system ($\sigma_{1,a} = \sigma_{L+1,a}$ for $a = \uparrow, \downarrow$). This does not correspond to periodic boundary conditions for the underlying electron system. The differences in boundary conditions, however, will not affect thermodynamic quantities like the specific heat.

For the classical counterpart in two dimensions (see Chapter 12) we considered a double-layer square lattice, consisting of $\uparrow$- and $\downarrow$-sublattices. Each local Hilbert space corresponding to a particular site of the lattice is indexed by an integer i, the sublattice is specified by the additional $a = \uparrow, \downarrow$. For vanishing on-site interaction ($U = 0$) the R-matrix is given by the product of vertex weights of the free-fermion six vertex model, $r(\lambda) = r_{\uparrow}(\lambda)r_{\downarrow}(\lambda)$, where

$$r_a(\lambda) = \frac{\cos(\lambda) + \sin(\lambda)}{2} + \frac{\cos(\lambda) - \sin(\lambda)}{2}\sigma_{1,a}^{z}\sigma_{2,a}^{z} + (\sigma_{1,a}^{+}\sigma_{2,a}^{-} + \sigma_{1,a}^{-}\sigma_{2,a}^{+}). \tag{13.74}$$

Hence r acts non-trivially in the product of local Hilbert spaces corresponding to sites 1 and 2.

Taking account of a non-vanishing U as done in [393] the following local vertex weights are found [393], (12.110)

$$\check{R}(\lambda, \mu) = \cos(\lambda + \mu) \cosh(h(\lambda, U) - h(\mu, U))\check{r}(\lambda - \mu)$$
$$+ \cos(\lambda - \mu) \sinh(h(\lambda, U) - h(\mu, U))\check{r}(\lambda + \mu)\sigma_{1,\uparrow}^{z}\sigma_{1,\downarrow}^{z}$$

where (12.108)

$$\sinh 2h(\lambda, U) := \frac{U}{4}\sin(2\lambda). \tag{13.75}$$

The L-operator is related to $\check{R}$ by (13.1). As shown in Appendix 12.A this R-matrix satisfies the Yang-Baxter relation for triple R matrices. The commutativity of the row-to-row transfer matrix is a direct consequence.

The R matrix and H are related by an expansion in small spectral parameters,

$$R_{1,2}(\lambda, \mu) = P[1 + (\lambda - \mu)H_{1,2}] + O(\lambda^2, \lambda\mu, \mu^2), \qquad (13.76)$$

where P denotes the permutation operator, $P(x \otimes y) = y \otimes x$. Note that there is a minus sign change in (13.76) in comparison to (13.3). We therefore follow Section 13.1 as closely as possible with the final replacement of the parameter τ by $-\tau$. Alternatively we may replace U by $-U$, i.e. we study $H(-U)$ which is equivalent to $-H(+U)$ upon a unitary transformation of the type of a sublattice gauge transformation, $c_{i,a} \to (-1)^i c_{i,a}$. Hence we find

$$Z = \lim_{L \to \infty} \mathrm{Tr}\, e^{-\beta H(U)} = \lim_{L \to \infty} \mathrm{Tr}\, e^{\beta H(-U)}. \qquad (13.77)$$

In order to calculate the partition function in the thermodynamic limit we further proceed as in Section 13.1

$$Z = \lim_{L \to \infty} \lim_{N \to \infty} \mathrm{Tr}\, [T(\tau)\overline{T}(\tau)]^{N/2}\big|_{\substack{\tau=\beta/N \\ U \to -U}}. \qquad (13.78)$$

In particular we introduce the QTM (13.9)

$$T^{\mathrm{QTM}}(\lambda, \tau) := \mathrm{Tr}_{\mathrm{aux}}\left[\bigotimes^{N/2} R(\lambda, -\tau) \otimes \widetilde{R}(\lambda, \tau)\right]. \qquad (13.79)$$

In the next section we will derive the BA equations by use of several of the techniques developed in Chapter 12 for the homogeneous row-to-row transfer matrix.

13.4 Diagonalization of the quantum transfer matrix

In this section we will diagonalize the QTM by the Quantum Inverse Scattering Method (QISM). At first glance, the diagonalization scheme for the QTM looks quite different from the row-to-row case. The QTM has a complicated inhomogeneous structure, seemingly demanding much more effort. Fortunately, this is not true. The crucial observation is, as remarked in the previous section, that QTMs share the same intertwining operator with the row-to-row transfer matrices. In view of the QISM, this results in identical operator algebras allowing for the diagonalization of the trace of the monodromy matrix. We note that we adopt periodic or twisted boundary conditions in the Trotter direction in order to account for the external magnetic field B and chemical potential μ, see the end of Section 13.1. Thus, the eigenvalue equation of the QTM involves the same combinations of 'dressing functions' in the terminology of the analytic Bethe ansatz as in the row-to-row case. One only has to replace the vacuum expectation values taking account of the inhomogeneity in the quantum space.

We define state vectors $|i\rangle, i = 1, \cdots, 4$ by

$$|1\rangle = |+, -\rangle, \quad |2\rangle = |+, +\rangle, \quad |3\rangle = |-, -\rangle, \quad |4\rangle = |-, +\rangle. \qquad (13.80)$$

A convenient vacuum in the present study is $|\Omega\rangle := |1, 4, 1, 4, \cdots, 1, 4\rangle$. Then the vacuum expectation values $\langle\Omega|T_{i,i}|\Omega\rangle$ read

$$\langle\Omega|T_{i,i}|\Omega\rangle = A_i \cdot e^{\beta\mu_i}, \quad i = 1, \cdots, 4,$$
$$A_i = [R_{i,1}^{i,1}(\lambda, -\tau)R_{4,i}^{4,i}(\tau, \lambda)]^{N/2}, \tag{13.81}$$

with $\mu_1, \mu_2, \mu_3, \mu_4 = \mu + B, 2\mu, 0, \mu - B$. Here μ is a chemical potential and B an external magnetic field. These fields merely lead to trivial modifications in Λ due to twisted boundary conditions for the QTM as in Refs. [101, 230, 231, 254].

Using the explicit expressions for R we obtain

$$\frac{A_1}{A_2} = \left[\frac{\omega_?(\lambda)\omega_2(\tau) - \omega_3(\lambda)\omega_1(\tau)}{\omega_2(\lambda)\omega_2(\tau) + \omega_3(\lambda)\omega_1(\tau)} \cdot \frac{\omega_2(\lambda)\omega_2(\tau) - \omega_3(\lambda)\omega_3(\tau)}{\omega_2(\lambda)\omega_2(\tau) + \omega_3(\lambda)\omega_3(\tau)}\right]^{N/2},$$

$$\frac{A_4}{A_2} = \left[\frac{\omega_1(\lambda)\omega_2(\tau) + \omega_2(\lambda)\omega_1(\tau)}{\omega_1(\lambda)\omega_2(\tau) - \omega_2(\lambda)\omega_1(\tau)} \cdot \frac{\omega_1(\lambda)\omega_2(\tau) + \omega_2(\lambda)\omega_3(\tau)}{\omega_1(\lambda)\omega_2(\tau) - \omega_2(\lambda)\omega_3(\tau)}\right]^{N/2},$$

$$A_2 = A_3 = \left[\cos^2\lambda\cos^2\tau\cos^2(\lambda - \tau)\cos^2(\lambda + \tau)\right]^N$$
$$\times \left(\frac{[\omega_1(\lambda)\omega_2(\tau) - \omega_2(\lambda)\omega_1(\tau)][\omega_3(\lambda)\omega_1(\tau) + \omega_2(\lambda)\omega_2(\tau)]}{\omega_1(\lambda)\omega_2(\lambda)\omega_1(\tau)}\right)^{N/2}. \tag{13.82}$$

Proceeding as in Section (12.6) we may derive the following expression for the eigenvalue

$$\Lambda(\lambda) = \left[e^{\beta(\mu+B)}\frac{A_1}{A_2} + e^{2\beta\mu}(-1)^m\prod_{\alpha=1}^{l}\frac{-1}{a(\lambda, \mu_\alpha)} + \prod_{j=1}^{m}\frac{1}{a(\lambda, \lambda_j)}\prod_{\alpha=1}^{l}\frac{-1}{a(\mu_\alpha, \lambda)}\right.$$
$$\left. + e^{\beta(\mu-B)}\frac{A_4}{A_2}\prod_{j=1}^{m}\frac{-1}{a(\lambda, \lambda_j)}\right](-1)^n A_2\prod_{j=1}^{m}\left(\omega_2(\lambda)\frac{\omega_2(\lambda_j)\omega_2(\lambda) + \omega_1(\lambda_j)\omega_1(\lambda)}{\omega_2(\lambda_j)\omega_2(\lambda) - \omega_1(\lambda_j)\omega_3(\lambda)}\right). \tag{13.83}$$

This is the analogy of (12.368).

Next we replace U by $-U$ as discussed in the last section. Furthermore we introduce the parameterizations of λ, τ in terms of x, w

$$e^{2x} = \tan\lambda, \qquad e^{2w} = \tan\tau, \tag{13.84}$$

and introduce the functions

$$z_\pm(x) := e^{2h(x)\pm 2x}, \quad 2h(x) := -\sinh^{-1}\left(\frac{U}{4\cosh 2x}\right). \tag{13.85}$$

In these notations the vacuum terms are expressed as

$$A_1/A_2 = \left(\frac{(1 - z_-(w)z_+(x))(1 - z_+(w)z_+(x))}{(1 + z_-(w)z_+(x))(1 + z_+(w)z_+(x))}\right)^{N/2}$$

$$A_4/A_2 = \left(\frac{(1 + z_-(w)/z_-(x))(1 + z_+(w)/z_-(x))}{(1 - z_-(w)/z_-(x))(1 - z_+(w)/z_-(x))}\right)^{N/2}$$

$$A_2 = A_3 = \left(\cos^2 \lambda \cos^2 \tau \cos^2(\lambda - \tau) \cos^2(\lambda + \tau) \right.$$

$$\left. \times\, e^{2h(w)} \left(\frac{1}{z_-(w)} - \frac{1}{z_-(x)}\right)\left(z_+(x) + \frac{1}{z_-(w)}\right)\right)^{N/2}.$$

The eigenvalue (13.83) takes the form

$$\frac{\Lambda(\lambda)}{A_2} = e^{\beta(\mu+B)}\frac{A_1}{A_2}\prod_{j=1}^{m} e^{2x}\frac{1 + z_j z_-(x)}{1 - z_j z_+(x)}$$

$$+ e^{2\beta\mu}\prod_{j=1}^{m} -e^{2x}\frac{1 + z_j z_-(x)}{1 - z_j z_+(x)}\prod_{\alpha=1}^{\ell} -\frac{z_-(x) - 1/z_-(x) - 2iw_\alpha + 3U/2}{z_-(x) - 1/z_-(x) - 2iw_\alpha + U/2}$$

$$+ \prod_{j=1}^{m} -e^{-2x}\frac{1 + z_+(x)/z_j}{1 - z_-(x)/z_j}\prod_{\alpha=1}^{\ell} -\frac{z_-(x) - 1/z_-(x) - 2iw_\alpha - U/2}{z_-(x) - 1/z_-(x) - 2iw_\alpha + U/2}$$

$$+ e^{\beta(\mu-B)}\frac{A_4}{A_2}\prod_{j=1}^{m} e^{-2x}\frac{1 + z_+(x)/z_j}{1 - z_-(x)/z_j}, \tag{13.86}$$

where we have set

$$z_j := z_-(\lambda_j), \quad 2iw_\alpha := 2i\nu(\mu_\alpha) - U. \tag{13.87}$$

The parameters $\{z_j\}, \{w_\alpha\}$ satisfy the Bethe ansatz equations,

$$e^{\beta(\mu-B)}\left(\frac{(1 + z_-(w)/z_j)(1 + z_+(w)/z_j)}{(1 - z_-(w)/z_j)(1 - z_+(w)/z_j)}\right)^{N/2}$$

$$= -(-1)^m \prod_{\alpha=1}^{\ell} -\frac{z_j - 1/z_j - 2iw_\alpha - U/2}{z_j - 1/z_j - 2iw_\alpha + U/2}, \tag{13.88}$$

$$e^{2\beta\mu}\prod_{j=1}^{m} \frac{z_j - 1/z_j - 2iw_\alpha + U/2}{z_j - 1/z_j - 2iw_\alpha - U/2} = -\prod_{\beta=1}^{\ell} \frac{2i(w_\alpha - w_\beta) - U}{2i(w_\alpha - w_\beta) + U}.$$

Here some remarks are in order:

(i) We have checked (13.86) for the largest eigenvalue against results from numerical diagonalizations of finite systems up to size $N = 6$. The leading state lies in the sector $m = N, \ell = N/2$. For the repulsive case and $\mu = B = 0$, all z_j's are on the imaginary axis, while all w_α's are real, cf. figures 13.10 and 13.11.

(ii) The free-fermion partition function is recovered in the limit $U \to 0$.

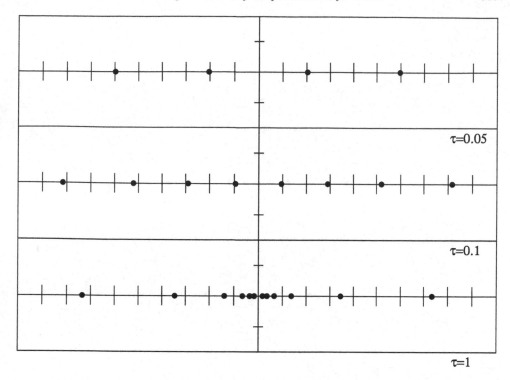

Fig. 13.10. Plots of the parameters z_j on the imaginary axis (horizontal lines) for $U = 8$, $N = 16$ and different temperatures $\tau = \beta/N = 0.05, 0.1, 1$. For the sake of clarity we have omitted the axes labels. The origin corresponds to 0, and marks are placed at integer values times i. In all three cases there are 16 parameters z_j, however only a subset is visible within the displayed window of absolute values less than 10. Note that for $\tau = 0.05$ all values of z_j are larger than 1, for $\tau = 0.1$ (1) two (six) values of z_j are less than 1.

(iii) Starting from another vacuum $|\Omega'\rangle = |2, 3, \cdots\rangle$, one obtains a different expression for $\Lambda(\lambda)$. The resultant one is actually identical to (13.83) after negating U and exchanging $B \leftrightarrow \mu$. This alternative formulation is in fact equivalent to (13.83) thanks to a partial particle-hole transformation.

The solution to the Bethe ansatz equations (13.88) corresponding to the largest eigenvalue shows a characteristic temperature dependence. For rather large values of the temperature T (i.e. small values of $\tau = \beta/N$) all z_j lie on the imaginary axis and have absolute values larger than 1. Upon lowering T (i.e. increasing τ) the parameters z_j decrease and converge towards the origin, see figure 13.10. In particular, at low temperatures a certain number of the z_j's have absolute values less than 1. A similar behaviour is shown by the w_α parameters on the real axis, see figure 13.11. We note that the motion of the z_j parameters has profound consequences. In the next section we will reparametrise the z_j's in terms of s_j parameters by use of a function $s(z)$ that is not one-to-one, i.e. $z(s)$ is double valued. Upon the action of $s(z)$, parameters z on the imaginary axis with $|z| > 1$ and those with $|z| < 1$ are mapped

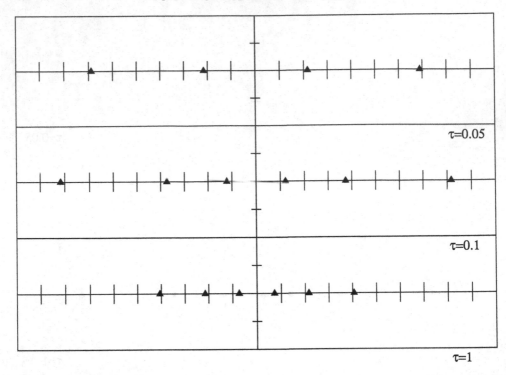

Fig. 13.11. Plots of the parameters w_α on the real axis (horizontal lines) for $U = 8$, $N = 16$ and different temperatures $\tau = \beta/N = 0.05, 0.1, 1$. For the sake of clarity we have omitted the axes labels. The origin corresponds to 0, and marks are placed at integer values. In all three cases there are 8 parameters w_α, however only a subset is visible within the displayed window of absolute values less than 10.

onto the same range of the real axis with $|s| > 1$. In this sense the motion of z_j parameters through the points $\pm i$ corresponds to a change of the new s_j from the first branch to the second branch (with a branch cut from -1 to $+1$). The set of parameters w_α will be kept and there is no branch cut in the w-plane.

13.5 Associated auxiliary problem of difference type

The thermodynamical information on the system is encoded in the solution to the Bethe ansatz equations (13.88) in the limit $N \to \infty$. For finite N it is possible to solve the Bethe ansatz equations numerically. However, for large N it is quite complicated to find the numerical solution even for the ground state. Furthermore, in the Trotter limit $N \to \infty$ the roots $\{v_k, w_k\}$ accumulate at infinity. This is similar to other models (Heisenberg model, t-J model) where the solutions of the Bethe ansatz equations of the QTM coalesce at the origin [100, 230–232, 253, 254]. This represents the main problem in analyzing the limit $N \to \infty$ directly on the basis of the Bethe ansatz equations. To overcome this difficulty one can express the solution of the Bethe ansatz equations by a system of non-linear integral equations in analogy with our treatment of the Heisenberg chain in Section 13.2.

The first problem to be overcome is the complicated structure of the Bethe ansatz equations (13.88). Introducing variables

$$s_j = \frac{1}{2i}\left(z_j - \frac{1}{z_j}\right),$$ (13.89)

we may recast equations (13.88) in a difference form (in the rapidities $\{s_j\}$, $\{w_\alpha\}$)

$$e^{-\beta(\mu-B)}\phi(s_j) = -\frac{q_2(s_j - iU/4)}{q_2(s_j + iU/4)},$$ (13.90)

$$e^{-2\beta\mu}\frac{q_2(w_\alpha + iU/2)}{q_2(w_\alpha - iU/2)} = -\frac{q_1(w_\alpha + iU/4)}{q_1(w_\alpha - iU/4)},$$ (13.91)

where we have defined

$$q_1(s) := \prod_j^m (s - s_j), \quad q_2(s) := \prod_\alpha^l (s - w_\alpha),$$ (13.92)

$$\phi(s) := \left(\frac{(1 - z_-(w)/z(s))(1 - z_+(w)/z(s))}{(1 + z_-(w)/z(s))(1 + z_+(w)/z(s))}\right)^{N/2},$$ (13.93)

$$z(s) := is\left[1 + \sqrt{(1 - 1/s^2)}\right].$$ (13.94)

Equations (13.90), (13.91) would be equivalent to (13.88) if the functions $\phi(s)$ and $z(s)$ were single-valued. However, these functions possess two branches.

The standard ('first') branch of $z(s)$ is defined by requiring that $z(s) \simeq 2is$ for large values of s, with a branch cut along $[-1, 1]$ (corresponding to values of z on the unit circle.) Hence the first branch of $z(s)$ maps the complex plane without $[-1, 1]$ onto the region of the complex plane outside the unit circle. Conversely the second branch of $z(s)$ maps the complex plane without $[-1, 1]$ onto the inner region of the unit circle.

Along the branch cut we find

$$z(x \pm i0) = ix \mp \sqrt{1 - x^2}, \quad x \in [-1, 1].$$ (13.95)

We will not often refer to the second branch of $z(s)$. We emphasize that $z(s)$ is in no way related to the functions $z_\pm$ defined in (13.85).

To summarize: equations (13.90), (13.91) are equivalent to (13.88) only if we specify on which Riemann sheet each parameter s_j in (13.90) lies. In figure 13.10 (and figure 13.11) we plot solutions z_j (and w_α) to (13.88) for three typical cases at high, moderate, and low temperatures. At high temperatures all z_j have absolute values larger than 1 with the corresponding parameters s_j lying on the first sheet, cf. figure 13.12 where the positions of s_j on the first sheet are depicted by squares. At low temperatures most of the z_j's have absolute values larger than 1, but there are some z_j with absolute values less than 1. The

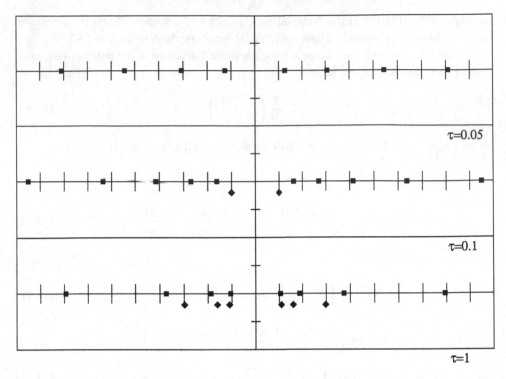

Fig. 13.12. Plots of the parameters s_j on the real axis (horizontal lines) for $U = 8$, $N = 16$ and different temperatures $\tau = \beta/N = 0.05, 0.1, 1$. The origin corresponds to 0, and marks are placed at integer values. In all three cases there are 16 parameters s_j, however only a subset is visible within the displayed window of absolute values less than 10. The depicted results correspond to those in figure 13.10 via the relation (13.89). Parameters s_j corresponding to z_j's with absolute values larger (smaller) than 1 are depicted by squares (diamonds). For the sake of clarity the symbols for s_j's lying on the second sheet are plotted slightly below the axis.

corresponding parameters s_j lie on the first as well as the second sheet, cf. figure 13.12 where the positions of s_j's that lie on the second sheet are depicted by diamonds.

We note that for $N \to \infty$ there are infinitely many rapidities on the first (upper) sheet and finitely many on the second (lower) sheet. The number of rapidities on the second sheet is increasing with decreasing temperature, resulting in a flow from the first to the second sheet, see figure 13.13.

The function $\phi(s)$ defined in (13.92) has two branches as well, which we denote by $\phi^+(s)$ and $\phi^-(s)$, respectively. The function $\phi^+(s)$ has a zero (pole) of order $N/2$ at the point s_0 ($-s_0$) defined by

$$z(s_0) := z_-(w), \qquad (2is_0 \simeq N/\beta \text{ for large } N). \qquad (13.96)$$

On the other hand, $\phi^-(s)$ has a zero (pole) of order $N/2$ at the point $-s_0 + iU/2$ ($s_0 - iU/2$).

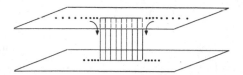

Fig. 13.13. Depiction of the flow of rapidities s_j from the first (upper) Riemann sheet to the second (lower) one as the temperature is decreased.

We note that the general expression (13.86) for the eigenvalue $\Lambda(\lambda)$ is quite complicated, but simplifies considerably at $\lambda = 0$ and $\tau \to 0$

$$\Lambda(\lambda = 0) = e^{\beta U/4}(1 + e^{\beta(\mu+B)})(1 + e^{\beta(\mu-B)})\tau^N \prod_{j=1}^{m} z_j. \tag{13.97}$$

At this point a comment on the difference type property of (13.91) is in order. These equations are 'Bethe ansatz equation compatible' with an auxiliary system in the following sense. Consider the auxiliary function

$$\Lambda^{\text{aux}}(s) := \lambda_1(s) + \lambda_2(s) + \lambda_3(s) + \lambda_4(s), \tag{13.98}$$

where the λ_j functions are defined by

$$\lambda_1(s) := e^{\beta(\mu+B)}\frac{\phi(s - iU/4)}{q_1(s - iU/4)}, \quad \lambda_2(s) := e^{2\beta\mu}\frac{q_2(s - iU/2)}{q_2(s)q_1(s - iU/4)},$$

$$\lambda_3(s) := \frac{q_2(s + iU/2)}{q_2(s)q_1(s + iU/4)}, \quad \lambda_4(s) := e^{\beta(\mu-B)}\frac{1}{\phi(s + iU/4)\,q_1(s + iU/4)}. \tag{13.99}$$

In this case, the condition of analyticity of $\Lambda^{\text{aux}}(s)$, i.e. the absence of poles, leads to (13.90), (13.91), which are the Bethe ansatz equations of the eigenvalue $\Lambda(\lambda)$ of the original QTM! An important qualitative difference between $\Lambda(\lambda)$ and $\Lambda^{\text{aux}}(s)$ is that while $\Lambda(\lambda)$ is analytic everywhere, i.e. on all branches, $\Lambda^{\text{aux}}(s)$ is analytic on the first (standard) branch, but may have singularities on the other (three) branches.

Let us illustrate this point by considering the first set of the Bethe ansatz equations (13.90). The latter set of equations arises from requiring that the zeroes $s_j + iU/4$ in the denominators of λ_1 and λ_2 are cancelled in the sum $\lambda_1 + \lambda_2$ (or equivalently that the zeroes $s_j - iU/4$ in the denominators of λ_3 and λ_4 are cancelled in the sum $\lambda_3 + \lambda_4$):

$$\frac{\lambda_1(s_j + iU/4)}{\lambda_2(s_j + iU/4)} = -1. \tag{13.100}$$

This condition is satisfied in the limit $N \to \infty$ for an infinite number of rapidities on the first branch of the function λ_1/λ_2, and for a finite number of rapidities on the second branch, see figure 13.13.

In figure 13.14 we show the distribution of zeroes, poles and the branch cut for the function $(\lambda_1 + \lambda_2)(s)$ at a relatively high temperature. Here all rapidities s_j satisfy Eqn. (13.100) on the first branch. Hence the cancellation of poles and zeroes happens entirely for the first

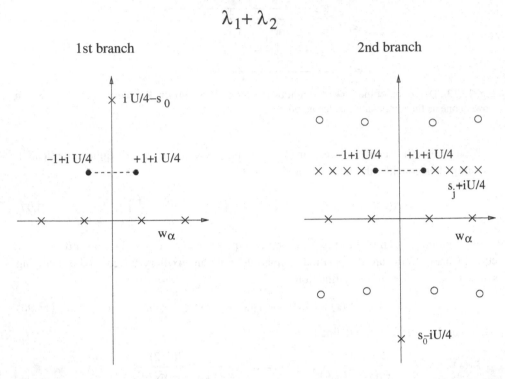

Fig. 13.14. Distribution of zeroes, poles and the branch cut of the function $(\lambda_1 + \lambda_2)(s)$ in the complex s-plane for zero magnetic field and chemical potential ($B = 0$, $\mu = 0$). Zeroes (poles) are depicted by open circles (crosses). Across the branch cut depicted by a dashed line, the transition from the first to the second branch and back is possible. All poles are of simple order except those at $iU/4 - s_0$ and $s_0 - iU/4$ which are of order $N/2$. The zeroes of the function (on the second branch) are not of further importance to our analysis. Here, for not too low temperatures, the poles with imaginary part $U/4$ occur on the second branch and are completely absent on the first branch. If the temperature is lowered, the poles on the second branch move through the branch cut onto the first branch.

branch of $(\lambda_1 + \lambda_2)(s)$, no cancellation takes place on the second branch. Therefore all poles with imaginary part $U/4$ occur on the second branch and are completely absent on the first one. In figure 13.15 we show the distribution of zeroes, poles and the branch cut for the function $(\lambda_3 + \lambda_4)(s)$. Last but not least, we illustrate the distribution of zeroes, poles and branch cuts for the function $(\lambda_1 + \lambda_2 + \lambda_3 + \lambda_4)(s)$, cf. figure 13.16. As we have two cuts of the type discussed above there are in total four branches. We are mostly interested in the case of the first (standard) branch.

As a final comment on the singularities of the introduced functions we note that there is no non-zero winding number of the functions $(\lambda_1 + \lambda_2)(s)$, $(\lambda_3 + \lambda_4)(s)$, $(\lambda_1 + \lambda_2 + \lambda_3 + \lambda_4)(s)$ around their branch cuts. The reason is simple: the total number of poles on the 1st branch is equal to N and the asymptotic behaviour of the functions is $1/s^N$. Of course, for lower temperatures isolated singularities may change from one branch to the other via the

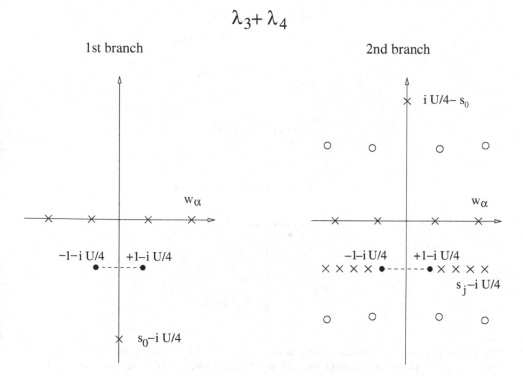

Fig. 13.15. Distribution of zeroes, poles and the branch cut of the function $(\lambda_3 + \lambda_4)(s)$ in the complex s-plane. The notation is identical to that of figure 13.14. Here the patterns look like the complex conjugates of those in figure 13.14. This is strictly true for zero magnetic field and chemical potential ($B = 0$, $\mu = 0$) and still holds approximately for finite fields.

branch cuts. Still the above reasoning remains correct if the loop surrounds the branch cut as well as the emerging singularities.

The construction (13.98), (13.99) is at this point purely mathematical; however, it will be the starting point of the derivation of integral equations in the next section.

13.6 Derivation of non-linear integral equations

In this section we are concerned with the derivation of well posed integral equations equivalent to the nested Bethe ansatz equations for the largest eigenvalue of the QTM for $U > 0$. (The case $U < 0$ is simply obtained via a particle-hole transformation, see Section 2.2.4.) We introduce a set of auxiliary functions satisfying a set of closed functional equations which later on are transformed into integral form.

At first glance a treatment strictly following the one for the Heisenberg chain would appear to be possible [255]. In fact, functions like $\lambda_1(s)/\lambda_2(s)$, $\lambda_2(s)/\lambda_3(s)$ etc. may lead to a closed set of equations. However, the analytic properties, i.e. distribution of poles and

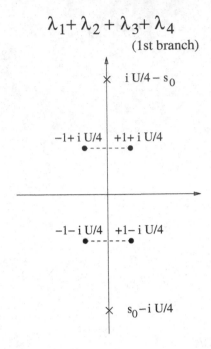

$$\lambda_1 + \lambda_2 + \lambda_3 + \lambda_4$$
(1st branch)

Fig. 13.16. Distribution of zeroes, poles and the branch cut of the function $(\lambda_1 + \lambda_2 + \lambda_3 + \lambda_4)(s)$ in the complex s-plane for zero magnetic field and chemical potential ($B = 0$, $\mu = 0$) and relatively high temperature. The notation is identical to that of figure 13.14 and figure 13.15.

branch cuts, are such that the integration contours are no longer straight lines. This causes severe problems in a numerical study and obstructs the analytical investigation of low-temperature properties. Some problems in such a direct approach are already encountered at moderate temperatures in a study of the BA equations for finite N. In figure 13.17 the functions λ_1/λ_2 and λ_2/λ_3 are shown for argument $s(z) + iU/4$ (with imaginary z) and argument s (real), respectively. Note that the phases of these functions are monotonic for λ_1/λ_2, but non-monotonic for λ_2/λ_3, which in the latter case complicates the determination of the relevant zeroes. The parameters w_α are to be determined from λ_2/λ_3 or equivalently from $(\lambda_1 + \lambda_2)/(\lambda_3 + \lambda_4)$ which shows the advantage of a monotonical varying phase.

The following explicit expressions of the functions $\mathfrak{b}$, $\overline{\mathfrak{b}}$, $\mathfrak{c}$, $\overline{\mathfrak{c}}$ turn out to be very useful

$$\mathfrak{b} = \frac{\overline{l_1} + \overline{l_2} + \overline{l_3} + \overline{l_4}}{l_1 + l_2 + l_3 + l_4}, \qquad \overline{\mathfrak{b}} = \frac{l_1 + l_2 + l_3 + l_4}{\overline{l_1} + \overline{l_2} + \overline{l_3} + \overline{l_4}},$$

$$\mathfrak{c} = \frac{l_1 + l_2}{l_3 + l_4} \cdot \frac{\overline{l_1} + \overline{l_2} + \overline{l_3} + \overline{l_4}}{l_1 + l_2 + l_3 + l_4 + \overline{l_1} + \overline{l_2} + \overline{l_3} + \overline{l_4}}, \tag{13.101}$$

$$\overline{\mathfrak{c}} = \frac{\overline{l_3} + \overline{l_4}}{\overline{l_1} + \overline{l_2}} \cdot \frac{l_1 + l_2 + l_3 + l_4}{l_1 + l_2 + l_3 + l_4 + \overline{l_1} + \overline{l_2} + \overline{l_3} + \overline{l_4}},$$

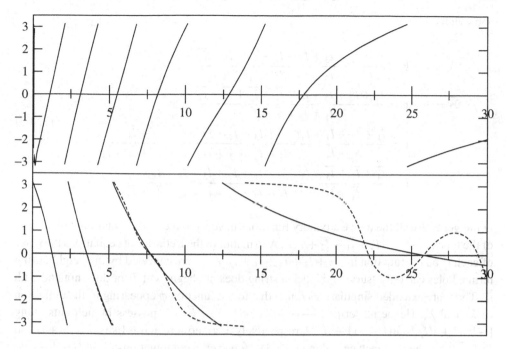

Fig. 13.17. Illustration of various functions for the case $U = 4$, $N = 16$ and $\tau = 0.1$. Upper panel: depiction of the phase of $-\frac{\lambda_1}{\lambda_2}(s(z) + iU/4)$ for z on the positive imaginary axis (horizontal line). Zeroes of this function correspond to the Bethe ansatz rapidities z_j. Lower panel: depiction of the phase of $-\frac{\lambda_2}{\lambda_3}(s)$ (dashed line) and of $-\frac{\lambda_1+\lambda_2}{\lambda_3+\lambda_4}(s)$ (solid line). Some (all) of the zeroes of the dashed (solid) line correspond to the Bethe ansatz rapidities w_α.

where the functions l_j and $\overline{l}_j$ are closely related to the λ_j defined in (13.99)

$$l_j(s) = \lambda_j(s - iU/4) \cdot e^{2\beta B} \phi^+(s)\phi^-(s),$$
$$\overline{l}_j(s) = \lambda_j(s + iU/4). \tag{13.102}$$

The main observation in connection with the functions defined in (13.101) is based on elementary facts of the theory of complex functions. In particular any analytic function on the complex plane is entirely determined by its singularities, i.e. poles and branch cuts, as well as its asymptotic behaviour at infinity. Below we will show that the singularities of $\log \mathfrak{b}$, $\log \mathfrak{c}$ etc. on the *entire complex* plane are exhausted by the singularities of $\log(1 + \mathfrak{b})$, $\log(1 + \mathfrak{c})$ etc. in the vicinity of the *real axis*.[2] Furthermore, all functions involved exhibit constant asymptotics for finite N. Hence there exists a suitable integral representation of $\log \mathfrak{b}$, $\log \mathfrak{c}$ etc. in terms of $\log(1 + \mathfrak{b})$, $\log(1 + \mathfrak{c})$ etc. The latter functions will be

---

[2] The relevant singularities are distributed exactly on the real axis for vanishing external fields. For this case the subsequent treatment can be taken literally. For finite external fields B, μ, deviations from the real axis occur. The following reasoning still applies *mutatis mutandis*.

abbreviated by

$$\mathfrak{B} = 1 + \mathfrak{b} = \frac{l_1 + l_2 + l_3 + l_4 + \overline{l_1} + \overline{l_2} + \overline{l_3} + \overline{l_4}}{l_1 + l_2 + l_3 + l_4},$$

$$\overline{\mathfrak{B}} = 1 + \overline{\mathfrak{b}} = \frac{l_1 + l_2 + l_3 + l_4 + \overline{l_1} + \overline{l_2} + \overline{l_3} + \overline{l_4}}{\overline{l_1} + \overline{l_2} + \overline{l_3} + \overline{l_4}},$$

$$\mathfrak{C} = 1 + \mathfrak{c} = \frac{l_1 + l_2 + l_3 + l_4}{l_3 + l_4} \cdot \frac{l_3 + l_4 + \overline{l_1} + \overline{l_2} + \overline{l_3} + \overline{l_4}}{l_1 + l_2 + l_3 + l_4 + \overline{l_1} + \overline{l_2} + \overline{l_3} + \overline{l_4}},$$

$$\overline{\mathfrak{C}} = 1 + \overline{\mathfrak{c}} = \frac{\overline{l_1} + \overline{l_2} + \overline{l_3} + \overline{l_4}}{\overline{l_1} + \overline{l_2}} \cdot \frac{l_1 + l_2 + l_3 + l_4 + \overline{l_1} + \overline{l_2}}{l_1 + l_2 + l_3 + l_4 + \overline{l_1} + \overline{l_2} + \overline{l_3} + \overline{l_4}}.$$

(13.103)

Quite generally all the above auxiliary functions have a product representation with factors of the type $\ldots + l_3 + l_4 + \overline{l_1} + \overline{l_2} + \ldots$. As a matter of the Bethe ansatz equations the poles of each l_j and $\overline{l_j}$ function in $\ldots + l_3 + l_4 + \overline{l_1} + \overline{l_2} + \ldots$ are canceled by the neighbouring terms. Poles can only 'survive' if such a string does not begin with l_1 or does not end with $\overline{l_4}$. There are extended singularities (cuts) due to the function ϕ appearing in the definition of λ_1 and λ_4. Hence all terms $l_1 + l_2 + \ldots$ and $\ldots + \overline{l_3} + \overline{l_4}$ possess branch cuts along $[-1, 1] + iU/2$ and $[-1, 1] - iU/2$, respectively. Furthermore, terms like $\ldots + l_3 + l_4$ and $\overline{l_1} + \overline{l_2} + \ldots$ have branch cuts along $[-1, 1]$. However in combinations $\ldots + l_4 + \overline{l_1} + \ldots$ the branch cut due to the ϕ function disappears, because

$$l_4(s) + \overline{l_1}(s) = e^{\beta(\mu+B)} \frac{\phi^+(s) + \phi^-(s)}{q_1(s)}, \tag{13.104}$$

and $\phi^+(s) + \phi^-(s)$ is analytic everywhere as a crossing of the line $[-1, +1]$ results into a simple exchange $\phi^+(s) \leftrightarrow \phi^-(s)$ leaving the sum invariant.

Inspecting the function $\lambda_1 + \lambda_2 + \lambda_3 + \lambda_4$ more closely we find poles of order $N/2$ at $s_0 - iU/4$ and $iU/4 - s_0$ where s_0 is defined in (13.96). In addition we find zeroes and branch cuts on the lines $\text{Im}(s) = \pm U/4$ which we write as

$$\log[\lambda_1(s) + \lambda_2(s) + \lambda_3(s) + \lambda_4(s)] \equiv_s -\frac{N}{2} \log[(s - s_0 + iU/4)(s + s_0 - iU/4)]$$
$$+ L_-(s + iU/4) + L_+(s - iU/4), \tag{13.105}$$

where $\equiv_s$ denotes that left- and right-hand sides have the same singularities on the entire plane, and $L_\pm$ are suitable functions possessing the desired singularities and being analytic otherwise. These functions can be constructed quite explicitly by contour integrals of the type (13.31). The functions $L_\pm$ are obtained by choosing $\mathcal{L}$ around the axes $\text{Im}(s) = \pm U/4$ and setting $f(v) \to L_\pm(s)$, $\mathfrak{A}(w) \to (\lambda_1 + \lambda_2 + \lambda_3 + \lambda_4)(w)$ and $v - w \to s - w \pm iU/4$. Alternatively, we can write the functions as $L_\pm = k \circ l_\pm$ where the function k and the symbol $\circ$ are defined below in (13.109), (13.110) and $l_\pm(s) = (\lambda_1 + \lambda_2 + \lambda_3 + \lambda_4)(s \pm iU/4)$. Note however, that we do not need these explicit expressions for our reasoning. From (13.105) we find the following singularities

$$\log[l_1(s) + l_2(s) + l_3(s) + l_4(s)] \equiv_s -\frac{N}{2}\log[(s - s_0)(s + s_0 - iU/2)]$$
$$+ \log[\phi^+(s)\phi^-(s)] + L_-(s) + L_+(s - iU/2),$$
$$\log[\overline{l}_1(s) + \overline{l}_2(s) + \overline{l}_3(s) + \overline{l}_4(s)] \equiv_s -\frac{N}{2}\log[(s + s_0)(s - s_0 + iU/2)]$$
$$+ L_-(s + iU/2) + L_+(s). \tag{13.106}$$

From this, and (13.101, 13.103) and the identity

$$\phi^+(s)\phi^-(s) = \left[\frac{(s - s_0)(s + s_0 - iU/2)}{(s + s_0)(s - s_0 + iU/2)}\right]^{N/2} \tag{13.107}$$

we find the singularities

$$\log \mathfrak{b}(s) \equiv_s L_-(s + iU/2) + L_+(s) - L_-(s) - L_+(s - iU/2),$$
$$\log \mathfrak{B}(s) \equiv_s -L_-(s) + \text{rest},$$
$$\log \overline{\mathfrak{c}}(s) - \log \overline{\mathfrak{C}}(s) \equiv_s L_-(s) - L_+(s) + \text{rest}, \tag{13.108}$$

where 'rest' indicates singularities not located on the real axis.

Next we introduce the notation

$$(g \circ f)(s) = \int_{\mathcal{L}} g(s - t)f(t)dt \tag{13.109}$$

for the convolution of two functions g and f with contour $\mathcal{L}$ surrounding the real axis at infinitesimal distance above and below in anticlockwise manner. From Cauchy's theorem we find for any function f analytic above and below the real axis

$$(k \circ f)(x \pm i0) = (k \circ f)(x) + f(x \pm i0), \text{ where } k(s) = \frac{1}{2\pi i}\frac{1}{s}, \tag{13.110}$$

and x is real (see also (13.34) and (13.35)). For further convenience we introduce the functions

$$K_1(s) = k(s - iU/4) - k(s + iU/4) = \frac{U}{4\pi}\frac{1}{s^2 + (U/4)^2},$$
$$\widehat{K}_1(s) = K_1(s + iU/4), \quad \overline{K}_1(s) = K_1(s - iU/4), \tag{13.111}$$
$$K_2(s) = k(s - iU/2) - k(s + iU/2) = \frac{U}{2\pi}\frac{1}{s^2 + (U/2)^2},$$

which will play the role of integral kernels. From (13.108), (13.110), (13.111) we find

$$[K_2 \circ \log \mathfrak{B} + \overline{K}_1 \circ (\log \overline{\mathfrak{c}} - \log \overline{\mathfrak{C}})] \equiv_s L_-(s + iU/2) + L_+(s) - L_-(s) - L_+(s - iU/2). \tag{13.112}$$

Upon comparing (13.108), (13.112) we conclude

$$\log \mathfrak{b}(s) = K_2 \circ \log \mathfrak{B} + \overline{K_1} \circ (\log \overline{\mathfrak{c}} - \log \overline{\mathfrak{C}}) + \text{const}, \qquad (13.113)$$

as both sides are complex functions with identical singularities. For a proof of the identity we consider the difference function which is entire, i.e. analytic on the entire complex plane. Furthermore the difference function is bounded, hence it is constant. The constant is computed from considering the asymptotic behaviour at $s \to \infty$. The function $\mathfrak{b}(s)$ (13.101) due to (13.102) has the limiting behaviour of $1/[e^{2\beta B} \phi^+(s) \phi^-(s)]$ which simply gives $e^{-2\beta B}$. As the integrals on the r.h.s. of the last equation turn to 0 we have

$$\text{const} = -2\beta B. \qquad (13.114)$$

For the derivation of the second type of integral equations we define an intermediate set of auxiliary functions

$$\mathfrak{t} = \frac{l_1 + l_2}{l_3 + l_4}, \qquad \mathfrak{T} = 1 + \mathfrak{t} = \frac{l_1 + l_2 + l_3 + l_4}{l_3 + l_4},$$
$$\overline{\mathfrak{t}} = \frac{\overline{l_3} + \overline{l_4}}{\overline{l_1} + \overline{l_2}}, \qquad \overline{\mathfrak{T}} = 1 + \overline{\mathfrak{t}} = \frac{\overline{l_1} + \overline{l_2} + \overline{l_3} + \overline{l_4}}{\overline{l_1} + \overline{l_2}}. \qquad (13.115)$$

Quite similar to the above reasoning we find

$$\log[\lambda_1(s) + \lambda_2(s)] \equiv_s -\frac{N}{2} \log(s + s_0 - iU/4) + L(s - iU/4) - \log q_2(s),$$

$$\log[\lambda_3(s) + \lambda_4(s)] \equiv_s -\log \phi(s + iU/4) - \frac{N}{2} \log(s + s_0 + iU/4)$$
$$+ L(s + iU/4) - \log q_2(s) \qquad (13.116)$$

with a suitable function $L(s)$. From this we find the singularities

$$\log \mathfrak{t}(s) \equiv_s \frac{N}{2} \log \frac{s + s_0}{s + s_0 - iU/2} + \log \phi(s) - L(s) + L(s - iU/2)$$
$$\log \overline{\mathfrak{B}}(s) + \log \overline{\mathfrak{T}}(s) \equiv_s -L(s) + \text{rest}, \qquad (13.117)$$

where 'rest' again indicates singularities not located on the real axis. Hence we conclude

$$\log \mathfrak{t}(s) = \beta(\mu + B) + \frac{N}{2} \log \frac{s + s_0}{s + s_0 - iU/2} + \log \phi(s) - \overline{K_1} \circ (\log \overline{\mathfrak{B}} + \log \overline{\mathfrak{T}}). \qquad (13.118)$$

The constant on the r.h.s. has been determined similarly to above from the limit of $\mathfrak{t}(s)$ which is straightforwardly found from (13.99) to be $\exp(\beta(\mu + B))$.

Next we deform the integration contour for $\log \overline{\mathfrak{B}}$ in (13.118) from a narrow loop around the real axis to a wide loop consisting of the two horizontal lines $\text{Im}(s) = \pm \alpha$, with $0 < \alpha \leq U/4$. The corresponding convolution is denoted by '$\square$'

$$\overline{K_1} \circ \log \overline{\mathfrak{B}} = \overline{K_1} \square \log \overline{\mathfrak{B}} - \log \overline{\mathfrak{B}}, \qquad (13.119)$$

and the additional contribution is due to the residue of $\overline{K_1}$, see (13.110). Taking into account (13.101, 13.115) we find

$$\log \mathfrak{c} = \log \mathfrak{t} - \log \overline{\mathfrak{B}}, \qquad \Delta \log \overline{\mathfrak{T}} = \Delta \log \overline{\mathfrak{C}}, \qquad (13.120)$$

where $\Delta f(x) = f(x + i0) - f(x - i0)$ denotes the discontinuity along the real axis. Therefore, (13.118) turns into

$$\log \mathfrak{c}(s) = \beta(\mu + B) + \frac{N}{2} \log \frac{s + s_0}{s + s_0 - iU/2} + \log \phi(s) - \overline{K_1} \square \log \mathfrak{B} - \overline{K_1} \circ \log \overline{\mathfrak{C}}. \qquad (13.121)$$

Lastly, we perform the limit $N \to \infty$ in the above equations yielding

$$\log \mathfrak{b} = -2\beta B + K_2 \square \log \mathfrak{B} + \overline{K_1} \circ (\log \overline{\mathfrak{c}} - \log \overline{\mathfrak{C}}),$$
$$\log \mathfrak{c} = -\beta U/2 + \beta(\mu + B) + \log \phi - \overline{K_1} \square \log \mathfrak{B} - \overline{K_1} \circ \log \overline{\mathfrak{C}}, \qquad (13.122)$$
$$\log \overline{\mathfrak{c}} = -\beta U/2 - \beta(\mu + B) - \log \phi + \widehat{K_1} \square \log \mathfrak{B} + \widehat{K_1} \circ \log \mathfrak{C},$$

where the equation for $\log \overline{\mathfrak{c}}$ has been derived in analogy to the one for $\log \mathfrak{c}$, and for $N \to \infty$ the function ϕ takes the simplified form

$$\log \phi(s) = -2\beta i s \sqrt{1 - 1/s^2}. \qquad (13.123)$$

yielding at the branch cut

$$\log \phi(x \pm i0) = \pm 2\beta \sqrt{1 - x^2}, \qquad x \in [-1, 1]. \qquad (13.124)$$

We want to point out that the function $\mathfrak{b}$ will be evaluated on the lines $\text{Im}(s) = \pm \alpha$ (notably with $\alpha = U/4$). The functions $\mathfrak{c}$ and $\overline{\mathfrak{c}}$ need only be evaluated on the real axis infinitesimally above and below the interval $[-1, 1]$. Also the convolutions involving the '$\mathfrak{c}$ functions' in (13.122) can be restricted to a contour surrounding $[-1, 1]$ as these functions are analytic outside.

Finally, we want to comment on the structure of the equations determining the thermodynamical properties of the Hubbard model. In contrast to long-range interaction systems [161, 240] we have to solve a set of subsidiary equations (13.122) for the 'distribution functions' $\mathfrak{b}$, $\mathfrak{c}$, and $\overline{\mathfrak{c}}$ before evaluating the free energy (13.154). Obviously, the dynamics of the elementary excitations of the nearest-neighbour systems are more involved than those of [161, 240] which may be viewed as 'free particles with exclusion statistics'.

13.7 Integral expression for the eigenvalue

Here we turn to the derivation of expressions for the largest eigenvalue of the QTM (13.97) in terms of the above auxiliary functions. We write $\sum_j \log z_j = \sum_j \log z(s_j)$ (13.89) as a Cauchy integral of the function

$$f(s) = \log z(s) \left[\log (1 + l_4/l_3(s)) \right]', \qquad (13.125)$$

where the s_j are precisely the zeroes of both branches of $(1 + l_4/l_3(s))$ on or close to the real axis. Therefore, we use a contour $\mathcal{L}_0$ surrounding the s_j in anticlockwise manner. The s_j are not located on the branch cut of $\log z(s)$ from -1 to 1, hence $\mathcal{L}_0$ consists of two disconnected parts. (For vanishing external fields these contours are loops around $]-\infty, -1]$ and $[1, \infty[$, respectively. In the general case they are appropriately deformed.) For not too low temperatures the z_j corresponding to a particular s_j are calculated by use of the first branch of $\log z(s)$. This is no longer possible at lower temperatures, so for the general case we must write

$$2\pi\mathrm{i} \sum_j \log z(s_j) = \underbrace{\int_{\mathcal{L}_0} f(s)\big|_{\text{1st branch}} ds}_{=:\,\Sigma_1} + \underbrace{\int_{\mathcal{L}_0} f(s)\big|_{\text{2nd branch}} ds}_{=:\,\Sigma_2}, \tag{13.126}$$

where the first and second term on the right hand side, Σ_1 and Σ_2, will be separately evaluated below.

13.7.1 First integral expression in terms of auxiliary functions

The function $l_4(s)/l_3(s)$ for $s \to \infty$ behaves like a rational function with finite limiting value $\exp(\beta(\mu - B))$ and next-leading asymptotics of order $\mathcal{O}(1/s)$. As we also have $z(s)$ of order $\mathcal{O}(s)$ we find the asymptotics of $f(s)$ as $\mathcal{O}(\log s/s^2)$. Therefore we are allowed to add two large 'semi-circles' to the contour $\mathcal{L}_0$ without changing the integral expression of Σ_1. Next we deform the integration contour by leaving the value of the integral unchanged. Due to Cauchy's theorem we may do so as long as the contour is not moved over singularities of $f(s)$ which result from a branch cut along the interval $[-1, +1]$ (depicted by a dashed line in figure 13.18), and poles resulting from zeroes and poles of the expression $1 + l_4/l_3(s)$ (depicted by open circles and crosses). Ultimately we find a contour consisting of three separate parts, cf. figure 13.19. Contour (a) consists of a path (a_1) from $-\infty$ to -1, a loop (a_2) around the interval $[-1, +1]$ and a path (a_3) back to $-\infty$.

Note that (a_1) and (a_3) are inverse to each other, but do not lead to a cancellation in the integral as the integrand shows a jump from (a_1) to (a_3), because $\log z(s)$ jumps by $-2\pi\mathrm{i}$. We find

$$\int_{(a_1,a_3)} \log z(s) \left[\log\left(1 + l_4/l_3(s)\right)\right]' ds = \int_{(a_3)} (-2\pi\mathrm{i}) \left[\log\left(1 + l_4/l_3(s)\right)\right]' ds$$

$$= -2\pi\mathrm{i} \log\left[1 + l_4/l_3(s)\right]\Big|_{s=-1}^{s=-\infty}. \tag{13.127}$$

Here we like to point out that a similar expression will be encountered below where the function l_4/l_3 will be evaluated on its second branch. Fortuitously, the values at the points -1 and $-\infty$ are independent of the particular branch (eventually leading to cancellation of these terms).

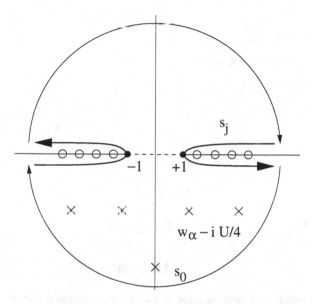

Fig. 13.18. Distribution of zeroes and poles of the function $1 + l_4/l_3(s)$. Zeroes (poles) are depicted by open circles (crosses) and are located at s_j ($w_\alpha - iU/4$). A pole of order $N/2$ is located at s_0. The integration contour $\mathcal{L}_0$ is indicated by thick solid lines. We can add two large 'semi-circles' with radius R depicted by thin solid lines without changing the integral as the integrand vanishes like $\mathcal{O}(\log R/R^2)$.

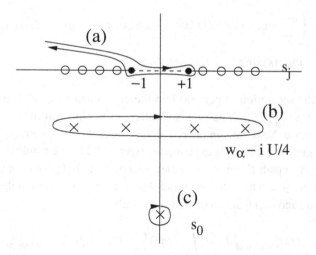

Fig. 13.19. Integration contour equivalent to that of the previous figure. There are three separate parts: (a) starting at $-\infty$, encircling the interval $[-1, +1]$ in clockwise manner, returning to $-\infty$, (b) loop surrounding the parameters $w_\alpha - iU/4$, (c) small circle arround s_0.

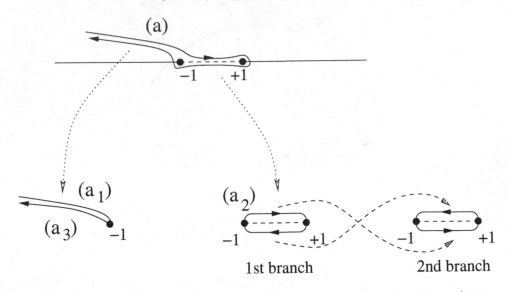

Fig. 13.20. Decomposition of the integration contour (a) into two parts. The function $f(s)\big|_{\text{1st branch}}$ on the upper/lower part of the loop takes values identical to $f(s)\big|_{\text{2nd branch}}$ on the lower/upper part of the loop.

The integrals along the parts (b) and (c) can be explicitly evaluated yielding

$$\Sigma_1 = +2\pi i \left(\sum_\alpha \log z(w_\alpha - iU/4) + \frac{N}{2} \log z(s_0) - \left[\log\left(1 + l_4/l_3(s)\right) \right]\Big|_{s=-1}^{|s=-\infty} \right)$$
$$+ \int_{(a_2)} \log z(s) \left[\log\left(1 + l_4/l_3(s)\right) \right]' ds. \tag{13.128}$$

In figure 13.20 the separation of path (a) into the components (a_1, a_3) and (a_2) is shown. The integral of $f(s)\big|_{\text{1st branch}}$ along (a_2) is identical to the integral of $f(s)\big|_{\text{2nd branch}}$ along (a_2) in reversed sense. Next we join the remaining integrals in Σ_1 with the one in Σ_2. The 'surgery' of integration contours is explained in figure 13.21. The resulting integral is over $f(s)\big|_{\text{2nd branch}}$ along a path $\mathcal{L}$ surrounding the real axis in anti-clockwise manner from $-\infty$ to $+\infty$ and back to $-\infty$ where the integrand shows a jump. In addition there are integrals over the paths (a_1) and (a_3) that can be done explicitly

$$\int_{(a_1,a_3)} f(s)\big|_{\text{2nd branch}} ds = \int_{(a_3)} (2\pi i) \left[\log\left(1 + l_4/l_3(s)\big|_{\text{2nd branch}}\right) \right]' ds$$
$$= 2\pi i \log\left[1 + l_4/l_3(s)\big|_{\text{2nd branch}} \right] \Big|_{s=-1}^{|s=-\infty}$$
$$= 2\pi i \log\left[1 + l_4/l_3(s) \right] \Big|_{s=-1}^{|s=-\infty}. \tag{13.129}$$

This calculation resembles that of (13.127). The different sign is due to the jump of

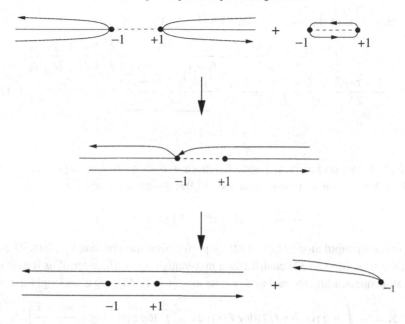

Fig. 13.21. The remaining integral in (13.128) and Σ_2 can be joined to a simple contour consisting of two lines above and below the real axis. The integrand is strictly $f(s)|_{2\text{nd branch}}$. Note that this integrand is analytic on the contour from -1 to $-\infty$, then from $-\infty$ to $+\infty$ and back to -1, but there is a discontinuity at -1. We can move this discontinuity from -1 to $-\infty$. This procedure involves an integral of $f(s)|_{2\text{nd branch}}$ along the contour (a_1, a_3).

$\log z|_{2\text{nd branch}}$ by $+2\pi \text{i}$ which is ultimately a consequence of

$$z|_{2\text{nd branch}} = -1/z|_{1\text{st branch}}. \qquad (13.130)$$

Hence (13.129) cancels exactly the corresponding term in (13.128), namely (13.127), yielding

$$\Sigma_1 + \Sigma_2 = +2\pi \text{i} \left(\sum_\alpha \log z(w_\alpha - \text{i}U/4) + \frac{N}{2} \log z(s_0) \right) + \int_{\mathcal{L}} f(s)|_{2\text{nd branch}} ds. \qquad (13.131)$$

Next we want to show that the r.h.s. of (13.131) is practically identical to

$$\Sigma := \int_{\mathcal{L}} [\log z(s - \text{i}U/2)]' \log \mathfrak{C}(s) ds + \int_{\mathcal{L}} [\log z(s)]' \log \frac{1 + c + \bar{c}}{\bar{c}} ds, \qquad (13.132)$$

which will complete our derivation. The necessary calculations are done by use of the explicit definitions of the involved functions (13.103)

$$\mathfrak{C} = \frac{l_1 + l_2 + l_3 + l_4}{l_3 + l_4} \cdot \frac{l_3 + l_4 + \bar{l}_1 + \bar{l}_2 + \bar{l}_3 + \bar{l}_4}{l_1 + l_2 + l_3 + l_4 + \bar{l}_1 + \bar{l}_2 + \bar{l}_3 + \bar{l}_4},$$

$$\frac{1 + \mathfrak{c} + \bar{\mathfrak{c}}}{\bar{\mathfrak{c}}} = \frac{\bar{l}_1 + \bar{l}_2 + \bar{l}_3 + \bar{l}_4}{\bar{l}_3 + \bar{l}_4} \cdot \underbrace{\frac{l_3 + l_4 + \bar{l}_1 + \bar{l}_2}{l_3 + l_4}}, \tag{13.133}$$

$$= 1 + \frac{l_4}{l_3}\bigg|_{\text{2nd branch}}$$

where the last fraction involving l and $\bar{l}$ functions was simplified by explicit use of (13.99) and (13.102). From the last expression and (13.99) the asymptotics can be easily read off

$$\frac{1 + \mathfrak{c} + \bar{\mathfrak{c}}}{\bar{\mathfrak{c}}} \rightarrow \left(1 + e^{\beta(\mu+B)}\right)\left(1 + e^{\beta(\mu-B)}\right). \tag{13.134}$$

We begin the manipulations of (13.132) by performing integrations by parts. The second term on the r.h.s. of (13.132) contributes a non-vanishing 'surface term' as $\log z(s)$ shows a jump after surrounding the real axis, not so $\log z(s - iU/2)$, $\log \mathfrak{C}$ and $\log(1 + \mathfrak{c} + \bar{\mathfrak{c}})/\bar{\mathfrak{c}}$:

$$\Sigma = -\int_{\mathcal{L}} \log z(s - iU/2)[\log \mathfrak{C}(s)]' ds - \int_{\mathcal{L}} \log z(s) \left[\log \frac{1 + \mathfrak{c} + \bar{\mathfrak{c}}}{\bar{\mathfrak{c}}}\right]' ds$$

$$+ 2\pi i \log\left[\left(1 + e^{\beta(\mu+B)}\right)\left(1 + e^{\beta(\mu-B)}\right)\right]. \tag{13.135}$$

Note that in the integral over $\mathfrak{C}$ only the first 'fraction of l functions' has to be kept, the second such term is analytic along the entire real axis and hence drops out according to Cauchy's theorem (note that the factor $\log z(s - iU/2)$ is analytic!). The only non-vanishing contribution to the first integral on the r.h.s. of (13.135) is

$$-\int_{\mathcal{L}} \log z(s - iU/2) \left[\log \frac{l_1 + l_2 + l_3 + l_4}{l_3 + l_4}\right]' ds \tag{13.136}$$

where the zeroes and poles of the fraction $(l_1 + l_2 + l_3 + l_4)/(l_3 + l_4)$ are depicted in figure 13.22. We deform the integration contour $\mathcal{L}$ as explained in the figure caption and obtain a contour with three separate parts shown in figure 13.23. The contribution of the first and third path to the integral in (13.136) can be explicitly given in terms of the surrounded poles. The contribution of the second path being identical to $\mathcal{L} + iU/2$ (in reversed sense) can be reformulated by a shift of the variable of integration from s to $s + iU/2$ leading to a replacement of the l_j functions by $\bar{l}_j$ functions:

$$(13.136) = +\int_{\mathcal{L}} \log z(s) \left[\log \frac{\bar{l}_1 + \bar{l}_2 + \bar{l}_3 + \bar{l}_4}{\bar{l}_3 + \bar{l}_4}\right]' ds$$

$$+ 2\pi i \left(-\frac{N}{2} \log z(-s_0) + \sum_\alpha \log z(w_\alpha - iU/4)\right) \tag{13.137}$$

Next, we insert this result into (13.135) and see that the integral term on the r.h.s. of (13.137) cancels one part of the contribution of the second integral in (13.135), i.e. that over

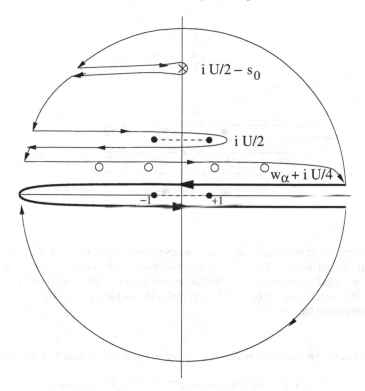

Fig. 13.22. Illustration of the singularities of $(l_1 + l_2 + l_3 + l_4)/(l_3 + l_4)$: branch cuts are depicted by dashed lines, zeroes by open circles and the pole by a cross. The integration contour $\mathcal{L}$ is depicted by thick solid lines. To the lower half-plane we add a large semi-circle with radius R depicted by a thin solid line. This path does not contribute to the integral (13.136) as the integrand asymptotically vanishes like $\mathcal{O}(\log R/R^2)$. We also add a path depicted by a thin solid line to the upper part of the complex plane. This path has no contribution as it can be closed to a point without touching any of the singularities. Finally, we deform the contour by letting the lower part consisting of the semi-circle and a straight line shrink to a point, and the upper semi-circle is dropped as its contribution vanishes. The remaining contour encircles the zeroes $w_\alpha + iU/4$, the branch cut $[-1, +1] + iU/2$, and the pole $iU/2 - s_0$ in clockwise manner, see figure 13.23.

$(1 + \mathfrak{c} + \bar{\mathfrak{c}})/\bar{\mathfrak{c}}$, which can be seen from the explicit form (13.133).

$$\Sigma = 2\pi i \log\left[\left(1 + e^{\beta(\mu+B)}\right)\left(1 + e^{\beta(\mu-B)}\right)\right]$$

$$+ 2\pi i \left(-\frac{N}{2} \log z(-s_0) + \sum_\alpha \log z(w_\alpha - iU/4)\right)$$

$$- \int_{\mathcal{L}} \underbrace{\log z(s) \left[\log\left(1 + \frac{l_4}{l_3}(s)\right)\Big|_{\text{2nd branch}}\right]'}_{\longrightarrow \, -f(s)|_{\text{2nd branch}}} ds. \qquad (13.138)$$

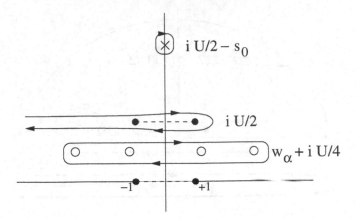

Fig. 13.23. Depiction of the three remaining contours around the zeroes $w_\alpha + iU/4$, the branch cut $[-1, +1] + iU/2$, and the pole $iU/2 - s_0$ in clockwise manner. The integrals over the first and third contour can be explicitly evaluated. Note that the second contour cannot be closed to a loop around $[-1, +1] + iU/2$ as the integrand in (13.136) contains the factor $\log z(s - iU/2)$ with a non-zero winding number along the contour.

In order to understand the last transformation we only have to replace in the last integral

$$\log z(s) = \log z(s)|_{\text{1st branch}} = \pi i - \log z(s)|_{\text{2nd branch}} \tag{13.139}$$

where the constant πi drops out as the winding number around the branch cut is zero and the remaining terms recombine into f, cf. (13.125). Comparing (13.138) with (13.131) we find

$$\Sigma_1 + \Sigma_2 = \Sigma + 2\pi i \left(N \log z(s_0) - \log \left[\left(1 + e^{\beta(\mu+B)} \right) \left(1 + e^{\beta(\mu-B)} \right) \right] \right), \tag{13.140}$$

where we have dropped a term $2\pi i \cdot N \cdot \log(-1)$ as it does not contribute to $\log \Lambda$. Inserting (13.126) into (13.97) by respecting (13.140) and (13.132) we are left with

$$2\pi i \log \Lambda = 2\pi i \left(\beta U/4 + N \log \tau + N \log z(s_0) \right)$$
$$+ \int_{\mathcal{L}} \left[\log z(s - iU/2) \right]' \log \mathfrak{C}(s) ds + \int_{\mathcal{L}} \left[\log z(s) \right]' \log \frac{1 + \mathfrak{c} + \bar{\mathfrak{c}}}{\bar{\mathfrak{c}}} ds. \tag{13.141}$$

This is the final result for the largest eigenvalue of the QTM in the case of finite Trotter number N. The constant on the r.h.s. may be simplified by use of (13.92), $\tau = \beta/N$, and (13.96), (13.84), (13.85). In the limit $N \to \infty$ we find

$$2\pi i \log \Lambda = -2\pi i \beta \frac{U}{4} + \int_{\mathcal{L}} \left[\log z(s - iU/2) \right]' \log \mathfrak{C}(s) ds$$
$$+ \int_{\mathcal{L}} \left[\log z(s) \right]' \log \frac{1 + \mathfrak{c} + \bar{\mathfrak{c}}}{\bar{\mathfrak{c}}} ds. \tag{13.142}$$

13.7.2 Alternative integral expressions

Next we like to give two alternative expressions for the integrals contributing to (13.141) and (13.142). First we separate on the r.h.s. of (13.142)

$$I := \int_{\mathcal{L}} [\log z(s)]' \log(1 + \mathfrak{c} + \bar{\mathfrak{c}}) ds$$

$$+ \int_{\mathcal{L}} [\log z(s - iU/2)]' \log \mathfrak{C}(s) ds - \int_{\mathcal{L}} [\log z(s)]' \log \bar{\mathfrak{c}} ds. \quad (13.143)$$

Noting $\bar{\mathfrak{c}} = \bar{\mathfrak{t}}/\mathfrak{B}$ and performing integration by parts on the contribution by $\bar{\mathfrak{t}}$ we get

$$\int_{\mathcal{L}} [\log z(s)]' \log \bar{\mathfrak{t}}(s) ds = -2\pi i \beta(\mu + B) - \int_{\mathcal{L}} \log z(s) [\log \bar{\mathfrak{t}}(s)]' ds, \quad (13.144)$$

where the constant is obtained from the 'surface term' and the asymptotic behaviour $\bar{\mathfrak{t}}(\infty) = \exp(-\beta(\mu + B))$. In the last expression the integrand possesses several singularities: there is a branch cut at $[-1, +1]$ with integration contour $\mathcal{L}$ around it. There is another branch cut at $[-1, +1] - iU/2$, and poles due to a zero and a pole of $\bar{\mathfrak{t}}(s)$ of order $N/2$ at $s_0 - iU/2$ and $-s_0$, respectively. We blow up the integration contour $\mathcal{L}$ and find (cf. figure 13.24) the equivalent contours $\mathcal{L} - iU/2$ (in reversed sense) and clockwise loops around the poles

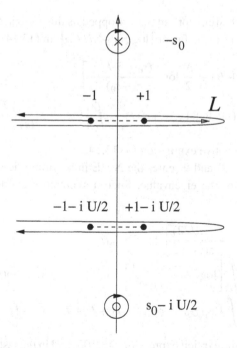

Fig. 13.24. Depiction of the singularities of $\bar{\mathfrak{t}}$, the integration contour $\mathcal{L}$ and its equivalent contours $\mathcal{L} - iU/2$ (in reversed sense) and clockwise loops around $s_0 - iU/2$ and $-s_0$.

yielding

$$\int_{\mathcal{L}} [\log z(s)]' \log \bar{\mathfrak{t}}(s)\, ds = -2\pi \mathrm{i} \left[\beta(\mu + B) + \frac{N}{2} \log \frac{z(s_0 - \mathrm{i}U/2)}{z(-s_0)} \right]$$
$$+ \int_{\mathcal{L}-\mathrm{i}U/2} \log z(s) [\log \bar{\mathfrak{t}}(s)]'\, ds. \qquad (13.145)$$

We use $\bar{\mathfrak{t}}(s) = 1/\mathfrak{t}(s + \mathrm{i}U/2)$ and again we perform an integration by parts

$$\int_{\mathcal{L}-\mathrm{i}U/2} \log z(s) [\log \bar{\mathfrak{t}}(s)]'\, ds = -\int_{\mathcal{L}-\mathrm{i}U/2} \log z(s) [\log \mathfrak{t}(s + \mathrm{i}U/2)]'\, ds$$
$$= -\int_{\mathcal{L}} \log z(s - \mathrm{i}U/2) [\log \mathfrak{t}(s)]'\, ds = \int_{\mathcal{L}} [\log z(s - \mathrm{i}U/2)]' \log \mathfrak{t}(s)\, ds, \qquad (13.146)$$

where in the last line no 'surface term' appears. The last integral is of the same type as the one in (13.143) involving $\mathfrak{C}$. Combining these terms we find

$$\frac{\mathfrak{C}}{\mathfrak{t}} = \frac{l_1 + l_2 + l_3 + l_4}{l_1 + l_2} \cdot \frac{l_3 + l_4 + \bar{l}_1 + \bar{l}_2 + \bar{l}_3 + \bar{l}_4}{l_1 + l_2 + l_3 + l_4 + \bar{l}_1 + \bar{l}_2 + \bar{l}_3 + \bar{l}_4}$$
$$= \frac{1}{\mathfrak{B}} \frac{l_3 + l_4 + \bar{l}_1 + \bar{l}_2 + \bar{l}_3 + \bar{l}_4}{l_1 + l_2} \qquad (13.147)$$

where the last ratio of l-functions may be dropped as this ratio is analytic in the neighbourhood of the real axis as is the factor $\left[\log z(s - \mathrm{i}U/2)\right]'$ in (13.143). We therefore find

$$I = 2\pi \mathrm{i} \left[\beta(\mu + B) + \frac{N}{2} \log \frac{z(s_0 - \mathrm{i}U/2)}{z(-s_0)} \right] \qquad (13.148)$$
$$+ \int_{\mathcal{L}} [\log z(s)]' \log(1 + \mathfrak{c} + \bar{\mathfrak{c}})\, ds - \int_{\mathcal{L}} \left[\log \frac{z(s - \mathrm{i}U/2)}{z(s)} \right]' \log \mathfrak{B}(s)\, ds,$$

yielding the first alternative expression to (13.142).

Next, we note that $\mathfrak{B}$ and $\bar{\mathfrak{B}}$ enter the NLIE in a symmetric way, not however the last integral expression for the eigenvalue. Such a symmetric expression can be derived by considering

$$\int_{\mathcal{L}} \left[\log \frac{z(s + \mathrm{i}U/2)}{z(s)} \right]' \log \bar{\mathfrak{B}}(s)\, ds$$
$$= -\int_{\mathcal{L}} \left[\log \frac{z(s + \mathrm{i}U/2)}{z(s)} \right]' \log(\bar{l}_1 + \bar{l}_2 + \bar{l}_3 + \bar{l}_4)(s)\, ds \qquad (13.149)$$
$$- \int_{\mathcal{L}} \left[\log z(s) \right]' \log(l_1 + l_2 + l_3 + l_4 + \bar{l}_1 + \bar{l}_2 + \bar{l}_3 + \bar{l}_4)(s)\, ds,$$

where we have used the explicit expression (13.103) and in the last integral we have dropped the contribution due to $z(s + \mathrm{i}U/2)$ leading to an analytic integrand. The first integral on the r.h.s. is treated by deformation of the path $\mathcal{L}$ illustrated in figure 13.25. The integrals

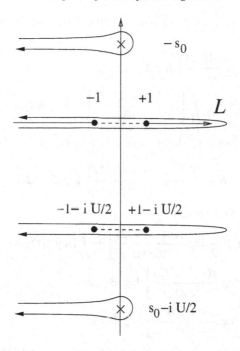

Fig. 13.25. Depiction of the integration contour $\mathcal{L}$ and its equivalent contours $\mathcal{L} - iU/2$ (in reversed sense) and contours from $-\infty$ to $s_0 - iU/2$ (respectively $-s_0$) and back to $-\infty$.

along the paths from $-\infty$ to the pole $s_0 - iU/2$ (respectively $-s_0$) of $\overline{l}_1 + \overline{l}_2 + \overline{l}_3 + \overline{l}_4$ and back to $-\infty$ yield

$$\int_{-s_0}^{-\infty} \left[\log \frac{z(s + iU/2)}{z(s)} \right]' 2\pi i \, ds = 2\pi i \log \left[\frac{z(-s_0)}{z(-s_0 + iU/2)} \right]$$

$$\int_{s_0 - iU/2}^{-\infty} \left[\log \frac{z(s + iU/2)}{z(s)} \right]' 2\pi i \, ds = 2\pi i \log \left[\frac{z(s_0 - iU/2)}{z(s_0)} \right] \qquad (13.150)$$

which cancel out. The remaining integral is

$$\int_{\mathcal{L}} \left[\log \frac{z(s + iU/2)}{z(s)} \right]' \log(\overline{l}_1 + \overline{l}_2 + \overline{l}_3 + \overline{l}_4)(s) ds$$

$$= -\int_{\mathcal{L} - iU/2} \left[\log \frac{z(s + iU/2)}{z(s)} \right]' \log(\overline{l}_1 + \overline{l}_2 + \overline{l}_3 + \overline{l}_4)(s) ds$$

$$= -\int_{\mathcal{L}} \left[\log \frac{z(s)}{z(s - iU/2)} \right]' \log \frac{(l_1 + l_2 + l_3 + l_4)(s)}{e^{2\beta B} \phi^+(s)\phi^-(s)} ds$$

$$= 4\pi i \beta B + \int_{\mathcal{L}} \left[\log \frac{z(s - iU/2)}{z(s)} \right]' \log(l_1 + l_2 + l_3 + l_4)(s) ds, \qquad (13.151)$$

where the $\phi^+(s)\phi^-(s)$ terms drop out after a short calculation. Inserting this into (13.149) we find

$$\int_{\mathcal{L}} \left[\log \frac{z(s+iU/2)}{z(s)} \right]' \log \overline{\mathfrak{B}}(s) ds$$

$$= -4\pi i \beta B - \int_{\mathcal{L}} \left[\log \frac{z(s-iU/2)}{z(s)} \right]' \log(l_1 + l_2 + l_3 + l_4)(s) ds$$

$$- \int_{\mathcal{L}} [\log z(s)]' \log(l_1 + l_2 + l_3 + l_4 + \overline{l_1} + \overline{l_2} + \overline{l_3} + \overline{l_4})(s) ds$$

$$= -4\pi i \beta B + \int_{\mathcal{L}} \left[\log \frac{z(s-iU/2)}{z(s)} \right]' \log \mathfrak{B}(s) ds. \tag{13.152}$$

Applying this to (13.148) we obtain

$$I = 2\pi i \left[\beta \mu + \frac{N}{2} \log \frac{z(s_0 - iU/2)}{z(-s_0)} \right] + \int_{\mathcal{L}} [\log z(s)]' \log(1 + \mathfrak{c} + \overline{\mathfrak{c}}) ds$$

$$- \frac{1}{2} \int_{\mathcal{L}} \left[\log \frac{z(s-iU/2)}{z(s)} \right]' \log \mathfrak{B}(s) ds$$

$$- \frac{1}{2} \int_{\mathcal{L}} \left[\log \frac{z(s+iU/2)}{z(s)} \right]' \log \overline{\mathfrak{B}}(s) ds. \tag{13.153}$$

This is the second alternative formula for the eigenvalue of the QTM. We collect the two alternative expressions (13.148) and (13.153) in the limit $N \to \infty$.

$$2\pi i \log \Lambda = 2\pi i \beta (\mu + B + U/4) + \int_{\mathcal{L}} [\log z(s)]' \log(1 + \mathfrak{c} + \overline{\mathfrak{c}}) ds$$

$$- \int_{\mathcal{L}} \left[\log \frac{z(s-iU/2)}{z(s)} \right]' \log \mathfrak{B}(s) ds,$$

$$= 2\pi i \beta (\mu + U/4) + \int_{\mathcal{L}} [\log z(s)]' \log(1 + \mathfrak{c} + \overline{\mathfrak{c}}) ds \tag{13.154}$$

$$- \frac{1}{2} \int_{\mathcal{L}} \left[\log \frac{z(s-iU/2)}{z(s)} \right]' \log \mathfrak{B}(s) ds$$

$$- \frac{1}{2} \int_{\mathcal{L}} \left[\log \frac{z(s+iU/2)}{z(s)} \right]' \log \overline{\mathfrak{B}}(s) ds.$$

These formulas are of particular importance to our further numerical and analytical treatment.

13.8 Numerical results

For the numerical treatment of equations (13.122), (13.154) we rewrite them in terms of usual convolutions of functions of a real variable

$$K * f = \int_{-\infty}^{\infty} K(x - y) f(y) \, dy. \tag{13.155}$$

For the functions (13.101) evaluated on the contours involved in (13.122), (13.154) we use the notation $\mathfrak{b}^\pm$, $\mathfrak{c}^\pm$ and $\overline{\mathfrak{c}}^\pm$

$$\mathfrak{b}^\pm(x) = \mathfrak{b}(x \pm iU/4), \quad \mathfrak{c}^\pm(x) = \mathfrak{c}(x \pm i0), \quad \overline{\mathfrak{c}}^\pm(x) = \overline{\mathfrak{c}}(x \pm i0). \tag{13.156}$$

Furthermore, we introduce the following relations:

$$\mathfrak{B}^\pm := 1 + \mathfrak{b}^\pm, \qquad\qquad \overline{\mathfrak{B}}^\pm := 1 + 1/\mathfrak{b}^\pm,$$
$$\mathfrak{C}^\pm := 1 + \mathfrak{c}^\pm, \qquad\qquad \overline{\mathfrak{C}}^\pm := 1 + \overline{\mathfrak{c}}^\pm, \tag{13.157}$$
$$\Delta \log \mathfrak{C} := \log(\mathfrak{C}^+/\mathfrak{C}^-), \qquad \Delta \log \overline{\mathfrak{C}} := \log(\overline{\mathfrak{C}}^+/\overline{\mathfrak{C}}^-), \quad \text{etc.}$$

Thus (13.122) is written in the form

$$\log \mathfrak{b}^+ = -2\beta B - K_{2,0} * \log \mathfrak{B}^+ + K_{2,U/2} * \log \mathfrak{B}^- - K_{1,0} * \Delta \log(\overline{\mathfrak{c}}/\overline{\mathfrak{C}}),$$
$$\log \mathfrak{b}^- = -2\beta B - K_{2,-U/2} * \log \mathfrak{B}^+ + K_{2,0} * \log \mathfrak{B}^- - K_{1,-U/2} * \Delta \log(\overline{\mathfrak{c}}/\overline{\mathfrak{C}}),$$
$$\log \mathfrak{c}^\pm = \Psi_c^\pm + K_{1,-U/2} * \log \overline{\mathfrak{B}^+} - K_{1,0} * \log \overline{\mathfrak{B}^-} + K_{1,-U/4} * \Delta \log \overline{\mathfrak{C}} \pm \tfrac{1}{2}\Delta \log \overline{\mathfrak{C}},$$
$$\log \overline{\mathfrak{c}}^\pm = \overline{\Psi}_c^\pm - K_{1,0} * \log \mathfrak{B}^+ + K_{1,U/2} * \log \mathfrak{B}^- - K_{1,U/4} * \Delta \log \mathfrak{C} \pm \tfrac{1}{2}\Delta \log \mathfrak{C},$$
$$\tag{13.158}$$

where

$$\Psi_c^\pm = -\beta U/2 + \beta(\mu + B) + \log \phi_{\pm 0},$$
$$\overline{\Psi}_c^\pm = -\beta U/2 - \beta(\mu + B) - \log \phi_{\pm 0}, \tag{13.159}$$

and we have used the notation f_α for a function f with shift of the argument by $i\alpha$

$$f_\alpha(x) = f(x + i\alpha),$$

and $K_{n,\alpha} := (K_n)_\alpha$ where the first index specifies the function and the second one specifies the shift of the argument. In particular $\phi_{\pm 0}$ denotes the function ϕ evaluated on the real axis from above/below. Notice that the convolutions of $K_{1,\pm U/4}$ with $\Delta \log \mathfrak{C}$ and $\Delta \log \overline{\mathfrak{C}}$ are determined by Cauchy's principal value. Remember that these functions vanish outside the interval $[-1, 1]$.

Similarly, from (13.142) and (13.154) we obtain two different relations for the eigenvalue

$$\log \Lambda = -\beta U/4 - \int_{-1}^1 \mathcal{K}_{+0} \log[(1 + \mathfrak{c}^+ + \overline{\mathfrak{c}}^+)(1 + \mathfrak{c}^- + \overline{\mathfrak{c}}^-)/(\mathfrak{c}^+ \overline{\mathfrak{c}}^-)]\, dx$$

$$- \int_{-1}^1 \mathcal{K}_{-U/2} \log[(1 + \mathfrak{c}^+)/(1 + \mathfrak{c}^-)]\, dx,$$

$$= \beta(\mu + B + U/4) - \int_{-1}^1 \mathcal{K}_{+0} \log[(1 + \mathfrak{c}^+ + \overline{\mathfrak{c}}^+)(1 + \mathfrak{c}^- + \overline{\mathfrak{c}}^-)]\, dx$$

$$= + \int_{-\infty}^\infty [(\mathcal{K}_{-U/4} - \mathcal{K}_{U/4}) \log \mathfrak{B}^+ - (\mathcal{K}_{-3U/4} - \mathcal{K}_{-U/4}) \log \mathfrak{B}^-]\, dx \tag{13.160}$$

with

$$K(s) = \frac{1}{2\pi i}[\log z(s)]' = \left(2\pi i\, s \sqrt{1 - 1/s^2}\right)^{-1}, \tag{13.161}$$

and at the branch cut we find

$$K(x \pm i0) = \mp \left(2\pi \sqrt{1 - x^2}\right)^{-1}, \qquad x \in [-1, 1]. \tag{13.162}$$

The branch of $\mathcal{K}$ is fixed by the requirement $\mathcal{K}(s) \simeq 1/(2\pi i s)$ for large s and $\mathcal{K}_\alpha$ is the related function with shifted argument. By means of the relation

$$\Delta \log \bar{\mathfrak{c}} = -\Delta \log \phi + \Delta \log \mathfrak{C}, \tag{13.163}$$

the first equation of (13.158) turns into

$$\log \mathfrak{b}^+ = \Psi_{\mathfrak{b}}^+ - K_{2,0} * \log \mathfrak{B}^+ + K_{2,U/2} * \log \mathfrak{B}^- - K_{1,0} * \Delta \log(\mathfrak{C}/\overline{\mathfrak{C}}),$$
$$\log \mathfrak{b}^- = \Psi_{\mathfrak{b}}^- - K_{2,-U/2} * \log \mathfrak{B}^+ + K_{2,0} * \log \mathfrak{B}^- - K_{1,-U/2} * \Delta \log(\mathfrak{C}/\overline{\mathfrak{C}}), \tag{13.164}$$

where

$$\Psi_{\mathfrak{b}}^+ = -\beta U - 2\beta B + \log \phi_{U/4} - \log \phi_{-U/4},$$
$$\Psi_{\mathfrak{b}}^- = -\beta U - 2\beta B + \log \phi_{-U/4} - \log \phi_{-3U/4}. \tag{13.165}$$

For the sake of completeness rather than for further applications we mention the results for finite Trotter number N. All equations above hold true after the replacement of the 'driving functions' ψ by (see also (13.121))

$$\Psi_{\mathfrak{b}}^+ = -2\beta B + \log \phi_{U/4} - \log \phi_{-U/4} - \frac{N}{2} \log \frac{x - s_0 + 3iU/4}{x - s_0 - iU/4},$$
$$\Psi_{\mathfrak{b}}^- = -2\beta B + \log \phi_{-U/4} - \log \phi_{-3U/4} - \frac{N}{2} \log \frac{x - s_0 + iU/4}{x - s_0 - 3iU/4},$$
$$\Psi_{\mathfrak{c}}^\pm = +\beta(\mu + B) + \log \phi_{\pm 0} + \frac{N}{2} \log \frac{x + s_0}{x + s_0 - iU/2},$$
$$\overline{\Psi}_{\mathfrak{c}}^\pm = -\beta(\mu + B) - \log \phi_{\pm 0} + \frac{N}{2} \log \frac{x - s_0}{x - s_0 + iU/2}, \tag{13.166}$$

where s_0 is defined in (13.96). These relations for finite Trotter number N have been used for a comparison of the results of the integral equations with a direct treatment based on the Bethe ansatz equations of Sections 13.4 and 13.5. Thus it was possible to ensure the accuracy (10^{-6}) of our numerics based on iterations and fast Fourier transform.

Next we present our numerical results for various physical quantities and discuss them in terms of the elementary spin and charge excitations, i.e. 'spinons' and 'holons' (plus gapped excitations based on 'doubly occupied sites'). Note that at half-filling the system possesses a charge gap such that the holons do not contribute at low temperatures. Furthermore, the hopping integral of the kinetic energy has been set to $t = 1$.

In figure 13.26 the temperature dependence of the specific heat is shown for densities $n = 1, 0.8$, and 0.5. For half-filling ($n = 1.0$) the specific heat shows one pronounced temperature maximum for lower values of the interaction U. For stronger U this maximum splits into a lower and a higher temperature maximum which are due to spin and (gapped)

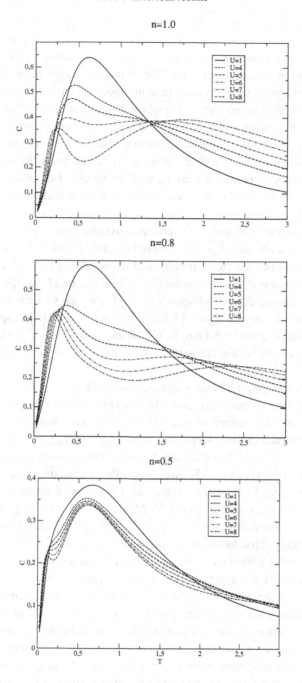

Fig. 13.26. Specific heat c (in units of k_B) versus T (in units of t/k_B) for particle densities $n = 1$, $n = 0.8$ and $n = 0.5$.

charge excitations, respectively. (These findings agree largely with those of [397].) The picture remains qualitatively true for small dopings ($n = 0.8$), however now the lower temperature peak receives contributions by gapless charge excitations, hence some weight is shifted from higher to lower temperatures. The situation changes quite drastically for fillings $n \approx 0.5$. Here a pronounced maximum in the specific heat is located at a temperature of about $T \approx 0.6$ which seems rather insensitive to the interaction. This is explained by the irrelevance of the onsite interaction at sufficiently large temperatures, because of the low-particle density. In addition, we find a maximum at very low temperatures which depends very sensitively on U as well as on the particle density n. In order to clarify the origin of this additional structure the variation of the specific heat with n is shown in figure 13.27 for $U = 8$. Decreasing the particle density n from half-filling ($n = 1$) to lower values ($n \approx 0.8$) the lower-temperature maximum increases. This picture is changed drastically below $n \lesssim 0.8$. Here the lower-temperature maximum and its location are suppressed for lower n and a shoulder at a slightly higher temperature develops into a clear maximum. This new structure in the specific heat is located at about $T \approx 0.6$ and quite independent of U as already mentioned. We interpret this maximum to be of 'charge' type. The complex behaviour at intermediate densities $0.5 \lesssim n \lesssim 0.7$ is due to a crossover of the 'spin' and 'charge' maxima, see also figure 13.32. For densities $n \approx 1$ the 'spin' maximum is located at finite temperature with finite height whereas the 'charge' maximum is located at very low temperature with small height. For densities close to $n \approx 0$ the situation is reversed.

In figure 13.28 and figure 13.29 the magnetic susceptibility χ is presented. Again we begin our discussion with the half-filled case which is known to correspond to the Heisenberg spin chain with interaction strength of order $O(t^2/U)$. Indeed, we observe a Heisenberg-like temperature dependence of the susceptibility with χ_{max} and T_{max} scaling with U and $1/U$ in the range of $U = 4, \ldots, 8$. Upon doping this behaviour remains qualitatively and quantitatively unchanged even for $n = 0.5$. Quite generally, the location T_{max} is shifted to lower temperatures, see figure 13.28. The maximal value χ_{max} decreases for decreasing particle density from $n = 1$ to $n \approx 0.8$, cf. figure 13.29. Below the value $n \lesssim 0.8$ the maximum χ_{max} increases for further decrease of the particle density. This behaviour is qualitatively explained by partially filled bands of charge carriers with spin. If the chemical potential moves away (towards) a band edge, the susceptibility decreases (increases). Somewhere inbetween the lower and the upper edge the minimal value is taken.

In contrast to χ the charge susceptibility κ ($= \partial n/\partial \mu$, i.e. compressibility) shows a more interesting dependence on the particle density n, see figure 13.30 and figure 13.31. At half-filling κ shows the expected exponentially activated behaviour in particular $\kappa = 0$ for $T = 0$ due to the charge gap. For any doping this behaviour is changed completely showing a finite value at zero temperature consistent with a partial filling of the lower Hubbard band. For density $n = 0.5$ we observe two different structures at low temperature similar to the case of the specific heat. The lower temperature 'spin' peak resembles the structure in the susceptibility χ, whereas the 'charge' maximum at slightly higher temperature is caused by the single-particle motion of the bare electrons. The charge susceptibility has a

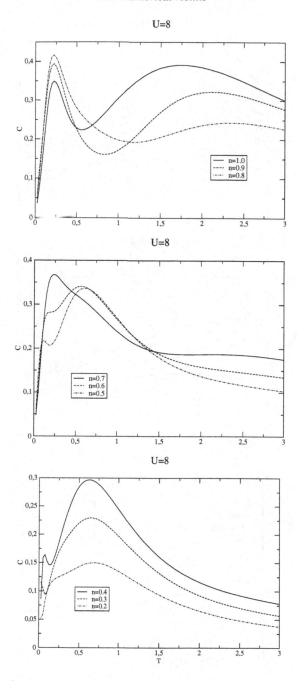

Fig. 13.27. Specific heat c (in units of k_B) versus T (in units of t/k_B) for fixed $U = 8$.

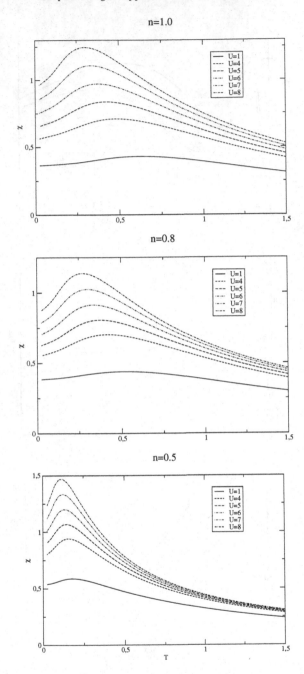

Fig. 13.28. Magnetic susceptibility χ (in units of μ_B^2/t) versus T (in units of t/k_B) for $n = 1, n = 0.8$ and $n = 0.5$.

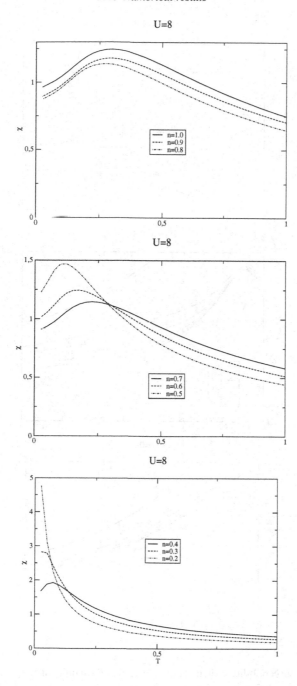

Fig. 13.29. Magnetic susceptibility χ (in units of μ_B^2/t) versus T (in units of t/k_B) for fixed $U = 8$.

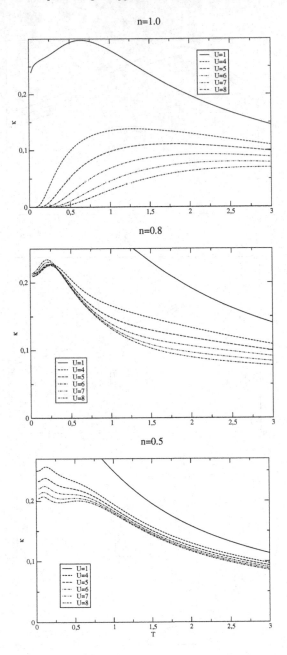

Fig. 13.30. Charge susceptibility κ (in units of $1/t$) versus T (in units of t/k_B) for particle densities $n = 1$, $n = 0.8$ and $n = 0.5$.

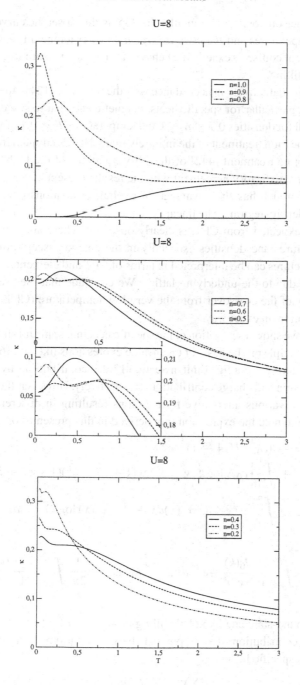

Fig. 13.31. Charge susceptibility κ (in units of $1/t$) versus T (in units of t/k_B) for fixed $U = 8$.

singular dependence on doping. The smaller the doping the closer the curves are to the case $n = 1$ at high temperatures and *the more divergent* at lower temperature, see first graph of figure 13.31. This, of course, is exactly the behaviour of a system exhibiting a Mott-Hubbard transition at half-filling.

Our findings are qualitatively in accordance with the results[3] of [241,469] for the dopings treated therein. In particular for specific heats, magnetic and charge susceptibilities the results compare well for densities $0.7 \le n \le 1$ and temperatures $T \ge 0.1$, giving independent support to the truncation treatment of the infinitely many NLIEs adopted in [241,469]. Very recently, an elaborate treatment [442] of the TBA equations of the Hubbard model [435] yielded extremely satisfactory agreement with the results presented here.

The present approach has the advantage of explicit evaluations over extremely wide temperature and density regions. The linear dependence of the specific heat on T at very low temperatures, as expected from CFT, is clearly observed. There are additional structures at lower temperatures and densities especially in the charge susceptibility as mentioned above. These structures can be interpreted in terms of CFT and elementary excitations with finite band width due to the underlying lattice. We conclude that the presented approach allows for a study of the crossover from the very low temperature (CFT) to the very high temperature region in an exact way.

In figure 13.32 we show a separation of the specific heat into spin and charge components. This is done in principle on the basis of eigenvalue expressions like (13.160). As motivated by the study of the strong-coupling limit in Section 13.9.1, contributions by $\mathfrak{b}$ and $\mathfrak{c}$ functions are interpreted as spin and charge contributions, respectively. However, the procedure is not unique as we have various alternative formulations resulting in different separations. In particular we like to note the expression (not derived in this presentation)

$$
\begin{aligned}
\log \Lambda = &-\beta(e_0 - U/4 - \mu) \\
&+ \int_{-1}^{1} \left[c_0 \Delta \log \mathfrak{C}/\overline{\mathfrak{C}} - \mathcal{K} \log(1 + \mathfrak{c}^+ + \overline{\mathfrak{c}}^+)(1 + \mathfrak{c}^- + \overline{\mathfrak{c}}^-) \right] dx \\
&+ \int_{-\infty}^{\infty} c_2(x) \log \mathfrak{B}^-(x) dx + \int_{-\infty}^{\infty} c_1(x) \log \overline{\mathfrak{B}}^+(x) dx,
\end{aligned}
\tag{13.167}
$$

with

$$
c_0(x) = \frac{1}{2\pi} \int_{-\infty}^{\infty} \frac{J_0(k)}{1 + e^{U|k|/2}} e^{ikx} dk, \qquad c_{1,2}(x) = \frac{1}{2\pi} \int_{-\infty}^{\infty} \frac{J_0(k)}{1 + e^{\mp Uk/2}} e^{ikx} dk.
\tag{13.168}
$$

Here e_0 is the ground-state energy at half-filling as given in [298] and the additional $\mathfrak{b}$ and $\mathfrak{c}$ terms represent contributions due to spin and charge excitations. In figure 13.32 we show the results for the specific heat

$$
c = T \left(\frac{\partial S}{\partial T} \right)_\mu + T \left(\frac{\partial n}{\partial T} \right)_\mu \left(\frac{\partial \mu}{\partial T} \right)_n,
\tag{13.169}
$$

---

[3] In [241,469] notice the factor 4 in the definition of the interaction parameter U.

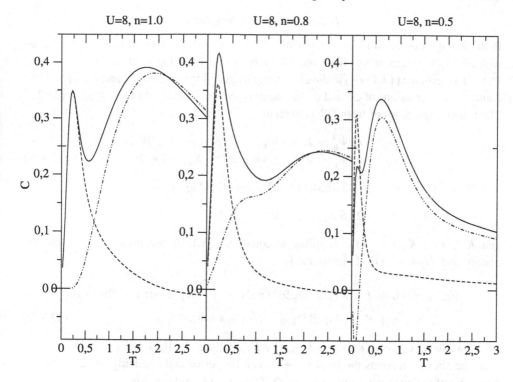

Fig. 13.32. Separation of specific heat (solid) in spin (dashed) and charge components (dashed-dotted).

where we have applied the separation based on (13.167) to the temperature derivatives of S and n. Note the functional form of the spin part is rather independent of the doping. Upon small doping the charge contribution develops a low-temperature shoulder which disappears for larger dopings. We would like to warn about the formal 'separation' of spin and charge that it may give rise to artificial results. For instance, at high (low) temperatures the 'partial specific heats' show negative values whereas the total specific heat, of course, is always positive. In Section 13.9.3 the spin-charge separation is treated properly at low temperatures and arbitrary particle density via an involved interplay of the various degrees of freedom rather than by a superficial interpretation of formulas.

13.9 Analytical solutions to the integral equations

In the previous sections we have derived non-linear integral equations for the largest eigenvalue of the QTM yielding directly the free energy of the Hubbard model at finite temperatures $T = 1/\beta$. For arbitrary temperatures and densities the integral equations can be solved only numerically. However, in some limiting cases analytical results can be derived and relations obtained which permit a comparison to known analytical results.

13.9.1 Strong-coupling limit

In the strong-coupling limit $U \to \infty$ at half-filling ($\mu = 0$) the Hubbard model is expected to reduce to the Heisenberg chain. Indeed, in the strong-coupling limit we find that $\mathfrak{c}^{\pm}$ and $\overline{\mathfrak{c}}^{\pm}$ tend to zero, see (13.159), yielding large negative driving terms for $\mathfrak{c}^{\pm}$ and $\overline{\mathfrak{c}}^{\pm}$ in (13.158). Hence all contributions of $\mathfrak{C}^{\pm}$ and $\overline{\mathfrak{C}}^{\pm}$ can be dropped on the r.h.s. of (13.164) and (13.122). The remaining equations from (13.164) read

$$
\begin{aligned}
\log \mathfrak{b}^{+} &= \Psi_{\mathfrak{b}}^{+} - K_{2,0} * \log \mathfrak{B}^{+} + K_{2,U/2} * \log \mathfrak{B}^{-}, \\
\log \mathfrak{b}^{-} &= \Psi_{\mathfrak{b}}^{-} - K_{2,-U/2} * \log \mathfrak{B}^{+} + K_{2,0} * \log \mathfrak{B}^{-},
\end{aligned}
\tag{13.170}
$$

and $\Psi_{\mathfrak{b}}^{\pm}$ are obtained from (13.165), (13.123), and (13.111) as

$$
\Psi_{\mathfrak{b}}^{+} = -2\beta B + 2\pi \beta \, K_{1,0}, \qquad \Psi_{\mathfrak{b}}^{-} = -2\beta B + 2\pi \beta \, K_{1,-U/2},
\tag{13.171}
$$

and $K_{j,\alpha}(x) = K_j(x + i\alpha)$. According to equation (13.160) and dropping an irrelevant energy shift $U/4$, the QTM eigenvalue is

$$
\begin{aligned}
\log \Lambda &= \beta B + \int_{\infty}^{\infty} K_{1,0}(x) \log \mathfrak{B}^{+}(x) dx - \int_{\infty}^{\infty} K_{1,-U/2}(x) \log \mathfrak{B}^{-}(x) dx \\
&= \beta B + K_{1,0} * \log \mathfrak{B}^{+}|_{x=0} - K_{1,U/2} * \log \mathfrak{B}^{-}|_{x=0},
\end{aligned}
\tag{13.172}
$$

where we have replaced $(\mathcal{K}_{-U/4} - \mathcal{K}_{U/4})$ by K_1 etc., see (13.161), and (13.111).

In the above equations the limit $U \to \infty$ can be performed explicitly by rescaling the argument of the auxiliary functions $x \mapsto (U/4)x$. To this end we define

$$
\mathfrak{a}^{\pm}(x) := \mathfrak{b}^{\pm}\left(\frac{U}{4}x\right), \qquad \mathfrak{A}^{\pm}(x) := \mathfrak{B}^{\pm}\left(\frac{U}{4}x\right) = 1 + \mathfrak{a}^{\pm}(x)
\tag{13.173}
$$

which are inserted into the integral equations yielding

$$
\begin{aligned}
\log \mathfrak{a}^{+} &= \Psi^{+} - \widetilde{K}_{2,0} * \log \mathfrak{A}^{+} + \widetilde{K}_{2,2} * \log \mathfrak{A}^{-}, \\
\log \mathfrak{a}^{-} &= \Psi^{-} - \widetilde{K}_{2,-2} * \log \mathfrak{A}^{+} + \widetilde{K}_{2,0} * \log \mathfrak{A}^{-},
\end{aligned}
\tag{13.174}
$$

which follows from (13.173) by use of the general relation $(K * B)(\frac{U}{4}x) = (\widetilde{K} * A)(x)$ for $\widetilde{K}(x) := \frac{U}{4}K(\frac{U}{4}x)$ and $A(x) := B(\frac{U}{4}x)$. In (13.174) the following definitions were employed

$$
\begin{aligned}
\widetilde{K}_1(x) &= \frac{U}{4} K_1\left(\frac{U}{4}x\right) = \frac{1}{2\pi} \frac{2}{x^2 + 1}, \\
\widetilde{K}_2(x) &= \frac{U}{4} K_2\left(\frac{U}{4}x\right) = \frac{1}{2\pi} \frac{4}{x^2 + 4}, \\
\Psi^{+}(x) &= -\tilde{\beta}\tilde{h} + 2\pi \tilde{\beta} \, \widetilde{K}_{1,0}(x) = -\tilde{\beta}\tilde{h} + \frac{2\tilde{\beta}}{x^2 + 1}, \\
\Psi^{-}(x) &= \Psi^{+}(x - 2i),
\end{aligned}
\tag{13.175}
$$

and we have introduced the rescaled (reciprocal) temperature $\tilde{\beta} := (4/U)\beta$ and rescaled magnetic field $\tilde{h} := (U/2)B$.

By inspection, we see that equations (13.174) for $\mathfrak{a}^{\pm}$ are identical to (13.48) for $\overline{\mathfrak{a}}$ in the case of the Heisenberg model if we identify $\mathfrak{a}^{\pm}(x) = \overline{\mathfrak{a}}(x \pm i)$.

Finally, the eigenvalue (13.172) is written as

$$\log \Lambda = \tilde{\beta}\tilde{h}/2 + \tilde{K}_{1,0} * \log \mathfrak{A}^{+}|_{x=0} - \tilde{K}_{1,2} * \log \mathfrak{A}^{-}|_{x=0}, \tag{13.176}$$

which is identical to the Heisenberg model result (13.49) (with $v = 0$) upon the identification $\mathfrak{a}^{\pm}(x) = \overline{\mathfrak{a}}(x \pm i)$.

13.9.2 Free-Fermion limit

Let us consider the opposite limit $U \to 0$, that is the case of two independent free-fermion systems. The non-linear integral equations (13.158) simplify to an algebraic set of equations due to $K_{1,0}(x) \to \delta(x)$, $K_{1,\pm U/2}(x) \to 0$, $K_{2,\alpha}(x) \to \delta(x)$ if $|\alpha| \le U/2$, and $K_{1,\pm U/4}(x) \to \frac{1}{2}\delta(x)$ (in principal value integrals):

$$\log \mathfrak{b}^{\pm} = -2\beta B - \log \mathfrak{B}^{+} + \log \mathfrak{B}^{-} - (\tfrac{1}{2} \pm \tfrac{1}{2})\Delta \log(\overline{\mathfrak{c}}/\overline{\mathfrak{C}}),$$

$$\log \mathfrak{c}^{\pm} = +\beta(\mu + B) + \log \phi_{\pm} - \log \overline{\mathfrak{B}}^{-} + (\tfrac{1}{2} \pm \tfrac{1}{2})\Delta \log \overline{\mathfrak{C}}, \tag{13.177}$$

$$\log \overline{\mathfrak{c}}^{\pm} = -\beta(\mu + B) - \log \phi_{\pm} - \log \mathfrak{B}^{+} + (-\tfrac{1}{2} \pm \tfrac{1}{2})\Delta \log \mathfrak{C}.$$

With $\mathcal{K}_{-U/4}(x) \to \mathcal{K}_{-0}(x) = -\mathcal{K}_{+0}(x)$ for arguments $x \in [-1, +1]$ we find from (13.160)

$$\log \Lambda = -\int_{-1}^{1} \mathcal{K}_{+0} \log \frac{(1 + \mathfrak{c}^{+} + \overline{\mathfrak{c}^{+}})(1 + \mathfrak{c}^{-} + \overline{\mathfrak{c}^{-}})(1 + \mathfrak{c}^{-})}{\overline{\mathfrak{c}^{+}}\,\overline{\mathfrak{c}^{-}}\,(1 + \mathfrak{c}^{+})} \, dx. \tag{13.178}$$

The equations (13.177) can be solved by standard techniques. Alternatively, we may just use the Bethe ansatz expressions (13.101) and (13.99) that we started with. There, all q_1 and q_2 functions cancel or completely factor out as they have the same argument. Hence, only the 'trivial' function ϕ remains, however, care has to be applied. The relevant arguments have real part in $[-1, 1]$, but infinitesimal imaginary part of both signs with $\phi(x - i0) = 1/\phi(x + i0)$. Hence, we obtain the following expressions

$$\mathfrak{b}^{+} = \frac{\left[1 + e^{\beta(\mu+B)}\phi\right]\left[1 + e^{\beta(\mu-B)}/\phi\right]}{e^{2\beta B}\left[e^{\beta(\mu+B)}/\phi + e^{2\beta\mu} + 1 + e^{\beta(\mu-B)}/\phi\right]},$$

$$\mathfrak{b}^{-} = \frac{\left[e^{\beta(\mu+B)}/\phi + e^{2\beta\mu} + 1 + e^{\beta(\mu-B)}/\phi\right]}{e^{2\beta B}\left[1 + e^{\beta(\mu+B)}/\phi\right]\left[1 + e^{\beta(\mu-B)}\phi\right]}, \tag{13.179}$$

$$\mathfrak{c}^{+} = \frac{e^{\beta(\mu+B)}}{\phi}\frac{1 + e^{\beta(\mu-B)}\phi}{1 + e^{\beta(\mu-B)}/\phi}\frac{\mathfrak{b}^{+}}{1 + \mathfrak{b}^{+}}, \qquad \mathfrak{c}^{-} = \frac{e^{\beta(\mu+B)}}{\phi}\frac{\mathfrak{b}^{-}}{1 + \mathfrak{b}^{-}},$$

$$\overline{\mathfrak{c}}^{+} = \frac{1}{e^{\beta(\mu+B)}\phi(1 + \mathfrak{b}^{+})}, \qquad \overline{\mathfrak{c}}^{-} = \frac{\phi}{e^{\beta(\mu+B)}}\frac{1 + e^{\beta(\mu-B)}/\phi}{1 + e^{\beta(\mu-B)}\phi}\frac{1}{1 + \mathfrak{b}^{-}},$$

Lastly, we substitute $x = \sin k$ in the integration for the eigenvalue leading to

$$\log \Lambda = +\frac{1}{2\pi} \int_{-\pi}^{\pi} \log\left[1 + \exp\left(\beta(\mu + B + 2\cos k)\right)\right] dk$$

$$+ \frac{1}{2\pi} \int_{-\pi}^{\pi} \log\left[1 + \exp\left(\beta(\mu - B + 2\cos k)\right)\right] dk. \qquad (13.180)$$

This is the result expected from the well-known properties of ideal Fermi gases.

13.9.3 Low-temperature asymptotics

The low-temperature regime is the most interesting limit as here the system shows Tomonaga-Luttinger liquid behaviour. We will derive analytic expressions for the thermodynamics within our first principles calculations and confirm the field theoretical predictions. In particular we will show how the non-linear integral equations correspond to the known dressed energy formalism of the Hubbard model. This represents a further and in fact the most interesting consistency check of the field-theoretical picture of the 1d Hubbard chain.

For $T = 1/\beta \ll 1$ we can simplify the non-linear integral equations as follows. We adopt fields $B > 0$, $\mu < 0$, such that $\mathfrak{b}^- \to 0$, $\mathfrak{c}^\pm \to 0$ for $\beta \to \infty$ with uniform exponential convergence for all arguments (which is observed numerically). The property $\mathfrak{c}^\pm \to 0$ is a consequence of the third equation in (13.158) and (13.159) for negative μ. The behaviour $\mathfrak{b}^- \to 0$ can be understood from the second equation in (13.158). However, $\mathfrak{b}^+$ and $\overline{\mathfrak{c}^\pm}$ do not vanish.

We simplify the notation by use of

$$\mathfrak{b}(\lambda) := \mathfrak{b}^+(\lambda), \qquad \mathfrak{c}(k) := \begin{cases} 1/\overline{\mathfrak{c}^+}(\sin k), & \text{for } k \in \left[-\frac{1}{2}\pi, +\frac{1}{2}\pi\right], \\ 1/\overline{\mathfrak{c}^-}(\sin k), & \text{for } k \in \left[+\frac{1}{2}\pi, +\frac{3}{2}\pi\right]. \end{cases} \qquad (13.181)$$

and find from (13.158)

$$\log \mathfrak{b}(\lambda) = -\beta \varepsilon_s^0(\lambda) - \int_{-\infty}^{\infty} K_2(\lambda - \lambda') \log[1 + \mathfrak{b}(\lambda')] \, d\lambda'$$

$$+ \int_{-\pi}^{\pi} K_1(\lambda - \sin k') \cos k' \log[1 + \mathfrak{c}(k')] \, dk' \qquad (13.182)$$

$$\log \mathfrak{c}(k) = -\beta \varepsilon_c^0(k) + \int_{-\infty}^{\infty} K_1(\sin k - \lambda') \log[1 + \mathfrak{b}(\lambda')] \, d\lambda',$$

where $\varepsilon_s^0 = 2B$, $\varepsilon_c^0 = -\mu - U/2 - B - 2\cos k$ and all $\log \mathfrak{B}^-$ and $\log \mathfrak{C}^\pm$ have been dropped. These expressions are valid at low temperatures where the correction terms are exponentially small, i.e. corrections are of order $\mathcal{O}(\exp(-\text{const} \times \beta))$ with some positive real constant related to the charge gap.

Also the eigenvalue expressions simplify in a similar way. From (13.160) we find

$$\log \Lambda = -\beta U/4 - \int_{-1}^{1} \mathcal{K}_{+0}(x) \log\left[\left(1 + \frac{1}{\mathfrak{c}^{+}(x)}\right)\left(1 + \frac{1}{\mathfrak{c}^{-}(x)}\right)\right] dx$$

$$= -\beta U/4 + \frac{1}{2\pi} \int_{-\pi}^{\pi} \log[1 + \mathfrak{c}(k)] \, dk \qquad (13.183)$$

Next we note that the solutions $\log \mathfrak{b}$ and $\log \mathfrak{c}$ to (13.182) are analytic functions of order $O(\beta)$, or more precisely $\log \mathfrak{b} = -\beta \, \varepsilon_s$ and $\log \mathfrak{c} = -\beta \, \varepsilon_c$ with some analytic functions ε_s and ε_c. These functions will be investigated in quite some detail below. They are real, symmetric functions possessing zeroes $\pm\lambda_0$, $\pm k_0$ and have the properties

$$\varepsilon_s(\lambda), \ \varepsilon_c(k) < 0 \quad \text{for} \quad |\lambda| < \lambda_0, |k| < k_0$$
$$\varepsilon_s(\lambda), \ \varepsilon_c(k) > 0 \quad \text{for} \quad |\lambda| > \lambda_0, |k| > k_0. \qquad (13.184)$$

such that $\mathfrak{b}$ and $\mathfrak{c}$ show steep crossover behaviour at low temperatures

$$|\mathfrak{b}(\lambda)|, \ |\mathfrak{c}(k)| \gg 1 \quad \text{for} \quad |\lambda| < \lambda_0, |k| < k_0$$
$$|\mathfrak{b}(\lambda)|, \ |\mathfrak{c}(k)| \ll 1 \quad \text{for} \quad |\lambda| > \lambda_0, |k| > k_0. \qquad (13.185)$$

As a consequence, the functions $\log(1 + \mathfrak{b})$ and $\log(1 + \mathfrak{c})$ are no longer analytic at low temperature: for arguments smaller than λ_0 and k_0 these functions are identical to $-\beta \, \varepsilon_s$ and $-\beta \, \varepsilon_c$, for arguments larger than λ_0 and k_0 the functions are identical to 0! Nevertheless, the convolutions on the r.h.s. of (13.182) yield $(\beta \times)$ analytic functions.

Linearization

The slopes at the crossover points are steep, allowing for certain approximations to the integral equations (13.182). We split the contribution of a typical integral term in (13.182) into three pieces

$$\int_{-\infty(-\pi)}^{\infty(\pi)} K(x') \log(1 + f(x')) \, dx'$$

$$= \int_{|x'|<x_0} K(x') \log f(x') \, dx'$$

$$+ \underbrace{\int_{|x'|<x_0} K(x') \log\left(1 + \frac{1}{f(x')}\right) dx' + \int_{|x'|>x_0} K(x') \log(1 + f(x')) \, dx'}_{=: I},$$

$$(13.186)$$

where in our applications $f(x)$ is an even function like $\mathfrak{b}(\lambda)$ or $\mathfrak{c}(k)$ with $\log f$ of order $\mathcal{O}(\beta)$, x (x_0) corresponding to λ or k (λ_0 or k_0), respectively. The function $K(x')$ is a shorthand for $K_j(\cdots - \lambda')$ or $K_j(\cdots - \sin k') \cos k'$ with x' corresponding to λ' or k', respectively.

As the slope $(\log f)'(x_0)$ is of order $\mathcal{O}(\beta)$ and hence sufficiently steep at low temperatures, we can approximate $1/f(x)(|x| < x_0)$ and $f(x)(|x| > x_0)$ by $\exp(+(\log f)'(x_0)|x \pm x_0|)$ in the vicinity of the 'Fermi' surfaces $\pm x_0$ (note that $(\log f)'(x_0)$ is negative). For the last two integrals in (13.186) this linearization can be justified over the total integration range. Hence the last two terms in (13.186) reduce to

$$
I = 2[K(x_0) + K(-x_0)] \int_0^\infty \log\left(1 + e^{(\log f)'(x_0)x}\right) dx + o\left(\frac{1}{\beta}\right)
$$

$$
= [K(x_0) + K(-x_0)]\frac{-\pi^2}{6(\log f)'(x_0)} + o\left(\frac{1}{\beta}\right),
$$

(13.187)

where we have evaluated the integral $\int_0^\infty \log(1 + e^z)dz = \pi^2/12$.

The resultant equations are now given by linear integral equations over finite integration intervals

$$
\log \mathfrak{b}(\lambda) = \phi_b(\lambda) - \int_{-\lambda_0}^{+\lambda_0} K_2(\lambda - \lambda') \log \mathfrak{b}(\lambda') \, d\lambda'
$$

$$
+ \int_{-k_0}^{+k_0} K_1(\lambda - \sin k') \cos k' \log \mathfrak{c}(k') \, dk'
$$

(13.188)

$$
\log \mathfrak{c}(k) = \phi_c(k) + \int_{-\lambda_0}^{+\lambda_0} K_1(\sin k - \lambda') \log \mathfrak{b}(\lambda') \, d\lambda'.
$$

The driving terms read

$$
\phi_b(\lambda) = -\beta \varepsilon_s^0(\lambda) + \frac{\pi^2[K_2(\lambda - \lambda_0) + K_2(\lambda + \lambda_0)]}{6(\log \mathfrak{b})'(\lambda_0)}
$$

$$
- \frac{\pi^2[K_1(\lambda - \sin k_0) + K_1(\lambda + \sin k_0)]}{6(\log \mathfrak{c})'(k_0)} \cos k_0,
$$

$$
\phi_c(k) = -\beta \varepsilon_c^0(k) - \frac{\pi^2[K_1(\sin k - \lambda_0) + K_1(\sin k + \lambda_0)]}{6(\log \mathfrak{b})'(\lambda_0)}.
$$

(13.189)

Retaining the leading terms in the integral equations we find the following connection between auxiliary functions and the dressed energy functions:

$$
\log \mathfrak{b} = -\beta \, \varepsilon_s + O(1/\beta) \quad \text{and} \quad \log \mathfrak{c} = -\beta \, \varepsilon_c + O(1/\beta).
$$

(13.190)

For a comparison with [140, 141] note the different normalization of the chemical potential.

A similar linearization of (13.183) yields

$$
\log \Lambda = -\beta U/4 - \frac{\pi}{6(\log \mathfrak{c})'(k_0)} + \frac{1}{2\pi} \int_{-k_0}^{k_0} \log \mathfrak{c}(k) \, dk.
$$

(13.191)

For our further manipulations we note the (linear) integral equations for the root density functions ρ_s and ρ_c characterizing the ground state (6.12, 6.13) or in the notation of this chapter

$$\rho_s(\lambda) = -\int_{-\lambda_0}^{+\lambda_0} K_2(\lambda - \lambda') \rho_s(\lambda') \, d\lambda' + \int_{-k_0}^{+k_0} K_1(\lambda - \sin k') \rho_c(k') \, dk',$$

$$\rho_c(k) = \frac{1}{2\pi} + \cos k \int_{-\lambda_0}^{+\lambda_0} K_1(\sin k - \lambda') \rho_s(\lambda') \, d\lambda'. \tag{13.192}$$

Note that the kernel matrices for the integral equations (13.188), (13.192) are mutually transpose.

Dressed function formalism

The above equations (13.188) can be summarized in the form

$$l_i(x) = l_i^0(x) + \sum_j \int_{-X_j}^{X_j} K_{ij}(x - x') l_j(x') y_j(x') dx' \tag{13.193}$$

with the correspondence of l_1, l_2 with log $\mathfrak{b}$, log $\mathfrak{c}$, and X_1, X_2 with λ_0, k_0 etc. In particular we have

$$l_i^0(x) = -\beta \varepsilon_i^0(x) - \frac{\pi^2}{6} \sum_j \frac{K_{ij}(x - X_j) + K_{ij}(x + X_j)}{l_j'(X_j)} y_j(X_j). \tag{13.194}$$

The functions l_i (l_i^0) are referred to as dressed (bare) functions in the sense that the functions l_i are identical to the bare l_i^0 functions dressed by integral terms along (13.193). At low temperature the dressed l_i functions are related to the dressed energy functions ε_i

$$l_i = -\beta \, \varepsilon_i + \mathcal{O}(1/\beta). \tag{13.195}$$

The equations (13.192) for the density functions can be written as

$$\rho_i(x) = \rho_i^0(x) + y_i(x) \sum_j \int_{-X_j}^{X_j} K_{ij}^T(x - x') \rho_j(x') dx' \tag{13.196}$$

with ρ_1, ρ_2 corresponding to ρ_s, ρ_c and bare densities $\rho_1^0 = 0$ and $\rho_2^0 = 1/2\pi$. The integration kernel in (13.196) is the transpose of the one in (13.193)

$$(K^T)_{ij}(z) := K_{ji}(-z). \tag{13.197}$$

The expression (13.191) for the eigenvalue can be cast into the form

$$\log \Lambda = -\beta U/4 + \sum_j \left[-\frac{\pi^2}{3} \frac{\rho_j^0(X_j)}{l_j'(X_j)} + \int_{-X_j}^{X_j} \rho_j^0(x) l_j(x) dx \right]. \tag{13.198}$$

It is this expression we want to simplify by use of (13.193) and (13.196).

It is an elementary excercise to prove the identity

$$\sum_j \int_{-X_j}^{X_j} \rho_j^0(x) l_j(x) dx = \sum_j \int_{-X_j}^{X_j} \rho_j(x) l_j^0(x) dx, \tag{13.199}$$

i.e. the sum of integrals of bare ρ^0 times dressed l functions is identical to the sum of integrals of dressed ρ times bare l^0 functions. The l.h.s. of (13.199) appears in (13.198), however the r.h.s. of (13.199) is more tractable as we have the explicit expression (13.194) for the bare l^0 functions. Inserting (13.194) (after an exchange of the 'dummy' indices i, j) into the r.h.s. of (13.199) we obtain

$$\log \Lambda = -\beta U/4 - \sum_j \left(\beta \int_{-X_j}^{X_j} \varepsilon_j^0(x) \rho_j(x) dx + \frac{\pi^2}{3} \frac{\rho_j^0(X_j)}{l_j'(X_j)} \right)$$
$$- \sum_i \sum_j \left[\frac{\pi^2}{6} \frac{y_i(X_i)}{l_i'(X_i)} \int_{-X_i}^{X_i} [K_{ij}^T(X_i - x) + K_{ij}^T(-X_i - x)] \rho_j(x) dx \right]. \tag{13.200}$$

Next, the integrations and sum over j in the last line can be performed explicitly by use of (13.196). The last line turns into

$$- \sum_i \left[\frac{\pi^2}{3} \frac{\rho_i(X_i) - \rho_i^0(X_i)}{l_i'(X_i)} \right] \tag{13.201}$$

hence the eigenvalue is

$$\log \Lambda = -\beta U/4 - \beta \sum_j \int_{-X_j}^{X_j} \varepsilon_j^0(x) \rho_j(x) dx - \frac{\pi^2}{3} \sum_j \frac{\rho_j(X_j)}{l_j'(X_j)}. \tag{13.202}$$

At low temperature the l_j functions can be replaced by the dressed energies (13.195)

$$\frac{\pi^2}{3} \frac{\rho_j(X_j)}{l_j'(X_j)} = -\frac{\pi^2}{3} \frac{\rho_j(X_j)}{\beta \varepsilon_j'(X_j)} = -\frac{\pi}{6\beta} \frac{1}{v_j}, \tag{13.203}$$

where we have introduced the velocity v_j of the elementary excitation number j

$$v_j = \frac{\varepsilon_j'}{2\pi \rho_j} \bigg|_{X_j}. \tag{13.204}$$

Finally, the eigenvalue reads

$$\log \Lambda = -\beta \varepsilon_0 + \frac{\pi}{6\beta} \sum_j \frac{1}{v_j}, \tag{13.205}$$

where the groundstate energy is given by

$$\varepsilon_0 = U/4 + \sum_j \int_{-X_j}^{X_j} \varepsilon_j^0(x) \rho_j(x) dx. \tag{13.206}$$

The computational framework is rather general as the expressions hold for many models.

Applied to the Hubbard chain we find the free energy up to $\mathcal{O}(T^2)$ terms in the low-temperature expansion

$$f = \varepsilon_0 - \frac{\pi}{6\beta^2} \left(\frac{1}{v_c} + \frac{1}{v_s} \right). \tag{13.207}$$

This agrees with general expressions of conformal field theory for low temperatures, see Section 9.2.1, where the spin and charge channels contribute independently each with central charge $c = 1$ and sound velocities $v_{s,c} = \varepsilon'_{s,c}/2\pi\rho_{s,c}|_{\lambda_0,k_0}$. The groundstate energy is given in terms of the densities by

$$\varepsilon_0 = \int_{-k_0}^{+k_0} \rho_c(k)\,\varepsilon_c^0(k)\,dk + \int_{-\lambda_0}^{+\lambda_0} \rho_s(\lambda)\,\varepsilon_s^0(\lambda)\,d\lambda,$$

Here the trivial shift in the energy $U/4$ is omitted. We thereby conclude that our formalism completely recovers the correct contribution from spinon and holon excitations in the low temperature behaviour. This is a manifestation of spin-charge separation due to which each elementary excitation contributes independently to (13.207) where the velocities v_c and v_s typically take different values.

13.9.4 High-temperature limit

Finally, we consider the high-temperature limit $T \to \infty$ with B, U as well as $\beta\mu$ fixed. The auxiliary functions in (13.158) become constant and convolutions $K_{2,\pm U/2} * f$, $K_{n,0} * f$ with a constant f yield f, but $K_{1,\pm U/2} * f$ yields 0. From (13.158) we find

$$\log \mathfrak{b}^\pm = 0, \quad \log \mathfrak{c}^\pm = \beta\mu - \log 2, \quad \log \overline{\mathfrak{c}}^\pm = -\beta\mu - \log 2, \tag{13.208}$$

and from (13.142)

$$\log \Lambda = \log(1 + \mathfrak{c} + \overline{\mathfrak{c}})/\overline{\mathfrak{c}} = 2\log(1 + e^{\beta\mu}). \tag{13.209}$$

Thus, the free energy reads

$$f = -2\,T \log(1 + e^{\mu/T}) \quad \text{with} \quad \mu/T = \log \frac{n}{2 - n}, \tag{13.210}$$

where n is the particle density. We obtain the entropy

$$S = 2 \log \frac{2}{2 - n} - n \log \frac{n}{2 - n}, \tag{13.211}$$

as expected by counting the degrees of freedom per lattice site. Especially, at half-filling $n = 1$ this is equal to $S = \log 4$.

13.10 Conclusions

In this chapter, the QTM formulation of the thermodynamics of 1d quantum systems has been developed for the Heisenberg model and the Hubbard model. Several quantities of physical interest have been evaluated with high numerical precision and various limiting cases have been studied analytically.

As already noted above, we may consider as one of the most practical advantages of the present formulation the fact that one only has to deal with a finite number of unknown functions and non-linear integral equations among them. This does not only imply convenience, rather it opens a more fundamental understanding related to the particle picture of 1d quantum systems. For the Heisenberg model, the complex conjugate auxiliary functions play a role which seems to correspond to the elementary spinon excitations [86, 133, 227]. In our treatment of the Hubbard model we have shown that the complete thermodynamics are described by three independent functions $\mathfrak{b}$, $\mathfrak{c}$, $\overline{\mathfrak{c}}$, physically corresponding to spinons, and holons in upper and lower Hubbard bands. In the $T \to 0$ limit, these functions are shown to reduce to energy-density functions ('dressed energy functions') for such elementary excitations.

The two apparently different approaches, the combinatorial TBA and the operator-based QTM, are not at all independent! In the latter approach there are several quite different ways of analysis of the eigenvalues of the QTM. This is very well understood for the case of the Heisenberg model. In the standard (and most economical) way a set of just two coupled non-linear integral equations (NLIE) is derived. Alternatively, an approach based on the 'fusion hierarchy' leads to a set of (generically) infinitely many NLIEs [253, 281] that are identical to the TBA equations, though a completely different reasoning has to be applied!

For the Hubbard model this explicit relation has not yet been established. It is still an open problem how to derive the TBA equations of the Hubbard model from the largest eigenvalue of the QTM. We leave the investigation of these questions as an interesting future problem.

Obviously, our formulation can be extended to the evaluation of the asymptotics of correlation functions, such as spin-spin correlation lengths etc. For the Heisenberg model the reader is referred to [253, 254, 258]. In the case of the Hubbard model, there is only preliminary (numerical) work on correlation lengths at finite temperature [459] and [465] (the latter on the basis of the formulation of Section 13.4).

Appendices to Chapter 13

13.A Derivation of TBA equations from fusion hierarchy analysis

In this appendix we present a treatment of the eigenvalue problem (13.23) different from that in Section 13.2. For all integers $j = 0, 1, 2, \ldots$ we define the objects

$$T_j(v) := q(v - (j+1)\mathrm{i})q(v + (j+1)\mathrm{i}) \cdot \sum_{l=0}^{j} \frac{e^{(2l-j)\beta h/2}\varphi(v + (2l-j)\mathrm{i})}{q(v + (2l-j-1)\mathrm{i})q(v + (2l-j+1)\mathrm{i})},$$

(13.A.1)

which are analytic functions due to the Bethe ansatz equations (13.28). In order to understand the cancellation of poles of the individual summands we consider the case $j = 1$ leading to the function (indeed identical to (13.24))

$$T_1(v) := q(v - 2\mathrm{i})q(v + 2\mathrm{i}) \cdot \left[\frac{e^{-\beta h/2}\varphi(v - \mathrm{i})}{q(v - 2\mathrm{i})q(v)} + \frac{e^{+\beta h/2}\varphi(v + \mathrm{i})}{q(v)q(v + 2\mathrm{i})} \right],$$

(13.A.2)

which is a meromorphic function with possible poles at the singularities of the function in square brackets. These singularities are of two types: either we have singularities due to the zeroes of $q(v \mp 2\mathrm{i})$ which are canceled by the prefactor in front of the square bracket, or the singularities are due to zeroes of $q(v)$ occurring in the first as well as second term in square brackets and cancel each other thanks to the BA equations.

The analyticity of (13.A.1) comes about in a very similar way. The only possible poles might arise due to zeroes of the functions $q(v - (j+1)\mathrm{i})$, $q(v - (j-1)\mathrm{i})$, $\ldots$, $q(v + (j-1)\mathrm{i})$, $q(v + (j+1)\mathrm{i})$ occurring in the denominators of the summands in (13.A.1). Again, the zeroes of the first and the last function are precisely cancelled by the prefactor of the sum in (13.A.1). The 'intermediate poles' cancel pairwise: the function $q(v + (2l - j + 1)\mathrm{i})$ with $l = 0, \ldots, j - 1$ appears in two terms of (13.A.1), namely the summands corresponding to labels l and $l + 1$. The sum of these two terms is identical to the square bracket in (13.A.2) (with a shift of the argument by $(2l - j + 1)\mathrm{i}$ and a common constant factor $e^{(2l-j+1)\beta h/2}$) and hence the poles cancel as in (13.A.2).

557

Next, we aim at the derivation of the following (quadratic) functional relation for $j = 1, 2, 3 \ldots$

$$T_j(v + i)T_j(v - i) = \varphi(v - (j + 1)i)\varphi(v + (j + 1)i) + T_{j-1}(v)T_{j+1}(v). \quad (13.A.3)$$

For the proof we introduce the compact notation

$$\lambda_j(v) := \frac{e^{j\beta h/2}\varphi(v + ji)}{q(v + (j - 1)i)q(v + (j + 1)i)}, \quad (13.A.4)$$

by use of which (13.A.1) is equivalent to

$$T_j(v) = q(v - (j + 1)i)q(v + (j + 1)i)\sum_{l=0}^{j}\lambda_{2l-j}(v). \quad (13.A.5)$$

From this we find

$$\frac{T_j(v - i)T_j(v + i)}{q(v - ji)q(v + ji)q(v - (j + 2)i)q(v + (j + 2)i)} = \sum_{l=0}^{j}\sum_{l'=0}^{j}\lambda_{2l-j}(v - i)\lambda_{2l'-j}(v + i)$$

$$= \sum_{l=0}^{j}\sum_{l'=0}^{j}\lambda_{2l-j-1}(v)\lambda_{2l'-j+1}(v), \quad (13.A.6)$$

where in the last step we have used that in $\lambda_{\ldots}(\ldots)$ a shift of the argument by i is equivalent to a shift of the index by 1 (with a constant factor of $e^{-\beta h/2}$). Rearranging the sums we find

r.h.s.(13.A.6)

$$= \left[\sum_{l=0}^{j+1}\lambda_{2l-j-1}(v) - \lambda_{j+1}(v)\right]\left[\sum_{l'=0}^{j-1}\lambda_{2l'-j+1}(v) + \lambda_{j+1}(v)\right]$$

$$= \sum_{l=0}^{j+1}\lambda_{2l-j-1}(v)\sum_{l'=0}^{j-1}\lambda_{2l'-j+1}(v) + \lambda_{j+1}(v)\left[\lambda_{-j-1}(v) + \lambda_{j+1}(v)\right] - [\lambda_{j+1}(v)]^2$$

$$= \frac{T_{j-1}(v)T_{j+1}(v)}{q(v - ji)q(v + ji)q(v - (j + 2)i)q(v + (j + 2)i)} + \lambda_{-j-1}(v)\lambda_{j+1}(v). \quad (13.A.7)$$

From (13.A.6), (13.A.7) and the identity

$$q(v - ji)q(v + ji)q(v - (j + 2)i)q(v + (j + 2)i)\lambda_{-j-1}(v)\lambda_{j+1}(v)$$
$$= \varphi(v - (j + 1)i)\varphi(v + (j + 1)i), \quad (13.A.8)$$

(13.A.3) follows immediately.

We may rewrite (13.A.3) by introducing a function Y_j as the ratio of the two terms on the r.h.s.

$$Y_j(v) = \frac{T_{j-1}(v)T_{j+1}(v)}{\varphi(v - (j + 1)i)\varphi(v + (j + 1)i)}, \quad j = 1, 2, \ldots. \quad (13.A.9)$$

Hence (13.A.3) takes the form

$$T_j(v + i)T_j(v - i) = \varphi(v - (j + 1)i)\varphi(v + (j + 1)i)[1 + Y_j(v)], \qquad (13.A.10)$$

for all $j = 1, 2, 3, \ldots$ and also $j = 0$ if we use

$$T_0(v) = \varphi(v), \qquad Y_0(v) := 0. \qquad (13.A.11)$$

From (13.A.9) and (13.A.10) we immediately obtain for all $j = 1, 2, 3, \ldots$

$$Y_j(v + i)Y_j(v - i) = [1 + Y_{j-1}(v)][1 + Y_{j+1}(v)], \qquad (13.A.12)$$

which is the final set of functional relations.

The strategy for solving these relations is: (i) solve (13.A.12) for the functions Y_j, (ii) by use of the solutions Y_j solve (13.A.10) for T_j. Quite generally, as in Section (13.2) above, we transform the multiplicative functional equations into (nonlinear) integral equations by first taking the logarithm of the functional equations and then the Fourier transform. In this manner and as $Y_j(v)$, $j = 2, 3, \ldots$ has neither pole nor zero in $-1 \le \mathrm{Im}x \le 1$, we find

$$\ln Y_j(v) = \mathbf{s} * [\ln(1 + Y_{j-1}) + \ln(1 + Y_{j+1})], \quad j \ge 2, \qquad (13.A.13)$$

where $*$ denotes convolutions and $\mathbf{s}$ is the function

$$\mathbf{s}(v) := \frac{1}{4 \cosh \pi x/2}. \qquad (13.A.14)$$

For $Y_1(v)$ the functional equation reads explicitly

$$Y_1(v - i)Y_1(v + i) = 1 + Y_2(v), \qquad (13.A.15)$$

and looks slightly simpler, however care has to be taken as Y_1 has zeroes at $\pm(1 - \tau)i$ and poles at $\pm(1 + \tau)i$. We set

$$Y_1(v) = \left[\frac{\tanh\frac{\pi}{4}(v + (1 - \tau)i)}{\tanh\frac{\pi}{4}(v + (1 + \tau)i)}\right]^{N/2} \tilde{Y}_1(v), \qquad (13.A.16)$$

which defines a function $\tilde{Y}_1(v)$ that is analytic and non-zero in the domain comprising the strip $-1 \le \mathrm{Im}(v) \le +1$ and satisfies the same inversion identity as $Y_1(v)$

$$\tilde{Y}_1(v - i)\tilde{Y}_1(v + i) = 1 + Y_2(v). \qquad (13.A.17)$$

This equation can now be cast in integral form

$$\ln \tilde{Y}_1(v) = \mathbf{s} * \ln(1 + Y_2), \qquad (13.A.18)$$

and for Y_1 we obtain

$$\ln Y_1(v) = \frac{N}{2} \log\left[\frac{\tanh\frac{\pi}{4}(v + (1 - \tau)i)}{\tanh\frac{\pi}{4}(v + (1 + \tau)i)}\right] + \mathbf{s} * \ln(1 + Y_2). \qquad (13.A.19)$$

Since $\tau = \beta/N$, the first term on the r.h.s. of (13.A.19) has a well defined limit for $N \to \infty$

$$\log \left[\frac{\tanh\frac{\pi}{4}(v + (1-\tau)\mathrm{i})}{\tanh\frac{\pi}{4}(v + (1+\tau)\mathrm{i})} \right]^{N/2} = -N\tau\mathrm{i}\frac{\mathrm{d}}{\mathrm{d}x}\log\tanh\frac{\pi}{4}(v + \mathrm{i}) = -\beta\frac{\frac{\pi}{2}}{\cosh\frac{\pi}{2}x}. \quad (13.A.20)$$

Therefore, in the Trotter limit $N \to \infty$ we have

$$\ln Y_1(v) = -\beta\frac{\frac{\pi}{2}}{\cosh\frac{\pi}{2}x} + \mathbf{s} * \ln(1 + Y_2). \quad (13.A.21)$$

Using equations (13.A.21), (13.A.13) together with the asymptotic behaviour with respect to large indices

$$\lim_{j \to \infty} \frac{\ln Y_j(v)}{j} = \beta h \quad (13.A.22)$$

we have a closed set of NLIEs.

Finally, from (13.A.10) we obtain for the largest eigenvalue $\Lambda(v) \equiv T_1(v)$ of the QTM

$$\ln \Lambda(v) = \ln T_1(v) = -\beta e(v) + \mathbf{s} * \ln(1 + Y_1), \quad (13.A.23)$$

where $e(v)$ is some β-independent function with $e(v = 0)$ being the groundstate energy. For the free energy per lattice site we find

$$\beta f = \beta e - \int_{-\infty}^{\infty} \mathbf{s}(x) \ln(1 + Y_1(x)) dx. \quad (13.A.24)$$

The equations (13.A.21), (13.A.13), (13.A.22) and (13.A.24) are completely identical to the TBA equations upon identifying $Y_j(x) \equiv \eta_j(x)$.

There are non-linear integral equations for a finite number of auxiliary functions 'interpolating' between 2 and ∞. These equations are particularly useful for the study of higher spin $SU(2)$ models [424].

13.B Derivation of single integral equation

From (13.A.3) we find for $j = 1$

$$T_1(v - \mathrm{i})T_1(v + \mathrm{i}) = \varphi(v - 2\mathrm{i})\varphi(v + 2\mathrm{i}) + \varphi(v)T_2(v), \quad (13.B.1)$$

where we have used $T_0(v) = \varphi(v)$. We want to solve for $T_1(v)(= \Lambda(v))$ which is a polynomial of degree N. The methods of Section 13.2 and appendix 13.A were based on the solution of multiplicative functional equations. In Section 13.2 the eigenvalue $T_1(v)$ was found by an ansatz satisfying the desired asymptotics and the location of zeroes (there are no singularities).

Here we want to apply a different strategy. We define a suitable function $u_1(v)$ as the ratio of $T_1(v)$ and a polynomial $\psi(v)$ of degree N with zeroes of order $N/2$ at points $\pm v_0$

soon to be specified

$$u_1(v) := \frac{T_1(v)}{\psi(v)}, \qquad \psi(v) := [(v - v_0)(v + v_0)]^{N/2}. \tag{13.B.2}$$

For this function the equation (13.B.1) turns into

$$u_1(v - i)u_1(v + i) = b(v) + u_2(v), \tag{13.B.3}$$

with

$$b(v) := \frac{\varphi(v - 2i)\varphi(v + 2i)}{\psi(v - i)\psi(v + i)},$$

$$u_2(v) := \frac{\varphi(v)}{\psi(v - i)\psi(v + i)} T_2(v). \tag{13.B.4}$$

For our further analysis we need the properties:

> (i) the points $\pm(v_0 - i)$ be zeroes of $\varphi(v)$,
> (ii) the points $\pm(v_0 - 2i)$ be far from the zeroes of $u_1(v)$. $\qquad$ (13.B.5)

For this setting we obtain from (13.B.3)

$$u_1(v + i) = \frac{b(v)}{u_1(v - i)} + \frac{u_2(v)}{u_1(v - i)}, \tag{13.B.6}$$

where $v_0 - i$ is a singularity of the l.h.s., but no singularity of the 2nd term on the r.h.s. as $u_2(v)$ is analytic due to property (i), $u_1(v - i)$ is non-zero due to (ii). If C is a suffiently narrow path surrounding 0 once in counterclockwise manner, the integral

$$\frac{1}{2\pi i} \oint_{C+v_0-i} \frac{1}{v - w} \frac{b(w)}{u_1(w - i)} dw, \tag{13.B.7}$$

defines, for v outside the contour $C + v_0 - i$, a meromorphic function with asymptotics 0 and only singularity at $v_0 - i$ being identical to that of the r.h.s. of (13.B.6), i.e. identical to the singularity of $u_1(v + i)$ at $v_0 - i$. Analogously we find

$$\frac{1}{2\pi i} \oint_{C-v_0+i} \frac{1}{v - w} \frac{b(w)}{u_1(w + i)} dw \tag{13.B.8}$$

defines a meromorphic function with 0 asymptotics and only singularity at $-v_0 + i$ identical to that of $u_1(v + i)$. Hence

$$\frac{1}{2\pi i} \oint_{C+v_0-i} \frac{1}{v - i - w} \frac{b(w)}{u_1(w - i)} dw + \frac{1}{2\pi i} \oint_{C-v_0+i} \frac{1}{v + i - w} \frac{b(w)}{u_1(w + i)} dw \tag{13.B.9}$$

is a meromorphic function with the same singularities of $u_1(v)$ (namely at $\pm v_0$). The function in (13.B.9) can be written by a shift of the variable of integration

$$\frac{1}{2\pi i} \oint_C \left[\frac{1}{v - v_0 - w} \frac{b(w + v_0 - i)}{u_1(w + v_0 - 2i)} dw + \frac{1}{v + v_0 - w} \frac{b(w - v_0 + i)}{u_1(w - v_0 + 2i)} \right] dw. \tag{13.B.10}$$

From this and a comparison of the asymptotic behaviour, which for $u_1(v)$ is the constant $2\cosh(\beta h/2)$, we obtain the identity

$$u_1(v) = 2\cosh(\beta h/2)$$
$$+ \frac{1}{2\pi i} \oint_C \left[\frac{1}{v - v_0 - w} \frac{b(w + v_0 - i)}{u_1(w + v_0 - 2i)} + \frac{1}{v + v_0 - w} \frac{b(w - v_0 + i)}{u_1(w - v_0 + 2i)} \right] dw.$$

$$(13.B.11)$$

Finally, we have to show that v_0 exists such that (13.B.5) is satisfied. Part (i) is satisfied by the choice $v_0 = i\tau$ or $v_0 = i(2 - \tau)$, but only in the latter case also (ii) is satisfied. Therefore

$$v_0 = \begin{cases} i(2 - \tau) = i\left(2 - \frac{\beta}{N}\right), & \text{for finite } N, \\ i2, & \text{for } N \to \infty. \end{cases}$$

$$(13.B.12)$$

We note that the resulting function $u_1(v)$ is identical to the function $\Lambda(iv, \tau)$ defined in (13.23). The function $u_2(v)$ is given by

$$u_2(v) = \frac{T_2(v)}{[(v - i(3 - \tau))(v + i(3 - \tau))]^{N/2}}.$$

$$(13.B.13)$$

More interestingly, the function $b(v)$ occurring in (13.B.11) takes the explicit form

$$b(v) = \begin{cases} \left[\frac{(v - i(1 + \tau))}{(v - i(1 - \tau))} \frac{(v + i(1 + \tau))}{(v + i(1 - \tau))} \right]^{N/2}, & \text{for finite } N, \\ \exp\left(i\frac{\beta}{v + i} - i\frac{\beta}{v - i} \right), & \text{for } N \to \infty. \end{cases}$$

$$(13.B.14)$$

The integral equation in the limit $N \to \infty$ reads

$$u_1(v) = 2\cosh(\beta h/2) + \frac{1}{2\pi i} \oint_C \left[\frac{b(w + i)}{v - 2i - w} + \frac{b(w - i)}{v + 2i - w} \right] \frac{1}{u_1(w)} dw. \quad (13.B.15)$$

14

The Yangian symmetry of the Hubbard model

In this chapter we will reveal another piece of the algebraic structure of the Hubbard model. As was first observed by Uglov and Korepin [462], the Hubbard Hamiltonian on the infinite line is invariant under the action of the direct sum of two so-called Yangian quantum groups, extending the rotational and the η-pairing su(2) symmetries we encountered earlier. Following [172] we shall address the issue in a more general context. We present two pairs of fermionic representations of the Y(su(2)) Yangian quantum group which commute with the trigonometric [162] and hyperbolic [40, 41] versions of a Hubbard Hamiltonian with non-nearest-neighbour hopping. In both cases the two representations are also mutually commuting, hence can be combined into a representation of Y(su(2))⊕Y(su(2)). The generators of the Yangian symmetry of the ordinary Hubbard model (with nearest-neighbour hopping) and of a number of other interesting models like the Haldane-Shastry spin-chain [194, 394] are obtained as special cases of our general result.

14.1 Introduction

Quantum groups were introduced by Drinfeld [107, 109]. His original intention was to put what we called the Yang-Baxter algebra into the mathematically more conventional context of Hopf algebras. The Yangians are special quantum groups. Their representation theory [80, 81] is intimately related to the classification of integrable quantum systems with rational R-matrices.

Later [57] it became apparent that Yangians also play an interesting role as additional hidden symmetries of integrable systems, and moreover [58], that Yangians are part of the symmetry algebra of such well studied integrable systems as the nearest-neighbour Heisenberg model. These symmetries had been overlooked for a long time, since for the models with nearest-neighbour interactions they are incompatible with periodic boundary conditions and for this reason do not combine with the familiar Bethe ansatz methods.

14.2 The variable-range-hopping Hamiltonian

We shall consider here a variant of the Hubbard model with more general hopping amplitudes:

$$H = \sum_{j,k} t_{jk} c^{\dagger}_{j,a} c_{k,a} + u \sum_j (1 - 2n_{j\uparrow})(1 - 2n_{j\downarrow}), \tag{14.1}$$

where $t_{jk} = t_{j-k}$ is a function of the difference of the site indices and

$$t_n = \begin{cases} -i\,\text{sh}(\kappa)\text{sh}^{-1}(\kappa n) & \text{for } n \neq 0, \\ 0 & \text{for } n = 0. \end{cases} \tag{14.2}$$

The Hamiltonian (14.1) is hermitian if and only if the hopping matrix is, $t_{jk} = \bar{t}_{kj}$. Setting $\kappa = a + ib$ in the definition (14.2) the requirement of hermiticity of the the Hamiltonian is seen to be equivalent to $a = 0$ or $b = m\pi$, $m \in \mathbb{Z}$. Moreover, the case $b = m\pi$ is easily recognized to be gauge equivalent to the case $b = 0$. We therefore restrict ourselves to purely real or purely imaginary values of κ. More precisely, our choices for κ are $\kappa = i\pi/N$ for a finite lattice of N sites ('trigonometric case'), and $\kappa > 0$ for an infinite lattice ('hyperbolic case'). The energy scale has been chosen such as to give hopping amplitudes of absolute value 1 between neighbouring sites. The summation indices run from 0 to $N - 1$ in the trigonometric case, and over all integers in the hyperbolic case. The thermodynamic limit of the trigonometric model and the limit $\kappa \to 0$ of the hyperbolic model coincide. In both cases t_{jk} turns into $-i/(j - k)$. The model is then called the $1/r$-Hubbard model or, for reasons that will become clear immediately, the chiral Hubbard model. In the limit $\kappa \to \infty$, the hyperbolic model turns, up to a local gauge transformation described below, into the nearest-neighbour Hubbard model.

It may be interesting to notice, that the trigonometric and hyperbolic hopping amplitudes can be interpreted as q-deformed $1/r$-hopping. The notion of q-deformation is defined by $r_q = (q^r - q^{-r})/(q - q^{-1})$. Setting q equal to e^κ the hopping amplitudes become $t_{jk} = -i/(j - k)_q$. The trigonometric case corresponds to q being the N-th root of unity, the hyperbolic case to $q > 1$ (see figure 14.1).

In order to understand the physical meaning of the above kind of hopping amplitudes, one has to consider the dispersion relation of the free model ($u = 0$) [40, 158, 162, 172]. For the trigonometric case we obtain

$$\varepsilon(p) = \sum_{n=1}^{N-1} t_n e^{ipn} = \frac{N}{\pi} \sin\left(\frac{\pi}{N}\right)(\pi - p), \tag{14.3}$$

where $p = 2\pi(m + 1/2)/N$, $m = 0, \ldots, N - 1$. This yields $\varepsilon(p) = \pi - p$ in the thermodynamic limit. The dispersion relation (14.3) is linear in the first Brillouin zone, the model is chiral. It contains only left-moving particles. The physically most interesting point about this chiral model is the appearance of a Mott transition at finite $u > 0$ [161, 162]. In the

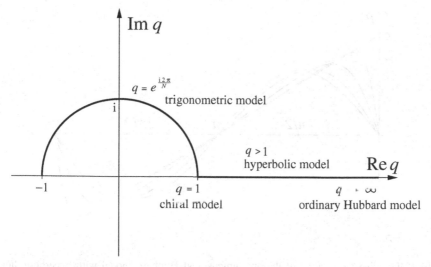

Fig. 14.1. Parameter space of the variable range hopping amplitude $t_{jk} = -i/(j-k)_q$. Points on the thick line correspond to hermitian Hamiltonians.

hyperbolic case the dispersion relation reads

$$\varepsilon(p) = \sum_{n=1}^{\infty} t_n e^{ipn} = 2\mathrm{sh}(\kappa) \sum_{n=1}^{\infty} \frac{\sin(pn)}{\mathrm{sh}(\kappa n)}. \tag{14.4}$$

The last expression is easily recognised as being, up to a redefinition of scales, the logarithmic derivative of the Jacobi theta function ϑ_4 (see [479]). As a function of κ it interpolates between the sinusoidal dispersion relation of the of the nearest-neighbour model and the saw-tooth-shaped dispersion relation of the $1/r$ model (see figure 14.2).

The local gauge transformation $c_{j,a} \rightarrow e^{i\varphi_j} c_{j,a}$, φ_j real, does not alter the canonical anticommutation relations between the Fermi operators. The electron densities $n_j = n_{j,\uparrow} + n_{j,\downarrow}$ are invariant under this transformation, hence the interaction part of the Hamiltonian (14.1) is invariant as well. This means that we can always use a local gauge transformation to modify the hopping term to our convenience. The modified model will be completely equivalent to the original one. Consider the example $\varphi_j = j\pi$. This transformation introduces a factor of $(-1)^{j-k}$ into the expression for the hopping amplitudes and shifts the dispersion relations by a half period. Using this transformation our conventions meet the conventions of Gebhard and Ruckenstein [162]. To recover the nearest-neighbour Hubbard model in its familiar form (2.31), it is not sufficient to consider $\kappa \rightarrow \infty$. In addition we have to apply a gauge transformation with $\varphi_j = j\pi/2$. This transformation removes the factor of 'i' in front of the hopping amplitude, changes the hopping amplitude to an even function, and shifts the dispersion relation by a quarter period. Hence, the quadratic bottom of the sinusoidal band is shifted to $p = 0$.

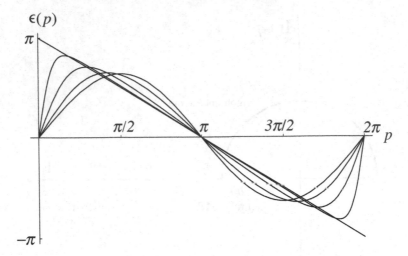

Fig. 14.2. The dispersion curve of the one-dimensional Hubbard model with hyperbolic hopping amplitudes. As a function of $q = e^{\kappa}$ the dispersion curve interpolates between the linear dispersion curve $\varepsilon(p) = \pi - p$ of the chiral model and the sinusoidal dispersion curve of the nearest-neighbour model.

For our Hamiltonian (14.1) the role of the Shiba transformation (2.59) is taken by

$$c_{j\downarrow} \to c_{j\downarrow}^{\dagger} . \tag{14.5}$$

This transformation leaves every Hamiltonian of the form (14.1) with antisymmetric hopping matrix invariant. However, the operator of the total spin is mapped to a new independent operator. Thus, the occurrence of two su(2) symmetries is generic for Hubbard models with antisymmetric hopping amplitudes. The ordinary nearest-neighbour Hubbard model has two su(2) symmetries, because it is gauge equivalent to the model with antisymmetric hopping amplitudes obtained as the limit $\kappa \to \infty$ of the hyperbolic version of our Hamiltonian (14.1). We shall see below that the additional generators of the Yangian symmetry as well are doubled by the transformation (14.5).

14.3 Construction of the Yangian generators

We shall now present an ad hoc construction of the generators of the Yangian symmetry of the Hamiltonian (14.1). The generators will be written as sums involving the local current operators S_{jk}^0, S_{jk}^α introduced in Chapter 2 (see (2.73)). The true usefulness of this formulation will become apparent to those readers who try to verify the commutation relations between the generators of the Yangian (Yangian Serre relations). The calculation considerably simplifies if one uses the algebra of the current operators instead of the elementary commutation relations between the Fermi operators.

The algebra of the current operators is rather rich. First of all we have the commutators derived in Chapter 2 (see equation (2.74)):

$$[S^0_{jk}, S^0_{lm}] = \delta_{kl} S^0_{jm} - \delta_{mj} S^0_{lk} , \tag{14.6a}$$

$$[S^0_{jk}, S^\alpha_{lm}] = \delta_{kl} S^\alpha_{jm} - \delta_{mj} S^\alpha_{lk} , \tag{14.6b}$$

$$[S^\alpha_{jk}, S^\beta_{lm}] = \delta^{\alpha\beta} \left(\delta_{kl} S^0_{jm} - \delta_{mj} S^0_{lk} \right) + i\varepsilon^{\alpha\beta\gamma} \left(\delta_{kl} S^\gamma_{jm} + \delta_{mj} S^\gamma_{lk} \right) . \tag{14.6c}$$

However, there are more, 'anticommutator-like' relations which are easily derived in a similar way as the relations (14.6),

$$S^\alpha_{jk} S^\alpha_{lm} + S^0_{jk} S^0_{lm} + 2S^0_{jm} S^0_{lk} = 4\delta_{kl} S^0_{jm} + 2\delta_{lm} S^0_{jk} , \tag{14.7a}$$

$$S^0_{jk} S^\alpha_{lm} + S^0_{lm} S^\alpha_{jk} + S^0_{lk} S^\alpha_{jm} + S^0_{jm} S^\alpha_{lk} = \delta_{jk} S^\alpha_{lm} + \delta_{lm} S^\alpha_{jk} + \delta_{lk} S^\alpha_{jm} + \delta_{jm} S^\alpha_{lk} , \tag{14.7b}$$

$$S^\alpha_{jk} S^\beta_{lm} + S^\beta_{jk} S^\alpha_{lm} + S^\alpha_{jm} S^\beta_{lk} + S^\beta_{jm} S^\alpha_{lk} = \delta^{\alpha\beta} \left(S^0_{jm} (2\delta_{lk} - S^0_{lk}) + S^\gamma_{jm} S^\gamma_{lk} \right) , \tag{14.7c}$$

$$-i\varepsilon^{\alpha\beta\gamma} S^\beta_{jk} S^\gamma_{lm} - S^0_{jm} S^\alpha_{lk} + S^0_{lk} S^\alpha_{jm} = 2\delta_{lk} S^\alpha_{jm} + \delta_{jk} S^\alpha_{lm} - \delta_{lm} S^\alpha_{jk} . \tag{14.7d}$$

Equations (14.6) and (14.7) generate a long list of succeedingly less general relations by systematically equating all possible combinations of site indices. Generating this list one may find convenient to introduce S^0_j as a short-hand notation for the particle number operator S^0_{jj}.

When written in terms of the current operators the Hamiltonian (14.1) assumes the form

$$H = \sum_{j,k} t_{jk} S^0_{jk} + 2u \sum_j \left((S^0_j - 1)^2 - \tfrac{1}{2} \right) . \tag{14.8}$$

Since the particle number $\hat{N} = \sum_j S^0_j$ is conserved, only the term $(S^0_j)^2$ is relevant in the interaction part of the Hamiltonian. The other terms could be removed by a shift of the chemical potential. Similar to the case of the nearest-neighbour model in Chapter 2 we retained them here to make obvious the invariance of H under the transformation (14.5).

Let us forget temporarily our definition (14.2) and consider the Hamiltonian (14.8) with antisymmetric but otherwise unspecified hopping matrix, $t_{jk} = -t_{kj}$. For $\alpha = x, y, z$ we define the operators

$$J^\alpha = \tfrac{1}{2} \sum_{j,k} \left[(f_{jk} + h_{jk}(S^0_j + S^0_k - 2)) S^\alpha_{jk} + 4g_{jk} \varepsilon^{\alpha\beta\gamma} S^\beta_j S^\gamma_k \right] , \tag{14.9}$$

where g_{jk} and h_{jk} are odd functions, and $f_{jj} = g_{jj} = h_{jj} = 0$ by convention. Using equations (14.6) and (14.7) the reader may verify that H commutes with J^α if and only if the following functional equations for the coefficients are satisfied,

$$t_{jk} = h_0 h_{jk}, \tag{14.10a}$$

$$(g_{jl} - g_{kl})h_{jk} = \tfrac{1}{2}h_{jl}h_{kl}, \quad j \neq k \neq l \neq j, \tag{14.10b}$$

$$iu f_{jk}/h_0 + g_{jk}h_{jk} = -\tfrac{i}{4}\sum_l h_{jl}h_{kl}, \quad j \neq k, \tag{14.10c}$$

$$\sum_l (f_{jl}h_{kl} - f_{lk}h_{lj}) = 0. \tag{14.10d}$$

Here h_0 is a free parameter which fixes the scale for J^α.

The only solutions to the equations (14.10) correspond to the cases of trigonometric and hyperbolic hopping amplitudes (14.2) under consideration. In the trigonometric case we find

$$f_{jk} = 0, \quad g_{jk} = \tfrac{1}{2}\operatorname{ctg}(\pi(j-k)/N), \quad h_{jk} = i\sin^{-1}(\pi(j-k)/N), \tag{14.11}$$

whereas the solutions in the hyperbolic case are

$$f_{jk} = \frac{\operatorname{sh}(\kappa)(j-k)}{2u\operatorname{sh}(\kappa(j-k))}, \quad g_{jk} = \tfrac{1}{2}\operatorname{cth}(\kappa(j-k)), \quad h_{jk} = i\operatorname{sh}^{-1}(\kappa(j-k)). \tag{14.12}$$

The parameter h_0 has to be real in order for J^α to be self adjoint. Choosing $h_0 = -\sin(\pi/N)$ in the trigonometric case and $h_0 = -\operatorname{sh}(\kappa)$ in the hyperbolic case we get back to our original definition (14.2) of the hopping amplitudes. It is an unexpected fact that J^α does not depend on u in the trigonometric case, where therefore (see Section 14.4) hopping part and interaction part of the Hamiltonian separately commute with J^α.

What is the nature of the operators J^α? They are of different kind than the conserved operators generated by the logarithm of the transfer matrix of the Hubbard model. In fact it turns out that the operators J^α combined with the spin operators S^α (see equation (2.66)) generate a representation of Drinfeld's $Y(su(2))$ Yangian [107], i.e., the spin operators S^α and the conserved quantities J^α satisfy the relations

$$[S^\lambda, S^\mu] = c_{\lambda\mu\nu}S^\nu, \tag{14.13a}$$

$$[S^\lambda, J^\mu] = c_{\lambda\mu\nu}J^\nu, \tag{14.13b}$$

$$\begin{aligned} [[J^\lambda, J^\mu], [S^\rho, J^\sigma]] + [[J^\rho, J^\sigma], [S^\lambda, J^\mu]] = \\ -4\delta(a_{\lambda\mu\nu\alpha\beta\gamma}c_{\rho\sigma\nu} + a_{\rho\sigma\nu\alpha\beta\gamma}c_{\lambda\mu\nu})\{S^\alpha, S^\beta, J^\gamma\}, \end{aligned} \tag{14.13c}$$

where $\delta = -1$ in the trigonometric case, $\delta = 1$ in the hyperbolic case, and the further abbreviations

$$c_{\lambda\mu\nu} = i\,\varepsilon^{\lambda\mu\nu}, \tag{14.14a}$$

$$a_{\lambda\mu\nu\alpha\beta\gamma} = c_{\lambda\alpha\rho}c_{\mu\beta\sigma}c_{\nu\gamma\tau}c_{\rho\sigma\tau}, \tag{14.14b}$$

$$\{x_1, x_2, x_3\} = \frac{1}{6}\sum_{i\neq j\neq k\neq i} x_i x_j x_k \tag{14.14c}$$

have been used. Equations (14.13a) are nothing but the su(2) commutation relations between the spin operators. Equations (14.13b) mean that the J^α transform like a vector representation of su(2), and are easily confirmed for our J^α. (14.13c) is called the Yangian Serre relation. Since both, (14.13b) and (14.13c), are homogeneous, we could have introduced a deformation parameter, say $h^2 \neq 0$, on the right-hand side of the Yangian Serre relation (14.13c). Because this parameter merely fixes the scale of J^α and has no deeper physical meaning, we suppressed it here. Equations (14.13) have been confirmed by direct calculation [172]. We have to warn the reader though, that the calculation is rather lengthy. Before we comment on its details we formulate one more result.

Under the transformation (14.5) the generators S^α, J^β transform into an independent set of generators S'^α, J'^β of another representation of the Y(su(2)) Yangian. The two representations mutually commute, hence can be combined into a representation of the direct sum Y(su(2))⊕Y(su(2)). Their mutual commutativity is non-trivial. It depends on the specific form of the hopping amplitudes t_{jk} and on the functional equations (14.10). It thus may be claimed as being of 'dynamical origin'. Of course, applying a local gauge transformation with parameters $\varphi_j = j\pi/2$ to the operators $-S'^\pm$ they turn into $\eta^\pm$ while S'^z turns into η^z. Thus, one of the Yangian representations extends the spin representation of su(2) while the other one extends the η-spin representation.

It seems rather hard to verify the validity of the Yangian Serre relation for our operators J^α in the original formulation formulation (14.13c). We may use the following simplification instead: let

$$K^\alpha = -i\varepsilon^{\alpha\beta\gamma}[J^\beta, J^\gamma] - 4\delta(S^\beta)^2 S^\alpha. \tag{14.15}$$

Then a short but slightly tricky calculation (see appendices 14.A.5 and 14.A.7) shows that (14.13c) can be replaced by the equation

$$[J^\alpha, K^\beta] + [J^\beta, K^\alpha] = 0. \tag{14.16}$$

The left-hand side of (14.16) has a property that turns out to be very useful in practical calculations. It is traceless. Assume we are given an operator J^α, and we do already know that it transforms as a vector representation of su(2). Then this knowledge assures the identity $[J^\alpha, K^\alpha] = 0$. It is therefore sufficient to show that the left-hand side of equation (14.16) is proportional to $\delta^{\alpha\beta}$. This is a severe simplification, since the symmetrization of the commutator produces numerous terms proportional to $\delta^{\alpha\beta}$, which can be neglected according to the above argument. In our case the explicit expression for K^α is

$$K^\alpha = \frac{1}{2} \sum_{j,k,l} \left\{ 8(8 g_{jk} g_{jl} - \delta) S_j^\beta S_k^\beta S_l^\alpha + 2 A_{jl} A_{lk} S_{jk}^\alpha - 8i A_{jk}(g_{jl} - g_{kl}) S_{jk}^0 S_l^\alpha \right.$$
$$\left. + 4 A_{jk}(g_{jl} + g_{kl}) \varepsilon^{\alpha\beta\gamma} S_{jk}^\beta S_l^\gamma + i h_{jl} \varepsilon^{\alpha\beta\gamma} (A_{jk} S_{jk}^\beta - A_{kj} S_{kj}^\beta)(S_{jl}^\gamma - S_{lj}^\gamma) \right\}, \tag{14.17}$$

where $A_{jk} = f_{jk} + h_{jk}(S_j^0 + S_k^0 - 2)$. To verify (14.16), one has to use the following relations among the coefficients f_{jk}, g_{jk}, h_{jk} in addition to their defining functional equations above.

$$f_{jk}(g_{jl} - g_{kl}) = \frac{i}{2}(f_{jl} h_{kl} - f_{kl} h_{jl}), \quad j \neq k \neq l \neq j, \tag{14.18a}$$

$$g_{jk} g_{jl} + g_{kl} g_{kj} + g_{lj} g_{lk} = \delta/4, \quad j \neq k \neq l \neq j, \tag{14.18b}$$

$$4 g_{jk}^2 + h_{jk}^2 = \delta, \quad j \neq k. \tag{14.18c}$$

The homogeneity of the lattices has not been used in the verification of the Yangian Serre relation in the bulk. However, it is necessary to guarantee the commutativity of J^α with the Hamiltonian. This situation is similar to the case of the Yangian symmetric spin chains. Therefore the existence of a Yangian symmetric long-range Hubbard Hamiltonian on an inhomogeneous lattice was conjectured in [172]. In analogy to the spin chain case [199] the generator of its Yangian symmetry might be constructed by adding 'potential terms' to the second order Yangian generator K^α, equation (14.15).

We would like to emphasize that Yangian symmetry does not imply integrability. Nevertheless, it seems likely that the models considered here are exactly solvable, and are special cases of a more general exactly solvable non-Yangian-symmetric model with elliptic hopping amplitudes, which is known to have additional conserved operators of non-Yangian type [210]. A proof of the 'integrability' of this model would provide the basis for an understanding of the Haldane-Shastry spin chain and the nearest-neighbour Hubbard model on a common ground. At our present state of knowledge these models appear rather dissimilar. The integrability of the Haldane-Shastry chain has been shown by exploiting a mapping to a related dynamical model [446], whereas the integrability of the nearest-neighbour Hubbard model is based on Shastry's R-matrix (12.120). We expect that a proof of the integrability of the non-nearest-neighbour Hubbard models would reveal a more generic structure.

14.4 Special cases

Our representation (14.9) of the generators J^α of the Y(su(2)) Yangian contains two free parameters κ and u in the hyperbolic case. This offers the possibility to consider various limiting cases. Another possibility comes from considering the restriction to the spin chain space of states with every lattice site occupied by precisely one electron. As we shall see below this restriction makes sense in the trigonometric as well as in hyperbolic the case.

We have shown above that the ordinary Hubbard Hamiltonian is a special case of the Hamiltonian (14.1) with hyperbolic hopping amplitudes and is recovered in the limit $\kappa \to \infty$ through a local gauge transformation $c_{j,a} \to i^j c_{j,a}$. Let us apply the same manipulations to the first-level generator J^α of the representation of the Yangian. For $\kappa \to \infty$ the solutions (14.12) of the functional equations (14.10) have the limits $f_{jk} = \delta_{|j-k|,1}/2u, g_{jk} = \text{sign}(j - k)/2, h_{jk} = 0$. We thus obtain

$$J^\alpha = \frac{i}{4u} \sum_j \left(S^\alpha_{jj+1} - S^\alpha_{jj-1} \right) - 2 \sum_{j<k} \varepsilon^{\alpha\beta\gamma} S^\beta_j S^\gamma_k \tag{14.19}$$

for $\alpha = x, y, z$. These operators are the generators of the Yangian symmetry of the Hubbard model on the infinite lattice. The Yangian generators E_1, F_1 and H_1 of Uglov and Korepin [462] are linear combinations of these operators, $E_1 = 2u(J^x + iJ^y)$, $F_1 = 2u(J^x - iJ^y)$ and $H_1 = 4uJ^z$.

Keeping κ fixed but sending u to infinity we obtain another interesting limiting case.[1] Let us define

$$J^\alpha_0 = \frac{1}{2} \sum_{j,k} \left[h_{jk}(S^0_j + S^0_k - 2)S^\alpha_{jk} + 4g_{jk} \varepsilon^{\alpha\beta\gamma} S^\beta_j S^\gamma_k \right], \tag{14.20}$$

$$J^\alpha_1 = \frac{u}{2} \sum_{j,k} f_{jk} S^\alpha_{jk}. \tag{14.21}$$

Then J^α_1 is independent of u and

$$J^\alpha = J^\alpha_0 + \frac{1}{u} J^\alpha_1. \tag{14.22}$$

This formula also holds for the trigonometric case where $f_{jk} = 0$. Clearly, S^α and $J^\beta_0, \alpha, \beta = x, y, z$, generate a representation of $Y(su(2))$. But what is the corresponding $Y(su(2))$-invariant Hamiltonian?

We note that

$$[S^\alpha, P_0] = [J^\alpha_0, P_0] = 0, \tag{14.23}$$

where $P_0 = \prod_j (1 - n_{j\uparrow} n_{j\downarrow})$ is the familiar projection operator (see appendix 2.A and Chapter 12.3.5) onto the space with no doubly occupied lattice site. The equations (14.23) mean that S^α and J^α_0 leave the space with no doubly occupied lattice site invariant. They are consequences of the local identities

$$[S^\alpha_j, P_0] = [(S^0_j + S^0_k - 2)S^\alpha_{jk}, P_0] = 0 \tag{14.24}$$

which follow from (14.6), (14.7). Let us consider the Hamiltonian $H = T + 4uD$, where

$$T = \sum_{j,k} t_{jk} c^\dagger_{j,a} c_{k,a}, \qquad D = \sum_j n_{j\uparrow} n_{j\downarrow}. \tag{14.25}$$

---

[1] This limiting case was explained to us by V. Inozemtsev to whom we are deeply grateful.

This Hamiltonian commutes with J^α, since the Hamiltonian (14.1) does and since $[J^\alpha, \hat{N}] = 0$. Hence,

$$[H, J^\alpha] = [T, J_0^\alpha] + 4u[D, J_0^\alpha] + \frac{1}{u}[T, J_1^\alpha] + 4[D, J_1^\alpha] = 0 \tag{14.26}$$

for all $u \in \mathbb{R}$. We conclude that $[D, J_0^\alpha] = 0$, which proves our claim above that in the trigonometric case hopping part and interaction part of the Hamiltonian separately commute with J^α. We further conclude that $[T, J_0^\alpha] = 4[J_1^\alpha, D]$ and thus, using (14.23) and $P_0 D = D P_0 = 0$,

$$[P_0 T P_0, J_0^\alpha] = P_0[T, J_0^\alpha]P_0 = 4P_0[J_1^\alpha, D]P_0 = 0. \tag{14.27}$$

This means that the t-0 Hamiltonian

$$H_{t-0} = P_0 \sum_{j,k} t_{jk} c_{j,a}^\dagger c_{k,a} P_0 \tag{14.28}$$

with trigonometric [476] or hyperbolic hopping amplitudes (14.2) commutes with the Yangian generators J_0^α. Again by (14.23) it also commutes with the spin operators S^α. Therefore the trigonometric and hyperbolic t-0 models are Yangian invariant.

We have yet another option to further specialize the generators J_0^α. Equation (14.23) means that S^α and J_0^α leave the space with a fixed number of doubly occupied sites invariant. Moreover, $[J_0^\alpha, \hat{N}] = [S^\alpha, \hat{N}] = 0$. It follows that the restrictions of J_0^α and S^α to the 'spin chain space of states', say $\mathcal{H}_S$, where every lattice site is occupied by exactly one electron, form a representation of the $Y(su(2))$ Yangian. Now, since $[(S_j^0 + S_k^0 - 2), S_{jk}^\alpha] = 0$ and since $S_j^0 + S_k^0 - 2$ annihilates $\mathcal{H}_S$, the restriction of J_0^α to $\mathcal{H}_S$ is

$$J_S^\alpha = 2 \sum_{j,k} g_{jk} \varepsilon^{\alpha\beta\gamma} S_j^\beta S_k^\gamma. \tag{14.29}$$

The restriction of H_{t-0} onto $\mathcal{H}_S$, however, vanishes identically. A non-trivial Hamiltonian commuting with J_S^α has to be construct by independent means. Using the simple ansatz

$$H_S = \sum_{j,k} s_{jk} S_j^\alpha S_k^\alpha \tag{14.30}$$

with an even function s_{jk} vanishing at $j = k$ we obtain the commutator

$$[H_S, J_S^\alpha] = 4i \sum_{j \neq k \neq l \neq j} \left[s_{jk}(g_{lj} - g_{lk}) + s_{jl}(g_{kj} - g_{kl}) \right] S_j^\alpha S_k^\beta S_l^\beta + 4i \sum_{j \neq k} s_{jk} g_{kj} S_j^\alpha (2 - S_k^0) S_k^0. \tag{14.31}$$

For this expression to be zero both sums have to vanish independently. The vanishing of the square bracket under the first sum is easily seen to be equivalent to $s_{jk} = c|t_{jk}|^2$, where t_{jk} is the hopping amplitude (14.2) and c is an arbitrary constant. Inserting this result into the second sum we see that it only vanishes when restricted to the spin chain space of states, where we can replace S_k^0 by 1 and S_j^α by $\frac{1}{2}\sigma_j^\alpha$. On the spin chain space of states the Hamiltonian H_S reduces to the well studied (for a review see [195]) Haldane-Shastry

Hamiltonian [194, 394] or to its hyperbolic counterpart [209]. The Yangian symmetry of these models was discovered and discussed in [59, 196].

Remark. We finally wish to point out the following subtlety: In the hyperbolic case the limit $\kappa \to \infty$ of the Yangian generators J_S^α yields

$$J_S^\alpha = \sum_{j \neq k} \text{sign}(j - k)\, \varepsilon^{\alpha\beta\gamma} S_j^\alpha S_k^\beta. \tag{14.32}$$

As we have seen above, these operators (together with the corresponding spin operators) generate a representation of $Y(su(2))$ commuting with the Hamiltonian

$$H_{t-0} = -P_0 \sum_j (c_{j,a}^\dagger c_{j+1,u} + c_{j+1,a}^\dagger c_{j,a}) P_0 \tag{14.33}$$

of the usual t-0 model (compare [179]). One might naively think that J_S^α, equation (14.32), would also commute with $\sum_j S_j^\alpha S_{j+1}^\alpha$ and that for this reason the t-J Hamiltonian

$$H_{t-J} = P_0 \left[-\sum_j (c_{j,a}^\dagger c_{j+1,a} + c_{j+1,a}^\dagger c_{j,a}) + J \sum_j S_j^\alpha S_{j+1}^\alpha \right] P_0 \tag{14.34}$$

would be Yangian invariant. This is, however, *not* the case. Due to (14.31) only the restriction of H_{t-J} to the spin chain space of states, which is equivalent to the Heisenberg Hamiltonian, commutes with J_S^α. This nicely relates to the fact that for general J the t-J model is not solvable by nested Bethe ansatz [377].

14.5 Conclusions

Following [172] we have constructed a pair of fermionic representations of the $Y(su(2))$ Yangian, one of them extending the spin representation of $su(2)$ the other one the η-spin representation. The two representations of the Yangian are simply related to one another by the Shiba transformation (14.5). It is was therefore sufficient to concentrate our discussion on the spin representation of the Yangian defined by the spin operators S^α and the additional 'first level generators' J^α, equations (14.9), (14.11) and (14.12). These generators commute with the Hamiltonian (14.1) which therefore can be called Yangian symmetric. For the Hamiltonian as well as for the generators J^α we have to distinguish between a trigonometric case which applies to a finite chain with periodic boundary conditions and a hyperbolic case which applies to an infinite chain. The hyperbolic case has two free parameters, the Hubbard interaction u and the decay length of the hopping κ. Thus, we could rather say we have obtained a two-parametric family of representations of the $Y(su(2))$ Yangian together with a two-parametric Yangian invariant Hamiltonian. This two-parametric family has several interesting limiting cases which were discussed in Section 14.4. Most important for us is the limit $\kappa \to \infty$ which, in conjunction with a gauge transformation, led us to the generators (14.19) of the spin representation of the Yangian of the usual Hubbard model.

This representation commutes with the Hubbard Hamiltonian on the infinite interval and is not compatible with periodic boundary conditions. Hence, there is no natural action of the first level Yangian generators on the Bethe ansatz eigenstates of Chapter 3 and Chapter 12. Yet, we shall see in Chapter 15 that the eigenstates of the Hubbard model one the infinite interval over the empty lattice vacuum can be constructed directly by a variant of the quantum inverse scattering method. These eigenstates are then shown to transform like tensor products of so-called evaluation representations of the Yangian.

In a number of appendices we shall present some background material about Yangians. Our account will stay elementary. We will concentrate on the specific example of the $Y(su(2))$ Yangian and will not say much about the general theory of quantum groups. We included proofs of several theorems which are not easily accessible in the literature. It is our hope that the appendix will provide a useful bridge for those readers who are familiar with the quantum inverse scattering method and would like to start studying the mathematical literature on quantum groups.

Appendices to Chapter 14

14.A Yangians

In this appendix we elaborate further on the mathematical structure of the class of models which have the same R-matrix $R(\lambda) = P \check{R}(\lambda)$ as the su(d)-XXX spin chain (see (12.49), (12.55)). For our convenience we write this R-matrix here as

$$R(\lambda) = \lambda + \eta P, \tag{14.A.1}$$

i.e., we set $\eta = ic$ and change the normalization as compared to (12.49). $R(\lambda)$ is often called the gl(d) invariant R-matrix.

14.A.1 Symmetrizers and antisymmetrizers

In the following we will need projection operators $P^{\pm}_{1,\dots,N}$ onto the completely symmetric and antisymmetric subspaces of N-fold tensor products of $\mathbb{C}^d$. The purpose of this section is to define these operators and to derive their characteristic properties.

We shall denote the canonical basis of $\mathbb{C}^d$ as before by $\{\mathbf{e}_a\}_{a=1}^d$. As we observed in appendix 3.B the formula

$$P(\mathbf{e}_{a_1} \otimes \cdots \otimes \mathbf{e}_{a_N}) = \mathbf{e}_{a_{P^{-1}(1)}} \otimes \cdots \otimes \mathbf{e}_{a_{P^{-1}(N)}} \tag{14.A.2}$$

where $P \in \mathfrak{S}^N$ and $a_j = 1, \dots, d$, defines a representation of the symmetric group acting on $(\mathbb{C}^d)^{\otimes N}$. Referring to this representation we define the operators

$$P^-_{1,\dots,N} = \frac{1}{N!} \sum_{P \in \mathfrak{S}^N} \text{sign}(P) \, P, \tag{14.A.3a}$$

$$P^+_{1,\dots,N} = \frac{1}{N!} \sum_{P \in \mathfrak{S}^N} P. \tag{14.A.3b}$$

If the operators $P^{\pm}_{1,\dots,N}$ act non-trivially on the factors $m, \dots, n = m + N - 1$ of multiple tensor products of spaces $\mathbb{C}^d$ we use the notation $P^{\pm}_{m,\dots,n}$.

Lemma 21. *Properties of the operators* $P^{\pm}_{1,...,N}$.

(i) Projection property. The operators $P^{\pm}_{1,...,N}$ *are projection operators,*

$$(P^{\pm}_{1,...,N})^2 = P^{\pm}_{1,...,N},$$ \hfill (14.A.4a)

$$P^-_{1,...,N}P^+_{1,...,N} = P^+_{1,...,N}P^-_{1,...,N} = 0.$$ \hfill (14.A.4b)

(ii) Determinant formula. Let $A \in \text{End}(\mathbb{C}^d)$, $N \le d$. *Then*

$$\text{tr}\big(P^-_{1,...,N} A^{\otimes N}\big) = \sum_{a_1 < \cdots < a_N} \sum_{P \in \mathfrak{S}^N} \text{sign}(P)\, A^{a_{P(1)}}_{a_1} \ldots A^{a_{P(N)}}_{a_N}.$$ \hfill (14.A.5)

This implies, in particular, for $N = d$ *that*

$$\text{tr}\big(P^-_{1,...,N} A^{\otimes d}\big) = \det A.$$ \hfill (14.A.6)

(iii) Recursion formulae.

$$P^-_{1,...,N+1} = P^-_{2,...,N+1} \frac{N - P_{1,N+1}}{N+1} P^-_{1,...,N},$$ \hfill (14.A.7a)

$$P^+_{1,...,N+1} = P^+_{1,...,N} \frac{N + P_{1,N+1}}{N+1} P^+_{2,...,N+1},$$ \hfill (14.A.7b)

where $P_{1,N+1}$ *is the familiar transposition operator interchanging the first with the* $(N+1)$*th factor in a multiple tensor product of spaces* $\mathbb{C}^d$.

(iv) Resolution of the recursion formulae. The recursion formulae (14.A.7) can be resolved as

$$P^-_{1,...,N+1} = \frac{1 - P_{N,N+1}}{2} \cdot \frac{2 - P_{N-1,N+1}}{3} \cdots \frac{N - P_{1,N+1}}{N+1} P^-_{1,...,N},$$ \hfill (14.A.8a)

$$P^+_{1,...,N+1} = P^+_{1,...,N} \frac{N + P_{1,N+1}}{N+1} \cdots \frac{2 + P_{N-1,N+1}}{3} \cdot \frac{1 + P_{N,N+1}}{2}.$$ \hfill (14.A.8b)

Proof. Point (i) is obvious from the definition (14.A.3).

Point (ii) is proven by direct calculation,

$$
\begin{aligned}
\text{tr}\big(P^-_{1,...,N} A^{\otimes N}\big) &= \text{tr}\big(P^-_{1,...,N}\, e^{b_1}_{a_1} \otimes \cdots \otimes e^{b_N}_{a_N}\big) A^{a_1}_{b_1} \ldots A^{a_N}_{b_N} \\
&= \frac{1}{N!} \sum_{P \in \mathfrak{S}^N} \text{sign}(P)\, \text{tr}\big(e^{b_1}_{a_{P^{-1}(1)}} \otimes \cdots \otimes e^{b_N}_{a_{P^{-1}(N)}}\big) A^{a_1}_{b_1} \ldots A^{a_N}_{b_N} \\
&= \frac{1}{N!} \sum_{P \in \mathfrak{S}^N} \text{sign}(P)\, A^{a_{P(1)}}_{a_1} \ldots A^{a_{P(N)}}_{a_N} \\
&= \sum_{a_1 < \cdots < a_N} \sum_{P \in \mathfrak{S}^N} \text{sign}(P)\, A^{a_{P(1)}}_{a_1} \ldots A^{a_{P(N)}}_{a_N}.
\end{aligned}
$$ \hfill (14.A.9)

Here we used the identity

$$\text{tr}\big(e^{b_1}_{a_{P^{-1}(1)}} \otimes \cdots \otimes e^{b_N}_{a_{P^{-1}(N)}}\big) = \text{tr}(e^{b_1}_{a_{P^{-1}(1)}}) \ldots \text{tr}(e^{b_N}_{a_{P^{-1}(N)}}) = \delta^{b_1}_{a_{P^{-1}(1)}} \ldots \delta^{b_N}_{a_{P^{-1}(N)}}$$ \hfill (14.A.10)

in the third equation and the fact that the matrix elements of A commute in the third and in the fourth equation.

Equation (14.A.7a) in point (iii) follows from

$$(N!)^2 \, P^-_{2,\dots,N+1}(N - P_{1,N+1})P^-_{1,\dots,N}$$

$$= N! P^-_{2,\dots,N+1} \sum_{\substack{P \in \mathfrak{S}^{N+1} \\ P(N+1)=N+1}} \mathrm{sign}(P)(N - P_{1,N+1})P$$

$$= \sum_{\substack{Q \in \mathfrak{S}^{N+1} \\ Q(1)=1}} \sum_{\substack{P \in \mathfrak{S}^{N+1} \\ P(N+1)=N+1}} \mathrm{sign}(PQ)Q(N - P_{1,N+1})P$$

$$= \sum_{\substack{Q \in \mathfrak{S}^{N+1} \\ Q(1)=1}} \left[\sum_{\substack{P \in \mathfrak{S}^{N+1} \\ P(N+1)=Q(N+1)}} N \, \mathrm{sign}(P)P \; + \sum_{\substack{P \in \mathfrak{S}^{N+1} \\ P(N+1)=Q(1)}} \mathrm{sign}(P)P \right]$$

$$= (N - 1)! \sum_{k=2}^{N+1} \sum_{\substack{P \in \mathfrak{S}^{N+1} \\ P(N+1)=k}} N \, \mathrm{sign}(P)P + N! \sum_{\substack{P \in \mathfrak{S}^{N+1} \\ P(N+1)=1}} \mathrm{sign}(P)P$$

$$= N! \sum_{P \in \mathfrak{S}^{N+1}} \mathrm{sign}(P)P \, . \tag{14.A.11}$$

The proof of (14.A.7b) is very similar and is left as an exercise to the reader.

The proof of (iv) relies on a simple induction argument using (14.A.7) and the properties $P^-_{2,\dots,N} P^-_{1,\dots,N} = P^-_{1,\dots,N}$ and $P^+_{1,\dots,N} P^+_{2,\dots,N} = P^+_{1,\dots,N}$ of the projection operators. $\qquad\square$

14.A.2 Quantum symmetric functions and a theorem of Kulish and Sklyanin

Using (14.A.5) we can represent the sum over the principle minors of rank N of the determinant of a monodromy matrix $T^{(cl)}(\lambda) \in \mathrm{End}(\mathbb{C}^d)$ belonging to a classical integrable system as $\sigma_N^{(cl)}(\lambda) = \mathrm{tr}\big(P^-_{1,\dots,N}\big(T^{(cl)}(\lambda)\big)^{\otimes N}\big)$. This function is easily seen to be the Nth order symmetric polynomial in the eigenvalues of $T^{(cl)}(\lambda)$ which is known to be a generating function of involutive integrals of motion (see [277] and references therein). One may ask, what is the quantum analogue of $\sigma_N^{(cl)}(\lambda)$? The answer is not known in the general case. For models with the R-matrix (14.A.1), however, Kulish and Sklyanin were able to construct such quantum analogue [277]. Their construction is based on equation (14.A.5) and will be presented below.

For any monodromy matrix $T(\lambda)$ representing the Yang-Baxter algebra with rational R-matrix (14.A.1) and for any $N = 1, \dots, d$ we define the 'quantum symmetric function'

$$\sigma_N(\lambda) = \mathrm{tr}\big(P^-_{1,\dots,N} \, T_1(\lambda)T_2(\lambda + \eta) \dots T_N(\lambda + (N-1)\eta)\big) \, . \tag{14.A.12}$$

Note that the classical limit ($\eta \to 0$) of $\sigma_N(\lambda)$ is $\sigma_N^{(cl)}(\lambda)$ and that $\sigma_1(\lambda)$ is the transfer matrix of the system.

Theorem 1. *(Kulish and Sklyanin [277]). The quantum symmetric functions $\sigma_N(\lambda)$ form a commutative family:*

$$[\sigma_N(\lambda), \sigma_M(\mu)] = 0, \tag{14.A.13}$$

for all $N, M = 1, \ldots, d$ and for all $\lambda, \mu \in \mathbb{C}$.

The higher quantum symmetric functions may generate additional conserved quantities which, because of the theorem, commute with the conserved quantities generated by the transfer matrix $t(\lambda) = \sigma_1(\lambda)$.

The remaining part of this subsection is devoted to the proof of the theorem. Since the proof is rather lengthy, we divided it into several lemmata.

Lemma 22.

$$P_{1,\ldots,N}^- \, T_1(\lambda) \ldots T_N(\lambda + (N-1)\eta) = T_N(\lambda + (N-1)\eta) \ldots T_1(\lambda) \, P_{1,\ldots,N}^-, \tag{14.A.14a}$$

$$P_{1,\ldots,N}^- \, T_N(\lambda - (N-1)\eta) \ldots T_1(\lambda) = T_1(\lambda) \ldots T_N(\lambda - (N-1)\eta) \, P_{1,\ldots,N}^-, \tag{14.A.14b}$$

$$P_{1,\ldots,N}^+ \, T_1(\lambda + (N-1)\eta) \ldots T_N(\lambda) = T_N(\lambda) \ldots T_1(\lambda + (N-1)\eta) \, P_{1,\ldots,N}^+, \tag{14.A.14c}$$

$$P_{1,\ldots,N}^+ \, T_N(\lambda) \ldots T_1(\lambda - (N-1)\eta) = T_1(\lambda - (N-1)\eta) \ldots T_N(\lambda) \, P_{1,\ldots,N}^+. \tag{14.A.14d}$$

Proof. We will prove equation (14.A.14a) by induction over N. The argument is based on the resolved recursion formula (14.A.8a) and on the elementary fact that

$$R_{jk}(-N\eta) = -\eta(N - P_{jk}). \tag{14.A.15}$$

Using the latter equation for $N = 1$ we find

$$P_{1,2}^- \, T_1(\lambda)T_2(\lambda + \eta) = -\frac{R_{12}(-\eta)}{2\eta} T_1(\lambda)T_2(\lambda + \eta)$$

$$= -T_2(\lambda + \eta)T_1(\lambda)\frac{R_{12}(-\eta)}{2\eta} = T_2(\lambda + \eta)T_1(\lambda) \, P_{1,2}^-, \tag{14.A.16}$$

and (14.A.14a) is proven for $N = 2$.

For the induction step we assume (14.A.14a) to be true for some $N \geq 2$. Inserting (14.A.15) into (14.A.8a) we obtain

$$P_{1,\ldots,N+1}^- = \frac{(-1)^N}{\eta^N (N+1)!} R_{N,N+1}(-\eta) \ldots R_{1,N+1}(-N\eta) \, P_{1,\ldots,N}^-. \tag{14.A.17}$$

Hence,

$$P^-_{1,\ldots,N+1} T_1(\lambda) \ldots T_{N+1}(\lambda + N\eta)$$

$$= \frac{(-1)^N}{\eta^N (N+1)!} R_{N,N+1}(-\eta) \ldots R_{1,N+1}(-N\eta)$$

$$\times T_N(\lambda + (N-1)\eta) \ldots T_1(\lambda) T_{N+1}(\lambda + N\eta) \, P^-_{1,\ldots,N}$$

$$= \frac{(-1)^N}{\eta^N (N+1)!} T_{N+1}(\lambda + N)\eta) \ldots T_1(\lambda)$$

$$\times R_{N,N+1}(-\eta) \ldots R_{1,N+1}(-N\eta) \, P^-_{1,\ldots,N}$$

$$= T_{N+1}(\lambda + N\eta) \ldots T_1(\lambda) \, P^-_{1,\ldots,N+1} \,, \tag{14.A.18}$$

and the proof of (14.A.14a) is complete. We used (14.A.17) and the induction hypothesis in the first equation, the Yang-Baxter algebra in the second equation, and again (14.A.17) in the third equation. The proof of the remaining equations is very similar. For the proof of (14.A.14b) and (14.A.14d) one has to take into account that $R_{12}(\lambda) = R_{21}(\lambda)$. $\square$

In preparation of the next lemma we define for fixed $M, N = 1, \ldots, d$

$$\tilde{R}^\ell(\lambda) = R_{M,M+\ell}(\lambda + (M-\ell)\eta) \ldots R_{1,M+\ell}(\lambda + (1-\ell)\eta), \quad \ell = 1, \ldots, N,$$

$$\tilde{R}_\ell(\lambda) = R_{\ell,M+1}(\lambda + (\ell-1)\eta) \ldots R_{\ell,M+N}(\lambda + (\ell-N)\eta), \quad \ell = 1, \ldots, M,$$

$$\mathcal{R}^{M+1,\ldots,M+N}_{1,\ldots,M}(\lambda) = \tilde{R}^1(\lambda) \ldots \tilde{R}^N(\lambda) \, P^-_{1,\ldots,M} \, P^-_{M+1,\ldots,M+N} \,. \tag{14.A.19}$$

Lemma 23.

$$\left(1 - P^-_{1,\ldots,M} P^-_{M+1,\ldots,M+N}\right) \mathcal{R}^{M+1,\ldots,M+N}_{1,\ldots,M}(\lambda) = 0 \,. \tag{14.A.20}$$

Proof. First of all we have

$$\tilde{R}^1(\lambda) \ldots \tilde{R}^N(\lambda) = R_{M,M+1}(\lambda + (M-1)\eta) \ldots R_{1,M+1}(\lambda)$$

$$R_{M,M+2}(\lambda + (M-2)\eta) \ldots R_{1,M+2}(\lambda - \eta)$$

$$\ldots$$

$$R_{M,M+N}(\lambda + (M-N)\eta) \ldots R_{1,M+N}(\lambda + (1-N)\eta)$$

$$= R_{M,M+1}(\lambda + (M-1)\eta) \ldots R_{M,M+N}(\lambda + (M-N)\eta)$$

$$R_{M-1,M+1}(\lambda + (M-2)\eta) \ldots R_{M-1,M+N}(\lambda + (M-1-N)\eta)$$

$$\ldots$$

$$R_{1,M+1}(\lambda) \ldots R_{1,M+N}(\lambda + (1-N)\eta)$$

$$= \tilde{R}_M(\lambda) \ldots \tilde{R}_1(\lambda) \,. \tag{14.A.21}$$

Since $R(\lambda)$ is a solution of the Yang-Baxter equation, $T_j(\lambda) = R_{j,n}(\lambda)$ is a representation of the Yang-Baxter algebra, and we can apply lemma 22,

$$P^-_{1,\ldots,M} R_{1,M+\ell}(\lambda + (1 - \ell)\eta) \ldots R_{M,M+\ell}(\lambda + (M - \ell)\eta)$$
$$= \tilde{R}^\ell(\lambda)\, P^-_{1,\ldots,M}\,, \quad \ell = 1,\ldots, N\,. \tag{14.A.22}$$

It follows that

$$\tilde{R}^\ell(\lambda)\, P^-_{1,\ldots,M} = P^-_{1,\ldots,M}\, \tilde{R}^\ell(\lambda)\, P^-_{1,\ldots,M}\,, \quad \ell = 1,\ldots, N \tag{14.A.23}$$

and by iteration

$$\tilde{R}^1(\lambda)\ldots \tilde{R}^N(\lambda)\, P^-_{1,\ldots,M} = P^-_{1,\ldots,M}\, \tilde{R}^1(\lambda)\ldots \tilde{R}^N(\lambda)\, P^-_{1,\ldots,M}\,. \tag{14.A.24}$$

Similarly, using $R_{j,n}(\lambda) = R_{n,j}(\lambda)$ we conclude with lemma 22 that

$$P^-_{M+1,\ldots,M+N} R_{\ell,M+N}(\lambda + (\ell - N)\eta) \ldots R_{\ell,M+1}(\lambda + (\ell - 1)\eta)$$
$$= \tilde{R}_\ell(\lambda)\, P^-_{M+1,\ldots,M+N}\,, \quad \ell = 1,\ldots, M\,. \tag{14.A.25}$$

Hence,

$$\tilde{R}_\ell(\lambda)\, P^-_{M+1,\ldots,M+N} = P^-_{M+1,\ldots,M+N}\, \tilde{R}_\ell(\lambda)\, P^-_{M+1,\ldots,M+N}\,, \quad \ell = 1,\ldots, M\,. \tag{14.A.26}$$

Iterating this equation and using (14.A.21) we arrive at

$$\tilde{R}^1(\lambda)\ldots \tilde{R}^N(\lambda)\, P^-_{M+1,\ldots,M+N} = P^-_{M+1,\ldots,M+N}\, \tilde{R}^1(\lambda)\ldots \tilde{R}^N(\lambda)\, P^-_{M+1,\ldots,M+N}\,. \tag{14.A.27}$$

Finally, the lemma follows from (14.A.24) and (14.A.27). $\qquad\qquad\square$

For the formulation of the last lemma needed in the proof of theorem 1 we define the operators

$$\mathcal{T}_{1,\ldots,M}(\lambda) = P^-_{1,\ldots,M}\, T_1(\lambda)\ldots T_M(\lambda + (M - 1)\eta)\,, \tag{14.A.28}$$

$$\mathcal{T}_{M+1,\ldots,M+N}(\mu) = P^-_{M+1,\ldots,M+N}\, T_{M+1}(\mu)\ldots T_{M+N}(\mu + (N - 1)\eta)\,, \tag{14.A.29}$$

where $M, N = 1,\ldots, d$. These operators satisfy a Yang-Baxter algebra like relation.

Lemma 24.

$$\mathcal{R}^{M+1,\ldots,M+N}_{1,\ldots,M}(\lambda - \mu)\, \mathcal{T}_{1,\ldots,M}(\lambda)\mathcal{T}_{M+1,\ldots,M+N}(\mu)$$
$$= \mathcal{T}_{M+1,\ldots,M+N}(\mu)\mathcal{T}_{1,\ldots,M}(\lambda)\, \mathcal{R}^{M+1,\ldots,M+N}_{1,\ldots,M}(\lambda - \mu)\,. \tag{14.A.30}$$

Proof. We first note that

$$\tilde{R}^\ell(\lambda - \mu)\, T_M(\lambda + (M - 1)\eta)\ldots T_1(\lambda)T_{M+\ell}(\mu + (\ell - 1)\eta)$$
$$= R_{M,M+\ell}(\lambda - \mu + (M - \ell)\eta)\ldots R_{1,M+\ell}(\lambda - \mu + (1 - \ell)\eta)$$
$$\times T_M(\lambda + (M - 1)\eta)\ldots T_1(\lambda)T_{M+\ell}(\mu + (\ell - 1)\eta)$$
$$= T_{M+\ell}(\mu + (\ell - 1)\eta)T_M(\lambda + (M - 1)\eta)\ldots T_1(\lambda)\tilde{R}^\ell(\lambda - \mu)\,. \tag{14.A.31}$$

Then, it follows that

$$
\mathcal{R}_{1,\dots,M}^{M+1,\dots,M+N}(\lambda - \mu)\, \mathcal{T}_{1,\dots,M}(\lambda)\mathcal{T}_{M+1,\dots,M+N}(\mu)
$$

$$
= \tilde{R}^1(\lambda - \mu)\dots\tilde{R}^N(\lambda - \mu)\, P_{1,\dots,M}^-\, \mathcal{T}_1(\lambda)\dots\mathcal{T}_M(\lambda - (M-1)\eta)
$$
$$
\times P_{M+1,\dots,M+N}^-\, \mathcal{T}_{M+1}(\mu)\dots\mathcal{T}_{M+N}(\mu + (N-1)\eta)
$$

$$
= \tilde{R}^1(\lambda - \mu)\dots\tilde{R}^N(\lambda - \mu)\, \mathcal{T}_M(\lambda - (M-1)\eta)\dots\mathcal{T}_1(\lambda)
$$
$$
\times \mathcal{T}_{M+N}(\mu + (N-1)\eta)\dots\mathcal{T}_{M+1}(\mu)\, P_{1,\dots,M}^-\, P_{M+1,\dots,M+N}^-
$$

$$
= \mathcal{T}_{M+N}(\mu + (N-1)\eta)\dots\mathcal{T}_{M+1}(\mu)\mathcal{T}_M(\lambda - (M-1)\eta)\dots\mathcal{T}_1(\lambda)
$$
$$
\times \mathcal{R}_{1,\dots,M}^{M+1,\dots,M+N}(\lambda - \mu)
$$

$$
= \mathcal{T}_{M+1,\dots,M+N}(\mu)\mathcal{T}_{1,\dots,M}(\lambda)\, \mathcal{R}_{1,\dots,M}^{M+1,\dots,M+N}(\lambda - \mu), \tag{14.A.32}
$$

and the proof of the lemma is complete. We used (14.A.4a) in the first equation, lemma 22 in the second equation, (14.A.31) in the third equation, and lemma 23 and lemma 22 in the last equation. □

We now proceed with the proof of theorem 2: $P_{1,\dots,M}^-$ and $P_{M+1,\dots,M+N}^-$ project on proper subspaces of $(\mathbb{C}^d)^{\otimes(M+N)}$. Therefore $\mathcal{R}_{1,\dots,M}^{M+1,\dots,M+N}$ has no inverse. However, $\mathcal{R}_{1,\dots,M}^{M+1,\dots,M+N}$ leaves the subspace

$$
\mathcal{H} = P_{1,\dots,M}^-\, P_{M+1,\dots,M+N}^-\, (\mathbb{C}^d)^{\otimes(M+N)} \subset (\mathbb{C}^d)^{\otimes(M+N)} \tag{14.A.33}
$$

invariant by lemma 23. The same is true for $\mathcal{T}_{1,\dots,M}$ and $\mathcal{T}_{M+1,\dots,M+N}$ by definition. Now $\left(P_{1,\dots,M}^-\, P_{M+1,\dots,M+N}^-\right)\big|_{\mathcal{H}} = \mathrm{id}\big|_{\mathcal{H}}$, and $\tilde{R}^\ell$ is invertible for $\ell = 1, \dots, N$. It follows that $\mathcal{R}_{1,\dots,M}^{M+1,\dots,M+N}\big|_{\mathcal{H}}$ is invertible and, by lemma 24, that

$$
\mathcal{R}_{1,\dots,M}^{M+1,\dots,M+N}\big|_{\mathcal{H}}(\lambda - \mu)\, \mathcal{T}_{1,\dots,M}\big|_{\mathcal{H}}(\lambda)\mathcal{T}_{M+1,\dots,M+N}\big|_{\mathcal{H}}(\mu)\left(\mathcal{R}_{1,\dots,M}^{M+1,\dots,M+N}\big|_{\mathcal{H}}\right)^{-1}(\lambda - \mu)
$$
$$
= \mathcal{T}_{M+1,\dots,M+N}\big|_{\mathcal{H}}(\mu)\mathcal{T}_{1,\dots,M}\big|_{\mathcal{H}}(\lambda). \tag{14.A.34}
$$

Therefore

$$
\mathrm{tr}_{\mathcal{H}}\left(\mathcal{T}_{1,\dots,M}\big|_{\mathcal{H}}(\lambda)\mathcal{T}_{M+1,\dots,M+N}\big|_{\mathcal{H}}(\mu)\right) = \mathrm{tr}_{\mathcal{H}}\left(\mathcal{T}_{M+1,\dots,M+N}\big|_{\mathcal{H}}(\mu)\mathcal{T}_{1,\dots,M}\big|_{\mathcal{H}}(\lambda)\right), \tag{14.A.35}
$$

where $\mathrm{tr}_{\mathcal{H}}$ denotes the trace on $\mathcal{H}$.

Let $\mathcal{H}^\perp$ be the orthogonal complement of $\mathcal{H}$. Then $\mathcal{H} \oplus \mathcal{H}^\perp = (\mathbb{C}^d)^{\otimes(M+N)}$ and there are orthonormal bases $\mathcal{B}$ and $\mathcal{B}^\perp$ of $\mathcal{H}$ and $\mathcal{H}^\perp$, such that $\mathcal{B} \cup \mathcal{B}^\perp$ is a basis of $(\mathbb{C}^d)^{\otimes(M+N)}$. Using lemma 22 we obtain

$$
\mathcal{T}_{1,\dots,M}(\lambda)\mathcal{T}_{M+1,\dots,M+N}(\mu)|b^\perp\rangle = \mathcal{T}_{M+1,\dots,M+N}(\mu)\mathcal{T}_{1,\dots,M}(\lambda)|b^\perp\rangle = 0 \tag{14.A.36}
$$

for all $|b^\perp\rangle \in \mathcal{B}^\perp$. We conclude that

$$
\begin{aligned}
\mathrm{tr}\big(&\mathcal{T}_{1,\dots,M}(\lambda)\mathcal{T}_{M+1,\dots,M+N}(\mu)\big) \\
&= \sum_{|b\rangle \in \mathcal{B}} \langle b|\mathcal{T}_{1,\dots,M}(\lambda)\mathcal{T}_{M+1,\dots,M+N}(\mu)|b\rangle \\
&\quad + \sum_{|b^\perp\rangle \in \mathcal{B}^\perp} \langle b^\perp|\mathcal{T}_{1,\dots,M}(\lambda)\mathcal{T}_{M+1,\dots,M+N}(\mu)|b^\perp\rangle \\
&= \mathrm{tr}_{\mathcal{H}}\big(\mathcal{T}_{1,\dots,M}\big|_{\mathcal{H}}(\lambda)\mathcal{T}_{M+1,\dots,M+N}\big|_{\mathcal{H}}(\mu)\big) \\
&= \mathrm{tr}\big(\mathcal{T}_{M+1,\dots,M+N}(\mu)\mathcal{T}_{1,\dots,M}(\lambda)\big).
\end{aligned}
\tag{14.A.37}
$$

Hence,

$$
\big[\mathrm{tr}\big(\mathcal{T}_{1,\dots,M}(\lambda)\big), \mathrm{tr}\big(\mathcal{T}_{1,\dots,N}(\mu)\big)\big] = 0,
\tag{14.A.38}
$$

which completes the proof of the theorem.

14.A.3 The quantum determinant

In this section we work out the properties of the 'highest' quantum symmetric function $\sigma_d(\lambda)$. We have already seen that this function turns into the determinant of the classical monodromy matrix in the classical limit $\eta \to 0$. We shall see below that even for non-zero η the function $\sigma_d(\lambda)$ has all the characteristic properties of the usual determinant. For this reason we call

$$
\det_q(T(\lambda)) = \sigma_d(\lambda - (d-1)\eta)
\tag{14.A.39}
$$

the quantum determinant of the monodromy matrix $T(\lambda)$. The shift in the spectral parameter by $(d-1)\eta$ has been introduced for later notational convenience.

The quantum determinant was first introduced in [213], where it was used to invert the monodromy matrix. Its most remarkable property is its commutativity with all the elements of the monodromy matrix. In other words, the quantum determinant is in the centre of the Yang-Baxter algebra. The reason why we discuss the quantum determinant here is that it also generates the centre of the Yangian quantum group $Y(gl(n))$ which (for $n = 2$) will be introduced in the following section.

Let us mention two alternative expressions for the quantum determinant of $T(\lambda)$. Using the totally antisymmetric tensor

$$
\varepsilon^{b_1,\dots,b_d}_{a_1,\dots,a_d} =
\begin{cases}
\mathrm{sign}(Q) & \text{if } \mathbf{a} = \mathbf{b}Q,\ Q \in \mathfrak{S}^d \text{ and } a_j \neq a_k \text{ for all } j \neq k \\
0 & \text{else}
\end{cases}
\tag{14.A.40}
$$

we can write

$$
\det_q(T(\lambda)) = \frac{1}{d!}\, \varepsilon^{b_1,\dots,b_d}_{a_1,\dots,a_d}\, T^{a_1}_{b_1}(\lambda - (d-1)\eta)\dots T^{a_d}_{b_d}(\lambda).
\tag{14.A.41}
$$

Another expression comes from the observation that $P^-_{1,...,d}$ projects on the one-dimensional subspace of $(\mathbb{C}^d)^{\otimes d}$ spanned by the vector

$$|-\rangle = \frac{1}{\sqrt{d!}} \sum_{P \in \mathfrak{S}^d} \text{sign}(P)\, \mathbf{e}_{P(1)} \otimes \cdots \otimes \mathbf{e}_{P(d)}. \tag{14.A.42}$$

This vector is of unit length, $\langle -|-\rangle = 1$. Hence, $P^-_{1,...,d} = |-\rangle\langle -|$ and therefore

$$\text{det}_q(T(\lambda)) = \langle -|T_1(\lambda - (d-1)\eta) \dots T_d(\lambda)|-\rangle. \tag{14.A.43}$$

Before coming to the general properties of the quantum determinant we would like to consider the important example of the fundamental representation of the Yang-Baxter algebra generated by (14.A.1). Recall that the corresponding fundamental L-matrix (see Section 12.1) has matrix elements $L^\alpha_\beta(\lambda) = R^{\alpha\gamma}_{\beta\delta}(\lambda)e^\delta_\gamma$.

Lemma 25. *The quantum determinant of the fundamental L-matrix associated with the rational R-matrix (14.A.1) is*

$$\text{det}_q(L(\lambda)) = I_d\,(\lambda + \eta) \prod_{n=1}^{d-1}(\lambda - n\eta). \tag{14.A.44}$$

Proof. [1] We are free to interpret the matrices $R_{n,d+1}(\lambda)$, $n = 1, \dots, d$, as matrices in the tensor product $(\mathbb{C}^d)^{\otimes d}$ with entries acting on a quantum space $\mathbb{C}^d$ carrying the label $d + 1$. With this interpretation in mind we find

$$L(\lambda - (d-1)\eta) \otimes \cdots \otimes L(\lambda) = R_{1,d+1}(\lambda - (d-1)\eta) \dots R_{d,d+1}(\lambda)$$

$$= (\lambda - (d-1)\eta + \eta P_{1,d+1}) \dots (\lambda + \eta P_{d,d+1})$$

$$= p(\lambda) + \sum_{j=1}^{d} p_j(\lambda)\, P_{j,d+1} + \sum_{\substack{j,k=1 \\ j<k}}^{d} p_{jk}(\lambda)\, P_{j,d+1} P_{k,d+1}$$

$$+ \cdots + p_{1,...,d}(\lambda)\, P_{1,d+1} \dots P_{d,d+1}, \tag{14.A.45}$$

where the coefficients in the sums over products of transposition operators on the right-hand side are polynomials in λ, e.g.

$$p(\lambda) = \prod_{n=0}^{d-1}(\lambda - n\eta). \tag{14.A.46}$$

We have to multiply equation (14.A.45) by $P^-_{1,...,d}$ from the left and have to take the trace with respect to the auxiliary spaces. Then the right-hand side of (14.A.45) simplifies thanks

---

[1] FG is indebted to Y. Komori for the communication of this proof.

to the identities

$$P^-_{1,\ldots,d} P_{j,k} = -P^-_{1,\ldots,d}, \quad j,k = 1,\ldots,d, \quad j \neq k,$$

$$\mathrm{tr}_{1,\ldots,d}\left(P^-_{1,\ldots,d}\right) = I_d, \qquad (14.\mathrm{A}.47)$$

$$\mathrm{tr}_{1,\ldots,d}\left(P^-_{1,\ldots,d} P_{j,d+1}\right) = \frac{I_d}{d}, \quad j = 1,\ldots,d,$$

where I_d is the $d \times d$ unit matrix. Using $P_{j_1,d+1} \cdots P_{j_n,d+1} = P_{j_2,j_1} \cdots P_{j_n,j_1} P_{j_1,d+1}$ and the identities (14.A.47) we obtain

$$\mathrm{tr}_{1,\ldots,d}\left(P^-_{1,\ldots,d} L(\lambda - (d-1)\eta) \otimes \cdots \otimes L(\lambda)\right)$$

$$= I_d\, p(\lambda) - \frac{I_d}{d}\left[-\sum_{j=1}^d p_j(\lambda) + \sum_{\substack{j,k=1 \\ j<k}}^d p_{jk}(\lambda) + \cdots + (-1)^d p_{1,\ldots,d}(\lambda)\right]$$

$$= I_d\left[p(\lambda) + \frac{1}{d}p(\lambda) - \frac{1}{d}p(\lambda - \eta)\right] = I_d(\lambda + \eta)\prod_{n=1}^{d-1}(\lambda - n\eta). \qquad (14.\mathrm{A}.48)$$

Note that the term in brackets on the right hand side of the first equation is up to a missing term $p(\lambda)$ equal to the right hand side of (14.A.45) with all the transposition operators replaced by minus one. This fact was used in the second equation.   □

Now we turn to a description of the properties of the quantum determinant in the general case.

Theorem 2. *Properties of the quantum determinant [277].*

(i) Invariance under similarity transformations. Let $A \in \mathrm{End}(\mathbb{C}^d)$, A invertible. Then

$$\det_q(AT(\lambda)A^{-1}) = \det_q(T(\lambda)). \qquad (14.\mathrm{A}.49)$$

(ii) Multiplicativity. For two commutative representations $T(\lambda|1)$, $T(\lambda|2)$, $[T_1(\lambda|1), T_2(\mu|2)] = 0$, the multiplication formula

$$\det_q(T(\lambda|1)T(\lambda|2)) = \det_q(T(\lambda|1))\det_q(T(\lambda|2)) \qquad (14.\mathrm{A}.50)$$

is valid.

(iii) Kramer's rule.

$$T^{-1}(\lambda) = \frac{d}{\det_q(T(\lambda))}\, \mathrm{tr}_{1,\ldots,d-1}\left(P^-_{1,\ldots,d} T_1(\lambda - (d-1)\eta)\ldots T_{d-1}(\lambda - \eta)\right).$$

$$(14.\mathrm{A}.51)$$

(iv) Commutativity.

$$[\det_q(T(\lambda)), T(\mu)] = 0 \qquad (14.\mathrm{A}.52)$$

for all $\lambda, \mu \in \mathbb{C}$.

Proof. (i) is left as an exercise to the reader.

(ii) is obtained as

$$\det{}_q(T(\lambda|1)T(\lambda|2))$$

$$= \langle -|T_1(\lambda - (d-1)\eta|1) \ldots T_d(\lambda|1)T_1(\lambda - (d-1)\eta|2) \ldots T_d(\lambda|2)|-\rangle$$

$$= \langle -|T_1(\lambda - (d-1)\eta|1) \ldots T_d(\lambda|1)|-\rangle\langle -|T_1(\lambda - (d-1)\eta|2) \ldots T_d(\lambda|2)|-\rangle$$

$$= \det{}_q(T(\lambda|1))\det{}_q(T(\lambda|2)). \tag{14.A.53}$$

Here we used the form (14.A.43) of the quantum determinant and the commutativity of the two representations in the first equation and the fact that $P^-_{1,\ldots,d} = |-\rangle\langle -|$ and lemma 22 in the second equation.

(iii) Using lemma 22 and the representation (14.A.43) of the quantum determinant we obtain

$$T_{1,\ldots,d}(\lambda - (d-1)\eta) = P^-_{1,\ldots,d}\, T_1(\lambda - (d-1)\eta) \ldots T_d(\lambda)$$

$$= |-\rangle\langle -|T_1(\lambda - (d-1)\eta) \ldots T_d(\lambda)|-\rangle\langle -|$$

$$= \det{}_q(T(\lambda))\, P^-_{1,\ldots,d} = P^-_{1,\ldots,d}\det{}_q(T(\lambda)). \tag{14.A.54}$$

Multiplying by $d\,(\det_q(T(\lambda)))^{-1}$ from the left and by $T_d^{-1}(\lambda)$ from the right and taking the trace with respect to the first $d-1$ auxiliary spaces we conclude that

$$\frac{d}{\det_q(T(\lambda))}\,\mathrm{tr}_{1,\ldots,d-1}\big(P^-_{1,\ldots,d}T_1(\lambda - (d-1)\eta) \ldots T_{d-1}(\lambda - \eta)\big)$$

$$= d\,\mathrm{tr}_{1,\ldots,d-1}\big(P^-_{1,\ldots,d}T_d^{-1}(\lambda)\big)$$

$$= \frac{1}{(d-1)!}\,\varepsilon^{a_1,\ldots,a_{d-1},b}_{a_1,\ldots,a_{d-1},a}\big(T^{-1}\big)^a_c(\lambda)\,e^c_b = T^{-1}(\lambda). \tag{14.A.55}$$

(iv) The assertion follows from lemma 24 with $M = d$ and $N = 1$. In order to see this we first note that $\mathcal{R}^{d+1}_{1,\ldots,d}(\lambda) = \tilde{R}^1(\lambda)P^-_{1,\ldots,d}$ (compare equation (14.A.19)). Using (14.A.54) and lemma 22 and lemma 24 we obtain

$$\mathcal{R}^{d+1}_{1,\ldots,d}(\lambda - \mu)T_{1,\ldots,d}(\lambda)T_{d+1}(\mu)$$

$$= \tilde{R}^1(\lambda - \mu)P^-_{1,\ldots,d}\,\det{}_q(T(\lambda - (d-1)\eta))T_{d+1}(\mu)$$

$$= R_{d,d+1}(\lambda - \mu + (d-1)\eta) \ldots R_{1,d+1}(\lambda - \mu)P^-_{1,\ldots,d}\det{}_q(T(\lambda + (d-1)\eta))T_{d+1}(\mu)$$

$$= P^-_{1,\ldots,d}R_{1,d+1}(\lambda - \mu) \ldots R_{d,d+1}(\lambda - \mu + (d-1)\eta)\det{}_q(T(\lambda + (d-1)\eta))T_{d+1}(\mu)$$

$$= T_{d+1}(\mu)\det{}_q(T(\lambda - (d-1)\eta))P^-_{1,\ldots,d}\tilde{R}^1(\lambda - \mu)P^-_{1,\ldots,d}$$

$$= T_{d+1}(\mu)\det{}_q(T(\lambda - (d-1)\eta))$$

$$\times P^-_{1,\ldots,d}R_{1,d+1}(\lambda - \mu) \ldots R_{d,d+1}(\lambda - \mu + (d-1)\eta). \tag{14.A.56}$$

Then we take the trace with respect to the first d auxiliary spaces of the fourth and the sixth line of this equation, making use of the formulae

$$\mathrm{tr}_{1,\dots,d}(AB_{d+1}) = \mathrm{tr}_{1,\dots,d}(A)\, B\,,$$
$$\mathrm{tr}_{1,\dots,d}(B_{d+1}A) = B\, \mathrm{tr}_{1,\dots,d}(A)\,, \tag{14.A.57}$$

which holds for any $A \in \mathrm{End}\big((\mathbb{C}^d)^{\otimes(d+1)}\big)$, $B \in \mathrm{End}(\mathbb{C}^d)$. The resulting equation is

$$\mathrm{det}_q(L(\lambda - \mu + (d-1)\eta))\,\mathrm{det}_q(T(\lambda - (d-1)\eta))T(\mu)$$
$$= T(\mu)\,\mathrm{det}_q(T(\lambda - (d-1)\eta))\,\mathrm{det}_q(L(\lambda - \mu + (d-1)\eta))\,, \tag{14.A.58}$$

where $\mathrm{det}_q(L(\lambda))$ is the quantum determinant of the fundamental L-matrix associated with the R-matrix (14.A.1). According to lemma 6 this quantum determinant is proportional to the unit matrix I_d, which completes the proof of (iv). $\qquad\square$

In the next section, where we introduce the Yangians $Y(\mathrm{gl}(2))$ and $Y(\mathrm{sl}(2))$, we shall need the quantum determinant for the case $d = 2$. For this case we would like to derive four more alternative expressions for the quantum determinant. Starting point of our derivation is lemma 3 with $N = d = 2$. Multiplying (14.A.14d) by $P_{1,2}^-$ and (14.A.14b) by $P_{1,2}^+$ from the left and using (14.A.4b) we obtain

$$P_{1,2}^-\big(T(\lambda - \eta) \otimes T(\lambda)\big)P_{1,2}^+ = 0\,, \tag{14.A.59a}$$
$$P_{1,2}^+\big(T(\lambda) \otimes T(\lambda - \eta)\big)P_{1,2}^- = 0\,. \tag{14.A.59b}$$

These are two 4×4 matrix equations. Picking out the inner 2×2 block of these equations, i.e., the intersections of the second and third rows and columns, we find, in particular,

$$\begin{pmatrix} 1 & -1 \\ -1 & 1 \end{pmatrix} \begin{pmatrix} A(\lambda - \eta)D(\lambda) & B(\lambda - \eta)C(\lambda) \\ C(\lambda - \eta)B(\lambda) & D(\lambda - \eta)A(\lambda) \end{pmatrix} \begin{pmatrix} 1 & 1 \\ 1 & 1 \end{pmatrix} = 0\,, \tag{14.A.60a}$$

$$\begin{pmatrix} 1 & 1 \\ 1 & 1 \end{pmatrix} \begin{pmatrix} A(\lambda)D(\lambda - \eta) & B(\lambda)C(\lambda - \eta) \\ C(\lambda)B(\lambda - \eta) & D(\lambda)A(\lambda - \eta) \end{pmatrix} \begin{pmatrix} 1 & -1 \\ -1 & 1 \end{pmatrix} = 0\,, \tag{14.A.60b}$$

where the four matrix elements of the monodromy matrix have been denoted $A(\lambda), \dots, D(\lambda)$. Each of these matrix equations is equivalent to a single equation for the elements of the monodromy matrix,

$$A(\lambda - \eta)D(\lambda) - C(\lambda - \eta)B(\lambda) + B(\lambda - \eta)C(\lambda) - D(\lambda - \eta)A(\lambda) = 0\,, \tag{14.A.61a}$$
$$A(\lambda)D(\lambda - \eta) + C(\lambda)B(\lambda - \eta) - B(\lambda)C(\lambda - \eta) - D(\lambda)A(\lambda - \eta) = 0\,. \tag{14.A.61b}$$

Using (14.A.61a) we infer that

$$A(\lambda - \eta)D(\lambda) - C(\lambda - \eta)B(\lambda)$$

$$= D(\lambda - \eta)A(\lambda) - B(\lambda - \eta)C(\lambda)$$

$$= \tfrac{1}{2}\big[A(\lambda - \eta)D(\lambda) - C(\lambda - \eta)B(\lambda) + D(\lambda - \eta)A(\lambda) - B(\lambda - \eta)C(\lambda)\big]$$

$$= \mathrm{tr}\big(P_{1,2}^-\big(T(\lambda - \eta) \otimes T(\lambda)\big)\big) = \det_q(T(\lambda)). \qquad (14.A.62)$$

Equation (14.A.61b), on the other hand, implies

$$A(\lambda)D(\lambda - \eta) - B(\lambda)C(\lambda - \eta)$$

$$= D(\lambda)A(\lambda - \eta) - C(\lambda)B(\lambda - \eta)$$

$$= \tfrac{1}{2}\big[A(\lambda)D(\lambda - \eta) - B(\lambda)C(\lambda - \eta) + D(\lambda)A(\lambda - \eta) - C(\lambda)B(\lambda - \eta)\big]$$

$$= \mathrm{tr}\big(\big(T(\lambda) \otimes T(\lambda - \eta)\big)P_{1,2}^-\big) = \det_q(T(\lambda)). \qquad (14.A.63)$$

We summarize the various expressions we have found for the quantum determinant of the gl(2) invariant R-matrix ((14.A.1) with $d = 2$) in the following lemma.

Lemma 26. *Equivalent forms of the gl(2) quantum determinant [407].*

$$\det_q(T(\lambda)) = \mathrm{tr}\big(P_{1,2}^-\big(T(\lambda - \eta) \otimes T(\lambda)\big)\big)$$

$$= \mathrm{tr}\big(\big(T(\lambda) \otimes T(\lambda - \eta)\big)P_{1,2}^-\big)$$

$$= A(\lambda - \eta)D(\lambda) - C(\lambda - \eta)B(\lambda)$$

$$= D(\lambda - \eta)A(\lambda) - B(\lambda - \eta)C(\lambda)$$

$$= A(\lambda)D(\lambda - \eta) - B(\lambda)C(\lambda - \eta)$$

$$= D(\lambda)A(\lambda - \eta) - C(\lambda)B(\lambda - \eta). \qquad (14.A.64)$$

Let us illustrate some of our results again with our chief example, the inhomogeneous spin-$\tfrac{1}{2}$-XXX chain (see Sections 12.1.6 and 12.1.7). The quantum determinant of the elementary L-matrix, (12.52) with $\alpha, \beta = 1, 2$, follows from lemma 6 by a change in notation and normalization,

$$\det_q(L_j(\lambda, v_j)) = \frac{\lambda - v_j - ic}{\lambda - v_j}. \qquad (14.A.65)$$

Here we suppressed the identity operator in quantum space on the right-hand side of the equation. The quantum determinant of the monodromy matrix $T(\lambda) = L_L(\lambda, v_L) \ldots L_1(\lambda, v_1)$ then follows by multiplicativity, equation (14.A.50),

$$\det_q(T(\lambda)) = \prod_{j=1}^{L} \frac{\lambda - v_j - ic}{\lambda - v_j}. \qquad (14.A.66)$$

Thus, for the fundamental representation the quantum determinant is just a number (multiplied by the identity operator) and generates no additional conserved quantities. We wish to emphasize, however, that the situation may be different for general non-fundamental representations, where the quantum determinant may be a non-trivial operator.

It follows from (14.A.66) and (14.A.51) that the monodromy matrix $T(\lambda)$ of the inhomogeneous spin-$\frac{1}{2}$-XXX chain is invertible for $\lambda \neq \nu_j, \nu_j + ic, j = 1, \ldots, L$, and that

$$T^{-1}(\lambda) = \prod_{j=1}^{L} \frac{\lambda - \nu_j}{\lambda - \nu_j - ic} \begin{pmatrix} D(\lambda - ic) & -B(\lambda - ic) \\ -C(\lambda - ic) & A(\lambda - ic) \end{pmatrix}. \tag{14.A.67}$$

14.A.4 The Yangians Y(gl(2)) and Y(sl(2))

Yangians associated with any simple Lie algebra $\mathfrak{a}$ were introduced by Drinfeld in his seminal article [107] which marked the advent of quantum groups. Of the different possible characterizations of the Yangians [107, 108, 247] (see also Chapter 12 of [82]) in terms of generators we shall choose the least general one [247] as a starting point of our account. It has the advantage of being closely related to concepts we are already familiar with, the Yang-Baxter algebra and the quantum determinant. This is the favoured approach to Yangians in the physics literature (e.g. [58, 195, 337]). For simple Lie algebras it is restricted to $\mathfrak{a} = \mathrm{sl}(n)$, but an extention to $\mathfrak{a} = \mathrm{gl}(n)$ is possible. We shall be rather modest and consider only the Yangians associated with sl(2) and gl(2) which will be enough for our purposes. Our presentation in this section closely follows the appendix of [337].

The Yangian Y(gl(2)) is the associative algebra (with unit e) generated by the entries of the matrix coefficients T_n in the formal asymptotic series

$$T(\lambda) = I_2 e + \eta \sum_{n=1}^{\infty} \frac{T_n}{\lambda^n} \tag{14.A.68}$$

modulo the constraints imposed by the defining relations

$$\check{R}(\lambda - \mu)\big(T(\lambda) \otimes T(\mu)\big) = \big(T(\mu) \otimes T(\lambda)\big)\check{R}(\lambda - \mu) \tag{14.A.69}$$

of the Yang-Baxter algebra with rational R-matrix $\check{R}(\lambda) = \eta + \lambda P$.

Inserting (14.A.68) into (14.A.69) and comparing order by order in λ and μ we obtain two infinite sets of relations for the T_m:

$$T_1 \otimes T_n - P(T_n \otimes T_1)P = P(T_n \otimes I_2 - I_2 \otimes T_n), \tag{14.A.70a}$$

$$T_m \otimes T_{n+1} - P(T_{n+1} \otimes T_m)P - T_{m+1} \otimes T_n + P(T_n \otimes T_{m+1})P$$
$$= \eta P(T_m \otimes T_n - T_n \otimes T_m) \tag{14.A.70b}$$

for $m, n \in \mathbb{N}$.

We will get deeper insight into the structure of the Yangian by expanding the T_n in a gl(2) basis consisting of the matrices I_2, σ^x, σ^y and σ^z,

$$T_n = J_{n-1}^0 I_2 + J_{n-1}^\alpha \sigma^\alpha, \quad n \in \mathbb{N}. \tag{14.A.71}$$

The set $\{I_2, \sigma^x, \sigma^y, \sigma^z\}$ is a gl(2) orthonormal basis with respect to the scalar product $\langle X, Y \rangle = \frac{1}{2}\mathrm{tr}(XY)$. Thus, $J_{n-1}^0 = \frac{1}{2}\mathrm{tr}(T_n)$ and $J_{n-1}^\alpha = \frac{1}{2}\mathrm{tr}(\sigma^\alpha T_n)$. The commutators between the J_n^0, J_n^α are obtained from (14.A.70) by use of the formulae

$$P = \tfrac{1}{2}(I_2 \otimes I_2 + \sigma^\alpha \otimes \sigma^\alpha), \tag{14.A.72a}$$

$$(\sigma^\alpha \otimes I_2)P = \tfrac{1}{2}(\sigma^\alpha \otimes I_2 + I_2 \otimes \sigma^\alpha - i\varepsilon^{\alpha\beta\gamma}\sigma^\beta \otimes \sigma^\gamma), \tag{14.A.72b}$$

$$(I_2 \otimes \sigma^\alpha)P = \tfrac{1}{2}(\sigma^\alpha \otimes I_2 + I_2 \otimes \sigma^\alpha + i\varepsilon^{\alpha\beta\gamma}\sigma^\beta \otimes \sigma^\gamma), \tag{14.A.72c}$$

$$(\sigma^\alpha \otimes \sigma^\beta)P = \tfrac{1}{2}\big(\sigma^\alpha \otimes \sigma^\beta + \sigma^\beta \otimes \sigma^\alpha + \delta^{\alpha\beta}(I_2 \otimes I_2 - \sigma^\gamma \otimes \sigma^\gamma)$$
$$+ i\varepsilon^{\alpha\beta\gamma}(\sigma^\gamma \otimes I_2 - I_2 \otimes \sigma^\gamma)\big), \tag{14.A.72d}$$

$$(\sigma^\alpha \otimes \sigma^\alpha)P = \tfrac{1}{2}(3\,I_2 \otimes I_2 - \sigma^\alpha \otimes \sigma^\alpha). \tag{14.A.72e}$$

To obtain an example of how this works take, for instance, the trace of equation (14.A.70a). Then, because of the invariance of the trace under cyclic exchange of matrices with commuting entries, the left-hand side of the equation turns into $[J_0^0, J_{n-1}^0]$. The right-hand side can be calculated by means of (14.A.72a). It vanishes since $\mathrm{tr}(I_2) = 2$ and $\mathrm{tr}(\sigma^\alpha) = 0$.

Proceeding in this spirit we arrive after a little algebra at the following set of equations,

$$[J_0^\alpha, J_n^0] = 0, \tag{14.A.73a}$$

$$[J_0^\alpha, J_n^\beta] = i\varepsilon^{\alpha\beta\gamma} J_n^\gamma, \tag{14.A.73b}$$

$$[J_m^0, J_n^0] = 0, \tag{14.A.73c}$$

$$[J_m^0, J_n^\alpha] - [J_n^0, J_m^\alpha] = 0, \tag{14.A.73d}$$

$$[J_m^0, J_{n+1}^\alpha] - [J_{m+1}^0, J_n^\alpha] = \tfrac{in}{2}\varepsilon^{\alpha\beta\gamma}(J_m^\beta J_n^\gamma - J_n^\beta J_m^\gamma), \tag{14.A.73e}$$

$$[J_m^\alpha, J_n^\beta] - [J_n^\alpha, J_m^\beta] = 0, \tag{14.A.73f}$$

$$[J_m^\alpha, J_{n+1}^\beta] - [J_{m+1}^\alpha, J_n^\beta] = i\eta\varepsilon^{\alpha\beta\gamma}(J_m^\gamma J_n^0 - J_n^\gamma J_m^0). \tag{14.A.73g}$$

These equations are equivalent to (14.A.70). They provide a characterization of the Yangian $Y(\mathrm{gl}(2))$ in terms of the generators J_n^0 and J_n^α.

It turns out that the generators are not all independent. (14.A.73g) can be used to derive a recursion relation expressing J_n^α in terms of J_m^0, J_m^β, $m \le n - 1$. For this purpose we first set $m = 0$ and replace n by $n - 1$ in (14.A.73g). Then we use (14.A.73b) on the left-hand side and multiply by $\varepsilon^{\delta\alpha\beta}$. The resulting equation is equivalent to the recursion relation

$$J_n^\alpha = \tfrac{1}{2i}\varepsilon^{\alpha\beta\gamma}[J_1^\beta, J_{n-1}^\gamma] + \eta(J_0^\alpha J_{n-1}^0 - J_{n-1}^\alpha J_0^0) \tag{14.A.74}$$

which determines J_n^α for $n = 2, 3, \ldots$

To get a complete recursive description of the algebra, which includes the J_n^0, we consider the asymptotic expansion of the quantum determinant,

$$\det_q(T(\lambda)) = A(\lambda)D(\lambda - \eta) - B(\lambda)C(\lambda - \eta) \tag{14.A.75a}$$

$$= D(\lambda)A(\lambda - \eta) - C(\lambda)B(\lambda - \eta) \tag{14.A.75b}$$

$$= 1 + \eta \sum_{n=0}^{\infty} \frac{a_n}{\lambda^{n+1}}. \tag{14.A.75c}$$

This equation defines the coefficients a_n. It enables us to express the a_n in terms of the J_n^0, J_n^α by inserting the formal asymptotic series (14.A.68) into (14.A.75a) and (14.A.75b).

From the comparison of (14.A.75c) and (14.A.75a) we obtain $a_0 = 2J_0^0$ and

$$a_n = T_{n+1\,1}^{\ \ 1} + \sum_{m=0}^{n} \binom{n}{m} \eta^{n-m} T_{m+1\,2}^{\ \ 2}$$

$$+ \sum_{l=0}^{n-1} \sum_{m=0}^{n-l-1} \binom{n-l-1}{m} \eta^{n-l-m} \left(T_{l+1\,1}^{\ \ 1} T_{m+1\,2}^{\ \ 2} - T_{l+1\,2}^{\ \ 1} T_{m+1\,1}^{\ \ 2} \right) \quad (14.A.76)$$

for $n \in \mathbb{N}$. If we use (14.A.75b) instead of (14.A.75a), then $T_{n\,1}^{\ 1}$ and $T_{n\,2}^{\ 2}$ are interchanged and so are $T_{n\,2}^{\ 1}$ and $T_{n\,1}^{\ 2}$. It follows that

$$a_n = \tfrac{1}{2}(T_{n+1\,1}^{\ \ 1} + T_{n+1\,2}^{\ \ 2}) + \sum_{m=0}^{n} \binom{n}{m} \eta^{n-m} \tfrac{1}{2}(T_{m+1\,1}^{\ \ 1} + T_{m+1\,2}^{\ \ 2})$$

$$+ \sum_{l=0}^{n-1} \sum_{m=0}^{n-l-1} \binom{n-l-1}{m} \eta^{n-l-m}$$

$$\times \tfrac{1}{2}\left(T_{l+1\,1}^{\ \ 1} T_{m+1\,2}^{\ \ 2} + T_{l+1\,2}^{\ \ 2} T_{m+1\,1}^{\ \ 1} - T_{l+1\,2}^{\ \ 1} T_{m+1\,1}^{\ \ 2} - T_{l+1\,1}^{\ \ 2} T_{m+1\,2}^{\ \ 1} \right). \quad (14.A.77)$$

Now the right-hand side is easily expressed in terms of J_n^0 and J_n^α, and we end up with

$$a_n = 2J_n^0 + \sum_{m=0}^{n-1} \binom{n}{m} \eta^{n-m} J_m^0 + \sum_{l=0}^{n-1} \sum_{m=0}^{n-l-1} \binom{n-l-1}{m} \eta^{n-l-m} \left(J_m^0 J_l^0 - J_m^\alpha J_l^\alpha \right).$$

$$(14.A.78)$$

This formula determines the coefficients of the asymptotic expansion of the quantum determinant in terms of J_n^0, J_n^α. Turning it around we obtain

$$J_n^0 = \tfrac{1}{2} a_n - \tfrac{1}{2} \sum_{m=0}^{n-1} \binom{n}{m} \eta^{n-m} J_m^0 - \tfrac{1}{2} \sum_{l=0}^{n-1} \sum_{m=0}^{n-l-1} \binom{n-l-1}{m} \eta^{n-l-m} \left(J_m^0 J_l^0 - J_m^\alpha J_l^\alpha \right)$$

$$(14.A.79)$$

for $n \in \mathbb{N}$. Using $J_0^0 = \frac{a_0}{2}$ and equations (14.A.74) and (14.A.79) we can recursively express J_n^0 and J_n^α in terms of J_0^α, J_1^α and the a_n. In other words, the Yangian Y(gl(2)) is generated by J_0^α, J_1^α, $\alpha = x, y, z$, and by the coefficients of the asymptotic expansion of the quantum determinant which belong to its centre. Setting all the a_n equal to zero we obtain the Yangian Y(sl(2)). Thus, the Yangian Y(sl(2)) is the associative algebra generated by entries of the matrices T_n in (14.A.68) modulo the relations (14.A.69) and the further constraint that $\det_q(T(\lambda)) = 1$. The Yangian Y(sl(2)) has only six independent generators, J_0^α, J_1^α, $\alpha = x, y, z$.

14.A.5 The deformed Serre relation for Y(sl(2)) and Y(gl(2))

We have seen in the preceding section that the Yangian Y(sl(2)) has only a finite number of independent 'elementary' generators: given J_0^α and J_1^β, the infinitely many remaining generators occurring in (14.A.73) are determined by the recursion relations (14.A.74) and (14.A.79). This raises the interesting question of whether we can define the Yangian entirely in terms of its elementary generators J_0^α, J_1^β by imposing a finite set of relations on them. This set must include the sl(2)-relations

$$[J_0^\alpha, J_0^\beta] = i\varepsilon^{\alpha\beta\gamma} J_0^\gamma \,, \tag{14.A.80a}$$

$$[J_0^\alpha, J_1^\beta] = i\varepsilon^{\alpha\beta\gamma} J_1^\gamma \tag{14.A.80b}$$

which are contained in (14.A.73b) and are the only relations in (14.A.73) which relate only J_0^α and J_1^β. It is clear on the other hand, that these two relations do not completely determine the structure of the Yangian, for they merely indicate that the J_0^α generate sl(2) as a subalgebra of Y(sl(2)) and that the J_1^β transform like a vector representation of sl(2).

Starting from (14.A.73) we shall derive an independent higher-order relation between the elementary Yangian generators J_0^α, J_1^β which is called the Yangian Serre relation. The Yangian Serre relation was introduced by Drinfeld [107]. Drinfeld could show that the Yangian Serre relation together with the two equations (14.A.80) uniquely determines Y(sl(2)) as a Hopf algebra (for the Hopf algebra structure of Y(sl(2)) see the next Section 14.A.6).

For our derivation of the Serre relations from (14.A.73) we shall need the explicit expressions for J_2^β, J_0^0 and J_1^0 as obtainable from (14.A.79) and (14.A.74),

$$J_2^\beta = \tfrac{1}{2i}\varepsilon^{\beta\gamma\delta}[J_1^\gamma, J_1^\delta] + \eta(J_0^\beta J_1^0 - J_1^\beta J_0^0), \tag{14.A.81}$$

$$J_0^0 = a_0/2\,, \quad J_1^0 = a_1/2 - \tfrac{\eta}{2}\big(a_0/2 + (a_0/2)^2\big) + \tfrac{\eta}{2}(J_0^\delta)^2\,. \tag{14.A.82}$$

Using (14.A.81) and (14.A.73b) we obtain

$$\varepsilon^{\rho\sigma\alpha}\varepsilon^{\lambda\mu\beta}[J_1^\alpha, J_2^\beta] = \big[[J_1^\lambda, J_1^\mu], [J_0^\rho, J_1^\sigma]\big] + \eta\varepsilon^{\rho\sigma\alpha}\varepsilon^{\lambda\mu\beta}[J_1^\alpha, J_0^\beta J_1^0 - J_1^\beta J_0^0]. \tag{14.A.83}$$

Now $[J_1^\alpha, J_2^\beta] = -[J_1^\beta, J_2^\alpha]$ due to (14.A.73f). Therefore the left hand side of (14.A.83) is antisymmetric under exchange of the ordered pairs of indices $(\lambda\mu)$ and $(\rho\sigma)$. Upon symmetrizing in these index pairs we obtain

$$\big[[J_1^\lambda, J_1^\mu], [J_0^\rho, J_1^\sigma]\big] + \big[[J_1^\rho, J_1^\sigma], [J_0^\lambda, J_1^\mu]\big]$$
$$= -\eta(\varepsilon^{\rho\sigma\alpha}\varepsilon^{\lambda\mu\beta} + \varepsilon^{\lambda\mu\alpha}\varepsilon^{\rho\sigma\beta})[J_1^\alpha, J_0^\beta J_1^0 - J_1^\beta J_0^0]. \tag{14.A.84}$$

This is already the deformed Serre relation for Y(sl(2)). In order to match Drinfeld's original formulation we have to eliminate J_0^0 and J_1^0 by means of (14.A.82).

The commutator on the right hand side of (14.A.84) is

$$[J_1^\alpha, J_0^\beta J_1^0 - J_1^\beta J_0^0] = J_0^\beta[J_1^\alpha, J_1^0] + [J_1^\alpha, J_0^\beta]J_1^0 - J_1^\beta[J_1^\alpha, J_0^0] - [J_1^\alpha, J_1^\beta]J_0^0$$
$$= J_0^\beta[J_1^\alpha, J_1^0] + \cdots = [J_1^\alpha, J_1^0]J_0^\beta + \dots \tag{14.A.85}$$

Here we used (14.A.82) to conclude that $[J_1^\alpha, J_0^0] = 0$. The dots denote terms which are antisymmetric in α and β. The second equation in the second line follows from (14.A.73a) which states, in particular, that $[J_0^\beta, J_1^0] = 0$. Alternatively, it can be derived using (14.A.82). Exploiting the fact that the combination of ε-tensors on the right hand side of (14.A.84) is symmetric in α and β we conclude that

$$\left[[J_1^\lambda, J_1^\mu], [J_0^\rho, J_1^\sigma]\right] + \left[[J_1^\rho, J_1^\sigma], [J_0^\lambda, J_1^\mu]\right]$$

$$= \eta(\varepsilon^{\rho\sigma\alpha}\varepsilon^{\lambda\mu\beta} + \varepsilon^{\lambda\mu\alpha}\varepsilon^{\rho\sigma\beta})\begin{cases} J_0^\beta[J_1^0, J_1^\alpha] \\ [J_1^0, J_1^\alpha]J_0^\beta \\ \frac{1}{2}\{J_0^\beta, [J_1^0, J_1^\alpha]\} \end{cases}$$

$$= -\frac{i\eta^2}{2}(\varepsilon^{\lambda\mu\nu}\varepsilon^{\rho\sigma\alpha} + \varepsilon^{\lambda\mu\alpha}\varepsilon^{\rho\sigma\nu})\varepsilon^{\nu\beta\gamma}\begin{cases} J_0^\alpha J_0^\beta J_1^\gamma + J_0^\alpha J_1^\gamma J_0^\beta \\ J_0^\rho J_1^\gamma J_0^\alpha + J_1^\gamma J_0^\beta J_0^\alpha \\ \frac{1}{2}\{J_0^\alpha, \{J_0^\beta, J_1^\gamma\}\} \end{cases} \qquad (14.A.86)$$

We used (14.A.82) and (14.A.73b) in the second equation. The large curly brackets indicate that we have three alternative expressions. In order to achieve complete symmetrization we use (14.A.73b) to rewrite the third expression on the right-hand side as

$$\frac{1}{2}\{J_0^\alpha, \{J_0^\beta, J_1^\gamma\}\} = \frac{1}{2}\left[J_0^\alpha, [J_0^\beta, J_1^\gamma]\right] + J_0^\alpha J_1^\gamma J_0^\beta + J_0^\beta J_1^\gamma J_0^\alpha$$
$$= \frac{1}{2}\left([J_0^\beta, [J_0^\alpha, J_1^\gamma]] - [J_0^\gamma, [J_0^\alpha, J_1^\beta]]\right) + J_0^\alpha J_1^\gamma J_0^\beta + J_0^\beta J_1^\gamma J_0^\alpha .$$
$$(14.A.87)$$

It follows that

$$\frac{1}{2}\varepsilon^{\nu\beta\gamma}\{J_0^\alpha, \{J_0^\beta, J_1^\gamma\}\} = \varepsilon^{\nu\beta\gamma}\left(J_0^\beta J_0^\alpha J_1^\gamma + J_1^\gamma J_0^\alpha J_0^\beta\right), \qquad (14.A.88)$$

and thus, inserting the latter result back into (14.A.86),

$$\left[[J_1^\lambda, J_1^\mu], [J_0^\rho, J_1^\sigma]\right] + \left[[J_1^\rho, J_1^\sigma], [J_0^\lambda, J_1^\mu]\right]$$
$$= -i\eta^2(\varepsilon^{\lambda\mu\nu}\varepsilon^{\rho\sigma\alpha} + \varepsilon^{\lambda\mu\alpha}\varepsilon^{\rho\sigma\nu})\varepsilon^{\nu\beta\gamma}\{J_0^\alpha, J_0^\beta, J_1^\gamma\}$$
$$= \eta^2(a_{\lambda\mu\nu\alpha\beta\gamma}c_{\rho\sigma\nu} + a_{\rho\sigma\nu\alpha\beta\gamma}c_{\lambda\mu\nu})\{J_0^\alpha, J_0^\beta, J_1^\gamma\}. \qquad (14.A.89)$$

Here we introduced the notation (14.14) for the fully symmetrized triple product $\{\cdot, \cdot, \cdot\}$ and the 'structure constants' $a_{\lambda\mu\nu\alpha\beta\gamma}$ and $c_{\lambda\mu\nu}$ in the last line.

Equation (14.A.89) is the Yangian Serre relation as introduced by Drinfeld [107]. (14.A.80) and (14.A.89) are compatible with a Hopf algebra structure (co-product, co-unit and antipode) which will be derived the following section.

To round off this section let us introduce another, basis independent form of the Yangian Serre relation which is frequently encountered in the literature (see e.g. [80–82]). For this purpose we introduce the notation

$$J_0(x) = \frac{1}{2}\text{tr}(xT_1) = x_\alpha J_0^\alpha , \qquad (14.A.90a)$$

$$J_1(x) = \frac{1}{2}\text{tr}(xT_2) = x_\alpha J_1^\alpha , \qquad (14.A.90b)$$

where $x = x_\alpha \sigma^\alpha \in \mathrm{sl}(2)$. Using (14.A.73b), we then rewrite the first equation (14.A.89) as

$$[[J_1^\lambda, J_1^\mu], [J_0^\rho, J_1^\sigma]] + [[J_1^\rho, J_1^\sigma], [J_0^\lambda, J_1^\mu]]$$
$$= -i\eta^2 \big((\delta^{\lambda\beta}\delta^{\mu\gamma} - \delta^{\lambda\gamma}\delta^{\mu\beta})\varepsilon^{\rho\sigma\alpha} + \varepsilon^{\lambda\mu\alpha}(\delta^{\rho\beta}\delta^{\sigma\gamma} - \delta^{\rho\gamma}\delta^{\sigma\beta})\big)\{J_0^\alpha, J_0^\beta, J_1^\gamma\}$$
$$= -\eta^2 \big(\{[J_0^\lambda, J_0^\mu], J_0^\rho, J_1^\sigma\} - \{[J_0^\lambda, J_0^\mu], J_0^\sigma, J_1^\rho\}$$
$$+ \{[J_0^\rho, J_0^\sigma], J_0^\lambda, J_1^\mu\} - \{[J_0^\rho, J_0^\sigma], J_0^\mu, J_1^\lambda\}\big) \tag{14.A.91}$$

and project it onto four $\mathrm{sl}(2)$ vectors x, y, w, z with components $x_\alpha, y_\alpha, w_\alpha$ and z_α. We obtain

$$[[J_1(x), J_1(y)], [J_0(w), J_1(z)]] + [[J_1(w), J_1(z)], [J_0(x), J_1(y)]]$$
$$= -\eta^2 \big(\{[J_0(x), J_0(y)], J_0(w), J_1(z)\} - \{[J_0(x), J_0(y)], J_0(z), J_1(w)\}$$
$$+ \{[J_0(w), J_0(z)], J_0(x), J_1(y)\} - \{[J_0(w), J_0(z)], J_0(y), J_1(x)\}\big) . \tag{14.A.92}$$

Similarly, the $\mathrm{sl}(2)$ relations (14.A.80) can be rewritten as

$$[J_0(x), J_0(y)] = J_0([x, y]), \tag{14.A.93a}$$
$$[J_0(x), J_1(y)] = J_1([x, y]). \tag{14.A.93b}$$

By virtue of (14.A.93a) $J_0(x)$ generates $\mathrm{sl}(2)$ as a Lie subalgebra of the Yangian. Since the map $x \to J_0(x)$ is bijective, one may therefore identify x with $J_0(x)$ as, for instance, in [80–82].

Remark. The Yangian Serre relation (14.A.89) holds for $Y(\mathrm{sl}(2))$ and also for $Y(\mathrm{gl}(2))$ as should be clear from our derivation. However, while the algebra $Y(\mathrm{sl}(2))$ is uniquely defined by the $\mathrm{sl}(2)$ relations (14.A.80) and the Yangian Serre relation (14.A.89), these relations define $Y(\mathrm{gl}(2))$ only modulo its non-trivial centre, generated by the quantum determinant. Saying that $Y(\mathrm{sl}(2))$ is uniquely defined by (14.A.80) and (14.A.89) means to say that the whole set of relations (14.A.73) follows from (14.A.80), (14.A.89) and the recursion relations (14.A.74) and (14.A.79). We are not aware of a direct proof of this fact in the literature. Clearly a proof should proceed by induction over n, but due to the rather complicated form of (14.A.79) this proof may be involved. Instead Drinfeld [107] used the 'natural' Hopf algebra structure of $Y(\mathrm{sl}(2))$ to be discussed in the following section in order to show its uniqueness as a 'quantized Lie bialgebra'.

14.A.6 The Hopf algebra structure of Y(sl(2))

Up to this point we have regarded the Yangian $Y(\mathrm{sl}(2))$ as an associative algebra, i.e., as a vector space endowed with an associative multiplication. We shall see below that much more structure is implicitly contained in our definition.

We observed in Section 12.1.5 that tensor products of representations of the Yang-Baxter algebra (e.g. products of elementary L-matrices) are again representations. This is one of the essential properties of the Yang-Baxter algebra. It allowed us to construct integrable, interacting L-site spin chains from elementary L-matrices describing a single site. In antic-ipation of a definition below, which is obtained by promoting this property from the level of representations to the level of algebras, we called it the co-multiplication property.

The elements of a monodromy matrix obtained by 'co-multiplication of elementary L-matrices' are

$$T_k^i(\lambda) = \left(L_L(\lambda) \dots L_1(\lambda)\right)_k^i = L_{j_{L-1}}^i(\lambda) \otimes L_{j_{L-2}}^{j_{L-1}}(\lambda) \otimes \cdots \otimes L_k^{j_1}(\lambda), \qquad (14.A.94)$$

where the tensor products on the right hand side are tensor products in quantum space. In former sections we were hiding these tensor products in our notation, either by using subindices, or the notation $L(\lambda|j)$ (see e.g. theorem 2) in order to indicate on which quantum space the L-matrix was acting non-trivially. In the following we shall partially adjust our notation to the conventions used in mathematics. We will write (multiple) tensor products of $Y(sl(2))$ explicitly and use indices for the description of the matrix structure in auxiliary space. Then (14.A.69), for instance, turns into

$$\check{R}_{jm}^{il}(\lambda - \mu) \, T_k^j(\lambda) T_n^m(\mu) = T_j^i(\mu) T_m^l(\lambda) \, \check{R}_{kn}^{jm}(\lambda - \mu). \qquad (14.A.95)$$

For brevity let us now write $A = Y(sl(2))$. We define a homomorphism of vector spaces (linear map) $\Delta : A \to A \otimes A$ by

$$\Delta(T(\lambda))_k^i = \Delta\left(T(\lambda)_k^i\right) = T_j^i(\lambda) \otimes T_k^j(\lambda). \qquad (14.A.96)$$

Using (14.A.95) we find that

$$\check{R}_{jm}^{il}(\lambda - \mu) \, \Delta\left(T_k^j(\lambda)\right) \Delta\left(T_n^m(\mu)\right) = \Delta\left(T_j^i(\mu)\right) \Delta\left(T_m^l(\lambda)\right) \check{R}_{kn}^{jm}(\lambda - \mu). \qquad (14.A.97)$$

Furthermore, the multiplication property (14.A.50) of the quantum determinant implies

$$\det{}_q(\Delta(T(\lambda))) = \left(\det{}_q(T(\lambda))\right)^2 = 1. \qquad (14.A.98)$$

Now (14.A.97) and (14.A.98) are precisely the equations that define the algebra structure of $A = Y(sl(2))$. This means that Δ is a homomorphism of *algebras*. In other words Δ induces the algebra structure of A onto the tensor product $A \otimes A$. Δ is called the co-multiplication or co-product.

It is not difficult to compute the images of the elementary generators J_0^α and J_1^β under the action of Δ. On the one hand we have

$$\Delta\left(T_k^i(\lambda)\right) = \left[\delta_j^i e + \eta \sum_{n=1}^{\infty} \frac{T_{nj}^i}{\lambda^n}\right] \otimes \left[\delta_k^j e + \eta \sum_{n=1}^{\infty} \frac{T_{nk}^j}{\lambda^n}\right]$$

$$= \delta_k^i e \otimes e + \frac{\eta}{\lambda}\left(T_{1k}^i \otimes e + e \otimes T_{1k}^i\right)$$

$$+ \frac{\eta}{\lambda^2}\left(T_{2k}^i \otimes e + e \otimes T_{2k}^i + \eta T_{1j}^i \otimes T_{1k}^j\right) + \dots, \qquad (14.A.99)$$

while, on the other hand, by linearity of Δ

$$\Delta\big(T_k^i(\lambda)\big) = \delta_k^i \Delta(e) + \frac{\eta}{\lambda}\Delta\big(T_{1k}^i\big) + \frac{\eta}{\lambda^2}\Delta\big(T_{2k}^i\big) + \dots \tag{14.A.100}$$

Comparing (14.A.99) and (14.A.100) we obtain

$$\Delta\big(T_{1k}^i\big) = T_{1k}^i \otimes e + e \otimes T_{1k}^i \,, \tag{14.A.101a}$$

$$\Delta\big(T_{2k}^i\big) = T_{2k}^i \otimes e + e \otimes T_{2k}^i + \eta T_{1j}^i \otimes T_{1k}^j \,. \tag{14.A.101b}$$

It follows again by linearity that

$$\Delta(J_0^\alpha) = \tfrac{1}{2}(\sigma^\alpha)_i^k \Delta\big(T_{1k}^i\big) = J_0^\alpha \otimes e + e \otimes J_0^\alpha \,, \tag{14.A.102a}$$

$$\Delta(J_1^\alpha) - J_1^\alpha \otimes e + e \otimes J_1^\alpha + \tfrac{\eta}{2}(\sigma^\alpha)_i^k T_{1j}^i \otimes T_{1k}^j$$
$$= J_1^\alpha \otimes e + e \otimes J_1^\alpha + i\eta\varepsilon^{\alpha\beta\gamma} J_0^\beta \otimes J_0^\gamma \,, \tag{14.A.102b}$$

where we used the fact that $J_0^0 = a_0/2 = 0$ and (14.A.72c) in (14.A.102b). Equation (14.A.97) implies, of course, that $\Delta(J_0^\alpha)$ and $\Delta(J_1^\beta)$ satisfy the sl(2) relations (14.A.80) and the Yangian Serre relation (14.A.89).

Next, we define a linear map $\varepsilon : A \to \mathbb{C}$, setting

$$\varepsilon\big(T_k^i(\lambda)\big) = \delta_k^i \,. \tag{14.A.103}$$

ε is called the co-unit.

With the aid of the co-unit we obtain (in a sense) the inverse of Δ, for we have

$$(\varepsilon \otimes \mathrm{id})\Delta\big(T_k^i(\lambda)\big) = \varepsilon\big(T_j^i(\lambda)\big) \otimes T_k^j(\lambda) = 1 \otimes T_k^i(\lambda) = T_k^i(\lambda) \,, \tag{14.A.104a}$$

$$(\mathrm{id} \otimes \varepsilon)\Delta\big(T_k^i(\lambda)\big) = T_j^i(\lambda) \otimes \varepsilon\big(T_k^j(\lambda)\big) = T_k^i(\lambda) \otimes 1 = T_k^i(\lambda) \,. \tag{14.A.104b}$$

In the last equations we identified as usual A with $\mathbb{C} \otimes A$ or $A \otimes \mathbb{C}$.

The images of the elementary generators J_0^α and J_1^β under the action of the co-unit are again easily obtained. Because of the linearity of ε we have $\varepsilon(\delta_k^i e) = \delta_k^i$ and $\varepsilon(T_{nk}^i) = 0$ for all $n \in \mathbb{N}$. Thus,

$$\varepsilon(J_0^\alpha) = \varepsilon(J_1^\alpha) = 0 \,. \tag{14.A.105}$$

With this trivial action on the generators ε is an algebra homomorphism $A \to \mathbb{C}$ which generates the so-called trivial representation in $\mathbb{C}$.

A vector space equipped with a co-unit ε and a co-multiplication Δ which satisfy the 'compatibility conditions' (14.A.104) is called a coalgebra. Thus, we have seen that the Yangian Y(sl(2)) has a natural coalgebra structure. It is an algebra and a coalgebra at the same time. Clearly the co-multiplication Δ defined in (14.A.96) is coassociative: $(\Delta \otimes \mathrm{id})\Delta = (\mathrm{id} \otimes \Delta)\Delta$.

Algebras and coalgebras are described as 'dual notions' in the mathematical literature (see e.g. [82]): the multiplication in an algebra naturally defines a linear map $\mu : A \otimes A \to A$ if one sets $\mu(a \otimes b) = ab$ for any two elements $a, b \in A$. Mathematicians call this map 'the multiplication'. They further introduce a linear map $\iota : \mathbb{C} \to A$ which formalizes the identification of a number $\lambda \in \mathbb{C}$ with the element $\lambda e \in A$, $\iota(\lambda) = \lambda e$. Then ι and μ satisfy relations similar (or dual) to the relations between ε and Δ. The associativity of the multiplication is, for instance, expressed as $\mu(\mu \otimes \mathrm{id}) = \mu(\mathrm{id} \otimes \mu)$. The idea of duality becomes more transparent when expressed in the language of commutative diagrams. For this issue the reader is referred to [82].

We are now prepared for the abstract definition of a Hopf algebra.

Definition. Hopf algebra.

(i) A Hopf algebra A (over $\mathbb{C}$) is both an associative algebra and a coassociative coalgebra.
(ii) The co-multiplication Δ and the co-unit ε are homomorphisms of algebras.
(iii) A is equipped with a bijective map $S : A \to A$, called the antipode, such that

$$\mu(S \otimes \mathrm{id})\Delta = \mu(\mathrm{id} \otimes S)\Delta = \iota \circ \varepsilon. \qquad (14.A.106)$$

We have seen so far that the Yangian Y(sl(2)) with co-unit (14.A.103) and co-multiplication (14.A.96) satisfies (i) and (ii). Can we find an antipode such that also (14.A.106) holds? To answer this question let us act with (14.A.106) on $T_k^i(\lambda)$. We obtain

$$S\big(T_j^i(\lambda)\big)T_k^j(\lambda) = T_j^i(\lambda)S\big(T_k^j(\lambda)\big) = \delta_k^i\, e. \qquad (14.A.107)$$

From the latter equation we conclude that we must necessarily have $S(T(\lambda)) = T^{-1}(\lambda)$. But, since $\det_q(T(\lambda)) = 1$, this inverse exists for the Yangian, and we obtain by (14.A.51)

$$S\big(T_k^i(\lambda)\big) = (T^{-1})_k^i(\lambda) = (\sigma^y T^t(\lambda - \eta)\sigma^y)_k^i. \qquad (14.A.108)$$

We have thus shown that the Yangian Y(sl(2)) has the structure of a Hopf algebra.

Let us calculate the action of the antipode S on the elementary generators J_0^α and J_1^β. First of all,

$$
\begin{aligned}
S\big(T_k^i(\lambda)\big) &= \Big(\sigma^y\Big[I_2 e + \frac{\eta}{\lambda - \eta}T_1^t + \frac{\eta}{(\lambda - \eta)^2}T_2^t + \dots\Big]\sigma^y\Big)_k^i \\
&= \delta_k^i\, e + \frac{\eta}{\lambda}(\sigma^y T_1^t \sigma^y)_k^i + \frac{\eta}{\lambda^2}(\sigma^y(T_2^t + \eta T_1^t)\sigma^y)_k^i + \dots \\
&= \delta_k^i S(e) + \frac{\eta}{\lambda}S\big(T_{1k}^{\;i}\big) + \frac{\eta}{\lambda^2}S\big(T_{2k}^{\;i}\big) + \dots \qquad (14.A.109)
\end{aligned}
$$

Comparing the last two lines we obtain

$$S\big(T_{1k}^{\;i}\big) = \big(\sigma^y T_1^t \sigma^y\big)_k^i \qquad (14.A.110a)$$

$$S\big(T_{2k}^{\;i}\big) = \big(\sigma^y(T_2^t + \eta T_1^t)\sigma^y\big)_k^i. \qquad (14.A.110b)$$

It follows that

$$S(J_0^\alpha) = \tfrac{1}{2}\mathrm{tr}\big(\sigma^\alpha \sigma^y T_1^t \sigma^y\big) = -J_0^\alpha \qquad (14.A.111a)$$

$$S(J_1^\alpha) = \tfrac{1}{2}\mathrm{tr}\big(\sigma^\alpha \sigma^y (T_2^t + \eta T_1^t)\sigma^y\big) = -J_1^\alpha - \eta J_0^\alpha. \qquad (14.A.111b)$$

As an exercise for the reader we propose to verify (14.A.106) for the elementary generators J_0^α and J_1^β.

Setting $S(T(\lambda)T(\mu)) = (T(\lambda)T(\mu))^{-1} = T^{-1}(\lambda)T^{-1}(\mu)$ we see that the antipode naturally extends to an antiautomorphism of A, i.e., $S(ab) = S(b)S(a)$ for any two elements $a, b \in A$. This is a property of the antipode which holds in any Hopf algebra [1].

Let us summarize: we have shown by identifying the structure maps $\Delta, \varepsilon, \mu, \iota$ and S that the Yangian $Y(\mathrm{sl}(2))$ has a natural Hopf algebra structure, and we have computed the images of the elementary generators J_0^α and J_1^β under Δ, ε and S.

Although it would provide a deeper insight into the algebraic structure of integrable systems we avoided Drinfeld's original way [107, 109] of introducing Yangians as 'quantizations of Lie bialgebras' (see also [82]). Drinfeld's original reasoning requires a background in algebra which is beyond the needs and the level of abstraction of this book. The intention of this appendix is to provide a bridge from the quantum inverse scattering method to the theory of quantum groups (the Yangians are special quantum groups), such that the interested reader is able to start studying the mathematical literature (e.g. the books [82, 310]) on his or her own.

Recall that we began our considerations with the Yang-Baxter algebra relations (14.A.69) imposed on the formal asymptotic series (14.A.68). We have shown that the algebra generated by the coefficients of the formal asymptotic series has a finite number of elementary generators J_0^α and J_1^β, $\alpha, \beta = 1, 2, 3$, if we impose the additional constraint $\det_q(T(\lambda)) = 1$. Then we have derived a set of relations (14.A.80) and (14.A.89) satisfied by the elementary generators and have finally worked out the Hopf algebra structure of the algebra.

Drinfeld reverses this reasoning. He defines the Yangian $Y(\mathfrak{a})$ of an arbitrary simple Lie algebra $\mathfrak{a}$ by prescribing the action of the Hopf algebra structure maps on a finite number of generators $(J_0(I^\alpha), J_1(I^\alpha)$ with $\{I^\alpha\}$ an orthonormal basis of $\mathfrak{a}$) and imposing appropriate 'constraints' (the deformed Serre relations) on these generators in such a way that the co-multiplication becomes a homomorphism of algebras. For $\mathfrak{a} = \mathrm{sl}(2)$ this means, in particular, that the Hopf algebra structure is defined by (14.A.102), (14.A.105) and (14.A.111) and the algebra structure by the $\mathrm{sl}(2)$ relations (14.A.80) and the Yangian Serre relation (14.A.89). In Drinfeld's approach the choice of the co-multiplication is primary. His choice is motivated by the 'quantization problem' for Lie bialgebras [107, 276] which is related to the problem of assigning a solution of the Yang-Baxter equation to every solution of the classical Yang-Baxter equation. A theorem of Drinfeld (theorem 1 of [107]) states that the Yangian $Y(\mathfrak{a})$ is unique as a quantization of the Lie bialgebra associated with $\mathfrak{a}$ (for details see [82, 107, 109]).

The representation theory of the Yangian $Y(\mathrm{sl}(2))$ has been worked out by Chari and Pressley. Their paper [80] is written on a moderate level of abstraction and should be rather easily accessible. The representation theory makes essential use of an alternative realization of the Yangians, again introduced by Drinfeld [108].

14.A.7 Some details of the verification of the Yangian Serre relation for the Hubbard model with variable range hopping

In this last section of the appendix we would like to relate the appendix with the main body of the chapter and add a few technical comments.

First of all we have to adjust the notation. Equation (14.13) follows from (14.A.80) and (14.A.89) by setting $\delta = -\eta^2/4$, $S^\alpha = J_0^\alpha$ and $J^\alpha = J_1^\alpha$. Moreover $K^\alpha = 2J_2^\alpha$, and (14.15) follows from (14.A.81) and (14.A.82) in the $Y(\mathrm{sl}(2))$ case $a_0 = a_1 = 0$. Our claim in the main text that we obtain an equivalent set of equations by replacing the Yangian Serre relation in (14.13) with (14.16) is easily proven by reversing the order of steps in the calculation in appendix 14.A.5 that led from (14.A.83) to (14.A.89).

Finally we wish to justify our statement that equations (14.13a) and (14.13b) imply $[J^\alpha, K^\alpha] = 0$. We prove the slightly stronger statement that (14.A.80)–(14.A.82) lead to $[J_1^\alpha, J_2^\alpha] = 0$:

$$
\begin{aligned}
[J_1^\alpha, J_2^\alpha] &= \left[J_1^\alpha, \tfrac{1}{2i}\varepsilon^{\alpha\beta\gamma}[J_1^\beta, J_1^\gamma] + \eta(J_0^\alpha J_1^0 - J_1^\alpha J_0^0) \right] \\
&= \eta[J_1^\alpha, J_0^\alpha J_1^0] \\
&= \tfrac{i\eta^2}{2}\varepsilon^{\alpha\beta\gamma}
\begin{cases}
J_0^\alpha \{J_0^\beta, J_1^\gamma\} \\
\{J_0^\beta, J_1^\gamma\} J_0^\alpha \\
\tfrac{1}{2}\{J_0^\alpha, \{J_0^\beta, J_1^\gamma\}\}
\end{cases} \\
&= i\eta^2 \varepsilon^{\alpha\beta\gamma} \{J_0^\alpha, J_0^\beta, J_1^\gamma\} = 0 .
\end{aligned}
\tag{14.A.112}
$$

Here we used the explicit form of J_0^0 and the Jacobi identity in the second equation, the explicit form of J_1^0 and (14.A.80) in the third equation and equation (14.A.88) (which was shown by exploiting (14.A.80)) in the fourth equation. The large curly brackets indicate that we have three equivalent expressions in the third equation.

15

S-matrix and Yangian symmetry in the infinite interval limit

In this chapter we carry out the thermodynamic limit on the level of the monodromy matrix introduced in Chapter 12. This means to change the strategy as compared to the Bethe ansatz solutions put forward in Chapters 3 and 12 which depended crucially on the use of periodic boundary conditions. The discreteness of the quasi momenta k_j in the Bethe ansatz wave function was due to the finite length L of the system. For infinite L there will be no Lieb-Wu equations which were our main tool for studying the Hubbard model in this book. Instead the commutation relations for the elements of the infinite interval monodromy matrix will utilized in the calculations shown below.

We basically follow the articles [335,336], where the quantum inverse scattering method, in the way as originally designed in [131,404,454], was applied to the Hubbard model. Our account will be restricted to the case of zero electron density. Quite generally, the quantum inverse scattering method, as originally conceived in the spirit of the 'inverse scattering theory' for classical non-linear evolution equations, is restricted to uncorrelated vacua (ground states) which limits the applicability of the method. Nevertheless, applying it to the empty band ground state of the Hubbard model we shall obtain valuable additional insights into its structure. We shall reveal the connection between Shastry's R-matrix and the Yangian symmetry discussed in Chapter 14. We shall construct creation and annihilation operators of elementary excitations. These form a representation of the Zamolodchikov-Faddeev algebra which means that their commutation relations provide the full bare S-matrix. We shall further see that the scattering states of elementary excitations transform as tensor products of evaluation representations under the Yangian.

15.1 Preliminaries

Before proceeding to the description of the infinite interval limit we have to adapt some of our conventions to this purpose. First of all the Hubbard Hamiltonian for the infinite interval is

$$H = -\sum_{j\in\mathbb{Z}}\sum_{a=\uparrow,\downarrow}(c_{j,a}^\dagger c_{j+1,a} + c_{j+1,a}^\dagger c_{j,a}) + u\sum_{j\in\mathbb{Z}}[(1-2n_{j\uparrow})(1-2n_{j\downarrow})-1], \quad (15.1)$$

where we subtracted 1 in the local interaction part in order to have $H|0\rangle = 0$. We also slightly modify the L-matrix, setting

$$\hat{\mathcal{L}}_j(\lambda) = i^{n_{j\uparrow}+n_{j\downarrow}} \mathcal{L}_j(\lambda, 0) \tag{15.2}$$

for $j \in \mathbb{Z}$, where $\mathcal{L}_j(\lambda, 0)$ on the right hand side of this equation is the L-matrix introduced in (12.227). The slight modification of the L-matrix induces a modification of the R-matrix which must be replaced with

$$\check{\mathcal{R}}(\lambda, \mu) = V \check{R}(\lambda, \mu) V^{-1}, \quad V = \mathrm{diag}(1, i, i, -1) \otimes I_4, \tag{15.3}$$

where $\check{R}(\lambda, \mu)$ is our original R-matrix (12.135). Then $\hat{\mathcal{L}}_j(\lambda)$ is a representation of the graded Yang-Baxter algebra with R-matrix (15.3) and grading $p(1) = p(4) = 0$, $p(2) = p(3) = 1$,

$$\check{\mathcal{R}}(\lambda, \mu)\big(\hat{\mathcal{L}}_j(\lambda) \otimes_s \hat{\mathcal{L}}_j(\mu)\big) = \big(\hat{\mathcal{L}}_j(\mu) \otimes_s \hat{\mathcal{L}}_j(\lambda)\big)\check{\mathcal{R}}(\lambda, \mu). \tag{15.4}$$

In preparation of the thermodynamic limit we introduce a monodromy matrix $\hat{T}_{mn}(\lambda)$ as

$$\hat{T}_{mn}(\lambda) = \hat{\mathcal{L}}_{m-1}(\lambda)\hat{\mathcal{L}}_{m-2}(\lambda)\ldots\hat{\mathcal{L}}_n(\lambda), \quad m, n \in \mathbb{Z}, \, m > n. \tag{15.5}$$

It follows that $\hat{T}_{mn}(\lambda)$ is another representation of the graded Yang-Baxter algebra (15.4).

15.2 Passage to the infinite interval

We shall see in what follows that the thermodynamic limit leads to drastic simplifications of the commutation relations between the elements of the monodromy matrix encoded in the structure of a simplified R-matrix derived from (15.3). The commutation relations will become simple enough to allow us to identify creation and annihilation operators of elementary excitations, generators of conserved quantities and symmetry operators.

In taking the thermodynamic limit one cannot proceed naively. Some contributions to the monodromy matrix which oscillate at large distances have to be treated appropriately, in close analogy with the classical case [134]. These oscillating contributions depend on the chosen vacuum which is characterized by the density of electrons ρ_N and by the magnetization density ρ_M. As a result of the thermodynamic limit we will obtain the finite energy excitations over this vacuum. In contrast to the case of the algebraic Bethe ansatz for the finite periodic system we will not be able any more to distinguish between a pseudo-vacuum, upon which all eigenstates of the transfer matrix are built by the action of creation operators, and the physical vacuum, which is the true ground state of the model. In general, both states will be characterized by different values of ρ_M, ρ_N and thus will be separated by an infinite energy difference in the thermodynamic limit.

The oscillating contributions to the monodromy matrix will be removed by splitting off the asymptotics of its vacuum expectation value for $m, -n \to \infty$, which therefore has to be known a priori. For this reason the method [131, 404, 454] is restricted to (asymptotically) uncorrelated vacua. In the case of the Hubbard model there are four possible choices: the empty band ($\rho_M = \rho_N = 0$), the completely filled band ($\rho_M = 0$, $\rho_N = 2$), and the half-filled band with all spins up ($\rho_M = 1$, $\rho_N = 1$) or all spins down ($\rho_M = -1$, $\rho_N = 1$). In the following we will restrict ourselves to the empty-band vacuum $|0\rangle$ which is defined by

$$c_{m,a}|0\rangle = 0, \quad m \in \mathbb{Z}, \ a = \uparrow, \downarrow . \tag{15.6}$$

Let us now describe the general method in which we closely follow [404]. We define the Hilbert space $\mathcal{H}$ of states of 'compact support' as the space of all finite linear combinations of vectors $c^{\dagger}_{m_1,a_1} \ldots c^{\dagger}_{m_N,a_N}|0\rangle$. The vacuum expectation value of the L-matrix will be denoted by

$$V(\lambda) = \langle 0|\hat{\mathcal{L}}_m(\lambda)|0\rangle. \tag{15.7}$$

$V(\lambda)$ is independent of m, since the vacuum is translationally invariant. Let us introduce

$$\tilde{\mathcal{L}}_j(\lambda) = V(\lambda)^{-j-1}\hat{\mathcal{L}}_j(\lambda)V(\lambda)^j , \tag{15.8}$$

$$\tilde{\mathcal{T}}_{mn}(\lambda) = V(\lambda)^{-m}\hat{\mathcal{T}}_{mn}(\lambda)V(\lambda)^n . \tag{15.9}$$

It is easy to see that the limits $\lim_{n \to -\infty}\langle x|\tilde{\mathcal{T}}_{mn}(\lambda)|y\rangle$ and $\lim_{m \to \infty}\langle x|\tilde{\mathcal{T}}_{mn}(\lambda)|y\rangle$ exist for all $|x\rangle$, $|y\rangle \in \mathcal{H}$. These weak limits determine a pair of operators

$$\tilde{\mathcal{T}}^+_m(\lambda) = \lim_{n \to +\infty} \tilde{\mathcal{T}}_{nm}(\lambda), \tag{15.10a}$$

$$\tilde{\mathcal{T}}^-_m(\lambda) = \lim_{n \to -\infty} \tilde{\mathcal{T}}_{mn}(\lambda) \tag{15.10b}$$

with asymptotics

$$\lim_{m \to +\infty} \tilde{\mathcal{T}}^+_m(\lambda) = \lim_{m \to -\infty} \tilde{\mathcal{T}}^-_m(\lambda) = I_4 . \tag{15.11}$$

Multiplying (15.5) from the left by $\hat{\mathcal{L}}_m(\lambda)$ or from the right by $\hat{\mathcal{L}}_{n-1}(\lambda)$, we obtain two recursion relations for $\hat{\mathcal{T}}_{mn}(\lambda)$, which induce a pair of recursion relations for $\tilde{\mathcal{T}}^+_m(\lambda)$ and $\tilde{\mathcal{T}}^-_m(\lambda)$. By use of the asymptotic condition (15.11) these are equivalent to the following pair of Volterra type 'integral equations' for $\tilde{\mathcal{T}}^{\pm}_m(\lambda)$,

$$\tilde{\mathcal{T}}^+_m(\lambda) = I_4 + \sum_{j=m+1}^{\infty} \tilde{\mathcal{T}}^+_j(\lambda)\,(\tilde{\mathcal{L}}_{j-1}(\lambda) - I_4), \tag{15.12}$$

$$\tilde{\mathcal{T}}^-_m(\lambda) = I_4 + \sum_{j=-\infty}^{m-1} (\tilde{\mathcal{L}}_j(\lambda) - I_4)\,\tilde{\mathcal{T}}^-_j(\lambda). \tag{15.13}$$

The above considerations imply the existence of the weak limit

$$\tilde{\mathcal{T}}(\lambda) = \lim_{m,-n \to \infty} \tilde{\mathcal{T}}_{mn}(\lambda) = \tilde{\mathcal{T}}^+_m(\lambda)\tilde{\mathcal{T}}^-_m(\lambda). \tag{15.14}$$

$\tilde{\mathcal{T}}(\lambda)$ is the 'regularized' monodromy matrix. As can be inferred from equation (15.12), or (15.13) respectively, $\tilde{\mathcal{T}}(\lambda)$ has the 'integral representation'

$$\tilde{\mathcal{T}}(\lambda) = I_4 + \sum_m (\tilde{\mathcal{L}}_m(\lambda) - I_4) + \sum_{m>n} (\tilde{\mathcal{L}}_m(\lambda) - I_4)(\tilde{\mathcal{L}}_n(\lambda) - I_4) + \ldots . \tag{15.15}$$

Note that $\langle 0|(\tilde{\mathcal{L}}_m(\lambda) - I_4)|0\rangle = 0$ by construction. Hence, $\langle 0|\tilde{T}(\lambda)|0\rangle = I_4$. Equation (15.15) can be taken as the definition of the monodromy matrix on the infinite line. Later we will use it in order to obtain the generators of the Yangian and to calculate the action of products of operator entries of $\tilde{T}(\lambda)$ on the vacuum $|0\rangle$.

All we have to do to set equation (15.15) at work for a concrete model is to calculate $\tilde{\mathcal{L}}_m(\lambda)$. This is easily done for the Hubbard model. Using (12.285) and (15.2) we find

$$V(\lambda) = \operatorname{diag}(e^{2h}, -\operatorname{tg}(\lambda), -\operatorname{tg}(\lambda), \operatorname{tg}^2(\lambda)e^{2h}), \tag{15.16}$$

and we conclude that

$$\tilde{\mathcal{L}}_m(\lambda) = V(\lambda)^{-m-1}\hat{\mathcal{L}}_m(\lambda)V(\lambda)^m$$

$$= \left(\begin{array}{cc}
(\mathrm{i}\operatorname{ctg}\lambda)^{-n_{m\uparrow}-n_{m\downarrow}} & \frac{\mathrm{i}e^{-h}}{\cos\lambda}(\mathrm{i}\operatorname{ctg}\lambda)^{-n_{m\downarrow}}c_{m\uparrow}^{\dagger}e^{-imk(\lambda)} \\[2mm]
-\frac{e^{h}}{\sin\lambda}(\mathrm{i}\operatorname{ctg}\lambda)^{-n_{m\downarrow}}c_{m\uparrow}e^{imk(\lambda)} & (\mathrm{i}\operatorname{ctg}\lambda)^{n_{m\uparrow}-n_{m\downarrow}} \\[2mm]
-\frac{e^{h}}{\sin\lambda}(\mathrm{i}\operatorname{ctg}\lambda)^{-n_{m\uparrow}}c_{m\downarrow}e^{imk(\lambda)} & \frac{\mathrm{i}}{\sin\lambda\cos\lambda}c_{m\uparrow}^{\dagger}c_{m\downarrow} \\[2mm]
\frac{1}{\sin^2\lambda}c_{m\downarrow}c_{m\uparrow}e^{im(k(\lambda)+p(\lambda))} & -\frac{e^{-h}}{\sin\lambda}(\mathrm{i}\operatorname{ctg}\lambda)^{n_{m\uparrow}}c_{m\downarrow}e^{imp(\lambda)}
\end{array}\right.$$

$$\left.\begin{array}{cc}
\frac{\mathrm{i}e^{-h}}{\cos\lambda}(\mathrm{i}\operatorname{ctg}\lambda)^{-n_{m\uparrow}}c_{m\downarrow}^{\dagger}e^{-imk(\lambda)} & \frac{1}{\cos^2\lambda}c_{m\downarrow}^{\dagger}c_{m\uparrow}^{\dagger}e^{-im(k(\lambda)+p(\lambda))} \\[2mm]
\frac{\mathrm{i}}{\sin\lambda\cos\lambda}c_{m\downarrow}^{\dagger}c_{m\uparrow} & \frac{\mathrm{i}e^{h}}{\cos\lambda}(\mathrm{i}\operatorname{ctg}\lambda)^{n_{m\uparrow}}c_{m\downarrow}^{\dagger}e^{-imp(\lambda)} \\[2mm]
(\mathrm{i}\operatorname{ctg}\lambda)^{-n_{m\uparrow}+n_{m\downarrow}} & -\frac{\mathrm{i}e^{h}}{\cos\lambda}(\mathrm{i}\operatorname{ctg}\lambda)^{n_{m\downarrow}}c_{m\uparrow}^{\dagger}e^{-imp(\lambda)} \\[2mm]
\frac{e^{-h}}{\sin\lambda}(\mathrm{i}\operatorname{ctg}\lambda)^{n_{m\downarrow}}c_{m\uparrow}e^{imp(\lambda)} & (\mathrm{i}\operatorname{ctg}\lambda)^{n_{m\uparrow}+n_{m\downarrow}}
\end{array}\right). \tag{15.17}$$

Here we have used the functions $k(\lambda)$ and $p(\lambda)$ introduced in (12.331), (12.332).

Next we will turn to the calculation of the commutation relations between the elements of $\tilde{T}(\lambda)$. Let

$$\mathcal{L}_m^{(2)}(\lambda,\mu) = \mathcal{L}_m(\lambda)\otimes_s\mathcal{L}_m(\mu), \tag{15.18}$$

$$\mathcal{T}_{mn}^{(2)}(\lambda,\mu) = \mathcal{T}_{mn}(\lambda)\otimes_s\mathcal{T}_{mn}(\mu). \tag{15.19}$$

We may apply the same regularization scheme we applied to $\mathcal{T}_{mn}(\lambda)$ also to the matrix $\mathcal{T}_{mn}^{(2)}(\lambda,\mu)$. We simply have to replace $V(\lambda)$ by

$$V^{(2)}(\lambda,\mu) = \langle 0|\mathcal{L}_m^{(2)}(\lambda,\mu)|0\rangle. \tag{15.20}$$

Note that $V^{(2)}(\lambda,\mu)$ is *not* just the tensor product $V(\lambda)\otimes_s V(\mu)$. There appear additional off-diagonal terms due to normal ordering of the operators. Following the same steps as in our discussion above we obtain a regularized tensor-product matrix

$$\tilde{T}^{(2)}(\lambda,\mu) = \lim_{m,-n\to\infty}V^{(2)}(\lambda,\mu)^{-m}\mathcal{T}_{mn}^{(2)}(\lambda,\mu)V^{(2)}(\lambda,\mu)^n \tag{15.21}$$

which satisfies $\langle 0|\tilde{T}^{(2)}(\lambda,\mu)|0\rangle = I_{16}$.

On the other hand, taking the vacuum expectation value of the local exchange relation (15.4) yields

$$\mathcal{R}(\lambda, \mu) V^{(2)}(\lambda, \mu) = V^{(2)}(\mu, \lambda) \mathcal{R}(\lambda, \mu), \tag{15.22}$$

and we conclude that

$$\mathcal{R}(\lambda, \mu) \tilde{T}^{(2)}(\lambda, \mu) = \tilde{T}^{(2)}(\mu, \lambda) \mathcal{R}(\lambda, \mu). \tag{15.23}$$

If $\tilde{T}_{mn}^{(2)}(\lambda, \mu)$ is defined in analogy with $\tilde{T}_{mn}(\lambda)$, with $V^{(2)}(\lambda, \mu)$ replacing $V(\lambda)$ in (15.9), then

$$\tilde{T}_{mn}(\lambda) \otimes_s \tilde{T}_{mn}(\mu) = U_m(\lambda, \mu)^{-1} \tilde{T}_{mn}^{(2)}(\lambda, \mu) U_n(\lambda, \mu), \tag{15.24}$$

where we have introduced the matrix

$$U_n(\lambda, \mu) = V^{(2)}(\lambda, \mu)^{-n} \big(V(\lambda)^n \otimes_s V(\mu)^n \big). \tag{15.25}$$

Let us tentatively assume that the limits

$$U_+(\lambda, \mu)^{-1} = \lim_{m \to \infty} U_m(\lambda, \mu)^{-1}, \quad U_-(\lambda, \mu) = \lim_{m \to -\infty} U_m(\lambda, \mu) \tag{15.26}$$

exist in some appropriate sense. Then, according to equation (15.24) $\tilde{T}_{mn}(\lambda) \otimes_s \tilde{T}_{mn}(\mu)$ has a weak limit for $m, -n \to \infty$. This limit may be identified with $\tilde{T}(\lambda) \otimes_s \tilde{T}(\mu)$,

$$\tilde{T}(\lambda) \otimes_s \tilde{T}(\mu) = U_+(\lambda, \mu)^{-1} \tilde{T}^{(2)}(\lambda, \mu) U_-(\lambda, \mu). \tag{15.27}$$

Inserting the above equation into (15.23), we arrive at the exchange relation for the monodromy matrix $\tilde{T}(\lambda)$ on the infinite interval,

$$\tilde{\mathcal{R}}^{(+)}(\lambda, \mu) \big(\tilde{T}(\lambda) \otimes_s \tilde{T}(\mu) \big) = \big(\tilde{T}(\mu) \otimes_s \tilde{T}(\lambda) \big) \tilde{\mathcal{R}}^{(-)}(\lambda, \mu), \tag{15.28}$$

where

$$\tilde{\mathcal{R}}^{(\pm)}(\lambda, \mu) = U_\pm(\mu, \lambda)^{-1} \mathcal{R}(\lambda, \mu) U_\pm(\lambda, \mu). \tag{15.29}$$

The calculation of the matrices $U_\pm(\lambda, \mu)$ is cumbersome but straightforward. Some of the technical steps involved are discussed in the appendix of [336]. Here we only note that there is no common domain of convergence for all matrix elements of $U_+(\lambda, \mu)$ and $U_-(\lambda, \mu)$. Therefore equation (15.28) has to be interpreted as a set of equations for the matrix elements with different domains of validity. The behaviour of the matrix elements at the boundaries of these domains is singular (see appendix of [336]) and it is only at these boundaries where the matrix elements of $U_+(\lambda, \mu)$ and $U_-(\lambda, \mu)$ may be distinct. In the following we stay away from these singular points. Then $U_{+ \gamma \delta}^{\alpha \beta}(\lambda, \mu) = U_{- \gamma \delta}^{\alpha \beta}(\lambda, \mu)$ and there is no difference between $\tilde{\mathcal{R}}^{(+)}(\lambda, \mu)$ and $\tilde{\mathcal{R}}^{(-)}(\lambda, \mu)$.

It is a nontrivial matter of fact that all the matrix elements of $U_\pm(\lambda, \mu)$ calculated from (15.25) and (15.26) turn out to be rational functions of the original Boltzmann weights $\rho_j(\lambda, \mu)$. Therefore the matrix elements of the new R-matrix $\tilde{\mathcal{R}}(\lambda, \mu) = \tilde{\mathcal{R}}^{(+)}(\lambda, \mu) = \tilde{\mathcal{R}}^{(-)}(\lambda, \mu)$ equally depend rationally on the Boltzmann weights. Numerous cancellations of terms occur such that the new R-matrix has only 18 non-vanishing elements as compared to the 36 non-vanishing elements of the original R-matrix. As in our treatment of the algebraic Bethe ansatz in Chapter 12 the structure of the commutation relations between the monodromy matrix elements encoded in (15.28) comes out more clearly when written in block form: we define 2×2 matrices $A(\lambda)$, $B(\lambda)$, $C(\lambda)$, $D(\lambda)$ by assuming that $\tilde{T}(\lambda)$ is of the form (12.272). Then applying the similarity transformation X defined in (12.279) to (15.28) we end up with

$$
\begin{pmatrix} \check{r} & 0 & 0 & 0 \\ 0 & 0 & \ell & 0 \\ 0 & \tilde{\ell} & 0 & 0 \\ 0 & 0 & 0 & \check{s} \end{pmatrix}
\begin{pmatrix} A \otimes \bar{A} & A \otimes \bar{B} & B \otimes \bar{A} & B \otimes \bar{B} \\ A \otimes \bar{C} & A \otimes \bar{D} & -B \otimes \bar{C} & -B \otimes \bar{D} \\ C \otimes \bar{A} & C \otimes \bar{B} & D \otimes \bar{A} & D \otimes \bar{B} \\ -C \otimes \bar{C} & -C \otimes \bar{D} & D \otimes \bar{C} & D \otimes \bar{D} \end{pmatrix} =
$$

$$
\begin{pmatrix} \bar{A} \otimes A & \bar{A} \otimes B & \bar{B} \otimes A & \bar{B} \otimes B \\ \bar{A} \otimes C & \bar{A} \otimes D & -\bar{B} \otimes C & -\bar{B} \otimes D \\ \bar{C} \otimes A & \bar{C} \otimes B & \bar{D} \otimes A & \bar{D} \otimes B \\ -\bar{C} \otimes C & -\bar{C} \otimes D & \bar{D} \otimes C & \bar{D} \otimes D \end{pmatrix}
\begin{pmatrix} \check{r} & 0 & 0 & 0 \\ 0 & 0 & \ell & 0 \\ 0 & \tilde{\ell} & 0 & 0 \\ 0 & 0 & 0 & \check{s} \end{pmatrix} . \quad (15.30)
$$

Here we suppressed the arguments of the various matrices. The 4×4 matrices $\check{r}$, ℓ, $\tilde{\ell}$ and $\check{s}$ depend on λ and μ through rational functions of the Boltzmann weights ρ_j introduced in (12.121). A, B, C and D depend on λ and $\bar{A}$, $\bar{B}$, $\bar{C}$ and $\bar{D}$ on μ. We have the explicit formulae

$$
\check{r}(\lambda, \mu) = \begin{pmatrix} 1 & 0 & 0 & 0 \\ 0 & \frac{1-\rho_3\rho_4}{\rho_4\rho_8} & -\frac{\rho_9\rho_{10}}{\rho_4\rho_8} & 0 \\ 0 & -\frac{\rho_9\rho_{10}}{\rho_4\rho_8} & \frac{1-\rho_3\rho_4}{\rho_4\rho_8} & 0 \\ 0 & 0 & 0 & 1 \end{pmatrix}, \quad
\check{s}(\lambda, \mu) = \begin{pmatrix} \frac{\rho_1}{\rho_4} & 0 & 0 & 0 \\ 0 & 0 & \frac{\rho_8}{\rho_4} & 0 \\ 0 & \frac{\rho_1}{\rho_7} & 0 & 0 \\ 0 & 0 & 0 & \frac{\rho_1}{\rho_4} \end{pmatrix},
$$

$$
\ell(\lambda, \mu) = \begin{pmatrix} \frac{-i\rho_1}{\rho_9} & 0 & 0 & 0 \\ 0 & 0 & \frac{i\rho_{10}}{\rho_4} & 0 \\ 0 & \frac{-i\rho_1}{\rho_9} & 0 & 0 \\ 0 & 0 & 0 & \frac{i\rho_{10}}{\rho_4} \end{pmatrix}, \quad
\tilde{\ell}(\lambda, \mu) = \begin{pmatrix} \frac{-i\rho_9}{\rho_4} & 0 & 0 & 0 \\ 0 & 0 & \frac{-i\rho_9}{\rho_4} & 0 \\ 0 & \frac{i\rho_1}{\rho_{10}} & 0 & 0 \\ 0 & 0 & 0 & \frac{i\rho_1}{\rho_{10}} \end{pmatrix}.
$$

$$\quad (15.31)$$

All information about the Hubbard model on the infinite line at zero density is contained in equations (15.30), (15.31) and (15.15). These equations will be studied in the following sections in order to determine and to classify the spectrum of elementary excitations of the Hubbard model. We wish to emphasize that the commutation relations encoded in (15.30) and (15.31) are much simpler as compared to the relations (12.280) which were the starting point for the algebraic Bethe ansatz calculation of Chapter 12. In fact, if it were not for the two diagonal elements $\frac{1-\rho_3\rho_4}{\rho_4\rho_8}$ of $\check{r}(\lambda, \mu)$ all commutation relations would reduce to the mere interchange of two factors along with the multiplication of some rational function of the Boltzmann weights.

15.3 Yangian symmetry and commuting operators

It is clear from the form of (15.30) that the 2×2 matrices $A(\lambda), \ldots, D(\lambda)$ generate subalgebras of (15.28). The understanding of the structure of these subalgebras will be the key for the interpretation of our results. Our first task will be to find again the Hubbard Hamiltonian which got lost during our passage to the infinite line. Before, in Chapter 12, we obtained it by expanding the logarithm of $\mathrm{str}(T(\lambda))$ in the vicinity of $\lambda = 0$. But this cannot work anymore, because the infinite line monodromy matrix $\tilde{T}(\lambda)$ is not analytic at $\lambda = 0$. Amazingly, the study of the subalgebra generated by $A(\lambda)$,

$$\check{r}(\lambda, \mu)\big(A(\lambda) \otimes A(\mu)\big) = \big(A(\mu) \otimes A(\lambda)\big)\check{r}(\lambda, \mu), \tag{15.32}$$

will provide an alternative generating function of commuting operators.

The matrix $\check{r}(\lambda, \mu)$ in equation (15.32) is exactly the R-matrix (12.302) introduced by Ramos and Martins [371] which played a key role in the construction of the algebraic Bethe ansatz for the Hubbard model. In our calculation above it appeared quite naturally and had not to be introduced ad hoc. From Chapter 12.6 we know already that the change of variables

$$v(\lambda) = -\mathrm{i}\,\mathrm{ctg}\,(2\lambda)\mathrm{ch}(2h) = -\sin k(\lambda) + \mathrm{i}u = -\sin p(\lambda) - \mathrm{i}u \tag{15.33}$$

transforms $\check{r}(\lambda, \mu)$ into the rational R-matrix of the XXX spin chain,

$$\check{r}(\lambda, \mu) = \frac{2\mathrm{i}u + (v(\lambda) - v(\mu))P}{2\mathrm{i}u + v(\lambda) - v(\mu)}. \tag{15.34}$$

Part of the general theory of the Yang-Baxter algebra connected with this R-matrix was developed in Chapter 12.1 and in the appendix to Chapter 14. We learned in particular that the coefficients J_n^0, J_n^α in the asymptotic expansion

$$A(\lambda) = I_2 + 2\mathrm{i}u \sum_{n=1}^{\infty} \frac{J_{n-1}^0 I_2 + J_{n-1}^\alpha \sigma^\alpha}{v(\lambda)^n} \tag{15.35}$$

generate a representation of the Yangian $Y(\mathrm{gl}(2))$ (see appendix 14.A.4) and that the elements in the centre of this algebra can be obtained in a similar way by expanding the quantum determinant,

$$\det{}_q(A(\lambda)) = A_1^1(\lambda)A_2^2(\check{\lambda}) - A_2^1(\lambda)A_1^2(\check{\lambda}) = 1 + 2\mathrm{i}u \sum_{n=1}^{\infty} \frac{a_{n-1}}{v(\lambda)^n}. \tag{15.36}$$

Here $\check{\lambda}$ is determined by the condition that $v(\check{\lambda}) = v(\lambda) - 2\mathrm{i}u$.

In our case the asymptotic expansions (15.35) and (15.36) can be calculated term by term from equation (15.15). Some care is necessary in the calculation since the limit $v(\lambda) \to \infty$ can be carried out in several different ways, only one of which leads to finite results for J_0^α, J_1^α. We have to take $\mathrm{Im}(\lambda) \to \infty$ and at the same time have to choose the correct branch of the solution of equation (12.109) which determines h as a function of λ. Solving (12.109) for

e^{-2h} we obtain

$$e^{-2h} = -u \sin(2\lambda) \pm \sqrt{1 + u^2 \sin^2(2\lambda)}. \tag{15.37}$$

In order to achieve convergence of the matrix elements $\tilde{T}^\alpha_\beta$ ($\alpha, \beta = 2, 3$) we have to choose the lower sign here. Then $e^{-2h(\lambda)}$ is approximately equal to $-2u \sin(2\lambda)$ for large positive values of $u \sin(2\lambda)$, and we obtain

$$e^{2h} = \frac{i e^{2i\lambda}}{u} + \mathcal{O}(e^{6i\lambda}), \qquad e^{ik(\lambda)} = -\frac{e^{2i\lambda} + 2e^{4i\lambda}}{u} + \mathcal{O}(e^{6i\lambda}),$$

$$e^{-ip(\lambda)} = \frac{e^{2i\lambda} - 2e^{4i\lambda}}{u} + \mathcal{O}(e^{6i\lambda}), \qquad \frac{1}{v(\lambda)} = \frac{-2i e^{2i\lambda}}{u} + \mathcal{O}(e^{6i\lambda}). \tag{15.38}$$

The leading terms in the series (15.15) are of order $e^{2i\lambda}$, $e^{4i\lambda}$, ... Thus, from the first two sums in (15.15), we get the expansion of the matrix $A(\lambda)$ up to the order $e^{4i\lambda}$, and the last equation in (15.38) yields the required expansion in $v(\lambda)^{-1}$ up to second order.

We obtain the following explicit expressions for the zeroth and first-level Yangian generators,

$$J^\alpha_0 = \sum_j S^\alpha_j, \tag{15.39a}$$

$$J^\alpha_1 = -\frac{i}{4} \sum_j \left(S^\alpha_{j\,j+1} - S^\alpha_{j\,j-1} \right) + 2u \sum_{j<k} \varepsilon^{\alpha\beta\gamma} S^\beta_j S^\gamma_k. \tag{15.39b}$$

Here we used again the notation introduced in Chapter 2.2.5. J^α_0 is equal to the operator S^α of the total spin, J^α_1 is up to a prefactor of $-u$ equal to the first-level Yangian generator obtained in Chapter 14 (see (14.19)). The prefactor is conventional and can be attributed to a different deformation parameter ($\eta = 2iu$ instead of $\eta = -2i$) in the Yangian Serre relation (see (14.13c), (14.A.89)).

The $\frac{1}{v(\lambda)}$-expansion of the quantum determinant of $A(\lambda)$ yields

$$a_0 = 0, \qquad a_1 = \frac{iH}{2}, \tag{15.40}$$

where H is the Hubbard Hamiltonian in the form (15.1). Thus, we have found the Hamiltonian among the operators in the centre of Y(gl(2)) and at the same time have shown in a completely different way than in Chapter 14 that the Hubbard Hamiltonian is Yangian invariant.

In Chapter 14 we have learned that the Shiba transformation (2.59) preserves the deformed Serre relations (14.A.89) but transforms the representation (15.39) into an independent representation connected with the η-pairing symmetry and commuting with the original one. The reason why we obtained only the representation connected with the rotations is that we performed the infinite interval limit with respect to the vacuum $|0\rangle$ which is rotationally invariant but breaks the invariance with respect to the non-Abelian gauge transformations generated by the η^α. A fully SO(4) invariant vacuum would be the singlet ground state at half-filling. Alas, at present we do not know how to perform an infinite interval limit with respect to such type of correlated ground state.

15.4 Constructing N-particle states

In this section we construct creation and annihilation operators of elementary excitations and study their commutation relations. We shall see that the elementary excitations over the empty band vacuum decay into two classes: electrons (with charge $-e$ and spin $\pm\frac{1}{2}$) and bound states (with charge $-2me$, $m \in \mathbb{N}$ and spin 0). General excitation are scattering states of elementary excitations. Our creation and annihilation operators are constructed in such a way that products of these operators generate normalized scattering states. The information about the S-matrix is then encoded in the commutation relations of the operators: the operators provide representations of the Faddeev-Zamolodchikov algebra whose 'structure constants' are the elements of the two-particle S-matrix.

We also obtain the commutation relations between the generators of the Yangian and the creation and annihilation operators of the elementary excitations. These commutation relations allow us to organize the scattering states into Yangian multiplets. It turns out, in particular, that scattering states of N electrons form a single Yangian multiplet of exponential, 2^N-fold degeneracy. This means that all states in the multiplet can alternatively be created by acting with the Yangian generators on a Yangian highest-weight state which is a linear superposition of non-interacting plane waves (all spins point upwards). The bound states, on the other hand, are all Yangian singlet.

15.4.1 Scattering states of electrons

The commutation relations of the monodromy matrix elements with the number operators $\hat{N}_\uparrow$ and $\hat{N}_\downarrow$, (12.275), remain valid in the infinite interval limit, i.e., after replacing $\mathcal{T}(\lambda)$ with $\tilde{\mathcal{T}}(\lambda)$. This is because $V(\lambda)$ commutes with $I_2 \otimes \sigma^z$ and $\sigma^z \otimes I_2$. Thus, the interpretation of the elements of $\tilde{\mathcal{T}}(\lambda)$ as creation, annihilation or particle number conserving operators remains the same as in Chapter 12.6.1. In particular, $C_1^1(\lambda)$ and $B_2^2(\lambda)$ add one up-spin electron to a state while $C_2^1(\lambda)$ and $B_2^1(\lambda)$ add one down-spin electron. $D_1^1(\lambda)$ and $D_2^2(\lambda)$ conserve the particle number.

The repeated action of operators $B_2^a(\lambda)$, $C_a^1(\lambda)$ on the vacuum produces N-particle eigenstates of the quantum determinant of $A(\lambda)$. This follows from the commutation relations in appendix 15.A.3. For small enough N the associated wave functions can be calculated from the 'integral representation' (15.15). They find their natural interpretation as unnormalized scattering states of N particles. In previous studies (e.g. [131, 366, 454]) it turned out that the standard normalization, with the amplitude of the incident wave equal to unity, could be obtained by introducing the operator analogue of the reflection coefficient of the corresponding classical inverse scattering problem. For this purpose the creation operators had to be multiplied by the inverse of generators of conserved quantities.

This idea was applied to the Hubbard model in [336], where the following two pairs of normalized creation operators were proposed:

$$F_a(\lambda)^\dagger = -ie^h \cos(\lambda) \, C_a^1(\lambda) D_1^1(\lambda)^{-1} \,, \tag{15.41a}$$

$$Z^a(\lambda)^\dagger = (-1)^{3-a} ie^{-h} \cos(\lambda) \, B_2^{3-a}(\lambda) D_2^2(\lambda)^{-1} \tag{15.41b}$$

for $a = 1, 2$ corresponding to spin-up and spin-down, respectively. The numerical prefactors have been determined by demanding that the one-particle states generated by $F_a(\lambda)^\dagger$ and $Z^a(\mu)^\dagger$ be normalized,

$$F_a(\lambda)^\dagger |0\rangle = \sum_m e^{-imk(\lambda)} c_{m,a}^\dagger |0\rangle \,, \quad Z^a(\mu)^\dagger |0\rangle = \sum_m e^{-imp(\mu)} c_{m,u}^\dagger |0\rangle \,. \tag{15.42}$$

Hereafter we assume that $k(\lambda)$ and $p(\mu)$ are real. Then the one-particle wave functions which can be read off from (15.42) are bounded for $m \to \pm\infty$ and hence describe physical excitations.

The behaviour of the monodromy matrix elements $\tilde{T}_\beta^\alpha(\lambda)$ under hermitian conjugation can be easily calculated from (12.277), (15.2), (15.9) and (15.16). The result is shown in appendix 15.A.1. It can be used to obtain the annihilation operators conjugated to $F_a(\lambda)^\dagger$ and $Z^a(\lambda)^\dagger$. Let $\lambda' = \frac{\pi}{2} - \lambda^*$ and $h' = h(\lambda')$. Then

$$F_a(\lambda) = (-1)^{3-a} e^{h'} \sin(\lambda') D_2^2(\lambda')^{-1} C_{3-a}^2(\lambda') \,, \tag{15.43a}$$

$$Z^a(\lambda) = -e^{-h'} \sin(\lambda') D_1^1(\lambda')^{-1} B_1^a(\lambda') \tag{15.43b}$$

for $a = 1, 2$.

The commutation relations between the normalized operators follow from (15.30). For $\lambda \neq \mu \pmod{2\pi}$ they are

$$F_a(\lambda)^\dagger F_b(\mu)^\dagger = -F_c(\mu)^\dagger F_d(\lambda)^\dagger \check{r}_{ab}^{cd}(\lambda, \mu) \,, \tag{15.44a}$$

$$F_a(\lambda) F_b(\mu)^\dagger = -F_c(\mu)^\dagger F_d(\lambda) \check{r}_{db}^{ca}(\mu, \lambda) \,, \tag{15.44b}$$

$$Z^a(\lambda)^\dagger Z^b(\mu)^\dagger = -\check{r}_{cd}^{ab}(\mu, \lambda) Z^c(\mu)^\dagger Z^d(\lambda)^\dagger \,, \tag{15.44c}$$

$$Z^a(\lambda) Z^b(\mu)^\dagger = -\check{r}_{ca}^{db}(\lambda, \mu) Z^c(\mu)^\dagger Z^d(\lambda) \,, \tag{15.44d}$$

$$F_a(\lambda)^\dagger Z^b(\mu)^\dagger = -Z^b(\mu)^\dagger F_a(\lambda)^\dagger \,, \tag{15.44e}$$

$$F_a(\lambda) Z^b(\mu)^\dagger = -Z^b(\mu)^\dagger F_a(\lambda) \,. \tag{15.44f}$$

These equations show that the operators $F_a(\lambda)$, $F_a(\lambda)^\dagger$ and $Z^a(\lambda)$, $Z^a(\lambda)^\dagger$ form (right and left) representations of the (graded) Faddeev-Zamolodchikov algebra [131, 182, 242–244, 420, 502] with two-particle S-matrix $\check{r}(\lambda, \mu)$. The grading is such that all operators are odd. The Faddeev-Zamolodchikov algebra guarantees by construction the factorization of the N-particle S-matrix into products of two-particle S-matrices. In the context of quantum field theory the Faddeev-Zamolodchikov algebra is usually treated as a more or less abstract means for handling the S-matrix. Here we are in the fortunate situation to have the explicit expressions (15.41), (15.43), (15.15).

The aforementioned equations allow us to calculate the action of the Faddeev-Zamolodchikov operators on the vacuum. With a growing number of particles this task becomes very cumbersome. However, the two-particle sector can still be worked out by hand. Using (15.15) and the commutation relations between the elements of the monodromy matrix we obtain

$$F_1(\lambda)^\dagger F_1(\mu)^\dagger |0\rangle = \sum_{n,m} c_{n,\uparrow}^\dagger c_{m,\uparrow}^\dagger e^{-ink(\lambda)} e^{-imk(\mu)} |0\rangle, \tag{15.45a}$$

$$F_2(\lambda)^\dagger F_2(\mu)^\dagger |0\rangle = \sum_{n,m} c_{n,\downarrow}^\dagger c_{m,\downarrow}^\dagger e^{-ink(\lambda)} e^{-imk(\mu)} |0\rangle, \tag{15.45b}$$

$$F_1(\lambda)^\dagger F_2(\mu)^\dagger |0\rangle = \sum_{n,m} c_{n,\uparrow}^\dagger c_{m,\downarrow}^\dagger \left[\theta(n \geq m) e^{-ink(\lambda)} e^{-imk(\mu)} \frac{v(\lambda) - v(\mu)}{v(\lambda) - v(\mu) + 2iu} \right.$$

$$\left. + \theta(n < m) e^{-ink(\lambda)} e^{-imk(\mu)} - \theta(n < m) e^{-imk(\lambda)} e^{-ink(\mu)} \frac{2iu}{v(\lambda) - v(\mu) + 2iu} \right] |0\rangle, \tag{15.45c}$$

$$F_2(\lambda)^\dagger F_1(\mu)^\dagger |0\rangle = \sum_{n,m} c_{n,\downarrow}^\dagger c_{m,\uparrow}^\dagger \left[\theta(n \geq m) e^{-ink(\lambda)} e^{-imk(\mu)} \frac{v(\lambda) - v(\mu)}{v(\lambda) - v(\mu) + 2iu} \right.$$

$$\left. + \theta(n < m) e^{-ink(\lambda)} e^{-imk(\mu)} - \theta(n < m) e^{-imk(\lambda)} e^{-ink(\mu)} \frac{2iu}{v(\lambda) - v(\mu) + 2iu} \right] |0\rangle, \tag{15.45d}$$

$$Z^1(\lambda)^\dagger Z^1(\mu)^\dagger |0\rangle = \sum_{n,m} c_{n,\uparrow}^\dagger c_{m,\uparrow}^\dagger e^{-inp(\lambda)} e^{-imp(\mu)} |0\rangle, \tag{15.45e}$$

$$Z^2(\lambda)^\dagger Z^2(\mu)^\dagger |0\rangle = \sum_{n,m} c_{n,\downarrow}^\dagger c_{m,\downarrow}^\dagger e^{-inp(\lambda)} e^{-imp(\mu)} |0\rangle, \tag{15.45f}$$

$$Z^1(\lambda)^\dagger Z^2(\mu)^\dagger |0\rangle = \sum_{n,m} c_{n,\uparrow}^\dagger c_{m,\downarrow}^\dagger \left[\theta(n \leq m) e^{-inp(\lambda)} e^{-imp(\mu)} \frac{v(\lambda) - v(\mu)}{v(\lambda) - v(\mu) - 2iu} \right.$$

$$\left. + \theta(n > m) e^{-inp(\lambda)} e^{-imp(\mu)} + \theta(n > m) e^{-imp(\lambda)} e^{-inp(\mu)} \frac{2iu}{v(\lambda) - v(\mu) - 2iu} \right] |0\rangle, \tag{15.45g}$$

$$Z^2(\lambda)^\dagger Z^1(\mu)^\dagger |0\rangle = \sum_{n,m} c_{n,\downarrow}^\dagger c_{m,\uparrow}^\dagger \left[\theta(n \leq m) e^{-inp(\lambda)} e^{-imp(\mu)} \frac{v(\lambda) - v(\mu)}{v(\lambda) - v(\mu) - 2iu} \right.$$

$$\left. + \theta(n > m) e^{-inp(\lambda)} e^{-imp(\mu)} + \theta(n > m) e^{-imp(\lambda)} e^{-inp(\mu)} \frac{2iu}{v(\lambda) - v(\mu) - 2iu} \right] |0\rangle. \tag{15.45h}$$

Note that the two-particle states (15.45a)–(15.45d) generated by $F_a(\lambda)^\dagger$ are in-states if $k(\lambda) < k(\mu)$ and out-states if $k(\lambda) > k(\mu)$. Moreover, they are normalized in the sense explained above. As for the operators $Z^a(\lambda)^\dagger$ we observe similar things. The two-particle states (15.45e)–(15.45h) are normalized in-states if $p(\lambda) > p(\mu)$ and normalized out-states if $p(\lambda) < p(\mu)$. These facts, together with the examples of other integrable models [366,454] lead us to the following conjecture:

Conjecture 1. *Provided that $k(\lambda_j)$ is real for $j = 1, \ldots, N$, the N-particle state*

$$F_{a_1}(\lambda_1)^\dagger \ldots F_{a_N}(\lambda_N)^\dagger |0\rangle \tag{15.46}$$

is a normalized in-state if $k(\lambda_1) < \cdots < k(\lambda_N)$ and a normalized out-state if $k(\lambda_1) > \cdots > k(\lambda_N)$.
Provided that $p(\mu_j)$ is real for $j = 1, \ldots, N$, the N-particle state

$$Z^{a_1}(\mu_1)^\dagger \ldots Z^{a_N}(\mu_N)^\dagger |0\rangle \tag{15.47}$$

is a normalized in-state if $p(\mu_1) > \cdots > p(\mu_N)$ and a normalized out-state if $p(\mu_1) < \cdots < p(\mu_N)$.

The proof of this conjecture seems difficult for general N since it seems to be unavoidable to use the series (15.15) and the explicit form (15.17) of $\tilde{\mathcal{L}}_m(\lambda)$.

We have constructed two pairs $(F_a(\lambda)^\dagger$ and $Z^a(\lambda)^\dagger, a = 1, 2)$ of normalized one-particle creation operators. Similar as in the construction of the algebraic Bethe ansatz eigenstates in Chapter 12.6, where we could have worked with the creation operators $B_2^a(\lambda)$ instead of $C_a^1(\lambda)$, not both pairs of Faddeev-Zamolodchikov operators are really needed for constructing multi-particle states. We may use the operator $F_a(\lambda)^\dagger$ only (or $Z^a(\lambda)^\dagger$ only). The reason is the following. From (15.42) we deduce that

$$Z^a(\lambda)^\dagger |0\rangle = F_a(\tilde{\lambda})^\dagger |0\rangle, \tag{15.48}$$

where $p(\lambda) = k(\tilde{\lambda})$. Hence, the action of a mixed product of operators $F_a(\lambda)^\dagger$ and $Z^a(\lambda)^\dagger$ on the vacuum can be expressed in the form (15.46) by use of (15.44e) and (15.48). In particular, one easily obtains

$$Z^{a_N}(\lambda_N)^\dagger \ldots Z^{a_1}(\lambda_1)^\dagger |0\rangle = (-1)^{\frac{N(N-1)}{2}} F_{a_1}(\tilde{\lambda}_1)^\dagger \ldots F_{a_N}(\tilde{\lambda}_N)^\dagger |0\rangle, \tag{15.49}$$

where $p(\lambda_j) = k(\tilde{\lambda}_j)$. The order of the operators is reversed when written in terms of $F_a(\lambda)^\dagger$ instead of $Z^a(\lambda)^\dagger$.

15.4.2 Action of the Yangian on scattering states of N electrons

Using the commutation relations contained in (15.30) of the submatrix $A(\lambda)$ of the infinite interval monodromy matrix with the remaining submatrices $B(\lambda)$, $C(\lambda)$, $D(\lambda)$ and the asymptotic expansion (15.35) we can calculate the commutators of the Yangian generators J_0^α and J_1^α with $B(\lambda)$, $C(\lambda)$ and $D(\lambda)$. The result is shown in appendix 15.A.2. From the definition (15.41) it then follows that

$$[J_0^\alpha, F_a(\lambda)^\dagger] = \tfrac{1}{2} F_b(\lambda)^\dagger \sigma_{ba}^\alpha, \tag{15.50a}$$

$$[J_1^\alpha, F_a(\lambda)^\dagger] = -\tfrac{1}{2} \sin k(\lambda) F_b(\lambda)^\dagger \sigma_{ba}^\alpha + u\, \varepsilon^{\alpha\beta\gamma} F_b(\lambda)^\dagger \sigma_{ba}^\beta J_0^\gamma, \tag{15.50b}$$

$$[J_0^\alpha, Z^a(\lambda)^\dagger] = \tfrac{1}{2} Z^b(\lambda)^\dagger \sigma_{ba}^\alpha, \tag{15.50c}$$

$$[J_1^\alpha, Z^a(\lambda)^\dagger] = -\tfrac{1}{2} \sin p(\lambda) Z^b(\lambda)^\dagger \sigma_{ba}^\alpha - u\, \varepsilon^{\alpha\beta\gamma} Z^b(\lambda)^\dagger \sigma_{ba}^\beta J_0^\gamma. \tag{15.50d}$$

These formulae induce the adjoint action of the Yangian on multi-particle scattering states [286, 337].

Taking into account that $J_0^\alpha |0\rangle = J_1^\alpha |0\rangle = 0$ we obtain the action of the Yangian on the one-particle sector as

$$J_0^\alpha F_a(\lambda)^\dagger |0\rangle = \tfrac{1}{2} F_b(\lambda)^\dagger \sigma_{ba}^\alpha |0\rangle, \tag{15.51a}$$

$$J_1^\alpha F_a(\lambda)^\dagger |0\rangle = -\tfrac{1}{2} \sin k(\lambda) F_b(\lambda)^\dagger \sigma_{ba}^\alpha |0\rangle. \tag{15.51b}$$

Since the action of J_1^α is $-\sin k(\lambda)$ times that of J_0^α, the representation is called the fundamental representation $W_1(-\sin k(\lambda))$ [80–82]. In the two-particle sector we obtain

$$J_0^\alpha F_a(\lambda_1)^\dagger F_b(\lambda_2)^\dagger |0\rangle = F_c(\lambda_1)^\dagger F_d(\lambda_2)^\dagger \tfrac{1}{2} \big[\sigma_{ca}^\alpha \delta_{db} + \delta_{ca} \sigma_{db}^\alpha \big] |0\rangle, \tag{15.52a}$$

$$J_1^\alpha F_a(\lambda_1)^\dagger F_b(\lambda_2)^\dagger |0\rangle = F_c(\lambda_1)^\dagger F_d(\lambda_2)^\dagger$$
$$\times \tfrac{1}{2} \big[-\sin k(\lambda_1) \sigma_{ca}^\alpha \delta_{db} - \sin k(\lambda_2) \delta_{ca} \sigma_{db}^\alpha + u\, \varepsilon^{\alpha\beta\gamma} \sigma_{ca}^\beta \sigma_{db}^\gamma \big] |0\rangle. \tag{15.52b}$$

This is a tensor product, $W_1(-\sin k(\lambda_1)) \otimes W_1(-\sin k(\lambda_2))$, of two fundamental representations with co-multiplication Δ defined by

$$\Delta(J_0^\alpha) = J_0^\alpha \otimes 1 + 1 \otimes J_0^\alpha, \tag{15.53a}$$

$$\Delta(J_1^\alpha) = J_1^\alpha \otimes 1 + 1 \otimes J_1^\alpha + 2u\, \varepsilon^{\alpha\beta\gamma} J_0^\beta \otimes J_0^\gamma. \tag{15.53b}$$

The representation is four-dimensional and irreducible since $k(\lambda_1)$ and $k(\lambda_2)$ are real. Due to the Yangian invariance of the Hamiltonian the four states in the Yangian multiplet are degenerate. The action of the su(2) sub-algebra associated with the spin degrees of freedom decomposes the Yangian multiplet into su(2)-triplet and su(2)-singlet, or, to turn it the other way round, the Yangian mixes spin singlet and spin triplet into a larger multiplet.

In a similar way the N-particle states $F_{a_1}(\lambda_1)^\dagger \ldots F_{a_N}(\lambda_N)^\dagger |0\rangle$ with $a_j = 1, 2$ transform under the Yangian as tensor product representations $W_1(-\sin k(\lambda_1)) \otimes \cdots \otimes W_1(-\sin k(\lambda_N))$. These representations are irreducible since the quasi-momenta $k(\lambda_j)$ are real [80]. However, they are not irreducible with respect to the action of the sub-algebra su(2). The 2^N N-particle states form a large degenerate multiplet with respect to the Yangian, but decompose into the usual spin multiplets under the restricted action of its su(2) sub-algebra.

The irreducibility of the N-particle multiplet leads us to the conclusion that we can construct all the N-particle states (15.46) by acting with Yangian generators J_0^α, J_1^α on the Yangian highest-weight state

$$F_1(\lambda_1)^\dagger \ldots F_1(\lambda_N)^\dagger |0\rangle . \tag{15.54}$$

The wave function of the highest weight state (15.54) must be of plane-wave form, since due to the Pauli principle the on-site interaction does not affect particles of like spin. Therefore, assuming that the state (15.54) is normalized (see Conjecture 1), we conjecture that the above state (15.54) is equal to the superposition of plane waves

$$c_\uparrow^\dagger(k(\lambda_1)) \ldots c_\uparrow^\dagger(k(\lambda_N))|0\rangle , \tag{15.55}$$

where $c_a^\dagger(k) = \sum_{j \in \mathbb{Z}} c_{j,a}^\dagger e^{-ijk}$. Thus, we have found an alternative method for constructing multi-particle scattering states. They can also be obtained by applying the Yangian to plane-wave states of the form (15.55). A similar situation was encountered in case of the Fermi gas with repulsive δ-function interaction [337] which is the proper continuum limit of the Hubbard model in the zero-density phase (see appendix 2.B).

Yangian representations of multi-particle states can of course also be constructed by use of $Z^a(\lambda)^\dagger$. The alert reader will have noticed the only small difference which is in the different signs in front of u in equations (15.50b) and (15.50d), leading to different definitions of the co-multiplication. Instead of (15.53) we obtain

$$\Delta'(J_0^\alpha) = J_0^\alpha \otimes 1 + 1 \otimes J_0^\alpha , \tag{15.56a}$$

$$\Delta'(J_1^\alpha) = J_1^\alpha \otimes 1 + 1 \otimes J_1^\alpha - 2u\, \varepsilon^{\alpha\beta\gamma} J_0^\beta \otimes J_0^\gamma . \tag{15.56b}$$

But this does not cause any harm. The order of the quasi-momenta in (15.49) is reversed in the multi-particle states expressed in terms of $F_a(\lambda)^\dagger$ compared to those expressed in terms of $Z^a(\lambda)^\dagger$. This corresponds to the reversed order of the tensor product $\otimes$ in the definition of the co-multiplication, which compensates the different sign in front of u in (15.53b) and (15.56b).

15.4.3 Free electron limit

The various parameters u, λ, h, v, k and p which we used so far are connected through the formulae (12.109) and (15.33). Thus, only two of them are independent. As a test of consistency of the results in this section let us consider the free fermion limit $u \to 0$. This limit is most conveniently taken for fixed v, for if we fix $v = v(\lambda)$ and $\bar{v} = v(\mu)$ in equations (15.45a)–(15.45h) and let u approach 0, we see that the products of operators $F_a(\lambda)^\dagger$ and $Z^a(\lambda)^\dagger$ act like products of creation operators of Bloch states $c_a^\dagger(k_0)$ on the vacuum. Here $k_0 = p_0$ is determined by the corresponding limit in equation (15.33),

$$\sin k_0 = -v . \tag{15.57}$$

λ and h are now dependent variables. Considering (12.109) and (15.33) for fixed v and small u we find the following solutions

$$\mathrm{i}\,\mathrm{ctg}\,\lambda = 1 + u(1 - v^2)^{-\frac{1}{2}} + \mathcal{O}(u^2)\,, \tag{15.58a}$$

$$\mathrm{e}^{2h} = \mathrm{i}(1 - v^2)^{\frac{1}{2}} - v + \mathcal{O}(u^2)\,. \tag{15.58b}$$

Using these equations and some standard trigonometric identities we can express all the functions of h and λ, which enter the definition of $\tilde{\mathcal{L}}_m(\lambda)$ (see (15.17)) in terms of v and u. Note that equations (15.58) are not the only possible solution of (12.109) and (15.33) for fixed v and small u. We chose the branches such that $\lim_{u\to 0} \tilde{\mathcal{L}}_m(\lambda)|_v = I_4$. For small u the odd elements of $\tilde{\mathcal{L}}_m(\lambda) - I_4$ are of the order of $u^{\frac{1}{2}}$ and the even elements are of the order of u. Thus, only the first sum on the right hand side of (15.15) contributes in order $u^{\frac{1}{2}}$ to the odd elements of $\tilde{T}(\lambda) - I_4$, and we obtain

$$C_a^1(\lambda) = \frac{\mathrm{i}\mathrm{e}^{-h}}{\cos\lambda} \sum_{m\in\mathbb{Z}} c_{m,a}^\dagger \mathrm{e}^{-\mathrm{i}mk_0} + \mathcal{O}(u^{\frac{3}{2}})\,, \tag{15.59a}$$

$$B_2^{3-a}(\lambda) = (-1)^a \frac{\mathrm{i}\mathrm{e}^h}{\cos\lambda} \sum_{m\in\mathbb{Z}} c_{m,a}^\dagger \mathrm{e}^{-\mathrm{i}mp_0} + \mathcal{O}(u^{\frac{3}{2}})\,, \tag{15.59b}$$

where $\mathrm{e}^{\pm h}/\cos\lambda = \mathcal{O}(u^{\frac{1}{2}})$. Since $D_{aa}(\lambda) = 1 + \mathcal{O}(u)$, $a = 1, 2$, it follows from the definitions (15.41) that

$$\lim_{u\to 0} F_a(\lambda)^\dagger\Big|_v = \lim_{u\to 0} Z^a(\lambda)^\dagger\Big|_v = c_a^\dagger(k_0)\,, \quad a = 1, 2\,. \tag{15.60}$$

The corresponding formulae for $F_a(\lambda)$ and $Z^a(\lambda)$ are true by hermitian conjugation. Equations (15.44a)–(15.44f) turn into the usual anticommutators between Fermi operators since

$$\lim_{u\to 0} \check{r}(\lambda, \mu)\Big|_{v,\bar{v}} = P\,, \tag{15.61}$$

where P is the permutation matrix. Hence, it is natural to interpret the Faddeev-Zamolodchikov algebra as a deformation with deformation parameter u of the canonical anticommutation relations between Fermi operators.

15.4.4 Bound states and scattering states of bound states

One of the delicate points in Bethe ansatz calculations is the question of completeness. For a lattice model in a finite volume completeness can in principle be established by counting the finitely many linearly independent Bethe ansatz eigenstates and possibly using symmetries of the model. This programme is not as easy to apply in practice as it may seem at first sight. It is neither easy to show the linear independence of the coordinate Bethe ansatz wave function, nor to determine the number of solutions of a typical set of Bethe ansatz equations. Proofs based on counting therefore usually rely on a number of further assumptions, the most important of which is that an appropriately formulated string hypothesis gives the correct number of (regular) Bethe ansatz states. Depending on the reader's taste such kind of proof may rather be called a consistency test. For the Hubbard model subject to periodic boundary conditions it was obtained in [123, 125] and is discussed at length in appendix B of Chapter 4.

Within our infinite chain formalism the question of completeness is even more difficult to answer. All we can offer here is that we construct the infinite-interval bound states that correspond to the k-Λ string states in the finite volume. As we have seen in the preceding section the operators $F_a(\lambda)^\dagger$ and $Z^a(\lambda)^\dagger$ create single electrons in scattering states. These correspond to solutions of the Lieb-Wu equations (3.95), (3.96) with real quasi momenta k_j. From the commutation relations of the infinite-interval monodromy matrix with the particle number operator we know that the only other candidate creation operator except for $B_2^a(\lambda)$, $C_a^1(\lambda)$ is the operator $D_2^1(\lambda)$ which adds two particles to the system. The action of $D_2^1(\lambda)$ on the empty band is easily calculated from (15.15),

$$D_2^1(\lambda)|0\rangle = -\tfrac{1}{\cos^2\lambda} \sum_{m,n} e^{-i(m+n)(k+p)/2 - i|m-n|(k-p)/2} c_{m\uparrow}^\dagger c_{n\downarrow}^\dagger |0\rangle . \tag{15.62}$$

We have to take into account here that $k(\lambda)$ and $p(\lambda)$ are not independent, but are connected by the constraint (15.33), $\sin k - \sin p = 2iu$. Furthermore, for the wave function in (15.62) to be bounded we need to have $\text{Im}(k + p) = 0$ and $\text{Im}(k - p) < 0$. With these conditions (15.62) describes precisely the bound state of two electrons obtained in Section 3.2.4 within the coordinate Bethe ansatz. In Section 4.2 we showed that this state can be interpreted as an exact k-Λ-2 string.

There are no operators that create more than two particles among the elements of the monodromy matrix. However, the string hypothesis suggests the existence of bound states of pairs. In order to obtain an idea of how to define the corresponding bound-state operators let us recall the string hypothesis. We will denote the spin rapidities of the coordinate Bethe ansatz by Λ_j and the charge momenta by k_j. According to Chapter 4 there are two types of string solutions of the Bethe ansatz equations:

(i) (Λ-string) m Λ_j's form a string configuration in which the real parts of the Λ_j's are identical, while the imaginary parts are arranged at equal spacing $2iu$. The centre of the string should be real.

(ii) (k-Λ-string) $2m$ k_i's and m Λ_j's form a string configuration. The values of the k_i's and Λ_j's are

$$
\begin{aligned}
k_1 &= \pi - \arcsin(\Lambda' + imu), \\
k_2 &= \arcsin(\Lambda' + i(m-2)u), \\
k_3 &= \pi - k_2, \\
&\;\;\vdots \\
k_{2m-2} &= \arcsin(\Lambda' - i(m-2)u), \\
k_{2m-1} &= \pi - k_{2m-2}, \\
k_{2m} &= \pi - \arcsin(\Lambda' - imu), \\
\Lambda_j &= \Lambda' + i(m+1-2j)u, \quad \Lambda' \text{ real}, \quad j = 1, 2, \dots, m .
\end{aligned}
\tag{15.63}
$$

We expect these solutions to represent exact bound states in the infinite interval limit. Since we are dealing with the zero density vacuum, there should be no spin excitations, and we do not have to consider the Λ-string here.

We shall now introduce an alternative construction of the bound state (15.62) which can be generalized to bound states of more particles. Using (15.15) and (15.17) it follows that the bound state (15.62) is proportional to $C_2^1(\lambda')C_1^1(\lambda'')|0\rangle$, if λ' and λ'' satisfy the following conditions;

$$p(\lambda') = \pi - k(\lambda'') \bmod 2\pi , \tag{15.64a}$$

$$k(\lambda'') = p(\lambda) \bmod 2\pi , \tag{15.64b}$$

$$k(\lambda') = k(\lambda) \bmod 2\pi . \tag{15.64c}$$

These are three conditions for three parameters $\lambda, \lambda', \lambda''$ which at first sight seems to violate the arbitrariness of λ. Yet, there is a redundancy in these equations. (15.64a) and (15.64b) imply that

$$p(\lambda') = \pi - p(\lambda) \bmod 2\pi , \tag{15.65}$$

which is compatible with (15.64c) by taking into account the constraint (15.33). Thus, we have obtained two possible 2-string creation operators, which are connected to each other by the relation

$$D_2^1(\lambda)|0\rangle = \frac{ie^{h(\lambda')+h(\lambda'')}\cos\lambda''\cos^2\lambda'\sin\lambda'}{\cos^2\lambda}$$
$$\times \frac{1 - e^{i(p(\lambda')-k(\lambda''))})(1 - e^{i(k(\lambda'')-k(\lambda'))})}{1 - e^{i(p(\lambda')-k(\lambda'))}} C_2^1(\lambda')C_1^1(\lambda'')|0\rangle . \tag{15.66}$$

λ, λ' and λ'' in this equation have to satisfy (15.64). Note that it follows from (15.30) that

$$C_2^1(\lambda')C_1^1(\lambda'') = -C_1^1(\lambda')C_2^1(\lambda'') . \tag{15.67}$$

We can now proceed with the general $2m$-string states. We conjecture that the creation operator of a k-Λ-$2m$-string can (up to an overall normalization factor) be expressed as

$$C^{(2m)}(\lambda_1, \ldots, \lambda_{2m}) = C_2^1(\lambda_1)C_1^1(\lambda_2)C_2^1(\lambda_3)C_1^1(\lambda_4)\ldots C_2^1(\lambda_{2m-1})C_1^1(\lambda_{2m}) , \tag{15.68}$$

where

$$k(\lambda_{2s}) + p(\lambda_{2s-1}) = \pi \bmod 2\pi , \tag{15.69a}$$

$$\sin k(\lambda_{2s-1}) = \sin k(\lambda_1) + 2(s-1)iu , \quad s = 1, \ldots, m . \tag{15.69b}$$

Following previous works [166, 167, 275, 278, 279, 421, 508] we shall call this operator a bound-state operator. The questions of the domain of existence of the term on the right hand side and of its analytic properties are rather delicate. An interpretation as a 'composite operator' was proposed in appendix C of [336]. Here we treat $C^{(2m)}(\lambda_i)$ rather formally and assume that the commutation relations of this operator with arbitrary elements of the infinite-interval monodromy matrix are given by (12.280) for all allowed values of spectral parameters. One can easily verify that the functions $\sin k(\lambda_i)$ in (15.69a) form the same configuration as in the k-Λ-string, if their centre

$$\zeta = \frac{1}{2m} \sum_{i=1}^{2m} \sin k(\lambda_i) = \sin k(\lambda_1) + (m - 2)iu \qquad (15.70)$$

is real.

We can normalize the bound-state operator by a similar method as used for the creation operators of electrons $F_a(\lambda)^\dagger$. Let

$$D^{(2m)}(\lambda_1, \ldots, \lambda_{2m}) = D_1^1(\lambda_1)D_1^1(\lambda_2)D_1^1(\lambda_3)D_1^1(\lambda_4)\ldots D_1^1(\lambda_{2m-1})D_1^1(\lambda_{2m}). \qquad (15.71)$$

Then we define a normalized bound-state operator as

$$F^{(2m)}(\lambda_1, \ldots, \lambda_{2m})^\dagger = C^{(2m)}(\lambda_1, \ldots, \lambda_{2m})D^{(2m)}(\lambda_1, \ldots, \lambda_{2m})^{-1}. \qquad (15.72)$$

Similar definitions of bound-state operators have been used before in the context of the XXZ-chain [275, 278, 279] and the Bose gas with attractive δ-function interaction [166, 167, 368, 508].

Formally using the commutation relations (15.30) for the elements of the monodromy matrix we obtain for our bound state operators

$$F^{(2m)}(\lambda_i)^\dagger F^{(2n)}(\mu_j)^\dagger = \frac{\zeta - \eta + (n + m)iu}{\zeta - \eta - (n + m)iu} \frac{\zeta - \eta + |n - m|iu}{\zeta - \eta - |n - m|iu}$$

$$\cdot \prod_{s=1}^{\min\{m,n\}-1}\left[\frac{\zeta - \eta + (n + m - 2s)iu}{\zeta - \eta - (n + m - 2s)iu}\right]^2 F^{(2n)}(\mu_j)^\dagger F^{(2m)}(\lambda_i)^\dagger, \qquad (15.73)$$

$$F^{(2m)}(\lambda_i)^\dagger F_a(\mu)^\dagger = \frac{\zeta - \sin k(\mu) + miu}{\zeta - \sin k(\mu) - miu} F_a(\mu)^\dagger F^{(2m)}(\lambda_i)^\dagger, \quad a = 1, 2, \qquad (15.74)$$

where ζ is the centre of the $2m$-string and η the centre of the $2n$-string. As before, we interpret these relations as Faddeev-Zamolodchikov algebra. This time particles without internal degrees of freedom are involved. Then the factor on the right-hand side of (15.73) is the S-matrix for the scattering of a bound state represented by a $2m$-string on another bound state with spectral parameters of a $2n$-string. Similarly, the factor on the right-hand side of (15.74) is the S-matrix describing the scattering of a $2m$-string by an electron. The bound-state bound-state S-matrix in (15.73) is of the same form as for the scattering of bound states of magnons in the XXX-chain [278, 279].

As for the transformation under the Yangian we can easily see that

$$[J_0^\alpha, F^{(2m)}(\lambda_i)^\dagger] = [J_1^\alpha, F^{(2m)}(\lambda_i)^\dagger] = 0, \qquad (15.75)$$

which follows from the commutation relations shown in appendix 15.A.2. We conclude that the k-Λ-$2m$-string states are Yangian singlets,

$$J_0^\alpha F^{(2m)}(\lambda_i)^\dagger|0\rangle = 0, \quad J_1^\alpha F^{(2m)}(\lambda_i)^\dagger|0\rangle = 0. \qquad (15.76)$$

15.5 Eigenvalues of quantum determinant and Hamiltonian

In the previous section we constructed independent sets of creation operators $F_a(\lambda)^\dagger$ and $F^{(2m)}(\lambda_1, \ldots, \lambda_{2m})^\dagger$ (see (15.41a), (15.72)). Using the commutation relations (15.A.6) and $\det_q A(\mu)|0\rangle = |0\rangle$ we conclude that the scattering states created by these operators are eigenstates of the quantum determinant of $A(\mu)$, e.g.,

$$\det_q(A(\mu))F_{a_1}(\lambda_1)^\dagger \ldots F_{a_N}(\lambda_N)^\dagger|0\rangle$$

$$= \left[\prod_{j=1}^{N} -\frac{\rho_1(\lambda_j, \mu)\rho_1(\lambda_j, \check\mu)}{\rho_9(\lambda_j, \mu)\rho_9(\lambda_j, \check\mu)} \frac{v(\lambda_j) - v(\mu) + 2iu}{v(\lambda_j) - v(\mu)}\right] F_{a_1}(\lambda_1)^\dagger \ldots F_{a_N}(\lambda_N)^\dagger|0\rangle, \quad (15.77)$$

where $\check\mu$ is defined by $v(\check\mu) = v(\mu) - 2iu$. Similar expressions follow for states created by $Z_u(\lambda)^\dagger$ and for states involving bound states. In order to calculate the corresponding eigenvalue of the Hamiltonian we expand (using (15.A.7)) the terms under the product up to second order in $v(\mu)^{-1}$,

$$-\frac{\rho_1(\lambda_j, \mu)\rho_1(\lambda_j, \check\mu)}{\rho_9(\lambda_j, \mu)\rho_9(\lambda_j, \check\mu)} \frac{v(\lambda_j) - v(\mu) + 2iu}{v(\lambda_j) - v(\mu)} = 1 + \frac{2u(\cos(k(\lambda_j)) + u)}{v(\mu)^2} + \mathcal{O}\left(\frac{1}{v(\mu)^3}\right),$$

$$(15.78)$$

and compare with (15.36), (15.40). We obtain

$$H\, F_{a_1}(\lambda_1)^\dagger \ldots F_{a_N}(\lambda_N)^\dagger|0\rangle = -2\sum_{j=1}^{N}\left(\cos(\lambda_j) + u\right) F_{a_1}(\lambda_1)^\dagger \ldots F_{a_N}(\lambda_N)^\dagger|0\rangle. \quad (15.79)$$

Applying a similar procedure to general states also involving bound states we reproduce the formula (4.52) which was obtained on the basis of the string hypothesis.

It follows from our considerations that the operator $\ln\big(\det_q(A(\mu))\big)$ has an 'additive spectrum', i.e., applying $\ln\big(\det_q(A(\mu))\big)$ to multiple scattering states produces a sum over terms which each depends only on λ_j, $j = 1, \ldots, N$. This observation was used in [336] to obtain a conjecture about how the higher conserved operators generated by $\ln\big(\det_q(A(\mu))\big)$ are related to the conserved operators [183, 187] previously constructed by hand.

15.6 Conclusions

This section was devoted to an algebraic study of the Hubbard model on the infinite interval. Our presentation closely followed [335,336], where the relation between Shastry's R-matrix and the Y(su(2)) Yangian was first explained. We saw how the Yangian acts on scattering states and we constructed explicit representations of the Faddeev-Zamolodchikov algebra providing us in a simple way with the bare S-matrix of the Hubbard model.

Appendices to Chapter 15

15.A Some useful formulae

This appendix contains a collection of formulae that could be useful for those readers who wish to verify the results of the present chapter.

15.A.1 Conjugation properties of the infinite interval monodromy matrix

The behaviour of the monodromy matrix elements under hermitian conjugation are obtained by combining equation (12.277) with the definitions (15.2) and (15.16), (15.17):

$$A(\lambda)^\dagger = \sigma^y A(\tfrac{\pi}{2} - \lambda^*)\sigma^y, \qquad (15.A.1)$$

$$B(\lambda)^\dagger = -i\sigma^y B(\tfrac{\pi}{2} - \lambda^*)\sigma^y, \qquad (15.A.2)$$

$$C(\lambda)^\dagger = -i\sigma^y C(\tfrac{\pi}{2} - \lambda^*)\sigma^y, \qquad (15.A.3)$$

$$D(\lambda)^\dagger = \sigma^y D(\tfrac{\pi}{2} - \lambda^*)\sigma^y. \qquad (15.A.4)$$

As in (12.277) the dagger in these equations means hermitian conjugation of the matrix elements but not of the 2×2 matrices.

15.A.2 Elements of the monodromy matrix under Yangian transformations

The commutation relations of the elements of the infinite-interval monodromy matrix contained in the submatrices $B(\lambda)$, $C(\lambda)$ and $D(\lambda)$ with the Yangian generators are obtained by extracting those of $A(\lambda)$ with these matrices from (15.30) and inserting the asymptotic expansion (15.35),

$$[J_0^\alpha, B(\lambda)] = -\tfrac{1}{2}\sigma^\alpha B(\lambda), \qquad (15.A.5a)$$

$$[J_1^\alpha, B(\lambda)] = \tfrac{1}{2}\sin p(\lambda)\sigma^\alpha B(\lambda) + u\,\varepsilon^{\alpha\beta\gamma}\sigma^\beta B(\lambda)J_0^\gamma, \qquad (15.A.5b)$$

$$[J_0^\alpha, C(\lambda)] = \tfrac{1}{2}C(\lambda)\sigma^\alpha, \qquad (15.A.5c)$$

$$[J_1^\alpha, C(\lambda)] = -\tfrac{1}{2}\sin k(\lambda)C(\lambda)\sigma^\alpha + u\,\varepsilon^{\alpha\beta\gamma}C(\lambda)\sigma^\beta J_0^\gamma, \qquad (15.A.5d)$$

$$[J_0^\alpha, D(\lambda)] = [J_1^\alpha, D(\lambda)] = 0. \qquad (15.A.5e)$$

15.A.3 Commutators involving the quantum determinant

The quantum determinant $\det_q(A(\mu))$ is in the centre of the Yang-Baxter algebra generated by $A(\lambda)$. The commutators with the remaining entries of the monodromy matrix $\check{T}(\lambda)$ can be calculated from (15.30),

$$\det_q(A(\mu))B_1^a(\lambda) = -\frac{\rho_9(\lambda, \mu)\rho_9(\lambda, \check{\mu})}{\rho_1(\lambda, \mu)\rho_1(\lambda, \check{\mu})} \frac{v(\lambda) - v(\mu)}{v(\lambda) - v(\mu) + 2iu} B_1^a(\lambda)\det_q(A(\mu)),$$

$$\det_q(A(\mu))B_2^a(\lambda) = -\frac{\rho_4(\lambda, \mu)\rho_4(\lambda, \check{\mu})}{\rho_{10}(\lambda, \mu)\rho_{10}(\lambda, \check{\mu})} \frac{v(\lambda) - v(\mu)}{v(\lambda) - v(\mu) + 2iu} B_2^a(\lambda)\det_q(A(\mu)),$$

$$\det_q(A(\mu))C_a^1(\lambda) = -\frac{\rho_1(\lambda, \mu)\rho_1(\lambda, \check{\mu})}{\rho_9(\lambda, \mu)\rho_9(\lambda, \check{\mu})} \frac{v(\lambda) - v(\mu) + 2iu}{v(\lambda) - v(\mu)} C_a^1(\lambda)\det_q(A(\mu)),$$

$$\det_q(A(\mu))C_a^2(\lambda) = -\frac{\rho_{10}(\lambda, \mu)\rho_{10}(\lambda, \check{\mu})}{\rho_4(\lambda, \mu)\rho_4(\lambda, \check{\mu})} \frac{v(\lambda) - v(\mu) + 2iu}{v(\lambda) - v(\mu)} C_a^2(\lambda)\det_q(A(\mu)),$$

$$\det_q(A(\mu))D_2^1(\lambda) = \frac{\rho_1(\lambda, \mu)\rho_1(\lambda, \check{\mu})}{\rho_9(\lambda, \mu)\rho_9(\lambda, \check{\mu})} \frac{\rho_4(\lambda, \mu)\rho_4(\lambda, \check{\mu})}{\rho_{10}(\lambda, \mu)\rho_{10}(\lambda, \check{\mu})} D_2^1(\lambda)\det_q(A(\mu)),$$

$$\det_q(A(\mu))D_1^2(\lambda) = \frac{\rho_9(\lambda, \mu)\rho_9(\lambda, \check{\mu})}{\rho_1(\lambda, \mu)\rho_1(\lambda, \check{\mu})} \frac{\rho_{10}(\lambda, \mu)\rho_{10}(\lambda, \check{\mu})}{\rho_4(\lambda, \mu)\rho_4(\lambda, \check{\mu})} D_1^2(\lambda)\det_q(A(\mu)),$$

$$[\det_q(A(\mu)), D_{11}(\lambda)] = [\det_q(A(\mu)), D_{22}(\lambda)] = 0, \tag{15.A.6}$$

where $a = 1, 2$ and $\check{\mu}$ is defined by $v(\check{\mu}) = v(\mu) - 2iu$. These equations involve only two ratios of Boltzmann weights. For the calculation of the eigenvalues of the Hamiltonian their asymptotic expansion in terms of $v(\mu)$ is needed:

$$-\frac{i\rho_1(\lambda, \mu)}{\rho_9(\lambda, \mu)} = 1 + \frac{iu}{v(\mu)} - \frac{(u - 2e^{-ik(\lambda)})u}{2v(\mu)^2} + \mathcal{O}\left(\frac{1}{v(\mu)^3}\right), \tag{15.A.7a}$$

$$\frac{i\rho_{10}(\lambda, \mu)}{\rho_4(\lambda, \mu)} = 1 + \frac{iu}{v(\mu)} - \frac{(u + 2e^{ip(\lambda)})u}{2v(\mu)^2} + \mathcal{O}\left(\frac{1}{v(\mu)^3}\right). \tag{15.A.7b}$$

16

Hubbard model in the attractive case

In this Chapter we discuss attractive case of the Hubbard model. We follow the papers [485] and [121].

We shall start with the Hamiltonian

$$H_0(u) = -\sum_{j=1}^{L}\sum_{\sigma=\uparrow,\downarrow}(c_{j,\sigma}^{\dagger}c_{j+1,\sigma} + c_{j+1,\sigma}^{\dagger}c_{j,\sigma}) + 4u\sum_{j=1}^{L}(n_{j,\uparrow} - \frac{1}{2})(n_{j,\downarrow} - \frac{1}{2}). \quad (16.1)$$

We shall consider the case of even L [number of sites of the chain]. The energy levels of the model are given by

$$E = -2\sum_{l=1}^{N_e}\cos k_l - 2uN_e + uL. \quad (16.2)$$

Here N_e is the number of electrons and k_l are the momenta of individual electrons. Later we shall add a magnetic field and chemical potential to the Hamiltonian.

In the attractive case $u < 0$ we can solve the model independently, using a technique similar to the one of the repulsive case. On the other hand we can use symmetries of the model to reduce the attractive case to the repulsive one. Let us mention relevant symmetries:

The model is invariant under the space reflection: $j \leftrightarrow L - j + 1$.

Partial particle-hole transformation is also important: $c_{j,\uparrow} \leftrightarrow c_{j,\uparrow}^{\dagger}$; $c_{j,\downarrow} \leftrightarrow (-1)^j c_{j,\downarrow}$.

Combination of these two can be represented as a nice unitary transformation:

$$W_1 c_{j,\uparrow} W_1^{-1} = c_{L-j+1,\uparrow}^{\dagger}; \quad W_1 c_{j,\downarrow}W_1^{-1} = (-1)^{L-j+1}c_{L-j+1,\downarrow} \quad (16.3)$$

Here W_1 is an involution $W_1^2 = I$. Here by I we mean the identify operator. The involution can be represented as an exponent of a sum of local operators: $W_1 = \exp[i\hat{S}_1] = W_1^{-1}$ and

$$\hat{S}_1 = \frac{\pi}{2}\sum_{j=1}^{L/2}(c_{j\uparrow} - c_{L-j+1\uparrow}^{\dagger})(c_{j\uparrow}^{\dagger} - c_{L-j+1\uparrow}) + (c_{j\downarrow} + ic_{L-j+1\downarrow}^{\dagger})(c_{j\uparrow}^{\dagger} - ic_{L-j+1\uparrow})$$

$$+ (2j-1)(c_{j\downarrow}c_{j\downarrow}^{\dagger} + c_{L-j+1\downarrow}c_{L-j+1\downarrow}^{\dagger}) \quad (16.4)$$

The involution changes the sign of the Hamiltonian:

$$W_1 H_0(u) W_1 = -H_0(u) \tag{16.5}$$

The operator of number of electrons and the third component of spin are transformed in the following way:

$$W_1 \hat{N} W_1 = L + \hat{N}_\downarrow - \hat{N}_\uparrow, \qquad W_1(\hat{N}_\uparrow - \hat{N}_\downarrow) W_1 = L - \hat{N} \tag{16.6}$$

The involution replaces charge and spin degrees of freedom.

Another involution $W_2 = W_2^{-1}$ can be represented as exponential of a boost operator: $W_2 = \exp[i\hat{S}_2]$, here

$$\hat{S}_2 = \pi \sum_{j=1}^{L} j \sum_{\sigma=\downarrow,\uparrow} n_{j,\sigma} \tag{16.7}$$

The Hamiltonian change the sign and repulsion is replaced by attraction:

$$W_2 H_0(u) W_2 = -H_0(-u), \tag{16.8}$$

The operator of number of electrons and the third component of spin does not change under W_2.

The product of both involutions act like this:

$$W_2 W_1 H_0(u) W_1 W_2 = H_0(-u); \qquad W_2 W_1 \hat{N} W_1 W_2 = L - (\hat{N}_\uparrow - \hat{N}_\downarrow);$$
$$W_2 W_1 (\hat{N}_\uparrow - \hat{N}_\downarrow) W_1 W_2 = L - \hat{N} \tag{16.9}$$

This product replaces repulsion by attraction in an intelligent way. It maps low-lying states into low-lying states. In 1983 F. Woynarovich found the action of these involutions on Bethe Ansatz [485]. We already presented eigenfunctions of the Hubbard Hamiltonian earlier in the book . These states parameterized by set of N_e momenta $\{k_l\}$ and another set of M spin rapidities $\{\lambda_\alpha\}$. Sometimes λ_α called spectral parameters. They satisfy Lieb-Wu equations [298] :

$$e^{ik_l L} = \prod_{\alpha=1}^{M} \frac{\sin(k_l) - \lambda_\alpha + iu}{\sin(k_l) - \lambda_\alpha - iu} \ , \ l = 1, \dots, N_e$$
$$\prod_{l=1}^{N_e} \frac{\sin(k_l) - \lambda_\beta + iu}{\sin(k_l) - \lambda_\beta - iu} = -\prod_{\alpha=1}^{M} \frac{\lambda_\alpha - \lambda_\beta + 2iu}{\lambda_\alpha - \lambda_\beta - 2iu} \ , \ \beta = 1, \dots, M . \tag{16.10}$$

Here N_e and M should belong to a fundamental region: $N_e \leq L$ and $M \leq N_e/2$.

We denote corresponding eigenfunction of the Hamiltonian by $|\{k_l\}, \{\lambda_\alpha\}, u\rangle$.

We can define operator of momentum $\hat{P}$ in a formal way. It is an operator with eigenfunctions $|\{k_l\}, \{\lambda_\alpha\}, u\rangle$ and eigenvalues $P = \sum_{l=1}^{N_e} k_l$.

Woynarovich found that the involutions W act on eigenstates in the following way:

$$W_2 |\{k_l\}, \{\lambda_\alpha\}, u\rangle = |\{k_l + \pi\}, \{-\lambda_\alpha\}, -u\rangle$$
$$W_1 |\{k_l\}, \{\lambda_\alpha\}, u\rangle = |\{k_g\}, \{\lambda_\alpha\}, u\rangle \tag{16.11}$$

Here $\{k_g\}$ are holes. Let us define these holes formally. The first of the Lieb-Wu equations can be considered as an equation for one k_l [at fixed λ_α]. It can be represented as an equation for the roots of a polynomial:

$$P(x) = x^L \prod_{\alpha=1}^{M}(x^2 - 2i(\lambda_\alpha + iu)x - 1)) - \prod_{\alpha=1}^{M}(x^2 + 2i(-\lambda_\alpha + iu)x - 1)) \quad (16.12)$$

Here $x_l = \exp[ik_l]$. The power of the polynomial is $L + 2M$, so it has $L + 2M$ roots. We use only N_e of them to construct the eigenfunction $|\{k_l\}, \{\lambda_\alpha\}, u\rangle$. We denote by $\{x_g\}$ the $L + 2M - N_e$ remaining roots. Corresponding k_g appear in the eigenfunctions after an action of W_1 involution. We call them holes. Please note that the new eigenfunction $|\{k_g\}, \{\lambda_\alpha\}, u\rangle$ still belongs to a fundamental region $L + 2M - N_e \leq L$ and $2M \leq L + 2M - N_e$. In Appendix 16.A we show that k_g satisfy Lieb-Wu equation with the same set of λ_α. In the same Appendix 16.A we shall also see that involutions act on the momentum in the following way:

$$W_2 \hat{P} W_2 = \hat{P} + \pi N_e, \quad \mathrm{mod}(2\pi)$$
$$W_1 \hat{P} W_1 = \pi(L + 1) + \pi M - \hat{P}, \quad \mathrm{mod}(2\pi). \quad (16.13)$$

Since L is even we can drop it from the right-hand side. Now we can combine both involutions into an equation:

$$W_1 W_2 \hat{P} W_2 W_1 = \pi(N_\uparrow + 1) - \hat{P}, \quad \mathrm{mod}(2\pi) \quad (16.14)$$

We shall call the product $W_1 W_2$ Woynarovich mapping. In the following sections we shall use it to construct the ground state and excitations in the attractive case, starting from the repulsive one.

16.1 Half-filled case

16.1.1 Ground state

Let us start our analysis of the ground state from the half filled case [no magnetic field]. In the repulsive case the ground state was invariant under Yangian symmetry [the eigenfunction of the ground state was annihilated by all generators of Yangian]. This high symmetry is more typical for quantum field theory then for condensed matter. This was the reason why we were able to solve all integral equations explicitly and get an expression for the ground-state energy in terms of special functions. This is the reason why the half-filled band is in the centre of the phase diagram. In the repulsive case for the ground state we had $N_e = L$ and $2M = N_e$. The product of two involutions $W_2 W_1$ (Woynarovich mapping) maps it to the ground state of the attractive case. We shall denote ground state in the infinite volume by $|gs\rangle$. The number of electrons and spin does not change $L + 2M - N_e = L$ and $2M = N_e$. In the repulsive case λ filled a symmetric interval. So the set of $\{\lambda\}$ does not change under Woynarovich mapping. Also the set of real $\{k_l\}$ is mapped into a set of complex $\{k_l\}$. In the attractive case the ground state is filled with bound states:

$$\sin k^{\pm} = \lambda \pm iu + O(\exp(-L)), \quad \text{Im}\lambda = 0 \tag{16.15}$$

This is a bound state of an electron with spin up and an electron with spin down. It is called $k - \lambda$ string. Root density of $\{\lambda\}$ in the attractive case is the same as it was in the repulsive:

$$\sigma(\lambda) = \frac{1}{2\pi} \int_{-\infty}^{\infty} d\omega e^{i\omega\lambda} \frac{J_0(\omega)}{2\cosh(u\omega)} \tag{16.16}$$

Here J_0 is a Bessel function. Specific energy for the ground state $e(u)$ is the same as in the repulsive case:

$$e(u) = \frac{E_{GS}}{L} = -|u| - \int_{-\infty}^{\infty} \frac{d\omega}{\omega} \frac{e^{-|\omega u|}}{\cosh u\omega} J_0(\omega) J_1(\omega). \tag{16.17}$$

Woynarovich matched the expressions for $e(u)$ from repulsive and attractive sides. He proved that both $e(u)$ and $de(u)/du$ are continuous across $u = 0$. Decomposition of $e(u)$ into Taylor series in u/π can be found in [439]:

$$\pi e(u) = -4 - 7\zeta(3)\left(\frac{u}{\pi}\right)^2$$
$$- \sum_{n=2}^{\infty} \left(\frac{(2n-1)(2^{2n+1}-1)\{(2n-3)!!\}^3}{2^{2(n-1)}(2n-2)!!}\right) \zeta(2n+1)\left(\frac{u}{\pi}\right)^{2n} \tag{16.18}$$

Here $\zeta(s)$ is Riemann zeta function

$$\zeta(s) = \sum_{n=1}^{\infty} \frac{1}{n^s}.$$

The series is divergent; it is actually an asymptotic series. This indicates a singularity at $u = 0$. The singularity might be related to charge and spin separation. At $u = 0$ excitation, electrons carry charge and spin. For other values of u charge and spin separates.

We also need to notice that coefficients of asymptotic series can be expressed in terms of the values of the Riemann ζ function at odd arguments and rational coefficients. The values of the Riemann ζ function at odd arguments are important objects of number theory; they are conjectured to be algebraically independent transcendental numbers. This has profound consequences for correlation functions of the XXX spin chain.

Under Woynarovich mapping spin and charge degrees of freedom change places. Spin $SU(2)$ interchanges with $\eta - SU(2)$. In the attractive case the ground state is also invariant under both $SU(2)$ [both spins are equal to zero]. It is a one-dimensional representation of each $SU(2)$. The ground state is also invariant under the whole Yangian of $SO(4)$, as it was in the repulsive case.

16.1.2 Excitations

The product of the two involutions $W_2 W_1$ maps exited states into exited states. Charge and spin degrees of freedom interchange. Charge and spin separate as they did in the repulsive case [120, 121].

Spin wave has a gap:

$$p_{sw}(k) = k - \int_0^\infty \frac{d\omega}{\omega} \frac{J_0(\omega)\sin(\omega \sin(k))e^{-|u|\omega}}{\cosh u\omega} \tag{16.19}$$

$$\varepsilon_{sw}(k) = 2|u| - 2\cos(k) + 2\int_0^\infty \frac{d\omega}{\omega} \frac{J_1(\omega)\cos(\omega \sin(k))e^{-|u|\omega}}{\cosh u\omega} \tag{16.20}$$

It does not carry electrical charge. It has spin 1/2. This is actually the spinon. The η 'spin' of these excitations is zero.

Charge-waves (the spinless charged carriers) have the dispersions

$$p_{cw}^p(\lambda) = \pi - \int_0^\infty \frac{d\omega}{\omega} \frac{J_0(\omega)\sin(\omega\lambda)}{\cosh u\omega} = \pi + p_{cw}^h(\lambda) , \tag{16.21}$$

$$\varepsilon_{cw}(\lambda) = 2\int_0^\infty \frac{d\omega}{\omega} \frac{J_1(\omega)\cos(\omega\lambda)}{\cosh u\omega} . \tag{16.22}$$

These excitations can be called holon and antiholon. The holon has electrical charge opposite to the electrical charge of an electron and the antiholon has electrical charge equal to the electrical charge of an electron. The expression for the momentum of holon and antiholon differs by π. These excitations are gapless. They have spin equal to zero. The η 'spin' of these excitation is $\frac{1}{2}$.

This is the complete list of all elementary excitations at half filled band in zero magnetic field. All other energy levels are scattering states of these elementary excitations [120, 121]. One can also calculate the scattering matrix of these excitations. Scattering matrix is the same as in the repulsive case, one should only relabel excitations [change charge and spin degrees of freedom].

Another important object is a Fermi velocity:

$$v = \frac{\varepsilon'_{cw}(\lambda)}{p'_{cw}(\lambda)}\bigg|_{\lambda\to\infty} \tag{16.23}$$

It was evaluated by M. Takahashi, see [439]:

$$v = 2\frac{I_1\left(\frac{\pi}{2|u|}\right)}{I_0\left(\frac{\pi}{2|u|}\right)} \tag{16.24}$$

Here $I_{0,1}$ are modified Bessel functions. Fermi velocity will be important for the description of correlation functions. We shall also use it in low-temperature thermodynamics.

16.2 The ground state and low-lying excitations below half filling

Let us introduce the chemical potential μ into the Hamiltonian:

$$H(u) = -\sum_{j=1}^{L} \sum_{\sigma=\uparrow,\downarrow} (c_{j,\sigma}^{\dagger} c_{j+1,\sigma} + c_{j+1,\sigma}^{\dagger} c_{j,\sigma}) + 4u \sum_{j=1}^{L} (n_{j,\uparrow} - \frac{1}{2})(n_{j,\downarrow} - \frac{1}{2}) - \mu \hat{N},$$

(16.25)

Here $\hat{N}$ is the operators of total number of electrons. For $\mu < 0$ the band is less than half filled. The ground state $|gs\rangle$ is still filled with $k - \lambda$ strings only

$$\sin k^{\pm} = \lambda \pm iu + O(\exp(-L), \quad \text{Im}\lambda = 0;$$

(16.26)

The energy of this string is

$$E(\lambda) = -4\text{Re}\sqrt{1 - (\lambda + iu)^2} + 4|u| - 2\mu$$

(16.27)

The momentum of the string is:

$$P(\lambda) = 2\text{Re} \arcsin(\lambda + iu), \quad P'(\lambda) = 2\text{Re}[1 - (\lambda + iu)^2]^{-1/2}$$

(16.28)

The root density of strings $\sigma(\lambda)$ satisfy the following integral equation:

$$\sigma(\lambda) + \int_{-\Lambda}^{\Lambda} K(\lambda, \nu)\sigma(\nu)d\nu = \frac{1}{2\pi} P'(\lambda)$$

(16.29)

Here

$$2\pi K(\lambda, \nu) = \frac{4|u|}{4u^2 + (\lambda - \nu)^2}$$

(16.30)

The full density of the ground state can be expressed as an integral of the root density:

$$D = \frac{N_e}{L} = 2 \int_{-\Lambda}^{\Lambda} \sigma(\lambda)d\lambda$$

(16.31)

Specific energy of the ground state is equal to

$$e(u) = \frac{E_{GS}}{L} = \int_{-\Lambda}^{\Lambda} E(\lambda)\sigma(\lambda)d\lambda$$

(16.32)

Zero value of chemical potential $\mu = 0$ corresponds to the half filled band. The negative value of chemical potential $\mu_c = 2|u| - 2\sqrt{1 + u^2}$ corresponds to an empty lattice. The density of the ground state $D = N_e/L$ monotonically depends on μ. Spin of the ground state is still equal to zero.

Let us briefly discuss excitations. The simplest excitation seems to be a hole. The energy of the hole can be denoted by $\varepsilon(\lambda)$. It satisfies an integral equation:

$$\varepsilon(\lambda) + \int_{-\Lambda}^{\Lambda} K(\lambda, \nu)\varepsilon(\nu)d\nu = -E(\lambda)$$

(16.33)

The function $\varepsilon(\lambda)$ has to vanish at the edges of integration $\varepsilon(\pm\Lambda) = 0$. This excitations has spin 0 and charge 2. The momentum of 'dressed' $k - \lambda$ string is:

$$-P(\lambda) - 2\int_{-\Lambda}^{\Lambda} \arctan\left(\frac{\lambda - \nu}{2u}\right)\sigma(\nu)d\nu. \tag{16.34}$$

If we map this excitation to repulsive case [by Woynarovich mapping] it will turn into a magnon [at half filled band in a magnetic field]. In the previous chapters we showed that the magnon is not an elementary excitation [below critical field]; a magnon is a scattering state of two spinons. We can apply a similar analysis here in the attractive case. It will show that the hole, which we considered above not to be an elementary excitation, consists of two holons. Dispersions of the holons can be obtained from dispersions of the spinons [in the repulsive case] by Woynarovich mapping.

16.3 Interaction with the magnetic field

Let us sum a magnetic field B interacting with the third component of spin:

$$H(u) = -\sum_{j=1}^{L}\sum_{\sigma=\uparrow,\downarrow}(c_{j,\sigma}^{\dagger}c_{j+1,\sigma} + c_{j+1,\sigma}^{\dagger}c_{j,\sigma})$$

$$+4u\sum_{j=1}^{L}(n_{j,\uparrow} - \frac{1}{2})(n_{j,\downarrow} - \frac{1}{2}) - \mu\hat{N} - B(N_{\uparrow} - N_{\downarrow}), \tag{16.35}$$

At small magnetic field the ground state is filled with $k - \lambda$ strings. These are bound states of an electron with spin up and another electron with spin down. If the magnetic field is strong enough it will break the pairs. The energy of one electron with spin up embedded into the ground state below half filling is:

$$E_e(k) = -2\cos k + 2|u| - \mu - B - 2\int_{-\Lambda}^{\Lambda} K(2\sin k, 2\lambda)\varepsilon(\lambda)d\lambda \tag{16.36}$$

Critical magnetic field makes this excitation gapless:

$$B_{c1} = -2 + 2|u| - \mu - \frac{|u|}{\pi}\int_{-\Lambda}^{\Lambda}\frac{\varepsilon(\lambda)}{u^2 + \lambda^2}d\lambda. \tag{16.37}$$

At half filled band the expression for critical magnetic field simplifies:

$$B_{c1} = -2 + 2|u| + 2\int_{0}^{\infty}\frac{d\omega}{\omega}\frac{J_1(\omega)e^{-|u|\omega}}{\cosh(u\omega)}. \tag{16.38}$$

At a larger magnetic field then critical free electrons with spin up will start filling up the ground state. This means that the ground state will consist of two Fermi spheres: one consists of $k - \lambda$ strings with real λ, another consists of real k [free electrons with spin up]. The reason why a broken pair is not replaced by one electron with spin up and another electron with spin down is because we are considering a grand canonical ensemble. We fix the magnetic field and a chemical potential, and then we compare energies in the sectors with different quantum numbers. Electrons with spin up have lowest energy because of the direction of the magnetic field. If we increase magnetic field even further then more $k - \lambda$ strings will be replaces by electrons with spin up. When the magnetic field passes the second critical value

$$B_{c2} = 2 + 2|u| \tag{16.39}$$

the ground state will be filled only with electrons with spin up; all $k - \lambda$ strings will disappear.

16.4 Phase diagram

In the attractive case the phase diagram can be obtained by 90 degree rotation from the repulsive case. In figure 16.1 we present only the fundamental part of the phase diagram. The rest of the phase diagram can be obtained by particle -hole transformation and by replacing electrons with spin up by electrons with spin down. In order to obtain our phase diagram in the attractive case we take the phase diagram from the repulsive case and replace B by $(-\mu)$ and μ by $(-B)$. Actually, ground-state energy $E_{GS}(\mu, B, T, u)$ has the following symmetry: $E_{GS}(\mu, B, T, u) = E_{GS}(-B, -\mu, T, -u) - \mu - B$. Let us comment on the phases.

Phase I is an empty lattice. Boundaries are given by the same equations as in the repulsive case. For example

$$\mu_c = 2|u| - 2\sqrt{1 + u^2}. \tag{16.40}$$

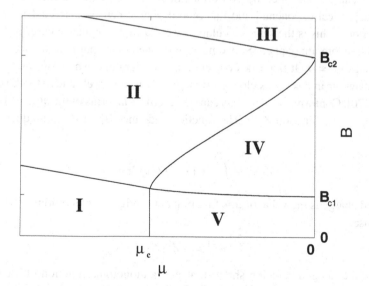

Fig. 16.1. Phase diagram of the attractive Hubbard model.

Phase II is less than half filled and fully polarized. The ground state is filled only with electrons with spin up and there are fewer electrons than lattice sites. The critical point is

$$B_{c2} = 2 + 2|u|. \tag{16.41}$$

Phase III is half filled and fully polarized. The number of electrons is equal to the number of lattice cites, all spins are up spins.

Phase IV is less than half filled and partially polarized. The ground state is filled with $k - \lambda$ strings [bound state of an electron with spin up and another electron with spin down] and real k [electrons with spin up].

$$B_{c1} = -2 + 2|u| + 2 \int_0^\infty \frac{d\omega}{\omega} \frac{J_1(\omega)e^{-|u|\omega}}{\cosh(u\omega)} \tag{16.42}$$

Phase V is less then half filled. It has no polarization [spin is equal to zero]. The ground state is filled only with $k - \lambda$ strings.

The origin $\mu = B = 0$ corresponds to the half filled band [no polarization]. This state is invariant under Yangian symmetry. This is the centre of the phase diagram.

The rest of the phase diagram can be restored by symmetry. The phase diagram is symmetric for free electrons and for the repulsive Hubbard model. Here the same symmetry holds.

16.5 Critical behaviour

At $T = 0$ some correlation functions decay algebraically, others exponentially. In this section we will be interested only in correlation functions, which decay algebraically. These correlations can be described by conformal field theory [Virasoro algebra]. Conformal dimensions [critical exponents] and the central charge can be extracted from the finite size corrections. This is the same calculation we did in the repulsive case; only now spin degrees of freedom are gapped. So we have only one conformal field theory with central charge equal to $c = 1$. It describes correlations of operators, which do not change spin. Critical behaviour in a sense is close to Bose gas. It is very well understood, see Chapter XVIII of [270]. Conformal dimensions can be described in terms of the dressed charge. It is a special value of a function $Z(\lambda)$. The function is defined by an integral equation, similar to (16.29):

$$Z(\lambda) + \int_{-\Lambda}^{\Lambda} K(\lambda, v)Z(v)dv = 1. \tag{16.43}$$

The dressed charge is the value of this function at the edge of integration $Z(\Lambda)$. Below we shall also use

$$\gamma^{-1} = 2Z^2(\Lambda). \tag{16.44}$$

Analysis of the integral equation shows that γ is a monotonic function of the density D. The integral equation for the dressed charge has been studied in detail in papers [65–67]. It was proved that γ changes in the interval:

$$\frac{1}{2} \leq \gamma \leq 1. \tag{16.45}$$

The γ approaches $\frac{1}{2}$ in the low density limit, it goes to 1 for the half filled band. It will be important to us that

$$\gamma \leq \frac{1}{\gamma}. \tag{16.46}$$

It turns into equality only for the half filled band. Let us consider examples of some correlations. Let us start with the superconducting correlation function. The formula, which describes the long distance asymptotic as $|x| \to \infty$ is:

$$\langle c^\dagger_{x,\uparrow} c^\dagger_{x,\downarrow} c_{1,\uparrow} c_{1,\downarrow} \rangle \sim \frac{1}{|x|^\gamma}. \tag{16.47}$$

Let us also mention the correlation function of local densities. Its asymptotic behaviour contains oscillations:

$$\langle c^\dagger_{x,\uparrow} c_{x,\uparrow} c^\dagger_{1,\uparrow} c_{1,\uparrow} \rangle - \langle c^\dagger_{1,\uparrow} c_{1,\uparrow} \rangle^2 \sim \frac{\cos(2\pi x D)}{|x|^{\frac{1}{\gamma}}}, \quad |x| \to \infty \tag{16.48}$$

Let us mention that the Green function $\langle c^\dagger_{x,\uparrow} c_{1,\uparrow} \rangle$ decays exponentially. Now we can compare asymptotic of the superconducting correlation function with density-density correlation. The inequality $\gamma \leq 1/\gamma$ shows that the superconducting correlation function decays slower than density-density correlation. This means that *super conducting correlations dominate*. More details on the conformal description of correlation functions in the attractive Hubbard model can be found in [65–67, 140]. Multi-point correlations also can be described by conformal field theory.

Entropy also can be described by conformal field theory. At zero temperature the ground state $|gs\rangle$ is unique and the entropy of the whole infinite ground state is zero. Nevertheless there is some entropy in a subsystem [part of the ground state]. Let us consider electrons present on some space interval $(0, x)$ in the ground state. They can be described by a density matrix:

$$\rho = \mathrm{tr}_\infty(|gs\rangle\langle gs|)$$

Here we trace out the degrees of freedom of electrons on the unification of intervals $(-\infty, 0) \cup (x, \infty)$. The entropy of the electrons on the interval $(0, x)$ can be defined as von Neumann entropy of the density matrix:

$$S(x) = -\mathrm{tr}_x \rho \ln \rho$$

Here we are taking the trace with respect to degrees of freedom of electrons on the interval $(0, x)$. It is difficult to calculate $S(x)$; nevertheless for large x it simplifies. In Appendix 16.B we show that the asymptotic can be described by conformal field theory:

$$S(x) \rightarrow \frac{1}{3} \ln x.$$

Notice that specific entropy $s = \lim(S(x)/x)$ as $x \rightarrow \infty$ vanishes according to the third law of thermodynamics. This result tells us that ground-state electrons from the interval $(0, x)$ can be in

$$n = \exp[S(x)] = x^{1/3}$$

different states $|x, j\rangle$ here $j = 1, \ldots, n$. Outside electrons, from the intervals $(-\infty, 0) \cup (x, \infty)$ also can be in n different states $|\infty, j\rangle$. The wave function of the ground state can be represented in the form:

$$|gs\rangle = \sum_{j=1}^{n} |x, j\rangle |\infty, j\rangle$$

Since n is large this describes entanglement of electrons from the interval $(0, x)$ with the rest of the ground state.

16.6 Thermodynamics

Partition function of the model is defined by

$$Z = \text{tr}[e^{-H(u)/T}] \tag{16.49}$$

In the thermodynamic limit ($L \rightarrow \infty$, $N_e \rightarrow \infty$, density $n = N_e/L$ fixed) the partition function can be asymptotically represented as :

$$Z = e^{\frac{-fL}{T}} \tag{16.50}$$

Here f is bulk free energy. We already described thermodynamics in the repulsive case in the frame of the Yang-Yang approach. Thermodynamics was described in terms of Takahahsi's equations. In the attractive case we can use the same equations; now the coupling constant u will take negative values in these equations, see [287, 288] .

  Another approach to thermodynamics in the attractive case is based on symmetries mentioned at the beginning of the chapter. One can use the symmetries to relate the bulk free energy in the attractive and repulsive cases:

$$f(\mu, B, T, u) = f(-B, -\mu, T, -u) - \mu - B. \tag{16.51}$$

This describes the bulk free energy in the attractive case.

  One can study analytical properties of f as a function of coupling constant u. The bulk free energy behaves differently from the the ground state energy $e(u)$. The bulk free energy f does not have a singularity at $u = 0$, see [21].

  Let us now discuss the entropy. Specific entropy s increases with temperature. So we shall discuss two limiting cases of small temperature and large temperature.

At small temperature entropy density [specific entropy] vanishes linearly, see [6]:

$$s = \frac{S}{L} = \frac{\pi T}{3v}. \tag{16.52}$$

Here v is Fermi velocity.

At very large temperature entropy also simplifies [because of a different reason]. Notice that we fixed the density $n = N_e/L$. We can express the specific entropy of the Hubbard model at infinite temperature s_∞ in terms of Boltzmann entropy:

$$S_B(p) = -p \ln p - (1 - p) \ln(1 - p)$$

Here p is a probability.

At infinite temperature different lattice sites are independent. If magnetic field $B = 0$ then half of the electrons have spin up another half have spin down. For an electron with spin up the probability of its presence in a lattice site is $p = n/2$. This means that a contribution of electrons with spin up to the specific entropy of the Hubbard model is $S_B(n/2)$. Electrons with spin down will give the same contribution to the entropy. So the total specific entropy of the Hubbard model at fixed density and infinite temperature is $s_\infty = 2S_B(n/2)$:

$$s_\infty = \ln 4 - n \ln n - (2 - n) \ln(2 - n), \qquad T = \infty, \qquad B = 0, \qquad n \text{ is fixed} \tag{16.53}$$

This agrees with the Takahashi equations. Let us emphasize again that entropy increases monotonically with temperature from its values at small temperature to its values at large temperature.

We can also study the entropy of electrons $S(x)$ on the space interval $(0, x)$. According to the second law of thermodynamics $S(x)$ is proportional to the length of the interval:

$$S(x) = sx \tag{16.54}$$

Now let us comment on correlation functions. At small temperatures the correlation function can be calculated by conformal mapping. Now they decay exponentially. Asymptotic of superconducting correlation is given by the following expression:

$$\langle c_{x,\uparrow}^{\dagger} c_{x,\downarrow}^{\dagger} c_{1,\uparrow} c_{1,\downarrow} \rangle \sim \exp[-\frac{\pi T \gamma}{v} x]. \tag{16.55}$$

Asymptotic of the correlation function of charge-density waves is:

$$\langle c_{x,\uparrow}^{\dagger} c_{x,\uparrow} c_{1,\uparrow}^{\dagger} c_{1,\uparrow} \rangle - \langle c_{1,\uparrow}^{\dagger}(0) c_{1,\uparrow}(0) \rangle^2 \sim \cos(2\pi n D) \exp[-\frac{\pi T}{\gamma v} x] \tag{16.56}$$

Some times nonlocal correlation are also interesting. An example is emptiness-formation probability $P(x)$. It is a probability that [because of fluctuations] there will be no electrons on a space interval of the length x. It is difficult to calculate, but large x asymptotic is simple:

$$P(x) \sim \exp\left(\frac{-\mathcal{P}x}{T}\right) \tag{16.57}$$

Here $\mathcal{P}$ is a pressure: $\mathcal{P} = -(f - u)$. This formula follows from Maxwell-Boltzmann statistics, because $\mathcal{P}x$ is work necessary to remove electrons from the interval $(0, x)$.

Appendices to Chapter 16

16.A Appendix A

The first of Lieb-Wu equations is equivalent to the equation $P(x) = 0$ with

$$P(x) = x^L \prod_{\alpha=1}^{M}(x^2 + (-2i\lambda_\alpha + 2u)x - 1)) - \prod_{\alpha=1}^{M}(x^2 + (-2i\lambda_\alpha - 2u)x - 1)) \quad (16.A.1)$$

The polynomial can be represented in the form $P(x) = \prod_{a=1}^{L+2M}(x - x_a)$. We divide the set of all roots into two subsets: $\{x_a\} = \{x_l\} \cup \{x_g\}$. Here $\{x_l\}$ describes the original electrons and $\{x_g\}$ describes the holes. We can represent the second Lieb-Wu equation in the form:

$$\prod_{l=1}^{N_e} \frac{x_l^2 + (-2i\lambda - 2u)x_l - 1}{x_l^2 + (-2i\lambda + 2u)x_l - 1} = -\prod_{\alpha=1}^{M} \frac{\lambda_\alpha - \lambda + 2iu}{\lambda_\alpha - \lambda - 2iu}. \quad (16.A.2)$$

Here λ belongs to the set $\{\lambda_\alpha\}$. We want to prove that $\{x_g\}$ satisfy the same equation:

$$\prod_{g=1}^{L+2N-N_e} \left(\frac{x_g^2 + (-2i\lambda - 2u)x_g - 1}{x_g^2 + (-2i\lambda + 2u)x_g - 1} \right) = -\prod_{\alpha=1}^{M} \frac{\lambda_\alpha - \lambda + 2iu}{\lambda_\alpha - \lambda - 2iu}. \quad (16.A.3)$$

It is equivalent to the following:

$$\prod_{a=1}^{L+2N} \left(\frac{x_a^2 + (-2i\lambda - 2u)x_a - 1}{x_a^2 + (-2i\lambda + 2u)x_a - 1} \right) = \prod_{\alpha=1}^{M} \left(\frac{\lambda_\alpha - \lambda + 2iu}{\lambda_\alpha - \lambda - 2iu} \right)^2$$

$$= \prod_{\alpha=1}^{M} \left(\frac{\lambda - \lambda_\alpha - 2iu}{\lambda - \lambda_\alpha + 2iu} \right)^2. \quad (16.A.4)$$

In order to prove this let us introduce roots of numerators and denominator. The roots of the numerator we shall define by:

$$x^2 + (-2i\lambda - 2u)x - 1 = (x - x_n^+)(x - x_n^-),$$
$$x_n^\pm = i\lambda + u \pm \sqrt{1 + (i\lambda + u)^2},$$
$$x_n^+ x_n^- = -1. \quad (16.A.5)$$

633

Similar definition for denominator is:

$$x^2 + (-2i\lambda + 2u)x - 1 = (x - x_d^+)(x - x_d^-),$$
$$x_d^\pm = i\lambda - u \pm \sqrt{1 + (i\lambda - u)^2},$$
$$x_d^+ x_d^- = -1. \tag{16.A.6}$$

We shall need to define logarithm as a function of x:

$$\ln\left(\frac{x^2 + (-2i\lambda - 2u)x - 1}{x^2 + (-2i\lambda + 2u)x - 1}\right) = \ln\left[\left(\frac{x - x_n^+}{x - x_d^+}\right)\left(\frac{x - x_n^-}{x - x_d^-}\right)\right]. \tag{16.A.7}$$

Here we make one cut from x_d^+ to x_n^+ and another cut from x_d^- to x_n^-. Let us present the identity; we want to prove in a logarithmic form:

$$\sum_{a=1}^{L+2N} \frac{1}{i}\ln\left(\frac{x_a^2 + (-2i\lambda - 2u)x_a - 1}{x_a^2 + (-2i\lambda + 2u)x_a - 1}\right) = \sum_{\alpha=1}^{M} \frac{2}{i}\ln\left(\frac{\lambda - \lambda_\alpha - 2iu}{\lambda - \lambda_\alpha + 2iu}\right) \quad \mathrm{mod}(2\pi). \tag{16.A.8}$$

Let us represent the left-hand side as a contour integral around all the roots of the polynomial $P(x)$:

$$\sum_{a=1}^{L+2N} \frac{1}{i}\ln\left(\frac{x_a^2 + (-2i\lambda - 2u)x_a - 1}{x_a^2 + (-2i\lambda + 2u)x_a - 1}\right)$$
$$= \frac{1}{2\pi i}\oint \frac{1}{i}\ln\left(\frac{x^2 + (-2i\lambda - 2u)x - 1}{x^2 + (-2i\lambda + 2u)x - 1}\right) d\ln P(x) \tag{16.A.9}$$

We can continuously deform the integration contour, enlarging it. The cuts of the integrand will contribute in the integral. The large circle will not contribute because the integrand decays asymptotically as $1/x^2$. Now it is convenient to represent the logarithm in the integrand in the form (16.A.7). The jump on the cut is $2\pi i$, so the integral is equal to :

$$\left(\int_{x_d^+}^{x_n^+} + \int_{x_d^-}^{x_n^-}\right)\frac{1}{i}d\ln P(x) = \frac{1}{i}\ln\frac{P(x_n^+)}{P(x_d^+)} + \frac{1}{i}\ln\frac{P(x_n^-)}{P(x_d^-)} \tag{16.A.10}$$

We can simplify the right-hand side, because the first product in the expression for the polynomial vanishes at x_d and the second product vanishes at x_n.

$$P(x_n) = x_n^L \prod_{\alpha=1}^{M}(x_n^2 + (-2i\lambda_\alpha + 2u)x_n - 1),$$

$$P(x_d) = -\prod_{\alpha=1}^{M}(x_d^2 + (-2i\lambda_\alpha - 2u)x_d - 1) \tag{16.A.11}$$

Now we can represent the right-hand side of (16.A.9), (16.A.10) in the form:

$$\frac{1}{i}\ln\frac{P(x_n^+)}{P(x_d^+)} + \frac{1}{i}\ln\frac{P(x_n^-)}{P(x_d^-)}$$

$$= \frac{1}{i}\ln\prod_{\alpha=1}^{M}\left(\frac{x_n^+ - (x_n^+)^{-1} - 2i\lambda_\alpha + 2u}{x_d^+ - (x_d^+)^{-1} - 2i\lambda_\alpha - 2u}\right)\left(\frac{x_n^- - (x_n^-)^{-1} - 2i\lambda_\alpha + 2u}{x_d^- - (x_d^-)^{-1} - 2i\lambda_\alpha - 2u}\right) \quad (16.A.12)$$

Here we used $x^+x^- = -1$, see (16.A.6) and (16.A.5). From the definition of $x^\pm$ we know that

$$x_n^\pm - (x_n^\pm)^{-1} = 2i\lambda + 2u,$$
$$x_d^\pm - (x_d^\pm)^{-1} = 2i\lambda - 2u \quad (16.A.13)$$

We can use this in order to simplify the right-hand side.

$$\frac{1}{i}\ln\frac{P(x_n^+)P(x_n^-)}{P(x_d^+)P(x_d^-)} = \sum_{\alpha=1}^{M}\frac{2}{i}\ln\left(\frac{2i\lambda - 2i\lambda_\alpha + 4u}{2i\lambda - 2i\lambda_\alpha - 4u}\right) \quad (16.A.14)$$

So we have proved equation (16.A.8) and equation (16.A.3). We have proved that the set $\{k_g\}$ satisfy Lieb-Wu equations with the same set of $\{\lambda_\alpha\}$. So we have completely described the involution W_1. It replaces electrons by holes $W_1\{k_l\} = \{k_g\}$.

Now let us calculate the total momentum

$$\exp[i\sum_{a=1}^{L+2M} k_a] = \prod_{a=1}^{L+2M} x_a.$$

It is equal to the coefficient at zero power of x in the polynomial, because $(-1)^{L+2M} = 1$. The coefficient is equal to

$$\exp[i\sum_{a=1}^{L+2M} k_a] = (-1)^{M+1}$$

This proves equation (16.13).

16.B Appendix B

Here we follow the argument of the paper [269]. Conformal field theory [51] is useful for the description of low-temperature behaviour of gapless models in one space and one time dimensions. We are interested in specific entropy s [entropy per unit length]. Let us start with specific heat $C = Tds/dT$. Low-temperature behaviour was obtained in [6]:

$$C = \frac{\pi T c}{3v}, \quad \text{as} \quad T \to 0 \quad (16.B.1)$$

Here c is a central charge of corresponding Virasoro algebra and v is Fermi velocity. We are more interested in s. We can integrate the equation and fix the integration constant from the third law of thermodynamics ($s = 0$ at $T = 0$). So for specific entropy we have the same low temperature behaviour:

$$s = \frac{\pi T c}{3v}, \quad \text{as} \quad T \to 0 \qquad (16.B.2)$$

We consider entropy of electrons on the interval $(0, x)$. The second law of thermodynamics states that the entropy is extensive parameter. So the entropy of electrons on the interval $S(x)$ is proportional to the size x:

$$S(x) = sx \quad \text{at} \quad T > 0. \qquad (16.B.3)$$

The laws of thermodynamics are applicable to a subsystem of macroscopic size, meaning large x. Here specific entropy s depends on the temperature. For small temperature the dependence simplifies, see (16.B.2):

$$S(x) = \frac{\pi T c}{3v}x, \quad x \gg \frac{1}{T}. \qquad (16.B.4)$$

Let us find how $S(x)$ depends on x for zero temperature. It is some function of the size x:

$$S(x) = f(x), \quad \text{at} \quad T = 0 \qquad (16.B.5)$$

Now let us apply the ideas of conformal field theory, see [6, 51] and also Chapter XVIII of [270]. We can arrive at small temperatures from zero temperature by conformal mapping $\exp[2\pi T z/v]$. It maps the whole complex plane of z without the origin to a strip of the width $1/T$. This replaces zero temperature by positive temperature T. The function f does not change, only its argument does. The conformal mapping results in a replacement of the variable x [argument of the function] by $[v/\pi T] \sinh[\pi T x/v]$. Now the entropy of a subsystem is given by the formula:

$$S(x) = f\left(\frac{v}{\pi T} \sinh\left[\frac{\pi T x}{v}\right]\right), \quad \text{at} \quad T > 0 \qquad (16.B.6)$$

So at positive temperature the entropy of a subsystem is described in terms of the same function with a different argument. In order to find the function f we should match two different expressions for asymptotic of $S(x)$ for positive temperature. For large x the formula (16.B.6) simplifies:

$$S(x) = f\left(\exp\left[\frac{\pi T(x - x_0)}{v}\right]\right), \quad Tx \to \infty. \qquad (16.B.7)$$

Here $\pi T x_0/v = -\ln(v/2\pi T)$. This formula should coincide with (16.B.4). Both represent the entropy of a subsystem for small positive temperatures. This provides an equation for f:

$$f\left(\exp\left[\frac{\pi T(x - x_0)}{v}\right]\right) = \frac{\pi T c}{3v}(x - x_0) \qquad (16.B.8)$$

This formula describes asymptotic of large x, so we have added $-x_0$ to the right-hand side. In the low-temperature case $x \gg x_0$. We are considering the region $x > 1/T$ and $x_0 \sim \ln(1/T)$ at $T \to 0$.

In order to solve the equation for f, let us introduce a new variable $y = \exp\left[\pi T(x - x_0)/v\right]$. Then the last equation turns into:

$$f(y) = \frac{c}{3} \ln y \qquad (16.B.9)$$

Thus we have found how the function f depends on its argument. The dependence will not change as we change the notation of the argument from y to x. So at zero temperature entropy of electrons containing on the interval $(0, x)$ is:

$$S(x) = \frac{c}{3} \ln x \quad \text{as} \quad x \to \infty, \quad T = 0 \qquad (16.B.10)$$

Let us remember that for the attractive Hubbard model $c = 1$, see [65].

17

Mathematical appendices

17.1 Useful integrals

In this appendix we list a number of identities that are useful for manipulating the TBA equations.

17.1.1 'Symmetric integration'

For any well-behaved function $f(x)$ we have

$$\int_{-\pi}^{\pi} dk \, \cos k \, f(\sin k) = 2 \int_{0}^{\pi} dk \, \cos k \, f(\sin k) = 0 . \tag{17.1}$$

The second identity is proved by substituting $k = \pi - k'$ and this implies that the first integral is zero as well.

17.1.2 Fourier transforms

For $a > 0$ we have

$$\int_{-\infty}^{\infty} \frac{dx}{2\pi} \, \exp(-i\omega x) \, \frac{2a}{a^2 + x^2} = \exp(-a|\omega|) , \tag{17.2}$$

$$\int_{-\infty}^{\infty} \frac{dx}{2\pi} \, \frac{\exp(-i\omega x)}{2 \cosh ax} = \frac{1}{4a \, \cosh(\omega\pi/2a)} , \tag{17.3}$$

$$\int_{-\infty}^{\infty} d\Lambda \, 2 \arctan(\Lambda/a) \, \exp(i\omega\Lambda) = -\frac{2\pi}{i\omega} \exp(-|a\omega|) . \tag{17.4}$$

17.1.3 Identities involving the integral kernels

Let us recall the definitions for the functions $s(x)$, $R(x)$, $a_n(x)$ and $A_{nm}(x)$:

$$s(x) = \frac{1}{4u \cosh(\pi x/2u)} = \frac{1}{2\pi} \int_{-\infty}^{\infty} d\omega \frac{\exp(-i\omega x)}{2 \cosh(\omega u)} , \tag{17.5}$$

$$R(x) = \int_{-\infty}^{\infty} \frac{d\omega}{2\pi} \frac{\exp(i\omega x)}{1 + \exp(2u|\omega|)}, \tag{17.6}$$

$$a_n(x) = \frac{1}{2\pi} \frac{2nu}{(nu)^2 + x^2}. \tag{17.7}$$

The following identities hold:

$$\int_{-\infty}^{\infty} dy \, s(x - y) \left[a_{m-1}(y) + a_{m+1}(y) \right] = a_m(x). \tag{17.8}$$

$$\sum_{n=1}^{\infty} \int_{-\infty}^{\infty} dy \, A_{kn}^{-1}(x - y) \, a_n(y - \sin k) = \delta_{k,1} \, s(x - \sin k). \tag{17.9}$$

$$\sum_{n=1}^{\infty} \int_{-\infty}^{\infty} dy \, A_{kn}^{-1}(x - y) \left(4\mathrm{Re}\sqrt{1 - (y - inu)^2} - 2n\mu - 4nu \right)$$
$$= \delta_{k,1} \int_{-\pi}^{\pi} dk \, 2\cos^2(k) \, s(x - \sin k). \tag{17.10}$$

$$\int_{-\infty}^{\infty} d\Lambda \, a_1(x - \Lambda) \, s(\Lambda - y) = R(x - y). \tag{17.11}$$

$$\int_{-\infty}^{\infty} d\Lambda \, a_1(x - \Lambda) \left[\delta(\Lambda - y) - R(\Lambda - y) \right] = s(x - y). \tag{17.12}$$

Equation (17.8) can be proved by Fourier transformation, (17.9) is a direct consequence of (17.8), and (17.10) follows from (17.13) and (17.9). Equations (17.11) and (17.12) are proved by Fourier transformation.

17.1.4 A list of useful integral identities

$$4\mathrm{Re}\sqrt{1 - (\Lambda - inu)^2} - 4nu = \int_{-\pi}^{\pi} \frac{dk}{\pi} \frac{\cos^2 k \, (2nu)}{(nu)^2 + (\sin k - \Lambda)^2}, \quad u > 0. \tag{17.13}$$

$$\int_{-\infty}^{\infty} \frac{d\mu}{2\pi} \frac{2a}{a^2 + (\lambda - \mu)^2} \frac{2b}{b^2 + (\mu - v)^2} = \frac{2(a + b)}{(a + b)^2 + (\lambda - v)^2}, a, b > 0. \tag{17.14}$$

$$2\mathrm{Re}\left[\arcsin(\Lambda + ia)\right] = \int_{-\infty}^{\infty} \frac{d\omega}{i\omega} J_0(\omega) \, \exp(-|a\omega| + i\omega\Lambda). \tag{17.15}$$

$$2\mathrm{Re}\left[\arcsin(\Lambda + ia)\right] = \int_{-\pi}^{\pi} \frac{dk}{2\pi} \theta\left(\frac{\Lambda - \sin(k)}{a}\right), \quad a > 0. \tag{17.16}$$

$$2\mathrm{Re}\frac{1}{\sqrt{1 - (\Lambda + ia)^2}} = \int_{-\pi}^{\pi} \frac{dk}{2\pi} \frac{2a}{a^2 + (\Lambda - \sin k)^2}, \quad a > 0. \tag{17.17}$$

$$\int_{-\infty}^{\infty} \frac{d\omega}{i\omega} \frac{\exp(i\omega x)}{1 + \exp(2u|\omega|)} = i \ln \left[\frac{\Gamma\left(\frac{1}{2} + i\frac{x}{4u}\right)\Gamma\left(1 - i\frac{x}{4u}\right)}{\Gamma\left(\frac{1}{2} - i\frac{x}{4u}\right)\Gamma\left(1 + i\frac{x}{4u}\right)} \right], \quad u > 0. \qquad (17.18)$$

17.1.5 Integrals involving Bessel functions

$$J_n(z) = \frac{1}{2\pi} \int_{-\pi}^{\pi} d\theta \, \exp(iz\sin\theta - in\theta). \qquad (17.19)$$

$$\int_{-\pi}^{\pi} dk \, \cos^2(k) \, \exp(i\omega\sin k) = \frac{2\pi \, J_1(\omega)}{\omega}. \qquad (17.20)$$

17.2 The Wiener-Hopf method

Consider a linear Fredholm integral equations of the type

$$f(x) = f^{(0)}(x) - \mathcal{K}_X * f(x). \qquad (17.21)$$

Here $\mathcal{K}_X$ is an integral operator which acts on a function $f(x)$ as

$$\mathcal{K}_X * f(x) = \int_{-X}^{X} dy \, K(x - y)f(y). \qquad (17.22)$$

For simplicity we assume that the both the kernel $K(x)$ of the integral operator and the 'driving term' $f^{(0)}(x)$ are even functions

$$K(x) = K(-x), \qquad f^{(0)}(-x) = f^{(0)}(x), \qquad (17.23)$$

which are defined on the entire real axis. The integration boundary X is supposed to be large but finite. The operator $1 + \mathcal{K}_\infty$ is non-degenerate and its resolvent $\bar{\mathcal{K}}$, defined as

$$(1 + \mathcal{K}_\infty)^{-1} \equiv 1 - \bar{\mathcal{K}} \qquad (17.24)$$

can be obtained e.g. by Fourier transformation. Following Yang and Yang [495] we now rewrite (17.21) as

$$(1 + \mathcal{K}_\infty) * f(x) = f^{(0)}(x) + \left\{ \int_{-\infty}^{-X} + \int_{X}^{\infty} \right\} dy \, K(x - y)f(y). \qquad (17.25)$$

Acting with $(1 + \mathcal{K}_\infty)^{-1}$ on both sides of (17.25) we arrive at

$$f(x) = f_\infty(x) + \left\{ \int_{-\infty}^{-X} + \int_{X}^{\infty} \right\} dy \, \bar{K}(x - y)f(y), \qquad (17.26)$$

where $\bar{K}$ is the kernel of the integral operator $\bar{\mathcal{K}}$ introduced in equation (17.24) and f_∞ is the solution of (17.21) for $X = \infty$. In many cases an explicit expression for f_∞ can be derived by Fourier transformation. Using that $f(x) = f(-x)$ and shifting the variables in

(17.26) by introducing $g(z) = f(X + z)$ we obtain

$$g(z) = f_\infty(X + z) + \int_0^\infty dz' \, \bar{K}(z - z') f(z')$$

$$+ \int_0^\infty dz' \, \bar{K}(2X + z + z') f(z'). \tag{17.27}$$

Assuming that the kernel $\bar{K}(x)$ vanishes sufficiently fast with $x > 0$, we may solve the integral equation (17.27) by a rapidly converging expansion [495]

$$g(z) = \sum_{n=0}^\infty g_n(x) . \tag{17.28}$$

The functions $g_n(z)$ fulfil linear integral equations of the form

$$g_n(z) = g_n^{(0)}(z) + \int_0^\infty dz' \, \bar{K}(z - z') g_n(z') , \tag{17.29}$$

where

$$g_0^{(0)}(z) = f_\infty(X + z) ,$$

$$g_n^{(0)}(z) = \int_0^\infty dz' \, \bar{K}(2X + z + z') g_{n-1}(z') , \quad n \geq 1. \tag{17.30}$$

The resulting equations for the functions $g_n(x)$ are of Wiener-Hopf type and can be solved as follows (see e.g. [329]). We start by Fourier-transforming (17.29)

$$(1 - \bar{K}(\omega)) g_n^+(\omega) + g_n^-(\omega) = g_n^{(0)}(\omega) , \tag{17.31}$$

where

$$g_n^\pm(\omega) = \int dz \, \theta_H(\pm z) g_n(z) \exp(i\omega z) \tag{17.32}$$

provide a decomposition of $g_n(\omega)$ into a sum of two parts that are analytic in the upper and lower half planes, respectively ($\theta_H(z)$ is the Heaviside step function). The key to the solution of the Wiener-Hopf equation (17.31) is to decompose the kernel into factors $G^\pm$ that are analytic in the upper and lower complex ω-plane, respectively

$$1 - \bar{K}(\omega) = [G^+(\omega) G^-(\omega)]^{-1}, \qquad \lim_{\omega \to \infty} G^\pm(\omega) = 1 \tag{17.33}$$

Using such a factorization equation (17.31) becomes

$$[G^+(\omega)]^{-1} g_n^+(\omega) + G^-(\omega) g_n^-(\omega) = Q_n^+(\omega) + Q_n^-(\omega) \tag{17.34}$$

where $Q_n^\pm(\omega)$ are analytic in the upper and lower half planes respectively:

$$Q_n^+(\omega) + Q_n^-(\omega) = G^-(\omega) g_n^{(0)}(\omega). \tag{17.35}$$

Using the analytic properties of the functions involved we obtain the solution of equation (17.34)

$$g_n^+(\omega) = G^+(\omega)Q_n^+(\omega) ,$$

$$g_n^-(\omega) = \frac{Q_n^-(\omega)}{G^-(\omega)} . \tag{17.36}$$

For practical applications the following identities are useful

$$\int_0^\infty dz \, g_n(z) = g_n^+(\omega = 0), \quad g_n(z = 0) = -i \lim_{\omega \to \infty} \omega g_n^+(\omega) . \tag{17.37}$$

References

[1] E. Abe, *Hopf Algebras*, vol. 74 of Cambridge Tracts in Mathematics (Cambridge: Cambridge University Press, 1980).

[2] M. J. Ablowitz and H. Segur, *Solitons and the Inverse Scattering Transform* (Philadelphia: SIAM, 1981).

[3] M. Abramowitz and I. Stegun, eds., *Handbook of Mathematical Functions*, 8th edition (New York: Dover Publications, Inc., 1975).

[4] A. A. Abrikosov, L. P. Gorkov and I. E. Dzyaloshinski, *Methods of Quantum Field Theory in Statistical Physics* (New York: Dover, 1975).

[5] I. Affleck, Exact critical exponents for quantum spin chains, non-linear σ-models at $\theta = \pi$ and the Quantum Hall effect, *Nucl. Phys. B* **265** (1986) 409.

[6] — Universal term in the free energy at a critical point and the conformal anomaly, *Phys. Rev. Lett.* **56** (1986) 746.

[7] — Field theory methods and quantum critical phenomena, in E. Brézin and J. Zinn-Justin, eds., *Champs, cordes et phénomène critique* (Amsterdam: North-Holland, 1990), 563–640. Les Houches, Session XLIX, 28 June – 5 August 1988.

[8] — Field theory methods and strongly correlated electrons, in H.-C. Lee, ed., *Physics, Geometry and Topology*, vol. I (New York: Plenum Publishers, 1990). Proceedings of a NATO Advanced Study Institute and Banff Summer School in Theoretical Physics, held 14–25 August 1989, in Banff, Alberta, Canada.

[9] — Boundary condition changing operators in conformal field theory and condensed matter physics, *Nucl. Phys. B* (Proc. Suppl.) **58** (1997) 35.

[10] I. Affleck, D. Gepner, H. J. Schulz and T. Ziman, Critical behavior of spin S Heisenberg antiferromagnetic chains: analytic and numerical results, *J. Phys. A* **22** (1989) 511.

[11] I. Affleck and F. D. M. Haldane, Critical theory of quantum spin chains, *Phys. Rev. B* **36** (1987) 5291.

[12] I. Affleck and A. W. W. Ludwig, Critical theory of overscreened Kondo fixed points, *Nucl. Phys. B* **360** (1991) 641.

[13] — The Fermi edge singularity and boundary condition changing operators, *J. Phys. A* **27** (1994) 5375.

[14] I. Affleck and M. Oshikawa, On the field-induced gap in Cu Benzoate and other $S = 1/2$ antiferromagnetic chains, *Phys. Rev. B* **60** (1999) 1038.

[15] F. C. Alcaraz and R. Z. Bariev, Interpolation between Hubbard and supersymmetric t-J models: two-parameter integrable models of correlated electrons, *J. Phys. A* **32** (1999) L483.

[16] F. C. Alcaraz and M. J. Martins, Conformal invariance and the operator content of the XXZ model with arbitrary spin, *J. Phys. A* **22** (1989) 1829.

[17] P. W. Anderson, Infrared catastrophe in Fermi gases with local scattering potentials, *Phys. Rev. Lett.* **18** (1967) 1049.

[18] N. Andrei, Integrable models in condensed matter physics, preprint, cond-mat/ 9408101.

[19] N. Andrei and C. Destri, Dynamical symmetry-breaking and fractionalization in a new integrable model, *Nucl. Phys. B* **231** (1984) 445.

[20] N. Andrei, K. Furuya and J. H. Lowenstein, Solution of the Kondo problem, *Rev. Mod. Phys.* **55** (1983) 331.

[21] H. Araki, Gibbs states of a one-dimensional quantum lattice, *Comm. Math. Phys.* **14** (1969) 120.

[22] V. I. Arnold, *Mathematical Methods of Classical Mechanics, 2nd edition* (Berlin: Springer-Verlag, 1989).

[23] H. Asakawa and M. Suzuki, Boundary susceptibilities of the Hubbard model in open chains, *J. Phys. A* **29** (1996) 7811.

[24] — Finite-size corrections in the XXZ model and the Hubbard model with boundary fields, *J. Phys. A* **29** (1996) 225.

[25] N. W. Ashcroft and N. D. Mermin, *Solid State Physics* (New York: Holt, Rinehart and Winston, 1976).

[26] R. Assaraf, P. Azaria, M. Caffarel and P. Lecheminant, Metal-insulator transition in the one-dimensional SU(N) Hubbard model, *Phys. Rev. B* **60** (1999) 2299.

[27] A. Auerbach, *Interacting Electrons and Quantum Magnetism* (New York: Springer-Verlag, 1994).

[28] L. V. Avdeev and B. D. Dörfel, The Bethe ansatz equations for the isotropic Heisenberg antiferromagnet of arbitrary spin, *Nucl. Phys. B* **257** (1985) 253.

[29] — Solutions of the Bethe ansatz equations for XXX-antiferromagnet of arbitrary spin in the case of a finite number of sites, *Theor. Math. Phys.* **71** (1987) 528.

[30] O. Babelon, H. J. de Vega and C. M. Viallet, Analysis of the Bethe ansatz equations of the XXZ model, *Nucl. Phys. B* **220** (1983) 13.

[31] H. M. Babujian, Exact solution of the isotropic Heisenberg chain with arbitrary spin – thermodynamics of the model, *Nucl. Phys. B* **215** (1983) 317.

[32] H. M. Babujian, A. Fring, M. Karowski and A. Zapletal, Exact form factors in integrable quantum field theories: the sine-Gordon model, *Nucl. Phys. B* **538** (1999) 535.

[33] H. M. Babujian and M. Karowski, Exact form factors in integrable quantum field theories: the sine-Gordon model (II), *Nucl. Phys. B* **620** (2002) 407.

[34] — Sine-Gordon breather form factors and quantum field equations, *J. Phys.* A **35** (2002) 9081.

[35] D. Baeriswyl, D. K. Campbell and S. Mazumdar, in H. Keiss, ed., *Conjugated Conducting Polymers* (Berlin: Springer, 1992).

[36] D. Baeriswyl, J. Carmelo and A. Luther, Correlation effects on the oscillator strength of optical absorption: sum rule for the one-dimensional Hubbard model, *Phys. Rev.* B **33** (1986) 7247. Erratum: *Phys. Rev.* B **34** (1986) 8976.

[37] J. Balog and M. Niedermaier, Off-shell dynamics of the O(3) NLS model beyond Monte Carlo and perturbation theory, *Nucl. Phys.* B **500** (1997) 421.

[38] R. Bannister and N. d'Ambrumenil, Spectral functions of half-filled one-dimensional Hubbard rings with varying boundary conditions, *Phys. Rev.* B **61** (2000) 4651.

[39] P.-A. Bares, J. M. P. Carmelo, J. Ferrer and P. Horsch, Charge-spin recombination in the one-dimensional t-J model, *Phys. Rev.* B **46** (1992) 14624.

[40] P.-A. Bares and F. Gebhard, Asymptotic Bethe-ansatz results for a Hubbard chain with 1/sinh-hopping, *Europhys. Lett.* **29** (1995) 573.

[41] — Critical behavior of a one-dimensional Hubbard-model with 1/sinh hopping, *J. Phys.: Condens. Matter* **7** (1995) 2285.

[42] R. Z. Bariev, Two-dimensional ice type vertex model with 2 types of staggered sites. 1. The free energy and polarization, *Theor. Math. Phys.* **49** (1981) 1021.

[43] R. J. Baxter, Eight-vertex model in lattice statistics, *Phys. Rev. Lett.* **26** (1971) 832.

[44] — Partition function of the eight-vertex lattice model, *Ann. Phys.* (N.Y.) **70** (1972) 193.

[45] — *Exactly Solved Models in Statistical Mechanics* (London: Academic Press, 1982).

[46] — Completeness of the Bethe ansatz for the six and eight-vertex models, *J. Stat. Phys.* **108** (2002) 1.

[47] V. V. Bazhanov, Integrable quantum systems and classical Lie algebras, *Comm. Math. Phys.* **113** (1987) 471.

[48] G. Bedürftig, B. Brendel, H. Frahm and R. M. Noack, Friedel oscillations in the open Hubbard chain, *Phys. Rev.* B **58** (1998) 10225.

[49] G. Bedürftig and H. Frahm, Spectrum of boundary states in the open Hubbard chain, *J. Phys.* A **30** (1997) 4139.

[50] — Tunneling singularities in the open Hubbard chain, *Physica E* **4** (1999) 246.

[51] A. A. Belavin, A. M. Polyakov and A. B. Zamolodchikov, Infinite conformal symmetry in two-dimensional quantum field theory, *Nucl. Phys.* B **241** (1984) 333.

[52] F. A. Berezin and V. N. Sushko, Relativistic two-dimensional model of a self-interacting fermion field with nonvanishing rest mass, *Sov. Phys. JETP* **48** (1965) 865.

[53] B. Berg, M. Karowski and P. Weisz, Construction of Green's functions from an exact S matrix, *Phys. Rev.* D **19** (1979) 2477.

[54] B. Berg, M. Karowski, P. Weisz and V. Kurak, Factorized $U(n)$ symmetric S-matrices in two dimensions, *Nucl. Phys.* B **134** (1978) 125.

[55] B. Berg and P. Weisz, Exact S-matrix of the chiral invariant SU(N) Thirring model, *Nucl. Phys. B* **146** (1978) 205.

[56] A. Berkovich, Temperature and magnetic field dependent correlators of the exactly integrable $(1+1)$-dimensional gas of impenetrable fermions, *J. Phys. A* **24** (1991) 1543.

[57] D. Bernard, Hidden Yangians in 2D massive current algebras, *Comm. Math. Phys.* **137** (1991) 191.

[58] — An introduction to Yangian symmetries, *Int. J. Mod. Phys. B* **7** (1993) 3517.

[59] D. Bernard, M. Gaudin, F. D. M. Haldane and V. Pasquier, Yang-Baxter equation in long-range interacting systems, *J. Phys. A* **26** (1993) 5219.

[60] H. Bethe, Zur Theorie der Metalle. I. Eigenwerte und Eigenfunktionen der linearen Atomkette, *Z. Phys.* **71** (1931) 205.

[61] J. L. Black and V. Emery, Critical properties of two-dimensional models, *Phys. Rev. B* **23** (1981) 429.

[62] H. W. Blöte, J. L. Cardy and M. P. Nightingale, Conformal invariance, the central charge, and universal finite-size amplitudes at criticality, *Phys. Rev. Lett.* **56** (1986) 742.

[63] N. M. Bogoliubov, A. G. Izergin and V. E. Korepin, Critical exponents for integrable models, *Nucl. Phys. B* **275** (1986) 687.

[64] N. M. Bogoliubov, A. G. Izergin and N. Y. Reshetikhin, Finite-size effects and infrared asymptotics of the correlation functions in 2 dimensions, *J. Phys. A* **20** (1987) 5361.

[65] N. M. Bogolyubov and V. E. Korepin, The role of quasi-one-dimensional structures in high-T_c superconductivity, *Int. J. Mod. Phys. B* **3** (1989) 427.

[66] — Correlation functions of the one-dimensional Hubbard model, *Theor. Math. Phys.* **82** (1990) 231.

[67] — The mechanism of Cooper pairing in the one-dimensional Hubbard model, *Proceedings of the Steklov Institute of Mathematics* **2** (1992) 47.

[68] M. Born and K. Huang, *Dynamical Theory of Crystal Lattices* (Oxford: The Clarendon Press, 1954).

[69] C. Bourbonnais and D. Jerome, in P. Bernier, S. Lefrant and G. Bidan. eds., *Advances in Synthetic Metals, Twenty Years of Progress in Science and Technology* (New York: Elsevier 1999). Preprint, cond-mat/9903101.

[70] M. Brech, J. Voit and H. Büttner, Momentum distribution function of the one-dimensional Hubbard model – an analytical approach, *Europhys. Lett.* **12** (1990) 289.

[71] J. C. Campuzano, M. R. Norman and M. Randeria, Photoemission in the high-T_c superconductors, in K.H. Bennemann and J B. Ketterson eds., *Physics of Conventional and Unconventional Superconductors* (Berlin: Springer Verlag, 2003). Preprint cond-mat/0209476.

[72] J. Cardy and G. Mussardo, Universal properties of self-avoiding walks from 2-dimensional field theory, *Nucl. Phys. B* **410** (1993) 451.

[73]   J. L. Cardy, Conformal invariance and surface critical behavior, *Nucl. Phys. B* **240** (1984) 514.

[74]   — Conformal invariance and universality in finite-size scaling, *J. Phys. A* **17** (1984) L385.

[75]   — Operator content of two-dimensional conformally invariant theories, *Nucl. Phys. B* **270** (1986) 186.

[76]   — *Scaling and Renormalization in Statistical Physics* (Cambridge: Cambridge University Press, 1996).

[77]   J. M. P. Carmelo, P. Horsch, P. A. Bares and A. A. Ovchinnikov, Renormalized pseudoparticle description of the one-dimensional Hubbard-model thermodynamics, *Phys. Rev. B* **44** (1991) 9967.

[78]   J. M. P. Carmelo, P. Horsch and A. A. Ovchinnikov, Static properties of one-dimensional generalized Landau liquids, *Phys. Rev. B* **45** (1992) 7899.

[79]   W. J. Caspers, M. Labuz, A. Wal, M. Kuzma and T. Lulek, From asymptotic to finite Heisenberg chain – the evolution of Bethe solutions, *J. Phys. A* **36** (2003) 5369.

[80]   V. Chari and A. Pressley, Yangians and R-matrices, *L'Enseignement Mathématique* **36** (1990) 267.

[81]   — Fundamental representations of Yangians and singularities of R-matrices, *J. Reine Angew. Math.* **417** (1991) 87.

[82]   — *A Guide to Quantum Groups* (Cambridge: Cambridge University Press, 1994).

[83]   T. C. Choy and F. D. M. Haldane, Failure of Bethe-ansatz solutions of generalisations of the Hubbard chain to arbitrary permutation symmetry, *Phys. Lett. A* **90** (1982) 83.

[84]   T. C. Choy and W. Young, On the continuum spin-wave spectrum of the one-dimensional Hubbard model, *J. Phys. C* **15** (1982) 521.

[85]   R. Claessen, M. Sing, U. Schwingenschloegl, P. Blaha, M. Dressel and C. S. Jacobsen, Spectroscopic signatures of spin-charge separation in the quasi-one-dimensional organic conductor TTF-TCNQ, *Phys. Rev. Lett.* **88** (2002) 096402.

[86]   J. des Cloizeaux and J. J. Pearson, Spin-wave spectrum of the antiferromagnetic linear chain, *Phys. Rev.* **128** (1962) 2131.

[87]   S. Coleman, Quantum sine-Gordon equation as the massive Thirring model, *Phys. Rev. D* **11** (1975) 2088.

[88]   C. F. Coll, Excitation spectrum of the one-dimensional Hubbard model, *Phys. Rev. B* **9** (1974) 2150.

[89]   D. Controzzi and F. H. L. Essler, Dynamical density correlation function of 1d Mott insulators in a magnetic field, *Phys. Rev. B* **66** (2002) 165112.

[90]   D. Controzzi, F. H. L. Essler and A. M. Tsvelik, Optical conductivity of one-dimensional Mott insulators, *Phys. Rev. Lett.* **86** (2001) 960.

[91]   A. Damascelli, Z. Hussein and Z.-X. Shen, Angle-resolved photoemission studies of the cuprate superconductors, *Rev. Mod. Phys.* **75** (2003) 473.

[92]   P. Dargis and Z. Maassarani, Fermionization and Hubbard models, *Nucl. Phys. B* **535** (1998) 681.

[93]   R. Dashen and Y. Frishman, Four-fermion interactions and scale invariance, *Phys. Rev. D* **11** (1975) 2781.

[94]   T. Deguchi, Non-regular eigenstate of the XXX model as some limit of the Bethe state, *J. Phys. A* **34** (2001) 9755.

[95]   T. Deguchi, F. H. L. Essler, F. Göhmann, A. Klümper, V. E. Korepin and K. Kusakabe, Thermodynamics and excitations of the one-dimensional Hubbard model, *Phys. Rep.* **331** (2000) 197.

[96]   P. A. Deift and X. Zhou, A steepest descent method for oscillatory Riemann-Hilbert problems. Asymptotics for the MKdV equation, *Ann. of Math.* **137** (1993) 295.

[97]   G. Delfino and G. Mussardo, The spin-spin correlation function in the 2-dimensional Ising model in a magnetic field at $T = T_c$, *Nucl. Phys. B* **455** (1995) 724.

[98]   C. Destri and J. H. Lowenstein, Analysis of the Bethe ansatz equations of the chiral-invariant Gross-Neveu model, *Nucl. Phys. B* **205** (1982) 369.

[99]   C. Destri and T. Segalini, A local and integrable lattice regularization of the massive Thirring model, *Nucl. Phys. B* **455** (1995) 759.

[100] C. Destri and H. J. de Vega, New thermodynamic Bethe ansatz equations without strings, *Phys. Rev. Lett.* **69** (1992) 2313.

[101] — Unified approach to thermodynamic Bethe ansatz and finite size corrections for lattice models and field theories, *Nucl. Phys. B* **438[FS]** (1995) 413.

[102] P. Di Francesco, P. Mathieu and D. Sénéchal, *Conformal Field Theory* (New York: Springer Verlag, 1997).

[103] A. Doikou, L. Mezincescu and R. I. Nepomechie, Factorization of multiparticle scattering in the Heisenberg spin chain, *Mod. Phys. Lett. A* **12** (1997) 2591.

[104] — Simplified calculation of boundary S-matrices, *J. Phys. A* **30** (1997) L507.

[105] B. Doyon and S. Lukyanov, Fermion Schwinger's function for the SU(2) Thirring model, *Nucl. Phys. B* **644** (2002) 451.

[106] R. M. Dreizler and E. K. U. Gross, *Density Functional Theory: an Approach to the Quantum Many Body Problem* (Berlin: Springer-Verlag, 1990).

[107] V. G. Drinfel'd, Hopf algebras and the quantum Yang-Baxter equation, *Dokl. Acad. Nauk SSSR* **283** (1985) 1060.

[108] — A new realization of Yangians and quantized affine algebras, *Dokl. Acad. Nauk SSSR* **296** (1987) 13.

[109] — Quantum groups, *Proceedings of the International Congress of Mathematicians*, 798 (American Mathematical Society, 1987).

[110] V. N. Dutyshev, 2-Dimensional isotopic model of a fermion field with broken SU(2) symmetry, *Sov. Phys. JETP* **78** (1980) 1698.

[111] G. I. Dzhaparidze and A. A. Nersesyan, Magnetic-field phase transition in a one-dimensional system of electrons with attraction, *JETP Lett.* **27** (1978) 334.

[112] I. Dzyaloshinskii, Some consequences of the Luttinger theorem: The Luttinger surfaces in non-Fermi liquids and Mott insulators, *Phys. Rev. B* **68** (2003) 085113.

[113] E. N. Economou and P. N. Poulopoulos, Ground-state energy of the half-filled one-dimensional Hubbard model, *Phys. Rev. B* **20** (1979) 4756.

[114] R. Egger and H. Grabert, Friedel oscillations for interacting fermions in one dimension, *Phys. Rev. Lett.* **75** (1995) 3505.

[115] V. J. Emery, A. Luther and I. Peschel, Solution of the one-dimensional electron gas on a lattice, *Phys. Rev. B* **13** (1976) 1272.

[116] F. H. L. Essler, The supersymmetric t-J model with a boundary, *J. Phys. A* **29** (1996) 6183.

[117] F. H. L. Essler and H. Frahm, X-ray edge singularity in integrable lattice models of correlated electrons, *Phys. Rev. B* **56** (1997) 6631.

[118] — Density correlations in the half-filled Hubbard model, *Phys. Rev. B* **60** (1999) 8540.

[119] F. H. L. Essler and V. E. Korepin, Higher conservation laws and algebraic Bethe Ansätze for the supersymmetric t-J model, *Phys. Rev. B* **46** (1992) 9147.

[120] — Scattering matrix and excitation spectrum of the Hubbard model, *Phys. Rev. Lett.* **72** (1994) 908.

[121] — SU(2) × SU(2) invariant scattering matrix of the Hubbard model, *Nucl. Phys. B* **426** (1994) 505.

[122] F. H. L. Essler, V. E. Korepin and K. Schoutens, Complete solution of the one-dimensional Hubbard model, *Phys. Rev. Lett.* **67** (1991) 3848.

[123] — Completeness of the SO(4) extended Bethe ansatz for the one-dimensional Hubbard model, *Nucl. Phys. B* **384** (1992) 431.

[124] — Fine structure of the Bethe ansatz for the spin-$\frac{1}{2}$ Heisenberg XXX model, *J. Phys. A* **25** (1992) 4115.

[125] — New eigenstates of the one-dimensional Hubbard model, *Nucl. Phys. B* **372** (1992) 559.

[126] — New exactly solvable model of strongly correlated electrons motivated by high-T_c superconductivity, *Phys. Rev. Lett.* **68** (1992) 2960.

[127] — Exact solution of an electronic model of superconductivity, *Int. J. Mod. Phys. B* **8** (1994) 3205.

[128] F. H. L. Essler and A. M. Tsvelik, Weakly coupled one-dimensional Mott insulators, *Phys. Rev. B* **65** (2002) 115117.

[129] — Finite temperature spectral function of Mott insulators and charge density wave states, *Phys. Rev. Lett.* **90** (2003) 126401.

[130] M. Fabrizio and A. O. Gogolin, Interacting one-dimensional electron gas with open boundaries, *Phys. Rev. B* **51** (1995) 17827.

[131] L. D. Faddeev, Quantum completely integrable models in field theory, *Sov. Sci. Rev. Math. Phys. C* **1** (1980) 107.

[132] L. D. Faddeev and L. A. Takhtajan, Spectrum and scattering of excitations in the one-dimensional isotropic Heisenberg model, *Zap. Nauchn. Sem. LOMI* **109** (1981) 134. Translated in J. Soviet Math. **24** (1984) 241.

[133] — What is the spin of a spin wave?, *Phys. Lett. A* **85** (1981) 375.

[134] — *Hamiltonian Methods in the Theory of Solitons* (Berlin: Springer-Verlag, 1987).

[135] J.-P. Farges, ed., *Organic Conductors* (New York, Marcel Dekker, 1994).

[136] J. Favand, S. Haas, K. Penc, F. Mila and E. Dagotto, Spectral functions of one-dimensional models of correlated electrons, *Phys. Rev. B* **55** (1997) R4859.

[137] A. M. Finkel'shtein, Correlation functions in one-dimensional Hubbard model, *JETP Lett.* **25** (1977) 73.

[138] A. Foerster and M. Karowski, Algebraic properties of the Bethe ansatz for an spl(2,1)-supersymmetric t-J model, *Nucl. Phys. B* **396** (1993) 611.

[139] E. Fradkin, *Field Theories of Condensed Matter Systems* (Reading, Mass.: Addison Wesley, 1991).

[140] H. Frahm and V. E. Korepin, Critical exponents for the one-dimensional Hubbard model, *Phys. Rev. B* **42** (1990) 10553.

[141] — Correlation functions of the one-dimensional Hubbard model in a magnetic field, *Phys. Rev. B* **43** (1991) 5653.

[142] H. Frahm and M. P. Pfannmüller, On the Hubbard model in the limit of vanishing interaction, *Phys. Lett. A* **204** (1995) 347.

[143] H. Frahm and A. Schadschneider, Critical exponents of the degenerate Hubbard model, *J. Phys. A* **26** (1993) 1463.

[144] H. Frahm and N. A. Slavnov, New solutions to the reflection equation and the projecting method, *J. Phys. A* **32** (1999) 1547.

[145] H. Frahm and N.-C. Yu, Finite size effects in the integrable XXZ Heisenberg model with arbitrary spin, *J. Phys. A* **23** (1990) 2115.

[146] H. Frahm, N.-C. Yu and M. Fowler, The integrable XXZ Heisenberg model with arbitrary spin: construction of the Hamiltonian, the ground-state configuration and conformal properties, *Nucl. Phys. B* **336** (1990) 396.

[147] J. Friedel, The distribution of electrons round impurities in monovalent metals, *Philos. Mag.* **43** (1952) 153.

[148] A. Fring, G. Mussardo and P. Simonetti, Form factors for integrable Lagrangian field theories: the sine-Gordon model, *Nucl. Phys. B* **393** (1993) 413.

[149] S. Fujimoto and N. Kawakami, Exact multicritical properties of the multicomponent interacting fermion model with boundaries, *Phys. Rev. B* **54** (1996) 5784.

[150] H. Fujisawa, T. Yokoya, T. Takahashi, S. Miyasaka, M. Kibune and H. Takagi, Angle-resolved photoemission study of Sr_2CuO_3, *Phys. Rev. B* **59** (1999) 7358.

[151] T. Fujita, T. Kobayashi and H. Takahashi, Large-N behaviour of string solutions in the Heisenberg model, *J. Phys. A* **36** (2003) 1553.

[152] P. Fulde, *Electron Correlations in Molecules and Solids* (Berlin: Springer Verlag, 1991).

[153] F. B. Gallagher and S. Mazumdar, Excitons and optical absorption in one-dimensional extended Hubbard models with short- and long-range interactions, *Phys. Rev. B* **56** (1997) 15025.

[154] M. Gaudin, Un système à une dimension de fermions en interaction, *Phys. Lett. A* **24** (1967) 55.

[155] — Thermodynamics of the Heisenberg-Ising ring for $\Delta \geq 1$, *Phys. Rev. Lett.* **26** (1971) 1301.

[156] — Diagonalisation d'une classe d'hamiltoniens de spin, *Journal de Physique* **37** (1976) 1087.

[157] — *La fonction de l'onde de Bethe pour les modèles exacts de la mécanique statistique* (Paris: Masson, 1983).

[158] F. Gebhard, *The Mott Metal-Insulator Transition* (Berlin: Springer Verlag, 1997).

[159] F. Gebhard, K. Bott, M. Scheidler, P. Thomas and S. W. Koch, Optical absorption of non-interacting tight-binding electrons in a Peierls-distorted chain at half band-filling, *Phil. Mag. B* **75** (1997) 1.

[160] — Optical absorption of strongly correlated half-filled Mott-Hubbard chains, *Phil. Mag. B* **75** (1997) 47.

[161] F. Gebhard, A. Girndt and A. E. Ruckenstein, Charge-gap and spin-gap formation in exactly solvable Hubbard chains with long-range hopping, *Phys. Rev. B* **49** (1994) 10926.

[162] F. Gebhard and A. E. Ruckenstein, Exact results for a Hubbard chain with long-range hopping, *Phys. Rev. Lett.* **68** (1992) 244.

[163] A. Georges, G. Kotliar, W. Krauth and M. J. Rozenberg, Dynamical mean-field theory of strongly correlated fermion systems and the limit of infinite dimensions, *Rev. Mod. Phys.* **68** (1996) 13.

[164] B. Gerganov, A. LeClair and M. Moriconi, Beta function for anisotropic current interactions in 2d, *Phys. Rev. Lett.* **86** (2001) 4753.

[165] T. Giamarchi, Umklapp process and resistivity in one-dimensional fermion systems, *Phys. Rev. B* **44** (1991) 2905.

[166] M. Göckeler, The construction of bound state operators in the cubic Schrödinger field theory, *Z. Phys. C* **7** (1981) 263.

[167] — Quantum Gelfand-Levitan method for the cubic Schrödinger field theory with attractive coupling, *Z. Phys. C* **11** (1981) 125.

[168] A. O. Gogolin, A. A. Nersesyan and A. M. Tsvelik, *Bosonization and Strongly Correlated Systems* (Cambridge: Cambridge University Press, 1998).

[169] F. Göhmann, *The one-dimensional Hubbard model – exact solution and algebraic structure*, Habilitationsschrift, Universität Bayreuth (2001).

[170] — Algebraic Bethe ansatz for the gl(1|2) generalized model and Lieb-Wu equations, *Nucl. Phys. B* **620** (2002) 501.

[171] — Universal correlations of one-dimensional electrons at low density, in J. Harnad and A. R. Its, eds., *Isomonodromic Deformations and Applications in Physics*, vol. **31** of the CRM Proceedings and Lecture Notes Series, p. 131 (Providence, R.I.: AMS, 2002).

[172] F. Göhmann and V. Inozemtsev, The Yangian symmetry of the Hubbard models with variable range hopping, *Phys. Lett. A* **214** (1996) 161.

[173] F. Göhmann, A. R. Its and V. E. Korepin, Correlations in the impenetrable electron gas, *Phys. Lett. A* **249** (1998) 117.

[174] F. Göhmann, A. G. Izergin, V. E. Korepin and A. G. Pronko, Time and temperature dependent correlation functions of the one-dimensional impenetrable electron gas, *Int. J. Mod. Phys. B* **12** (1998) 2409.

[175] F. Göhmann and V. E. Korepin, The Hubbard chain: Lieb-Wu equations and norm of the eigenfunctions, *Phys. Lett. A* **263** (1999) 293.

[176] — Universal correlations of one-dimensional interacting electrons in the gas phase, *Phys. Lett. A* **260** (1999) 516.

[177] — Solution of the quantum inverse problem, *J. Phys. A* **33** (2000) 1199.

[178] F. Göhmann and S. Murakami, Algebraic and analytic properties of the one-dimensional Hubbard model, *J. Phys. A* **30** (1997) 5269.

[179] — Fermionic representations of integrable lattice systems, *J. Phys. A* **31** (1998) 7729.

[180] F. Göhmann and H. Schulz, The exact susceptibility of a Kondo spin-$\frac{1}{2}$ for ferromagnetic coupling and $T = 0$, *J. Phys. Condens. Matter* **2** (1990) 3841.

[181] F. Göhmann and A. Seel, Algebraic Bethe ansatz for the gl(1|2) generalized model II: the three gradings, *J. Phys. A.* **37** (2004) 2843.

[182] C. Gómez, M. Ruiz-Altaba and G. Sierra, *Quantum Groups in Two-dimensional Physics* (Cambridge: Cambridge University Press, 1996).

[183] M. P. Grabowski and P. Mathieu, Structure of the conservation laws in quantum integrable spin chains with short range interactions, *Ann. Phys. (N.Y.)* **243** (1995) 299.

[184] I. S. Gradshteyn and I. M. Ryzhik, *Tables of Integrals, Series, and Products* (London: Academic Press Inc., 1980).

[185] M. T. Grisaru, L. Mezincescu and R. I. Nepomechie, Direct calculation of the boundary S-matrix for the open Heisenberg chain, *J. Phys. A* **28** (1995) 1027.

[186] D. Gross and A. Neveu, Dynamical symmetry breaking in asymptotically free field theories, *Phys. Rev. D* **10** (1974) 3235.

[187] H. Grosse, The symmetry of the Hubbard model, *Lett. Math. Phys.* **18** (1989) 151.

[188] M. C. Gutzwiller, Effect of correlation on ferromagnetism of transition metals, *Phys. Rev. Lett.* **10** (1963) 159.

[189] F. D. M. Haldane, General relation of correlation exponents and spectral properties of one-dimensional Fermi systems: application to the anisotropic S = 1/2 Heisenberg chain, *Phys. Rev. Lett.* **45** (1980) 1358.

[190] — Demonstration of the 'Luttinger liquid' character of Bethe-Ansatz-soluble models of 1-D quantum fluids, *Phys. Lett. A* **81** (1981) 153.

[191] — Effective harmonic-fluid approach to low-energy properties of one-dimensional quantum fluids, *Phys. Rev. Lett.* **47** (1981) 1840.

[192] — 'Luttinger liquid theory' of one-dimensional quantum fluids: I. Properties of the Luttinger model and their extension to the general 1D interacting spinless Fermi gas, *J. Phys. C* **14** (1981) 2585.

[193] — Quantum fluid ground state of the sine-Gordon model with finite soliton density – exact results, *J. Phys. A* **15** (1982) 507.

[194] — Exact Jastrow-Gutzwiller resonating-valence-bond ground-state of the spin 1/2 antiferromagnetic Heisenberg chain with $1/R^2$ exchange, *Phys. Rev. Lett.* **60** (1988) 635.

[195] — Physics of the ideal semion gas: spinons and quantum symmetries of the integrable Haldane-Shastry spin chain, in A. Okiji and N. Kawakami, eds., *Correlation Effects in Low Dimensional Electron Systems* (Berlin: Springer Verlag, 1994).

[196] F. D. M. Haldane, Z. N. C. Ha, J. C. Talstra, D. Bernard and V. Pasquier, Yangian symmetry of integrable quantum chains with long-range interactions and a new description of states in conformal field-theory, *Phys. Rev. Lett.* **69** (1992) 2021.

[197] O. J. Heilmann and E. H. Lieb, Violation of noncrossing rule – Hubbard Hamiltonian for benzene, *Ann. N.Y. Acad. Sci.* **172** (1971) 584.

[198] W. Heisenberg, Über quantentheoretische Umdeutung kinematischer und mechanischer Beziehungen, *Z. Phys.* **33** (1925) 879.

[199] K. Hikami, Yangian symmetry and Virasoro character in a lattice spin system with long-range interactions, *Nucl. Phys. B* **441** (1995) 530.

[200] K. Huang, *Statistical Mechanics*, 2nd edition (New York: John Wiley & Sons, 1987).

[201] J. Hubbard, Electron correlations in narrow energy bands, *Proc. R. Soc. (London) A* **276** (1963) 238.

[202] — Electron correlations in narrow energy bands. 2. Degenerate band case, *Proc. R. Soc. (London) A* **277** (1964) 237.

[203] — Electron correlations in narrow energy bands. 3. Improved solution, *Proc. R. Soc. (London) A* **281** (1964) 401.

[204] — Electron correlations in narrow energy bands. 4. Atomic representation, *Proc. R. Soc. (London) A* **285** (1965) 542.

[205] — Electron correlations in narrow energy bands. 6. Connexion with many-body perturbation theory, *Proc. R. Soc. (London) A* **296** (1967) 100.

[206] — *Electron correlations in narrow energy bands*. V. A perturbation expansion about atomic limit, *Proc. R. Soc. (London) A* **296** (1967) 82.

[207] L. Hulthén, Über das Austauschproblem eines Kristalles, *Arkiv för Matematik, Astronomi och Fysik* **26A** (1938) 1.

[208] A. Ilakovac, M. Kolanovic, S. Pallua and P. Prester, Violation of the string hypothesis and the Heisenberg XXZ spin chain, *Phys. Rev. B* **60** (1999) 7271.

[209] V. I. Inozemtsev, On the connection between the one-dimensional $S = 1/2$ Heisenberg chain and Haldane-Shastry model, *J. Stat. Phys.* **59** (1990) 1143.

[210] V. I. Inozemtsev and R. Sasaki, Scalar symmetries of the Hubbard models with variable range hopping, *Phys. Lett. A* **289** (2001) 301.

[211] A. R. Its, A. G. Izergin, V. E. Korepin and N. Slavnov, Differential equations for quantum correlations functions, *Int. J. Mod. Phys. B* **4** (1990) 1003.

[212] A. R. Its, A. G. Izergin, V. E. Korepin and G. G. Varzugin, Large time and distance asymptotics of field correlation function of impenetrable bosons at finite temperature, *Physica D* **54** (1992) 351.

[213] A. G. Izergin and V. E. Korepin, A lattice model related to the nonlinear Schrödinger equation, *Dokl. Acad. Nauk SSSR* **259** (1981) 76.

[214] — The quantum inverse scattering method approach to correlation-functions, *Comm. Math. Phys.* **94** (1984) 67.

[215] A. G. Izergin, V. E. Korepin and N. Y. Reshetikhin, Correlation-functions in a one-dimensional Bose-gas, *J. Phys. A* **20** (1987) 4799.

[216] — Conformal dimensions in Bethe ansatz solvable models, *J. Phys. A* **22** (1989) 2615.

[217] A. G. Izergin and A. G. Pronko, Correlators in the one-dimensional two-component bose and fermi gases, *Phys. Lett. A* **236** (1997) 445.

[218] — Temperature correlators in the two-component one-dimensional gas, *Nucl. Phys. B* **520** (1998) 594.

[219] A. G. Izergin, A. G. Pronko and N. I. Abarenkova, Temperature correlators in the one-dimensional Hubbard model in the strong coupling limit, *Phys. Lett. A* **245** (1998) 537.

[220] J. D. Jackson, *Classical Electrodynamics*, 3rd edition (New York: Wiley, 1999).

[221] G. I. Japaridze, A. A. Nersesyan and P. B. Wiegmann, Exact results in the two-dimensional U(1) symmetric Thirring model, *Nucl. Phys. B* **230** (1984) 511.

[222] E. Jeckelmann, unpublished.

[223] — Dynamical density-matrix renormalization group method, *Phys. Rev. B* **66** (2002) 045114.

[224] E. Jeckelmann, F. Gebhard and F. H. L. Essler, Optical conductivity of the half-filled Hubbard chain, *Phys. Rev. Lett.* **85** (2000) 3910.

[225] M. Jimbo, A q-difference analogue of U($\mathfrak{g}$) and the Yang-Baxter equation, *Lett. Math. Phys.* **10** (1985) 63.

[226] J. Johnson, S. Krinsky and B. McCoy, Vertical-arrow correlation length in the eight-vertex model and the low-lying excitations of the X-Y-Z Hamiltonian, *Phys. Rev. A* **8** (1973) 2526.

[227] P. Jordan and E. Wigner, Über das Paulische Äquivalenzverbot, *Z. Phys.* **47** (1928) 631.

[228] G. Jüttner and B. D. Dörfel, New solutions of the Bethe ansatz equations for the isotropic and anisotropic spin-1/2 Heisenberg chain, *J. Phys. A* **26** (1993) 3105.

[229] G. Jüttner and A. Klümper, Exact calculation of thermodynamical quantities of the integrable t-J model, *Euro. Phys. Letts.* **37** (1997) 335.

[230] G. Jüttner, A. Klümper and J. Suzuki, Exact thermodynamics and Luttinger liquid properties of the integrable t-J model, *Nucl. Phys. B* **487** (1997) 650.

[231] — Thermodynamics of correlated electrons with bond-charge and Hubbard interaction in one dimension, *J. Phys. A* **30** (1997) 1881.

[232] — The Hubbard chain at finite temperatures: ab initio calculations of Tomonaga-Luttinger liquid properties, *Nucl. Phys. B* **522** (1998) 471.

[233] J. Kanamori, Electron correlation and ferromagnetism of transition metals, *Prog. Theor. Phys.* **30** (1963) 275.

[234] A. Kapustin and S. Skorik, Surface excitations and surface energy of the anti-ferromagnetic XXZ chain by the Bethe ansatz approach, *J. Phys. A* **29** (1996) 1629.

[235] M. Karbach, D. Biegel and G. Müller, Quasiparticles governing the zero-temperature dynamics of the one-dimensional spin-1/2 Heisenberg antiferromagnet in a magnetic field, *Phys. Rev. B* **66** (2002) 054405.

[236] M. Karowski and H. J. Thun, Complete S-matrix of the O(2N) Gross-Neveu model, *Nucl. Phys. B* **109** (1981) 61.

[237] M. Karowski, H. J. Thun, T. T. Truong and P. Weisz, On the uniqueness of a purely elastic S-matrix in (1+1) dimensions, *Phys. Lett. B* **67** (1977) 321.

[238] M. Karowski and P. Weisz, Exact form factors in 1+1 dimensional field theoretic models with soliton behaviour, *Nucl. Phys. B* **139** (1978) 445.

[239] Y. Kato and Y. Kuramoto, Fractional exclusion statistics for the t-J model with long range exchange and hopping, *J. Phys. Soc. Jpn.* **65** (1996) 1622.

[240] N. Kawakami, T. Usuki and A. Okiji, Thermodynamic properties of the one-dimensional Hubbard model, *Phys. Lett. A* **137** (1989) 287.

[241] N. Kawakami and S. K. Yang, Luttinger anomaly exponent of momentum distribution in the Hubbard chain, *Phys. Lett. A* **148** (1990) 359.

[242] I. M. Khamitov, Local fields in the inverse scattering method, *Theor. Math. Phys.* **62** (1985) 217.

[243] — Quantum field scattering theory for the nonlinear Schrödinger equation with repulsive coupling, *Theor. Math. Phys.* **63** (1985) 486.

[244] — Constructive approach to the quantum $(\cosh \varphi)_2$ model. I. the method of the Gel'fand-Levitan-Marchenko equations, *J. Soviet Math.* **40** (1988) 115.

[245] C. Kim, A. Y. Matsuura, Z. X. Shen, N. Montoyama, H. Eisaki, S. Uchida, T. Tohyama and S. Maekawa, Observation of spin-charge separation in one-dimensional $SrCuO_2$, *Phys. Rev. Lett.* **77** (1996) 4054.

[246] C. Kim, Z. X. Shen, N. Montoyama, H. Eisaki, S. Uchida, T. Tohyama and S. Maekawa, Separation of spin and charge excitations in one-dimensional $SrCuO_2$, *Phys. Rev. B* **56** (1997) 15589.

[247] A. N. Kirillov and N. Yu. Reshetikhin, The Yangians, Bethe ansatz and combinatorics, *Lett. Math. Phys.* **12** (1986) 199.

[248] A. N. Kirillov and F. Smirnov, Form factors in the SU(2)-invariant Thirring model, *Zap. Nauchn. Sem. LOMI* **164** (1987) 80.

[249] — A representation of the current algebra connected with the SU(2) invariant Thirring model, *Phys. Lett. B* **198** (1987) 506.

[250] — Form factors in the O(3) nonlinear sigma model, *Int. J. Mod. Phys. A* **3** (1988) 731.

[251] N. Kitanine, J. M. Maillet and V. Terras, Form factors of the XXZ Heisenberg spin-$\frac{1}{2}$ finite chain, *Nucl. Phys. B* **554** (1999) 647.

[252] A. Klümper, Free energy and correlation length of quantum chains related to restricted solid-on-solid lattice models, *Ann. Physik* **1** (1992) 540.

[253] — Thermodynamics of the anisotropic spin-1/2 Heisenberg chain and related quantum chains, *Z. Physik B* **91** (1993) 507.

[254] A. Klümper and R. Z. Bariev, Exact thermodynamics of the Hubbard chain: free energy and correlation length, *Nucl. Phys. B* **458** (1996) 623.

[255] A. Klümper and M. T. Batchelor, An analytic treatment of finite-size corrections of the spin-1 antiferromagnetic XXZ chain, *J. Phys. A* **23** (1990) L189.

[256] A. Klümper, M. T. Batchelor and P. A. Pearce, Central charges of the 6- and 19-vertex models with twisted boundary conditions, *J. Phys. A* **24** (1991) 3111.

[257] A. Klümper, J. R. Martinez, C. Scheeren and M. Shiroishi, The spin-1/2 XXZ chain at finite magnetic field: crossover phenomena driven by temperature, *J. Stat. Phys.* **102** (2001) 937.

[258] A. Klümper, A. Schadschneider and J. Zittartz, A new method for the excitations of the one-dimensional Hubbard model, *Z. Phys. B* **78** (1990) 99.

[259] A. Klümper and J. Zittartz, The eight-vertex model: spectrum of the transfer matrix and classification of the excited states, *Z. Phys. B* **75** (1989) 371.

[260] T. Koma, Thermal Bethe-ansatz method for the one-dimensional Heisenberg model, *Prog. Theor. Phys.* **78** (1987) 1213.

[261] R. M. Konik, Haldane gapped spin chains: exact low temperature expansions of correlation functions, *Phys. Rev. B* **68** (2003) 104435.

[262] R. M. Konik and A. A. W. Ludwig, Exact zero-temperature correlation functions for two-leg Hubbard ladders and carbon nanotubes, *Phys. Rev. B* **64** (2001) 115112.

[263] I. G. Korepanov, Tetrahedral Zamolodchikov algebra and the two-layer flat model in statistical mechanics, *Mod. Phys. Lett. B* **3** (1989) 201.

[264] — Tetrahedral Zamolodchikov algebras corresponding to Baxters's *L*-operators, *Comm. Math. Phys.* **154** (1993) 85.

[265] — Tetrahedron equations and the algebraic geometry, *Zap. Nauchn. Sem. POMI* **209** (1994) 137.

[266] V. E. Korepin, Direct calculation of the S-matrix in the Massive Thirring Model, *Teor. Mat. Fiz.* **41** (1979) 953.

[267] — Analysis of a bilinear relation for the six-vertex model, *Dokl. Acad. Nauk SSSR* **265** (1982) 1361.

[268] — Calculation of norms of Bethe wave functions, *Comm. Math. Phys.* **86** (1982) 391.

[269] — Universality of entropy scaling, *Phys. Rev. Lett.* **92** (2004) 096402.

[270] V. E. Korepin, N. M. Bogoliubov and A. G. Izergin, *Quantum Inverse Scattering Method and Correlation Functions* (Cambridge: Cambridge University Press, 1993).

[271] T. D. Kühner and S. R. White, Dynamical correlation functions using the density matrix renormalization group, *Phys. Rev. B* **60** (1999) 335.

[272] P. P. Kulish, Integrable graded magnets, *Zap. Nauchn. Sem. LOMI* **145** (1985) 140.

[273] P. P. Kulish and N. Yu. Reshetikhin, Generalized Heisenberg ferromagnet and the Gross-Neveu model, *Zh. Eksp. Teor. Fiz.* **80** (1981) 214.

[274] P. P. Kulish, N. Yu. Reshetikhin and E. K. Sklyanin, Yang-Baxter equation and representation theory: I, *Lett. Math. Phys.* **5** (1981) 393.

[275] P. P. Kulish and E. K. Sklyanin, Quantum inverse scattering method and the Heisenberg ferromagnet, *Phys. Lett. A* **70** (1979) 461.

[276] — Solutions of the Yang-Baxter equation, *Zap. Nauchn. Sem. LOMI* **95** (1980) 129.

[277] — Quantum spectral transform method – recent developments, in *Lecture Notes in Physics 151* (Berlin: Springer Verlag, 1982), 61–119.

[278] P. P. Kulish and F. A. Smirnov, Quantum inverse problem and Green functions for the Heisenberg ferromagnet, *Phys. Lett. A* **90** (1982) 74.

[279] — Anisotropic Heisenberg ferromagnet with a ground state of domain wall type, *J. Phys. C* **18** (1985) 1037.

[280] A. Kundu, Integrability and exact solution of correlated hopping multi-chain electron systems, *Nucl. Phys. B* **618** (2001) 500.

[281] A. Kuniba, K. Sakai and J. Suzuki, Continued fraction TBA and functional relations in XXZ model at root of unity, *Nucl. Phys. B* **525** (1998) 597.

[282] H. Kuzmany, *Solid-State Spectroscopy* (Berlin: Springer Verlag, 1998).

[283] L. D. Landau and E. M. Lifshitz, *Quantum Mechanics* (Oxford: Pergamon Press, 1977).

[284] A. LeClair, F. Lesage, S. Sachdev and H. Saleur, Finite temperature correlations in the one-dimensional quantum Ising model, *Nucl. Phys. B* **482** (1996) 579.

[285] A. LeClair and G. Mussardo, Finite temperature correlation functions in integrable QFT, *Nucl. Phys. B* **552** (1999) 624.

[286] A. LeClair and F. A. Smirnov, Infinite quantum group symmetry of fields in massive 2d quantum field theory, *Int. J. Mod. Phys. A* **7** (1992) 2997.

[287] K.-J.-B. Lee and P. Schlottmann, Thermodynamic Bethe-ansatz equations for the Hubbard chain with an attractive interaction, *Phys. Rev. B* **38** (1988) 11566.

[288] — Low-temperature properties of the Hubbard chain with an attractive interaction, *Phys. Rev. B* **40** (1989) 9104. See also comment by F. Woynarovich, *Phys. Rev. B.* **43** (1991) 11448, and reply, *Phys. Rev. B.* **43** (1991) 11451.

[289] F. Lesage, H. Saleur and S. Skorik, Form factors approach to current correlations in one-dimensional systems with impurities, *Nucl. Phys. B* **474** (1996) 602.

[290] E. H. Lieb, Exact solution of the *F* model of an antiferroelectric, *Phys. Rev. Lett.* **18** (1967) 1046.

[291] — Exact solution of the two-dimensional Slater *KDP* model of a ferroelectric, *Phys. Rev. Lett.* **19** (1967) 108.

[292] — Residual entropy of square ice, *Phys. Rev.* **162** (1967) 162.

[293] — Two theorems on the Hubbard model, *Phys. Rev. Lett.* **62** (1989) 1201.

[294] — The Hubbard model – some rigorous results and open problems, in D. Iagolnitzer, ed., *Proceedings of the XIth International Congress of Mathematical Physics, Paris 1994* (Cambridge, Mass: International Press, 1995), 392–412.

[295] E. H. Lieb and T. Kennedy, Proof of the Peierls instability in one dimension, *Phys. Rev. Lett.* **59** (1987) 1309.

[296] E. H. Lieb and W. Liniger, Exact analysis of an interacting Bose gas. I. The general solution and the ground state, *Phys. Rep.* **130** (1963) 1605.

[297] E. H. Lieb and B. Nachtergaele, The Stability of the Peierls Instability for Ring Shaped Molecules, *Phys. Rev. B* **51** (1995) 4777.

[298] E. H. Lieb and F. Y. Wu, Absence of Mott transition in an exact solution of the short-range, one-band model in one dimension, *Phys. Rev. Lett.* **20** (1968) 1445. Erratum: *Phys. Rev. Lett.* **21** (1968) 192.

[299] — Two dimensional ferroelectric models, in C. Domb and M. Green, eds., *Phase Transitions and Critical Phenomena, Vol. I*, (London: Academic Press, 1972), 331–490.

[300] — The one-dimensional Hubbard model: a reminiscence, *Physica A* **321** (2003) 1.

[301] J. Links, H.-Q. Zhou, R. M. McKenzie and M. D. Gould, Ladder operator for the one-dimensional Hubbard model, *Phys. Rev. Lett.* **86** (2001) 5096.

[302] A. W. W. Ludwig and K. J. Wiese, The 4-loop beta-function in the 2d non-abelian Thirring model, and comparison with its conjectured 'exact' form, *Nucl. Phys. B* **661** (2003) 577.

[303] S. Lukyanov, Free-field representation for massive integrable models, *Comm. Math. Phys.* **167** (1995) 183.

[304] — Form factors of exponential fields in the sine-Gordon model, Mod. *Phys. Lett. A* **12** (1997) 2911.

[305] — Low energy effective Hamiltonian for the XXZ spin chain, *Nucl. Phys. B* **552** (1998) 533.

[306] — Finite temperature expectation values of local fields in the sinh-Gordon model, *Nucl. Phys. B* **612** (2001) 391.

[307] S. Lukyanov and A. B. Zamolodchikov, Form factors of soliton-creating operators in the sine-Gordon model, *Nucl. Phys. B* **607** (2001) 437.

[308] M. Lüscher, Dynamical charges in the quantized renormalized massive Thirring model, *Nucl. Phys. B* **117** (1976) 475.

[309] A. Luther and I. Peschel, Single-particle states, Kohn anomaly, and pairing fluctuations in one dimension, *Phys. Rev. B* **9** (1974) 2911.

[310] Z.-Q. Ma, *Yang-Baxter Equation and Quantum Enveloping Algebras* (Singapore: World Scientific, 1993).

[311] Z. Maassarani, Exact integrability of the su(n) Hubbard model, *Mod. Phys. Lett. B* **12** (1998) 51.

[312] — Hubbard models as fusion products of free fermions, *Int. J. Mod. Phys. B* **12** (1998) 1893.

[313] Z. Maassarani and P. Mathieu, The su(n) XX model, *Nucl. Phys. B* **517** (1998) 395.

[314] A. H. MacDonald, S. M. Girvin and D. Yoshioka, t/U expansion for the Hubbard model, *Phys. Rev. B* **37** (1988) 9753.

[315] G. D. Mahan, *Many-Particle Physics* (New York and London: Plenum Press, 1990).

[316] J. M. Maillet and J. Sanchez de Santos, Drinfel'd twists and algebraic Bethe ansatz, in M. Semenov-Tian-Shansky, ed., vol. 201 of *Amer. Math. Soc. Transl., Ser. 2*,

(Providence, R.I.: Amer. Math. Soc., 2000), 137–178. L. D. Faddeev's Seminar on Mathematical Physics

[317] J. M. Maillet and V. Terras, On the quantum inverse problem, *Nucl. Phys. B* **575** (2000) 627.

[318] S. Mandelstam, Soliton operators for the quantized sine-Gordon equation, *Phys. Rev. D* **11** (1975) 3026.

[319] M. J. Martins and P. B. Ramos, The algebraic Bethe ansatz for braid-monoid lattice models, *Nucl. Phys. B* **500** (1997) 579.

[320] — The quantum inverse scattering method for Hubbard-like models, *Nucl. Phys. B* **522** (1998) 413.

[321] D. Mattis, New wave-operator identity applied to study of persistent currents in 1d, *J. Math. Phys.* **15** (1974) 609.

[322] J. B. McGuire, Study of exactly solvable one-dimensional N-body problems, *J. Math. Phys.* **5** (1964) 622.

[323] V. Meden and K. Schönhammer, Spectral functions for the Tomonaga-Luttinger model, *Phys. Rev. B* **46** (1992) 15753.

[324] E. Melzer, On the scaling limit of the 1d Hubbard model at half filling, *Nucl. Phys. B* **443** (1995) 553.

[325] W. Metzner and D. Vollhardt, Correlated lattice fermions in $d = \infty$ dimensions, *Phys. Rev. Lett.* **62** (1989) 324. Erratum: *Phys. Rev. Lett* **62** (1989) 1066.

[326] — Ground-state energy of the $d = 1, 2, 3$ dimensional Hubbard model in the weak-coupling limit, *Phys. Rev. B* **39** (1989) 4462.

[327] P. K. Mitter and P. H. Weisz, Asymptotic scale invariance in a Massive Thirring Model with $U(n)$ symmetry, *Phys. Rev. D* **8** (1973) 4410.

[328] Y. Morita, M. Kohmoto and T. Koma, Quasi-bound states of two magnons in the spin-1/2 XXZ chain, *J. Stat. Phys.* **88** (1997) 745.

[329] P. M. Morse and H. Feshbach, *Methods of Theoretical Physics* (New York: McGraw-Hill, 1953).

[330] J. Moser, 3 integrable Hamiltonian systems connected with isospectral deformations, *Adv. Math.* **16** (1975) 197.

[331] N. F. Mott, The basis of the electron theory of metals, with special reference to the transition metals, *Proc. Phys. Soc. A* **62** (1949) 416.

[332] — *Metal-Insulator Transitions* (London: Taylor and Francis, 1990).

[333] G. Müller, H. Thomas, H. Beck and J. C. Bonner, Quantum spin dynamics of the antiferromagnetic linear chain in zero and nonzero magnetic field, *Phys. Rev. B* **24** (1981) 1429.

[334] S. Murakami, New integrable extension of the Hubbard chain with variable range hopping, *J. Phys. A* **31** (1998) 6367.

[335] S. Murakami and F. Göhmann, Yangian symmetry and quantum inverse scattering method for the one-dimensional Hubbard model, *Phys. Lett. A* **227** (1997) 216.

[336] — Algebraic solution of the Hubbard model on the infinite interval, *Nucl. Phys. B* **512** (1998) 637.

[337] S. Murakami and M. Wadati, Connection between Yangian symmetry and the quantum inverse scattering method, *J. Phys. A* **29** (1996) 7903.

[338] G. Mussardo, Spectral representation of correlation functions in two-dimensional quantum field theories, preprint, hep-th/9405128.

[339] — Off-critical statistical models – Factorized scattering theories and bootstrap program, *Phys. Rep.* **218** (1992) 215.

[340] R. Neudert, M. Knupfer, M. S. Golden, J. Fink, W. Stephan, K. Penc, N. Motoyama, H. Eisaki and S. Uchida, Manifestation of spin-charge separation in the dynamic dielectric response of one-dimensional Sr_2CuO_3, *Phys. Rev. Lett.* **81** (1998) 657.

[341] M. den Nijs, Derivation of extended scaling relations between critical exponents in two-dimensional models from the one-dimensional Luttinger model, *Phys. Rev. B* **23** (1981) 6111.

[342] S. Nishimoto, E. Jeckelmann, F. Gebhard and R. Noack, Application of the density matrix renormalization group in momentum space, *Phys. Rev. B* **65** (2002) 165114.

[343] J. D. Noh, D. S. Lee and D. Kim, Origin of the singular Bethe ansatz solutions for the Heisenberg XXZ spin chain, *Physica A* **287** (2000) 167.

[344] M. Ogata and H. Shiba, Bethe-ansatz wave function, momentum distribution, and spin correlation in the one-dimensional Hubbard model, *Phys. Rev. B* **41** (1990) 2326.

[345] M. Ogata, T. Sugiyama and H. Shiba, Magnetic-field effects on the correlation functions in the one-dimensional strongly correlated Hubbard model, *Phys. Rev. B* **43** (1991) 8401.

[346] T. Ogawa, A. Furusaki and N. Nagaosa, Fermi-edge singularity in one-dimensional systems, *Phys. Rev. Lett.* **68** (1992) 3638.

[347] E. Olmedilla and M. Wadati, Conserved quantities for spin models and fermion models, *J. Phys. Soc. Jpn.* **56** (1987) 4274.

[348] — Conserved quantities of the one-dimensional Hubbard model, *Phys. Rev. Lett.* **60** (1988) 1595.

[349] E. Olmedilla, M. Wadati and Y. Akutsu, Yang-Baxter relations for spin models and Fermion models, *J. Phys. Soc. Jpn.* **56** (1987) 2298.

[350] L. Onsager, Crystal statistics. I. A two-dimensional model with an order-disorder transition, *Phys. Rev.* **65** (1944) 117.

[351] Y. Oreg and A. M. Finkel'stein, Resonance in the Fermi-edge singularity of one-dimensional systems, *Phys. Rev. B* **53** (1996) 10928.

[352] A. A. Ovchinnikov, Excitation spectrum of the one-dimensional Hubbard model, *Sov. Phys. JETP* **30** (1970) 1160.

[353] A. Parola and S. Sorella, Asymptotic spin-spin correlations of the $U \to \infty$ one-dimensional Hubbard model, *Phys. Rev. Lett.* **64** (1990) 1831.

[354] — Spin-charge decoupling and the Green's function of one-dimensional Mott insulators, *Phys. Rev. Lett.* **45** (1992) 13156.

[355] — Theory of hole propagation in one-dimensional insulators and superconductors, *Phys. Rev. B* **57** (1997) 6444.

[356] K. Penc, K. Hallberg, F. Mila and H. Shiba, Shadow band in the one-dimensional infinite-U Hubbard model, *Phys. Rev. Lett.* **77** (1996) 1390.

[357] — Spectral functions of the one-dimensional Hubbard model in the $U \to \infty$ limit: how to use the factorized wave function, *Phys. Rev. B* **55** (1997) 15475.

[358] K. Penc, F. Mila and H. Shiba, Spectral function of the 1D Hubbard model in the $U \to \infty$ limit, *Phys. Rev. Lett.* **75** (1995) 894.

[359] K. Penc and F. Woynarovich, Novel magnetic properties of the Hubbard chain with an attractive interaction, *Z. Phys. B* **85** (1991) 269.

[360] D.-T. Peng and R.-H. Yue, Yang-Baxter equation for the R-matrix of the 1-D SU(n) Hubbard model, *J. Phys. A* **35** (2002) 6985.

[361] M. Pernici, Spin and pairing algebras and ODLRO in a Hubbard model, *Europhys. Lett.* **12** (1990) 75.

[362] M. Pfannmüller and H. Frahm, Algebraic Bethe ansatz for gl(2|1) invariant 36 vertex models, *Nucl. Phys. B* **479** (1996) 575.

[363] V. L. Pokrovsky and A. L. Talapov, Ground state, spectrum, and phase diagram of two-dimensional incommensurate crystals, *Phys. Rev. Lett.* **42** (1979) 65.

[364] W. H. Press, S. A. Teukolsky, W. T. Vetterling and B. P. Flannery, *Numerical Recipes in C, The Art of Scientific Computing* (Cambridge: Cambridge University Press, 1992).

[365] R. Preuss, A. Muramatsu, W. von der Linden, P. Dieterich, F. F. Assaad and W. Hanke, Spectral properties of the one-dimensional Hubbard model, *Phys. Rev. Lett.* **73** (1994) 732.

[366] F. C. Pu, Y. Z. Wu and B. H. Zhao, Quantum inverse scattering method for multicomponent non-linear Schrödinger model of bosons or fermions with repulsive coupling, *J. Phys. A* **20** (1987) 1173.

[367] F. C. Pu and B. H. Zhao, Exact solution of a polaron model in one dimension, *Phys. Lett. A* **118** (1986) 77.

[368] — Quantum inverse scattering transform for the nonlinear Schrödinger model of spin-$\frac{1}{2}$ particles with attractive coupling, *Nucl. Phys. B* **275** (1986) 77.

[369] S. Qin, M. Fabrizio and L. Yu, Impurity in a Luttinger liquid: a numerical study of the finite size energy spectrum and of the orthogonality catastrophe exponent, *Phys. Rev. B* **54** (1996) R9643.

[370] P. B. Ramos and M. J. Martins, One-parameter family of an integrable spl(2|1) vertex model: Algebraic Bethe ansatz and ground state structure, *Nucl. Phys. B* **474** (1996) 678.

[371] — Algebraic Bethe ansatz approach for one-dimensional Hubbard model, *J. Phys. A* **30** (1997) L195.

[372] Y. Ren and P. W. Anderson, Asymptotic correlation functions in the one-dimensional Hubbard model with applications to high-T_c superconductivity, *Phys. Rev. B* **48** (1993) 16662.

[373] N. Yu. Reshetikhin, Calculation of the norm of Bethe vectors in models with SU(3) symmetry, *Zap. Nauchn. Sem. LOMI* **150** (1986) 196.

[374] N. Reshetikhin and F. Smirnov, Hidden quantum group symmetry and integrable perturbations of conformal field theories, *Comm. Math. Phys.* **131** (1990) 157.

[375] H. Saleur, A comment on finite temperature correlations in integrable QFT, *Nucl. Phys. B* **567** (2000) 602.

[376] A. Schadschneider and J. Zittartz, On the one-dimensional Hubbard model away from half-filling, *Z. Phys. B* **82** (1991) 387.

[377] P. Schlottmann, Integrable narrow-band model with possible relevance to heavy fermion systems, *Phys. Rev. B* **36** (1987) 5177.

[378] P. Schmitteckert and U. Eckern, Phase coherence in a random one-dimensional system of interacting fermions: a density-matrix renormalization-group study, *Phys. Rev. B* **53** (1996) 15397.

[379] K. D. Schotte and U. Schotte, Tomonaga's model and the threshold singularity of X-ray spectra of metals, *Phys. Rev.* **182** (1969) 479.

[380] H. Schulz, Hubbard chain with reflecting ends, *J. Phys. C* **18** (1985) 581.

[381] H. J. Schulz, Critical behavior of commensurate-incommensurate phase transitions in two dimensions, *Phys. Rev. B* **22** (1980) 5274.

[382] — Correlation exponents and the metal-insulator transition in the one-dimensional Hubbard model, *Phys. Rev. Lett.* **64** (1990) 2831.

[383] — Correlated fermions in one dimension, *Int. J. Mod. Phys. B* **5** (1991) 57.

[384] — Interacting fermions in one dimension: from weak to strong correlation, in V. J. Emery, ed., *Correlated Electron Systems*, Vol. IX (Singapore: World Scientific, 1993), 199–241. Jerusalem Winter School for Theoretical Physics, 30 Dec. 1991–8 Jan. 1992.

[385] H. J. Schulz, G. Cuniberti and P. Pieri, Fermi liquids and Luttinger liquids, in G. Morandi et al. eds., *Field Theories for Low-Dimensional Condensed Matter Systems*, (Berlin: Springer, 2000).

[386] A. Schwartz, M. Dressel, G. Gruner, V. Vescoli, L. Degiorgi and T. Giamarchi, On-chain electrodynamics of metallic $(TMTSF)_2X$ salts: observation of Tomonaga-Luttinger liquid response, *Phys. Rev. B* **58** (1998) 1261.

[387] J. Schwinger, Field theory commutators, *Phys. Rev. Lett.* **3** (1959) 296.

[388] H. Segur and M. J. Ablowitz, Asymptotic solutions and conservation laws for the nonlinear Schrödinger equation. I, *J. Math. Phys.* **17** (1976) 710.

[389] D. Sénéchal, An introduction to bosonization, preprint, cond-mat/9908262.

[390] D. Sénéchal, D. Perez and M. Pioro-Ladrière, Spectral weight of the Hubbard model through cluster perturbation theory, *Phys. Rev. Lett.* **84** (2000) 522.

[391] B. S. Shastry, Exact integrability of the one-dimensional Hubbard-model, *Phys. Rev. Lett.* **56** (1986) 2453.

[392] — Infinite conservation-laws in the one-dimensional Hubbard-model, *Phys. Rev. Lett.* **56** (1986) 1529. Erratum: 2334.

[393] — Decorated star triangle relations and exact integrability of the one-dimensional Hubbard-model, *J. Stat. Phys.* **50** (1988) 57.

[394] — Exact solution of an $S = 1/2$ Heisenberg antiferromagnetic chain with long-ranged interactions, *Phys. Rev. Lett.* **60** (1988) 639.

[395] B. S. Shastry and B. Sutherland, Twisted boundary conditions and effective mass in Heisenberg-Ising and Hubbard rings, *Phys. Rev. Lett.* **65** (1990) 243.

[396] H. Shiba, Magnetic susceptibility at zero temperature for the one-dimensional Hubbard model, *Phys. Rev. B* **6** (1972) 930.

[397] — Thermodynamic properties of the one-dimensional half-filled-band Hubbard model. II, *Prog. Theor. Phys.* **48** (1972) 2171.

[398] N. Shibata, K. Ueda, T. Nishino and C. Ishii, Friedel oscillations in the one-dimensional Kondo lattice model, *Phys. Rev. B* **54** (1996) 13495.

[399] M. Shiroishi, H. Ujino and M. Wadati, SO(4) symmetry of the transfer matrix for the one-dimensional Hubbard model, *J. Phys. A* **31** (1998) 2341.

[400] M. Shiroishi and M. Wadati, Decorated star-triangle relations for the free-fermion model and a new solvable bilayer vertex model, *J. Phys. Soc. Jpn.* **64** (1995) 2795.

[401] — Yang-Baxter equation for the R-matrix of the one-dimensional Hubbard model, *J. Phys. Soc. Jpn.* **64** (1995) 57.

[402] — Bethe ansatz equation for the Hubbard model with boundary fields, *J. Phys. Soc. Jpn.* **66** (1997) 1.

[403] — Integrable boundary conditions for the one-dimensional Hubbard model, *J. Phys. Soc. Jpn.* **66** (1997) 2288.

[404] E. K. Sklyanin, Quantum version of the method of inverse scattering problem, *Zap. Nauchn. Sem. LOMI* **95** (1980) 55.

[405] — The quantum Toda chain, in *Lecture Notes in Physics 226* (Berlin: Springer Verlag, 1985), 196–233.

[406] — Boundary conditions for integrable quantum systems, *J. Phys. A* **21** (1988) 2375.

[407] — Quantum inverse scattering method. Selected topics, in M.-L. Ge, ed., *Quantum Group and Quantum Integrable Systems (Nankai Lectures in Mathematical Physics)*. (Singapore: World Scientific, 1992), 63–97.

[408] — Separation of variables – new trends, *Prog. Theor. Phys. Suppl.* **118** (1995) 35.

[409] — Generating function of correlators in the sl(2) Gaudin model, *Lett. Math. Phys.* **47** (1999) 275.

[410] E. K. Sklyanin and L. D. Faddeev, Method of the inverse scattering problem and quantum nonlinear Schrödinger equation, *Dokl. Acad. Nauk SSSR* **244** (1978) 1337.

[411] — Quantum mechanical approach to completely integrable models of field theory, *Dokl. Acad. Nauk SSSR* **243** (1978) 1430.

[412] S. Skorik and H. Saleur, Boundary bound states and boundary bootstrap in the sine-Gordon model with Dirichlet boundary conditions, *J. Phys. A* **28** (1995) 6605.

[413] N. A. Slavnov, Calculation of scalar products of the wave functions and form factors in the framework of the algebraic Bethe ansatz, *Teor. Mat. Fiz.* **79** (1989) 232.

[414] F. Smirnov, A general formula for soliton form factors in the quantum sine–Gordon model, *J. Phys. A* **19** (1986) L575.

[415] — Proof of identities which arise in the calculation of form-factors in the sine-Gordon model, *Zap. Nauchn. Sem. LOMI* **161** (1987) 98.

[416] — Quantum groups and generalized statistics in integrable models, *Comm. Math. Phys.* **132** (1990) 415.

[417] — Reductions of the sine-Gordon model as a perturbation of minimal models of conformal field theory, *Nucl. Phys. B* **337** (1990) 156.

[418] — A new set of exact form factors, *Int. J. Mod. Phys. A* **9** (1994) 5121.

[419] — Counting the local fields in the sine-Gordon theory, *Nucl. Phys. B* **453** (1995) 807.

[420] F. A. Smirnov, *Form Factors in Completely Integrable Models of Quantum Field Theory* (Singapore: World-Scientific, 1992).

[421] K. Sogo and M. Wadati, Boost operator and its application to quantum Gelfand-Levitan equation for Heisenberg-Ising chain with spin one-half, *Prog. Theor. Phys.* **69** (1983) 431.

[422] B. Sutherland, Exact solution of a two-dimensional model for hydrogen-bonded crystals, *Phys. Rev. Lett.* **19** (1967) 103.

[423] — An introduction to the Bethe ansatz, in *Lecture Notes in Physics 242* (Berlin: Springer Verlag, 1985), 1–95.

[424] J. Suzuki, Spinons in magnetic chains of arbitrary spins at finite temperatures, *J. Phys. A* **32** (1999) 2341.

[425] J. Suzuki, Y. Akutsu and M. Wadati, A new approach to quantum spin chains at finite temperature, *J. Phys. Soc. Jpn.* **59** (1990) 2667.

[426] J. Suzuki, T. Nagao and M. Wadati, Exactly solvable models and finite size corrections, *Int. J. Mod. Phys. B* **6** (1992) 1119.

[427] M. Suzuki and M. Inoue, The ST-transformation approach to analytic solutions of quantum systems. I. General formulations and basic limit theorems, *Prog. Theor. Phys.* **78** (1987) 787.

[428] A. Swieca, Solitons and confinement, *Fortschr. Phys.* **25** (1977) 303.

[429] M. Takahashi, Magnetization curve of the half-filled Hubbard model, *Prog. Theor. Phys.* **42** (1969) 1098.

[430] — Ground state energy of the one-dimensional electron system with short range interaction. I, *Prog. Theor. Phys.* **44** (1970) 348.

[431] — On the exact ground state energy of Lieb and Wu, *Prog. Theor. Phys.* **45** (1971) 756.

[432] — One-dimensional electron gas with delta-function interaction at finite temperature, *Prog. Theor. Phys.* **46** (1971) 1388.

[433] — One-dimensional Heisenberg model at finite temperature, *Prog. Theor. Phys.* **46** (1971) 401.

[434] — Thermodynamics of the Heisenberg-Ising model for $|\Delta| < 1$ in one dimension, *Phys. Lett. A* **36** (1971) 325.

[435] — One-dimensional Hubbard model at finite temperature, *Prog. Theor. Phys.* **47** (1972) 69.

[436] — Low-temperature specific heat of one-dimensional Hubbard model, *Prog. Theor. Phys.* **52** (1974) 103.

[437] — Half-filled Hubbard model at low temperature, *J. Phys. C* **10** (1977) 1289.

[438] — Correlation length and free energy of the S = 1/2 XYZ chain, *Phys. Rev. B* **43** (1991) 5788. See also *Phys. Rev. B.* **44** (1991) 12382.

[439] — *Thermodynamics of One-Dimensional Solvable Models* (Cambridge: Cambridge University Press, 1999).

[440] Simplification of thermodynamic Bethe ansatz equations, in A. K. Kirillov and N. Liskova, eds., *Physics and Combinatorics* (Singapore: World Scientific, 2001), 299–304.

[441] M. Takahashi and M. Shiroishi, Integral equation generates high-temperature expansion of the Heisenberg chain, *Phys. Rev. Lett.* **89** (2002) 117201.

[442] — Thermodynamic Bethe ansatz equations of one-dimensional Hubbard model and high-temperature expansion, *Phys. Rev. B* **65** (2002) 165104.

[443] M. Takahashi, M. Shiroishi and A. Klümper, Equivalence of TBA and QTM, *J. Phys. A* **34** (2001) L187.

[444] L. Takhtajan, The picture of low-lying excitations in the isotropic Heisenberg chain of arbitrary spins, *Phys. Lett. A* **87** (1982) 479.

[445] L. A. Takhtajan and L. D. Faddeev, The quantum method of the inverse problem and the Heisenberg XYZ model, *Usp. Mat. Nauk* **34** (1979) 13.

[446] J. C. Talstra and F. D. M. Haldane, Integrals of motion of the Haldane-Shastry model, *J. Phys. A* **28** (1995) 2369.

[447] — Dynamical $T = 0$ correlations in the S = 1/2 one dimensional Heisenberg antiferromagnet with $1/r^2$ exchange in a magnetic field, *Phys. Rev. B* **54** (1996) 12594.

[448] V. Tarasov and A. Varchenko, Asymptotic solutions of the quantized Knizhnik-Zamolodchikov equation and Bethe vectors, *Amer. Math. Soc. Transl. (2)* **174** (1996) 235. Hep-th/9406060.

[449] V. O. Tarasov, Structure of quantum L-operators for the R-matrix of the XXZ model, *Teor. Mat. Fiz.* **61** (1984) 163.

[450] — Irreducible monodromy matrices for the R-matrix of the XXZ model and local lattice quantum Hamiltonians, *Teor. Mat. Fiz.* **63** (1985) 175.

[451] — Algebraic Bethe ansatz for the Izergin-Korepin R-matrix, *Theor. Math. Phys.* **76** (1988) 793.

[452] H. Tasaki, The Hubbard model – an introduction and selected rigorous results, *J. Phys.: Condens. Matter* **10** (1998) 4353.

[453] M. G. Tetel'man, Lorentz group for two-dimensional integrable lattice systems, *Sov. Phys. JETP* **55** (1982) 306.

[454] H. B. Thacker, Exact integrability in quantum field theory and statistical systems, *Rev. Mod. Phys.* **53** (1981) 253.

[455] T. T. Truong and K. D. Schotte, Inhomogeneous eight-vertex system and the one-dimensional Fermi gas, *Phys. Rev. Lett.* **47** (1981) 285.

[456]   — Quantum inverse scattering method and the diagonal-to-diagonal transfer matrix of vertex models, *Nucl. Phys. B* **220[FS8]** (1983) 77.

[457]   O. Tsuchiya, Boundary *S*-matrices for the open Hubbard chain with boundary fields, *J. Phys. A* **30** (1997) L245.

[458]   O. Tsuchiya and T. Yamamoto, Boundary bound states for the open Hubbard chain with boundary fields, *J. Phys. Soc. Jpn.* **66** (1997) 1950.

[459]   K. Tsunetsugu, Temperature dependence of spin correlation length of half-filled one-dimensional Hubbard model, *J. Phys. Soc. Jpn.* **60** (1991) 1460.

[460]   A. M. Tsvelick and P. B. Wiegmann, Exact results in the theory of magnetic alloys, *Adv. Phys.* **32** (1983) 453.

[461]   A. M. Tsvelik, *Quantum Field Theory in Condensed Matter Physics* (Cambridge: Cambridge University Press, 1995).

[462]   D. B. Uglov and V. E. Korepin, The Yangian symmetry of the Hubbard model, *Phys. Lett. A* **190** (1994) 238.

[463]   G. V. Uimin and G. V. Fomichev, Excitations in the one-dimensional Hubbard model, *Sov. Phys. JETP* **36** (1973) 1001.

[464]   Y. Umeno, Fermionic R-operator and algebraic structure of 1d Hubbard model: its application to quantum transfer matrix, *J. Phys. Soc. Jpn.* **70** (2001) 2531.

[465]   Y. Umeno, M. Shiroishi and A. Klümper, Correlation length of the 1d Hubbard model at half-filling: equal-time one-particle Green's function, *Europhys. Lett.* **62** (2003) 384.

[466]   Y. Umeno, M. Shiroishi and M. Wadati, Fermionic R-operator and integrability of the one-dimensional Hubbard model, *J. Phys. Soc. Jpn.* **67** (1998) 2242.

[467]   — Fermionic R-operator of the Fermion chain model, *J. Phys. Soc. Jpn.* **67** (1998) 1930.

[468]   T. Usuki, N. Kawakami and A. Okiji, Charge susceptibility of the one-dimensional Hubbard model, *Phys. Lett. A* **135** (1989) 476.

[469]   — Thermodynamic quantities of the one-dimensional Hubbard model at finite temperatures, *J. Phys. Soc. Japan* **59** (1990) 1357.

[470]   H. J. de Vega and F. Woynarovich, Method for calculating finite size corrections in Bethe ansatz systems – Heisenberg chain and 6-vertex model, *Nucl. Phys. B* **251** (1985) 439.

[471]   A. Virosztek and F. Woynarovich, Degenerated ground states and excited states of the S = 1/2 anisotropic antiferromagnetic Heisenberg chain in the easy axis region, *J. Phys. A* **17** (1982) 2985.

[472]   J. Voit, Charge-spin separation and the spectral properties of Luttinger liquids, *Phys. Rev. B* **47** (1993) 15472.

[473]   — One-dimensional Fermi liquids, *Rep. Prog. Phys.* **58** (1995) 977.

[474]   — Dynamical correlation functions of one-dimensional superconductors and Peierls and Mott insulators, *Europhys. Jour. B* **5** (1998) 505.

[475]   M. Wadati, E. Olmedilla and Y. Akutsu, Lax pair for the one-dimensional Hubbard model, *J. Phys. Soc. Jpn.* **56** (1987) 1340.

[476] D. F. Wang, Q. F. Zhong and P. Coleman, Gutzwiller-Jastrow wave functions for the 1/*r* Hubbard model, *Phys. Rev. B* **48** (1993) 8476.

[477] Y. Wang and J. Voit, An exactly solvable Kondo problem for interacting one-dimensional fermions, *Phys. Rev. Lett.* **77** (1996) 4934.

[478] Y. Wang, J. Voit and F.-C. Pu, Exact boundary critical exponents and tunneling effects in integrable models for quantum wires, *Phys. Rev. B* **54** (1996) 8491.

[479] E. T. Whittaker and G. N. Watson, *A Course of Modern Analysis*, 4th edition (Cambridge: Cambridge University Press, 1963), ch. 21.

[480] P. B. Wiegmann, Phase transitions in two-dimensional systems with commutative symmetry group, *Sov. Sci. Rev. Ser. A* **2** (1980) 43.

[481] F. Woynarovich, Excitations with complex wavenumbers in a Hubbard chain: I. States with one pair of complex wavenumbers, *J. Phys. C* **15** (1982) 85.

[482] — Excitations with complex wavenumbers in a Hubbard chain: II. States with several pairs of complex wavenumbers, *J. Phys. C* **15** (1982) 97.

[483] — On the eigenstates of a Heisenberg chain with complex wavenumbers not forming strings, *J. Phys. C* **15** (1982) 6397.

[484] — On the $S^z = 0$ excited states of an anisotropic Heisenberg chain, *J. Phys. A* **15** (1982) 2985.

[485] — Low-energy excited states in a Hubbard chain with on-site attraction, *J. Phys. C* **16** (1983) 6593.

[486] — Spin excitations in a Hubbard chain, *J. Phys. C* **16** (1983) 5293.

[487] — Finite-size effects in a non-half-filled Hubbard chain, *J. Phys. A* **22** (1989) 4243.

[488] F. Woynarovich and H. P. Eckle, Finite-size corrections and numerical calculations for long spin 1/2 Heisenberg chains in the critical region, *J. Phys. A* **20** (1987) L97.

[489] — Finite-size corrections for the low-lying states of a half-filled Hubbard chain, *J. Phys. A* **20** (1987) L443.

[490] F. Woynarovich and P. Forgacs, Scaling limit of the one-dimensional attractive Hubbard model: the half-filled band case, *Nucl. Phys. B* **498** (1997) 565.

[491] — Scaling limit of the one-dimensional attractive Hubbard model: the non-half-filled band case, *Nucl. Phys. B* **538** (1999) 701.

[492] F. Woynarovich and K. Penc, Novel magnetic properties of the Hubbard chain with an attractive interaction, *Z. Phys. B* **85** (1991) 269.

[493] C. N. Yang, Some exact results for the many-body problem in one dimension with repulsive delta-function interaction, *Phys. Rev. Lett.* **19** (1967) 1312.

[494] — η pairing and off-diagonal long-range order in a Hubbard model, *Phys. Rev. Lett.* **63** (1989) 2144.

[495] C. N. Yang and C. P. Yang, One-dimensional chain of anisotropic spin-spin interactions. II. Properties of the ground-state energy per lattice site for an infinite system, *Phys. Rev.* **150** (1966) 327.

[496] — Thermodynamics of a one-dimensional system of bosons with repulsive delta-function interaction, *J. Math. Phys.* **10** (1969) 1115.

[497] C. N. Yang and S. C. Zhang, SO_4 symmetry in a Hubbard model, *Mod. Phys. Lett. B* **4** (1990) 759.

[498] K. Yosida, *Theory of Magnetism* (Berlin: Springer-Verlag, 1996).

[499] R. Yue and T. Deguchi, Analytic Bethe ansatz for 1-D Hubbard model and twisted coupled XY model, *J. Phys. A* **30** (1997) 849.

[500] V. P. Yurov and Al. B. Zamolodchikov, Correlation functions of integrable 2d models of the relativistic field theory; Ising model, *Int. J. Mod. Phys. A* **6** (1991) 3419.

[501] A. M. Zagoskin and I. Affleck, Fermi edge singularities: Bound states and finite-size effects, *J. Phys. A* **30** (1997) 5743.

[502] A. B. Zamolodchikov and Al. B. Zamolodchikov, Factorized *S*-matrices in two dimensions as the exact solutions of certain relativistic quantum field theory models, *Ann. Phys. (N.Y.)* **120** (1979) 253.

[503] — Relativistic factorized S-matrix in two dimensions having O(n) isotopic symmetry, *Nucl. Phys. B* **133** (1978) 525.

[504] — Massless factorized scattering and sigma models with topological terms, *Nucl. Phys. B* **379** (1992) 602.

[505] Al. B. Zamolodchikov, Thermodynamic Bethe ansatz in relativistic models: scaling 3-state Potts and Lee-Yang models, *Nucl. Phys. B* **342** (1990) 695.

[506] — Mass scale in the sine-Gordon model and its reductions, *Int. J. Mod. Phys. A* **10** (1992) 1125.

[507] H.-Q. Zhou, Quantum integrability for the one dimensional Hubbard open chain, *Phys. Rev. B* **54** (1996) 41.

[508] Y. K. Zhou, The construction of the eigenstates for nonlinear Schrödinger model with supermatrices and attractive coupling, *Z. Phys. C* **39** (1988) 215.

[509] J. M. Ziman, *Principles of the Theory of Solids*, 2nd edition (Cambridge: Cambridge University Press, 1972).

[510] A. A. Zvyagin, Theory of current states in the Hubbard chain, *Sov. Phys. Solid State* **32** (1990) 905.

Index